THE NATURE OF CONSCIOUSNESS THE STRUCTURE OF REALITY

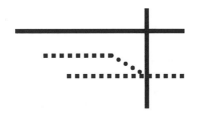

Theory of Everything Equation Revealed
Scientific Verification and Proof of Logic God Is

JERRY DAVIDSON WHEATLEY

Research Scientific Press
Phoenix AZ
2001

Research Scientific Press
P.O. Box 50132
Phoenix Arizona 85076-0132

ISBN   0-9703161-0-0
LCCN 00-191570

P-CIP Data
Wheatley, Jerry D., 1943–
Main entry under title:
        The nature of consciousness — the structure of reality:
    theory of everything equation revealed—
    scientific verification and proof of logic god is
        Includes bibliographic references and index.
        xxx, pp 770., 6 x 9 inches
        1. Reality, Structure of. 2. God Proof
        3. Physics—Philosophy.
        4. Consciousness. 5. Unified Field Theory.
        6. Metaphysics.
        7. Philosophy and Science.   8. Religion and Science.
        I. Title
        QC6.4.R42W48   530/ .01–dc21

Figures 12:3 and 12:4 Based on work by Roger Penrose.
Adapted with permission from Roger Penrose.

Printing: 9 8 7 6 5 4 3 2 1
Printed in the United States of America
on alkaline paper

TABLE OF CONTENTS

CAPTIONS

PREFACE

- Goal of science
- Author's work
- Book's utility

This book represents a revolution in scientific understanding. It will change how we perceive our world. The skeptic asks how is that possible? This is an expected question: one that furnishes us with a reasonable beginning.

In 1970 when 26 years of age, the author was a teacher of chemistry and the physical sciences. One day a student flipped ahead in his text and posed a question. The author gave a simple but scientifically definitive answer. After toying a week with the answer, he also realized it inferred several consequential facts not immediately implied in the question.

He constructed a large diagram to illustrate his reasoning. It incorporated not only the answer to the student's question but also explained many things not apparently connected to the original question.

The first diagram led to other diagrams. The diagrams show how things interrelate. Simply stated: the author has "delineated the structure of reality."

Project Idea: the Problem—the Solution

The long-term goal of science is to understand everything in one conceptual scheme. It has been called "the theory of everything." It answers the most basic of questions. What is reality? How can it be ultimately understood?

The philosopher B. Spinoza thought a systematic arranging of all true concepts could represent the structure of reality. He supposed the whole of reality is God. The implication is that we would know God if we could discern the whole structure of reality.

Some scientists think that a single principle must represent the theory of everything. This project offers an excellent candidate for that principle. Mindful of Ockham's Razor, the principle must be very simple; indeed, it is. A child can understand it. Yet, it must also explain everything; it does so quite handily. Let's not get ahead of ourselves.

Reality can be represented as a "map." The map has two basic aspects. There is consciousness and there is that of which we are conscious—namely, experience.

There is an extraordinary relation between consciousness and experience. How is consciousness connected to experience? This question is unanswerable until we fully understand experience.

What is experience? We experience things. Things, indeed everything, comprise the structure of reality. The structure of reality is a "mapping" of the relationship of everything. We need to understand the structure of reality before we can determine the nature of consciousness.

One of our first questions addresses "What" we experience. Do we experience things as they are in themselves? Or, do we merely experience things indirectly? For example, You may perceive (identify) a table, but most people agree that the table also exists independently from its perception.

Bertrand Russell has called the "patches of color" we experience sense data or simply, phenomenon. Immanuel Kant has called the table "in itself" the noumenon. From this reasoning we obtain an explanatory construct that immediately allows us to demarcate a reality divided into "our consciousness" and "what we are conscious of," therefore, what we experience.

There has been a problem in scientific methodology. Scientific knowledge is ideally an inductive endeavor. Yet, in experimental practice the method is deductive. Experiments are designed to verify hypotheses. Expected experimental results verify the hypothesis. The results are considered to be a deductive consequence of the hypothesis. This is putting the "cart before the horse." This method does not produce inductive understanding. It is a poor way to generalize. Experimenters rationalize: the more we determine what is not the case, the closer we will be to a solution. However, this is not reflected in most experiments. This is expected when there is not sufficient knowledge to do otherwise. Experimentation is ineffective in building unifying concepts.

The real problem surfaces when considering how inductive statements are proven. This has been a concern of philosophy and logic. There is a related problem. Every experimental solution poses more questions. Where does it end?

Why does the methodology problem persist? The reason is not reality: it is what it is. The problem is a shortcoming in man's inductive reasoning ability. Yet, now and then a theorist constructs a unifying concept that explains things very generally, but in an abstract way. Newton and Einstein's work are examples.

All things interconnect. An important task for scientists is to discover functional links between different phenomena. They seek to understand the underlying order behind everything. On a highway map, we find dots connected by lines. They correspond in reality to cities and highways. To understand the relationship between things, we also resort to mapping.

Constructing a map of reality is like piecing together a jigsaw puzzle. Each puzzle piece represents a statement of scientific fact. Confirming a statement as fact is one thing: to understand it is quite another.

We understand by establishing connections between one fact and others. In puzzle terms: We gain understanding by seeing how puzzle pieces fit together—how they interrelate. The "piecing process" demands redefining many existing scientific concepts and words in noncontradictory terms.

Only in understanding do we grasp the actual meaning of a particular fact. To understand is to be conscious of the relationship between things.

The author is a concept-research scientist. He has developed a systematic process (a refined-scientific methodology) which enables noncontradictory alignment of facts into generalized categories. The categories are grouped into "encompassing" concepts. Like Chinese boxes, there are some puzzles within puzzles just as there are different levels of generality found in science. For example, Einstein's Special Relativity groups the concepts of space and time into the enveloping conceptualization called space-time.

Reality is not experienced as space-time. We experience reality as a separate space and time. The lesson is reality is structured in a nonapparent way. Its structure; however, can be illustrated by diagrams.

Although too complex to explain here, we will state a little about how what we experience can be generalized into encompassing concepts. How do we know how to connect one fact, or puzzle piece, to another? Physics principles tell us how to connect puzzle pieces into map sections or diagrams representing parts of reality. Applying this tactic is guided by two major considerations.

First, diagrams correspond to structure found in nature. Arrangement of facts relates levels of scientific conceptualization.

Second, conceptualizations in diagrammatic form must be coherent. There should be no contradictions. Puzzle pieces or facts must fit together.

Given the first consideration, the second one assures that diagram coherency represents the "uniformity of nature." Uniformity or order is implicit in the fact we associate one thing with another; the very nature of knowledge infers it. Uniformity also implies problems are solvable only if we sufficiently understand reality.

What Has Been Accomplished

The author studied contemporary knowledge. He particularly focused on physics, the behavioral sciences, biology, philosophy, religion and logic. He intently studied these subjects to develop an acceptable way of explaining the meaning of the diagrams in terms easily discernible by knowledgeable people in those fields. The meaning of the diagrams must corroborate evidential facts, but not necessarily their present day interpretation.

Many present theories are reinterpretable using the context of the diagrams. The author did much of this himself. For example, he explains the so called "left-right split-brain" phenomenon using the diagrams.

Accepted-physical theories support the diagrams. There are also a few untested and not as well-known hypotheses that correlate with some of the diagrams. Experimental findings do not contradict this book's interpretations.

How were the diagrams pieced together? Contemporary scientific methodology is practically useless for constructing and verifying nonphysical theories. For example, one area that had been virtually immune to rigorous-scientific methodology is the mind. Examples supporting this statement abound. Psychologists attribute behavior to certain ill-defined concepts such as childhood experiences, innate drives, and contingencies of reinforcement. The starting point for any reasonable explanation of behavior must be the premise "people behave in such a way to become more secure." The problem is to find a way to do this.

A refined methodology was used to probe how the mind functions. This methodology was also applied to other difficult areas of science. The following suggests how the methodology developed.

After understanding the first diagram, the author realized it answered one of five basic questions. They are: Where? How? When? What? and Why? The first diagram answered the "where question" at the highest level it can be asked. It took many more diagrams and many years to interconnect the answers of the five questions. The diagrams are "frames of reference" from which we can visualize reality to more easily understand it. The diagrams helped the author develop a new and highly refined-scientific methodology. It sounds simple here, but it was not simple in practice.

Like all revolutionary-scientific theories, this work supplements those that precede it. Like those before, it simplifies explaining disparate phenomena at the expense of being more complex. Yet once studied, it is not so difficult to understand. Its diagrams, coupled with refined technique, enabled a scientific assault on long standing questions in not only science and behavior but also philosophy, logic, and religion.

There is an interesting relation between religion and science. It is one not envisioned by most theologians or scientists. Religion "culminates" in science. The meaning of this statement will become clear as one studies this book.

The author experienced little difficulty in elucidating reality's structure except for some minor problems. It has been a matter of patience and thirty years of full-time work. This book is the product of: extraordinary experiences, extraordinary research and extraordinary inductive reasoning ability.

Problems typically indicate there is something in reality we do not understand. The trick is to learn as much as possible and solve the problem by inductive conceptualization.

Solutions were found by using a step by step process. When a cause was not easily discerned, the author applied the refined methodology. How? In considering one cause there are initially two extreme possibilities. Either the possibility is generally correct, or it is not.

In considering higher-level concepts: one possibility, as a generality, is usually a "play it safe" deductive consequence of the other possibility. One of the possibilities is quickly eliminated because it does not cohere with other facts, or it does not correspond to anything in reality.

If reality does not outright contradict the remaining plausible cause, the task becomes one of either modifying diagrammatic references (which is rare) or redefining the cause so that it coheres with the diagrams. Definitions must cohere; puzzle pieces must fit together.

The following is an example of the noncontradictory aspect of the refined methodology.

Situation: people often say, if you really want to do something, you will do it!

Question:

(A) Main question: Is "wanting" a cause of doing something? In everyday life people behave as though they can do what they want. For instance, an individual wants ice cream. He gets ice cream, and believes he got it because (cause and effect) he wanted it.

(B) First probing question: Is wanting a necessary condition for doing something? We look for possible situations, which contradict either a yes or no answer. Consider: An individual turns around only to encounter a small projectile coming in his direction. What does he do? With no thought of wanting, he automatically blinks.

Answer: Wanting had nothing to do with his blinking! The answer is no. Wanting is not the cause of his blinking.

(C) Second probing question: Is wanting a sufficient condition for doing something? An individual crawling in the middle of a desert is extremely thirsty. He wants water.

Answer: Wanting does not produce water. Here again the answer is no! Wanting is not a sufficient condition for doing something.

Analysis: What have we learned and what does it mean? The water must be available and the individual must be able to drink it. Wanting does not cause what an individual does. Yet, no reasonable person will deny there is often a connection between the wanting and the having of what one wants. So we redefine the connection in terms other than direct cause and effect. We say the person must "translate his wants into action." This calls for two circumstances. He must have the physical ability, and the water must be accessible.

The author does not here elaborate on this aspect of behavior. However, properly understanding behavior offers clues about the structure of reality. For example, if there are barriers between wanting and doing, what does that say about free will? This analysis suggests free will is limited.

All diagrams unify as one-all-encompassing generality—the concept "Reality." All diagrams collectively represent reality's structure. Structural integration is the idea behind physics grand unification theories: supergravity, super-symmetry, super-string theories, and M-theory. Note: Although these theories may eventually unify the four major forces of nature under extreme conditions, they do not explain how reality is conceptually composed as a structure! This book supplies the structure.

As pertinent-experimental results (such as verifying proton decay) accumulate, they can be interpreted by the structure depicted in this book. When gravitation is "unified" with the other three forces, two questions remain. First: What does it all mean?

After determining meaning, there is another question. Why? The hypothesis "all forces unify under extreme conditions of temperature and pressure," even if verified, does not tell us enough. This book answers these other questions. Here is a hint how this is done:

> Experimental results finally verify one of the mathematically unifying hypotheses. The Ultimate Principle, suggested in this book, yields a unique concept for the theory of everything.

> What happens next? If not for this project, scientists would still need to interpret the unification hypothesis regarding its meaning. The problem of induction remains!

Note: What people believe is not the issue! What matters is that what we say has meaning. Real meaning is determined by concept coherency. The diagrammatic concepts must unify completely and consistently. If they unify via the ultimate principle, the diagrams probably represent the "structure of reality." Physicists are seeking a concept that unifies the quantum nature of things with relativity. The ultimate principle should seemlessly connect them.

At the turn of the twentieth-century, Max Planck found an equation describing light spectrum emitted from a black-body. The equation suggested black-body radiation is emitted and absorbed in discrete units of energy. Consider its consequences. The whole development of quantum mechanics had its birth in Planck's breakthrough. Scientists are still debating the meaning of quantum phenomena. Quantum theory is finally understood by interpreting the diagrams in this book using solutions to problems concerning logical completeness and consistency. The ultimate principle is an equation that connects relativity to quantum theory.

Lesson: We cannot know the real meaning of any particular "thing" until we understand everything generally. This infers the whole structure of reality needs to be understood. Why? Because each thing relates somehow to everything else.

> Diagrams represent the structure of reality. The diagrams provide a general context for practically any evidential fact or theory: including the unifying hypotheses.

The significance of this project is apparent.

The book's final two sections "interpret" the diagrams to definitively answer and settle major questions. The final section dispels any lingering doubts about the significance of the structure presented in this book. The power of scientific explanation presented overcomes the most difficult philosophical and religious questions. The diagrams are pieced together by scientific knowledge. The diagrams clarify scientific understanding.

Some Uses for the Project

The diagrams provide a context to understand things. The first diagram answers a major question about behavior. Why do we behave the way we do? We earlier stated that "wanting" alone has no direct cause and effect relation to "doing" something. We now realize: Though a person "wants" to help others, does not mean he is able.

People have problems because they encounter situations in reality they do not understand. In this sense one can only help others when he understands what those in need of help do not know.

Many people confuse understanding with familiarity. Decisions for handling a problem based on ignorance help no one. To the contrary, ignorance usually compounds the problem.

One cannot help someone merely because he "wants" to help. This is especially true if the problem involves behavior. Behavior has not been well understood. Why do we behave the way we do? A person can effectively help others only when he understands the deep-rooted causes of behavior. Then it is a matter of correcting the behavioral problem by manipulating the circumstances (or causes) to alter their effect on the problem individual.

The diagrams provide a means to understand problems confronting people. If enough damage has not already rendered the situation irrevocable, proper causal understanding can pull the individual out of his predicament. There is a bonus. Understanding how a person got himself into the predicament can also prevent him from repeating it. The basic benefit of knowledge is to help people.

Why do we have problems? Problems "force" the mind to change and grow. Review: Reality gives us problems to solve. We learn and mentally grow. We learn ways to help those less fortunate than ourselves. We cannot help our-fellow man until we first help ourselves to understand.

There has been no "psychological" standard by which to understand behavior. The author has developed a novel way to explain behavior. Diagrams explain human behavior differently from any present-day theory. For example, while B.F. Skinner's behaviorism adequately explains habits of daily living, it practically ignores consciousness.

Asserting the environment is the main cause of "why we do what we do" is not a sufficient explanation for behavior. Most behavioral scientists have come to grips with this fact. They have yet to postulate a theory that incorporates consciousness without colliding with strongly contested issues like determinism. The diagrams point to a noncontradictory way of explaining environmental influences on behavior. One diagram lists several "intelligence functions," which describes the workings of awareness without appealing to the "hidden-inner man."

What other behaviors do the diagrams explain? They explain the source of personal memory. They explain how a person can be comatose yet not brain-dead. They explain sleep and sleep-walking. They explain motivation.

What are some problems resolved by correlating physics with the diagrams? According to the CPT theorem (charge-parity-time) there must be some type of time reversal. Although the equations describing particle motion are temporally symmetric, entropy does not permit time reversal as in a movie run backwards.

We note this problem in the following question. How can there be time reversal without running afoul of the causality paradox? Physicists have yet to present a noncontradictory explanation. The diagrams easily handle that question too. Another major physics question is where is the missing antimatter? Certain principles suggest the quantity of antimatter must equal that of matter. To explain Maxwell's equations, physicists postulated a magnetic analog of the electron—the magnetic monopole. Where is it? Reasonable solutions are offered for these questions as well.

What is the proper interpretation of quantum mechanics in the sense first proposed by the Einstein-Podolsky-Rosen Paradox? They asked if the quantum theory of reality is complete. The diagrams suggest two solutions. One has been experimentally verified. The word "solution" here means noncontradictory explanation—one not invalidated by either self incoherence or noncorrespondence to reality.

Science concerns how phenomena interrelate. The grand idea of science is not to find evidence to support "somebody's theory," but to let the "cards fall where they may." Evidential facts speak for themselves. Like puzzle pieces, facts fit together only one way. There is but one reality.

The structure of reality solves problems originally posed by philosophers. One such question ponders whether there is a world apart from our awareness (of it). Is the "external world" merely a projection of consciousness? These questions, though raked over by philosophers, have not been settled scientifically.

Science is predicated on the idea things are determinable. Because scientific methodology is also deterministic, it has trouble addressing issues such as "free will." Is the "free will—determinism" issue solvable? This book offers a noncontradictory solution to this problem. These are a few of the questions resolved by understanding the structure of reality.

Leibniz thought a universal-logical system could serve as a standard against which disputes could be compared and settled. Can the structure of reality settle conflicts? Specific cases best demonstrate this use of the diagrams. The resolution of such cases, however, is describable in general terms. Here is how.

Each side of the conflict is asked to state their case. They are instructed to clarify what their case means. The reasons behind this are simple. All situations have causal factors. The real problem is not what they believe. The problem is solvable because of the following. There is something about reality these defendants do not understand or acknowledge.

Meaning pertains to the order of things: to the relationship between one thing and another. To save face, eventually the contestants are forced to adhere to one position. This is accomplished by maneuvering them using the following questions. Then, is that what you mean or do you mean this? They say, no, not that! Such tactics finally draws out "their" meaning in explicit terms. This step locks them into a position—backed into a corner so to speak. At this stage of questioning their arguments begin to crumble because reality contradicts "their" arguments. Credibility begins to suffer. The defendants can no longer fall back on their "honor" or reputation.

The conflict is not yet resolved. The contestants have already committed themselves to explicit statements describing their case. In the second part of questioning, the defendants are asked to state "why" they support that position. Maneuvering as before, the contestants are again locked into a group of explicit statements defining their defense. They probably cannot support their case with reality-based reasons.

The difference between beliefs and realistic reasons is revealed in the following two questions. Why do YOU say that? And, Why IS that? In simple terms, the "YOU" in the first question refers to one's beliefs. It represents what one does not know. The "IS" in the second question pertains to Reality.

The actual conflict is not between the two contestants, but between them and reality. They "just" believe it is between them. The real problem is their "personal" interpretation of reality. How can science help? Science concerns how things really are.

In the third stage of questioning the contestants may yet insist their case is valid. Most of their arguments are challenged in the first stage of questioning. That narrows their arguments. Arguments are further restricted in the second stage. Once the defendants' arguments are explicitly formulated, the task is handed over to reality. Most likely their arguments do not refer to reality. That is the problem!

At the onset, if the contestants agree to an honest (realistic) settlement and someone with sufficient knowledge and understanding mediates the dispute, then it is a matter of having the absurdities in their "personal" positions brought to the fore. Although only one contestant could be "in the right," usually both factions by now would have contradicted themselves.

How does this method work? If the contestants are realistically honest, they can probably manage the first why-question about order. However, if their reasons are self-serving instead of realistic, they will not get by the second why-question. Asking them to explain why they answered the first why-question, shows their "real" intentions. Usually, people are not cognizant of the real reasons for their decisions and actions.

The methodology proves the reasons are not what the contestants originally stated. Their only recourse is to say, but that is not what we meant! Since they had already locked themselves into explicit statements, their arguments are nullified because they did not know what they were talking about! Most people do not "know" why they do what they do. The mediator resolves the case by using realistic solutions. Why people behave the way they do is fully explained in this book.

What is the point? Someone in a position of knowledge and understanding is not always present to mediate conflicts. This is where the "structure of reality" helps. After asking, "why do YOU say that?" it becomes readily apparent that the question must instead be: "Why IS that?" That question is quickly satisfied by saying, "because that IS the way things are!" This question is answered only by reality. The structure of reality is an aid to settle disputes. Why? It explains everything. This project's significance is obvious.

How are people to be helped without a mediator who really understands the structure of reality? One does not have to understand the whole structure to find a reasonable solution. The solution is a deductive consequence of reality's structure. Only the structure is needed! It is a noncontradictory explanation of reality. The structure is logically complete. If it were otherwise, science would contradict it. If the structure is complete, any answer must be either explicitly or categorically derivable from it.

Because the structure of reality is complete and logically tight, it can be programmed into a computer. Any situation is solvable by "inputting" the contestants' problem into the programmed computer. It is simple. There is only one reality and one basic solution to every question. There is no reason such a computer program will not work. The major obstacle is supplying the proper premises. This book provides the premises. We pursue knowledge to help people. In helping others, we also help ourselves.

Doubting Project Claims

Revolutionary-scientific advances occur when independent thinkers examine "old problems" in a new way. Newton connected the "sideways force" of "celestial bodies" with that experienced by a falling apple and derived his three laws of motion. Newton's ideas held universally until inherent contradictions in his theory (e.g., planet Mercury's orbit) were explained away by Einstein. Einstein's General Theory of Relativity describes gravitation as the curvature of space-time. Einstein's theory not only supersedes Newton's theory, but, as a generality, encompasses it as well.

Each time a revolutionary theory is published, it meets the indignation of the "old guard." They are often intimidated by the boldness and imaginative effort needed to produce a new theory. Many big names in physics have more often than not embarrassed themselves by their public comments when a new theory is striving for respectability. Yet, some of those very people are the ones saying we must reexamine the old problem in a new way! For example, scientists at the turn of the century knew Maxwell's equations, coupled with Michelson-Morley's experiments, strongly contradicted the concept of absolute space and time. Though aware of apparent contemporary problems, the old guard clung to the familiar. Einstein replaced the notion of absolute space-time with relative motion. The lesson is simple. New solutions inevitably have consequences, which can be tested. Recognizing consequences via inductive reasoning inaugurates new ways of understanding.

The next revolution in science is not happening where millions of dollars are spent refurbishing high-energy accelerators. Discovering more elementary particles will not enhance understanding.

There have been many scientific breakthroughs: many in theory and many in experimental results. History attests that most scientific advances happen through the efforts of lone individuals. On the efforts of those few, follow other scientists who refine the original understanding.

Well, here we are. What supports the diagrams in this book as representing the structure of reality? The physics is already there. Most of the puzzle pieces (facts) are "on the table!" Using higher-level physics principles, the author has put the "given pieces" together—forming the diagrams. Though there were some "missing pieces," he had enough of the puzzle together to determine what they were. By supplying missing pieces the author has found solutions that push back the frontier of scientific understanding.

Every question has an answer. The first why-question that can be asked of anything, concerns the order of things—how one thing relates to another (how puzzle pieces fit together). We have stated that experimental methods use a deductive approach. Experimentalists ask what is this? What does it mean? A deduction is only as valid as its premises. The overriding problem is inductive—posing the question why is this? Simply, what value is a deduction if its premises are not verified? This project not only resolves why that problem exists, but answers it. This is the crux of the whole problem arising out of present-day scientific methodology.

Many scientists suppose that someday there would be sufficient knowledge that someone will be able to conceptually generalize it into grand patterns. These patterns will inevitably be hierarchical diagrams representing scientific understanding. They will represent the structure of reality. Once that is accomplished, the task of refining the understanding begins. That time is now.

Despite this book, readers should continue to doubt. This is why. To doubt is to question. To question is to seek knowledge and understanding. Question everything in this book. The idea is to acquire enough facts to minimize doubt—not to eliminate it. The more noncontradictory evidence one accumulates the more one approaches its true meaning.

Officially, the author claims nothing except Reality must speak for itself. It is difficult, if at all possible, to explain without making a claim. A statement, before establishing factuality, is a claim.

There are four possibilities.

- No claim—no verification or proof.
- A claim—but no verification or proof.
- A claim—with verification or proof.
- No claim—but verification or proof.

The first possibility is a logical "filler." The second possibility is very common. It is the bane of religion. The third possibility is the providence of science. The fourth possibility suggests verification and proof is already present in the Platonistic sense.

What more is there to say about doubt? There is always a point reached where reasonable people will acknowledge evidential facts. What can be said for instance if someone asserts he is not sitting on a certain chair when a roomful of people honestly state otherwise? There is "no problem" in "the Rest of Reality." A person encounters problems when he either does not understand or does not acknowledge what is real. Problems arise from incorrect inference.

Speaking as scientists, we do not make absolute statements because such statements are not amenable to reinterpretation through new evidence. It is not for man to say what is true about reality, but for reality to say what is true about itself. Our task, as scientists, is to find ways (such as experiments, noncontradictory reasoning) to cajole reality into yielding its secrets.

Some Notes

The author has extensively explored the last frontier—knowledge itself. This book represents the "expanse of knowledge." The preface mentions this last frontier—the one you will study. Let's call it the "mind expanse." Studying is one thing—reality another. To enter the mind expanse, you must do more than study: you must reason inductively. The phrase "mind expanse" is a euphemism for the product of inductive reasoning—knowledge itself.

The mind is every bit as important as is the physical world. Without mind, there is no awareness of material things. The mind can deduce the nature of experience by inferring the functionality of consciousness. That is the ultimate-scientific goal realized in this project.

The unrealized potential of mind infers the mind expanse. The following question suggests it. Why do "why-questions" evade scientific scrutiny? Why-questions are easy to formulate. Why are they so difficult to answer?

Scientists have been unable to systematically venture into the mind expanse. Scientists are mentally no different from other people. Their inductive-inferential abilities are nonfunctional. Why? This book answers and explains that question. Remember: Just because a person wants to do something does not mean he is able. The book also describes the refined methodology needed to explore the mind expanse.

Because the mind expanse is nonphysical, the methodology of empirical science is practically useless to understand it. To explore the mind expanse one uses a "refined" scientific methodology. One enters by reasoning inductively. There is one guiding question. It is: "How can this (whatever) be explained?" Understanding enables one to "move about" in the mind expanse. It is not unlike finding one's way in total darkness. You "mentally" learn where everything is located.

A few theorists have metaphorically envisioned the mind expanse. Plato called it the "Ideal": the world of forms. It takes one beyond the everyday world of "Whatness." It takes one into the world of "Whyness."

The last scientific frontier awaited methodical exploration. Serious-truth seekers have ventured ahead of their time. Philosophers "mapped" expectations. Scientists learned how things interrelate. Yet, the knowledge of scientific history does not prepare one for the journey because "things are not what they seem."

However, the legacy of scientific thought endures as "landmarks" in the mind expanse. These landmarks are guideposts for deeper exploration.

Few people understand the last-scientific frontier. That creates a problem for the reader. It makes understanding more difficult. The reader will be confused without proper methodology and guidance. Go slowly. Learn the methodology. Question everything stated in this book. An easy way to begin studying is to read the first time to get the general idea. (Or, at least read the "Key Points" at the end of each chapter. In the final Section, the Conclusion, Key Points follow each heading.) Then return to the beginning to seriously study.

The information in this book is presented step by step. Study the scientific basis for the presented structure. Ask probing questions. Why this—why that? Try to refute what is stated. If lies cannot be easily found in this book, most of its statements may be true.

Questioning is the first step into the mind expanse. One explores by seeking noncontradictory explanations. Ask, "How can this be explained?" The refined methodology enables one to obtain answers from reality.

One does not blaze a trail by continuing to follow in the footsteps of others. This project is conceptually innovative, challenging, and bold. These characteristics describe those individuals who forged ahead and revolutionized scientific thought.

Studying this book will instill a greater understanding and appreciation of reality. A critical examination of the knowledge represented in this book should enable the reader to methodically enhance his or her abilities to cope with everyday life.

INTRODUCTION

Perspective

- Knowledge, questions, and pursuit of truth
- Bringing order to knowledge
- Organized knowledge helps understand reality
- Tools to understand Reality

?
REALITY

Chapter 1

Man, Knowledge, And Reality

"People are not justified in limiting knowledge
because of ignorance, but because of ignorance, are
justified in not limiting knowledge. Any question posed
herein, the reader could just as well have asked."

People always have searched the unknown. Despite much effort, reality has
eluded man's search for self-knowledge, self-understanding, and self-meaning.
This book describes that knowledge, that understanding, and that meaning. In
short, this book explicates the nature of reality.

This book is a journey in self-discovery. There is one basic question most
individuals ask: Who am I? For some, it is a fleeting thought. For others, it fires
a consuming passion to learn. We are sure of one thing. If there is no meaning to
life, there is no satisfactory answer. However, if life has meaning, it is because
reality has purpose. To find our purpose requires an understanding of reality.

What is reality? What is knowledge? Reality can be defined as That which is, or
All that is actual or ultimately undeniable. Although a very simple definition, it
is suitable for this project. It follows—knowledge (or truth) corresponds to
"That which is." Knowledge is that information which represents reality.

Understanding is obtained by clarifying the connection between different
conceptual levels. Connections form patterns, and patterns are well suited for
explaining the structure of reality. In understanding reality, man finds real
meaning. That meaning should also reveal the purpose of man's existence.

Many generalizations in this exposition, though well defined, are fully
understood only in the context of the whole structure. The structure makes it
easier to understand the meaning of abstract-scientific concepts as they emerge
from the orderly generalization of supporting evidential facts. Many
explanations were only completely developed after finally understanding reality
as a whole.

People are not justified in limiting knowledge because of ignorance, but because
of ignorance, are justified in not limiting knowledge. For many people,
knowledge is attainable. The fortunate ones are not compelled by "outside"
influences to limit their knowledge, although some outside pressure is often
there, albeit in unrecognizable form.

Knowledge is acquired by asking questions. There is more to questioning than
most people realize. The method of questioning simplifies addressing complex
"why-what" relationships.

Is reality ultimately intelligible? If it is, another no less important question is
whether man is capable of understanding it. This reminds us of Albert Einstein's
statement: "The most incomprehensible thing about the world is that it is
comprehensible."

We begin by focusing on man's nature. Some think that man is what man can do. So what are his abilities? He is consciously aware and capable of observation. He experiences activity, sensation, apprehension, perception, cognition, imagination, and memory. From a "higher level" of understanding however, man will be defined not by what he can do, but by how he fits into the structure of reality.

Information Systems And People Classification

Information is categorized into two systems: knowledge and belief. Knowledge is information that represents and therefore corresponds to reality. Belief is information not connected to reality. Beliefs represent the individual apart from reality. Unsubstantiated judgments (labeled thoughts, opinions, and feelings) also are not classifiable as knowledge because they too correspond to an individual's presumptions of reality and not to reality itself. Just as there are two types of information, there are two types of people. People are classifiable by the information they use. One is egocentric; the other is realistic.

The egocentric (i.e., ignorant) person tells what he thinks about reality, instead of being realistic and explicitly explaining things about it. His attention centers upon himself. He emphasizes his perceptions. This is indicated in his conversation with other people. He consistently talks in the first person: I think, I believe, and I want.

The realistic (anti-egocentric) person talks about reality and therefore of what he knows. His attention focuses upon reality. He speaks from reality's perspective. He hypothesizes and refers to "evidence" and considers possible "whyness-whatness" relationships. He seldom proffers deductive conclusions.

Knowledge of Relationships

Just as there are structural connections between phenomena (actually—noumena) there are relational connections (representing structure) between facts. There are also connections between an observer and that observed.

Observations pose questions for the observer. Often, in asking questions, an individual is admitting he does not understand something. This is the first step in acquiring knowledge.

There are only a few basic questions to ask of any given phenomenon. First, one can ask *What* something is. He can expand by asking: *how*, *when*, and *where*, that something is. Finally, he can pose the enveloping question of *why* that something is.

The first four questions are adequately addressed with present scientific methodology. However, when the question is asked why something is, most people, including scientists, either do not know how or are unable to adduce realistic answers through reasoning. The result is a lack of understanding.

Most people fill this void with beliefs and opinions. In not admitting ignorance the individual loses the justification (or reason) for seeking knowledge by asking questions. He is doomed to think his beliefs are true. He probably fails to realize the inherent contradiction in that idea.

To question reality, an individual needs some standardizing principles and methods. The operative principles of logic and an adaptation of the scientific method are cornerstones for constructing knowledge. Deductive premises are not assumed self-evident but are conditional upon the final establishment of truth for validity. Inductive generalization is used to erect "ordered" concepts amenable to scientific scrutiny. Order is realized in patterns and diagrams. These are used to build models. We use models to understand.

Standards

Major standards used in acquiring understanding together with an accompanying brief description are:

(1) Uniformity

- All realistic induction presupposes uniformity in nature. It ascribes to reality various forms of regularity such as similarity, consistency, and continuity. Uniformity relates order in knowledge.

  Without uniformity empirical investigation would be untenable. It establishes the initial premise for organizing thoughts. This principle is not assumed *a priori*: its validity is accepted conditionally. Its truth depends upon the coherency of the whole system of knowledge.

(2) Identity

- The principle of identity asserts that a phenomenon is what it is; or that a statement is true if it correspondingly represents reality.

(3) Contradiction

- The principle of contradiction asserts no phenomenon can both be and not be at the same time; or that a statement cannot simultaneously represent and not represent reality. It suggests reality is not self-contradictory.

(4) Excluded Middle

- The principle of excluded middle asserts phenomena must either be or not be; or that any statement is either true or false. If a statement is true, it corresponds to something in reality. If it is false, it does not correspond to anything in reality. There is no middle ground.

(5) Correspondence

- Correspondence stresses the actual relationship between knowledge and phenomena. Knowledge is the symbolic representation of phenomena. This relation is initially contingent upon the empirical confirmation of physical theories. It is finally contingent upon the coherency of all knowledge.

  Relevant physical evidence serves as the (corresponding) basis for acquiring and building knowledge.

(6) Symbols, Description, and Definition

- The basic building blocks of knowledge are symbols that represent and correspond to phenomenal referents.

- Symbols acquire significance by description of a referent's properties (or relational characteristics) in a definition. These stated characteristics constitute a referent's description.

- A definition involves an expression to be defined and a defining expression that states the meaning of some symbol or term.

  Symbols, descriptions, and definitions are the functional building blocks of knowledge.

(7) Meaning and Context

- An explanation of a symbol infers two meanings. An extensional meaning referentially denotes a class of phenomena having similar characteristics or properties. Individual properties connotatively relate the intentional meaning of a symbol. Objectively ascribed properties aid in building concepts.

  Symbols also find meaning in context from accompanying ideas or purpose. It would be difficult to define explicitly every word used in this book. Since understanding is an individual matter, meaning should be construed in the context of accompanying ideas.

(8) Concepts

- A concept is the referential meaning of a symbol. A symbol's definition may imply an underlying class concept inferring some relevancy to other phenomena. Connecting properties by similar meaning widens generalizations obtained from interpretive experience. These abstractions may also enhance the explication of meaning itself.

  Concepts are universal representations of phenomena. They constitute the basic universal referential standards of knowledge.

(9) Hypothesis, Theory, Verification, and Proof

- A tentative scientific explanation is called a hypothesis. It is usually a generalization that conditionally explains some type of relation between phenomena.

- A wide hypothetical generalization having some supportive evidence is called a theory. Theories are hypothetical, although pertinent experiments may support their truth.

- It is the uncertainty of when there is sufficient evidence to warrant verification that complicates the factuality of a theory.

- Verification implies an acceptance of some hypothesis or theory within the context of experiment, but does not itself constitute proof.

Scientists use hypotheses to represent possible explanations of various phenomena. Hypotheses and theories are the keys used in objectively advancing knowledge and science. Logical proof is the final goal of science.

(10) Generalities and Particulars; Deduction and Induction

- An evidential fact is usually a generalized representation of particular observations or experimental results (indicating some type of recurring regularity). Observations may be transposed into a universal statement. Its truth may be supposed in an encompassing generality. The immediate justification for such inference is based on the principle of uniformity.

- An inference may be deductive or inductive, depending on the direction of specificity or generality.

- Deduction involves a closed system with inference from universals to subsumed universals or particulars.

- Induction involves an open system, which projects evidential facts into ever enveloping universals or encompassing concepts.

Induction makes possible greater integration by higher level classification.

(11) Class, Order, and Systems

- The categorizing, of similarly related facts, inductively leads to a general classification of various phenomena.

- These groupings, through common class association, lend themselves to some specific arrangement or order.

- Arrangement of interrelated classes constitutes a system. If abstract systems are properly built upon corresponding structure (established by principles of physics) of related phenomena, they will be coherent in keeping with the principle of uniformity.

Systems infer a greater connection of facts ordered through the higher association of different classes.

(12) Phenomenal Class Relevancy: Intraphenomenal and Interphenomenal

- Class relevancy defines the scope of descriptive and explanatory inquiry. Phenomenal class relevancy concerns generalities representing phenomena. It ascribes (relevant) meaning by classifying concepts. It represents categorically what one is attempting to understand and explain.

    There are two aspects of phenomenal relevance. The first "defines" by classification. The other extends the class by broadening its "level of inclusion." Each type confers meaning by establishing connections within or beyond a given generality. Phenomenal revelancy sets the defining limits of a class. We are concerned with two types of phenomenal class relevancy: intraphenomenal and interphenomenal.

- Intraphenomenal relevancy refers to the classifying characteristics of a phenomenon. It defines something by classifying it. Also, it excludes things not sharing the characteristics of membership inclusion. Ascribed properties of an intraphenomenal class are restricted by a superclass set (next higher class generality) and by (its) subclass sets (if each component is a subordinate class). Intraphenomenal relevancy focuses on a thing's class and its defining properties.

An example of intraphenomenal relevance is the periodic table of atomic elements. Functionally, each element is distinguished by its own set of properties. Yet, every element is classified as "atomic." The class "atom" limits what can be denoted as atomic elements. Here, we are not concerned with atomic construction. Each phenomenon (e.g., this hydrogen atom and that oxygen atom) can be conceptually universalized to represent a whole class of similar phenomena (i.e., all atoms). Intraphenomenal relevancy implies class unity on a set (atomic) of properties (that characterize the different atoms). In a typical hierarchical system, it pertains to the "horizontal delineation" of defining properties.

- Interphenomenal relevancy distinguishes class level. It recognizes class differences by scope of generality. It implies a categorical separation of two or more phenomena (or two or more classes of phenomena). Conceptually, each class represents different levels of generality. In a hierarchical system interphenomenal relevancy refers to the vertical association between classes, i.e., the superclasses, and subclasses. In set theory interphenomenal relevancy distinguishes one class level from others. For example, atoms combine forming molecules "on the high end." Molecules are of a different level than its constituent elements (the atoms). Molecules exhibit different defining properties than those of atoms. At the low end: atoms are characterized differently from their components (nucleons and electrons).

Interphenomenal relevance can be used to describe whyness-whatness associations. In this use "whyness" denotes a higher generality of a "whatness." Here the word "explanation" is a valid substitute for whyness. Many whatnesses (particularizations) may have one whyness (explanation) that becomes a constituent whatness for the next higher level of generalization. Atoms are explained by molecules.

Phenomenal relevancy orders phenomenal classes into hierarchical arrangement. Intraphenomenal relevancy defines a class by a thing's properties or characteristics. Interphenomenal relevancy describes relationships between different levels of generalization.

Cause and (then) effect is a limited attempt to explain the relevancy of things. All events have a reason for their state or condition. The reason can be elucidated by careful classification of phenomenal characteristics into various levels of generality. Phenomenal relevancy sets patterns in building knowledge. Phenomenal relevancy extends "explanation" beyond the limiting idea of cause and (then) effect.

Phenomenal relevancy provides the scientific context for understanding. It uses a classificatory system to understand how things (phenomena) interconnect.

(13) Coherency

- The coherency principle involves the relevancy of knowledge. Class definition constrains meaning. Various phenomena have similar characteristics that classify them as generalizations. As generalization (of classes) broadens, knowledge becomes more abstract and losses its immediate correspondence with observable phenomena. At that point, the criterion for verification often involves more than the usual means of (empirical) confirmation (statement correspondence to reality). When this happens, verifiability becomes increasingly dependent upon the coherency of the knowledge system as a whole. The system must be consistent.

(14) Synthesis and Unity

- The coherent integration (synthesis) of associated systems may constitute an ordered interphenomenal all-inclusive unit.

- This unit, since it is representatively all-inclusive, would be the "coherent knowledge representation of a correspondingly structured reality." Its inherent structural relations are governed by the (logical) standards used in interpreting the scientifically determined interphenomenal connections that "represent" reality's structure.

Although evidential facts are important, it is the interrelating of facts that is most significant in synthesizing knowledge. It is the synthesis of relations between generalizations that yields the knowledge of the whole of reality (as a unit) in terms that our awareness can understand.

Key Points:

- Reality is "that which is."

- Knowledge corresponds to and therefore represents reality.

- Ask questions to seek truth.

- There are five basic questions: why, what, where, how, and when.

- There is no reason to suppose reality is anything but uniform. So, we expect concepts representing reality to be consistent. There should be no contradiction in understanding reality.

- Higher level concepts are generalities that "categorically" define universal inclusions.

- Phenomenal relevancy relates things categorically. Intraphenomenal relevancy represents a class and its defining characteristics. Interphenomenal relevancy represents the connection between classes.

- Testing the interconnection of all classes for consistency should yield an "all embracing" classificatory synthesis of knowledge. Complete knowledge synthesis should lead to a unit structure representation of reality.

- Standards without a methodology to gain deeper knowledge are useless. We need to develop a methodology capable of applying these standards to obtain a coherent and complete understanding of reality.

Chapter 2

Methodology

"People should not fear what they do not understand;
people should fear because they do not understand."

Methodology infers many things. There is the rigorous methodology of science and logic. Scientific verification is premised on seriously seeking truth about reality on its terms. Logic concerns coherency of relationships. If applied to knowledge, logic helps establish scientific generalities on principles that interconnect subsumed evidential facts. Before considering what constitutes a serious methodology, let's briefly look at a false methodology. A false methodology is no methodology at all. Yet, a false methodology often pretends to be a serious guide for seeking truth.

Non-Methodology

What is believing? People believe beliefs. Beliefs are accepted on faith. Faith is the methodology of believing. Faith is the uncritical acceptance of information as fact. It is the most insecure and untrusted method of determining information validity. Faith always misleads (in seeking truth). It is an anti-reality method. It is anti-reality because it pertains to what an individual supposes *a priori* to be true without appropriate evidence. Faith results from not seriously seeking truth.

Beliefs often reinforce an individual's fantasies to a point where he loses his ability to continue learning about reality. The more he thinks "his" beliefs are "right" (or true) the more he isolates himself from reality awareness. The egocentric person usually increases support for a believed premise by deductively inferring from it. Yet in doing that he waives researching the validity of the original premise. A person can believe anything, but there is only one reality, and therefore, only one set of information called knowledge that corresponds to it.

It is not difficult to comprehend an argument is only as valid as its weakest premise. Belief is not a valid premise for any inferences about reality.

Man has searched for meaning in many ways. How can man find real meaning when he is defined not by what he believes or does, but by how he fits into the design of reality? First, he must undergo a change of attitude and adopt critical standards in acquiring knowledge. Along with previously listed standards, he must adopt a personal scientific-like method of studying reality. He should hold all incoherent unsubstantiated information in abeyance.

Knowledge

The first goal is developing a method that minimizes mistakes in acquiring knowledge. Any information not "explainable without contradiction" must be suspected of error. One problem encountered by an individual attempting to change his method of inquiry, is trusting an unfamiliar method. It is necessary to explain phenomena without contradiction. This recognition should ease one's insecurities about adapting a non personal method. One need not "attach" to beliefs. An individual can invest personal security in a critical method of inquiry.

As one becomes more experienced and adept at analyzing possible explanations of phenomena, his confidence in the method improves. Self-security is not a problem because the basic method is not changed—only refined. It is the type of information acquired, which is altered.

One can capitalize on information already gleaned through science for a beginning understanding of reality. The more one is critically honest, the more he becomes aligned with the scientific method and the knowledge of science. Slowly he distinguishes belief from knowledge. Formerly held beliefs give way to knowledge.

Understanding entails awareness of meaning by recognizing relations. An individual must be familiar with all relations that correspondingly represent phenomenal structure before he can understand reality.

A logical argument is only as good as the method used to establish its premises. The first step of admitting ignorance becomes the personal factual premise for acquiring knowledge. One knows because he can empirically substantiate or explain without contradiction (what is purported to be fact). This is the personal explanation principle (PEP). That is, one cannot assert the evidential truth of a proposition until after showing (verifying) that it corresponds to reality. One needs to be realistic while seeking understanding through PEP, for his mental processes to develop to the inductive-intelligence level needed to synthesize knowledge to higher levels of generalization. One must be able to explain everything generally before explicitly understanding anything specifically. PEP commits one to a policy of limiting premature deductions.

One must make full use of both deductive and inductive techniques to obtain a synthesis of knowledge. The unceasing study of ever smaller bits (by "reduction")—deduction of a "jigsaw" puzzle—is counterproductive to placing all the pieces together (by induction). We need to start somewhere. There now is sufficient knowledge to piece together the puzzle of reality.

Induction leads the way in piecing evidential facts into relations used to ferret out causes. Ideas that contradict experimental evidence should be withheld, but not abandoned. Ideas that conflict with the coherency of well-established facts (within any class or generalization level) should be retained for further consideration. Questionable ideas often have significance at other levels of generality.

In one sense to understand is to know the relationship between different fields of study. It takes specialists to determine what the individual puzzle pieces are, but it is the ardent generalist who puts it together. There is an axiom that specialists learn more and more about less and less until eventually, they know everything about nothing. Generalists learn less and less about more and more until eventually, they know nothing about everything. The trick is to strike a balance bridging the two extremes. Knowledge correctness stems from the work of the specialists; understanding comes from the work of the generalist.

An individual is defined by his experiences. "You are what you know." If he acts upon beliefs, his world develops around fantasies. If he consistently uses knowledge his mind adapts to seek more knowledge. As a person becomes more realistically truthful, he becomes more clearly defined by relative fact. As his (factual) knowledge increases to the point of understanding the structure of reality, his place and meaning in the structure begin to unfold. He not only will understand his part in reality, but will also be able to explain "why" he is what he is. He becomes identified as one with reality. This event completes one's transformation from an egocentric being to a realistic being. From a personal perspective this book represents man's evolved meaning as revealed in the progressively unfolding structure of reality.

Concepts: Relating and Understanding

When an individual totally immerses himself in some subject or project, in a way it becomes him. He may become very knowledgeable by exploring its fine points. With sufficient information and improved methods he fashions the tools needed to push him beyond the frontier of present understanding. This also applies to science.

Any broad-minded change or accelerated addition of information to a discipline often is accompanied by a radical change in the method used to acquire it. If there is more to reality than the so-called physical world then traditional empirical methods must be supplemented to acquire any significant knowledge of a nonphysical world. This implies the method will develop from someone's increased ability to reason (inductively). This book explains how this ability evolves. Some of our ideas are represented in a simplified diagram. (See Figure 2:1.)

The diagram represents a physical object and how we can understand it. Dealing with unfamiliar phenomena is more laborious and complex than already having a diagram to aid understanding. Synthesized ideas only come after carefully establishing inter-phenomenal connections of many factual concepts.

The physical object in Figure 2:1 is ascribed with many properties. Borrowing from philosophy we say an object is characterized by an essence: one that is not directly knowable by empirical examination. However, it is revealed by understanding reality as a whole. The essence manifests as an attribute that, in the Kantian sense, is the noumenon or the thing-in-itself. The noumenon can be indirectly known through inductive inferences drawn from experiments. Properties of noumena are deductively surmised from noncontradictory considerations.

The phenomenon is what appears to us. We are acquainted directly with phenomena through the apprehension of sense data (in our minds). Structure is built inductively by interconnecting various phenomena by linking their ascribed properties in relationships. Relations between phenomena were culled under the cloak of phenomenal relevancy.

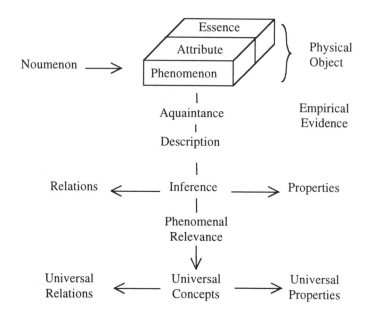

Figure 2:1. Physical Object. Ascribed Properties and Concept Formation.

Cross inferences of relations and properties lead to a categorization of evidential facts that, when expanded as generalities, are used to erect universal concepts representing the structure of reality.

Whyness-Whatness Association

We can ask specific questions about any phenomenon. We ask these questions in the form of What? How? When? Where? and Why? The how, when, and where questions pertain to specific aspects of the what-question. Whatness is manifested as howness, whenness, and whereness. Although the three basic what-questions are structurally equivalent, in practice they are not "information equivalent." The what-question considers a phenomenon as a particularity or as an effect. The why-question pertains to an explanation or reason for the what.

The tactic of seeking truth by addressing questions in what-why formulation, enabled methodological delineation of the structure of reality. The why-what formulation is used in this book to explain the structure of reality. Let's call it the "why-what paradigm." It was obtained by applying the five basic questions to the idea represented by interphenomenal relevancy. It gives us the means to construct a multi-level explanatory structure. The why-what paradigm forms the "shell" structure of reality.

Because uniformity is assumed in nature, the use of whyness-whatness associations—based on scientifically determined properties—establishes structures that link phenomena. When properly done, it simplifies organizing a coherent representation of reality. There are three types of whyness-whatness associations.

Type I Whyness-Whatness Association

Type I whyness-whatness associations are often expressed as "cause and (then) an effect." In reality the whyness factor is extrinsic to the whatness factor. Type I whyness-whatness associations describe interactions between specific phenomena. (Type I may also represent the interaction between phenomena as a class concept if the relation describes a regularity or physical law of nature.)

Here, one phenomenon is spatially or temporally separate from the other phenomenon. One phenomenon acts upon the other phenomenon. Example: when it rains (the whatness factor) it is because of clouds (the whyness factor).

This is cause and effect. The whatness factor is the effect; the whyness factor is the cause. Analysis indicates a problem in this common interpretation. In the above example, clouds, although a necessary condition, are not a sufficient condition for producing rain. (Clouds may be present and produce no rain.) Rain may become a cause of flooding. Flooding can cause destruction—another sequential effect. Yet, if it rains over the ocean, the same amount of rain may be no problem (no flooding effect) at all.

Labeling contributing factors as causes, creates a problem in terminology. One problem is that the "cause" is also changed along with the "effect." In the rain example the clouds lose moisture. This type "cause" is but one of many contributing factors in a sequence of events.

In other cases the whatness factor may seem (temporally) coincident with the whyness factor. Again, closer analysis reveals something else. Example: in a contained volume of gas, the pressure (the effect) increases as the temperature increases (the cause). At the micro-level, even this reduces to a sequence. Brownian motion (the micro-cause of temperature) of the gas molecules bouncing off the container's walls produces the increase in pressure. In micro-time the molecules act in an incalculable sequence producing the "overall" effect.

Justifiably or not, the cause and effect idea often describes change. That is acceptable if restrictions in the use of the concept are noted. We observe in both above examples that the contributing factors belong to the same class (has the property of tardyon physicality). These types of phenomena all act as contributing factors. All exist in time. These factors give the appearance of causation. Contributing factors are causes in the sense that without them the final state (or effect) does not occur. Based on the idea—there is no change without cause—it does describe what happens to something. In that respect Type I whyness-whatness association does embody the familiar notion of cause and effect.

Type II Whyness-Whatness Association

Type II whyness-whatness associations are based on interphenomenal relevancy. This association is expressed by "hierarchical" arrangement. The whatness factor is a more specific case of the whyness factor. The whatness factor is a subsumption of the whyness factor. Conceptually, the whatness factor and whyness factor belong to different levels of generality.

An example is the different levels of generality invoked by ordering a specific animal into a classifying scheme. First, a question. How does man (as an animal) fit into the animal kingdom? We can answer that question by seeking how man relates to "higher" levels of animal classification. (See Table 2:1.)

Whyness Factor	Whatness Factor	Characteristics
Phenomenal Class	Divergence	Defining Character
matter	animal [life form]	oxygen consumer
animal	*Chordata* [phylum]	bodily segmented, notochord
Chordata	*Vertebrata* [subphylum]	developed backbone, kidneys, brain
Vertebrata	*Mammalia* [class]	body hair, lactation, 4-chamber heart
Mammalia	*Eutheria* [subclass]	placenta
Eutheria	Primates [order]	tree dweller, large brain
Primates	*Anthropoidea* [suborder]	good vision, complex brain
Anthropoidea	*Hominoidea* [super family]	tailless, no cheek pouches
Hominoidea	*Hominidae* [family]	
Hominidae	Homo sapiens [genus, species]	erect, highly developed cerebrum

Table 2:1 Classification of Man

Categorically, scientists divide physical life into two kingdoms—Plantae, and Animalia. Man is classified as an animal.

What is the general plan for classification? The taxonomic idea is based on the following levels. Kingdom—Phylum—Class—Order—Family—Genus, and Species. As we ascend the hierarchy, there is greater inclusion. As we descend, there is greater specificity. Some of these levels are extended to allow more categories. Taxonomists now have broadened the original two categories (plants and animals).

Each hierarchical level is defined by delimiting characteristics. Refining the characteristics, we eventually reach Man (Homo sapiens). This category includes every man and woman that ever lived. At this level we jump "reductively" from general abstraction and denote real individuals.

There is more to this hierarchical system than that suggested in Table 2:1. Answering how man fits in with other animals by categorizing him in a table does not tell the whole story.

Most categories include more variation and divergence than the one listed. To see how any one subclass fits in with all other classes, we could list the others. However, there is a better way. An extensive "flow chart" shows all branches of the hierarchy. (See Figure 2:2 b.) Here we start with a high level of generality and work toward revealing all the particularized subclasses included under each superclass.

Each class is but one branch (of its superclass). For example, we find that the Order Primate has more than one suborder. Lemuroidea and Anthropoidea are additional suborders. Anthropoidea includes some "super families." One includes the rhesus monkey; another categorizes the spider monkey.

The super family Hominoidea includes two families. One is the higher ape: e.g., orangutan, chimpanzee and gorilla. The family Hominidae has only one genus and species—Homo sapiens.

As the animal hierarchy (structure) is elucidated, we more clearly understand our place in the animal kingdom. Why did we choose this example? This very idea carries over to the effort to fully understand ourselves. If we can delineate the whole structure of reality, we may have a basis for ascertaining the very meaning of our existence. Meaning is derived from understanding relationships.

Type III Whyness-Whatness Association

Type III whyness-whatness associations are also based on the structure inferred by interphenomenal relevancy. This association expresses "cause- (and simultaneous) effect." In reality the whyness (cause) is intrinsic to the whatness (effect). There are other ways of saying this.

> The whyness factor is coincident in space and time with the whatness factor.

> The whatness factor is produced and maintained by the whyness factor.

> The whatness factor is a direct manifestation of the whyness factor.

> The whyness factor is an essence; the whatness factor is an attribute.

Example: physical matter is a direct manifestation of electromagnetic energy.

Whyness-Whatness Modes

The initial problem of piecing together the structure of reality is not already having all the pieces (or facts) at hand. That means we have to place together the available pieces into some type of meaningful arrangement. That preliminary arrangement is called a mode. A mode is a way of looking at something diagrammatically.

We use modes to delineate whyness and whatness qualities that are scientifically ascribed to objects or classes of objects. Just as there are three types of whyness-whatness associations, there are three basic modes for depicting phenomenal relationships. Type I whyness-whatness associations are expressed using extrinsic modes. Type II whyness-whatness associations are expressed with hierarchical modes. Type III whyness-whatness associations are expressed using intrinsic modes.

Modes constitute an understanding of a phenomenon (or class of it) by ascribing it with whyness and whatness properties using the five basic questions. Again, conceptually, whyness relates cause and explanation (or high level generality). Whatness relates an effect and particularity (or low level generality) as whereness, whenness, or howness.

Extrinsic Modes

We use extrinsic modes to visualize the interaction between very particularized classes of objects. An extrinsic mode includes a whyness factor and a whatness factor. The whatness factor has three aspects—the howness aspect, the whenness aspect, and the whereness aspect. We use extrinsic modes to describe changes. (See Figure 2:2 a.)

Hierarchical Modes

We use hierarchical modes to visualize the structure created by interconnecting different levels of generality. (See Figure 2:2 b.) Animal classification exemplifies how structure is formed from hierarchical modes. For example: If we are asked to supply the specific classes of the order Primate (superclass in Figure 2:2 b) we can answer with three suborders (classes). We may further specify the suborders by listing their superfamilies (subclasses).

Also, if we were asked to supply the next more general example of the genus and species Homo sapiens (subclass) we would list the family Hominidae (class). Again, if we were asked to give the next "higher" general classification of the family Hominidae, we would list the superfamily Hominoidea (superclass).

Intrinsic Modes

Intrinsic modes depict the immanent relationship between two phenomena. (See Figure 2:2 c.) Type III whyness-whatness association efficaciously asserts that objects or phenomena have a dual nature. One is "whatness": defined as an intrinsic whatness factor (of the Type III whyness-whatness association). The other is "whyness": defined as an intrinsic whyness factor (of the Type III whyness-whatness association).

The duality expresses a simple idea. First, Type III whyness-whatness association is the widest form of a Type II whyness-whatness association. Second, in a narrower sense, Figure 2:2 b shows that many class levels exhibit both a whyness factor and a whatness factor. It all depends on direction of association. If the class is associated with the next higher level of generality, it is a whatness factor. If it is associated with the next lower level of generality, it is considered to be a whyness factor.

Modes expressing whyness-whatness associations show the functional aspect of a specific phenomenon. We address the functional nature of a phenomenon (or class of it) as whatness (when questioning by logic—therefore "what" is this?—we answer deductively). We address the functional nature of a phenomenon as whyness (when questioning by reason—therefore "why" is this?—we answer inductively).

When asking what-questions (about some character or property) we expect deductive answers. When asking why-questions, we expect inductive answers.

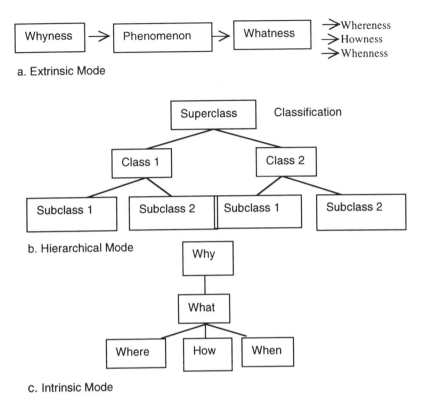

a. Extrinsic Mode

b. Hierarchical Mode

c. Intrinsic Mode

Figure 2:2. Classification Modes.

Deduction and Induction

Any phenomenon (or class) considered in the context of its relation to other phenomena (at any level of generality) is describable in one of the modes showing a whyness-whatness association. Whyness factors and whatness factors are most useful in understanding relationships between phenomenal properties when ascribed in terms answering the five basic questions—why, what, when, where, and how.

Deductive and inductive methods are most useful in understanding relationships between whyness factors and whatness factors when phenomenal properties are ascribed in terms supplying answers to the five basic questions. The more universalized the ascribed properties (of the phenomenon) the more its nature is characterized by whyness. The more particularized the properties, the more its nature is characterized by whatness.

Despite this simplicity, we are not directly aware of whyness factors. We only understand whyness factors by associating one phenomenon with other phenomena in whyness-whatness associations.

However, we can follow mental operations. The mind can reflect evidential facts in a hierarchical arrangement of properties that link phenomena together.

Conceptually moving in the direction of greater specificity exemplifies deduction. Deduction follows inferences from whyness to whatness. Conceptually moving in the direction of greater generality typifies induction. Induction follows inferences from whatness to whyness.

Functionally, whatness factors and whyness factors share an inverse relation and are oppositely directed. The more a phenomenon relates to other phenomena based on "what" questions, the easier it can be addressed and understood through the inductive process. The more it relates to other phenomena based on "why" questions, the easier it can be addressed and understood through the deductive process. However, there are practical limits to this ideal.

If we consider the most generalized concept of all (reality itself) we can only move in a deductive direction from it. Here, we are limited at the high end. (There is nothing beyond this concept whereto we could induct which is not already a subclass of this concept).

Remembering practical limitations, let's take another look at using induction and deduction on modes. Inductive or deductive operations can be performed on any mode (whether it is characterized predominantly by whatness or whyness). As we have seen, no matter which direction we perform an inference operation, we find there is no real limit to questioning. There is only a limit to what can be addressed by questioning.

It is easier to acquire information about a phenomenon using deduction (called "logicing") on whyness factors (which are quite generalized). Similarly, it is easier to use induction (called reasoning) on whatness factors (which are particularized).

Which operation best serves understanding? With sufficient knowledge, one recognizes the generalization level of confronting concepts. (Again, refer to the different levels of generality in Figure 2.2–b.) That is comforting to know when we are studying something that has already been laid out before us. But, how does one begin when only having a vague notion of what he is seeking to understand?

Although deductive methods are valuable in identifying properties, it is inductive methods that lead to synthesizing properties that are meaningful. Deductive methods break down "what" we experience into fragments (of whatness characterization) amenable to scientific study. These fragments (or pieces of experience) are ascribed with various properties. We develop concepts (of whyness) from careful universalization of these properties. The structure represented, using phenomenal relationship, is inferred by inductively reintegrating these pieces of information through a system that interrelates concepts (such as whyness-whatness association).

The modal system is most useful in explaining the structure of reality. We easily understand that whatness factors are deductive consequences of whyness factors. The three aspects of the whatness factor (the whereness, the whenness, and the howness aspects) are subsumed under the whatness factor and are therefore, deductive consequences of it.

Personal

Before embarking on our epistemological journey, we briefly describe a methodology tailored for the reader. Though simpler, it parallels that used by the author.

People usually interpret (judge) what they do not understand. This is human nature. There are many obstacles to understanding. To understand this book, the reader must counter the tendency to judge statement meaning. Here we describe a useful approach to understand this project. First, some preliminary statements.

Learning

There are three basic ways to learn. All three ways are useful for the serious truth seeker. We learn from schooling. We learn from personal experience. We learn from reasoning.

Most educated people have been "schooled." Schooling often involves teachers who have developed some expertise in various subjects. With the aid of teachers we learn language, grammar, and reading skills by studying books or learning devices. Books, journals, and papers are excellent sources to gain a wide perspective on many subjects.

From books and schooling we learn acceptable terminology for things and concepts. This enables us to communicate with others who have acquired expertise on a given subject. Generally, the more schooling the better. A "student" can also learn much on his own. Teachers are very helpful. Guidance saves time.

Memorizing is often associated with schooling. Students can learn much in school yet use their mind very little. Memorizing does not itself engender understanding. Our goal is to do more than study books. We, as truth seekers, need to understand. There is more to understand than books and journal articles.

We also need to consider personal experiences (with reality). Things we experience are not necessarily what they seem. If our experiences were straight forward, they would not confuse us.

We sometimes ask what does this experience mean? We constantly search for meaning. To understand is to ascertain meaning. To ascertain meaning means we must go beyond our study of books, journals and what our teachers taught us. By applying what we learned from books to our experiences, we can figure out some meaning. This is the tricky part. We can easily, and more often than not, misconstrue the meaning of things we experience. We have to do more than experience things to understand them.

We need to use our mind. We can reason to understand what we have studied to better understand our experiences. We can use terminology learned from schooling; we can ask guidance from people with expertise. Yet, each person must use his or her mind to infer meaning to gain understanding.

Consider the following examples. "Alfred" studies books about automobiles. He reads about how they function. He studies how automotive parts interact to produce transportation. He could study books for many years about automobiles. Many people might consider him to be an automotive expert. Many people could come to him for advice about cars. He may answer their questions with aplomb. Yet, those seeking answers may not even suspect that Alfred never encountered a real automobile: that he did not learn anything about cars by "working" on them—repairing them to working order. This "expert" represents someone who has much "formal" schooling, but no "practical" experience.

Consider "Allen." He cannot read. He learns about cars by "working" on them. He became an expert by experience. People come to him for advice. He has little difficulty giving mechanical advice as long as someone does not point out the problem using a book.

In both examples, we might have someone who is an expert on automotive structure yet might not understand how parts composing the structure function in higher level systems (e.g., electrical and cooling systems). This example, though simple, may strain the reader's assessment of the point. The point is that a person can recite what he has memorized from books. Yet, he might lack "hands on contact" with the reality (of the situation) represented in the books he has studied. Another person might work on cars extensively, has no book learning, and yet be an expert. Obviously, the second person would need much time to learn about cars, especially if he has not completed an apprenticeship in automotive technology. There is a decided difference between learning from books and directly learning from reality. In both cases we may have someone who still does not fully understand what he has learned.

Learning takes time and much experience. It is easiest to gain understanding by books and by direct experience.

Reasoning how automotive parts interrelate and function in higher-level systems (i.e., electrical, cooling, motor, drive train, for example) enables one to solve problems he has not encountered in either books or by apprenticeship. To push beyond book learning and learning by direct experience, one needs to reason with the knowledge he has acquired.

Reasoning enables the expert to better understand what he has learned (by book or by experience). Reasoning is the center of the learning process. The serious seeker of truth will always learn something new.

Reasoning not only carries us beyond books, it pushes us beyond physical experience. Reasoning inductively leads us to understand why things happen. Reasoning, pursued to the extreme, may lead us to an understanding of the meaning of life itself.

There is life. Our main question is what is the meaning to life? How can life be explained? We seek truth to answer this question. To obtain answers, we need to understand the role of explanation. One begins the quest for understanding by asking, "How can a given statement be explained?" What does this mean?

What is an Explanation?

This is a "strange" world. Things are often perceived "backward" from the way they really are. If we experience "whatness" things as they are in themselves there would be little disagreement among people about what is real. We learn from experience things are not "what" they seem.

Countering the deceptive nature of appearance we find each question suggests two answers. One answer concerns the phenomenal aspect of a "whatness" object. This is appearance. We identify objects based on appearance through deduction. Identifying an object is not sufficient to understand it. Appearance does not directly tell us about the thing-in-itself. Another answer represents the visually hidden "thing in-itself." Philosophers call this "form" of whatness the noumenon. We comprehend things in themselves through inductive reasoning. The deeper aspect of whatness things is different from their appearance.

Immediate deductions based on appearance are often "backward" from the truth about the thing-in-itself. Asking why something appears the way it does, is to inquire about its deeper nature. To ask why is to seek an explanation. A simple explanation is a reason or "whyness" that accounts for the phenomenal experience of whatness objects.

This book explains phenomena. When studying, consider the following:

> Approach each statement as though you asked for a clarification of meaning. Ask how does this relate to other closely allied information?

> Information is presented step by step as though you were continually asking for more information to better understand. Ask why is there a connection? Does the succession of concepts explain at a deeper level?

> Question every statement. Ask what does this mean? How can it be explained? Why is it? To understand things, questions need to be answered. Again, things are not what they seem. A major theme in this book is that appearances oppose truth. Do not make unnecessary deductions. Ask whyness questions.

You, the reader, might have involved yourself in things other than "what" they "appear" to be. This is the right time for serious truth seekers to begin learning. Use the methodology described to assess the meaning of your experiences.

Problems, Solutions, and Knowledge

We are trying to learn the true nature of reality. Before studying explanations of reality, there are preliminary considerations. Do not become complacent in your search to understand. Do not shy from difficult situations. There is no reason to be afraid of reality. The only thing to fear is ignorance itself—ignorance about things.

Problems are for solving. Problems generate the need to understand greater things. Facing a situation acknowledges a problem exists. Problems offer opportunity to grow mentally and cognitively. Recognizing problems opens the gateway to understanding.

People experience fear because they do not really understand. By understanding, we slowly but automatically become more secure. Complete understanding confers real security.

Sailors once feared if they ventured too far they would fall off the earth. They did not understand how the earth could be anything but flat. Today we are "educated"—we know better. But do we know better?

People look back in time and say our ancestors were ignorant and backward. What applies to earlier times also applies to today and tomorrow. We ask of the ignorance of contemporary man. We suppose that people of the future will be more knowledgeable. Just as we have done, they will look back to our time and say we were ignorant. How present man behaves speaks for itself. There is still wide spread killing, maiming, stealing, and a general disregard for others. These atrocities indicate the populace continues to have insufficient knowledge to do otherwise.

What applies to society as a whole, also applies to individual members of that society. In not understanding, we are all like children. Surely, you have noticed children (and adults) sometimes do foolish things. Why? What is more important is—why do you see their efforts as foolish? To you they have just not learned enough to not be foolish. It is relative. Children do not see themselves as foolish. They just do not know any better. However, by "adult" standards they are foolish.

Where is the line drawn between behaving foolishly and behaving wisely? Just as children do not see themselves as foolish, is it not possible you do not see yourself as foolish? In many respects, we are all like children when it comes to what we do not understand. When you finish studying this book—you may realize how little you really do understand. You may not be as intellectually mature as you suppose.

Consider the following example. A second-grade student does not understand higher mathematics. He lives in a (mental) world limited by what he has experienced—limited by what he knows and understands. He knows little and understands even less. Generally, his world consists primarily of "familiar everyday experiences." Although the child has yet to understand, say tensor calculus, does not mean there is no such mathematical specialty nor does it mean it is devoid of meaning.

> Tensor calculus is a higher mathematics developed by a friend of Einstein et al. It enabled Einstein to formulate gravity as a curvature of space and time.

Not understanding now does not preclude the possibility a child may someday, with sufficient guidance, become an expert in mathematics. Yet, he continues on, usually making decisions and "acting" as though he knows and understands everything. The following explains the problem. The child's world, to him, is everything he has experienced to date. He is familiar with "regular" everyday experience. The child's error is twofold. He does not differentiate between familiarity and understanding. Granted, people are familiar with everyday experience. Familiarity is not understanding. The child does not consider the possibility he really does not understand "everything." His "mental" world will not develop as he matures if he does not methodically question the real world in which he lives.

The lesson is clear. One does not quickly become an expert in higher mathematics unless he masters the fundamentals to gain that understanding. One does not become an expert on reality overnight. Mastering any subject is a step by step process.

Problems are to be solved. Having memorized a book or subject does not mean a person is an expert "in reality." If a student thoroughly studies this book, he still will not "know" the reality. One understands by using a methodology that "connects" him to reality.

Questioning itself does not gain understanding. Anyone can question. People sometimes question others to rebuke them. Serious questioning needs to be followed by explanatory answers.

Security invested in familiarity without understanding stunts mental growth. Many people are aware of Plato's "allegory of the cave" in his work *The Republic*. People are comfortable with the familiar. Many people reach a point where they get in a rut from which they no longer learn anything new about the real world. When this happens to you, reality may start "kicking you" to learn new things about itself. How does this happen? Life's difficulties often are the most educational. The bigger the obstacle—the greater the lesson.

In facing difficulties it may seem you are "mentally falling apart" or "losing control." You probably never really "had it together": never had control. You just believed you did. If you really had it together—you could not lose it! The obvious contradiction is—if you had it together—how do you account for problems arising from applying a supposedly adequate methodology to gain understanding? Inadequate methodology creates its own set of problems. "The idea is not to get your act together, but to get it together by getting rid of the act."

People react to "trying" situations in one of two ways:

> Hide from the "real challenge." People "turn" to religion or some organization providing unverified and unexplained answers.

> Face situations by questioning its meaning and why it is happening. Do this by studying and understanding the scientific method until you can comfortably incorporate it into your life. Try answering why you are confronted with difficult situations.

Is it not wiser to trust a method that determines truth on reality's terms instead of trusting one's "foolish method" using beliefs?

Problems "push" us to learn and understand the world in which we find ourselves. The more we understand reality the more it helps us. No one is taken unwillingly from "Plato's cave of ignorance" to have knowledge forced upon them. There is "real" knowledge. This book outlines knowledge and its understanding in "explanatory form."

This book describes many concepts that the "backward people" of today do not understand. Many concepts are not easily explainable. Study how each concept is scientifically verified and logically proven. Concepts that are not contradicted probably represent the way things really are.

There can be no incoherence or contradictions in information (knowledge) representing reality. If there is just one reality all that exists is a part of this one reality. All its parts, like pieces of one big jigsaw puzzle, fit together just one way. Parts coherently intermesh. This is not unlike a road map representing real roads. If there is a map error, though minor, you might not reach your destination. Constantly question whether you tread the intended path. If you cease questioning, you are not even on the path.

People believe when they do not know. Not knowing causes fear. Do not fear what you do not know, fear because you do not know. There is a difference. The "unknown" tends to cause insecurity. People avoid the unknown—avoid facing it. Interestingly enough, instead of avoiding the things that cause insecurity, if people would face the unknown, they might begin to understand it. With real understanding they would no longer have "reason" to fear and hence be insecure. The very thing causing them to be insecure, is the very thing, which if understood, would bring them real security. This is another perspective on why we encounter problems.

All unknowns exist in this one reality. The goal is to find a method to know and understand.

Example, those backward sailors of yesteryear believed the earth was flat. They just did not understand otherwise. Surely the earth appears flat. There were apparent contradictions. As ships sailed farther, they seemed to sink below the horizon. Today we know they went far enough that earth's curvature intervened between shore observers and the ship. The invention of the telescope enhanced this observation.

How could Magellan's or Drake's voyage around the world be explained when their ships did not reverse course? Knowledge accrued. A "pattern" formed. It was "explainable" by supposing earth was spherical and, after Newton, that physical bodies are attracted to earth by a "force" called gravity.

All things interrelate. Comparing is the initial process in seeking explanation. The next step incorporates a given thing into a higher-level concept (explanation). Once a connection (between a phenomenon and its explanation) is made, the next challenge is to broaden the contextual basis of understanding. Let's elaborate.

Context

The idea is to develop the explanatory question "how can this be explained," into a methodology capable of answering it.

Hierarchical structures are context intensive. We understand statements by context. Context is bidirectional. We learn context by specific examples. This is low-level context. Context defines or categorizes specific examples. Universal context is grasped by higher-level generalizations of which the idea is a more particularized example. The idea is to build an all-inclusive-hierarchical structure. Observe how concepts (described in this book) are hierarchically defined.

The structure in this book offers a high-level context for (understanding). Each chapter is a unit covering a particular subject. Each unit is "defining" information and supplies a certain level of context. Observing how subjects interrelate produces rudimentary higher-level understanding.

If serious about learning, continue studying. Study science for an overview. Study particular-scientific disciplines. The more you study science, the more you will learn of the evidential basis for the low-level context described in this book.

Study logic to better understand. Attempt to "contradictorily" refute statements in this book. Next, study philosophy to learn the "big questions." Observe how constructural conceptualization answers the "big questions." The structure of reality helps understand high-level context. Its immediate-lower levels are highly generalized concepts.

There are many questions to ask. Are major religious questions amenable to scientific scrutiny? Study religions to see how highly-generalized-scientific concepts correlate with major ideas of religion. Ask whether the presented structure verifies major religious concepts.

With elementary understanding, continue to the next section (of higher-level structure) that envelops that understanding. It is a deliberate and step by step process. Try to refute every major step in the assembling of this structure. If a step is refutable, it probably does not correlate with reality. This is real learning. Now let's be more specific.

Definition, Interpretation, and Assertion

Socrates used a dialectic now called the Socratic method. It is easily used in personal exchanges of dialogue. It is useful for avoiding misinterpretation. When a teacher defines a word or statement, the student can question it. By questioning the student learns. The presenter defines each step of his "explanation." The questioner attempts to refute each step by suggesting possible errors in reasoning.

The student suggests reasons the teacher's definitions may be invalid. Typically, a student's questioning reflects a lack of understanding. The teacher explains why the student's concepts are based on faulty reasoning.

Learning the definition of ignorance is the first step to knowing. If the student cannot explain his assertions, he probably does not understand. If a student merely "believes" what he states to be correct, the teacher suggests the student has no scientific or logical basis for his assertions. The teacher then explains how the student's statement leads to contradictions. The student learns he does not really know of what he asserts. One knows a statement to be true only when he can verify it by reality. We are developing a methodology to understand reality. If we understand, we can explain.

It is easier to explain in person than by written word. In person, teaching is tailored to a student's level of knowledge and understanding. The author cannot individually teach every person.

The written word has advantages. A student can pace study to his ability. Interesting enough, if a student seriously seeks truth, his intellectual ability increases.

Study this book as though you are an "objective outsider." Be scientific. Do not allow "feelings and emotions" distort what you are trying to understand. Read once to get the general idea (for context) then reread with context in mind.

Wisdom and Foolishness

While studying, carefully consider the following:

> The "Fool's Fatuity." Do not commit the error of "using what you do not know" (beliefs—what you think) to gain an understanding of what you are trying to learn (know). Do not judge with preconceptions what you will be reading. Do not say, "this statement is not true." But also, do not say it is true. One cannot learn by using ideas (concepts) not scientifically verified nor contradictorily defined. One does not learn about reality by this "method." Human nature hinders the search for truth.

> The "Wise Man's Wisdom. " Ask, "what is the meaning of these statements?" Why is this? Is there any evidence or reason (in what is already known and hence scientifically verifiable or noncontradictorily defined) which either supports or contradicts a given (hypothetical) statement's truth? Learn by using verified concepts. If not verified, a concept should be defined noncontradictorily. If properly defined, it should directly correlate to other verified concepts. This method leads to real learning. The Wise Man's Wisdom is a variation of the personal explanation principle (PEP).

These definitions are criteria for distinguishing foolishness from wisdom. The difference becomes more discernible as you study this book.

To understand statements in this book, forget personal ideas about things; instead, further study science. You can stay on the path to greater knowledge and understanding. There is no need to be foolish. This book explains simple basic concepts, which, if understood, should get you on the path. The following focuses on the methodology needed to stay on the path.

The Path

The Path is applying the "methodology" to experience. Getting on the path to greater knowledge and understanding is easy. Staying on the path is very difficult, not because of the methodology, but because of distractions. One usually "stumbles" onto the path by trying to understand everyday experiences by concepts amenable to contemporary-scientific verification.

Initially, the mind is not developed sufficiently to handle problems other than those of everyday experience. Routinely solving everyday problems sets you on the path. Getting on the path does not mean you will follow it. If you reason to solve and understand the "meaning" of everyday experiences, you begin to travel along the path.

The mind develops by reasoning to understand meaning. The Wise Man's Wisdom suggests one cannot get ahead of himself because of the following: To understand greater things (ever enveloping generalities) the truth seeker must constantly refine the methodology used to ascertain verifiability. Be "open minded." These are the basics:

A.   Analyze each "interesting" phenomenon. Break it down into simple parts. Here is a lengthy examination of this point.

Science excels at analysis. We wish to understand life. Analysis is our first task. Can analysis yield a satisfactory explanation of life? We ask particularly, can life forms be explained by biochemistry? Does biochemistry adequately explain manifest "form?" If we answer yes, then we have another problem.

This is a general pattern associated with any attempt to explain a given system by denoting its parts. The problem, called reductionism, is that parts do not fully explain the whole.

For example, let's say we have already "reduced" our understanding of life forms to bio-molecular structures like DNA. DNA is "genetic material." This implies DNA "encodes" life's structure and functions. The encoding process suggests reductive mechanisms are responsible for life itself. If we accept this "explanation" of life, we might state genetic material explains life by "blue-printing" the structure of life forms.

Life forms are composed of molecules, but do molecules explain life? Particularly, does molecular structure explain higher-level functions? Let's test the answer. We ask how can genetic material explain life when we can also ask for an explanation of DNA? We reduce DNA molecular structure to its parts—atomic structure. Let's simplify our task. Let's drop the idea of function—that our form (physical body) has the function of life. Our reductive answer (DNA) now stands in need of a "deeper" explanation.

We ask specifically, can atomic "energy fields" explain the "form" of DNA? Is it not the structure of DNA that defines it? More generally: do atomic-energy fields "explain" the form of manifested structures? In the same reductive direction, we again ask how is atomic structure itself explained? Do parts (energy packets) explain their arrangement into higher-level forms?

The same shape (form) can be copied in different substances. Consider a pet cat. We have artistic representations of the cat-form in plastic, aluminum, iron, and wood, and so on. Although model representations do not exhibit life, they do replicate form. Formness is what we are trying to explain. Although model content differs, its form is the same.

Energy quanta are an inadequate explanation of form. Reductive mechanisms only "appear" to explain substantial manifestation. A plastic-formed cat seems to be explained by reductive analysis. It is made of polymer chains. We ask, does the reductive structure explain how higher-level cat-forms come into the arrangement defining it? Does reductive analysis yield a satisfactory explanation? Let's continue our analysis.

A pure metal, like iron, has an immediate very low-level-reductive form—atomic structure. An atom of iron is not directly visible to our senses. Anyway, an iron atom cannot be shaped into the form of a cat. We already discern a problem in this approach. It is unreasonable to suppose low-level structures can explain higher level form. Let's dig deeper. With the benefit of modern science, we can further analyze atomic structure.

Physicists have determined a nucleus (which defines atomic identity) is composed of lepto-quarks. Let's make our point. Do lepto-quarks explain atomic form? If so, how? Let's generalize our question. Do lower-level structures explain higher level forms? Let's continue our analysis.

Do lepto-quarks explain the difference between a plastic cat and one constituted (made?) of iron? Here we address a difference not of form, but of content (actually, the form of content at a lower level). This evades our direct question. At the "bottom level" our variously constituted cat forms are all composed of the same "energy stuff" (lepto-quarks and "quantum packets of energy"). If there is no difference between a lepto-quark in one atom and another, how can this very low-level structure explain any form? If this is an unsatisfactory explanation, how can it explain the difference between a cat modeled of plastic and iron, let alone explaining the difference between a "live" cat and a dead one?

It is easier to use examples directly experienceable instead of invisible elements. However, there is a significant benefit in considering low level parts as a possible explanation. All lepto-quarks, within their limits, are indistinguishable. We ask how can that which in itself is all reductive sameness, explain distinguishable form? The only possibility is that it is imbued with even higher intelligence than higher-level structures. If that is the case, how is it to be explained? If mere high-level forms cannot be explained by lower-level structures, then neither can those forms, imbued with life functions, be explained by lower-level structures. Our analysis reaches a dead end. Reductive analysis yields an unsatisfactory answer to the question "how can life forms be explained."

Energy quanta do not "explain" form! Energy quanta neither explain nor suggest a reason for existence—regardless of form. Therefore, it is not energy packets that do "the forming." Rather, somehow energy packets are "created into form!" Because reductionism fails to explain form, we must seek and refine the methodology to gain a deeper understanding.

Higher-level forms are not explained satisfactorily by their structural composition. It is how low-level parts are arranged into higher level form that needs explaining. We conclude if energy packets cannot explain life forms, then neither can DNA. It is DNA that needs explaining. And, it needs to be explained by means other than by reductive mechanisms.

Obstacles appear to hinder truth seekers. An interesting "side" question is: "why has man deduced (via reduction) the proper constituency of form but has not found a satisfactory explanation of form?" This is one of the main questions addressed in this book. A little about it is mentioned here. If we can explain form, we may also be on the path to finding an answer to existence itself. Is there meaning to man's existence? Because this question has not been settled, we suppose, if there is an answer, there must be obstacles in the way to learning what it is. If we recognize some of the obstacles, we may be able to overcome them and continue our journey along the path to greater knowledge and understanding.

One obstacle is the mind. In this existence, the mind primarily functions deductively. Scientists do well at "reducing" (specific) phenomena to their lowest common denominator (energy packets) but have trouble "piecing" the "reduced" parts into higher-level concepts.

This problem arises again in cosmological physics. Some cosmologists (Paul Davies, Parker, and others) have concocted the Anthropic Principle. They tried to rectify the idea that the "form" of the physical universe (for example that it expands) happens in a way that allows the universe to exist. And, that it exists in such a way we can observe it. This is a tautological explanation. We are speaking of the nature of existence, i.e., that form exists. Let's get to the other basic ideas about our methodology.

B. Ascertain meaning of form. Represent part-forms in the "form" of knowledge. Hypothesize and generalize knowledge of parts into categorical wholes. Therefore, classify according to similarities and differences.

Scientists struggle with this because their minds do not function inductively. This is where "order" is determined. An excellent example is the taxonomic classification of plants and animals. Taxonomy, though a categorization of life forms, does not explain life.

C. In examining order or "pattern," see whether it suggests some type of underlying cause or reason for phenomenal form.

Reason explains phenomena. By sufficiently understanding the nature of cause, we come to realize the reason for phenomenal form. Reason must correlate with higher-level wholes. We find lower-level explanations do not satisfactorily explain higher-level wholes. We are left with the only other option: Higher-level wholes must explain lower-level structures (or "lesser" form). The whole explains the parts.

D. Do not believe you have finally found the answer. Do not become "security attached" to "what" you have learned to date. Be objective. There is more to learn.

Do not rest on contemporary understanding if big questions are left unanswered.

E.   Be patient yet persistent. It is a slow step by step process.

Do not get ahead of yourself. The path to greater knowledge and understanding is an arduous journey. Staying on the path calls for adherence to the Wise Man's Wisdom.

Do your homework; learn what is already scientifically understood. If you apply these ideas, you are ready to understand greater things. Once you understand the methodology, only obstacles stand in the way. Obstacles force us to refine the methodology. Refining methodology allows you to further traipse along the path.

You, the reader, may not trust the methodology enough to begin your journey to greater knowledge and understanding. Not doubting the methodology is also a problem. To doubt is to question. Question the methodology. Try to find an error in its use. If you find no reason to distrust the methodology; then, tentatively accept it and apply it. It works. The idea is to apply the methodology to the obstacles. When we finally overcome an obstacle, we learn that our problem was a shortcoming in our methodology. There is no problem with reality. Most problems can be traced to an inadequately developed methodology used to obtain understanding. This occurs because there is a lack of mental development. Yet, it is by solving problems that the mind develops.

Attachment

People become distracted from seeking truth and the security it brings. What are some obstacles to seeking truth? Human nature is a hindrance. For example, getting married is often a means to escape understanding greater things. That is a selfish "reason" for marrying anyone. Yet that is why people usually marry—to have someone take care of them—that is—their security. The logical absurdity in this is shown in the following. Each person marries the other for security. The question is how can a man have security in a woman, if the woman also marries the man for security? More succinctly, how can an individual have security in someone who is so insecure that they cling onto the other for security?

Insecurity causes people try to control their spouse by dictating or by physical mistreatment. They use their "security-blanket" spouse for their own security. This is the need to be taken care of: wanting to be wanted. This is why security dependency on anything other than oneself does not work.

Instead of pushing to understand insecurity (people just do not know enough) people hang security-wise (attach) onto something. They substitute one security-blanket for another. A security blanket is no substitute for real security. Why do people turn to religion? For security. That is a selfish reason. When one avenue of security dependency fails, others are adopted. "Outside" security blankets will fail.

The mind needs a proper methodology to seek truth. Using proper methodology—real knowledge is obtainable. The answer is not "out there!" The answer is found in the inner self—by using one's mind. Knowledge is "in you." You cannot lose it. If obstacles do not pressure you, you have no reason to learn enough to be responsible.

Reality constantly tests you. You experience pressure, conflict, and bewildering situations. These are symptoms of obstacles hindering progress along the path to greater knowledge and understanding. But these very experiences are reasons to understand. Instead of seeking "why" this (testing) happens, most people look for another "outside" security blanket.

One moves little along the Path to knowledge unless he accepts responsibility for himself. Only by accepting responsibility for one's existence, does mentality develop to a higher level. No one changes unless reality applies pressure.

"Hanging on to" someone or something (for security) mutes progress in obtaining knowledge. Seeking knowledge is painful. One becomes realistic by being self-responsible. The path of least resistance is not the path to greater knowledge and understanding. One can have a "real" relationship with someone without being sidetracked from seeking truth.

Marriage is not the only obstacle to security. Many people hang on (attach) to others for security. Many follow religious leaders. Leaders are followed because they know supposedly what they are doing. Do people apply the Wise Man's Wisdom and seriously question their leaders?

The self does not develop when it relies for security on things outside itself! In this existence, to become realistic is to be pulled from (the attachment to) things and groups (i.e., their beliefs). A corollary of a law of physics is "you do not get something for nothing." On the path to greater knowledge and understanding, you may suffer some. You suffer when pulled away from security attachment. These things (groups and things outside yourself) are not you (i.e., your mind).

What is the test of attachment? Simple: avoid the thing, which you unwittingly attached and observe what happens. (Observing and testing are the hallmarks of science.) Do you experience a withdrawal? Do you feel insecure? The error is a common one. As long as you "feel" secure you believe you really are secure. Again, you cannot trust your "own" wisdom, feelings, beliefs, and preconceptions as to what is real. The alcoholic consumes alcohol and feels secure. Prayer and church attendance make people feel secure. Meditation helps others. Prove these activities make you really secure. Ask a question. Do these activities give you knowledge about reality? If you say yes, your next task is to verify and prove it. Remember: truth and proof are inseparable.

Suppositions

Suppositions are a beginning to truth seeking. Try to verify what you suppose. Most people do not verify what they suppose because they are too insecure to question things.

What is the nature of the self? The self is defined by those beliefs that define one's identity. If one is a "Christian," he will not question Christianity. Even religious concepts fall into the category of everyday experience. They too should be questioned.

Big questions are not easily answered. High-level questions cannot be answered until all lower-level questions are solved and understood. If one does not understand the meaning of everyday experience, he will not develop the refined methodology needed to tackle the big questions.

The author does not suppose truth. He explains by using enveloping concepts. These concepts are used to answer philosophical questions. The answers are then correlated with religious concepts. The author continually sought the deeper meaning of existence by eliminating contradictions in what he already understood. He applied the Wise Man's Wisdom. He learned about "forms." Does form have structure? How can structure be explained?

If concepts presented in this book are true, it is because reality asserts that truth. Concepts can be verified experimentally. Higher-level concepts are established by logical coherency. Coherency is sought by reasoning to eliminate contradictions.

Asserting expertise and being an expert are not synonymous. Anyone can claim expertise. Many people are experts in some discipline. Real experts need not claim anything. One should be leery of those who make claims and argue about their expertise yet produce no real verification or noncontradictory explanation of what they suppose is true.

Experience does not confer understanding. Familiarity is not understanding. Some people can do interesting things, but that proves little. Some people say, just experience this and you will understand. Touching a hot stove is different from understanding pain.

If a person understands something, they can explain it (without contradiction). The physical world is experienced through the physical body. The mind questions. If we ask whether soda exists (in a refrigerator) we need not argue the case. Just ask someone to open the refrigerator door. Either there is or is not soda in the refrigerator. Scientific verification is letting reality speak for itself.

The idea is to understand things at a deeper level. There is no problem with what we experience. The problem is interpreting what we experience. In seeking truth we focus on definitions. Do they correctly represent reality?

There should not be any contradictions in a definition if the statement defines what is real. Feelings and emotions often fool people into believing they are experiencing something that is not what they suppose. The appearance of something often misrepresents what it really is (in itself).

The problem is the real self (one's mind) is not very well developed (when it inhabits a physical body). Most situations we face are not immediately understandable. Why a phenomenon occurs is found by mental effort. Ask what does this mean? How can it be explained? The mind develops by seeking to understand.

Either you control the situation, or automatically the situation controls you. The only way to control any situation is to understand it. If you do not understand, you will be controlled.

Key Points:

- We conceptually understand by interrelating things.

- Concepts categorize knowledge according to phenomenal similarities and differences.

- Interphenomenal relevancy suggests constructing a multi-level explanatory structure. We call it the "why-what paradigm." The why-what paradigm "classificatorily" configures the five basic questions: why, what, where, how, and when. This construct conceptually represents the "shell" structure of reality. The why-what paradigm superimposes explanatory meaning upon interphenomenal relevancy. The why-what paradigm generates an explanatory system.

- The mental process of deduction "directionally" follows inferences from whyness factors to their whatness particularizations.

- The mental process of induction synthesizes pieces of experience into unified conceptualizations.

- Aside from memory, we learn from schooling, personal experience, and inductive reasoning.

- Problems push us to understand: to explain.

- People often deceive themselves about the "appearance" of things.

- Scientific explanations answer why (something) is. Each explanation entices us to ponder greater problems.

- It is foolhardy to use what one does not know to gain deeper understanding. This is embodied in the Fool's Fatuity.

- It is wise to build a conceptual understanding of things upon what has been explicitly defined and verified. This ideation is represented in the Wise Man's Wisdom.

- The Path to knowledge and understanding is obtained by applying a carefully reasoned methodology to experience. Define and test statements to determine whether they correctly represent reality.

- The operative methodological question is "how can 'this' be explained?"

- Do not become "security" attached to ideas or things. Develop a methodology by inductive reasoning and continually test its efficacy. Do not let feelings and emotion have undue influence upon the quest to seek deeper truth.

The standards listed in Chapter One are arranged to reflect the intellectual shift from deduction to induction used to delineate the structure of reality.

The standards and modal system (of ascribing class characteristics) together lend themselves very well in building a converging hierarchical arrangement of whyness factors and whatness factors.

The modal system was not used to acquire the knowledge of reality. That was only accomplished by intensely studying science, and reasoning from well-established evidential facts. How the available pieces of fact fit together yielded the rest of the information. The interrelating of phenomena merely took a form that is easily characterized by ascribing class properties that can be explained using the modal system to describe whyness-whatness associations.

The trick has been one of intermeshing all phenomena by coherently representing them as general concepts. The concepts took shape by connecting scientific generalities with the five basic questions (where, how, when, what, and why). Functionally, phenomena are denoted by the level of universalization of their ascribed properties. This becomes more apparent as we proceed. Whyness-whatness modes (categorically) show how phenomena (or phenomenal classes) "fit in" with other phenomenal classes much as how a jigsaw puzzle piece fits in with other puzzle pieces.

Understanding is the product of perceiving relationships. Meaning is found in relationships. We describe relationships in their most elementary form—that of diagrams. Scientific constructs can represent the structure of reality.

One preliminary question remains. How are you, the reader, to approach this book? To rely on the self is to use the mind. Essentially, the self is the mind. If you study this book, question every statement. Accept responsibility for yourself. Refine your methodology to gain a deeper understanding of reality. Apply it introspectively. Examine everything as though you are an objective outsider. You are on the threshold of real-mental development and real learning.

Knowledge begets more knowledge. The more you learn about reality—the more your mind develops. The more it develops—the more you will learn. The idea is to learn enough to become self-secure.

How do scientists learn about things? They break things down into parts to see how they are put together. They determine structure. Then they examine the parts to learn how they function. This is the same tactic employed to understand reality. Determine structure and function.

Descriptions support the structure in this book. Some areas of study are described for purposes of "categorical inclusion." Therefore, each area of study mentioned is categorically explained by higher-level structures. Quarks are explained "categorically" by nuclear structure, which in turn is explained by atomic structure. The idea is to supply a sense of how things interrelate: both with other closely connected phenomena and with how each category correlates with the hierarchical structure elucidated in this book. Chapter and Section arrangements correspond to their categorical inclusion in a hierarchy.

To gain an understanding of reality is to look beyond oneself. Behavior needs to be well understood so that one's behavioral quirkiness and ignorance do not distort what one is attempting to learn about reality. That is the idea behind the Wise Man's Wisdom. Therefore, chapters five through eight describe basic behavior. A clear understanding of behavior puts us on the Path to delineate the deeper structures of reality. Once on the Path, physics principles become the essential clues to the structure of reality. Let's begin our journey into the mind expanse.

Section Two

REALM OF WHATNESS

Perspective

- Where we are: universal location: our place in the scheme of things
- How we are: what it means: understanding behavior
- When we are: meaning of the present
- The realm of experience

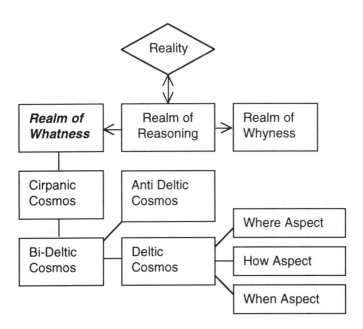

Figure 3:1. Perspective: Whatness

(Normal type— indicates topic to be discussed. Italicized-bold— indicates current topic of discussion. Normal-bold — indicates topic previously discussed.)

Part A: Cosmos

Reality is All-inclusive and therefore, absolute. It has no extrinsic characteristics and hence no extrinsic relations. There is no external reason to reality. It cannot be understood relative to something outside itself. It just is. Reality can only be understood by delineating its internal structure. If there is a supernal reason to what we find in reality, it too must refer to some basic aspect of the structure of reality. The basis for our structuring is the why-what paradigm.

Reality can be described by juxtaposed whatness and whyness factors in a graduated modal system. Reality *altogether* is designated as the answer to the widest generality of a "What" question that can be asked or "What is All?" This is an interrogative representation of the widest class mode.

Structure is understood by interrelating properties ascribed to classes of phenomena. Class properties are intrinsic aspects of the widest class mode—the "All mode." The All mode can be resolved into composite Whatness and Whyness modes. The intrinsic whyness mode pertains to the widest noncircular why-question that can be asked. The answer to this why-question denotes the Realm of Whyness. It relates the supernal reason for "everything." The intrinsic whatness mode is split into two sub-whatness modes (an oversimplification later to be rectified) which in physics' terminology is designated the Cosmos and Anti-Cosmos. Cosmic structures collectively denote the Realm of Whatness. Everything is represented as some aspect of the Realm of Whatness. Between the Realm of Whatness and the Realm of Whyness is a factor called the Realm of Reasoning. It is a connecting factor entwining the Whyness and Whatness factors.

The Realm of Whatness is composed of a few substructures. The highest-level substructure is the Cirpanic Cosmos. The Cirpanic Cosmos is composed of the Bi-Deltic Cosmos represented for every possible temporal instance it inhabits. A Deltic Cosmos and Anti-Deltic Cosmos constitute the Bi-Deltic Cosmos.

The Deltic Cosmos, as a whatness mode, is further divided into three aspects corresponding to the Where, How, and When questions. It is composed of the three universes: the Where Aspect, the How Aspect, and the When Aspect.

The Where Aspect defines "whereness." It suggests categorical placement in spatiality. The How Aspect concerns order. Order is a logical intelligence attribute of whyness (or reason). The When Aspect relates change or time (temporal flow).

These aspects are described in the sequence they were inductively integrated into the more encompassing mode (Deltic Cosmos) of which they are a function. Let's get to work and examine how the higher-level structures are built from the lower-level structures.

Where Aspect

Perspective

- Where Aspect defines three Universes
- Universe of Perfect Whatness is physical matter
- Universe of Imperfect Whatness is deductive mental function
- Universe of Imperfect Whyness is inductive mental function
- Higher level mental functions decipher greater structural "layers" of reality
- Structural levels are represented conceptually
- Each conceptualization is a partial self realization
- The Where Aspect is the first aspect of the Deltic Cosmos structure

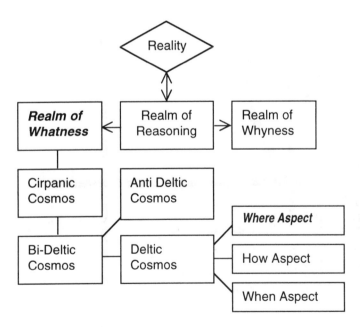

Figure 3.2. Perspective: Where Aspect.

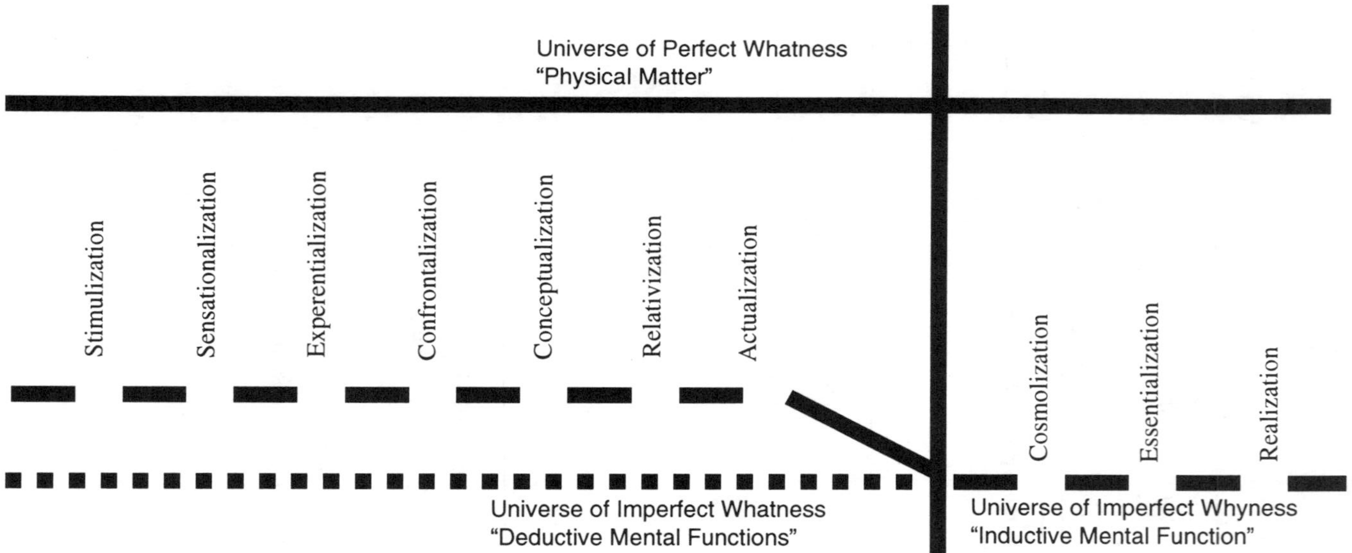

Universe of Perfect Whatness
"Physical Matter"

Universe of Imperfect Whatness
"Deductive Mental Functions"

Stimulization

Sensationalization

Experientialization

Confrontalization

Conceptualization

Relativization

Actualization

Universe of Imperfect Whyness
"Inductive Mental Function"

Cosmolization

Essentialization

Realization

This is an abbreviated version of the original Where Aspect Diagram. Its basic structure is a cross. Three universes are related. The Universe of Perfect Whatness occupies the second quadrant. The Universe of Imperfect Whatness lies in the third quadrant. The Universe of Imperfect Whyness is represented in the fourth quadrant. The diagram lists different levels of conceptualization. Each concept represents different levels of awareness processed by either the deductive or inductive inferential function.

Figure 3:3. Where Aspect Diagram.

Chapter 3

Where Aspect

"The things causing insecurity, are the very things
that, if understood, would lead to real security."

The Where Aspect of the widest intrinsic whatness mode entails "where" a
being (individual) having the characteristic of awareness (mind) is located
(conceptually) in reality.

The Where Aspect is divided (for analytical purposes) into three different
universes based on phenomenal relevancy. Each higher level of
conceptualization represents a greater delineation of reality's structure. Each
level (representing structure) not only recognizes greater understanding, but also
better defines man's role in reality.

Each conceptual level has its own relevant properties and relations that
characterize it as a phenomenal class. Intraphenomenal relevancy differentiates
one conceptual class from another (within one intrinsic unit). In practice there is
no sharp boundary from one level to another. Each encompassing concept
merges with others.

A seeker of truth cannot understand a conceptual stage until he reaches a high
level of "universalizing" many generalities. These generalities lead to an
ordering of concepts into coherent phenomenally relevant arrangements. The
Where Aspect Diagram is an example.

One's mental processes (of inferring) must shift from deducting to inducting to
understand how "high level" concepts (of enveloping generalities) are
constructed. Why someone experiences induction is addressed in the latter
chapters of this book.

Although the three main universes are easily identified through their
intraphenomenal properties, they are categorized and named according to
interphenomenal-class relevancy. This scheme interrelates the universes as
aspects of the Whole (of reality) by using the modal system. This also
exemplifies the various levels of inferring (deductive and inductive) needed to
construct an understanding of reality's structure. (Using "Whatness" and
"Whyness," with deduction and induction, expresses this relevancy.)

Some brief definitions:

The major category "matter" (or the class of physical composition) is
denoted the Universe of Perfect Whatness. Each person has a physical body
in this existence. The body is denoted the "individual" Universe of Perfect
Whatness.

A major class of mental function—"deducting"—is called the Universe of
Imperfect Whatness. In the physical existence a person's mind functions
deductively. "What" we experience is an "imperfect" representation of the

physical world. A person's deductive processes we denote "individual" Universe of Imperfect Whatness.

Another major class of mental function—"inducting" is called the Universe of Imperfect Whyness. The inductive reasoning process is an "imperfect" representation of the Realm of Whyness. A person's inductive processes are designated the "individual" Universe of Imperfect Whyness.

Where Aspect Diagram

The Where Aspect Diagram describes the Where Aspect of the widest intrinsic whatness mode. This diagram also characterizes the answer to the widest "where" question that can be asked. It represents the "where" of reality as structurally answered through the whatness and whyness modal system.

The diagram is divided into quadrants. The upper right quadrant is not relevant for describing the whatness aspect of cosmic structure. It represents the Cosmos as an essence (i.e., the Realm of Reasoning). It is described later.

The upper left quadrant represents the Universe of Perfect Whatness. All physical matter (atoms, molecules and any "solid" substance) is categorized as the Universe of Perfect Whatness.

The lower left quadrant denotes the Universe of Imperfect Whatness. All individual minds functioning deductively are categorized as the Universe of Imperfect Whatness.

The lower right quadrant corresponds to the Universe of Imperfect Whyness. All minds functioning inductively are grouped in the Universe of Imperfect Whyness.

It should be noted: Although all three universes collectively represent a third of the immediate cosmic structure, they do not function apart from the other two aspects (the When Aspect and How Aspect).

Let's proceed with an elaboration of the diagram's significance. Most emphasis is directed to an understanding of Universe of Imperfect Whatness and Universe of Imperfect Whyness. How the "where concepts" interrelate by varying degrees of deductive and inductive inferring (reasoning needed to assimilate them) is discussed in Chapter 4: the "How Aspect."

Universe of Perfect Whatness

Matter has been studied thoroughly. So much is understood about the properties of matter that little is mentioned here. Consequently, this book does not elaborate the gross phenomenal properties of the Universe of Perfect Whatness. We are concerned, however, with the interphenomenal relevancy that categorically describes material properties.

Concepts of material things infer spatial extensionality. Things do exist in space. Unnecessarily "mixing" different levels of generality (referring different hierarchical structures) is confusing. Things are not space. Things also "exist" in time. Things are not time. The Universe of Perfect Whatness does not denote space or time, but only physical matter (rest mass).

Relations between (rest) mass and time depend on the principle of the constancy of the light speed and the principle of relativity. The time factor is considered an aspect itself aside from mass. A similar separation is implied in the scientific problem of defining time in a manner not conflicting with Special Relativity and symmetry-relating experiments. Especially noteworthy is the theoretical attempt to construct a time-reversal theory that does not contradict the ideas of charge conjugation and parity in elementary physics. The idea is to erect a structure not conflicting with acceptable physics. The structure should also better explain contemporary physics.

The Principle of Relativity (and its verification) dealt a final blow to any idea of ever understanding something except through a proper standardizing of interphenomenal relations. This is epitomized by considering space (distance) and time as two "aspects" of some "one-thing" (space-time). Yet, there is more to cosmic structure than matter-time. One could, for example, concoct a theoretical conjunction of mind-time. However, such a construct is not an aid to understanding. An extenuated dilemma is avoided by considering space and time in a separate context from both matter and mind although they all operate in concert.

Our task is to analyze phenomena to learn how things "fit together." This is relevancy. We then piece together these "parts" into higher-level structures to understand how things work (or function). If we can properly define things by function, we should be able to address purpose.

We stress again that the Universe of Perfect Whatness pertains to a mass-cosmic aspect without space or time. There is also a difference between the Universe of Perfect Whatness and what is known about it. What the Universe of Perfect Whatness is itself cannot be known directly. It can be inferred by the mind. Inference concerns the Universe of Imperfect Whatness and the Universe of Imperfect Whyness.

Universe of Imperfect Whatness

The Universe of Imperfect Whatness represents the undeveloped mind (compared to the inductive ability of the developed mind). This mind functions deductively. It uses the premise of "sense data" obtained through stimuli generated by Universe of Perfect Whatness. This mental function is responsible for behavior usually associated with human nature. Human nature reflects the interplay of the deductive function with feelings and emotions. Examination of the "deductive" mental function is deferred until discussion of the "How Aspect." For the present we describe only what this mind uses for security. It acquires knowledge of physical reality.

Striving for self-security is reflected in the question: "Who am I?" The more an individual can answer this question by using induction premised on physical reality (Universe of Perfect Whatness) the more he acquires real meaning in answering it. This meaning only comes from understanding how man "fits" into the design of reality. Design relates man's phenomenal relevancy. To understand design calls for an extensive knowledge of the Universe of Perfect Whatness. And that is the task of Universe of Imperfect Whatness.

The mind can be "awakened" by its interaction with the Universe of Perfect Whatness. The individual (as a mind) can study and understand "whatness" by conceptually ascribing and aligning properties. Concepts form depicting relations. Relations reveal relevancy.

One's reasons for understanding reality must be unselfish. If need for understanding is self-motivated, it becomes a desire. That creates conflict. We observe conflict when one's attention is not fully focused on reality because his reason for doing so is self-serving. "What he does" is directed away from himself but "Why he does it" points to himself. This creates a gap between what one is doing and why he is doing it. It is a gap not easily closed.

The individual Universe of Imperfect Whatness (deductive function) centers attention upon itself instead of focusing on the surrounding world. This is indicated by the following type of statement: "*I want* to do this!" However, if someone uses the "personal explanation principle," his attention shifts from what he is doing to why he is doing it. The reasons for one's behavior is reflected by "why" questions. Reasons (causes and explanations) are intangible to most individuals who mentally infer through the deductive mode.

Order (in the material world) is the conceptual basis for understanding of man's place in the scheme of things. The trick is to erect high-level concepts that further define man. Each conceptual level is an "ordered" class unit of phenomenal relevancy. Each level represents a partial self-realization. Here the focus of one's perceptions is reality-oriented. (Note: The idea the self can be conceptually defined is found in the work of the American humanist-psychologist C. Rogers.)

Conceptual levels (pivotal points of the Where Aspect Diagram) of understanding the self (via encompassing realizations) are:

(1) Self-Stimulization

The environment stimulates life. Animals and plants are physically affected at this level. A plant leans toward sunlight (heliotropism). Plant roots grow toward water. A Venus fly-trap closes on a victim. These are all examples of an organism's contact with its environment.

(2) Self-Sensationalization

The mind becomes aware of physical stimulation. The self is cognizant of sense data. Awareness develops. This level differentiates plants from animals.

(3) Self-Experientialization

The mind recognizes patterns of sense data. Memory develops and self-identity evolves.

(4) Self-Confrontalization (self-confrontal realization)

The mind has already encountered its environment through the cognitive reception of stimuli. The mind is confronted by "something outside itself." From this "something" emerges sensory-images. The mind attempts to give meaning to these images by comparing their similarities and differences. At first these perceptions are vague and inaccurate representations of reality itself. Interaction with one's environment enables the mind to acquire the notion of "what something is."

This level of "learning" probably begins before physical birth, and ends with the individual having the simple idea he is somehow different from his environment. "Thinking" (the deductive function) is very weak at this stage. This level of experience prepares the mind for development.

Self-confrontalization culminates in the individual having differentiated himself from the outside world. An individual may consider only his body to be the "self" because it is through the body that his mind interacts with its surroundings. One's survival or security also depends on his environment. The completion of self-confrontalization serves as the initial basis for developing self-concepts.

(5) Self-Conceptualization (self-conceptual realization)

This level concerns "thoughts" about experiences. Ideas develop from one's ability to cope with and manipulate his surroundings. He ascribes "properties" to what he experiences and develops perceptions about them. Encountering similar experiences, he more easily remembers a typifying representation for future referral. He develops perceptual references as guides that facilitate the handling of his ever-widening encounters with reality. He slowly defines himself by what he can do with his environment. This is another way of saying he is self-defined by his perceptions of reality.

The self develops concepts that relate a personal interpretation of his experiences with reality. These (conceived) ideas, however, may hinder the search for real security. They may psychologically block what an individual could do. The real meaning of experience is subverted when one incorrectly ascribes properties to "perceptual objects." This happens when one misleads himself about the reasons for something. Not understanding does not justify using beliefs to characterize something. Believing shows lack of understanding. Misconstrued concepts hinder an individual's search for real security. (More about this later.) Using beliefs to characterize experience stunts the mind and hinders understanding beyond (the level of) self-conceptualization.

Concepts should correlate directly with experiences of reality: not to beliefs (about reality). The mind naturally adjusts to the need to understand. Slowly, the mind reasons (inductively). Initially, inductive development is practically unnoticeable.

Once an individual begins defining himself through reality by ascribing properties, he acquires a self-concept that automatically enables his mind to expand and further define itself. This also increases one's real security.

At this stage, conceptual levels parallel J. Piaget's idea of cognitive development. Piaget's stages represent development of abilities: The ones listed here refer to conceptual levels produced as a function of distinct intelligence levels (these are described in Chapter 4: the "How Aspect"). A properly completed self-concept is the basis for initiating self-relativization.

(6) Self-Relativization (self-relative realization)

This level defines the individual by constructing relations of scientifically ascribed properties. Recall: Understanding is acquired by knowing interrelations of ascribed properties. By recognizing relations the individual can answer low-level questions of causes pertaining to certain observed properties. Meaning is next recognized by interrelating properties through phenomenal relevance (concerning other properties). This "order of properties" can answer initial questions of reason (why-questions).

Ordering of properties is the first major development of self-relative realization. People need not adhere to beliefs about something when it can be immediately explained by its interaction with other phenomena. For example, something may change another thing's properties to the extent it is "recharacterized." A changed condition implies some type of interconnection. Recognizing this relevancy in a why-question propels the individual to ask for yet a higher reason (whyness). These higher-level questions are not easily answered. Science, philosophy, logic and religion form the context for structuring an answer.

Despite these obstacles, there is no justification for someone to substitute beliefs for knowledge. We will learn that beliefs distort the very process of gaining real understanding. Beliefs produce conflict of attention.

Much learning and reasoning is needed to understand phenomena via higher-level relations. High-level relations enable a realistic interpretation of personal experiences. One should consider the possibility of never obtaining real security with these methods. This realization makes one critically honest and pushes him to sharpen his methods of inquiry.

The individual eventually arrives at the frontier of knowledge by studying science. Reality has not been completely understood. How does a seeker of truth proceed beyond his present knowledge? He knows too much to accept information on faith. He is left with the sole choice of accepting self-intellectual responsibility. His only recourse is to invest security in a method of inquiry. He must refine his methodology to acquire greater understanding (to satisfy the need for "real security").

One's (real) security infers how much he understands about the real world in which we all live. When an individual has exhausted science for significant relations representing low-level understanding (usually denoting the physical world) he has no recourse but to rely on his own (methodology). He is pointed in the direction of reality. He is intellectually ready to find his way. He is ready to be self-actualized by reality.

(7) Self-Actualization (self-actual realization)

The individual earlier defined himself by understanding the Universe of Perfect Whatness through science. Science does not, however, do much for personal security, which—instead of depending on knowledge—depends on the method of acquiring it.

The scientific method becomes personal and eventually so does the knowledge that it acquires. The individual's higher-level-why questions remain unanswered even though he has a substantial scientific background. This increases self-doubt and insecurity. Doubt arises from the individual's questioning of where he has (conceptually) already been. He considers the possibility he erred in accepting the responsibility for his security. Placing security in a method of inquiry does not assure success. Doubt merely points out nondirectionality in his present knowledge. Knowledge has no guidelines to tell the person what to do with it. So what benefit is knowledge? There is no benefit from knowing if there is no security in knowing. Surely, there must be more security in knowing than not knowing.

At this stage the individual has (probably unknowingly) overcome indoctrination. His real security is tentatively dependent on what he will do. He knows he cannot turn back to more naive methods of satisfying the natural need for security. What the individual has already learned is maybe not what one would surmise. By reasoning (inductively) he has learned what is not the case—what is not true. Possible answers to why-questions might have been narrowed down but not answered. The wider questions of reason are not immediately implicit in the interrelations of the Universe of Perfect Whatness phenomena. These are answerable only by developing the higher-level-mental functions of induction.

I become more personal in what follows. I wrote a letter to a twin brother, who at the time, was serving in the military and living in another country. I had earlier written him that I was giving-up the study of dentistry. The letter describes my latest understanding about life and its meaning. Most of the realizations were hastened by the death of another brother.

The brother who died had been very close. I later realized I had betrayed myself by investing some of my security in him. It involved more than my respect for him. When an individual is unsure of himself, as anyone would be who does not have complete understanding, it seems natural to seek some type of security outside himself. This is natural and expected because a person is not born with understanding. Knowledge must accumulate to a point where one can trust it and so have security in it. That point is reached when one reasons explanatory knowledge of reality. Such awareness signifies the inception of "individualization."

When some of my outside security passed away (with the death of my brother) I was severely anguished for three months. I was anguished more by the questions death posed than by death itself. Especially noteworthy is the question concerning the meaning of life. I was struggling to regain a foothold and continually attempting to "face the facts" despite the consequences. Now instead of dealing with the outside world (the Universe of Perfect Whatness) through science, it was I face to face with reality itself.

My twin brother eventually brought this letter home. Later, I realized it summarized my self-actualization. It was dated Thursday, February 22, 1968.

The Letter

I have searched many years for a meaning to life in this world, as I perceived it. Alternately I played it safe and created my (own) meaning if the search proved fruitless. But, a funny thing happened while prodding along in logic: a strange thing took possession of me—myself. I mean to say that what happened is that I accepted my thoughts so strongly that my self-security came not from what I had hoped to invest it in (dentistry) but from what I actually thought itself. You know there are things that you do understand and things of which you have never heard. Yet, there are some things you barely comprehend and other things that you are acquainted from which you find your beliefs, religion and philosophy. What I use is a practical philosophy—an applicable thought, reason, solution type of thing that enhances one's sagacity to limits beyond oneself. As an analytical tool it is unequaled, unbiased and judiciously scrutinizing. I have realized a goal that maybe should not have been fulfilled. For confrontation with oneself is what is sought, whether you deceive yourself into thinking your goal is something else. If you do not confront your reasons to live, you may be better off. I can tell you now this is probably true.

I am not positive (for I know little of taught religion or philosophy) but I may be an existentialist. I need to read about it to find out. Yet what I use is my own and I suppose it would have to be to dispose self-security in thought.

The problem now manifests itself. Since fourth grade I had intended to work in dentistry. I planned around it and had my future security invested in it. It was a goal to be sought. I had some security by supposedly knowing what I would be doing.

The security is still there but stronger and is now invested in a "philosophy of knowing." Notice that knowing cannot be destroyed. Knowledge is you. I have therefore saved myself from anything like mere existence and lastly even death itself.

I have always thought rather deeply. After being thrown into the "cruel world," I spent day in and out philosophizing and reasoning just to gain more understanding. It all led to some tentative conclusions. I will not tell you much but only some of my premises and subsequent conclusions.

O.K.—Here goes:

I. What is *the meaning of life?*

> This is the main question. Well, I have searched it out. You may deceive yourself, but the answer may be that *there is no meaning to life.* This is the basic "play-it-safe" fact.

II. Consequences of previous question and answer:

A. This is why there are religions—so people will be spurned to live a natural lifetime through an acceptance of existence on faith. Notice the drawback. If they believe this religion, they are not real individuals and only exist, for truth comes not by faith but through thought and reason.

B. That is why these non-individuals have "created" a God—to save themselves. (That is not to say there is no God, but that if there is one, it is not the one they worship.) You may ask, to save themselves from what? I will tell you—from the probable fact "there is no meaning to life." It just is! You see: they create "their God" (in their own image and expectations) as a cover to hide from this fact. Yet that answer for me is just enough reason to continue living. When faced with a possible meaningless and therefore cruel death—what other choice do you have but to try making your own meaning?

III. Because there is probably no meaning to life and yet you intend to live a lifetime—you ask:

A. *How then can I live life to its fullest?*

O.K.—You set this as the goal. There is only one answer if you stick to your (play-it-safe) premises.

B. *Since "being" (as opposed to existing) is only in thoughts, then "living" is applying these thoughts to the "existing."* That which exists is your physical self and surroundings. Now I said my philosophy is an applicable one. Maybe you now see what I mean.

Now my corollary premises for applying thought to existence to give *being* or a *meaning to life* begins with one that will keep you in an objective world and enable you to deal with realistic situations.

a. *As an individual—I am insignificant.* If you accept this corollary premise of the evidential basic (play-it-safe) fact that there is no meaning to life, it will enable you to cope with any situation objectively and unemotionally. With this acceptance you can set out to make a meaning in life by using your thoughts—your own reason. In this lies your individuality.

b. You may argue what I have stated is a mere construct! Yet, you can see it has to be, because of the "play-it-safe" answer to the main question (that there might be no meaning to life). In it also lies the ugliness of it all—confrontation of yourself with reality.

c. *You have meaning only in what you do for other people.* This suggests your meaning lies with other people and not yourself. So you should not step on other people's toes. Every time you do, you would detract yourself back to nothingness.

d. *This means you live the Golden Rule—not by it.* You help others and then by the "Rule" they can help you (define yourself). Yet, you cannot ask them to help you since you are nothing except what they "make" you.

You may finally ask what is the difference between living an application of this personal construct and that practiced by a very religious person? Here is an analogy. These constructed thoughts are like an inventor's attitude toward his invention. He invents something and continues to improve it. Although fascinated by his achievements, he continually improves his invention. An "individual" also continually updates his thoughts and constructs.

Now a religious person or non-individual may use these already created inventions or thoughts and may even say they are the greatest. Yet these inventions or thoughts are not his. He does not know the inventor's reasons for them and is usually content to accept them on faith. He even lives a faked existence because he is not an individual defined by reason.

It is not easy to attain meaning in living life to the fullest. The ugly fact of real meaninglessness remains and returns when you step outside your constructed meaning. That "Cruel World" you have often heard is recognized when you understand the real meaninglessness of life. The only way to alleviate the suffering of insecurity is to construct meaning based on evidential facts.

The trick is to not let others know about life's meaninglessness, but to keep it to yourself. Let them be happy in their ignorance. I know the real meaning of humbleness. You cannot be humbler than to know you are really nothing in the face of reality. You must help others find happiness while you are still alive, for when you are dead and gone—what else is there? Play it safe! Why take a chance on there being an afterlife? Be the best person that you can be—now!

IV. Conclusions

Now I will mention some consequences of the "play-it-safe" fact that life is meaningless. One thing does prevail; it is because of life itself. One thing does make some sense: That is—propagation of mankind. Have a hand in creating another human being so someone else may have an opportunity to master the predicament of life.

You can make life as good, or as bad as you want, but it remains your responsibility. Please do something decent with it while maintaining it within the boundaries of the Golden Rule.

You see: I am letting FATE ride. What matters to me, is that I am an individual although only as a construct. I know myself as well as one can. I am self-secure in thought. I could talk to you for many days on how I apply this philosophy. But, I only intended today to give you some whys. It is now ten minutes to four, and I have written almost two hours. I think I will go to bed now. Oh! One thing on death. Some people say that death is for the better when one has already suffered enough. This, however, evades the point of life. The idea is to do something good for others while you are still alive. Then when you are on your death bed, you will not have any regrets about how you have lived your life. So remember, when your life flashes in front of you—the test is the Golden Rule. That is one of our brother's contributions. He set a good example. I therefore do not see his death as a vain one. The better example you are, the better example you set. Remember that, when someday you want your kids to be good.

The letter highlights some noteworthy characteristics of self-actualization. This developmental stage finds unfounded beliefs supplanted by evidential facts. These facts are supported by explanation. Most explanations are; however, tentative. They are offered as a "play-it-safe" type of induction (hypothetical generalization). Tentative explanations are narrowed. Contending explanations are refined (made more specific) but are sufficiently general to encompass the noncontradictory reason being sought.

Another characteristic of this level is personal anxiety. It arises because the individual shoulders most of the responsibility for his life. He has reached a point of letting reality (the reality of the situation) suggest what to do. This is suggested in the letter when the author says he is "letting fate ride." Let's examine the consequences.

Conflict arises from deductively wanting security from without while inductively attempting to acquire security from within through knowledge. The more the individual inducts knowledge the more he defines himself through reason. This also implies his actions are determined by reason. Although his actions are increasingly supported by inductive reason, his personal security lags behind. One's security is still dominated at this stage by his deductive mental functions. This is easy to understand. He has an idea where he has been; he does not know where he is going.

This is a crucial period. A person's deductive function is slowly giving into something other than itself—something that is anti-human nature, hence the conflict. The mind continues developing inductively. This ability enables the individual to learn about deeper aspects of reality. Conflict between emotional security (feelings) and intellectual understanding continues unabated. At this stage the mental inductive function is not sufficiently developed to explain why this is all happening.

This is a time of change. The individual is not aware of the meaning of what is happening. He does not know what lies ahead. He faces predicaments with one strengthening characteristic—his honesty with reality. The more honest he is—the more he learns. Honesty and learning complement each other. The more he learns—the more honest he becomes. He soon realizes he has traded dishonest deductions (mostly beliefs) for honest inductions (representing reality). Yet, knowledge obtained through induction is tentative at best. Note again: This conceptual level is not unlike self-actualization—developed by many religious leaders and psychologists—especially C. Rogers. The difference lies in our usage of the word self-actualization. For us, it pertains to the greater development and defining of the individual in a way that is verifiable by reality. This is an individual defined by reality. This is different from someone defining himself by his social identity.

After living a couple of years using truthful constructs, the individual is ready for a unique experience. He is readied for self-cosmic realization. This stage signals the beginning command (security wise) of the self through inductive reason.

Universe of Imperfect Whyness

The Universe of Imperfect Whyness represents an advanced functioning of the mind. It is the mind functioning inductively. The individual Universe of Imperfect Whyness is subfunctional before actualizing this level of awareness: viz, it functions intuitively. In that capacity it acts as a reality-oriented conscience. Note: On the original Where Aspect diagram (not shown here) there are responses opposite (below) the Universe of Imperfect Whatness conceptual levels. These responses are vague reasons. They suggest striving for self security via inductively-reasoned generalities.

By this stage, the individual's self-importance (in the face of reality) is minimized. His significance becomes even more dependent on understanding reality. Only traces of egocentrism remain. Yet, eliminating these last remnants of self interest is the most difficult. This happens because the Universe of Imperfect Whatness dominates behavior through "wanting" (desire). Elimination of desire results from higher-level reasoning. However, until information is verified (as knowledge) conflict continues. Security is affected. Doubt is not erased because two opposing functions (deductive and inductive) strive for answers to big questions. The first big question is "what is the meaning to life?" It is answered in one of two ways. One can determine his meaning (through deduction) or allow himself to be defined by reality (through induction). Asking about the meaning to life indicates the individual is being security-responsible for his life.

A transformation of the individual occurs at this level. He becomes mentally different (in awareness) from what he has been. He realistically inducts knowledge and begins obtaining definitive answers. Although the inductive answers are very general, they are precise because they are reasoned into patterned arrangements. One generality explains many effects. Patterns exhibit coherency. Coherency dictates the preciseness of inductive explanation. Coherency also eliminates contradictions and produces an intelligible relevancy of phenomena. The inductive function's (the Universe of Imperfect Whyness) first major task is to arrange self-concepts into patterns not unlike the Where

Aspect Diagram. This pattern reflects a beginning understanding of cosmic structure. This level initiates self-cosmic realization.

(8) Self-Cosmolization (self-cosmic realization)

This level signifies an onslaught of unique experiences. Initially, the individual does not, however, "see" the underlying meaning because of the effect on his security. Remember: He (the author) has earlier avowed to let FATE direct his life. In practice this means one accompanies the "flow of experience." More precisely, it suggests the individual has sharpened his methodology for whatever happens. These experiences happen to him—he does not make them happen. This stage surely starts innocently enough.

How did the author's realization of cosmic structure begin? During this time the author was a teacher of the physical sciences. On February 24, 1970, a student flipped ahead in his text and asked a then seemingly innocent question. He asked: "What is color?" The author replied:

Color is a manifestation of light. An observed object is not the color you see. You only see a reflection or appearance of that object. The object's real color is: *All color but the color you see.* If the object appears white, then it is actually black. If it appears blue then the object itself is yellow. Yellow is all color except blue. Yellow and blue are complementary colors. Green and magenta, red and cyan, are complementary.

Reasoning differentiates between the actual object and its appearance. Reasoning also suggests an individual is only directly acquainted with the appearance of reality and not with reality itself. It also implies reality can be indirectly known through reasoning.

Mindful that I. Newton, the English physicist and mathematician, refracted "white light" (in 1666) into components using a prism, let's look at an example tending to verify what has been stated. What happens when someone takes a photograph? For the sake of simplicity, first consider black and white film. White light (sunlight) strikes the object (with "whatness" properties) which to us appears white (an "imperfect" representation). Ideally speaking, all visible wavelengths impinging upon the surface are reflected from the object. A representative portion of the reflected rays enters the lens of the camera striking the film. The reflected light darkens grains on the film and the object's image looks black. This is the "negative." The appearance of the film image is a negative of the appearance of the object. In appearance it directly represents the actual color of the real object. When a "positive" is developed from the negative (film) we again get an image of the object as it appears to us. The same reasoning applies to other colors. The complementary color of the apparent object's color appears (neglecting the "orange mask") on the negative. We can say the negative (in appearance) represents the actual color of the object.

Reasoning dictates there is color only because of light. Without light there is no color. Consider sunlight shining on a "black" car. The car's surface becomes very hot because of the quantity of energy absorbed because of increased molecular-kinetic activity. By contrast, a "white" car is cool because it reflects most of the sun's impinging energy.

Objects absorb certain light rays (various colors or wavelengths of electromagnetic energy) and reflect the rest. That, which is reflected, acts as an initial stimulus for vision. It is this appearance or "sight" that the individual uses as the immediate premise (sense datum) for his perceptions (for purposes of identification).

Perceptions depend on what an individual infers from similar experiences. These references include what the individual has been told about such experiences. This poses an obvious question. Can an individual rely on what others say for that knowledge that corresponds to reality (and therefore, truth)? What if he has been blinded by indoctrination and all he "knows" is only what others have told him? In the author's case this problem had already been eliminated. He had overcome indoctrination. He paid little heed to information lacking explanation. This is itself an acknowledgment of ignorance. Yet this very confession is what allows one to continue learning. Knowledge is acquired by inductive reasoning.

Seven days after the student had asked his question, the author drew a diagram based on his answer. It is now called the Where Aspect Diagram. It is not surprising, in retrospect, the color question would lead to the Where Aspect Diagram. "Light" enables us to learn of the surrounding world. The color question and answer to how the diagram was originally structured will shortly be described. For the moment, we digress and supply a little background information about ideas.

When an idea is conceived, it is generally simple and vague. A construct must be elaborated with details to amplify meaning. It must also be coherent and not contradict already known facts if it is to disclose real meaning. The following descriptions describe the author's initial perspective on structuring the diagram.

Appearances or visual acquaintances of the Universe of Perfect Whatness are linked to our minds through perception. Perceptions infer the recognition of sense data, but do not lend themselves directly to the meaning or significance of sensory acquaintance. It has been stated that meaning is referentially ascribed by assigning properties and establishing relevancy.

Deductive consequences of appearances are represented in the lower left quadrant of the Where Aspect Diagram. For example, that something "looks white" is a deductive inference of its appearance (visual image). The visual image (the whiteness) links the mind to the physical world. The lower right quadrant represents the "reasoned knowledge" that the object's "unseen" noumenal color is actually black. The "unseen noumenal blackness" is represented by the upper left quadrant.

The nature of sight should be understood by the whole process of visual acquaintance. This tacitly assumes an understanding of what happens after light has reached the eyes' sensory receptors. This is not, however, our immediate concern.

The Where Aspect Diagram is erected using inferences about color. The slanted "bridge" connecting the lower left quadrant to the lower right quadrant represents a leap from one mental function to another (deductive to inductive). This also implies the two functions, at this stage, operate on nearly the same level of awareness. This means the individual Universe of Imperfect Whatness is functioning on a very high level while the individual Universe of Imperfect Whyness is just beginning to assert itself. The bridge represents a transition. It is a leap made by reasoning: a "leap by induction."

Although an inductive bridge connects the individual Universe of Imperfect Whatness with the individual Universe of Imperfect Whyness, it is the color paradigm that ties the phenomenal relevancy of color into the coherent unit depicted in the Where Aspect Diagram.

Key Points:

- The Where Aspect is an aspect of the Deltic Cosmos.

- The Where Aspect is structurally composed of three universes (two are real; one is a pseudo-universe).

- The Universe of Perfect Whatness is the universe of physical matter. It is constituted of material whatness things; it is real.

- The Universe of Imperfect Whatness is the mind functioning deductively. When a mind "inhabits" a physical body, it (the mind "itself") is "asleep." This means the mind is not functioning on its own. It functions with respect to the physical world. In this state, it functions deductively. It asks: what is this; what is that? It deduces answers based upon sense data gotten from sensory experience. It is a pseudo-universe. It is actually the Universe of Imperfect Whyness in a subfunctional state: it is an "asleep" mind.

- The Universe of Imperfect Whyness is the universe of all minds. A self-functioning mind reasons knowledge by induction. It asks "why?" It gets answers by inductive reasoning. A mind functioning inductively is one that is "awaken." Typically, this does not happen while the physical universe "dominates" the mind (when the mind "inhabits" a physical body).

- The Where Aspect Diagram relates a series of concepts that correlate with the intelligence functions (these are described in the next chapter). It demarcates the structure of our universe. The Where Aspect Diagram resulted from a student's question about color. The diagram is based on the author's answer to that question.

Additional Information

The author could not fully appreciate the diagram's significance when it was first revealed. Originally the universes were not named through the modal system. The updated nomenclature was adopted when the author recognized the benefit of using interphenomenal relevancy to show the connection between the distinct Where Aspect structures. The Universe of Perfect Whatness was called the "infinite universe." The Universe of Imperfect Whatness was called the "relative universe" and the Universe of Perfect Whyness was called the "absolute universe." The original nomenclature did not reflect much knowledge beyond the Where Aspect Diagram. Understanding the Where Aspect Diagram in the wider context of cosmic structure reveals its full significance. The inadequacy of the original designations conflicted with a later understanding of higher-level structures. The words absolute and infinite designate other aspects of reality.

Experiences happened to the author. They did not concur with what he would have willed or preferred. He did not coax the physics student to ask the color question. Yet, it was this question at a specific time in his conceptual understanding that inaugurated a series of unique experiences. It is not the intent here to expound upon those unusual events. Our main purpose here is to inform about the knowledge obtained because of these experiences.

"Apparently coincidental" experiences pose entangling questions. How could all these events be so timely? The more unusual the experiences, the more unlikely they could be coincidences. Did some type of intervening super-intelligence cause these events? Are all events predetermined? These are crucial questions. Obtaining noncontradictory answers required: added years of studying science (and other pertinent subjects) greater mental inductive ability, and more unique experiences. These interesting occurrences finally led to the coherent-knowledge realizations in this book.

The individual Universe of Imperfect Whatness wants surety. These were the waning days of the author's Universe of Imperfect Whatness' vain struggle for security. The deductive function cannot deliver real security because it does not reason explanations. When the author appropriately answers (by induction) the question of whether there is free will (or determinism) his last "desire" for self security should subside. There is no reason for someone to desire something that he has. We are a mind. A mind is aware of experience. It infers meaning by ascribing knowledge. When we acquire (have) knowledge, the desire for it is extinguished.

If the author had inducted knowledge on an assumption of determinism, he would have been placing the responsibility for acquiring knowledge on something outside his awareness. This would be foolish because he would be using an unsafe premise. Insufficient knowledge should be recognized for what it is. If one cannot explain what he knows, he must proceed with safe premises. (This is another example of using the personal explanation principle.)

Honesty is associated with the method of acquiring knowledge. Knowledge, once obtained, becomes a source of security. Acquiring knowledge is successful only to the extent one has developed his methodology. Here the reader should clearly understand why the seeker of truth invests security in a method. This prevents the individual from becoming security dependent upon unverified ideas. If the author placed responsibility for knowledge on something other than his self, it would eliminate the very reason (insecurity) which prompts development of his mental abilities. This is something one learns by hindsight. At any given time, one does not know that mental functions are amenable to greater development.

The author responded to his unique experiences through reasoning. In formulating the big questions, one first considers answers to lesser questions that are associated with formulation of the bigger questions. Reasoning raises many questions. Does an individual ask to be born? Does one determine when and where he is born? Does he ask to be what he is—physically and mentally? The best answer to all these questions would have to be an uncontested NO. These answers directly suggest we, individually, do not have untethered free will.

Next question: Are our actions and thoughts totally determined? Purpose to existence would be irrelevant if that were the case. We hypothesize there is a balance between extremes. If we have free will, it is limited. If there is a supernal Being, its powers are limited to the extent we have free will. The question is why do we find ourselves in this predicament? We are back to our main question. What is the meaning to life? We have tentatively eliminated the possibility there is unbridled free will. Also, we have dismissed the idea we have no free will. To answer these questions is to understand reality at a deeper level. Our search for real security continues.

Learning is a product of "what" we experience. Personal experiences also depend upon where we live. Many of our early views were a product of acculturation. An individual often becomes sociologically satiated with an ethnocentric perspective (of reality). Learning is affected by the ability to assimilate experience. This ability is determined by our mental and physical capabilities. Again, these capabilities are not sole conditions of our choice. Further, if some condition seems to be result of our choice, why did we so choose? Determinism permeates an analysis of our usual understanding of experience.

Unless we can answer why we do what we do, we are in no position to support a doctrine of unlimited free will. Can we really do what we want to do? What does "wanting" really mean? These questions need to be answered, or we have unsettled questions affecting self-security.

Simple reasoning shows there is no absolute free will. The question of determinism is entwined with universal causation. This implies all events have a necessary cause. Cause is descriptively represented by explanation. If there is an ultimate reason, induction should unravel higher explanations in a hierarchy of causes. The author's personal experiences and coherency of the diagrams, implicate some type of orchestrating-super intelligence. The task becomes one of seeking to understand this supernal Being. It is only in understanding this intelligence, and its purpose, that the seeker of truth can find his place (purpose) in the scheme of things. The first task is to understand the "scheme of things." And that entails delineating the structure of Reality. This book is the result of that effort. This book defines how man "fits" into reality's structure.

We have surveyed our first major structure in the mind expanse. We represented our findings with the Where Aspect Diagram. It is based on the "color" paradigm. It defines and relates the universes. The physical universe is called the Universe of Perfect Whatness. The mental universe is the Universe of Imperfect Whyness. When the Universe of Imperfect Whyness becomes "encapsulated" in a physical body, it generates a pseudo-universe called the Universe of Imperfect Whatness.

We categorized knowledge into three subclasses. Each class represents one aspect of an enveloping "whatness." The whatness is called the Cosmos. Its intrinsic classes are the Where Aspect, the When Aspect, and the How Aspect.

The author's unique experiences resulted in this coherent knowledge. In this context, certain experiences could not have been otherwise and still adduce this knowledge. The experiences imply man does not happen to life, rather, life happens to man.

By the end of July 1970, the author experienced the cessation of desire. It reflected a security shift from deductive methodology to one of inductive methodology. This occurrence also signaled the end of self-cosmic realization.

The advent of self cosmolization extends one's depth of knowledge beyond the Where Aspect Diagram. The How Aspect and When Aspect complete this level of understanding. Understanding this enveloping structure begins self-essentialization (which is addressed in Section Three called the "Realm of Reasoning").

How Aspect

Perspective

- Defines levels of order
- Functionally: Each level implies intelligence
- Relates intelligence functions to reality characterizations
- Intelligence functions correlate with Where Aspect concepts
- The How Aspect is the second aspect of the Deltic Cosmos structure

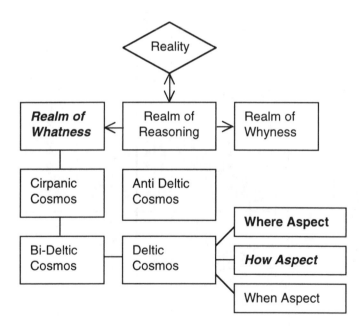

Figure 4:1. Perspective: How Aspect.

Realm of			Column 1 Reality Intelligence Levels		Column 2 Corresponding Ascribed Structural Orders		Column 3 Where Aspect Concepts
Universe		Universe	Functions in degrees of Whatness & Whyness	Functional Premises (Derived)	Develops Recognition of	Culminates in Recognition of	Self-Understanding Relating Design of Realty
	Whatness	Universe of Perfect Whatness	Whatness	Nonvital Substance	Mass (matter)	Plants	Matter
			Whatness	Vital Substance Plant	Plants	Animals	Stimulization
			Whatness	Vital Substance Animal	Animals	Sense Data	Sensationalization
	Mental Functions	Universe of Imperfect Whatness	Deducting	Extrinsic Perception	Sense Data	Identity	Experientialization
			Deducting	Extrinsic Perception	Indentity	Properties	Confrontalization
			Deducting	Extrinsic Whatness	Properties	Relationships	Conceptualization
			Deducting	Intrinsic Whatness	Relationships	Individualization	Relativization
		Universe of Imperfect Whyness	Inducting	Intrinsic Whyness	Individualization	Realm of Whatness	Actualization
			Inducting	Extrinsic Whyness	Realm of Whatness	Realm of Reasoning	Cosmolization
			Inducting	Intrinsic Whyness	Realm of Reasoning	Realm of Whyness	Essentialization
			Inducting	Intrinsic Perception	Realm of Whyness	Reality	Realization
	Reasoning		Whyness-Whatness	Intrinsic Conceiving			
	Whyness		Whyness	Consciousness			

Represents the intelligence of reality. It denotes the order of reality in a hierarchical sense. The intelligence levels are manifested in different levels of whatness and whyness. The are listed in column 1. Each intelligence level corresponds either to aspects of reality' structure or to the structure itself. Each intelligence level is recognized through a certain level of understanding. The understanding or awareness is listed in column 3.

Table 4:1. How Aspect Diagram.

Chapter 4

How Aspect

> "There was not only a design to what was occurring,
> but what was occurring was a revelation of design. It
> is difficult to know a difference unless one has
> experienced a difference."

The How Aspect of the widest intrinsic whatness mode refers to "levels of order" in reality. It relates functional intelligence, which is an implied-manifestation of reason (whyness). Mental functions are a part of the How Aspect.

Mental functions correlate with understanding levels of structural order. Each mental-intelligence level corresponds to a specific conceptual level (denoted in the Where Aspect Diagram). Intelligence levels engender comprehension of functional order using categories of whatness, deducting, inducting, and whyness.

Where Aspect concepts are quite specific, but their correlated How Aspect intelligence levels are not. Some vagueness is reflected in the merging of whatness and whyness in each level. For example, the How Aspect intelligence level "inducting intrinsic whatness" infers both whyness (because of inducting function) and whatness (because of the whatness premise). Even in the Universe of Perfect Whatness some whyness is implied because of "arrangement" of physical components (exemplified by atoms and molecules). Order implies reason. Similarly, the Realm of Whyness is imbued with some characteristic of whatness. For example, the wholeness function also has a particulate aspect. (More about this later.) Intelligent levels are arranged to explain the conceptual levels of self-understanding.

Sense experience represents an indirectly knowable "whatness" (the physical world) called the Universe of Perfect Whatness. The mental function immediately responsible for deduction is the Universe of Imperfect Whatness. Functional levels are defined by inductive methods.

The classificatory scheme used in the How Aspect Diagram is implied by the "color paradigm." It is the initial basis for separating levels of conceptual understanding (the concept levels) into different universes (viz, Universe of Perfect Whatness, Universe of Imperfect Whatness, and Universe of Imperfect Whyness). Each conceptual level of understanding was later associated with the intelligence level most appropriately representing it. Take a quick look at the How Aspect Diagram. Because the conceptual levels (column 3) are knowledge representations of structure (column 2) obtained through the individual's mental functions (column 1) the resultant concepts (column 3) are partial self-realizations. For example, it is the individual (self) recognizing sense data through extrinsic perceptions by which one experiences the physical world.

Concepts (generated by the intelligence levels) compose the Where Aspect Diagram. In the How Aspect Diagram conceptual levels of understanding are listed by their inclusion in higher-level generalizations. The intelligence functions (column 1) are also grouped by "how" the conceptual levels (listed in column 3) are inferentially derived (see the left margin of How Aspect Diagram). The How Aspect delineates the shifting emphasis of methodology needed to comprehend higher-level generalities.

Man's mental-inferential functions reveal levels of structural order in reality. Each mental function is developed by nurturing ideas culled from one's experiences. Induction synthesizes these ideas into arrangements. The final coherent-structural order (column 2) is recognized in its "knowledge" representation (column 3) corresponding to the structure of reality.

The How Aspect (column 1) is graduated on an intermixing scale of whatness and whyness. The extremes of the scale are an extension of the range of our potential mental abilities (its functions) yielding an intelligence continuum for the whole of reality.

How Aspect Diagram

The intelligence continuum (column 1) is sectioned into three major categories representing the three realms of reality: Whatness, Reasoning, and Whyness. The Realm of Whatness is divided into three intelligence groups (the universes). Each universe is divided into functional levels. Each intelligence level (or function) in an organizational (hierarchical) sense, evolves into and is enveloped by the next higher level. Each level is a more specific functional instance of the level that subsequently encompasses it.

A listing of intelligence levels (or functional orders):

Universe of Perfect Whatness

(a) Whatness—Nonvital Substance

Although nonliving matter is the lowest form of intelligence, it is nonetheless complex. Man is most familiar with material substance. Chemists study molecular interaction; high energy physicists delve into nucleonic structure; astronomers scan cosmic space.

Only imagination limits what can be learned. The more we observe—the more we learn of order in complexity. Though man is quite familiar with material order, he does not understand why there is order.

Order is imbued with an intelligence of its own; one that has evaded scientific understanding. This book defines this underlying intelligence that "guides" not only the ordering of cosmic structure, but everything within it.

(b) Whatness—Vital Substance: Plants

Some groups of molecules are endowed with life. It is difficult to ascertain where nonvital substance ends, and physical life begins. There is a continuum of order that defies attempts to demarcate the living from the nonliving. Borderline cases delimit resolution. Where does nonliving end and life begin? Where is the boundary between plants and animals? The continuum suggests one life form evolves into others.

Plants are thought to have arisen and diversified from nonvital substance. Structural similarities suggest plants interrelate. Organization of cellular tissue into specialized organs implicitly infers an underlying order of intelligence: an intelligence that guides life's sustaining activities. This manifested intelligence probably directs the evolutionary processes. Philosophers have suspected this factor and have called it the "life force."

(c) Whatness—Vital substance: Animals

Animals constitute the highest order of material life. At the lower end are organisms, which are structurally simpler than some plants. At a higher level are some subaverage humans who are less capable of dealing with reality than some "lower" primates. It is difficult to say specifically that one particular form in physical or mental capabilities has a decided functional superiority over another.

Generally, the "average" or normal individual can represent the species. A relative intelligence function is then assignable to organisms based on their mental capability. The life-form continuum suggests a diverging evolutionary hierarchy. This categorization implicates some type of inherent intelligence that guides evolution.

Animals have a mind. Mental development parallels physical development. Higher-level organisms, like man, develop complex inferential abilities. One function is an organism's ability to receive stimuli relating events of a "physical (whatness) nature." The individual becomes aware of the "extrinsic world" through the experience of sense data.

Sense data induce mental development. These ideas intimate purpose in evolutionary design. This hints eventual development of a species, which comes to know the reasons for its existence and its nature. Though this suggests a quasi-teleological explanation for life, it should not be construed as the sole reason for life. Teleology must also be explained.

Universe of Imperfect Whatness

(d) Extrinsic Perception

Extrinsic perceptions constitute the link between the individual's mind and his surrounding world. Extrinsic perceptions connect the Universe of Perfect Whatness (the physical) and the individual Universe of Imperfect Whatness (mental deductive functions). Though extrinsic perception depends on both the physical and the mental, it is categorically grouped with the latter. There would be no extrinsic perceptions without awareness.

Perceptions indicate that sense data are the "substance" of experience. Recognizing something depends on the individual's familiarity with similar sensory images.

Visual phenomena are an example of sense data. Visual phenomena indirectly "represent" objects-in-themselves (the noumena). Noumena are only knowable by their appearance. An object's appearance is the phenomenon. Phenomena constitute sense data.

To experience sensations the sense receptor must operate properly, and the causal factor stimulating the sense receptor must be present. An individual must not only be able to see, but a light source must be present for him to experience a visual sense datum of extrinsic origin. This idea generally applies to all the senses.

What we see or sense becomes the basis for intellectual manipulation. Mental pursuit indigenously attempts to define sense data by associating them with recollections of similar experiences. Perception is the unique identification of sensory experience. Some discernment must occur before the mind can infer anything about sense data. Immediate visual identification is practically automatic.

Extrinsic perception ascribes sense data with identity. Once identity is established, the mind adds meaning by abstraction. The mental functions construct inferences. Elaboration of relevancy is the first step to understanding. This process develops the higher intelligence levels.

This intelligence level produces awareness through one's phenomenal encounters representing the "outside world." It results not only in sensory recognition, but also establishes an identity image for subsequent reference. Perception of sense data functions as the backbone of knowledge. Extrinsic perceptions intermediate the confrontation between inferential-mental functions and phenomenal experience.

(e) Deducting Extrinsic Perception

Deducting extrinsic perception is usually the first inferential-mental function performed on sensory experience. It is a personal confrontation with phenomena. This level is premised on a phenomenon's immediate identifying import (denoted extrinsic perception). It is the first mental assigning of abstract significance. It explicitly relates the separation of identity from that identified—the sense datum. Note: There is first an automatic associative identification by recognition (extrinsic perception) pertaining to sensory imprinting. Second, there is a not so automatic abstract associative identification through recollection. The first depends on the latter once recollective processes become fully operative. Thingness identity becomes the abstract basis of knowledge.

Mental functions (of the intelligence continuum) are derivatively named according to succeeding envelopment of previous premises. The inferential product of deducting extrinsic perceptions is called extrinsic whatness. Extrinsic whatness represents ascribed properties. Properties, after being inductively rearranged, emerge as the next enveloped premise used by the next higher-intelligence function. Each ensuing premise is more generalized and abstract. As the individual functionally ascends through the higher-intelligence levels of the Universe of Imperfect Whatness, he deductively infers more from premises that have been inductively changed. These changes result from greater generalization of previously constructed premises.

This intelligence level, when fully operative, initiates phenomenal representation by abstract characterization. These abstractions lead to formation of reality-corresponding ideas. Abstraction initiates conceptualization.

(f) Deducting Extrinsic Whatness

Deducting extrinsic whatness is an entirely abstract mental function. It induces conceptual development. It develops deductions on premises already abstracted from sense data. Having established identity, the individual signifies phenomena by ascribing properties. Properties become a detached form of identification. Properties clarify the knowledge representation of phenomenal objects. Properties are assigned through a deductive mode; hence, the labeling of this mental function as "deducting."

Deductive inferences are postulated from extrinsic whatnesses (things). We see an object and say it is a table. Tableness is a concept: an idea. The concept "table" serves as an identifying premise. It represents the "categorization" of every structure called a table. The premise is the means to identify structure.

These premises represent phenomenal effects that are recognized by ascribed properties and characteristics. Ascribed properties are mentally refined by repeated matching to other phenomena exhibiting similar characteristics. Similar properties are then inductively categorized and extended to cover many sense objects. This leads to classifying phenomena by recognizing common attributes. This germinates familiarity through similarity of phenomenal effects.

Before assimilating an understanding of relations, one must be familiar with properties (used in establishing them). Although textbook science is second hand information, it provides the background reference. Personal references govern interpretation of experience. If someone is a theologian, he probably assesses experience through religious concepts. If someone is a scientist, he assesses experience using scientific concepts.

Scientifically ascribed properties lend realistic significance to phenomenal experience. They save the individual much time in developing sound ideas needed in coping with problems presented by reality.

It is no accident that problems continually happen. These "unwanted" occurrences pressure the individual to learn from direct experience.

Learning about reality is one of "give and take." Problems result from not understanding. If an individual analyzes problems with scientific like methods, he may learn enough that similar problems do not recur. Although science is about the only reliable indirect source of descriptive information concerning reality, it does not represent the complete knowledge of reality. Science has been the best source for indirect knowledge that stimulates reasoning. It also provides the best methodology for the beginner seeker of truth

Phenomena are logically grouped by similarities and differences (of properties). This classifying process suggests a basis for establishing relations between phenomena. This intelligence level culminates with a meaningful understanding of reality.

(g) Deducting Intrinsic Whatness

Deducting intrinsic whatness is the most complex mental function classifiable by deductive methods. It represents the pinnacle of Universe of Imperfect Whatness intellectual capability. Although this level is served by a deductive function, it depends on many inductive generalizations for premises. There is almost as much inductive synthesis of particulars in this level as there is deductive inferring from generalizations.

Intrinsic whatness is an inductively modified premise obtained from deducting extrinsic whatness. The premise concerns relations that have been inductively rearranged and grouped into more generalized patterns. Later these premises are incorporated into the modal system. Generalized premises are not well defined at this stage.

This mental function refines relevancy of phenomenal effects. Refinement reveals additional behavioral characteristics suggesting phenomena have intrinsic properties. These are noumenal properties that schematically connect disparate noumenal effects with a common cause or explanation. We recognize these effects in our modal construction as intrinsic whatnesses.

Noumenal effects refer to an object's nonapparent properties. Examples of noumenal effects are ("ascribed") chemical properties. These are ascertained by "reductively" understanding a substance's phenomenal effects. This level generates deeper (relational) meaning by recognizing order and design. Atomic structure is an example of a "deeper" reductive nature.

Relations between phenomena are (categorically) described by inter-phenomenal relevancy. This concept is the basis of modern science. Although science ascribes properties and relations, it is the method used in their determination that provides the seeker of truth with the tool to unravel deeper levels of reality. Greater mental development is not realized without a very knowledgeable understanding of scientific methodology.

Deducting intrinsic whatness is the mental function separating egocentric behavior from realistic behavior. The dismissal of one's self importance signals one's acceptance of realism. This intelligence level results in a basic realization of the self by understanding phenomenal relations. This level contributes to a deeper understanding of phenomena.

Universe of Imperfect Whyness

(h) Inducting Intrinsic Whatness

Inducting intrinsic whatness is an intelligence level developed by adopting scientific methodology to seek answers to deeper questions underlying reality. It involves the initial development of the "real self"—the individual Universe of Imperfect Whyness. The individual's previously nonfunctioning (inductive) mind slowly activates toward functional awareness. It plays an increasing role in the search for meaning and security. Its activation corresponds to self-actualization of the "real individual" or individualization. Individualization involves restructuring previously learned concepts. It enables direct "mental" contact with reality. There is also an adjustment of methodology. Inferential methods change from deduction to induction.

Most pretenses and facades disappear with direct exposure of the awakening mind to reality. The awakening occurs only after the individual has practically exhausted his deductive abilities in attempting to find security. He searches for meaning by understanding reality at a deeper level. (A "deeper level" does not necessarily pertain to the more "reductive" aspects of reality.)

Many relations seem to lack causal explanation. Questions concerning "cause" are not answered by merely ordering effects into simple relations. Coherency suggests itself in the arrangement of many relations—denoting a pattern. Individualization is premised on understanding "relational significance."

Individual Universe of Imperfect Whyness develops by a heightened awareness generated by the increase in inductive ability. This mental function seeks security through questioning and answering "why" something is or is not. It seeks security in reasoning itself. Reasoning is entwined with the methodology.

An early sign of inductive ability is intuition. Creative ability: in the arts, in poetry, in problem solving, are all expressions of intuition. A mind functioning intuitively is one not functioning inductively. It is an unawakened mind. The mind can synthesize ideas even when it is subfunctional. Only a functioning mind eliminates contradictions in generalities.

An example realization of this intelligence function is the "Letter" (described in the Where Aspect: subheading: self-actualization). The letter exemplifies the search for security when one arrives at the frontier of knowledge. This level finds the individual becoming acutely aware of the significance of reason (cause and effect) in the search for meaning. He recognizes his inability to explain (what he has already learned). He couples insight to the personal explanation principle (PEP). He begins to construct hypothetical explanations. He continues to study contemporary science.

This intelligence level attempts to inductively pull causal realizations from established concepts. This reflects an ignorance of actual causes. The anxiety of not knowing is heightened, but draws the individual into a deeper confrontation with reality. He recognizes ignorance is caused by one's lack of explanatory knowledge. Self security demands the seeker of truth continues along the road to greater knowledge and understanding.

This intelligence level recognizes knowledge to be a personal responsibility. The seeker does not depend on what others say concerning information validity. He now understands knowledge must be coherent. If a phenomenon cannot be explained without contradiction it must be considered tentative, misinterpreted or inextricable by contemporary scientific knowledge. The individual also applies these ideas in interpreting personal experience. The idea is to use scientific knowledge to build a logical (coherent) understanding of reality.

This intelligence level has performed its basic function when the individual has transferred all externally placed security onto himself. How does he do this? Security is transferred from unverified ideas to a methodology. This is acceptance of "real" self responsibility. This is scientific honesty.

This intelligence function is operatively superseded by the next-higher level leading to self-cosmolization.

(i) Inducting Extrinsic Whyness

Inducting extrinsic whyness is an intelligence level that assimilates knowledge into coherent units (patterns) which begin unraveling cosmic structure. The individual Universe of Imperfect Whyness is brought to full functionality. Knowledge is premised on concepts previously abstracted by inducting intrinsic whatness. This level is a complete Universe of Imperfect Whyness process in both function and derived premise.

It is challenging to translate specific experiences of extrinsic perception into meaningful abstractions using science and logic. Ascribed properties are continually reinterpreted to systematically organize phenomena (phenomenal and noumenal effects) into different (encompassing) arrangements. With the advent of self-cosmolization, realizations of various structural aspects become aligned into specific patterns. One recognizes the conceptual distinction of three major phenomenal classes—the Universe of Perfect Whatness, Universe of Imperfect Whatness, and Universe of Imperfect Whyness. This pattern has been diagrammatically presented and named the Where Aspect. This aspect (conceptually) reveals man's place (hence "where") within cosmic structure.

Lower level intelligence levels are subservient to the highest functioning level. Lower levels assimilate scientific understanding. Higher levels use this knowledge to interpret the many unique experiences encountered in self-cosmolization. These experiences have abstract meaning. The individual Universe of Imperfect Whyness considers reasons that can explain meaning. The seeker asks whether the meaning has personal significance.

Unique Universe of Perfect Whatness experiences are clues for answering deeper questions. The critical questions are usually those that have evaded resolution. And, they are probably questions that the seeker has spent much time contemplating. Many scientists and mathematicians have had this experience. After spending much time learning and trying to solve some problem, the solution comes unexpectedly and rapidly.

A simple communication between an individual and the rest of reality is the exchange between mind and Universe of Perfect Whatness. Although deducting and inducting are practically the only direct functional activity of awareness, all intelligence levels are involved in personal experiences. For example the Universe of Perfect Whatness, though not a mental component, is represented as sense data. There is an indirect systematic communication between all intelligence functions. Communication, however, often calls for an intermediating factor.

Cosmic structure determines noumenal effects. Noumena generate phenomenal effects. Each noumenal effect functions as a component, working harmoniously with all noumena. Noumenal effects intermediate phenomenal cause and its effect (sense data). In this view, cosmic structure (as noumenon) acts as the "phenomenal cause."

Separating phenomenal and noumenal effects is not always a simple matter. An oversimplification is phenomenal effects are directly represented in extrinsically derived sense data while noumenal effects are not. Though this idea is applied easily to a specific instance of Universe of Perfect Whatness and its corresponding apprehended sensation, it is more difficult to broaden this idea to analyze the functional aspects of Universe of Imperfect Whatness and Universe of Imperfect Whyness. These minds (viz, deductive and inductive functions) are understood categorically only by considering "phenomenal cause" to be an enveloping concept of cosmic structure.

Many noumenal effects lead to their sensory counterpart—phenomenal effects. Noumenal effects are products of interaction between noumenal properties. This is typified by the "action and reaction" in a collision of material bodies. The bodies are affected lawfully according to their inherent characteristics (whether the physical properties are understood). It involves not only the interaction between colliding bodies, but also the interaction of properties within each body. Both cases implicate noumenal properties. Phenomenal effects may follow noumenal involvement. The noumenal event (if there is one) is inferred from the individual's sensation of the phenomena.

Stimuli intermediate noumenal effects and possible phenomenal effects. Light intermediates the observation with that observed. Auditory stimulation is generated by sound (vibration of air molecules).

Noumenal effects are not always mentally represented as sense data. Unrealized phenomenal effects are "potential." (Realized phenomenal effects are extrinsic perceptions.) An individual's sensory apparatus may be impaired, or he may not be mentally receptive to stimuli. He may not even be present to be stimulated. Phenomenal effects depend upon sensory receptors. Phenomenal effects do not exist in themselves. They do not exist separately from noumenal effects (Universe of Perfect Whatness) and the mind.

The interrelation between cosmic structure (composed of noumena) and phenomenal effects is used in establishing the concept of Type I Whyness Whatness Association. This concept, disregarding potential phenomenal effects, requires a communicative relation between a mind (extrinsic perceptions) and the Universe of Perfect Whatness (noumena). Type I Association does not apply to Universe of Imperfect Whatness and Universe of Imperfect Whyness because they are a collective end in themselves and do not propagate an extended chain of "apparent" (phenomenal) events. Type I Associations are closely linked to "observed change."

Deductive and inductive functions respond to "whatness" and "whyness" (events and reason—explanation). According to Type I Association, cosmic structure is simplified in the modal system as extrinsic whyness (refer to Chapter 2).

Inducting extrinsic whyness, initiates basic knowledge of cosmic structure. The first structure realized is the enveloped tripartite structure relating "Whatness" in terms of Where, How, and When. This realization is the basis for using the (hierarchical) modal system as a paradigm to describe conceptual understanding of cosmic structure. This structure is the Deltic Cosmos.

When an individual continues to ask questions about "what" he already knows (recalls) he realizes there is more to understand. And so there is a "higher" level enveloping structure that synthesizes an understanding incorporating the Cosmos and its Anti-cosmos into an assemblage called the Bi-Deltic Cosmos. Finally, a Cirpanic-Cosmos arrangement fully incorporates the Realm of Whatness. These other structures are detailed later in their respective chapters. This intelligence level ends with the advent of self-essentialization.

(j) Inducting Intrinsic Whyness

Inducting intrinsic whyness is a high-level inferential intelligence level. It leads to knowledge of a structure referring to a "higher reality." Though implicated in the design of the Cirpanic Cosmos, it is a structural "unit in itself." It is called the Realm of Reasoning.

When an individual recognizes this structure by how it relates to the Realm of Whatness, he has epistemologically ascended to the conceptual level of self-essentialization. The necessity of this structure arises because cosmic design does not explain itself. This is easy to understand because a question can be asked of the cosmos that its structure cannot answer. That question is—"why?"

If the Realm of Whatness is ascribed the status of "attribute," the Realm of Reasoning is its "essence." This association relates another type of cause and effect connection. In the (intrinsic) modal system, from the perspective of phenomena, the Realm of Whatness (the effect) would be called "intrinsic whatness." Using the same idea, the Realm of Reasoning (the cause) "connects" the intrinsic whatness to its intrinsic whyness. Intrinsic whyness is the cause of noumenal "whatness." Type III Whyness-Whatness Association is described by coupling intrinsic whyness to whatness.

Again, a noumenon may be characterized by an effect or "whatness." From the phenomenal view, it would be called intrinsic whatness. When a noumenon is ascribed an inherent cause or "whyness," it is called intrinsic whyness. Such a causal connection establishes what we have denoted Type III Whyness-Whatness Association (see Chapter 2). This action is conceptually akin to the explanatory analysis of intra-phenomenal relevancy as described by intrinsic modes.

The Realm of Reasoning is described in Chapter 13. This intelligence function fulfills its purpose when the individual Universe of Imperfect Whyness begins to recognize an even "higher realm."

(k) Intrinsic Perception

Intrinsic perception is a noninferential intelligence function. It serves as a premise for inferences, or a representative product of inferences. It may correlate or associate with inferences, but is not itself an inferential process. It does not involve deducting or inducting. It is a center of visualization originating from a deeper aspect of the self. Most people are familiar with one product of this intelligence level—dreaming. When this level collaborates with the inferential functions, especially the inductive processes, it generates "imagination." Though operating in conjunction with all mental-inferential functions, this level is classified with the Universe of Imperfect Whyness (for reasons discussed in Chapter 9: the "When Aspect").

This intelligence level has the optimum task of visually representing the various levels of generalization obtained by mental inference. With recognition of the structural unit called the Realm of Whyness, the individual is equipped with the knowledge needed to elucidate the whole structure of reality. When one has visualized the interrelationship of all generalized components or classificatory levels, he recognizes the whole structure of reality. His individuality or self is definable in this totality of knowledge. Understanding is complete; his security is complete. He has conceptually fulfilled Self-Realization. However, as with all enveloping levels, he must continually refine and simplify the knowledge describing the "whole." Although this intelligence level culminates in the total recognition of reality, it has yet another feature without which experiences could not be correlated. This is the phenomenon of recollection.

Each individual is thought to possess a "memory" consisting of "stored" information. When mental functions collaborate with intrinsic perception, there results an encoding of experiential understanding. Consider a specific sense datum (such as a table) in the "field" of an extrinsic perception. Say you now experience the visual sense datum, which is a Universe of Perfect Whatness table. Now close your eyes. The original sense datum leaves a cognitive impression (the identified table) which can be mentally re-represented as an image or "intrinsic perception." The image is usually not as vivid as the original sense datum, but is nonetheless a representation of it. Identity is established between the extrinsic sense datum and its intrinsic perceptive representation.

Later the individual could be asked to "recall" the specific table. He may visualize a similar image. Remembering is a two-step process, encoding and extraction of information.

An inference from a premise can also be remembered. Experiences are recalled in representative visualization. Even language information is usually remembered by visualization. Anything one can experience can be remembered.

The manner in which an individual identifies an image often affects how he remembers it. People can compare present experiences to those of the past. Past experiences are references for interpreting present experience. Many past experiences become conceptual references, which function in extrinsic perception (especially identification).

Intrinsic perception envelops all mental functions. It enables one to unify the inferential products of other intelligence levels by interrelating them through a mental facility that visually recalls them. Reality is experienced in piecemeal. Intrinsic perception coordinates experience with cognitive order. This results in coherent understanding.

How does one remember? The externally derived sense datum (or extrinsic perception) leaves a central impression: a cognitive image encoded into memory. Encoding is automatic. One is seldom aware of the associative processes involved. By increasing deductions (thinking about it) about a sense datum, one can more easily recall it. The impact or selectivity of a particular sense datum (culled from the extrinsic perception—the field view which includes the object—in our example: the table) also affects recall.

How does memory figure in recollection? Recollection poses some legitimate questions about amnesia. Consider a hypothetical case where someone forgets who he is for a couple of years, and then suddenly remembers. How does this relate to an admission that during the interim he still experienced dreams (intrinsic perceptions) and could make inferences but could not remember his former life? One could argue since the amnesiac's intrinsic perception was still operative that his memory had somehow malfunctioned. This suggests the agent of one's memory is separate from intrinsic perception (which supposedly conveys remembrances to consciousness). It is arguable there must be a memory bin or agency that not only stores representations of former experiences, but is also a "server" from which stored information can be retrieved. (Note: Correlation does not necessarily establish a cause and effect connection.) How can amnesia be explained, considering other intact abilities, if memory is not a separate agency from intrinsic perception?

Let's face the dilemma of the common view of memory. A recalled image is an intrinsic perception. The pertinent question: Where is information stored when not being used? This is the crux of the problem. The common view suggests a storage bin or memory. Postulating is fine, only why is a storage bin needed to explain recall? Obviously, the common view calls for it. Many people may not be aware of the presupposition that (deductively) infers a memory storage bin. We briefly note this hidden premise: It is the "belief" in free will.

A memory bin is practically necessary to sustain the hypothesis of self-volition. Otherwise, it becomes difficult to explain free will. What causes one to remember? The common view suggests "a person determines at 'will' what he wants to recall." Thus the will or "desire" to recall becomes the reason for information retrieval. This presents a dilemma that can be succinctly stated as a question. Why did someone desire to remember in the first place? This question is not answered by stating the desire and postulating a memory (information storage bin). Such is the nature of causes. Desire and memory do not answer why we recall. Neither do they explain memory storage. Is there an alternative possibility why we remember?

If some things are determined (over what an individual wills) the agency of recalling need not reside in the individual. To ask why we experience recollection is to ask why there is recollection. This is different from asking where recollections are stored when not being used. When we point out the source of extrinsic perception, we will also be pointing out the immediate source (or cause) of intrinsic perception and hence recollection. The subject of this determining source will be discussed in the section on the Realm of Reasoning.

The intelligence level of "intrinsic conceiving" is described in chapters 13 and 15 and Section Six. The intelligence level "consciousness" is described in chapters 20 and 21.

Key Points:

- The How Aspect is the "intelligence" aspect of the Deltic Cosmos. It reflects order. Order implies intelligence.

- Intelligence levels are defined by function. At the "low end" are material things and physical life forms. In the middle are mental functions—deductive and inductive. At the high end is consciousness.

- There are two inferential functions. Deductive functions define sensory data obtained from the physical "whatness" universe. Inductive functions answer "why" questions by explanation.

- Extrinsic perception identifies whatness things. It connects the mental functions to whatness experience. Intrinsic perception provides a memory data bank from which present experience is compared to past experiences. It connects mental functions to consciousness.

- Consciousness is the whyness for everything. Intrinsic perception connects consciousness to experience.

We have retreated from the mind expanse to map what we learned about intelligence levels. The How Aspect Diagram is the result. We learned the mind can function deductively and inductively. Deduction and induction define the mechanisms of awareness. These functional mechanisms must also explain everyday behavior. Behavior has not been well understood. The next three chapters explain how the mechanisms of awareness affect and explain behavior.

Chapter 5

Behavior

"People behave in a way to become more secure."

Introduction

Behavior and Interacting Systems

Behavior is a broad subject. Psychology is the usual study of behavior. Behavior is a product of many factors: some extrinsic and some intrinsic. Behavior is characterized extrinsically by observation: the basis of empirical science. We can extrinsically observe an individual's physical activity. Behavior can be intrinsically described by studying: memory, the learning process, inner emotion, perception and the cognitive processes. Many of these phenomena are studied by administering performative tests. Mechanisms of sensory perception can be understood by studying anatomy and physiology. Cognitive processes are studied by correlating neurophysiological events with verbal description of mental images. Although such studies elucidate how we experience things, they offer little about why we experience things. The empirical approach does not explain the cognitive processes themselves.

Knowing behavior is different from understanding behavior. To understand behavior is to understand experience. Understanding experience involves more than the study of individual behavior. For example, sociology concerns behavioral interaction (between individuals). The cognitive processes are best defined by understanding logic (deductive and inductive)—not just to know logic—but to understand it purposively.

Heredity is a prominent factor in understanding behavior. To understand physical maturation, we look beyond anatomy, histology and physiology. We study genetics. We study biogenetics to understand the mechanisms of heredity. Yet these disciplines, and genetics, are understood by studying biochemistry. Biochemistry is understood by studying more chemistry. Chemistry is understood by studying physics. This compilation of peripheral subjects offers a clue of what we must do to understand behavior. Finally, behavior will be understood when we can answer the question bearing on the meaning to life. Answering that question should also explain behavior—why we do what we do. Our knowledge of reality must bridge the gap between understanding individual behavior and learning the meaning to life.

To understand behavior requires studying many subjects. Many scientists have been especially interested in how the mind develops. Mental maturation is "mirrored" in observable behavior. To obtain a deeper understanding of behavior, however, pushes us beyond the boundaries of psychology. We wonder why our intellect matures in the way that it does? The point is to classify "cognitively" many seemingly disparate phenomena. We do this by studying many subjects to learn how they all interrelate.

To understand is to interrelate phenomena. Let's begin by keeping it simple. Categorically, behavior is described often by overt and covert activities.

Covert and Overt Behavior

There are many types of behavior. Most can be classified as either covert or overt. An individual may be observed "doing" something. What one does overtly is what another person can observe through sense data. One "sense" observable is speaking and gesturing. These often infer something about an individual's covert behavior. Overt behavior is studied with relative ease. This project is concerned more with understanding covert behavior and intrinsic processes.

Universe of Perfect Whatness, Universe of Imperfect Whatness, and Universe of Imperfect Whyness

The essence of Beingness is interaction and change. Phenomenal change reflects noumenal interaction. The Individual's Universe of Imperfect Whatness (mind functioning deductively) reacts to the Universe of Perfect Whatness (experienced as sense data). The mind can also affect change in the environment (Universe of Perfect Whatness).

Mental functions (deductive and inductive) operate upon the environment through the brain and body. Conversely, the physical world operates upon the mind through the body and brain.

Behavior can be analyzed by understanding how the individual (as a mind) interacts with his environment. The mind (deductively) assesses its experience with the physical world. Our task is not only to observe "what" the mind does in this interaction, but also to understand "why" man behaves as he does. We build concepts to understand.

Basic Natures of Man

Classification: Basic Concepts

Each person behaves differently. Behavior can be classified according to why people are motivated. More specifically: Behavior can be explained by a person's security-orientation.

The classifying system used here is not one of specific categories. It is a spectrum with any position representing a varying composite of two extremes. One extreme (or goal) is represented simply by the knowledge of reality. The other extreme represents fantasy (opposed to the knowledge of reality). Fantasy represents the lack of knowledge. A person's behavior can be anywhere on the spectrum.

Draw a horizontal line representing the spectrum. Wherever the individual is placed, his behavior points him toward one of the two extremes. He is either realistic or egocentric. He is realistic if his behavior directs him to understand reality for purposes of obtaining security. His behavior is egocentric if his ignorance is sustained by bolstering and protecting his beliefs.

The extreme positions of the spectrum are a polarity of opposites. Although people are motivated by the same need for security, they may be oriented toward either pole in their drive to fulfill that need. An individual cannot simultaneously be directed to both extremes without conflict. If an individual is oriented toward reality and the corresponding knowledge it represents, he is directed away from self-fantasy. Conversely, if an individual is generally oriented away from reality and its knowledge, he is directed toward fantasy. Consider the following figure.

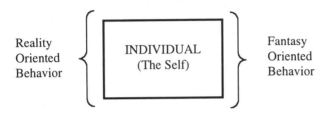

Reality Oriented Behavior { INDIVIDUAL (The Self) } Fantasy Oriented Behavior

Figure 5:1. Behavior Orientation

An individual primarily oriented toward reality to satisfy the need for security is realistic. An individual primarily oriented toward self-fantasy to satisfy the need for security is egocentric. In a motivational sense an individual's actions result from his security needs being directed toward one of the opposing poles.

Describing one's security orientation suggests his motivation. However, a description of either extreme does not represent one's entire behavior. Most people are egocentric and realistic at different times. They often experience conflict in directing their behavior toward either fantasy or reality. Generally, an individual is either realistic or egocentric. Although one's behavior is often a composite of these extremes, it can be categorized by his security predisposition of acting toward just one pole.

Egocentric Behavior

Egocentric behavior results in an individual who is directed to fantasy in fulfilling his need for security. He unintentionally adheres to the premise that he can trust his judgments about the meaning of his experiences. This does not mean he is cognizant of why he clings to ideas that he cannot explain. He does not question why he believes (what he does). To question why is to seriously question what he believes. That causes doubt, and doubt undermines security. An egocentric individual does not invest security in ideas that to him seem questionable and hence uncertain. He builds a belief system impregnable to self doubt.

Egocentric individuals think or judge their ideas (or beliefs) are right or true. It usually is unimportant to them if someone else, or even reality, poses possible contradictions in what they believe. Although challenged, they adhere to beliefs because of their security dependency upon them. If one has no meaning in life—he usually has no reason to continue living. So one holds onto whatever security he already has. One's meaning in life is tied to his security.

People are not born with knowledge intact. Knowledge is a by-product of experience. This suggests an obvious question. How does an individual acquire his ideas? More specifically, why is someone pulled toward fantasy to fulfill the need for security? First, we recognize people are born self-centered. Much parental-like attention is needed to assure infant survival. The caretaker assumes responsibility for the child's welfare. The child is not capable of being self-responsible. An individual begins life not being self-responsible. This sets up a process that continues into adulthood unless checked by reasoning. As the individual matures, he adopts methods of his contemporaries. He arrogates "acceptable social rules of conduct." These rules guide his behavior. He adjusts his behavioral code so that he is not socially rejected. He imitates others and is socially accepted. He obtains added security in being accepted.

People are socially indoctrinated by unquestioningly adopting prevailing codes of conduct. This hinders developing real-personal security. Deeper insecurities eventually surface. When this happens, people begin searching for answers. Unanswered questions cause doubt. Doubt creates "intellectual" insecurity.

There are two ways to seek security by asking questions. The first is to ask questions and accept answers from "experts." This is the most common. Religious organizations adeptly provide "ready made" answers. Accepting answers without further question does little to develop the mind. The second way is to continue doubting answers until possible contradictions are eliminated. This is very rare. This method marks a serious seeker of truth.

When questioning concerns "why" something is or is not, well-intentioned members of society often provide a different type of answer. They say you have to "believe" and have "faith." People have said they cannot live without faith. So what happens? Many individuals place their security in religion. By doing this they no longer need to question or doubt because they believe they are being "taken care of" by some benevolent superior Being—a Being which by practice or by definition cannot be questioned. It seems easy enough. All one has to do is be "obedient" to a set of "ready made rules" and he can bypass being responsible for himself. He has unwittingly shifted the responsibility for his life onto an unverified or imaginary Being. His security needs are satisfied because he now "feels secure." He no longer has to question his experiences or the meaning of his existence. His indoctrination is reflected in his imitation of other peoples' answers. His life becomes an imitation. That which he does not know nor understand is accepted on faith. It appears to work for the individual because he "feels" secure in his faith and beliefs. The egocentric individual lives a life of custom instilled in him since birth.

Occasionally, every individual is confronted by situations he does not understand. Even the most "religious" of individuals cannot account for all their experiences. Egocentric people adjust to "problems" through a variety of unrealistic methods. To protect security investment in certain beliefs, they often react defensively.

A belief is a defense mechanism that shields the individual from facing reality. What is a belief? A belief is a statement accepted as true without verification.

He accepts its truth and disregards whether it represents some aspect of reality. He decides in his mind what is true. He defines reality using his standards. This is why he is egocentric. He is the measure of what constitutes truth. He centers reality around himself. He comes first; all else comes second.

Beliefs are directed by security needs expressed through desire or want. The individual desires security. Much of man's activity is directed toward the goal of self security. When an egocentric individual feels secure, it is often experienced as the lack of frustration or anxiety. Frustration is experienced because of a conflict of motives or obstacles blocking the path to some goal. Anxiety accompanies frustration and is usually abated by eliminating sufficient doubt. This process contributes to individual homeostasis. We later see the "wanting" of some goals leads to its own set of problems.

The egocentric individual behaves to eliminate doubt by not doubting. Frustration and anxiety are often stilled through belief. The more one eliminates doubt using belief, the more he retreats into self fantasy. This is a self-feeding theme in egocentric people. The more self-centered one becomes, the quicker he unrealistically eliminates the nuisance of problems and the attending consequence of frustration and anxiety. In many religious individuals the only clue of insecurity is observing their frustration and anxiety. These characteristics become exaggerated in the extreme as self-righteousness and anger. Egocentric people excuse themselves by not doubting the justification for their views. Defense mechanisms permit the individual to defend his acts and allow the continual system building of beliefs. The self-centered individual shows a semblance of self security. A measure of egocentric success in self security is recognized in the individual's nonquestioning and total elimination of doubt.

Defense mechanisms take many forms. Projection is most common. A belief is a projection of what an egocentric individual asserts is true. He projects the truth of a statement onto reality. He slowly develops attitudes predisposing him to respond to challenges of reality by withdrawing from the questions they pose. When his beliefs are confronted by reality, he rationalizes by withdrawing defensively into security-satisfying fantasies. He builds up a repertoire of defenses. Why should he question his methods? Almost everyone else shows the same behavior. His methods are not uncommon. He pledges allegiance to his social group to allay self doubt. Society reinforces egocentric behavior. Such behavior is not the most effective way to cope with reality.

Realistic Behavior

Realistic behavior is exhibited in an individual who is directed to reality to fulfill the need for security. He does not trust what he "believes" to be true. He does not necessarily accept society's teachings as true. He realizes what is real, cannot be contradicted. He also understands knowledge must be coherent. He acknowledges coherency represents the unity (or consistency) in nature. He accepts the tentativeness of factual statements. He recognizes every statement must be questioned until all possible contradictions are resolved.

Although someone may become reality oriented, he obviously did not begin life being realistic. The realist begins like any other self-centered individual.

Eventually, he must "break" with the unquestioned teachings of society. He is first oriented toward reality by methods that direct questioning to his surrounding world. If he continues studying his environment, he needs to improve his methods of assessing truth. He develops critical methods like those used in science. Learning about the environment from a scientific perspective enables the realist to develop and refine his methodology.

The reality-oriented individual can save much time learning by examining scientifically discerned truth. He can quickly acquire an understanding of his surroundings by capitalizing on the work of others. He soon realizes that many scientists are "truth seekers" in their occupation, but not in their personal life. They are scientists in name only. Their personal security is not oriented toward reality. These "scientists" do not experience conflict between their personal life and doing science because they maintain a separation between science and their personal beliefs.

Solely studying science is not sufficient to become realistic. The seeker of truth must be able to explain effects using causes. When one can uncover an understanding of effects using his intellect, he is on his way to becoming a scientist in his personal life. His mind develops by reasoning inductively. His security becomes invested in his method of acquiring knowledge. Although one cannot trust the truth of statements, one can trust the method used in ascertaining verity. The idea statements are absolutely true is replaced by the idea verity is contingent on how well evidence supports its truth.

The scientific method must be refined if the individual is to understand higher-level causes. Experimentation in the nonphysical world of causes, occurs within the mind—the mind expanse. This is where the personal explanation principle comes into play. It is the first major prerequisite in using the scientific method. This principle acts as a sieve to distinguish between acceptable (possible) causes and those lacking merit. The personal explanation principle (PEP) helps order acceptable cause into generalities comprising hierarchical patterns.

What criteria distinguish higher level generalities representing reality? First, bits of information composing a generality cannot contradict its supporting evidence. Second, the generality should explain the supporting evidence as an effect or more specific instance of the generality. The closer the specifics correspond to reality, the more easily they align into hierarchical patterns. When sufficient bits of information are aligned into higher-level generalities, they categorically represent an entire group of evidential facts.

Categorical representation must be coherent. Coherency is easily displayed in diagrams. A diagram's logical construction characterizes phenomenal relevancy—representing encompassing concepts.

When an individual sufficiently understands reality to the point where he can assemble disparate knowledge into enveloping constructs, he begins acquiring personal security by being realistic. The overriding question the realistic individual asks is "how can this be explained?" Every answer is again subjected to the same question. Each time, a higher-level answer is sought. With each success, the individual becomes less egocentric and more realistic (in his knowledge).

Reality-oriented people do not experience frustration and anxiety, as do egocentric people. The realist experiences concerns while learning how to learn. He has intellectual problems to solve, but realizes the foolishness and counterproductive nature of anger. Contrast this with the egocentric individual who experiences anger because his desire of reaching some goal is thwarted. Anyway, each solution is followed by other problems. These problems need deeper comprehension. This is achieved by inductively building more encompassing explanations.

The realist and anti-realist (egocentric person) are opposite in how they react to "what" they experience. Where fantasy-oriented individuals react to problems by building defense mechanisms, reality-oriented individuals react by arranging pertinent facts inductively. The realist recognizes why problems occur. This enables him (by refining his methodology) to face similar but more complex problems. Both individuals build information systems to alleviate doubt.

The anti-realist or egocentric individual builds a structure immune to doubt. He eliminates doubt through "greater faith." His methods allow the return of similar problems. The same reason for not understanding happens in different form. He builds a fantasy, composed of statements shielded from doubt. The realistic individual formulates (hypothetical) concepts based on questioning. He tends to scientifically eliminate possible contraction (lies or untruths) by understanding reality at a deeper level.

What are the prospects of a reality oriented individual succeeding in obtaining real-personal security? Success depends on how well he develops the ability to interrelate evidential facts. The looming question is, whether he can continue seeking truth until all highly-ordered generalities are answered. After many experiences, he realizes one cannot understand anything until he sufficiently understands everything in a methodically generalized construct. A dawning realization of fulfilling this goal calls for a hierarchical structure connecting cause and effect using encompassing (explanatory) concepts. If he succeeds, the individual could transfer personal security from his successful methodology to the knowledge it acquires. Once started down the road to greater knowledge and understanding, there is no turning back. One cannot become more ignorant, only more knowledgeable.

Reassessing methodology is important. The realist continually refines his methods to meet the challenge of advancing scientific knowledge. Experience is also a teacher. From reasoning causal relationships there is usually obtained a generalized by-product allowing the individual to reappraise and adjust his methods of inquiry to deal effectively with "whyness" questions. Being self-critical enables the truth seeker to abstractly probe deeper into reality.

We observe two types of behavior. One is egocentric; one is realistic. The egocentric individual is security oriented toward fantasy. Because a person is born ignorant, he is initially defined by fantasy. The egocentric individual is self-security oriented. The realistic individual is security oriented toward reality.

The major defining characteristic of the egocentric person is his believing. The defining characteristic of the realist is his seeking of truth. We need to examine believing and knowing to better understand behavior.

Believing and Knowing

Two essential concepts are needed to understand behavior—believing and knowing. Believing and knowing are methods of assessing verity (truth of statements).

Believing is the dominant covert characteristic of egocentric individuals. Believing is a method. (Technically, it reflects a lack of method.) In a simple sense, believing is the personal acceptance of a statement or concept to be true without sufficient supporting evidence. In a more strict sense, it is the acceptance of a general truth even when there is much supporting evidence. The question, not easily answered, is when there is sufficient noncontradictory support (for a statement's truth). The problem recalls the idea one cannot understand anything until everything is understood as a whole. The personal explanation principle leads to this realization.

A wise individual does not assert the truth of any statement. Rather, he allows the evidence or verifying material to speak for (or assert) itself. The strict sense allows for error resulting from the use of insufficient noncontradictory support. Defining belief on verification raises the question of how verification is defined. Does verification represent reality?

Knowing is a characteristic of realistic individuals. Knowing, in a naive sense, is the personal acceptance of an idea (or statement) as true because of supposed verification. There is still the looming question of proper verification. People can use words (like verification) inappropriately. Examples are statements attributed to authority.

Authority appears in many guises. It may be an individual of high social standing, or a person of high religious status. Authority is also found in books of unquestioned nature such as religious texts. The only real authority is reality itself.

Common use suggests there is a middle ground between believing and knowing. This probably happens because we cannot properly understand anything except in the context of the whole of reality. This caveat includes improper use of verification. How is the word verification used? Anyone can be confused in the middle of not knowing and knowing. This is expected. Seeking explanations enables the individual to better define verification and refine his methodology in assessing truth. Yet the realist recognizes there cannot be a middle ground. He sees the middle ground as a "convenient crossover" between believing and knowing. The middle ground reflects a lack of methodological refinement. It becomes an excuse for failure. We have seen this confusion in logic itself. Some say the law of excluded middle is invalid.

The realist's methods shift from one of verifying statements representing "whatness" to one representing "whyness." The refinement is exhibited in a developing ability to eliminate contradictions in higher-level generalities. The methodology becomes one of using noncontradictory technique. This is akin to placing pieces (evidential facts) of the puzzle (of reality) into proper place. Final placement is dictated by the whole of reality. A place for everything and everything in its place. This reflects uniformity in nature. Regularity and uniformity are supposed, otherwise, scientific knowledge would be impossible. This is why there cannot be any middle ground. If there were, truth cannot be separated from falseness.

Logic should be followed as far as it can take us. The trick is one of understanding logic by solving problems pertaining to high-level concepts representing reality. Notice this is "backward" from using logic when it is not understood sufficiently to understand reality.

Certain problems in logic represent a lack of understanding certain deep-seated problems concerning reality. Use a high-level understanding of reality to see whether it reveals how the middle ground can be circumvented. Certain problems in logic need to be resolved to understand higher-level concepts that enmesh space and time. (This problem is later addressed in this book.) We cannot properly understand reality using logic if it appears to have limitations. Rather, if logic appears to have limitations, it indicates a lack of understanding reality at a very deep level. Understanding reality can be used to correct the deficiency in logical understanding. (We resume this chore in Chapter 18.)

Scientists have trouble clearly discerning the difference between believing and knowing. Literature on the subject shows this confusion. It is simple. A statement about anything is either true or false (is not true). There is no allowance in reality for a middle ground.

Believing is predicated on self-centered premises. Beliefs come from the self (fantasy). We observe this when someone says "I believe." The "I" represents the self. Believing does not represent reality.

Knowing is directed to reality. Knowledge is constituted of statements corresponding to reality. Reality does not represent the self. Knowledge represents reality. As the individual acquires knowledge, he becomes defined more by reality.

We are striving to explain how mind relates to behavior. How can we apply our behavioral spectrum exhibiting the polarity of egocentrism at one extreme and realism at the other extreme? We described (conceptualized) behavior using egocentrically and realistically oriented goals, viz, fantasy represented by beliefs and reality represented by knowledge. Although each individual is not the whole of reality, the whole should be understandable. Before continuing, let's reiterate and refine our understanding.

How can someone be sure whether he knows or merely believes? The safe assumption is to say one does not know until he fully understands reality as a whole. Besides this extensive and difficult assessment, in practical terms, one should not assert truth at all. One only has to point to reality and say, this is what reality says! The question is, does the observer correctly interpret what he experiences? As long as his security is invested in a method used to acquire knowledge, this question will not distort obtaining real knowledge. No individual can assert what truth is, he can only present evidence and look for contradictions in the concepts he uses to explain reality.

Because the middle ground resides in the individual's methods (or lack of it) he must find a way around the confusion it generates. Here, we are talking about ascertaining a definitive distinction between believing and knowing. The egocentric individual deals with it by not dealing with it. The realist can apply the personal explanation principle. In practice, he would have to offer an explanation to support defining the difference. Otherwise, he must say he does not have a clue to the difference resulting from insufficient knowledge. Contrast this with the believer who asserts truth of any statement to which he has attached security. He uses the term believing and yet cannot define it! He does not seriously define belief because to do so would undermine personal security. He is unaware of doing this. This is an affect of social indoctrination. Not only is he not born with knowledge, social conditioning hampers his effort to seek truth.

A definitive boundary is established between believing and knowing. The egocentric individual either believes or believes not. He cannot prove (verify) his asserted concepts either way. Further, because of his vested interest in his security-created fantasy, he is not inclined to question his methods. His methods become incapable of the refinement needed to tackle complex concepts—ones entailing higher-level generalities.

Let's review the condition for knowing. There is no simpler way, but to technically state "knowing" is ability to define information premised on an explanation that defies contradiction. Verification rests on meaning. Meaning depends on context supplied by noncontradictorily built structure. The completed structure represents the whole of reality. The realistic individual strives to understand the whole to attain real security in knowing his place in the scheme of things. Surely, is not an individual a "part" of the whole?

Let's be more explicit in defining our terms. The realistic individual either knows or knows not. The realist must (be able to) verify either way what he asserts as evidentially true (about reality). He must (be able to) explain his espoused statements without any contradiction. That is easy, but how does he verify what he does not know? The verification is that he cannot verify. Not knowing is a self admission of not being able to verify. To "know not" is verified by lack of verification. This is negative verification.

Each explanation engenders another onslaught of "why" questions. Seeking answers to these questions and the ever sharpening of mental abilities (to answer the questions) develops the mind. Later we examine just how these abilities manifest themselves in the individual. For now we concentrate on explaining with specific examples how a statement can both be a belief and a fact. Explicit examples are used to explain. With this understanding, it is not difficult to see there is no middle ground between believing and knowing.

Consider the following statements:

Person A states: The earth approximates the shape of a
sphere.

Person B states: The earth approaches the shape of a
plane.

Most educated people would say the statement of "person A" is true and
therefore constitutes a parcel of knowledge.

Most people would also agree Person B's statement is false.

At one time the most educated of people would have said the earth is flat. If
such a person could be asked "why" he so supposes, what would he say? He
might elaborate "if the earth is not flat, people would fall off it!" Today, we
would say this individual is ignorant of the truth and lacked sufficient
knowledge to say anything about earth's shape. We easily discern the problem
turns on the question of knowing when one has sufficient knowledge to assess
the truth of higher level questions. The shape of the earth is like a higher level
question because its direct determination has, until recently, been beyond the
capability of everyday experience. Its truth is not obvious.

Who is to say what constitutes sufficient knowledge? Better, what are the
criteria for determining sufficient knowledge? Knowledge is established when
(its) truth is assessed within an encompassing concept. Its truth cannot contradict
already established knowledge. It is not this simple in practice. If we could
somehow observe the earth from the moon, we could directly and easily see the
solution. Here the wider context is one that includes earth and its surroundings.

There are other ways to ascertain the truth of earth's shape, but they call for
ingenuity. Indirect evidence can also be used to ascertain truth. This is
circumstantial evidence. Both statements of the individuals (Persons A and B)
can be tested within the context of a wider generality. If the earth is spherical,
what can it explain? Sailors of yesteryear noticed as ships sailed from port the
ocean surface intervened between them and land structures (tall buildings and
mountains). This would be perplexing if the earth were flat. It is easily explained
if the earth is spherical.

A spherical earth also explains other evidential facts. What circumstances can
explain the sun traversing the sky? Information accumulates. How can it be that
Magellan sails in one direction and returns in another? A perplexing problem is
solvable if a concept can explain it with minimal conflict. Phenomenal
description cannot conflict with higher-level concepts that explain it. There is a
difference between apparent conflict and real explanation. Apparent conflict
results from lack of noncontradictory explanation.

Now to our specific problem. How can a statement be both fact and belief? Let's
use our example. If a person long ago said the earth is spherical, is this a belief
or fact? Obviously, we now say it is a true statement. Then how can it be a
belief? The answer is simple. That statement is a belief for someone like our
ancient questioner, who asserts truth but cannot verify it. The same statement is
an evidential fact to an astronaut. The difference is one of having sufficient
knowledge or experience to warrant assessing the statement as true. Yes, a

statement may be true. The central question is how can a given statement be verified?

Whether a statement functions as a belief or evidential fact depends on the individual's methods used in ascertaining its truth. What does this mean? Man does not experience reality, as it is itself. Instead, he experiences the "whatness" physical world indirectly. There is a separation between man (that is—his awareness) and the phenomenon (reality) he experiences. We learned this from the color paradigm.

Although believing and knowing are covert behaviors, they do have overt expression. We relate these expressions by the concepts of "wanting and having."

Wanting and Having

Overt behavior is an expression of covert behavior. Overt behavior, in a phenomenal sense, represents those actions of an individual that another person could observe or experience through sensory apparatus as extrinsic perceptions. Covert behavior represents mental activity that another person cannot observe. In the noumenal sense covert behavior is intrinsic mental activity. Intrinsic behavior is implicated in motivation.

Gaining a deeper understanding of behavior calls for an idea connecting overt and covert behaviors. Phenomenal behavior, i.e., overt, has been successfully studied and classified by psychologists. However, a real understanding of behavior hinges on understanding how covert behavior explains overt behavior. Motivation, a covert behavior, is a basic consideration in classifying behavior. The problem is easily stated as a question. "Why do we do what we do?" We begin with some generalities and use the classifying scheme of egocentric and realistic behaviors.

The egocentric individual is self-centered. Reasons (or motivations) for his overt behavior are directed to his self. The realistic individual is not self-centered. Reasons for his overt behavior are directed away from his self.

What motivates the egocentric individual? If an egocentric person is asked why he did something (overt behavior) he usually answers (if he attempts any reply) with a statement premised on belief. If further asked about his premise—he often responds with statements similar to the following. "I do not know" or "just because!" The question is why is he so motivated?

People experience the physical world (Universe of Perfect Whatness) through their extrinsic perceptions. Why do people not understand what they perceive? Perception is one thing, cognition another. Perception implies meaning while cognition suggests knowing. The egocentric individual misinterprets meaning when he incorrectly assigns properties to his perceptions. Describing a deductive cognitive function does not adequately explain motivation.

The egocentric person is not motivated by reality. He fashions concepts that do not represent reality when he uses deductive functions. Fantasies do not connect to reality, but may indicate one's wishful perspective toward reality.

Strong wishes become desires and wants. The egocentric individual views his percepts through his desires. He derives deductions about the world using his percepts and desires as premises. Desires distort the understanding of reality. Desire often imbues sense data with erroneous meaning. The egocentric individual makes reality to be what he wants. He constructs his personal version of reality.

Man begins life ignorant of reality. Fabricating constructs based on belief maintains this separation between man and reality. Wanting is the basic covert characteristic of egocentric people. Desire permeates the overt activity of egocentric people. How is desire explained?

All individuals are motivated by their security orientation. Egocentric people are oriented toward self-beliefs and fantasies to satisfy the inherent need for security. This general need is reflected in a constant drive to reassure their beliefs. Egocentric people desire to find truth and certainty, but ignorance perpetuates itself. Egocentric people do not know how to seek truth and understanding. If they did, they would not be ignorant or egocentric.

People behave in a way to become more secure. This is the primary-motivational factor. It answers the question "why do we do what we do?" The next question is "why is that?" This question will not be addressed until we learn more about (the whatness of) behavior.

Desire functions inconspicuously. Desire allows an egocentric individual to suppose truth in mental conviction. Strong desire brings psychological security to ignorant people. For example, the previous statement often "offends" egocentric people. Instead of questioning its truth and possibly gaining a deeper understanding of their own nature, they unconsciously "hide" from the truth by protecting their beliefs. Being offended is one way of hiding.

Egocentric people do not understand the shortcomings of their own nature. Whatever the intensity, beliefs are incapable of satisfying the need for real security. People believe when they do not know. Consequently, beliefs need to be reinforced. When an egocentric person begins doubting his beliefs, he modifies them to maintain his security.

What motivates the realistic individual? If asked why he did something (overt behavior) he answers with a reality-oriented reason (of cause or explanation). He does not answer "just because" or "I wanted to!" Again, the question is "why is he so motivated?" He is motivated to understand.

Real explanations are reality-depicting generalizations. For the realist they exhibit a link to the physical world of his actions. He reacts to his experiences (extrinsic perceptions) by forming general concepts that "explain" those experiences. He asks what does this mean? An exchange develops between his two mental functions (deductive and inductive) and the physical world.

The realist directs his covert mental activities (deducting and inducting) to understand reality as a whole. His mind relates to reality using symbolic representation. (Knowledge represents reality.) What is there for him to desire? Not anything! Having knowledge is all one can have. He is one with reality—on its terms. He is initially guided by the interaction between the physical and the mental. He does not decide what is truth. He relates to reality as it is.

Security develops in step with the attainment of knowledge. If the realist attains complete understanding, he is not found wanting. Seeking real security is the goal. If the realist attains real security, the "wanting it" automatically vanishes. Knowledge is truth. Complete understanding thrusts one toward certainty in what he does (overt behavior). He knows where he stands in relation to reality. He has all one (a mind) can have—knowledge. For consciousness, ultimately, is all one is. And consciousness shows itself as a function that obtains information. When that information is knowledge—instead of fantasy—one has done all he can do in fulfilling the need for security. Having (knowledge) is a basic covert characteristic of realistic individuals. Realistic people are motivated by security orientation to understand evidential facts in the widest possible context.

Let's examine wanting and having. It is simple. If an individual has (having) something, there is no reason for him to want it. For why would a person want something he already has? Conversely, a person only wants something when he does not have it.

Look at Table 5:1 about phenomenal causes. We use security as the topic. If one has security (a) he has no reason to want it (d). If he does not have security (b) he may want it (c). In a phenomenal sense wanting (c) is correlated to not having (b) and having (a) is correlated to not wanting (d).

The realistic individual speaks about what he has. Either he has it, or he does not have it. The egocentric individual talks about what he does not have (and therefore what he wants). The former speaks of reality (a and b) while the latter is unknowingly enmeshed in fantasy (c and d).

	Positive Statement	Negative Statement
Realistic Individual	(a) I Have Security	(b) I Have Not Security
Egocentric Individual	(c) I Want Security	(d) I Want Not Security

Table 5:1. Individuals Defined by Behavioral Orientation.

"Wanting" characterizes egocentric people. They are oriented toward fantasies (or beliefs) for security. Belief is information they "unknowingly" want (desire) to be true. By perceiving beliefs as fact they fool themselves into accepting false premises to satisfy their need for security. Belief indicates a desire (or want) for truth. Since one's security is often a matter of life or death, one fights to keep those beliefs that guard the sense of his well being. This means egocentric people are against reality because it attests beliefs and knowledge are functionally opposite.

Because the egocentric individual centers his attention more on what he wants instead of on what he has, he meanders through life on faith and hope. He has faith that his beliefs are true and hopes his wants will be met. He is a fool for faith and a slave to wants (desire). With desire, faith, and hope he builds his world of fantasy. Because of his selfish need for security, he becomes even more egocentric and further separated from reality.

What happens when an egocentric individual does not get what he wants? When a wanted objective is not obtained, the individual experiences frustration or anxiety. If he had obtained his objective, the desire would fade, and he would be happy. Such is naturally not the case because the dynamics of desire (wanting) usually operate unconsciously. Security orientation for the egocentric individual is practically automatic. In this existence, one is naturally egocentric by design. And this means people are unhappy not only by falling short of their goals, but also unhappy because one's mind is incarnated in a physical body. This is contrary to the mind functioning independently. (While "inhabiting" a physical body, the mind functions with respect to (depends upon) the physical world.)

Natural conflict develops between wanting and having. It is the nature of egocentrism that "wanting" be maintained. After obtaining some objective, the egocentric person eventually "wants not to have (it)." Conflict is exhibited when the egocentrist obtains the wanted objective and then wants not to have it.

Happiness is associated with wanting and having in the form of getting. The egocentrist believes he would be happy if he gets what he wants. For a while he is happy. Later he will desire to be rid of what he has. The egocentric individual is supposedly unhappy when he has not obtained some objective. (Unhappiness results from not getting according to expectation or desire.) Happiness and unhappiness are emotional measures of one's status in goal obtainment. Ironically, egocentric people often measure success in life by their happiness.

It is often heard certain couples "cannot live with each other and cannot live without each other." This phenomenon results from wanting that which one does not have and not wanting (wanting not) that which one does have. The conflict between wanting and having recurs in egocentric people because desire (or wanting) is maintained. This is the "Ping-Pong" effect. Yet, reasoning shows the foolishness in wanting something just because one does not have it. There is absurdity in "not wanting" when one does "have." The same reasoning not only applies to happiness and unhappiness, but also to what one intends to do.

If the egocentric individual wants to do something it means at least he has not yet done it. It means at best that he cannot do it. If someone wants to do something because he has not or cannot do it, it means if he did or could, he would not have to want to. Note: Although someone may own a physical object (in the sense of security) he does not really have it. He can have but a conscious representation of it. If one has trouble understanding this, then consider that death ends all pretenses one can own anything but his self. This is why all one can have is information.

Only with knowledge can one have real security. If there is an "afterlife" and people, in some sense, maintain their identity, all they will have is that information pertinent to their identity. Identity is based on experiences. Experiences shape the self. Note again: Wanting and having is defined through the perspective of security. This is why wanting and having is primarily a covert (or psychological) phenomenon.

Wanting often incapacitates people. Just as belief inhibits one's search for evidential facts, so it is that wanting often prevents one from having or doing. Desire lies at the very core of egocentric behavior. Desire is maintained by not attempting to have or do that which one wants. When one does try to obtain (have) what he wants, he is often satisfied by his attempt although his goal of having was not met.

Obtaining or doing what one wants, does not mean that what he obtained or did, occurred because of his desires. Real cause and effect only apply if people have free will. The question is—why did he want what he wanted (the objective) and why did he do what he did? The answer for what he did is not found in his desire. If it were, we can ask why did he originally so desire. That is unanswerable without settling the issue of determinism and free will (see Chapter 16).

The case can be made where someone desires something but in the end does not have (obtain) it. This clearly shows that desire for something does not itself cause the effect of having (it). This is no less true when the individual does end up having what he wanted.

The realistic individual becomes closer to reality because of his knowledge (of it). He lives not by faith, but by trust. One cannot trust himself (his beliefs)—only reality. He can trust that which is and that which shall be. He does not experience happiness or unhappiness because he does not have security invested in what he does not have. He does however experience peace of mind because he is (self) secure in the knowledge of reality.

Wanting or having indicates a person's security orientation toward either reality or fantasy. Just as wanting and having are opposites so it is that egocentric and realistic people are opposite in their motivation. One directs attention to himself and fantasy while the other directs it away from himself and toward reality. If a person consistently says he wants or believes something—he is drifting toward fantasy for his security. If he consistently says he knows (via the personal explanation principle)—he is directed toward reality.

There is a type of covert activity that is not so intimately involved with security attainment. A "preference" is nonsecurity oriented. It is the liking for some objective that is of no consequence to one's drive for security. Whether a preference is obtained or not has little bearing on one's well being. No anxiety or frustration results. Because a realistic individual subjugates himself to reality—his individuality survives as preferences.

Hate and Love

Wanting and having have been explained. These concepts can be expanded to include the wider context of sociology. We begin by posing a question. How can wanting and having explain social interaction? First, note social interaction is a more specific instance of the individual's interaction with his surroundings—the physical world (Universe of Perfect Whatness). To simplify, we use a concept embracing a dichotomy of extremes. Any behavior falling between the extremes is explained by the interplay of the extremes.

Wanting and having lead to the sociological phenomenon of hate and love. For explanatory purposes we say there are two general categorical objectives for wanting and having. Wanting results in self-fantasies. Having leads to knowledge of reality.

It is meaningless when someone wants something he already possesses (have). Desire (want) for knowledge is automatically eliminated when one has knowledge. Knowledge connects consciousness to the "rest of reality." Knowledge is the primary characteristic of consciousness. Wanting knowledge (of reality) signifies one does not have it (that is: the knowledge connection to reality).

Fantasy is the lack of connection to reality. Fantasies or beliefs, in an interesting and twisted way, do not represent anything. Beliefs represent what one wants, i.e., what one does not have.

Believing one has something, when he does not, generates fantasy. Wanting indicates fantasy. People are egocentric because they use beliefs for premises from which they ascertain (deduct) meaning (about their perceptions). Sometimes perceptions are images of other people. The individual deals with other people just as he does with any of the outside world (phenomenally represented in his perceptions). Overt behavior is usually based—not on understanding the world—but on what one thinks or believes about the world. One's covert behavior determines the meaning of his overt behavior, which is directed toward other people.

Let's examine the meaning of an egocentric male telling a woman he loves her. The first word many people associate with love (besides sex) is caring.

> Sex is not love because one can have sex without love and love without sex. The male says he cares for one special person above all others. Yet this is an improbable definition, as we will shortly determine.

What could the pursuer say if asked why he loves her? He may say he very much likes her. But, we are not discussing preferences and nonsecurity-oriented behavior. He may say he loves her just because he does. Such an explanation is no explanation at all. An egocentric individual cannot explain what he means by love. The personal explanation principle applies. If he cannot explain without contradiction, it probably means what he perceives as love is really not love (what he supposes).

We often observe selfish behavior when a man "loses" control over his spouse. Let's say a woman leaves her husband and he "wants" her not to leave. His desire (want) becomes magnified and overtly exhibited. We see his anger; he threatens her. Clearly, this does not indicate unselfishness.

People usually are friendly when there are no problems. Problems test relationships. Problems reveal the real basis of a relationship. Egocentric people do not experience real love because they do not have the necessary knowledge and understanding.

Many people suppose a special desire for someone is a form of love. Again, why does the individual want (love) the other person? We are getting closer to the truth. How can desire be real love (based on reality) if it is the causal trademark of egocentric love? Again, desire is associated with self-fantasies. Man does not love (want) a woman for herself. Man wants woman for himself and vice versa. It is he who wants. Desire comes from the individual and is for the individual.

The reason for the egocentric individual's love for another is his desire for security. Each wants the other person (e.g., a woman) for (because of) him or her self. They want each other for themselves because the reason for their love is themselves. They care for their security needs over their care for the other person. Egocentric (self-centered) people are selfish by reason because of desire (love). Can love emerge from selfishness? It cannot. It is selfishness (or desire) that contradicts this form of behavior as real (unselfish) love.

Just as the egocentric person believes when he does not know, he wants (loves) what he does not have (love object). Because he does not (have) love, he is by his nature (of wanting) incapable of experiencing it. It is not love he experiences. It is something else. More is explained later.

How is the term hate commonly used? When an egocentric person hates someone (or something) he usually implies he has a strong dislike or disdain for someone (or something). Yet, another person may "love" or like the object of this person's hate. Therefore, the object cannot be the cause of the hate despite what the hater believes. The cause of the supposed hate must reside in the individual who hates. He (the hater) cares not for the hated object. Yet, when the individual (the hater) is asked why he hates the object (of hating) he often says something like the following: because, or I just do!

Either the reason for hating is in the object of the hating or in the hater: it cannot be both. We have already established the reason must be in the individual who is doing the hating. It is the fault of the individual who hates and not the fault of the hated object. Why does he supposedly hate? First, we simplify and state that he wants not (to have) the object of his hating. The hated object is probably a threat to his sense of security.

Obviously, he cannot explain his supposed hate. It may be this supposed hate is not really hate at all. It may be something else other than what he believes. We will return to this subject soon.

We have seen what egocentric people supposedly experience as love and hate may not really be what they believe. One does not know if he cannot explain without contradiction what he states to be true. That a person believes he "loves or hates" someone is not the point. Hate and love must be defined in a way not contradicted by reality. To define love and hate is to understand them in a wider context. We must appeal to a higher-level generality that embraces these concepts. Logically, it must be an inductive concept revealing a deeper aspect of reality.

How is hate and love realistically defined? Let's lay a foundation from which
we will explain. Look at the diagram entitled Real Hate (Supposed Love):
Figure 5:2 A.

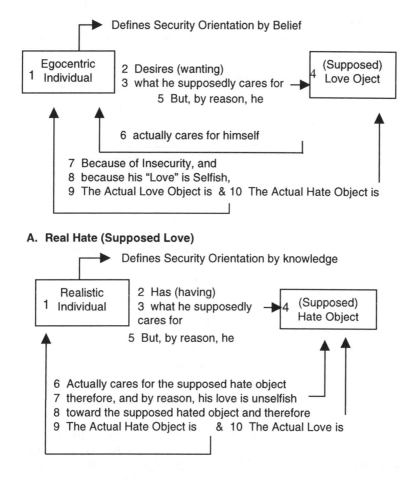

A. Real Hate (Supposed Love)

B. Real Love (Supposed Hate)

Figure 5:2. Hate and Love

The common denominator of hate and love is caring. How caring is directed
determines whether it leads to hate or love. Again, we look to the noumenal
reason (covert behavior) to explain the phenomenal effect (overt behavior) as
one denoting real hate or real love.

We explained the basic psychological drive of egocentric people is wanting
(knowledge). We stated self-centered people are security oriented toward beliefs
in their futile attempt to fulfill their wants. They use other people to satisfy their

need for security. Security-need is the reason for their apparent caring of another person.

Which is cared for most, the self or the object of caring? The apparent object of caring arises from the security condition of the one who cares. If the cause of caring is in the self (from the beliefs of the one who cares) the supposed caring is based on a selfish reason. The supposed love appears unselfish, but its cause is selfish.

There is a conflict between the cause of this "love" and its effect. No wonder egocentric people find difficulty relating to other people. They experience a constant conflict between supposedly wanting and having the object of their caring. This results in a "Ping-Pong" effect.

When a boy is attracted to a certain girl, but does not have her, he wants her. He construes this as love. He marries the girl. Now he has that which he once wanted. Because love is associated with wanting, in his mind, he now wants not to have her: he now loves her not. So he abuses her, treats her very differently than he had before marriage. He effectively is trying "to have her not." The wanting continues. He begins by wanting that which he has not, and finishes wanting not to have that which he has. Wanting is maintained. That is the nature of egocentric behavior. Reminder: having something physical (another person) is impossible. An individual is a mind. A mind can only have knowledge. Attempting to own anything other than knowledge is futile. Real security can only come from knowledge.

Love is caring directed toward a love object. What one cares for most by reason defines the actual love object. There are only two categories for the objects of wanting and having—self-fantasies or reality. To avoid contradiction we state the person either cares for his self or all else (the rest of reality including the specific object of the caring). If he cares by reason for himself—above the caring of the supposed love object—then he has love of himself. He cares less for the focus of his attention. This is selfish love.

There are (security-wise) only two choices. If one actually loves himself; then, he must actually hate the supposed object of his love. (We later explain more.) Because of their nature (wanting) egocentric people are incapable of caring for (love) another person (or for that matter—the rest of reality). Supposed caring is another attempt of the egocentrists to rid themselves of insecurity.

That which one cares for less by reason defines the actual hate object. Hate in the egocentric individual is real caring directed away from the supposed love. This reflects that the actual love object is the self and the supposed love object is actually the hate object. Real love depends on understanding. Look at the diagram entitled Real Love (Supposed Hate) Figure 5:2 B). Realists are oriented toward reality to fulfill the need for security. We explained the basic motive of realists is to have (knowledge). The realistic individual can only "really" care for another person when he is able to help that person. The realist only can help when he understands what the person in need does not understand.

One can only care to the extent he has knowledge and understanding. One is unable to really care if he lacks knowledge. He may "want" to care for the health of his child, for example, but unable to help. Surely, if he could help (actually care-love) wanting (to help) would not be a factor.

Security is a natural outgrowth of knowledge. The realistic individual, by reason (of having knowledge) can actually care (love) because his security is invested in reality. One can trust reality. Love (caring) is a natural outgrowth of this trust. The realistic individual acquires security by trusting reality. Having security ends the want of security. Instead of depending on others for security, the realists can offer security to others by explaining what they do not understand. People become secure when they understand. One attains complete security when he understands All of reality.

The realistic individual cares unrequitedly. He has (having) unselfish love. The actual love object is all else except the self (including any other specific instance of reality). There are only two categories: the "self" (represented by self fantasies) and "reality"—the rest of reality, represented by knowledge. This leaves two major factors to consider: One is that knowledge of reality is required for real love. The other is that the individual is either security oriented toward or away from reality.

One is either capable of real love, or he is not. Someone is security directed toward himself when he is security oriented away from reality. Problems are difficult to solve when the self is defined by fantasy.

Definitions of love that exclude these ideas are contradictory. For example: Saying someone cares for one person more than he does for others indicates it cannot be unselfish (real) love. It would be selfish love. The reason is simple. He categorically equates all else (and all people excepting himself). One person cannot be singled out from the "rest of reality." One cannot say, this is the individual I love above all else. The choice is between the Self and the Rest of Reality. Circumventing that choice is to not face reality on its terms. Without the necessary understanding, real love is elusive. One can only care (love) to the extent he has the knowledge to help (care). Any less than the necessary understanding of reality to help others, renders one's "caring" ineffective.

One has real hate for himself when he really loves that which is other than himself. So, he lives his life for others despite himself. In an indirect way his personal meaning in life is affected by how well he (really) helps other people. He must be as honest (realistically) with them as they can gently tolerate. The catch is he cannot impose himself on other people. They must seek his help.

Once the realist understands reality, he attributes problems to some "thing" in the "rest of reality" that he does not understand.

Problems arise when the realistic individual attempts to help an egocentric person (using knowledge). Egocentric people have trouble accepting real help—it is contrary to their nature. They act as though they want help, but either have a problem accepting it, or, if accepted, fail to apply it. Egocentric people usually only want the attention obtained from seeking help and not the help itself (because they depend on other people for security). The test is whether they apply a given solution.

It is not enough to apply a solution—people need to understand it. That is the problem. Some people understand but cannot apply a given solution because their insecurities surface to block the help they need. Others can apply a solution, but do not understand what they are doing. They may see the solution work if they persevere. They experience an increase in their security level by just being persistent.

The realistic individual realizes the difference between himself and self-centered people. They are opposite by nature. The egocentric person obtains more beliefs to bolster his security. He does not do this by intention. This is his nature. The more falseness (unverified statements) he constructs, the more he must shore up his beliefs. His security depends on "solidifying" his ideas (fantasies). Eventually, he erects impenetrable barriers between himself and the rest of reality. He listens only to those who agree with him. He does not seriously question his own beliefs. Nor does he allow others to question them. He does not doubt his beliefs. Yet, doubt is the main reason the serious truth seeker reasons to more deeply understand.

The egocentric person, by nature, is anti-knowledge. Seeking truth is contrary to his nature. He is defined by a set of beliefs. Beliefs constitute his identity. Acquiring knowledge changes one's identity. Obtaining knowledge destroys ignorance. An ignorant person "feels" threatened by ideas other than his own. Ignorance carries its own baggage of insecurity. Any experience threatening one's security is shunned. There is no real security in fantasy.

An egocentrist only cares for his self (self-fantasies). He "means good," but his efforts are usually worthless. Egocentric people cannot accept and apply help because it is contrary to their nature. The realist is socially out of place and time. However, when a reality oriented individual has a problem—the realistic individual can help. The realistic individual lives by a simple axiom that effectively relates him to any other individual. He cares only if they care. If they do not care—his caring is wasted.

Let's take a general view of egocentric love. What the egocentric individual believes to be love—is really hate. When he wants to love, the only choice he has is to hate. He cannot have what he wants. He cannot do what he wants. His nature does not allow it. Conversely, what he believes is hate must really be love. Although a bold statement, this is correct reasoning. It is explainable via security orientation.

How can supposed hate actually be love? We use an idealized example. The idea of actual caring is to engender real love in the actual love object. Real love is an outgrowth of understanding. It develops by learning about reality. If a person is supposedly hated by an egocentric individual, the supposed hated object cannot continue being security-dependent on the supposed hater. Security-wise, the supposed hated object's only recourse is to turn away from the egocentric individual. That leaves the hated object to turn to the "rest of reality" (all except himself and the hated object). Refer to Chapter 8, Figure 8:1.

The supposedly hated individual can learn and understand from a relationship with the "rest of reality." Again, the outgrowth of understanding reality is real love. Hence, actual love can develop in the supposed hated object (person) because he is kept from obtaining security from the one who supposedly hates him. He is forced to turn to reality to find out why. This is an idealized construct. It is based on two major premises. One is the supposed hated object would not turn to someone else for security. The other is we assume he would be directed toward reality for his security. Yet the idea of supposed hate being real love is not contradicted by reality. Remember, terms and concepts should not be defined by what people believe and want, but by a correspondence to reality.

It should not be surprising that egocentric love and realistic love are opposites. They have opposite motivations. Egocentric love is based on believing; realistic love is based on knowing. Believing and knowing are opposites. Believing is based on wanting; knowing is based on having. Wanting and having are opposites. Wanting indicates selfishness; having suggests unselfish.

Key Points:

- Behavior is explainable by one's security orientation. One's security is categorically directed in one of two opposing directions—the "Self" or "the rest of reality."

- Egocentric behavior results from security attachment to beliefs. Beliefs arise from ignorance (of reality) and (by default) define the self via fantasy.

- Beliefs are statements accepted as true by an individual who cannot properly verify its correspondence to reality. People believe when they do not know.

- Realistic behavior is exhibited when an individual allows reality to "speak for itself." He allows reality to verify whether a statement is true or not. The realist is defined via reality.

- An individual "knows" a statement is true when it is verified by reality. Knowing is only as valid as reality's assurance of its truth.

- Believing and knowing are opposites.

- Wanting and having are opposites. People want when they do not have. Wanting contributes to egocentric behavior.

- Having eliminates wanting. Having knowledge engenders realistic behavior and eliminates ignorance.

- People "believe" when they do not "have" knowledge. When one "has" (have) knowledge—there is no reason to "want" it. Consciousness cannot "have" things. It can only "have" a knowledge representation of things.

- "Wanting" can lead to hate. A person "having" knowledge is capable of love.

- Love is that which one cares for most—either the "self" or the rest of reality. If the individual cares for his or her self over the rest of reality, then he hates reality. If he cares more for others (they exist in the "rest of reality") then he hates his or her self. Hate and love are opposites.

Why do not people just reason inductively and develop the capability to acquire knowledge and be able to love? This question can be answered on more than one level. All levels will be considered, but for now, we continue to explain behavior in the context of everyday events.

Chapter 6

Mind Control

"Before getting to the truth, lies must be gotten out of
the way. Were we given a mind only not to use it?"

The self (mind) not only controls but also can be controlled. Here, we speak in
the sense of how life is experienced daily. At this level of generality, we do not
address whether man has free will or not. This awaits later sections.

People are confronted everyday by advertising. We are enticed to want
something we did not know we needed. We watch television, listen to radio and
see billboards. Sales people are taught how to overcome objections of
prospective buyers. With personal defenses down, another sale is made. Did the
buyer need the product? Will the product live up to the sales pitch? On both
sides information is distorted, and facts are ignored.

We watch political campaigners deceive and make promises that should not fool
anyone. Countries become embroiled in wars in which people die but never
know why. Propaganda besieges us from all directions.

The problem does not lie with the deceivers but with the deceived. Why are we
so easily duped and how do we protect ourselves from deceptive practices? The
answer to the first question is simple. We are not born with complete
knowledge. The answer to the second question is to acquire knowledge and
understanding.

Infants mimic parents. Success in society presumably depends on how well we
imitate those around us: those who often set our expectations. In school we find
the more we memorize the better student we become. By nature (being ignorant)
we are insecure in our decisions and unsure of our actions. Memorization is not
understanding.

We are counseled, guided and molded by the social setting. We are deemed
normal only if we easily fit in with the cultural milieu. Although individualism
is encouraged, it is not preferred over social consensus. Individual-analytical
reasoning is not encouraged except in academia.

Why are people so gullible? Our first example represents someone who is
isolated and is strongly dependent. We proceed from the "hard sell" to examples
of lesser control where a person seemingly could choose otherwise in what they
accept as true.

Brainwashing

The phenomenon of brainwashing became widely known from events
surrounding the Korean War. Brainwashing is a form of mind control. Many
individuals, most often American air force personnel and foreign missionaries,
were captured and placed under guard as prisoners of war. R. Lifton, an
American psychologist, interviewed such individuals soon after their release.

Years later he conducted follow-up studies. He describes the steps used in this form of mental coercion.

The captive's sense of identity is assaulted. He is humiliated, stripped of clothing, chained and mistreated. Defenses are weakened; self-identity begins to unravel. The prisoner loses his hold on self-security. He is made to feel guilty and ashamed of his beliefs. He is worn down by incessant talk sessions: kept in a standing position for hours at a time, placed on irregular schedules and allowed very little sleep. He is told old friends have betrayed him and are cooperating with officials. He soon realizes he is at the total mercy of his captors. He fears for his life and nears a breaking point. Anxiety and nervousness overwhelm his physical-mental integration. His identity collapses. He has trouble distinguishing the real from the not so real. He may hallucinate and even drift into and out of consciousness. He lives in total alienation, a fearful nightmare leading to self-destruction.

Guards and interrogators begin to show sporadic bursts of leniency. The prisoner is allowed some rest. His medical needs are given attention. The unexpected kindness is an advance on what the captive could expect if he totally cooperates with officials. Cooperate or perish. Previous incessant talks, in which the prisoner was accused of conspiring against the people (through alleged germ warfare—if a pilot, or social corruption—if a missionary) become a way out.

By confessing to the accusations the prisoner receives better treatment and improves his chances of survival. He is told to write down how activities before capture, support official accusations. Meaning is twisted. Former experiences become cloaked in the captors' rhetoric. The prisoner is informed that officials know of specific things he has done against the people. He willingly does what is expected. In exchange for kindness, he feels compelled to confess. The prisoner may even invent imagined experiences to please his captors. He comes to "understand the people's viewpoint." He is reeducated.

He is redefined as an individual. The ideological change calls for emotional support. He is treated well and gratefully does what is expected. He writes and rewrites a final statement of confession to crimes against the people.

It is said even the hardiest of individuals have a breaking point. When the alternative is death, there is little choice but to cooperate. In follow-up interviews Lifton found that ex-prisoners, although insisting they were not brainwashed, still espoused the ideology of their former captors. Such behavior is not totally unexpected considering the circumstances surrounding their captivity.

These cases, coupled with historical incidences suggesting apathy of supposedly honorable people, spawned research into the role of authority. For example, many people found it difficult to understand why German Christians stood by while millions of Jews were annihilated during the Second World War. Furthermore, why did officers carry out orders to have them killed?

Submission to Authority and Group Conformity

In the early 1960's psychologists, such as L. Berkowitz and A. Buss, studied aggression using a machine called the electro-shock generator. S. Milgram, another American psychologist, used a similar device in his studies on how people react to authority. The meaning of authority here is not that of someone who has expertise in some field. Milgram's use of the word pertains to an individual asserting some right to control the behavior of others.

Authority arises from sociological settings. Some people lead and give orders; others follow and obey orders. This is the basis of social structure. Milgram asked to what extent would people obey an order that increasingly opposes their own desires or moral values.

Milgram set up an experiment with three participants, the authority or "experimenter," the subject or "teacher," and the victim or "learner." Unknown to the teacher the other two participants were confederates working for the laboratory. As part of the act a drawing is held to see who plays what role. It is predetermined the outside participant is to be the teacher.

The experiment uses an electric-shock generator. The machine has thirty switches. The first switch labeled "slight shock" represents a 15 volt electrical jolt. Shocks are administered in increments of fifteen volts up to 450 volts. The last switch is labeled "danger—severe shock."

The object of the experiment is to study the amount of shock the teacher-subject is willing to administer to the victim-learner. The subject, however, is told the experiment is designed to determine the effects of punishment on the learner's ability to remember pairs of associated words.

The typical setting has the learner sitting in an enclosed "cage" seemingly wired to the shock generator controlled by the teacher-subject. The teacher can see the learner through a glass window. The experimenter is near the subject. The learner has previous instructions on how to perform his duties. When he misses a word association, the teacher pushes a button. Unbeknownst to the teacher no shock is given to the learner. However, before beginning the experiment a sample 45 volt shock is given to the teacher to convince him of its authenticity.

Soon after the experiment begins the learner, according to plan, misses a word pairing; the teacher pushes a button. At 75 volts the learner tells of his "heart problem" and asks to be let out. At 195 volts he says he is having heart trouble and repeatedly asks to be let out. At 210 volts he tells the experimenter to get him out, he has had enough. The "teacher" typically questions about continuing with the experiment. When needed the experimenter states, "You have no other choice—the experiment must go on." The experimenter later augments that by saying he takes full responsibility for what may happen to the learner. At 300 volts the learner refuses to give any more answers to the test. The experimenter informs the teacher to treat unanswered problems as incorrect ones. Despite verbal resistance and ever increasing tension the teacher often continues to administer all shock levels although the learner is soon screaming in agonizing pain. The tension experienced by the teacher indicates the buildup conflict between what he thinks he should do and the orders he is to follow.

Milgram conducted variations of his experiment. In most of those experiments, while in the presence of authority, the "teacher" fails to initiate any plan that countermands orders though many believed the victim was being seriously harmed. Milgram found people deemed good-upright citizens in everyday life would carry out a destructive act irrespective of their conscience as long as they perceived themselves not responsible.

There is a fine line between submissiveness and cooperation. As a footnote, experiments were conducted where the subject is told he alone is responsible for what happens to the learner-victim. In those cases there was a significant reduction in the extent to which teacher-subjects obeyed.

Milgram has also carried out experiments to check the tendency to conform to group consensus. It is a variation of an earlier experiment by S. Asch that sought to determine how a person reacts in judging different lengths of lines after first hearing the wrong answers of others. He set up six enclosed booths. Booths one through five were unoccupied. However, when the subject of the experiment walks into the "lab" room he sees five coats lined up on a coat rack. Being number six he enters the last booth. He is provided with headphones.

The object of the experiment is to observe how the "subject" reacts when he hears the answers of the other "participants" given first. There are two tone bursts. Tone A is longer than tone B. The "subject" is to compare and tell which is shorter. Through the headphones he first hears the responses (actually a recording) of the other five "participants," who all announced incorrectly that tone A is the shorter. Subjects most often pick the same tone, as did the other participants. Here, we find individuals submitting not to a particular authority but to a noncoercive group.

Discussion: People carry out orders despite their conscience for fear of being uncooperative with authority. In a social setting, identity with a group assures acceptance by the group. It also means conversely that an individual concomitantly surrenders his individuality and becomes an unwitting participant in group activities.

The phenomenon recalls the idea of safety in numbers. The absurdity of adhering to group conformity (without questioning) is shown by the fact there is often no distinctive reason upon which group members focus for guidance. The Vietnam War provided a poignant reminder of what can happen in such circumstances.

Many people supported the war because they thought American soldiers were fighting for their country. Yet the United States is not located in Southeast Asia. Many thought their fellow man would not be fighting unless there was a good reason. No valid reason was ever given. What about those involved in the actual fighting? Although no reason (such as self defense) was ever given, most soldiers joined in the action of their peers and did what was expected. They followed the orders of authority. For many the price for following orders was death.

Relaxation, Suggestibility, and Compliance

We have just described how easily behavior is molded by either authoritarian or public pressure. In the early stages of brainwashing the subject is usually aware of what is happening. Self-esteem and personal references are openly assaulted. The subject becomes a different person in his behavior and beliefs. Although individuals may resist rational argument in everyday settings, under pressure they succumb to irrational doctrine.

We briefly examine a mild form of mind control usually associated with the word "hypnosis." The word itself has connotations of mysticism. (Symptoms such as epilepsy, convulsion, and hysteria were once associated with it.) Little was understood about hypnotism until recently.

Hypnosis is not easily defined. It is considered a state of relaxation in which suggestion is uncritically accepted. It is a narrowing of one's attention to the point of not noticing certain sensory activity. Under hypnosis, sight, hearing, taste and pain can be effectively diminished. For instance, although there is no difference physiologically, sense activity is perceived differently.

In everyday life people have experiences of: dwindling attention while driving a long monotonous road, daydreaming in class, or simply entranced by watching the undulating action of the ocean tide. Some individuals often become so engrossed in reading a book they seem oblivious to their surroundings. Some athletes concentrate so much on performing that they do not hear the roar of the crowd. They may suffer severe injuries and not notice pain.

How is the hypnotic state induced? The subject is directed to relax. The hypnotist begins a routine to focus the subject's attention. A common technique involves the repetition of phrases such as "I want you to listen carefully to my voice, think of nothing—nothing—nothing, your eyes are closing. You are relaxed and breathing heavy and deeper—deeper, count from ten to one. As you count, you are going into a deep sleep—deep sleep—much deeper." It usually takes around eight minutes or more to produce the desired state. How deep a subject is "under" can be determined by the handclasp test where the subject is told he cannot unclasp his hands and then told to try. The subject is told the harder you try, the harder it is to pull them apart.

Can a hypnotized subject carry out instruction contrary to his personal preferences? It has been suggested that if the task were carefully planned, that many individuals would do it. (This reminds us of Milgram's experiments.)

Some practitioners give a post hypnotic suggestion to make it easier to re-hypnotize. Some people are very quickly hypnotized just by being in the presence of the hypnotist.

The study of hypnosis raises other interesting questions. Just what role does hypnotist's suggestion and subject's expectation play in hypnosis? People do like to relax. They also like to be directed in what to do. People often readily submit to someone in control (the authority). Some hypnotists assert there is no unusual state of consciousness only that the subject does what is expected when relaxed. Some say all they have to do is establish rapport and make the subject comfortable to increase suggestibility. Subjects will do what is expected to please the individual giving the suggestion.

Some scientists have critically examined the phenomenon of hypnosis and have demystified much of it. W. Edmonston has conducted experiments in hypnosis and has extensively researched its literature. He surmised hypnosis is not an altered state of awareness. He describes the change induced by hypnosis as nothing more than relaxation. Even if a subject appears not to be relaxed physically, he is relaxed mentally.

Edmonston found if subjects are told not to relax that they could not be hypnotized. However, the fact they complete that directive is itself instructive. Edmonston noted the same physiological changes (for example EEG [electroencephalograph] determination of an increase in alpha-brain waves) that occur under hypnosis also appear during relaxation.

G. Wagstaff has intensely studied hypnosis and concluded many people comply with the directives of the hypnotist because they are expected to. Studies have shown it is easier to carry out orders than suffer social embarrassment or guilt. In one interesting experiment a group of subjects are told beforehand only unintelligent and psychopathic people cannot be hypnotized. After hypnotic induction many said they were hypnotized. Another group is informed the experiment is a test of gullibility and imagination. Most of these subjects end up saying they were not hypnotized although the same procedures were used.

Perceived competency of the hypnotist also affects results. If the hypnotist informs the group that he is new at hypnotizing, fewer subjects are hypnotized than when he is perceived as more experienced. Here again, the role of authority influences results.

Other studies revealed many subjects were not sure if they were hypnotized or whether they acted on their own volition. So why do people say they were hypnotized? It has been suggested that they act according to "demand characteristics." They act out a role they believe to be suitable to the situation. Once the procedure is started, the subject becomes ever more obligated to play the role. This is self-persuasion. The hypnosis effect arises from situations where an individual passively complies to suggestions while relaxed. The next question is how does man function in everyday life?

The Need to be Taken Care of (Wanting to be Wanted)

Experiments have been conducted to study the effect of security attachment. In 1950 R. Spitz reported a study involving the placement of infants in strange rooms—without their mothers. He called the fear they experienced "eighth-month anxiety." He found babies exhibited this fear as early as five months and as late as twelve months. These types of experiments are limited for obvious reasons.

In 1959 H. Harlow and R. Zimmermann studied the effects of "contact comfort" on Rhesus monkeys. The experimenters constructed two wire-mesh surrogate mothers. To one surrogate was added a nipple for supplying milk. To the other was added a soft-cloth covering, but no nipple. Although the infant monkey received nourishment from the wire-surrogate mother, it did not obtain contact comfort from it. When not "feeding," it consistently clung to the cloth surrogate mother. If the cloth surrogate is removed from the cage, the young monkey, having known no other "mother," would grimace and crouch in the corner. When the cloth surrogate is reintroduced the infant would quickly rush and cling to it. Harlow and Zimmermann found that clinging to the cloth surrogate was more important than nourishment.

Security attachment in humans is important. Babies cling to mothers: in times of duress, in strange surroundings, and in the presence of strangers. That "need to be taken care of" carries over into adult life. Although social adjustment supplants the literal need for adults to cling to others, we still find that most people cling to others in a figurative or psychological sense.

Typically, security transference occurs. When some children are separated from their mothers, they search for and develop other sources of solace (or security). Here is an obvious example. Cigarette smoking supplants the sucking of a mother's nipple. The use of cigarettes becomes a partial transference of the need for security. Smoking becomes a "substitute security blanket." Tobacco is also physiologically additive.

People's security blankets take many forms. Some people develop a security orientation toward things of which they have no control. Keep in mind: Throughout the following discussion, the meaning of behavior is found in the search for self-security (either deliberate or unintentional).

We arrive at the not so surprising premise to understand behavior. "People behave in a way to become more secure." This does not mean a person is necessarily aware why he does what he is doing. Nor does it mean a person will always act to obtain "real security."

We first discuss a major means for obtaining personal security—close-intimate relationships. We present an analogy representing physically how a person reacts psychologically in seeking security. In the following, a woman seeks to increase her chances of survival and be more secure.

The scenario is a small boat with three people. A storm destroys the boat, and it sinks. Each individual now faces only two external objects of possible dependency (the other two people). The lives of three people are in serious jeopardy. They have no "life jackets" and nothing else (no "crutches") to hang on to. One man (man A) is the average athletic type and a good swimmer. The other (man B) can swim but is not in good physical condition. He is not a very secure object for the others. The third individual is the woman-subject who is also in poor physical condition. She has enjoyed a long relation (maybe married) with man B. However, lately she has been attracted to man A. We find the woman harboring doubts about her present security-blanket man B.

"Everyone wants to be taken care of." The three individuals tread water. After a short time, the woman is close to physical exhaustion—indeed an insecure situation. Struggling, she swims to "hang on" (security-wise) to man B who is having some trouble himself. To her consternation she soon discovers he is not as strong (secure) as she thought. He is not strong enough to keep her afloat. Her struggling greatly weakens him. Hanging on to another adds weight (or responsibility) to the other individual. They both begin going under. They are practically fighting each other to save themselves. The woman quickly abandons man B for man A because it appears to her that man A is stronger. She feels more secure in her chances of survival. Man A is perceived stronger because he had no added weight (responsibilities) causing him any problem.

Struggling to survive in water represents what happens often in relationships. Each individual (most probably egocentric) marries the other for security. This poses a problem. How can the woman have real security in a man who is insecure to the point of placing his security in her? (In the analogy they end up clinging to each other.)

The reason people behave the way they do is seldom recognized by those involved. Rather, people "react" to the situation they face in a way to better their immediate well being. They act in their self-interests (the major behavioral premise). Yet, it is those reasons that most people marry. It is also those reasons that later destroy the relationship.

One person says he, or she, "fell out of love" with the other. He is unaware of the real reason the relationship disintegrated. The meaning of this is quite simple. Each person married ostensibly for love. Though the woman once said she loved the man it really only meant she "wanted" him for her personal security. Therefore, saying I love you, means I want you. That is, I want this person to take care of me (the reason). This is what happens. Like the woman in the water, when she had obtained what she wanted, the basis for the wanting disappears. The reason she wanted him was that she did not have him (as a security blanket). That is, she did not have him taking care of her (saving her in the water). The wanting continues because of human nature. Since she already has what she wanted—she can now only want not to have what she has. She pushes him away (in the analogy she swims to someone else). Two things happen. The more a person is an individual—the more they want to be taken care of by something else. The complementary idea is—the more a person is taken care of (by someone other than themselves) the more they want to be an individual.

Why cannot the woman have real security in her husband? We are a mind-consciousness. We can only "have" knowledge. A basic characteristic of mind is the ability to infer knowledge and gain understanding. A mind cannot have anything physical. Hence, positing security in a relationship (to someone else) or anything external to one's mind fails to provide real security. In our analogy the woman cannot keep herself from drowning without help. So it is with other people in the water. The yo-yo effect comes into play. Supporting the woman, man A could tire and man B is then perceived by the woman to be more secure. She goes back and forth. This sustains her tendency to want.

The point is that the woman could not take care of herself and hence was not self-secure. She had not developed the ability to control her well being. She married for the wrong reasons. She married for what she lacked in herself (not having control) and hence needed someone to take care of her. If she had developed enough understanding to control herself, she would be free and capable of marrying someone for what she liked about the person despite her shortcomings. In the "treading water" analogy she could have developed the ability and physical conditioning needed to save herself without burdening others.

The general idea is people by nature (egocentric) lack self-control and attempt to control someone else. People gauge their controlling ability not by how well they can care for (control) themselves, but by how well they can get (control) someone else to take care of them. (The subject of control is discussed in Chapter 8: subheading titled "You and the Reality of the Situation.") The treading water analogy suggests someone can control someone else for a while. (The woman could sustain herself in the water without help for a short time.) The common tendency is people couple to someone just as insecure as they are.

We next examine some other objects upon which people develop fixed dependencies. Remember the basic idea. The thing upon which someone allows a security dependency is also that which controls him or her.

Difference Between Brainwashing and Submission to Authority

In each individual we see a certain factor at work. People do not have enough understanding and knowledge to cope with reality. So, they trust someone else to be responsible for them. It may be someone who even understands less than they do. (An individual can only understand someone else to the extent he understands himself.) We have already seen this factor at work in the phenomenon of marriage. It often happens one member of the relationship will lose his or her grip on his or her security-blanket spouse. Lacking a security blanket, self-identity disintegrates. This phenomenon also characterizes brainwashing. It is evident many people have a nervous breakdown. Eventually, such individuals are forced to look elsewhere for someone to hang onto (or to believe in). After many disappointments, the individual wants something incapable of disappointing his or her need for security.

What fulfills the need for security? Having found no real person who meets the need, insecure people often begin looking for ideas supposedly incapable of being found unworthy of being a security blanket: something that cannot be dislodged from a position of authority. The individual has unknowingly been led down a narrowing corridor: one leading in a direction of ever increasing dependency on something other than the self (the mind).

Sometimes insecure people join religious cults. Some say cult members are brainwashed. What difference is there between someone who has been authentically brainwashed (e.g., a prisoner of war) and those who have joined a religious cult? Let's first examine living conditions.

The subject is deprived of proper nutrition and lacks adequate sleep. If he does what he is told, he is taken care of. His remaining individuality (exemplified by seeking truth—asking questions) is surrendered and replaced by an attitude of not doubting authority. The authority is incapable by definition or remoteness to being questioned. The subject often recites certain phrases repeatedly until he espouses "their truth" as his own.

What distinguishes religious conversion from brainwashing? The only difference is in the treatment by authorities. The subject abides by the authority. He acts in concert with the new order—the carefully orchestrated communal effort. His former identity is isolated; then broken down. He becomes a different person: one defined by a new set of beliefs. He soon wholly identifies with the group.

The subject walked unwittingly into a cult that outsiders unabashedly declare has brainwashed the subject. If the subject is asked if he had been brainwashed—he denies it. He may say he is a new person: one that markedly contrasts with his old self. The means used to achieve the change varies, but the result is the same. The subject is brainwashed. He accepts certain statements as true but cannot offer any proof.

What is the difference between religious cults and main stream religion? People in main stream religion assert certain statements to be true yet offer no proof. For example, statements from an authority might be excerpted from a religious book or quoted from some figurehead. Some subjects say their source is inerrant, yet they offer no proof.

Converts give testimony about how their life has changed: that they are "born again" and made over into a different person. All because they have found the "answer." What is the answer? It is the authority one accepts on faith (without proof). One that is inaccessible and immune to questioning. How can a supposed answer be verified if it by definition cannot be questioned? More than once a week these people attend "brainwashing sessions" (place of worship) to have their beliefs strengthened. This is behavioral reinforcement.

Subjects become entwined in a web—in an ever narrowing mind trap. It protects them from self-doubt and the probing questions of outsiders. They surrender their life to unproved statements and lies in exchange for a security blanket—an authority shielded from not only themselves—but from seekers of truth. No matter how members of cults or mainstream-religious organizations are inculcated, the result is the same. The difference lies only in the degree or type of treatment received.

Brainwashing by mainstream religious organizations is usually deceptively subtle and cleverly done. Yet the techniques are the same. Instead of leaders proving their positions, they instill fear: fear of death, fear of hell, unless the subject accepts "their way." Their interpretation of holy books is the only one that can save them. Join and be saved.

Psychological mistreatment coerces cooperation. (Some groups inflict physical punishment upon their subjects.) Some joiners begin to doubt themselves and their ability to discern fantasy from reality. They surrender themselves completely to "the organization." Watch an evangelist psychologically beat people into submission.

Those who have no control over themselves, control those around them. Many churches were founded by individuals who had so little self-security that they attempted to control others. The more successful they are, the more convinced they become of their authority to fulfill their mission.

The history of religion is saturated with the blood of self-righteousness. The Inquisition, the crusades, and thousands of witch trials attest to this.

In the midst of all this, however, have been individual seekers of truth (many were also religious). Galileo is an example. He turned his telescope skyward. He announced his findings. They supported the truth of the Copernican theory of earth motion. People like Martin Luther and Catholic officials condemned him. They pointed to scripture to prove him wrong. Like most other people, they worshiped the Bible and not what is represented by truth. Their ultimate authority extends no further than scripture from their holy book. The Bible was their protection and security blanket. This is not to say those who do brainwashing, do so intentionally. They were brainwashed themselves and were merely dutifully passing it on.

People are defined by their security blanket. To the extent they identify with it, they are defined by it. They abandon their only chance to seek truth in exchange for something that could be a lie.

Is there safety in numbers or could thousands of people be in error? There should be some normal people in any large group. Some people should be sane enough and wise enough not to mislead the rest of us. Are there such people?

Will Normal People Please Stand Up

There is practically no difference between brainwashing of war prisoners and what happens to mainstream-religious converts. A reasonable question is "can a normal person be defined?" Most people cling to external-security blankets (religious or otherwise). Are there behavioral traits that distinguish normal people from those considered not normal?

The local group or cultural setting largely defines normality, in a sociological sense. The mores and beliefs of the group are usually the standards by which normality is measured. That does not help us much. If someone is noticeably different from the norm, he becomes a social deviant and most probably labeled not normal. Someone is thought to "have a problem" when he espouses statements not generally accepted by the group. This happens anyway whether he can prove the truth of what he says. He is often considered deviant. Again, Galileo is an example.

Can professional psychologists define normality? Can they differentiate between sane and insane people? (Sanity is more a legal term than a psychological term. We consider sane to be synonymous with normality.)

Studies were conducted in the early 1970's; the results published in *Science*. The article, by D. Rosenhan, was tilted "On Being Sane in Insane Places." It provided information on the questions just posed.

It was believed certain behavioral characteristics could be used to classify individuals as either normal or abnormal. Another view held that most behaviors are determined by the setting in which the individual is found. Therefore, observed characteristics reflect more of the observer's percepts than it does of those observed.

The experiment involved eight pseudo-patients who were admitted to twelve mental hospitals. There were three psychologists, one psychiatrist, a pediatrician, a psychology graduate student, a housewife, and a painter. None had a previous history of mental problems. Each pseudo-patient was instructed to get out of the hospital without help.

They phoned the hospital and complained of hearing voices. Once admitted, they behaved their normal self. If asked by staff how they felt they would reply "fine." They stated they no longer exhibited the symptoms.

During the experiment none of the participants were detected by the staff as being a pseudo-patient though they were soon openly taking notes of their experiences. Only some other patients thought the pseudo-patients did not belong there.

After a patient is labeled schizophrenic there is little he can do to remove the stigma. Much of a pseudo-patient's everyday behavior was interpreted by the staff to coincide with a symptom of their diagnosed condition. For instance, one pseudo-patient was bored and paced the hall when a staff member presumptuously asked: "Are you nervous?" The staff never considered whether the confining environment, or even something they instigated could have affected the behavior observed in the pseudo-patients.

The experiment shows a line cannot be drawn in separating normal from abnormal behavior. That is not to say there are not individuals needing some form of psychological help, only that trained professionals cannot adequately define ordinary individuals as normal. The lesson is that many people put labels on things and then "think" (deduct) they understand.

We have already seen from Milgram's experiments that insecure people have a natural inclination to submit to some outside authority. This tendency, along with social pressures, often leads people on a one way trip. One where, step by step, they are pushed along by deceptively subtle brainwashing. Individuality is eroded and eventually surrendered in exchange for some outer security blanket with which they identify. That does not mean they are defined realistically by their assertions merely because they say so. They are defined by their assertions if what they identify with, is scientifically proven by them to be what they purport. We shortly examine the meaning of this statement.

Key Points:

- Individuals either control situations (in the "rest of reality") or, by default, situations control the individual (self).

- One can only control situations if he or she understands.

- "Brainwashing" occurs when the self (accepts or) asserts statement truth without verification by reality.

- Psychological experiments show people willingly conform to social pressure. Social conformity explains many inhuman and heinous acts.

- People conform because they have a "need to be taken care of." It is easier to depend upon someone else (or "their ideas") for security than to rely on one's own thought processes (to discern true statements from false ones).

- The concept of normality is muddled because there has not been a definitive understanding of behavior.

The conceptual role of definition, as it pertains to science, needs to be stated explicitly. Conceptualization automatically poses the question whether there is a purpose for man's perceptual abilities (i.e., awareness). Man defines things he experiences. We need to examine definitions as they pertain to scientific understanding.

Chapter 7

Science and Definitions

"There is a crucial difference between a statement
being true and whether it is known to be true."

In science there must be means for testing any given hypothesis. A hypothesis is
an explicitly defined statement or concept that is possibly true. It is an
unverified-scientific concept. A hypothesis can represent the physical or
nonphysical.

Physical hypotheses are verified empirically. Verity is established when
correspondence links a hypothesis to "what" it represents.

Nonphysical hypotheses pertain to abstract concepts describing behavior and
higher-level generalizations. They are "tested" by coherency. They should not
be contradicted by other well-established concepts.

Terms and concepts must be clearly defined in science. Otherwise, it is difficult
to do any testing. Testing really concerns definitions. Testing is often arduous
and slow because definitions must be refined during the verification process. As
a seeker of truth, we are skeptical of the completeness of any given definition.

Understanding does not spring from projecting preconceptions. Defining things
by belief creates semantical problems. We learned about believing and knowing.
We need to better understand how they interrelate.

Let reality define itself (through experimentation and reason) and the problem of
semantics practically becomes a mute issue. Scientific methodology is used to
seek reality based answers. If all the big questions have been already
satisfactorily answered, then it is difficult to explain why those questions are
still being asked!

The denotative meaning of a word is determined by its correspondence to
something in reality. Semantical difficulty arises because someone has not
sufficiently understood how the word corresponds to reality.

The idea is to be able to define a term by reality. This is the way of science. All
people have this "ONE REALITY" in common. It is the final arbiter of what is
true! Once understood, there should not be any semantical difficulties. So how
are terms to be defined? An example showing the conceptual basis for
ascertaining definitions, follows:

Consider a specific nounal word. There are objects in reality to which the word
corresponds. For example, there is the general word "car." There are specific
objects in reality to which that general concept refers. Elaborating: There are
phrases (showing relationship between words) which are understandings
(relation by formula). In this example, we have the formula (or demonstrative
statement) "my car." There is a connection between the owner and a certain car.
To know whether this is true, a skeptic would ask for evidence (title
registration?) showing ownership.

The point is to approach an elimination of skepticism. To do this one must strive to exhaustively eliminate all possible explanations for the situation being examined. To help end any remaining skepticism, you have to dig further. Using the above example, ask yourself: Is this really the car's registration? (Make a phone call to the appropriate governmental agency to investigate.) We seek truth to know that a given statement represents "what is real." That is the ideal. In practice the idea is to "proceed according to the evidence."

When you can use your mind and search for truth, there is no reason to accept any statement as possibly true without proper-supporting evidence. Is there any reason not to accept what the evidence suggests?

Are you skeptical? One is not a skeptic because of belief. Religious people are skeptical of most religious organizations, but exempt their own affiliation from skepticism. This is no different from someone who is skeptical of other people's beliefs, but not skeptical about their own ideas.

There is no problem in reality. It is the way that it is! (A tautological point.) No, the problem seems to be the individual. The individual is also a part of reality. Therefore, the problem is not the individual. The problem is how the individual ascertains verity.

The problem narrows to the methodology the individual uses to ascertain verity. How are true statements distinguished from false ones? However, there is a crucial problem that will shortly be addressed.

What is the test of truth? First, it is prudent to refine the definition of truth. Truth is that information (statements) which corresponds to, and therefore represents reality. Reality is the test of truth. This definition generally suffices for testing statements corresponding to the immediate-physical world. Extreme skepticism aside, this is an acceptable "working definition."

> It is not this simple because of contradictions for naive realism: reality is not just what appears to us. The ubiquitous bent-stick example suffices. A straight stick poked into water appears bent.

Truth in the higher-level-classificatory definitions must also meet the criteria of noncontradiction (within the classificatory system). General concepts must be coherent—self-consistent. This test requirement is met by eliminating all contradictions. We still have a crucial problem. Let's pinpoint the problem.

A Crucial Difference

For us, the problem is not so much what is true; but rather, how do we (as individuals) "know" whether a given statement is true? (This is the crucial problem.) What does this mean? What does knowing mean? Knowing means one can prove by reality that a given statement corresponds to some thing or structure in reality. Let's restate the question. Do we know to what in reality a statement corresponds? Our problem and attention must focus on whether we "know" if a statement is true.

We know when we can verify that a statement represents some aspect of reality. A statement may be true. Yet, if you cannot verify by showing it represents reality, you are not warranted in saying you know (that it is true).

There is a crucial difference between a statement being true and your knowing the statement to be true. If you say a given statement is true (implying you know) but cannot prove it by reality, it is a lie that you know (it is true). If you say you know something, but cannot verify it (by reality) then you are a liar (because it is not true you know it, although the statement may be provable by someone else).

It is reality that we can directly know (by experience—knowledge by acquaintance: B. Russell). (However, knowing does not mean we understand.) All we can know about statements (propositions) is whether they are true or not. The common mistake is confusing the word (or statement) with the object (or situation) it represents. This is reification (or concretization). There is the word car, then there is the object car. People confuse statements with the reality they represent. This is the problem of belief. It is a statement, which a person (fantasizingly) projects onto reality. In short: This is the real meaning of the often heard phrase "you create your own reality."

> This phrase is a cornerstone of the "New Age" movement. It is not new age at all, but rather, old age (Hinduism, Buddhism, Chinese I Ching, for example) recycled. Sorry, the new age, if any deserves that nomenclature, is science.

If each individual creates reality there would be as many realities as there are people who fantasize (that their beliefs represent reality). If this were the case science would be impossible—and we know it is a very successful endeavor—hence the contradiction.

Science concerns knowledge. Do not confuse science with its application—technology and engineering. Applying knowledge does help verify or refine given scientific concepts.

A statement is true because it represents some aspect of reality. If you cannot verify it yourself, you are not warranted in saying you know that it is true (you are only warranted in saying you "know not") until you learn how to verify the given statement. Then what is the meaning of saying one "does not know" something to be true?

Proving a Negative

"Knowing not" is verified when one cannot produce the proof. Yes, this is a general example of proving a negative—the proof of "knowing not." The proof that someone "knows not" is that he cannot produce the evidence that demonstrates the truth of the statement (being considered). People are fooled because they believe if one does not know that it shows a lack of knowledge. To the contrary, it is the beginning of knowledge.

One knows he does not know when he cannot verify (or prove). Not to know this, is not to understand science (and logic). We learn from reality by verifying our ideas. Ideas are produced by reasoning. If someone does not reason to extricate themselves from fantasy, they probably never will. It is like physical exercise—no one can do it for you. Fantasy results from not verifying ideas.

Scientific methodology narrows possible (empirical) explanations by experimentation. It is easier to prove what is not true (i.e., prove a negative) than to prove what is true. To prove a statement as not true only takes one contradictory instance.

To prove a (general) statement as (positively) true calls for an exhaustive examination of every specific correspondence to which it refers. For example, it is easy to prove not all swans appear white by just finding one black (or other colored) swan. It is impossible to prove all swans appear white by examining every swan that ever lived (there are black swans). This is where the problem of (logical) induction emerges. However, at the higher levels of classificatory structures, this is not a problem (in a closed-hierarchical system). (B. Russell attempted such a structure—for logic—in his "theory of types.")

He who says, you cannot prove a negative probably has not discerned the difference between a statement being true and whether he knows it to be true. This is why there is science. As individuals, we do not necessarily experience reality, as it is itself. If we did, there would be no differences even in opinion.

The appearance is different from the reality. That was the purpose of asking about a thing's color. If we experience reality just as it is itself (the noumena) as Kant would say, there could only be one answer. There are two answers. Only one of which explains the other. That an object is in itself blue, for example, explains why it appears yellow (which is the complementary color of blue). (The underlying reality contributes to the appearance of things.) The problem of induction implies that proving positives is practically impossible. Science proceeds by proving what is not the case (not true). This process establishes inductive hypotheses as more probable.

Testing Definitions for Truth

Before one can ascertain which statements are true, he must first get lies out of the way. From what we learned above, it is simply a matter of not making personal assertions of truth, but waiting until sufficient evidence is accrued so reality can assert itself. We cannot honestly state what the big-picture-jigsaw puzzle (representing reality) is all about until after it is completed. Semantical difficulties indicate one does not understand reality.

How evidential facts (pieces of the jigsaw puzzle) fit together yields understanding (of the jigsaw pieces). It is not for man to assert truth, but for reality to demonstrate it.

How do we test for truth? We make conjectures: general statements that help us further classify phenomena by their similarities and differences. (An example is the classificatory system of plants and animals.) Then we develop tests (experiments) to determine if we correctly conceived of a covering generality that "categorically" explains the phenomena grouped under the proposed concept.

The idea is to find a single all encompassing concept under which every phenomenon can be subsumed. Every positive statement will be included within this "all-embracing system." That is the goal of science: to explain how everything can be derived deductively from one principle. Many scientists doubt it can be done. Stephen Hawking, the physicist, figures it will be done. So what does all this searching for truth have to do with us? It helps to understand and define the nature of the self.

Key Points:

- Scientists explicitly define statements or concepts as hypotheses to test against reality (for truth).

- Physical hypotheses are tested empirically. Nonphysical hypotheses are tested by their coherency with scientifically established concepts.

- Reality is the final arbiter of what is true. There is a crucial difference between a statement's truth and whether someone knows (can verify) its truth.

- Negatives are provable. If the solution is "not this or that," then it is otherwise (the negative).

- One obtains truth (knowledge) by suspending his or her judgments and beliefs. With scientific help, truth is knowable.

Chapter 8
Defining the Self

"A person either controls situations, or is controlled by
situations."

You and "the Reality of the Situation"

How can individuals be defined in functional terms? It was previously stated
there is your awareness or "self" and there is that of which you are aware—the
"rest of reality." But, you are not aware of the "whole" rest of reality: only a
particular series of events in the rest of reality. Any of those series of events, in
the rest of reality, we call the "reality of the situation."

We usually deal with these events conceptually. We have narrowed our focus to
two things: "you" (the self as you really are) and the "reality of the situation."
Either you control the reality of the situation or the situation controls you.

Each situation (you come in contact) poses a potential problem. These problems
afford opportunity to move along the road to intellectual maturity: the road
leading to knowledge and understanding.

Human nature (ignorance and egocentricity) limits a mature understanding of
things (i.e., physical world). This is so, unless some "outside pressure" emerges,
forcing one to seek truth (to alleviate suffering).

You can change or act in one of two ways. Each way is a method leading in
opposite directions. Each way yields dramatically different results. One is a
destructive process; the other is a creative process.

Controlled by the Situation and Self-Destruction

Most individuals tend to "destroy" themselves in one way or another. People
consistently behave (react to situations) in a way halting their progress on the
road to greater knowledge and understanding. (Only going that route can the real
self be defined.) Excessive use of alcohol, drugs, gambling, unnourishing
relationships, and religion, are ways to run from reality. Anything used as an
antidote to suppress anxiety induced by not understanding (some situation)
hinders solving problems presented by the situation. Treating symptoms does
not correct the cause of the problem.

Let's look at some brief examples. A clear example is the alcoholic: "man with a
problem." It typically begins innocently enough. A man stops after work to
drink a couple of beers. Drinking is a way to avoid going home to face a
situation: perhaps a deteriorating home life. The man has lost control over the
situation. He does not understand it, and it controls him. His family probably
despises him and treats him badly. This compounds the problem. For him family
life is no longer worth the effort. Not understanding the situation (insecurity
caused by lack of understanding behavior) he does not know what to do nor
know how to correct it. He increasingly takes refuge in his already established
escape pattern—drinking alcohol.

We have described what psychological studies have shown. People naturally look for something to be responsible for themselves: something in a position of authority (or somebody to take care of them). By default they allow themselves to be controlled by some "situation" because they refuse to assume the responsibility in understanding it.

They are no longer responsible for their own well being. In the example the man's former security-blanket—the family—has been transferred to alcohol. His whole life has been a constant shift by kind or degree to something that is even less capable of taking care of him. His dependency narrows. His life focuses on one thing—alcohol. He surrenders self-control. He exists at the mercy of the situation—alcohol. What began with a couple of beers turns into a progressive nightmare of dependency.

Depending on something outside oneself (the situation) gives the individual an excuse to be less responsible. He slowly surrenders control over himself and merely becomes a "victim of circumstances" (a sufferer of the situation).

If you have a security problem like this, unless a change occurs, you live and will die at the mercy of the situation that controls you. If you are in this or any similar situation you become defined by your dependency—your security-blanket. It becomes your authority figure: your God.

The price paid for security dependency is destruction of self (identity). You cease to exist as an "individual" because you become increasingly defined by the situation-fixation holding you captive. When you finally realize what happened, you may be too weak (insecure) to do anything about it.

Another example is a business man who slowly loses control over his business. He might have defined himself as a successful entrepreneur. The business goes sour with creditors at the door. It could have been caused by tough economic conditions. It might have been no direct result of the man's business acumen. That is the peril of allowing a situation to become a security-blanket. Something outside the self can always be taken away. With business success swept away, the man—by definition—is swallowed up in its wake.

Another example is the rejected lover. Some people live for the one they love. When the object of their love (the supposed love object) leaves them, they do not function well. Life seems not worth living because their rationale for living is gone.

The problem was there all along. It takes the disappearance of the love object for the real problem to surface. The problem is generated when the subject did nothing to stop his lover from also becoming his security-blanket. One possibility is suicide. Death by one's hands is total realization of self-destruction.

Because of ignorance most people do not perceive themselves as they really are (see following heading: "The Three You"). People are unaware of what they do not perceive or do not understand. They do not realize self-destruction is occurring. Deciding not to seek help is a choice itself. Not decreasing alcoholic consumption endangers one's health and well being.

Most people tend to self-destruct. There are many forms. Most cases of self-destruction are understood by observing what provokes insecurity. A person's defensiveness often indicates insecurity. It may be about a personal relationship, money matters, job, or religion. If an individual's security-blanket is challenged by verbal assault or by attempt at removal, he becomes belligerent or "withdraws."

Although not as revealing as subjects people avoid, a lot can be learned from how they handle topics they will discuss. Many people behave as though they are knowledgeable in subjects beyond their areas of expertise. When subject to probing questions people often change the topic of conversation. Sometimes they project onto the questioner demeaning declarations implying he does not understand. They offer no explanation to support their charge.

Recall the analogy of the three-boat-wrecked people. Their plight would not have developed if they were responsible for their own lives. How does one become self-responsible?

Controlling Situations and Self-Creation

Let's reiterate the basis for understanding. There are two things to consider. One is the "reality of the situation" and the other is your "self." You (the self) can allow a situation to control you, or you can control the situation.

In self-destruction, something outside yourself (by default) assumes responsibility for you. In self-creation, you assume responsibility for your "self." Recall: There is no problem in the situation. The problem resides in the individual. He either does not understand the situation, or he does not apply the understanding.

Solving a problem is a two-stage process. The first stage involves understanding the problem. Acknowledge the problem and recognize its consequences (determine its meaning). For example, driving a speeding automobile is hazardous. Why is this a problem? Is the automobile speeding too fast for conditions? The immediate answer is (in our example) the driver has relinquished responsibility for his actions and lost control of the situation.

It is an error to suppose (or believe) anything outside one's self will assume responsibility for the self. Consider the typical-personal relationship. Between you and another person, it will not be yourself who will be responsible. Each person clings to the other. Neither person is capable of real responsibility.

Everything has a cause. Understand the problem's cause. Understanding leads to solution. Knowing the solution is not enough. The second stage is applying the solution to the reality of the situation. Many people falter at this stage. To apply solutions is to take responsibility. Many people are not secure enough to be responsible.

There are difficulties in correcting problems. One difficulty occurs from using an inappropriate methodology to understand the problem. Another difficulty is people knowing the cause of their problem and solution, but doing nothing about it. They may not have enough resolve (self-security) or confidence in the solution to apply it. They need help in reestablishing self-control.

The more self-control gained by experience, the more one is ready for other problems. If today's problems are successfully handled, there is a greater chance of solving tomorrow's problems. Failure often suggests a refinement for the methodology used to gain understanding.

As one acquires knowledge, he learns from many "reality of the situations." Each situation is like a puzzle piece. The more those situation-puzzle pieces "come together," the more he understands.

As the seeker of truth travels the road of knowledge, he becomes acutely aware of the "rest of reality." This process is an ongoing defining of the individual, a continuous "creation of the self." His self identity becomes "redefined" by reality.

Basic training for self-creation culminates in "self-relativization" (See headings: "Deducting Intrinsic Whatness" in Chapter 4, "Self-Relativization" in Chapter 3). This level of understanding is the foundation for "self-individualization." It engenders actualization of the self through self-creation (See "Self-Actualization" in Chapter 4). Because self-creation is an individual process, the author can use his experiences as an example.

While a fourth-grade student, the author decided to work toward entering a health profession. For that reason he studied science.

After graduating from college he entered dental school. Dentistry was not to be an end in itself, but a means to help people. Though not recognized, a problem had long been in the making. A fourth grader does not understand that people need more help than a specialized health profession can provide. Dentistry was incapable of helping the human condition. He had already pondered the meaning to life. Realizing a health career would not provide the best means for him to help others, the author withdrew from his original goal of dentistry.

The author had no particular plans. Something happened that changed his life. A close brother died. That death brought anguish. He was not bothered by the death itself, for his brother had lived an unselfish life: no regrets here. Leaving school and the brother's death were consolidated into a pivotal question. "What is the meaning to life?" That question pressures the self to seek an answer. If he does not obtain a reasonable answer, he realizes death means the end of life. (Refer to "The Letter" in Chapter 3.)

The author recognized the "meaning to life" is the central question vexing existential philosophers. Many of them decided there is no meaning to life because they could not find any. Without restating details, how did he resolve the "existential dilemma?" He reasoned out a construct that became his raison d'être.

The construct emerged as the premise that restructured the author's meaning to life. It meant he took sole responsibility for his actions. The construct concluded: because we are nothing in ourselves—we can be defined only by what we do for other people. Our meaning is defined by how well we help others. Now he had a "real" reason behind his original idea of helping people. The construct also implies the author had to minimize errors in reasoning. Why?: To not compound the problem of others while helping them.

The author had to understand more than those needing help, or he would be incapable of helping them. His reasoning abilities needed refinement. A formal education is not enough. Memorizing is one thing, learning from reality is another.

How does high-level learning develop? First, an analogy. An individual who trains physically becomes stronger. If he trains enough he can break records in certain physical feats. Not realizing it, the author had already found the key to "strengthening" his powers of reason. It enabled him to push beyond the limiting barriers of the (egocentric) self. It developed mental inferential abilities.

What does this mean and what are the barriers? The author no longer acted upon what he "wanted" to do. Instead, he acted upon the "reality of the situation." It was evident one can only obtain "real" meaning from helping others. The author was compelled by reacting to non-self circumstances, i.e., the "rest of reality," instead of "wants" emerging from the "self." He did not cling to anything outside his awareness or self.

Clinging to something produces security-fixation. Fixations do release emotional pressure caused by insecurity. Fixations effectively circumvent self-responsibility. Scapegoating hinders development of mental-awareness functions.

Security fixations act as buffers against reality. The most common form is described in the following example. If someone absolutely "thinks" a certain statement is true, he is not security-wise motivated to research and scientifically verify it. Though the opportunity for greater learning is in the "outside world," the egocentrist, by his very nature, is sheltered (from the outside world) because of ignorance. He is too insecure to do anything but cling to what is familiar. Familiarity is not understanding. The familiar needs to be understood.

What "training" pushes the mind to learn? Begin by not acknowledging any statement as true without scientific verification. Leave personal interpretations behind. No statements should be accepted as true on "personal" terms. It is not for man to decide what is true. Reality can "speak" for itself. Seeking truth is the exercise that develops the mind.

Instead of interpreting what something means, reason how it relates to other things. Reasoning generates understanding. All one can really "have" is knowledge. Reasoning directly accesses reality on its terms.

The author reached a stage where he could help people by understanding what in others was blocked by insecurity. Fixation is often the reason people cannot solve problems.

The author studied to understand contemporary science. He approached the frontier of knowledge. He was the "right person in the right place at the right time." Something happened which initiated the bringing together of all his accumulated "pieces" of knowledge into categorical subsumption.

The author began teaching science. He hoped to instruct his students how to use a refined scientific methodology to solve personal problems. One day one of his students asked "that question" which when scientifically answered suddenly thrusts him past the present frontier of knowledge. He entered the mind-expanse. He quit teaching and began a relentless pursuit of knowledge and understanding. Why? To better help people.

Recall Newton's third law, there is a price extracted one way or the other for what one does or does not do. If one elects to stay ignorant (again, no decision is a decision itself) the price is obvious. He will make more errors and endure the consequences of not having added choices greater knowledge brings to the decision making process.

A Representative Case History: Problem and Solution

Here is a simple case history using previously mentioned concepts as a basis to understand what happened. The case is simple because the subject's insecurity does not involve inhaled, ingested, or injected chemicals that foster addiction. It is a neurotic condition in which the subject behaves in a way to avoid responsibility.

People feel secure with what they are familiar. To assume added responsibility is to change. Change means venturing into unfamiliar territory. Many people reach a point when crossed that pushes them beyond the limiting effect of their past experiences. Any such challenge tests one's security. Here is one person's problem and its solution.

Reality of the situation: individual suffers from S-K syndrome. Individual with initials SK is not wise in handling money. He is very much in debt. He earns considerable money compared to the amount needed to meet his daily living expenses. When he acquires extra money that could pay previous debts, he uses it for a down payment on something new. The more he borrows the more in debt he becomes. Why does he do this when the solution is obvious and simple?

The simple solution: Instead of buying more items, he needs to use the extra money to begin paying off debts. To gain control over the situation he must temporarily deny himself. Why does not he do it? Printed here is a paraphrased copy of a letter sent to SK describing his situation and explaining the problem.

Letter to SK

> Analysis: SK, first it can be said either you control the situation or the situation controls you. What does this mean? Quite simply, either you initiate something (do something constructive) in the "rest of reality" or you remain passive and merely react to what happens in the "rest of reality."

> Remember, there is your "self" or awareness; then, there is that of which you are aware—the "rest of reality." Most people do no more than react to what transpires around them. Others take control. They initiate something

around them to help control the situation. Many of these people are successful although their decisions are often based on what they do not know, therefore based on what they think and believe. In that respect they are no different from those who succeed by applying an inaccurate understanding of the problem.

Refer to Figure 8:1, "The Self and the Rest of Reality." This diagram was inductively reasoned to explain and help SK solve his problem. It is an effective way to explain behavior.

Here again, we are talking about egocentric people on one hand and realistic people on the other hand. There are some major differences between the two types. They are opposites. The egocentrist derives solutions deductively. The realist solves problems inductively.

The egocentrist side steps responsibility for his behavior. He can say he is a victim of circumstances (therefore controlled by the situation). He does not see himself as he really is.

The realist controls his situation. He takes responsibility for his life and directs his activities (controls the various situations he encounters). His abilities are tested by the situations he faces. He more clearly sees himself as he really is. This is represented by the statement: If you do not play the game—you cannot lose—but you cannot win either! But the idea is to play the game. So the next thing is, you must know the rules!

If you play the game and do not know the rules you make decisions based on what you do not know. To win you should minimize the gamble. You must know why you do something before you do it, and why you say something before you say it. You must not only take control of what you do, but also take responsibility for its consequences.

Most people are afraid of what they do not know. Most people are afraid of failure. Fear prevents them from taking control of their situations. That almost seems forgivable because they really do not understand how reality is structured and therefore do not know why things happen the way they do.

There is a second reason people do not know why things happen the way they do. They do not know how to control the situation. They naturally develop a habit of not coping effectively with the world outside their "selves." They offer many reasons (really excuses) for not controlling the situation. They fear failure. Because they do not know nor understand the "rest of reality," they naturally doubt any course of action that could remedy their problem. They should doubt. To doubt and do nothing, does not solve anything. Questions have answers. To find answers calls for a systematic methodology.

The individual is redefined as he adopts new methods to understand "difficult" situations. There is a sense of security surrounding you the way you are now. It is a false security because you will not stay the same. The reason is the "rest of reality" is ever changing around you. If you do not begin controlling the situation around you, you will be controlled and molded into something you will not like.

Someday you will wake up and discover you do not like yourself. You lose self respect. You will not like the situation in which you will find yourself. This happens to many people; they eventually realize it. They just do not know where they went wrong. Many people know there was no problem when they entered adult life. A person's character is shaped by his experiences.

You really cannot know all the consequences of your decisions or actions. At first, it is a little scary to take control of the situation. But that is the challenge! It is what makes life worth living. You will know the consequences of not taking control—you will be controlled! So what will it be? Do nothing and you will be controlled and molded into something of which you have no control!

How do you take control of yourself? You got into the situation step by step. You get out of a situation step by step. With help from others, if needed, objectively analyze your present situation. Ascertain its meaning and understand its cause. What are its consequences? Then ask what would be better for you in the end. What can you do to control the situation? Next pick a "little" problem and take control of it. With one success you will be more confident to take on bigger problems. As pennies add-up to dollars and dollars add-up to hundreds, so it will be of every little situation you control. Each step slowly leads to success in self-control.

So SK, now you should understand why you have the problem. Now just take one debt you owe and pay it off without incurring more debt. The more you do this the easier it gets, and the more confident you will become. And notice—slowly you will develop responsibility by what you do. You see—thinking never got anything done! You must act! There is no reason to wait any longer because if you just stand by and do nothing—your situation will only deteriorate. From your experience you know this is true.

Now you should understand how you got into this situation. Originally, you might have tried controlling situations, but having failed a few times, you soon developed, by default, the habit of not taking control. It seemed to you that if you did not actively interject yourself into the situation to change it that you could not be held responsible for the consequences. In a phenomenal sense whether you fail by default or by trial, you cannot escape the responsibility for the consequences of a situation that could have been controlled. You will be changed eventually by these very same circumstances anyway!

There is a difference between failure in self-responsibility and failure because of extenuating circumstances. If you control circumstances (situations)—you do risk occasional failure. But, you can also learn by failure. Failure offers a second chance to gain understanding. But, there are those who are too afraid to change (because of insecurity) despite having repeatedly committed the same error.

Because of one's present state of insecurity he should seek to better his position. There is no staying the same because the situation around you is constantly changing and that will inevitably leave its mark on you. So if all you do is react to reality emotionally—you eventually become mentally crippled and incapable of controlling even simple situations.

You now understand the consequences of not taking control of your situation. Once you understand—it is simple. Remember the saying "a winner never quits and a loser never wins." Next, do something about the situation. You need to apply the solution despite yourself. That is, in spite of your immediate need for security.

End Letter

We have seen what happens to people who do not face reality. They foster habits (or the lack of them) which put them on the path to self-destruction. Instead of acquiring new and more refined methods for coping with reality they become more inhibited and less tolerant of situations they face. Their inhibiting habits spill over into other areas. How someone deals with one problem is a clue to his methodology.

We have also seen what happens during self-creation. The realistic person analyzes situations. He checks to see if there is any reason not to study his situation. If it appears investigating will hurt no one else, he should pursue. If there is no reason not to investigate what one does not know, then continue to learn despite personal desires or inhibitions. These obstacles are often the very things leading to new understanding.

A person's future should not be blinded by his past. One should continue despite his self. Although the "self" is defined by one's experiences, that should not preclude him from "becoming other than what he has been." This is how one matures and learns.

The self is changed by knowledge. That should be welcomed. To do otherwise is to live in the past. Again, fear is caused by what one does not know. So one should push forward and learn all he can.

It is not easy to push forward and allow change. The reasons go deeper than mere thinking. The question can still be asked—why do we become controlled by situations? Or, if we do apply a solution, why must it be carried out despite our "selves." Why does the realistic solution run counter to our nature?

We have seen egocentric and realistic behavioral types tend toward self-destruction and self-creation. Why do their generating processes of deduction and induction seem out of balance? Why does self-destruction prevail over self-creation? The next subheading supplies an answer at the phenomenal level. We turn our attention to studies of the brain to seek an answer.

Split-Brain Phenomenon, Mental Function, and Behavioral Types

The brain is anatomically divided into two halves called hemispheres. It has been known each hemisphere is associated with movement on contralateral sides

of the body. For example, when an individual suffers a stroke of the left
hemisphere, he incurs paralysis on the right side of his body.

Not all functions are equally divided between the two hemispheres; that is, in the
temporal and frontal lobes. M. Dax, a French physician, presented a paper in
1836 describing observations that aphasia (loss of speech) seemed linked to
damage of the left hemisphere. Many people who had suffered a stroke resulting
in paralysis on the right side of the body, also loss the ability for articulate
speech. Dax had not observed aphasia with any right-brain-damaged patients
and concluded that the left hemisphere controls speech.

P. Broca, a French surgeon, also studied functionally impaired patients. He
concluded by 1861 that speech is articulated from a particular cerebral location
in the left brain: now called Broca's area. In 1870 two Germans, G. Fritsch and
E. Hitzig, using electrodes on dogs, mapped out motor areas on the cerebral
cortex. About the same time J. Jackson, a British neurosurgeon, observed that
epileptic seizures began with twitches in one part of the body and spread to other
parts. He believed the fits were associated with certain areas in the motor cortex.
K. Wernicke, a German neurologist, in 1874 discovered that damage to the
posterior region of the left temporal lobe is associated with difficulty in
understanding speech. This area was subsequently named after him. His
discovery and others contributed to the idea of cerebral localization.

Accumulated evidence suggests the right brain is associated with musical ability
and spatial recognition. Some patients could sing although they had left-brain
damage and suffered speech impairment. In other cases right-brain-damaged
individuals sometimes exhibited a marked decrease in singing ability (amusia)
yet had no difficulty speaking. Jackson reported that damage of the right
posterior lobe of one patient rendered him incapable of recognizing faces (facial
agnosia) and objects. There seemed to be an asymmetry for certain functions.
These findings not only lend credence to localization but to cerebral dominance,
an idea espoused as early as 1868 by Jackson.

The two halves of the brain are connected to each other by over 200 million
fibers grouped into many bands or commissures. In humans the largest is called
the corpus callosum. It measures 3.5 inches long and a quarter of an inch thick.
In the 1930's surgeons cut this tract in a patient with a tumor between the
hemispheres. They were surprised when the patient apparently suffered no
consequences. It was discovered seizures in some epileptics decreased if there
were a tumor on these fiber tracts. Because of questionable success the
operations were stopped.

R. Sperry, an American psychobiologist, was trying to find where certain
knowledge is stored in the brain. Specifically, where are memories of certain
tasks found in the visual cortex? In 1953 he asked a graduate student, R. Myers,
to begin a series of experiments using cats.

Vision in one eye normally was conveyed to both hemispheres. Myers found
that visual information presented to one hemisphere could be rendered
inaccessible to the other. He cut the cat's optic chiasma—a pathway through
which visual information is transmitted from each eye to its contralateral

hemisphere. This procedure limits direct transmission from one eye to the ipsolateral (same side) hemisphere.

When the cat, with a patch placed over the right eye, is taught a visual discriminatory task, it still remembers after the patch is shifted to the other eye. This indicated information was still getting to the contralateral hemisphere.

Myers next severed the corpus callosum and the optic chiasma. He repeated the experiment. Again the cat learns the task with the patch over one eye. With the patch switched to the other eye, the cat reacts to the task as though it had never learned it. Learning the task again took just as long as it first did with the patch over the other eye.

For example, the cats could be taught to recognize a circle with one eye and one-brain half and taught to recognize a square with the other eye and other brain half. For a given figure, if the patch is switched to the other eye the cat no longer associated the task with the pattern. Findings suggested each half of the brain could function independent of the other. More studies showed animals did not exhibit any ill effects from the operation nor were there any apparent differences between the hemispheres.

Armed with these new findings, P. Vogel and J. Bogen reconsidered the use of commissurotomy for patients with intractable epilepsy. They performed a complete commissurotomy on a forty-eight-year old patient who had suffered shrapnel wounds during World War II. Before surgery nothing abated his seizures. After surgery he was freed from epilepsy. The veteran did not suffer any serious-personality changes.

In 1961 Sperry, associate M. Gazzaniga and Bogen began testing the veteran. Gazzaniga found when the veteran was given verbal commands, like "raise your arm," that he could only respond with the right side of his body. He occasionally experienced problems putting on clothes. While one hand would pull up his pants, the other might push them down. It was as though one side of the body was unaware of what the other was doing. They also found he could only read printed material in his right visual field. After the operation he could not write anything with his left hand. However, one day from a picture, the veteran copied a Greek cross with his left hand. When asked to do the same with his right hand he could do no better than draw a few disconnected lines.

Far from being unintelligent the right brain seemed to function well but in a nonverbal way. In other tests the veteran tries to duplicate with colored blocks what he sees in a diagram. Trying with his right hand to place blocks in the required pattern, he becomes hopelessly confused. He also had trouble with both hands. One hand puts the blocks together while the other hand pulls them apart. There is a struggle for control. However, he did well at a spatial task when only allowed to use his left hand.

It is known the human eye zigzags as it shifts from one point to another. This is called saccadic eye movement. It occurs in about a tenth to a fifth of a second. Knowing this, Sperry and Gazzaniga designed some tests similar to the earlier cat experiments.

They used an instrument called a tachistoscope. In front of the subject is placed a screen that is briefly illuminated from behind using a special projector. A dot divides the screen into left and right visual fields. A split-brain individual sees the left visual field with his right brain and the right visual field with his left brain. The patient is told to look at the dot. The symbols "?" and "$," separated by the dot, are projected on the screen. The image is flashed for only about a twentieth of a second; so, by the time the patient shifts his eyes, the symbols are no longer visible. When the patient was instructed to write what he had seen, he draws a "?" (a question mark). When asked what he drew, he replies "dollar sign" ($).

A pair of scissors was flashed to a patient's left brain in another experiment. When asked to pick from many objects hidden behind the screen, which matches what he had seen, he says he cannot do it. Yet, his left hand quickly picks up the scissors.

During the early 1970's J. Levy and C. Trevairthen designed experiments using chimerical figures. Chimerical figures are composites. The experimenter made many face-figures combining the right half of one person's face and left half of another. Some faces, for example, are composites of an old man and young woman. A tachistoscope is used, and the subjects are instructed to focus on the dot in the middle of the screen. A chimerical face is flashed. The task is to identify the face.

The subject can choose from many faces, two of which were used in making the chimerical figure. Split-brain patients, when instructed to point to the face seen, always picked the one represented in the left visual field. Pointing with either hand did not affect their choice. When asked to verbally pick which face was seen, they mention the face matching the one whose half was flashed to the right visual field. Although sometimes admitting confusion, they never seemed to detect or mention anything unusual about the chimerical pictures.

A young woman sits in front of the screen in another test. The experimenters put in a slide of a nude woman and flashed it to the subject's left visual field. She slowly smiles, acting obviously embarrassed, but, still thinks she had not seen anything. She laughs and is asked to explain why. She answers: "Oh that funny machine!"

What do these findings mean? These experiments and others strongly suggest the mind has a dual nature. As early as 1844 it was found that each hemisphere could function independent of the other. Some autopsies had revealed certain individuals had but one hemisphere, yet, when alive had functioned normally. Gazzaniga suggested that evidence shows hemispheric specialization is not complete until age ten. If damage to one side of the brain occurs before age four, the other side is thought to assume the tasks usually governed by the affected hemisphere.

A split-brain patient can do two tasks simultaneously. Sperry, Gazzaniga, Bogen, and others believe the two hemispheres, not only function independently, but process information differently.

In blindfold tests a split-brain patient is given the task of tactually sorting through a pile of various-shaped beads. He is instructed to place spheres into the upper tray and cylinders into the lower tray. The other hand does just the opposite. Not only does one mind make decisions (for instance in which tray to place spheres) but each mind is oblivious that the opposing mind is directing an undoing of what the other does. This is like two people carrying out simultaneous, but mutually contrary tasks.

An experiment using a tachistoscope flashes the word "h-e-a-r-t." The letters are separated enough so that the eye-cue dot separates the word into "he" and "art." The normal subject views the flash in his left visual field and then the right visual field. When asked what is seen he reports "heart." When the same test is given a split-brain patient one of two answers is given depending on how he is instructed. When asked to verbally name the word seen, he replies, "art." When instructed to choose (using his left hand) the card showing the word seen, he points to "he."

Some thinkers believe each hemisphere is a separate consciousness. This idea is probably derived from the fact memories of tasks, learned independently by one hemisphere, is not recalled by the other (when later exposed to the same task). (This was first evident in Myer's cat experiments.) Sperry said split-brain surgery seems to leave "...people with two separate minds; two separate spheres of consciousness."

The left brain controls the motor functions of the right hand. Ninety-five percent of all people are right-handed. Bogen found split-brain patients, although retaining the ability to write with the right hand, lose the ability to draw with it. Using the left-hand, one patient could copy a simple geometric figure but could not write a word. The two hemispheres seem to complement each other.

Going one step further, R. Ornstein, who has studied the split-brain phenomenon, believed the two hemispheres are specialized for different types of thought. He conducted experiments on normal (nonsplit brain) people. He gave problems to solve, which, in split-brain patients, would be expected to be handled by just one hemisphere. Using an electroencephalograph (EEG) Ornstein and D. Galin found that alpha-brain waves decrease in amplitude on the side expected to solve the assigned problem (amplitude decrease indicates increased activity). Verbal tasks increased left-brain activity; spatial tasks increased right-brain activity.

So how does each hemisphere process information? Previously mentioned experiments and others, says J. Levy, imply the right hemisphere is oriented toward solving visual (spatial) tasks in a holistic manner. (Note: visualizing solutions uses intrinsic perception.) The right brain synthesizes many features into patterns; it recognizes faces. It processes information in Gestalt form: looking for the "overall picture."

The right brain cannot totally express itself because of its subaware nature. When it expresses itself, it often does so through feelings and intuition. The right brain displays emotions—because of its lack of dominance over the left hemisphere. Sometimes incomplete expression takes the form of art. The right brain is considered the seat of creativity. Some think the right hemisphere produces imagination, daydreaming, and maybe dreaming itself.

It has been reported that right brain damage or its removal has resulted in cessation of dreaming. W. Penfield, a brain surgeon, reported in 1959 that electrical stimulation on the right brain, during surgery, caused the patient to experience dreamlike states in which they recalled memories or experienced visual illusions. (Note that individualization is a creative process. Refer to subheading "Controlling Situations and Self-Creation.)

Tests reveal the left hemisphere processes information in a linear-sequential manner. It operates analytically: chopping input into fragments. It discriminates perceptually in identifying sensory data. The left hemisphere is thought to deal with the "objective world." Because we live in a whatness world, it is not difficult to grasp why the left brain is considered dominant.

What else can be said about these findings? Bogen and others say many experiments suggest the human mind functions in two different if not opposing ways. At most the studies suggest there are two minds functioning independently. In 1949 J. Wada developed a procedure to anesthetize one hemisphere at a time. It is used to determine which side is verbal. (The left brain is verbally dominant in about 97% of the people.) Sodium amytal is injected into the left or right major neck artery.

In one experiment, after the left brain is anesthetized, patients were asked to feel an unseen object. In about five minutes the effect wears off. The patients could not verbally identify what they had felt. However, when allowed to choose from a group of objects, they picked the correct one. Each brain half seems able to perform, but in a different way. Even in cases of right hemispherectomy, patients seem to recover thinking they are normal and no different from before the operation.

It has been found the left brain responds more to the literal meaning of conversation. The right brain senses inflection of voice and connotative meaning. The left brain seems to live in a fantasy world. It rationalizes behavior produced by the right brain. This happens because right brain behavior is not easily verbalized. Maybe what psychologists call the "unconscious" is, in reality, memories buried in the nonverbal right brain. The right brain is also very much involved in the often subconscious search for self-security.

Some tests (such as the Welchsler Intelligence Scale for Children—Revised; WISC-R) have been used to assess individual's relative intelligence in both verbal and nonverbal abilities. Some children are very intelligent verbally, but show retarded ability nonverbally. Others are very intelligent in nonverbal tasks, but are unintelligent in verbal tasks. Hemispheric specialization goes a long way in explaining these phenomena.

Teachers are now able to test, measure, and categorically recognize special learning disabilities. Commonly, children can be given the proper attention needed to help compensate for their specific learning disability. For example, some children may have a problem in carrying out writing assignments. So, they develop poor writing abilities. If the same child is allowed to "critically" express his thoughts in story form, he may develop reasonable writing skills.

Challenging the concepts in this subheading to an extreme may explain even more bizarre behavioral phenomena. For instance, how can the genius-like talents of idiot-savants be explained? Some such people, although unable to care for themselves on a daily basis, can listen once to a long classical piece of music and replay it. They have never heard the piece before nor have had formal-piano training. How many years of "practice" is needed for most "intelligent" people to become a proficient pianist?

Other idiot-savants can name the day, in the past or future, on which a certain date will fall. Many of us could work out a formula to get the answer. Yet these geniuses quickly give the correct answer without knowing how they did it. Do these cases represent an idiot level of intelligence in one hemisphere of the brain and a genius level of intelligence in the other half of the brain?

The Three You

For categorical purposes, we say each individual is constituted of three identities. One is the "you" that you perceive yourself to be. It is usually a positive-upbeat definition. This definition often includes such terms as kind, truthful, trustworthy—all the qualities a Boy Scout would envy. You may define yourself as an unselfish-God-worshiping individual. That is how you see yourself. It is what you think and believe is true, not only of yourself, but also by the standards of the outside world.

There is you as other people see you. It varies considerably. It depends on your relationship to others. Is the other person a foe, a marriage partner, a lover, or teacher? It also depends on how the other person perceives himself.

Does either of these definitions correctly define you? Obviously not. So, what defines you as you really are? It is surely not how others see you, and it is probably not how you see yourself. It has already been inferred you (the self) can be defined in a functional sense as indicated in the following questions. How do you cope with reality? Are you realistic or egocentric? This question is closely tied with the "three you."

The situation as you see it.

The situation as others see it.

The situation as it is (in itself).

The first and third bear heavily on how well you handle a particular situation, or how well you consistently deal with reality (the rest of reality). You can be defined by how well you cope with the real world in terms recognizing a situation for what it is as opposed to how you may distort it in your mind. Note this is different from how man generally is defined. Man is defined by his place in the scheme of things—therefore, where he fits into the structure of reality—a major topic of this book. The next question is how do people function as individuals?

The Self and the "Rest of Reality"

Let's take some previously examined concepts and tie them together. A diagram better clarifies meaning. First, we set down some definitions and premises.

Reality is that which is. As a logical construct, Reality can be broken down into parts representing the relationship between an individual and reality. See Figure 8:1.

There is the Self; therefore, an individual's awareness, and there is that of which an individual is aware—the "rest of reality." In practice the self is only limitedly aware of certain features in the rest of reality.

Awareness concerns information. Information is composed of statements. Statements are classifiable by usage. For our immediate purpose, there are four varieties of statement usage. Recall there is a crucial difference in usage between a statement's truth and whether a person knows the statement is true (refer to subheading "A Crucial Difference" in the previous chapter.

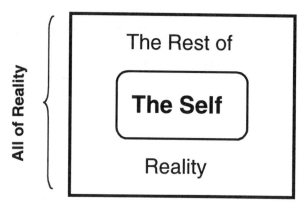

Figure 8:1. The Self and the Rest of Reality

Statements of Fact

Factual statements correspond to and therefore represent (or not) some aspect of reality. A statement is factual because of reality. Reality asserts the "truth" of the given statement. Factual statements are expressed in the positive or negative sense. Factual statements are either true or false.

A true positive statement is one that is verified as corresponding to something in reality (the "rest of reality"). A true-positive statement symbolically represents something in reality (i.e., the "rest of reality"). A true positive statement may also represent some situation in reality. Most true statements are stated in the positive sense. For example, there is water in the refrigerator. We often state it is a fact statement X is true, or it is a fact statement X is false.

A false statement is one that is proven not to correspond to anything in reality. For example, if there is no water in the refrigerator, then the statement "there is water in the refrigerator," is false. A false statement does not represent something in reality: it is not true. Most false statements are stated in the negative sense. For example, there is no water in the refrigerator.

True (factual) statements collectively constitute knowledge or truth in the broadest sense. We learn in the final sections that the meaning of true and false is not as simple as described here.

Statements of Belief

Belief statements are those an individual asserts are true, but has not shown (i.e., verified) they correspond to reality. Direct verification of belief statements is unknown to a believer. Believers assert statement truth although they cannot verify it by reality. Beliefs are also expressed in either the positive or negative sense.

A belief statement is one that an individual (self) asserts to be either true or false, but cannot verify it by reality. Someone may assert there is water in the refrigerator, but has not shown whether his assertion is mirrored by reality.

Because belief statements do not correspond to reality (the "rest of reality") they represent an individual's personal (self) views. Beliefs correspond to and therefore define the (egocentric) Self.

Statements of Opinion

Opinion statements are those that an individual considers to be true, but does not actually assert its truth. Someone may say, I think (meaning he does not really know) there is water in the refrigerator.

Statements of Hypotheses

Hypothetical statements are those considered, by evidence or supportive reasons, to be either true or false. Someone might guess there could be water in the refrigerator.

If the words, fact, belief, opinion, and hypothesis, are used in ways other than which can be categorized by the above definitions, they are being used incorrectly.

Assessment

An individual (self) can assert the truth of a statement only if he can prove it by reality. Reality does the asserting via evidence (or verification). The assertion is contingent only on the strength (or support) of evidence. The evidence may yet present a problem because the individual might have incorrectly interpreted its meaning. Meaning is not ascertained with finality until its reason or cause is discerned by categorical clarification. In knowledge terms—this is an explanation.

For example, a guest asks a friend if he has some cold water in the refrigerator. The guest means he would like a drink. The friend, a biochemist who also considers himself a philosopher of science, goes to the refrigerator and pulls out a sample of chemically contaminated water. He places it in front of his guest—who removes its cap. He detects a pungent odor. Holding it in the light he notices the water has a brownish tinge. The guest now knows this water is not meant for drinking. Continuing from an earlier discussion the biochemist makes his point. The lesson is to not predetermine a statement's "meaning."

The meaning of something is determined by examining the "reality of the situation" corresponding to the statement. Meaning is predicated upon reality, not upon what someone believes about reality. Evidence "speaks" for itself. The wise person shows how reality speaks for itself. This is the ideal of science. Reality-connected evidence is the difference between believing and knowing. Things are not so simple.

A statement can function as both fact and belief depending on usage. One individual may be able to verify a statement by reality—in which case the statement functions (in his use) as a fact. Another individual asserts its truth, but cannot verify it by reality. Here the statement functions as a belief. He does not know that the statement is true because he cannot offer sufficient evidence to support his claim. (Evidential) facts represent reality while beliefs represent the self. There is a difference between knowing a statement and knowing the reality the statement represents.

There are essentially two categories of statement usage:

> Knowledge systems, facts (truths) and perhaps hypotheses (which do not represent selves) are potentially factual in usage. This information corresponds to the "rest of reality" (see Figure 8:1 The Self and the Rest of Reality.).

> Belief systems, beliefs, opinions, and judgments. This information exclusively corresponds to the "self."

Given any information in statement form, either the statement is true (in which case it represents reality—or what is real) or it is not true (therefore false) in which case it does not represent reality. If information does not represent reality—then what could it represent? It represents the Self. This is belief.

When an individual asserts his beliefs are true—he does so without using a methodology that invokes an inquiry of reality. He fails to ascertain if reality supports his claim. To check concepts against reality is the essence of scientific methodology. One cannot trust his conceptions or constructs (in themselves) as a guide to understanding reality. When someone does not test, he effectively regards his self (through his beliefs) to be more significant than facts (which represent reality). He is saying his self (by asserting his beliefs to be true) is "right."

What is happening here? The individual is actually projecting himself onto reality. He projects certain statements, his beliefs, onto reality. In effect, he is telling reality what is true or real. This is foolish. It is for reality to tell the individual about itself. Such projections are defense mechanisms. They shield the individual from possible contradictions posed by reality.

A person who uses a method that shows (evidence-proving) the truth of a statement gives priority to reality. The substantiation of any statement is only as strong as the evidence supporting its truth. The "self" makes an assertion (of truth) by citing proof. That means the assertion arises from the proof (a part of the "rest of reality") and does not arise from the "self." It is for reality to say what is true. One who recognizes that reality decides what is true (using this method) is realistically honest. He represents what reality says when he speaks.

An individual knows (a statement) when he can verify his statements by reality. There is a categorical difference between this and someone saying "I only have to prove it to myself!" Such a person not only does not question his beliefs; he does not allow his beliefs to be seriously questioned by others. A serious seeker of truth only states what is "right" (or truthful) to the extent reality supports what he states. A realistically honest individual asks reality to verify what is true. He understands reality is the criterion by which truth or knowledge is defined. Again, knowledge is that information that corresponds to (reality) and therefore represents reality.

The idea is to develop a method that is "realistic": to develop a method by which reality can convey what is true about itself. Verification is the method of reality. However, if an individual adheres to what he believes to be true—how can reality convey its truth to the individual? It cannot. Reality is all around us; one does not have to go looking for it. The problem is not reality. Reality is—what it is. The problem is that most people have not developed a realistic method by which to gain an understanding of reality. If a person believes he knows the truth, but cannot in reality verify it, then he is dishonest with the truth.

What is understanding? Understanding involves more than mere facts. An individual understands a fact when he can connect it to other facts. The more he can interrelate facts—the more he gains understanding. Learning involves the reasoned interrelating of facts. Compare this method to memorization.

Facts can be likened to pieces of a jigsaw puzzle. The placing of specific facts into relationships with other facts is like placing together pieces of a puzzle. To complete the puzzle—the pieces must fit together. So it is with reality. Facts relate to each other in one way. This is true, for there is but ONE reality. This concession is based on the idea we observe order in everything. It is not this simple, but at this stage there is not anything that blatantly contradicts the jigsaw analogy.

To understand reality is to know how all the more general facts fit together. This is an arduous task. Conceding uniformity in nature (that it exhibits order) allows one to say there cannot be any contradiction in a given-factual statement. Well-established-evidential facts cannot be easily contradicted. If we find that a true statement is contradicted, it means we have not sufficiently connected it to other established facts, or that the "established facts" have not been correctly understood.

Misinterpretation is always a problem. Misinterpretation occurs when the "Self" projects its beliefs onto reality. In puzzle terms, either the puzzle piece (or fact) does not fit-in with the other puzzle pieces (already pieced together) or we have the right puzzle piece, but are trying to place it into the wrong puzzle. If much of the puzzle is pieced together, then the problem is that we are trying to place a piece, which does not belong (a statement misinterpreted as true or miscategorized).

The pieces (evidential-factual statements) must be verified. One cannot use unverified statements to erect the structure (puzzle) of reality. If a puzzle piece does not fit into the structure of reality, it probably represents someone's personal (belief) system.

A serious seeker of truth questions every statement, and checks it against reality. A realistically honest person states only what reality demonstrates (verifies). This is testing—to ascertain if a statement is true. Statements representing the physical world are tested by empirical methods. Experimentation is the hallmark of physical science. If one cannot verify a statement as fact, he does not "know" what he claims. To verify a statement as true is to test its truth by experimentation.

There are two directions. Either you are honest with reality (the "rest of reality") or you are honest only with your SELF—which excludes the "rest of reality."

Some Guide Lines:

There is no such thing as a problem in reality. Reality (in terms of knowledge) is noncontradictory and coherent. There is no problem in the "rest of reality." We assume coherency because we observe uniformity in nature. We find no reason to contradict that assumption. Hence, we also assume knowledge is ultimately noncontradictory.

Therefore, problems exist only in the self. It means there is something that the self does not understand about what is real (in reality) and about what is true (information that corresponds to reality).

If the problem exists in your SELF then it concerns the methods you use to ascertain verity (what is true).

What you believe has no direct connection to what is real or true. Like our previous example of the water in the refrigerator, it is unimportant what you believe about it. It is unimportant whether you believe or believe not that "there is drinkable water in the refrigerator." Either there is or is not potable water in the refrigerator.

What you do know does have a connection to reality. For example, you know, given there is water in the refrigerator, that it could be fit to drink. Develop a method for testing the truth of that proposition—get reality to reveal the truth about it. You could sit in a chair as long as you like and believe or not believe the water in the refrigerator is good. Truth is not revealed by belief. Believing is no method at all. It is acceptance of a statement as true without verification. Seeking truth is simple. Open the refrigerator door; look at the water. Smell it; examine it. Conduct tests to determine if there is any question about the water's quality.

What are consequences of not proving what you say? Failing to verify (what you state) is failure to make a deliberate connection to reality. "Proving" something to yourself means little except you consider your (egocentric) SELF to have precedence over reality.

We live in reality. Without the proper mental connection to reality, we cannot properly define ourselves via reality. Without the proper connection to reality, conflict inevitably develops and poor mental health results. The proper connection between our SELVES (mental) and the physical world can lead to proper mental health.

Question everything—question the truth of every statement. Reality and its truth can withstand any questioning. Beliefs are recognized for what they are. A belief is a statement one asserts is true, but cannot verify. It arises from self ignorance. This poses a question. Why were we not taught correctly? The following chapters begin to answer this question.

Key Points:

- Problems are opportunities to learn and understand.

- It is against the nature of the (deductive) mind to solve problems realistically.

- If one does not control himself in the face of some situation, he can be controlled by the situation.

- The way to control oneself is through reasoning. Solve problem; apply solution. The mind develops only if it is pressured to solve problems by (inductive) reasoning.

- Split-brain experiments suggest the brain processes information in two ways. This book suggests these processes correspond to the deductive function and the inductive function.

- There are three ways to define the self. There is how one perceives himself. There is how one is perceived by others. And there is how the self is defined by the scheme of things.

- Ultimately, man is defined, not by what he believes, but by how he "fits into" the structure of reality. There is the "self" and there is the "rest of reality." We need to understand reality before man can be understood and defined (by the "rest of reality").

- Factual statements represent some aspect of reality. Belief statements do not define the self by reality. To be defined by reality means the self must learn how reality supports factual statements.

The How Aspect Diagram lists intelligence levels from low nonliving physical objects up to the "Realm of Whyness." Intelligence levels lead one to understand the structure of reality (the "Realm of Whatness"). The structures can be conceptualized with knowledge. A mind directly acts with information and indirectly acts upon what that information represents.

Using the color paradigm, we established that the mind is distinguishable from the physical world. This helps understand the mind-body problem. At least it is settled for the level of generality dealt with here. The mind, though distinct from the brain, is "manifested" in the brain. (Note that absence of mind manifestation in brain [in an otherwise, physiologically healthy brain] may explain some cases of coma.)

It seems each half of the brain is itself a mind. One mind, representing verbal abilities and verbal memory, is found most often in the left brain. The other, the nonverbal mind and its memory, is found usually in the right brain. Because we only "perceive" the physical world, much of behavior—in which we are interested—is not observed. Most behaviors need to be inferred from what we do see.

We began with the idea behavior results from the interaction between these two minds and the physical world. Depending how people react to situations in the world—by security orientation—we classified the basic natures of man as either egocentric or realistic. The prominent covert characteristic of egocentric behavior is "believing." It precipitates from the mental processes of deduction. The most covert characteristic of realistic behavior is "knowing." It develops from induction. In applying these two categories of information, people perform the overt characteristic of "wanting and having." We also showed how wanting and having generates the sociological concepts of hate and love.

With the basic concepts of behavior behind us, we delved into defining man in more precise terms. We saw how the self can be defined by its relation (or lack of it) to reality. We showed how man reacts to situations ranging from strict coercion to the subtle settings of everyday life.

We found individuals could be more specifically defined by how they react to situations. Either an individual controls situations or they control him. Being controlled by situations produces habits of inhibition. Under these conditions the self slowly destructs and regresses to an infantile-like dependency on something outside its "self."

The inductively active "mental" individual is pressured into assuming a more dominating role over the physical world. Controlling situations calls for a greater understanding of reality. This allows the "self" to acquire more knowledge that continually redefines the "self." In striving toward higher levels of awareness the individual becomes more inductively aware. This process creates the "real" self—the self recognized as an individual.

The real self emerges as a "mental" individual—one oriented toward reality for security and fulfillment. Although the author's concept of the two minds was originally derived from the structure of reality, he also discovered this idea has an experimental basis. One half of the brain processes information in a deductive mode. The other half processes information in an inductive mode. The author suggests that split-brain experiments indicate the left brain is the usual seat of the deductive process. Similar experiments point to the right brain as the usual seat of the inductive processes.

We learn man is definable by how he reacts to situations.

Egocentric behavior predominates in society. Why are there not more realistic people? The first problem is that most people are also egocentric and do not realize the actual circumstances under which the two behaviors operate. There are two basic TIMES when each behavioral type is dominant. The reasons are discussed in the next chapter—"The When Aspect."

We have set up a logical construct that defines man by relating him to the rest of reality. There is the Self and there is the "Rest of Reality." We can fill in the details after we fully understand reality's structure.

We again venture back into the mind expanse to continue our surveying.

When Aspect

Perspective

- Defines time flow
- Time flow is an abstract conceptualization of change
- There is a two-phase time cycle
- Time flows "forward" during physical universe dominance
- Time flows "backward" during mental universe dominance
- The When Aspect is the third aspect of the Deltic Cosmos structure

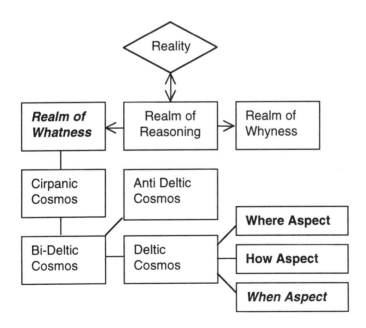

Figure 9:1. Perspective: When Aspect

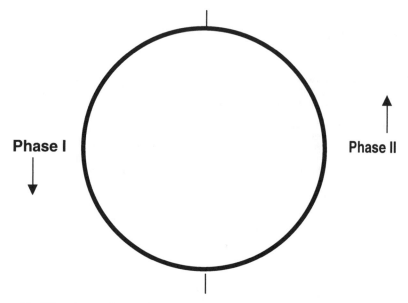

Phase I

Phase II

The When Aspect diagram is composed of two phases: Phase I and
Phase II. Time flows "forward" in Phase I. Time flows "backward" in
Phase II. Phase I represents tardyon-physical matter dominance. Phase
II represents tachyonic-mental dominance.

Figure 9:2. When Aspect Diagram

Chapter 9

When Aspect

"We know the past; we do not know the future. For
everything there is time; for nothing there is no time."

The When Aspect is the third and final component of the Widest Intrinsic
Whatness mode. It represents every temporal-manifestation of the Where
Aspect. The usual categorization of the When Aspect falls under the heading of
TIME. One of our important tasks is to understand time itself.

What is time? It is usually considered an intangible component of the world in
which we live. Time has been subject to repeated philosophical discussion. This
has resulted in the not uncommon notion time by nature is incapable of clear
definition.

Time will not be defined until it is completely and scientifically understood.
This chapter gathers some useful concepts about time. From that idea, we later
"springboard" to a proper understanding of time itself.

Without defining time, Greek philosophers believed we only dip into a specific-
temporal instance only once. The temporal "nowness" (the "here and now") we
experience is the "conscious-present." The conscious present was thought to be
the same for everyone. Similarly, it was believed this specious present flowed
into the past at the same rate for all possible (inertial) observers. Instances not
yet experienced were thought to comprise the future for other possible
observers.

Scientists have offered many practical definitions of time. They interpret most
physical phenomena in temporal terms. Concepts of energy; for example, force,
acceleration and dynamic hypotheses are understandable in relation to time.
Time itself has remained an enigma despite scientific progress.

Mathematical treatment of time has yet to yield the proper higher-level context
required to fit dynamic phenomena into the scheme of things. The best
conceptualized mathematical understanding relates time to space. Because of
this, time itself cannot be defined until space is better understood. Though much
has been learned, proper-categorical conceptualization of time has eluded
scientific understanding.

Anyway, our task is to begin a logical treatment of time. To understand is to be
able to relate one phenomenon to another. How can time be conceptualized to
gain a proper understanding? Time must be associated with an enveloping
concept that subsumes time as it is presently understood—viz, "the flow of
time." We ask how can the flow of time be explained?

The needed "enveloping concept" has eluded understanding. We know this
because many outstanding questions are unanswered. For example, how did time
originate? Has it always been? A proper conceptualization of time should
answer these questions.

What has hindered understanding time? Many people have exhibited the same mystical attitude toward time, as have philosophers. Even scientists must begin an understanding of time with logical concepts. They have done this, but have not advanced beyond the mathematical treatment of time. To understand time we must first understand the physicist's formulation of space-time.

Let's briefly examine what has been learned. What are some of the better approaches in defining time? Scientists believe time was co-created with space. Can time be defined without space? The answer is no. We address a proper logical treatment of space-time in the final chapters.

Time is perceived in different ways. There is "psychological" time. This is time as we "experience" it. This is different from measuring it with a "clock." Experiencing a phenomenon does not automatically confer understanding. There is time as physicists define it (in a practical sense): that most phenomena are defined against "measured" time. The measuring of time (using clocks) is something man superimposes onto time. The measuring of time, though useful, sheds little light on conceptually understanding time. Time itself is not a measurement. The measurement of time is something we do to exert some control in coping with it. The idea of "clocking" time is a helpful beginning.

We easily see why the present understanding of time is inadequate. The common notion time is a measurable duration implies a beginning and an end. This is derived from the once prevalent idea of absolute time—that time is constant and itself unchanging. The idea that rate change of time flow is "unchanging" helps us not at all. What could it mean if time is unchanging?

The distinction between the past, present, and future was thought to be the same for all observers. A future event in one frame of reference was thought to be a future event in any other. It is naive to suppose an understanding of time is reflected in one's experience of time. Perception of time is influenced by cultural attitudes and beliefs. Many culturally induced temporal concepts have been corrected by scientific understanding.

Time is definable in both a phenomenal and noumenal sense. Let's examine this approach. Phenomenal time is tied to the unidirectional idea of time. It is derived from immediate experience—appearances. The phenomenal sense of time is imbued with a superimposition of the metrical definition. This is because time can be gauged using a clock.

In a metrical sense, time is one dimensional. A temporal interval can be measured and represented by a one-coordinate system. We concede time "appears" to flow just one way and at the same rate for all observers. This is how time is experienced. However, the idea time is absolute and "flows" at an unchanging rate is an example of culturally induced thinking.

The passing of events does not explain time. Rather, it is the passing of events that needs explaining. Time is unobservable but for the change of events. In this view time is a construct used to understand events. Defining time by the passing of events (by measuring or clocking) adds little to understanding. The more significant question is why is there a passing of events?

The phenomenal sense of time confuses the understanding of time. Stripping time of its clock-like connotations should suggest a noumenal definition. To understand time in a noumenal sense is to look beyond the customary notions of time. The "customary" idea time is clock-like is so embedded in our perception of time that it is difficult to see how else time could be understood. The clock notion does not immediately suggest a higher-level conceptualization of time. The "clocking" of time is an appeal to reductive definition, and confuses understanding. It is fruitless to define time in terms of itself or by "redefinition." To correctly define time is to relate it to a proper understanding of space.

Cause and effect is entwined in the notion of unidirectionality. The present is thought to divide the past from the future. The present (even at any given time) is also thought to separate cause from effect. We observe the phenomenal effect of the force of one body acting upon another. We live everyday life based on such (customary) expectations. Experience reinforces the percept cause always precedes effect. It was deemed impossible for effect to precede cause. Usually, only science fiction writers entertain such ideas.

The contemporary scientific conceptualization of time as space-time has yet to yield the hoped for stepping stone to better understand time. If we only experience a phenomenon from a limited perspective, we are usually stifled in "getting the big picture." We need to grasp "context" before gaining understanding.

The unidirectionality of cause and (then) effect provides a context from which to begin an analysis of light. Time has not been understood because its directionality was thought to be unchangeable. Other problems surfaced when temporal reversibility is considered. Let's examine those problems and do some reasoning.

Most scientists agree there is temporal unidirectionality. They point to the concept of entropy—that the universe becomes more disordered with the passing of time—to support the concept of unidirectionality. Entropy is usually cited as the reason time cannot go backwards in the world as we experience it—the (physical) world in which we live. Entropy offers the best stepping stone to understand time.

Entropy does not seem to apply to every level of reality. Elementary-particle interactions seem to evade the label of temporal unidirectionality. We could play a movie of certain particle interactions. Even experts could not say whether the movie was shown as it was filmed or run backwards. Using entropy to ascertain temporal directionality is useless in these elementary interactions. The emission of certain particles in "forward" time can also be interpreted as the absorption of their antiparticles in "reverse" time. Here, temporal directionality is arbitrary and cause and effect can be switched without any resulting logical problems.

We initially understand time as we experience it in everyday life—exhibiting unidirectionality. We have learned about time in the micro-world—defying directionality. We can also guess there may be yet another sense of time measured against reality in the grandest sense. That is the idea we are seeking—a context enveloping time. The concept of space-time is a step in the right direction. Space-time provides the wider context needed to formulate a proper noumenal understanding of time. How can space-time be explained?

The noumenal sense of time implies a property hidden from the world we experience. Phenomenal interpretations of time usually suggest change occurs "in time." This idea suggests time itself can be characterized apart from the reality of events. And that means time itself (noumenal) cannot be fully understood by events (things in motion).

The everyday sense of time is linked to things in motion. Because we observe things in motion, we say there is a flow of time. The micro-world undermines defining time in an everyday sense. That leaves us to ponder the essence of time in some grander sense. This greater sense should explain the "flow of time."

If time is to be understood, we need to wait until we sufficiently understand reality's structure at a "higher" (conceptual) level. We must understand space itself apart from how we experience it. We do concede that "what" we experience are "things in space." Things change. This idea must carry over into a noumenal understanding. We develop a proper understanding of space in the final chapters.

For now, let's consider a not so grandiose understanding of time. Let's stay within the confines of time as we phenomenally experience it. That leaves measurable time and psychological time. Both recognize time as a passing of durational instances. We use this idea to explain "variation"—that things change. Even if we discern no variation, we know from kinetics, that "things" are in constant motion (at least at the micro level and cosmic macro level).

Stripped of everyday conventionalism (therefore measured or clock time) we define time as "change." Change is pegged to events: change in the condition of things.

The general notion of "the present dividing the past from the future" is an excellent point to begin our examination of how things change. This raises two questions. How and why do things change? Only the first aspect of this question is answered in this Section. The "conscious present" can be viewed as a passing of an individual's future into his past. We say this because awareness does not operate through some instantaneous present, but through a continual duration. (If things were instantaneously presented—and therefore, for no duration—we could not "experience" things. Further, an infinity of real instances [no duration regardless how small] would be no different from one instant.) Past, present, and future must be explained using a noumenal definition of time.

Each individual experiences the "conscious present" differently—in a personal way. People watching the same thing, for example, a car accident, do not share the same view.

Time, measured by change in the physical world, is a projection of the conscious present—a conscious-present continuum. We do not experience the whole present, only a representative part.

Our cognizance of time or change depends on the phenomenon of recollection. The temporal continuum is connected somehow to our memories since we experience the past only through the conscious present of recollection. Memory is needed to connect present awareness to the past. Intrinsic perception is intimately bound to the conscious present. Our noumenal understanding of time must also include the "idea" of memory. Memory links one thing to another. We need to incorporate consciousness, memory, and experience, in our understanding of time. Change should be ultimately definable in terms of consciousness, memory, and experience. These ideas are explained in the final Section.

"Changes" on the continuum of time vary because the ideas of past, present, and future have no absolute significance. This is Einstein's contribution. Yet, if the rate of change is the phenomenal definition of time, then time cannot be defined in the customary sense—that time is uniformly unidirectional. The unidirectionality of time applies to inertial frames of reference. There are many possible frames of reference, hence the problem. We must look beyond this snag in understanding time. The greater problem is explaining why time is unidirectional. What is it that suggests time is unidirectional? We have pieced together certain considerations bearing on an understanding of time. We are ready to look at the bigger picture. Before doing so, we must tidy up our concepts. Let's be explicit.

When Aspect Diagram

Many thinkers suppose things "exist" in time. This suggests time is distinct from things (existing in time).

It has been thought events (things "in time") change because of time. Some philosophers even believed things change because of the "flow of time." We concede the concept of time emerges from the idea of change. But, we also know things are not time. This, however, does not mean that we can categorically divorce things and time. It may be there is no time (flow) without things. Instead of things depending on the prevailing notion of an all pervasive temporality, maybe time flow arises only because of things, but how?

If things do not experience change because of time, then there must be some other explanation for the fact things change. Whatever the explanation, it also must explain time flow. The deeper answer to this question is more than this chapter is designed to handle. We revisit the deepest questions about time and answer them in the final Section.

For now we suppose a thing does not change itself. (This is probably the biggest reason change has been traditionally attributed to time.) This means there may not be time flow apart from things changing. Therefore, time (flow) arises because of change. Now we can ask what is there about things changing that is interpreted as time flow? Answer: motion. More specifically, we constantly experience motion and interpret that as change or time flow.

"Things in motion" is implicitly associated with the flow of time. No motion—no change—no time (flow). That is interesting. There is not anything that is not in motion on some level. That is not to say someone can sit in a quiet room, and there be no things moving. We know that things, though appearing still, are made of atoms, which are in constant motion. And, everything is in motion through space. Astronomical objects are receding in space. We have already stated that micro events (elementary interaction) and cosmic events (stellar and planetary) are in constant motion.

The idea change allows the emergence of time implies time is different from things existing in time. Though true, this is confusing because it does not tell us the most intimate nature of time, nor does it inform us that temporal flow emerges from things in motion. Yet, if this (that time flow emerges from motion) is correct, then there is also a deeper sense of time than temporal flow. But for now, we ask how does motion arise?

We know "things in motion" change. Because time, as we experience it (flowing—ever changing) is change, we state time (flow) does not exist apart from that (i.e., things) which is subjected to change. Rather, change produces the appearance of time flow. Things in motion constitute the noumenal definition of temporal flow. To define time (cosmically) is to describe where (relates the Where Aspect) change occurs and how (relates the How Aspect) events are ordered to constitute change (relates the When Aspect).

To sustain interest, let's ask another question about temporal flow. We would like to know why there is motion (instead of no motion). To answer that question, we must completely understand change. That understanding should lead to a full explanation of time. Before the end of this book, we will elucidate a complete understanding of time. Specifically, as a question, what is the most intimate connection between time and space?

The task of understanding things, change, and time flow, reduces to understanding existence itself. Why is there existence? Answering that question will also tell us why there is motion, and hence change, and hence time flow. We note for now that we are further along in our pursuit to understand time than one might guess. "Thingness" is associated with the Where Aspect. Change (order) is associated with the How Aspect. Time flow is represented by the When Aspect. In this treatment, we can understand why someone might confuse that things change (When Aspect) with the nature of the change (How Aspect—order). We now move onto the idea temporality is dimensional.

"Duration" provides an interesting approach to understanding time dimensionality. We experience time through "psychological" durations. (We grant time is described by a sequence of durational instances. Physicists refer to the minimal possible change as the Planck time.)

Durations have a beginning and an end. The notion of change suggests a beginning and end; therefore, the beginning and end of some change. We ask what happens during the duration? We observe the effect of entropy. We find it helpful to apply the idea of entropy to an understanding of duration apart from our direct (psychological) experience of it (duration).

We seek a "cosmic duration" defined by change. On a grand scale we suspect either entropy or negentropy defines change. Therefore, each cosmic duration should be "directionally" definable by entropy or negentropy. Although we observe the effect of entropy every day, we are usually not very cognizant of its meaning in a grander sense. Entropy must also be factored into understanding time at the most basic level. That again entails the utmost understanding of consciousness, memory and experience.

We link "experiencing change" to time. A key consideration is "we are conscious of experience itself"—this is embedded in the conscious present. A generalized conception of time must somehow explain a proper understanding of consciousness. We strive to understand consciousness and experience. Time should be explained by connecting consciousness to experience.

The flow of time can be extrapolated from our conscious present to any extreme—yielding a very short or very great duration. Again, the idea of beginning and end applies to any durational length. The temporal duration in which we now find ourselves is definable by entropy. Entropy dominates all physical interactions. An interesting question is "what is the greatest durational length dominated by entropy?" Answering such a question does not explain the origin of time. But, it should lead us in the proper "direction" to understand change, and therefore, time.

What is the greatest duration demarcated by entropy? When was the beginning of time, as we know it? Is there an end to this duration? Realistic answers must be without contradiction. Answering questions on this scale calls for a reexamination of the very definition of contradiction. If we fail to do this, we may overlook unwarranted assumptions. Although we have now set up some of the premises from which to better understand how time is created, we will only complete our understanding of time toward the end of this book when consciousness is fully explained.

If a certain beginning were cited, someone could ask "but what was there before the beginning?" Saying there was "nothing before the beginning" sheds little light on the nature of time. Such an answer is not an understandable explanation (unless nothingness itself has a definable function; it does) and explains little. These questions are beyond the scope of the present section. For now, let's define time by how it applies to our immediate world—our universe.

Time (or change) rate may appear faster or slower in different inertial systems. That has little bearing on the idea time, in "every day" experience, is unidirectional. To understand this we note change occurs by a sequence of instances. Instead of focusing on the rate of time, let's narrow our focus to the idea change happens in sequence (for any given observer).

Change occurs step by step; it is the sequence or order of change that holds the clue to understanding time itself. The following is an interesting question. Is there some type of end to a categorically defined series of sequences? For example, is there an end (a temporal end) to the physical world (duration) as we presently understand it? A similar tract of questioning is what was there before the beginning of time? Was there a time before the physical world? If so, how was it different from time during the physical world?

Surely, time did not bring itself into manifestation. And surely time is an aspect of manifestation—viz, an aspect of what we experience. Change did not create itself. If it did: how did it? How can time create itself prior to its existence? It is unwarranted and meaningless to say time creates itself after its existence. Therefore, we seek an explanation that supersedes time. It must be an explanation outside time. It must be a "timeless" explanation. Yet how can that be? Again, we are working toward these answers. Before answering these questions we must fully understand time as it is connected to our universe.

Narrowing our focus to the "sequence of things" we ask, what is the difference in the sequence of events during one categorical (durational type) description and another? Is there a difference in change before or after time, as we know it? If there is a difference, then it is a difference in change in the different categorical durations. For now we note generally that change involving the physical world, except some elementary particle interactions, is summed up by the phenomenon of entropy. The "forces of nature" bearing on physical matter are subject to entropy. The direction of time in the physical world is definable by entropy.

We are attempting to solve the problem of the "beginning of time." If we say there was a beginning of time but that it was preceded by no time, what would that mean? If there were no "type" of time preceding time as we know it (change in the physical world) how did time come into being? It sounds like a dilemma. We must first seek a "practical" answer—one in which we find no contradiction. Our practical solution hinges on understanding entropy. For example, why is there entropy? Our immediate answer must be viewed in the context of entropy. All change in our world (the physical) is dominated by entropy.

There is a simple solution to the problem of beginning and end of time. A two-phase-time cycle solves the dilemma. One phase arbitrarily represents "forward" time; the other phase represents "reverse" time. Physicists suppose time must "reverse" somehow.

A two-phase-time cycle allows an "effective" reversal without an actual reversal of what has already transpired. Time reversal occurs in a different phase. That is, there is a temporal reversal without a reversal of motion—as in a movie run backwards. In one phase we find entropy dominates (interactions). Entropy defines the directionality of time in the physical world. That is, change happens in a way that embodies a "breaking down" of matter or systems of matter. As expected, the "other phase" would involve a universe where negative entropy or negentropy dominates. Here, change is primarily directed toward a "coming together" or building up of parts into higher-level systems (or functions).

We find two factors implicated in change: that which in-itself is acted upon and ordered—the Where Aspect; that which in-itself constitutes order—the How Aspect. Can order on a universal level be better defined? On a grand scale we observe order is represented by entropy in the physical world. The When Aspect is change in-itself. Entropy dominates Phase I. Negentropy dominates Phase II.

We have now connected change to order. Entropy and negentropy refer to order—how things are ordered. In the physical world entropy defines the overall directionality of change. Time is change. Directionality concerns order. Order is represented as the How Aspect. With this understanding, we delve into what happens in each phase of the cycle.

Phase I

Before describing Phase I, let's clarify some Ideas. Recall, an animal or person is (usually in this existence) composed of entity units from the three major universes.

These are the three universes composing the Where Aspect. They are the Universe of Perfect Whatness (the physical world, here, a person's physical body) the Universe of Imperfect Whatness (the deductive function of mind) and the Universe of Imperfect Whyness (the inductive function of mind).

The first phase of the cycle is dominated by "whatness." Whatness occurs in the "forms" of the Universe of Perfect Whatness. Physical things have three-dimensional form.

The physical universe evidently originated more than twelve billion earth years ago. Many cosmologists believe the physical world resulted from a "Big Bang" giving birth to an expanding universe of "perfect whatness."

Matter is thought to have formed into aggregate clusters yielding galactic star systems familiar to astronomers. The earth was formed about 4.7 billion years ago. Nonvital substance developed into vital substance. This occurred between 3.5 and 3.2 billion years ago. Plants evolved and paved the way for complex-animal life forms. Animals developed minds. Nonvital substance and vital substance (generally represented by plants and animals) constitute the basic levels of the Universe of Perfect Whatness. (See Chapter 4: How Aspect Diagram.)

Many animals exhibit the ability to perceive their environment. This ability evolved into a mental intelligence that discerns change in the surrounding whatness. Cerebral intelligence (Universe of Imperfect Whatness) experiences the Universe of Perfect Whatness through the senses, but in an "imperfect way." Toward the end of Phase I, the Universe of Imperfect Whatness (mental development) evolves into a very high state of awareness in man (deductive inferential ability). This has been achieved through an evolutionary succession of vital Universe of Perfect-Whatness units (Universe of Perfect Whatness individuals—physical bodies). We appropriately ask why does this happen?

The Universe of Imperfect Whatness (deductive-inferential functions of the mind) makes deductive inferences based on extrinsic perceptions and recollections (memory is a phenomenon of intrinsic perception). The Universe of Imperfect Whatness (deductive minds) develops from an interaction with the Universe of Perfect Whatness (physical world). From a behavioral viewpoint, individuals in Phase I are egocentric. This is understandable. They need to focus on themselves to survive. They must acquire food, and water to sustain their Universe of Perfect Whatness (bodily) functions. Therefore, desire is necessarily maintained since a Universe of Imperfect Whatness unit must depend on a Universe of Perfect Whatness unit for survival (to sustain the deductive functioning of mind). The mind depends upon the body.

The condition of a Universe of Imperfect Whatness unit (deductive-mental function) is nonfunctional during sleep. During Phase I the mind functions only for the physical world. When sensory input ceases (for example when one "falls asleep") the mind has no sensory premises (extrinsic perceptions "of the physical world") from which to deductively infer anything. Sleep (the mind not functioning on its own) mimics the state of mental functioning when the mind is separated from a physical body (as happens after physical death). Even when separated from the body during Phase I, the mind is still unawakened—still not functioning in itself.

The Universe of Imperfect Whyness unit has minimal influence over the Universe of Perfect Whatness unit. However, there are times when the Universe of Imperfect Whyness unit exerts more control. Evidently, it can "take over," for example, when an individual is sleepwalking. In Phase I the Universe of Imperfect Whyness unit usually functions at a constant level. When the Universe of Perfect Whatness unit is resting, the Universe of Imperfect Whatness unit's functioning "falls" below functioning of the Universe of Perfect Whyness unit. Then the mind can influence the actions of the body.

The Universe of Imperfect Whyness (inductive mental function) is seemingly present throughout Phase I. Its condition is nonfunctional whether inhabiting a physical body or not. When the Universe of Imperfect Whyness unit (a mind) is associated with a body, however, it tends to act as a force to draw the Universe of Imperfect Whatness unit from mere extrinsic perception into inferential awareness. (See Chapter 4: How Aspect Diagram.) It does this by functioning in a default mode, i.e., deductively. A mind not functioning inductively functions by default deductively.

Perception

Extrinsic perception (best represented by vision) is usually fully operative in a vital Universe of Perfect Whatness unit during Phase I unless the individual has incurred some handicap limiting its sensing abilities. Eye vision is the most influential sense during Phase I.

Intrinsic perception, like the Universe of Imperfect Whyness inferential functions, only operates subfunctionally during Phase I. It usually yields only vague images. It does, however, serve as the functional memory. We encounter its presence as dreaming.

Life and Death

In Phase I there is a phenomenon commonly known as "life and death." Using man for instance, we look at how the various "universal" units are affected by the phase cycles.

What happens when man is born? His physical body develops from the union of egg and sperm. The body develops as it interacts with its surroundings. It takes in nutrients from its environment.

One's mind is usually passed on to him from his parents, just as is his body. The mind acquires information by interacting with the physical world; the deductive mental function is exercised. One's inductive abilities might also develop.

In normal life all three units (physical body, mental deductive and inductive inferential functions) are present and functional to some extent. If certain nerve pathways to or from the brain are blocked, the individual may be incapable of communicating with the "outside" world. This effect also results from the absence of a mind. Here, there is a body but no awareness is present. Medically, this is called brain-deadness.

What happens when death occurs? The body or Universe of Perfect Whatness unit ceases functioning. The Universe of Imperfect Whatness unit (mental deductive function) ceases to exist. Each unit might have been passed on to progeny. Death frees the mind from encapsulation in the physical universe.

Reincarnation

We address a subject that has been circumvented—reincarnation. Much evidence suggests the possibility of reincarnation. There have been many reports of someone remembering a past life. The idea is "past memories" are somehow passed on to another physical body. This is the "evidence" that some aspect of the individual is linked to someone who previously existed. If there is reincarnation it is quite possible that there are people who, in some sense, are more than merely an "offspring" of their physical parents.

If we grant the possibility of reincarnation, there is an obvious question. What is it that reincarnates? An individual only has two real aspects—the physical body and the mind. The reincarnating aspect is not the physical body—which leaves the mind as the only reasonable answer. The problem with this approach is that the mind must somehow survive physical death. For that to happen the mind must be categorically separate from the body.

The mind is more than a mere outgrowth of the physical brain. If the mind is an epiphenomenon that too must be explained (refer to Chapter 20: "Transcendent Function"). The deductive inferential functions of the mind "develop" via the brain. We understand from the How Aspect that the mind functions "by default" in the deductive mode while inhabiting a physical body. The mind operates in the physical world through the body's senses. Although the mind is functionally tied to the body's senses, it does not necessarily follow that the mind is born out of the physical world. To understand this is to view the mind's place in the scheme of things. To obtain this view is to better understand the structure of reality—the purpose of this project (Refer to Section Six, Interpreting Structure).

If mind is not a direct product of the brain, then we are left to wonder about the origin of the mind. Just as one's physical body is "given" birth from one's parents, so it is the mind is also given birth by one's parents. There is an exception. Instead of obtaining a mind from one's parents, an individual (physical body) receives a mind that previously existed: reincarnation.

Change from Phase I to Phase II

The cyclical nature of cosmic structure entails a "time" when a "phase-shift" occurs. This happens at the juncture of Phase I and Phase II. There must be two such "changes." The other occurs at the change from Phase II to Phase I. The change happens in a unique "jump" from the last Planck duration of Phase I to the first Planck duration of Phase II.

How would this occur? From physics, we know matter can annihilate into light. Matter can instantaneously disappear. Yet the changeover of the universe can only happen if there is some "higher level" intelligence governing the material world. We address this question later at a more appropriate time (See Chapter 20: and Chapter 21).

We justifiably ask whether time phases have a purpose. The structural purpose is addressed in the following section on the Bi-Deltic Cosmos. The phenomenal cause is to "split and diverge" the Universe of Imperfect Whyness. That is, to create many minds from one original mind (in Phase I). This is a result of entropy dominating the processing of "things" in Phase I: where things break ("split") apart.

If we "go back" in time, we eventually come to a point where few people, other wise known as minds, existed. Going further back, we reach a time when but one mind existed. This is based on the idea we get our mind from either our parents or someone who previously existed. Either way, there is a time before which no mind existed in (inhabited) the physical world. This again leaves one question opened. Where did that first mind come from? The answer to this question is addressed shortly. If we can also answer what was there before the beginning of time, as we know it, we may be able to determine the origination of the first mind.

What is the result of the entropic splitting and divergence of minds (in Phase I)? Many minds are produced. In the beginning there was just one mind, alone. This sounds like there is some "cosmic plan." Keep in mind, the question foremost on our minds is "how can this be explained?" In seeking the meaning to life, this is the question we are attempting to answer. How can life be explained? Is there meaning to life? We grapple again with this question in Chapter 21.

What evidence or circumstances support the idea there is a mind that is, in a sense, immortal? This question is only properly addressed by seeing how evidential facts fit together. Once these "pieces" of the puzzle are sufficiently placed together, certain patterns emerge (suggesting mind immortality).

How long does Phase I last? The answer is dated from the Big Bang to at least the present. Many events in the present suggest that the "change over" from Phase I to Phase II is near. We are ready for Phase II.

Phase II

Circumstances of Phase II are different from those of Phase I. Where Phase I is dominated by physical "whatness," Phase II is the cosmic "imperfect" representation of "whyness." Whyness is reason itself.

The Universe of Perfect Whatness, Universe of Imperfect Whatness and extrinsic perception are not present in Phase II. Instead, the Universe of Imperfect Whyness wholly occupies this half of the cycle. The Universe of Imperfect Whyness is the "awakened" mind—one functioning inductively. The predominant characteristic of Phase II is "knowledge" of reality (imperfect whyness). Phase II begins after Phase I ends.

Universe of Imperfect Whyness

Toward the end of Phase I there are trillions of Universe of Imperfect Whyness units. These are nonfunctioning minds. They do not function on their own, but only function with respect to the physical world. They function in terms of "whatness."

At the end of Phase I some of the nonfunctioning minds inhabit physical bodies (Universe of Perfect Whatness units). Some had incarnated Universe of Perfect Whatness units. So during Phase I these minds are nonfunctioning in themselves—they do not reason inductively, but only function with respect to the physical world.

In Phase II the minds begin functioning in themselves. They reason inductively—they are "awakened" minds. This is the "second birth." These minds seek knowledge about Reality. They become oriented toward reality for security. They become identified with reality by being defined by it. They become "realistic." In seeking truth, the real mind develops.

The product of reasoning (to an understanding of) whyness is a "figurative whyness" representation of Whyness itself. This whyness representation of reason itself is called "knowledge"—knowledge of "form," knowledge of structure. Not whyness itself, but rather its knowledge representation. One of our goals is to understand this eminent-perfect whyness.

We, as mental beings, are caught between full whatness—physical creation (our physical environment) and full whyness (reason itself—or Consciousness). Regardless which direction (deductive—pointing toward experience [perfect whatness] or inductive—pointing toward consciousness [perfect whyness]) we—as minds (fully functioning or not)—are found between the extremes (experience—whatness and consciousness—whyness). During Phase I we are mentally oriented toward whatness. During Phase II we are oriented toward whyness. These statements encapsulate the meaning inferred in the following statement. Varying degrees of whatness and whyness "conceptually" represent reality.

In Phase II the mind is inherently self-sufficient. Desire is practically eliminated because there is no "want" of security. From a behavioral perspective, individuals during Phase II are generally realistic. They do not need another universe to survive (function) as does the mind (universe of imperfect whyness units) during Phase I.

In Phase II the mind rests during periods of decreased activity. This rest is not unlike "daydreaming."

Perception

Perception is intrinsic during Phase II. Intraphenomenal relevancy places not only the effect, but also the cause of intrinsic perception within the Universe of Whyness unit. (See How Aspect Diagram.)

Intrinsic perceptive processes are "intrinsically" connected to the mind. That perception is intrinsic does not mean that the cause is (categorically) "in" the Universe of Perfect Whyness unit. It means there is no temporal delay between experiencing an intrinsic perception and its cause.

Intrinsic perception links the mind to whyness itself (just as extrinsic perception links the mind to the physical whatness world in Phase I). In Phase I there is a delay between something in the physical world (Universe of Perfect Whatness) and its sensory representation as an apprehended image (extrinsic perception). Intrinsic perception links the functioning mind to the Realm of Whyness. (More about this in a later section.)

In Phase II reason (or whyness) is perceived intrinsically (inductively inferred by the mind). Where the common denominator of experience (during Phase I) for the nonfunctioning mind is the physical-whatness world, the common denominator for the functioning mind (during Phase II) is the Realm of Whyness. The nature of this commonalty is that each individual shares in the singular source of experience. This confers similarity of experience.

Extrinsic perception (in Phase I) links the (in itself—nonfunctioning) mind to the physical world. Extrinsic perception is a representation of the "whatness" physical world. Intrinsic perception links the functioning mind to the Realm of Whyness where one experiences a facsimile of reason (or whyness)—as knowledge.

There is a difference how each perceptive process functions. Extrinsic perception functions in "serving up" whatness images as premises from which one builds a "repertory of familiarity" with the physical world. We deduce information about the physical world.

Intrinsic perception functions to "push" one "toward" an understanding of one's purpose in life. In the broadest sense this understanding encompasses life itself. We induce information about the whyness realm—why things are the way we find them. For example, why do we have minds and why do we experience things the way we do? In short, what is the meaning of life? Again, to obtain these answers also means we must understand time and how we fit into the scheme of things.

In Phase I intrinsic perception is subfunctional. We know its presence because of dreaming. Dreaming is intrinsic perception in a subfunctional mode. The main point is that intrinsic perception is present in Phase I. This apparently means it is an inherent functional addendum to the mind. A more in-depth explanation of intrinsic perception is later addressed.

In Phase II intrinsic perception becomes fully functional—not unlike extrinsic perception in Phase I. Intrinsic perception becomes an expression of inductive inferring. In Phase II this is the "form" of what we experience—knowledge.

We reason to experience why we exist. Without intrinsic perception the mind would have no means to direct its inferences in Phase II. Inferring implies something. Without "something" we would have no experience. "Something" (whatness) takes the form of extrinsic perception during Phase I. Somethingness takes the form of intrinsic perception during Phase II.

Intrinsic perception follows inductive reasoning. We reason inductively and then experience "patterns" of thought (hence its form). These "pictures" entail clues pointing to an understanding of why (this or that [something]).

Extrinsic perception is diametrically opposed to intrinsic perception. The dominant perceptive process in Phase I is functionally reversed from the dominant perceptive process during Phase II. In Phase I deductive inferences follow from extrinsic perception. We experience the physical world through extrinsic perception. We mentally "draw" deductive inferences from physical experience.

In Phase II intrinsic perceptions follow from inductive inferences. Even in Phase I when intrinsic perception is sub functional, we still notice the individual can "direct" his dreaming. The distinction between directing (with the mind) and what is directed is not easily noticed because the connection between inductive inferences and intrinsic perception is a seamless one. We find the same operative seamlessness in Phase I with extrinsic perception and deductive inferences.

Extrinsic perception is distinct from the physical world. Extrinsic perception seamlessly mirrors the physical world. In extrinsic perception we automatically identify "what" we experience. That extrinsic and intrinsic perceptions are diametrically opposed also falls in line with the idea deduction and induction are oppositely directed.

Life: Discussion

There are a few topics that should be mentioned about Phase II. There is no death in this phase of the cycle, at least in terms of how most people understand it. And people do not properly understand death. Life is immortal—the mind does not cease to be. Mostly, life is perpetual during this phase of the cycle. There is not anything (like an entity Universe of Imperfect Whatness unit in Phase I) in this phase, which the Universe of Imperfect Whyness unit (mind) can incarnate. There is no reincarnation in Phase II.

Change from Phase II to Phase I

Major changes do occur during Phase II. Before explaining, we reiterate their corollary conditions in Phase I. During much of Phase I there evidently is no Universe of Perfect Whatness life of any kind. It is only during the latter stages of Phase I that animal life emerges. Evidently all animals multiply and diversify from one simple creature. This is a key statement about Phase I. It was stated that this evolutionary development is preparatory for Phase II. It is the phenomenal reason for life in Phase I.

Reason dictates there be a counterpart to the above key statement about Phase II. What prepares Phase II for Phase I? There are some premises embedded in symmetry principles that point the way to understanding.

First, Phase II will last exactly as long as Phase I.

Second, desire must not be sustained during Phase II.

Third, the conditions in Phase II should inversely duplicate what happens in Phase I. Since there is a divergence of Universe of Imperfect Whatness units during the latter stages of Phase I, we expect a convergence of Universe of Imperfect Whyness units during the latter stages of Phase II. This idea follows from the understanding of entropy during Phase I. We expect negentropy plays a role in Phase II.

Fourth, just as there is an "evolutionary" increase of deductive abilities toward the end of Phase I, we expect a decrease of inductive abilities toward the end of Phase II. Such reasoning, among many other things, would explain the increase in "desire" during the latter stages of Phase II. The next question is "how can all these conditions be fulfilled in Phase II?"

The problem of means focuses on another question. How can desire be eliminated when two Universe of Imperfect Whyness units desire each other? Simply, the two units (minds) coalesce and become one. They unify. This converging process happens during the latter stages of Phase II. Eventually, there is only one Universe of Imperfect Whyness unit left. That individual becomes the historical representative of every Universe of Imperfect Whyness unit (mind) which formerly existed.

This unifying (converging) process is inversely similar to the "diverging" process found in Phase I, e.g., physical birth—increasing the number of Universe of Imperfect Whatness units.

At the instant of "changeover" the solitary Universe of Imperfect Whyness (let's call it Adam) unit becomes asleep. Its mind ceases functioning inductively. It no longer functions on its own. The formerly "mental being" later takes on physical form. It becomes subconscious—it becomes physical. The phenomenal purpose of Phase I begins again.

The physical is a representation of knowledge in "material form." Knowledge, which during Phase II was a property of consciousness, becomes represented in (a property of) physical form. Physical structure is a "memory" of the knowledge representing it. Just as knowledge represents things (every something) things also represent knowledge.

An automobile, for example, embodies the knowledge used to design (create) it. Knowledge can be "extracted" from seriously studying the automobile to "understand" it. Again, our leading question is "how can this be explained?

Here, how can automobile transportation be explained? We do this by "breaking down" the automobile to see how it all fits together. In this example, we can "explain how this functions" (transportation) by learning how its components interact to produce transportation. (We return to the automobile example in final chapters.)

The aforementioned mind (Adam) does not function again (on its own) until the latter stages of Phase I. Again, this is preparatory for the beginning of Phase II. When Phase II begins, time is again inverted. This is an endless two-phase cycle.

The endless two-phase time cycle does not mean every cycle occurs with the transpiring of the same events. However, the means of occurrence, must be mechanically consistent. Each phase develops for the phenomenal purpose of setting up the next phase. Consistency is found in the mechanisms of entropy and negentropy.

Key Points:

- The When Aspect represents time (flow). It is the third aspect of the Deltic Cosmos.

- Time is associated with change: there is not one without the other. Change occurs unidirectionally.

- Einstein's Special Theory of Relativity indicates time and space are intertwined. They have a commonality (a constancy—light velocity) in space-time. Einstein reasoned time (rate change of flow) and space (distance) are relative to the reference frame (or motion) of the observer. (There is no change without motion.) There is no motion without duration.

- Time duration is measured by clocks. The shortest duration is known as the Planck time. It is derived from an understanding of quantum processes.

- Duration suggests the possibility of a longest duration. If time flow commenced at the Big Bang, is there an end to time? In short, is there a longest duration? Is there an end to existence?

- If there is a longest duration that poses other questions. What was there before the duration? Simply, what was there before the beginning of time? Put another way. What was there before the beginning of whatever? Duration implies an end of time. If time ends, what then? The problem of "before and after" can be asked. There is an answer.

- Entropy suggests time cannot reverse its directionality. Yet, symmetry considerations in physics suggest somehow there must be a "reversal" to temporal directionality. Time directionality must somehow "flow the other way," but it (time directionality) must do so without reversing upon itself.

- The When Aspect Diagram illustrates a two phase time cycle. Time flow goes one way in Phase I. Time flow goes the "other way" in Phase II. Time flow reverses (directionality) in a different phase of the two part time cycle. There is a moment when temporal directionality changes. This happens at the beginning and ending of each phase. The end of one phase is the beginning of the other phase.

■ The When Aspect Diagram provides a noncontradictory answer to problems associated with unidirectionality, entropy and the beginning and ending of time. Forward time is negated by backward time in a different phase of the cosmic cycle.

How does the structure of reality "contextualize" time? Time exists as a conjunction of the Where Aspect (events) and How Aspect (order). Time comes into being, as an interaction of events and order. Events and ordering produce relative motion. Light also enters the equation. Relative motion and the constant velocity of light directly affect the rate of change. This means we must also understand how light fits into the scheme of things. There also would be no recognition of these evidential facts without awareness (of memory, for example). Obviously, consciousness and awareness must be factored into understanding time. But for awareness, time would have no meaning. To understand time is to view it within the context of the whole structure of reality. We are working toward that goal.

We have set forth some preliminary ideas to guide us toward a proper understanding of time. Cause preceding effect must be balanced by cause following effect. Our concept of change (as applied to the universe) can be found in entropy. Entropy suggests temporal unidirectionality. Entropy must be balanced somehow by negative entropy. Temporal directionality must also somehow be balanced. Forward time must be balanced somehow by backward time. Memory must also be incorporated into the understanding of time. All these concepts must be related to the conscious present. This brings up the idea of unconsciousness versus consciousness. Is there physics suggesting this pairing of ideas? When we study the next higher-level structure of reality we will better understand.

Our adventure into the mind expanse enabled us to map a two-phase time-cycle. Entropy and negentropy needed to be understood in terms of reality's structure. We asked how can entropy be explained? This question, along with other considerations, is answered by the When Aspect Diagram.

Our "construct" system-building continues. (The concepts of entropy and negentropy return in the final chapters to help solve some other problems.)

Completing an understanding of the cosmic cycle enables us to conceptually leap and view the structure of the cosmos in its entirety. We begin by collectively considering the three Cosmic Aspects of Where, How, and When. We do this in Part B called the Deltic Cosmos.

Part B: Deltic Cosmos

Perspective

- The integrated structure composed of the Where, How, and When Aspects
- Defines eventualities by where, how, and when they occur
- Deltic Cosmos is one half of the Bi-Deltic Cosmos structure

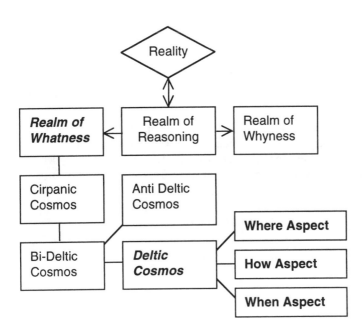

Figure 10:1. Perspective: Deltic Cosmos

166

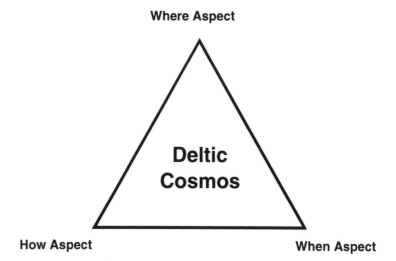

The Deltic Cosmos is composed of three Aspects: the Where Aspect, the How Aspect, and the When Aspect. The Where Aspect is composed of the universes. The How Aspect pertains to the order of things. The When Aspect represents the "forward" flow of time.

Figure 10:2. Deltic Cosmos Diagram.

Chapter 10
Deltic Cosmos

"The idea is not for people to get their act together,
but to get it together by getting rid of the act."

The Deltic Cosmos is the first major integrated structure of the Realm of Whatness. It is constituted by the Where Aspect, the How Aspect, and the When Aspect. (They were described in Part A.) The Deltic Cosmos results from the collective activity of these three aspects—"acting as a whole." This chapter also deals with change in total perspective—not just change-in-itself.

Change has three major features. There is change itself—denoted by the When Aspect. There is the object or entity that undergoes change—the Where Aspect. And there is an aspect that relates "changed-objects" (entities) to the "order" induced by change. Change happens because a difference in order has occurred. Order is represented by the How Aspect.

The Deltic Cosmos is a functional unit. It is the basic "whatness" unit relating cosmic process. No "Aspect" (Where, How, When) of the Cosmos exists or functions apart from the other two.

Deltic Cosmos Diagram

Because the Deltic Cosmos is a three in one structure, it is represented by a triangle. The term "Deltic" is used not only to denote basic cosmic structure, but also to distinguish it from its Anti-Cosmos. To keep this chapter simple, we look at large scale Cosmic properties exhibited during each phase of the When Aspect. Let's examine the overall properties of the Deltic Cosmos.

Cosmic Properties During Phase I

Phase I evolves around the macro-cosmic behavior of physical matter. "Behavior" refers to "change in order." So, we focus on Universe of Perfect Whatness and what happens via large-scale dynamics.

Particles of the Universe of Perfect Whatness (of physical matter) have a velocity less than light. In Special Relativity they are denoted "tardyons." Tardyons are always subluminal.

Let's consider what happens to a hypothetical-physical rod subjected to extreme theoretical conditions.

First, a rod will be seen as a tardyon in any subluminal Lorentz frame (of reference). It will always appear to have positive energy.

When a tardyon rod moves (at relativistic velocities) past an inertial observer, its dimension parallel to the direction of motion will contract as viewed by the inertial observer it passes. The observer photographs the rod as it passes in front of him. Upon examining the snapshot, he notices the object appears slightly rotated away from him. A Lorentz-contraction appears rotated because the observer (at any given instant) sees only by simultaneous arrival of light.

Second, the Principle of Relativity states there is no difference between a system considered at rest and one moving uniformly. To each inertial system it appears that the other system is the one undergoing a one-dimensional contraction.

Because the Principle of Relativity shows there is no preferred frame of reference, there is no significance to an object's size. Dimensionality or spatial extensionality is a variant when comparing different inertial systems. Relativity also implies the laws of physics are indifferent to the uniform motion of the observer.

Third, Relativity phenomena primarily occur because of the constancy of the velocity of light (represented by the letter "c"). The constant velocity of light is the connecting factor between space and time. This is a key clue to the structure of reality.

Another Universe of Perfect Whatness property is physical mass. Mass is a measure of a material body's resistance to acceleration. Proper (rest or inertial) mass is a measure of resistance using a system's own frame of reference.

Inertial forces were thought to arise from the universal distribution of matter. In the Machian sense the value of proper mass is associated with a system's relative motion to all matter in the physical universe (Universe of Perfect Whatness). These ideas are still subject to debate. Anyway, when a rod speeds past a system (observer) considered at rest, its mass (or better, energy) appears greater to the inertial observer than if it were measured in the rod's frame of reference. Its energy increases.

Four basic coordinates (Lorentz frame) are used to track movement of material objects. Three coordinates represent an object's location (using the object's center of mass) in space. One coordinate represents "placement" of an instant in the flow of time.

Tardyons (Universe of Perfect Whatness) are represented as a collection of spatial points extended along a temporal line. Space is three dimensional and isotropic (same energy everywhere) while time is one-dimensional and unidirectional.

A subluminal trip is "time-like." That is, much time passes, but little distance is traversed. This is the nature of Universe of Perfect Whatness—physical matter.

Momentum reflects an object's combined mass and velocity. The total momentum of a tardyon object at rest (in an inertial system) is zero. In such a system there are no added relativistic effects on energy and the object's mass will have its minimum value. An object's mass, in a proper frame of reference, is often correlated with an equivalent amount of energy by the famous Einstein equation $E = mc^2$.

For an object in relativistic motion, Minkowski represented mass as an energy-momentum 4–vector. Four coordinates identify the course of space-time curvature. The 4–vector measures the magnitude of mass. Mass remains unchanged in any coordinate transformation to other inertial systems. Mass is an invariant property.

An accelerating body accrues relativistic effects (compared to inertial observer). Energy and momentum increase. These effects need not be interpreted as an increase in mass. Rather, the usually interpreted change in mass can be attributed directly to an increase in kinetic energy since it takes increasing energy to increase speed. The compounding effects of the space-time through which the body accelerates produce the increase in energy.

Inertial mass is one concept, but relativistic mass more appropriately refers to the system's increased energy. Some added energy is translated into an object's momentum taking the form of kinetic energy.

A tardyon can have almost any speed. However, an object's velocity can never equal or exceed the velocity of light since a limitless amount of energy would be needed. This is the usual understanding. We later find there are cosmological reasons that light is a barrier to tardyon velocity.

The total energy of an object or system appears to change when there is a difference in the rate of time flow. Energy, represented by one vector, can be correlated with the one temporal dimension. The term time, as used in relativity, does not pertain so much to change-in-itself (the When Aspect) as it does to the rate of (temporal) change. Temporal rhythm (frequency) is directly associated with the varying energies of a system as viewed from different frames of reference. Energy is a variant property. The time factor, discussed in Chapter 9, is due for another look.

A tardyon rod not only exists as a part of space, but also time (the When Aspect—Phase I). As a rod moves past an inertial observer, its molecular and atomic motion "ticks" at a slower rate as determined by the stationary observer (when he compares it to a similar "clock" measured in his own system). When time slows, it dilates because there is a longer duration between ticks. The phenomenon of varying rates of change occurs because of the chosen vantage point one uses in calculating time. Time rate is a variant property.

Matter-energy and space-time are merged in the General Theory of Relativity in an attempt to explain gravity. A basic premise for General Relativity is the Equivalence Principle by which Einstein locally equated inertial mass with gravitational attraction.

Gravitational mass is a measure of the "force" acting upon matter from the gravitational field in which an object is immersed. The field is determined by the quantity and distribution of masses in the region.

Gravity has been considered the natural result of the interaction of mass with space. The presence of matter (mass) causes a distortion of space and time. This distortion is the curvature of space-time. The distortion is described as a gravitational field that determines the movement of an object placed within it. This does not mean space-time is altered, but that we must use distorted models to represent what happens to space-time in the presence of matter. Note: In a mathematical sense, the term curvature denotes the intrinsic structure of a manifold as defined by the parallel transport of a vector in a medium (here, a four-dimensional sphere).

When a tardyon is subjected to a gravitational field its energy increases, and it experiences lateral forces causing it to accelerate toward the source. Other relativistic phenomena occur. If a distant tardyon object emits light, it is redshifted because of a gravitational presence that dilates time—slowing time flow.

Measurements of space or time intervals have only relative significance. It is more meaningful to determine a system's coupled space-time interval because when a system's spatial dimension is contracted, it is accompanied by a corresponding dilation of time.

The space-time interval between two events disappears when only the velocity of light separates the events. That is, one event happens at the arrival of a light signal from the other event. Although space and time are translated one into another when considered from the perspective of different inertial observers, the coupled space-time interval remains invariant. That is the significance of the constancy of the velocity of light. Light velocity (c) is the constancy of proportionality between time (time flow) and space (spatial extensionality).

Again, note changes in space-mass are still attributable to the When Aspect because differences only occur when time (or change) is involved. The difference or result of change is represented by the How Aspect.

Let's digress: We must separate our perceptual experiences from the knowledge of natural processes. Although this forces a seemingly egocentric slant on our sensory experiences, it should not interfere with our inferential processes of obtaining knowledge. For example, though we nightly experience the presence of distant stars, we need not infer that we are seeing them as they presently are themselves.

Knowing something as it is, is more a matter of inferential knowledge than sensory experience. The idea is to understand "what" we experience. Relativity-wrought changes should be understood in this way. While relativity abstractly extends our knowledge, it does not enlarge our sphere of sensory experience.

Let's simplify what we have learned. In an inertial system all tardyons carry zero momentum and have minimum energy. As a tardyon rod is accelerated toward the velocity of light, it increases in energy and momentum. The increase of energy is supplied by what is causing the acceleration. It may, for example, be caused by "gravitational attraction" to some massive object.

From the perspective of General Relativity an object's increase in energy is caused by the space-time geometry in which the object is moving. The secondary effect of increased momentum results from the increase in energy.

Aside from the constant velocity of light the most crucial factor in relativity is the varying position of an observer's frame of reference. Differing relativistic values are attributable to the perspective from which the values are calculated. The previous statements are oversimplifications because the values or effects are inherent in the structure of reality. All forces are intrinsic and interactive. Extrinsic forces arise from our scheme to classify knowledge.

Preceding chapters (Part A) touched upon all three major Aspects. Properties of the Where Aspect's Universe of Perfect Whatness were especially delineated. The When Aspect was described. The How Aspect was implied in the order wrought by change. (It is through order we countenance change.) Properties of the Universe of Imperfect Whyness, though present during Phase I, are discussed shortly. With a basic understanding of Phase I properties, we address those of Phase II.

Cosmic Properties During Phase II

Scientists attempt to maintain symmetry in piecing together the structure of reality. Here we place puzzle pieces without direct empirical evidence. The Principle of Contradiction becomes useful in these circumstances. The idea is to maintain the symmetry allowed by certain equations while not contradicting well-established information when affixing meaning.

Astronomical studies and experimentation in high-energy physics have supported equations dealing with tardyon relativistic effects on space, time, energy, momentum, and mass. Tardyon physics pertains to only one side of the relativity equations. The other side involves what are now called superluminal frames of reference.

Interest in possible superluminal particles increased after 1962 when Bilaniuk, Deshpande, and Sudarshan (B-D-S) advanced their "Reinterpretation Principle." The principle is an example of using the logical Law of Contradiction. B-D-S said different observers need not agree on the interpretation or description of specific events, but must agree with the laws governing those events. That is, the laws must remain invariant when transformed from one frame of reference to another.

In subluminal frames all particles are customarily assigned positive energy. However, in superluminal frames some particles will appear to certain observers to have negative energy. B-D-S provided a way around this difficulty. They said a description can be used that always relates a superluminal particle. They reinterpret negative energy particles that appear to travel backward in time as positive energy particles traveling forward in time, but in the opposite direction.

Two different observers may not agree on the description of what happened but have to agree on the governing laws. A superluminal particle may appear positive to one observer yet negative to another in relative motion. Therefore, the interpretation of energy depends upon the velocity of the observer. It is necessary to reinterpret what occurred to avoid a negative-time interval between two events. For example, observer "A" may describe an event saying a particle, which he considers to be positive, was emitted first and then absorbed, while observer "B" interprets the same event as one in which a particle was absorbed first and then emitted.

Agreeing with the Reinterpretation Principle, observer A can accept observer B's interpretation if the particle is considered (from observer A's frame of reference) to be traveling backward in time with negative energy. These very conditions allow observer B to see the event as one that relates a positive energy particle going in the opposite direction (here, the roles of emission and absorption are switched). The two observers will see the same process, but in reversed order. These ideas work for any observers disputing what actually occurred.

The Reinterpretation Principle overcomes a problem of phenomenal causality. It changes the roles of cause and effect by describing what happened in reverse order. These ideas add meaning to the theory of superluminal particles. In the context of Relativity these theoretical superluminal particles are now called (after Feinberg) tachyons.

This project postulates tachyon physics describes Phase II mechanics. The Universe of Imperfect Whyness is the dominating universe during Phase II. We propose that the dominant universe during Phase II is constituted of tachyon "matter." It follows that Universe of Imperfect Whyness "particles" travel at a speed greater than the velocity of light. These are the superluminal particles. The terms "matter" and "particle," as used in this discussion, should not be construed to be physical although Universe of Imperfect Whyness properties, generally, will be treated as though they are physical. Doing this maintains a standard by which to assign properties.

There are two perspectives from which to describe tachyons. One is to relate what happens to tachyons when compared to tardyons under similar conditions; the other is to use symmetry and describe tachyons from an inverted view. We use both. A hypothetical tachyon rod will be subjected to varying circumstances. This enables us to describe superluminal properties of the Universe of Perfect Whyness.

We first compare tachyons by describing their behavior as directly compared to tardyon mechanics. The same basic principles of tardyon relativity also apply to tachyons. A tachyon will be observed to remain a tachyon in any superluminal frame.

As a tachyon rod increases in speed, there is a spatial Lorentz expansion. Remember, tachyons travel at velocity greater than light. There is no upper limit for tachyon velocity.

Another Universe of Imperfect Whyness property is tachyon mass. Recall that mass is a measure of an object's resistance to change. It measures an object's tendency to resist a change in energy caused by relativistic effects. This perspective helps free the individual from unnecessarily linking tachyon mass to physical mass.

As a tachyon object loses energy it accelerates. When it has lost all its translational energy, it attains infinite velocity. It would be everywhere on the super-circle of its trajectory (Bilaniuk and Sudarshan). The When Aspect also affects tachyon mechanics. As a tachyon object increases in speed, time contracts and speeds up. Because change is synonymous with the When Aspect, we say that as time contracts (rate of change increases) a tachyon in such a system accelerates. It is also possible tachyons are associated with some kind of universal wholeness function (see Chapters 20 and 21).

One might ask what all these statements mean for cosmic structure. To get a better idea, we present these ideas from a more symmetrical perspective.

A tardyon at rest or in uniform motion is symmetrically analogous to a tachyon traveling at infinite velocity. Because there is no absolute simultaneity, these values are contingent upon what observer is doing the measuring.

Tardyon mechanics is easily determined from an inertial observer or proper frame of reference. Tachyon phenomena can also be viewed from its proper frame of reference—one at infinite velocity. Note that infinite velocity is assigned by how it relates to a tardyon standard.

In 1969 L. Parker introduced the Extended Principle of Relativity stating that the laws of physics are the same in superluminal frames as they are in subluminal frames of reference. Therefore, superluminal particles behave as common tardyons relative to superluminal frames of reference, but take on the properties of tachyons when standardized by subluminal frames. These ideas are also incorporated into the Principle of Duality. It states that the choice of how to label these two systems is arbitrary. Labeling depends on how the observer defines his own system. In a structural sense, it may be both tardyon and tachyon proper frames coexist in the same way. Maybe further elimination of contradiction in reinterpretations will reveal this.

It is thought by some investigators, like C. Schwartz, that negative energy for some tardyon states does not present a problem. He says a free particle moving "in" or "out," relative to some interaction, should not be separate from the concepts of "charge," "momentum," or "energy." He continues that the charge of a tachyon particle should not be defined by superimposing the same ideas used in understanding tardyons. Schwarz's ideas are interesting because they do not require use of the "reinterpretation principle." With these ideas in mind, we look closer at the symmetrical interpretation of Phase II mechanics.

In an inertial system a tardyon has zero momentum and minimum (a non-zero finite value) energy. In a tachyon proper frame of reference the total energy is zero, and the momentum has a non-zero finite energy. In their proper frames of reference tardyons and tachyons have opposite energy-momentum characteristics.

Because energy and momentum are measurable, they are usually assigned real numbers. If this is done; space, mass, and time in a tachyon proper frame of reference take on imaginary (or complex) values. That is, those values are associated with the square root of a negative number. Since these quantities are unmeasurable from a tardyon standard no problem arises. That these are mathematical properties (for example, as opposed to perceptual properties) should not make tachyons seem any less real. Imaginary quantities have to be assigned to values of tachyon-proper frames to give them physical significance. This relates their properties to what we already understand about tardyons. Properties with imaginary values have nothing to do with how those properties would be perceived in their own system.

We have seen what befalls tardyons as their velocity approaches the velocity of light (c). The velocity of light is the dividing line that creates symmetry in relativity. We will describe what happens to tachyons as their rate of motion changes from a proper frame and (de)accelerates toward the speed of light. Tachyons approach (c) from above instead of below.

A tardyon object will spatially contract in the direction of motion as it accelerates and approaches (c). At (c) the one dimension disappears. Time flow ceases. Under the same conditions, time will dilate or slow until at (c) the temporal aspect disappears. When force is applied to a tachyon object, it slows relative to the direction of source. It would take a limitless amount of energy to decelerate a tachyon to the velocity of light. Tardyon-tachyon symmetry clearly emerges from these interpretations of the "other side of Einstein's equations for Special Relativity."

Superluminal time is isotropic while space is unidirectional. In a proper frame of reference tachyons can be represented as a collection of temporal points extended along a spatial line. A superluminal trip would be space-like. That is, much space would be covered in very little time.

A superluminal frame of reference can be found from which two simultaneous events seem to be instantaneously translated through space. Functionally, space-like and time-like intervals are interchanged in tachyon mechanics from what they are in subluminal frames. More explicitly, the temporal aspect in a subluminal frame is analogous to the spatial aspect in a superluminal frame. Subluminal spatiality, to a certain extent, corresponds to superluminal temporality.

What do these statements mean when making transformations between subluminal and superluminal frames? In keeping with a physical interpretation, one-dimensional subluminal time would have to translate into three-dimensional superluminal space. A similar problem arises when transforming subluminal space into superluminal time.

Some theoreticians, like Antippa, have developed various schemes to save a physical interpretation of tachyons. Antippa introduced a tachyon corridor in which superluminal space acts more like subluminal time. He does this by orienting superluminal space in a preferred direction moving at a preferred velocity. This concept is a tachyon analog of the temporal world line in subluminal frames.

Other ideas have been developed which completely symmetrize the interpretation of tachyons on a tardyon basis. Telli and Sutar have developed equations showing tardyons have real rest mass in subluminal frames and tachyons have real rest mass in superluminal frames. The reasoning is based on the idea that in subluminal frames tachyons behave like tardyons in superluminal frames. These ideas are fine and suggest tachyon properties by standardizing them on tardyon mechanics.

This book postulates the carrying agency of the Universe of Imperfect Whyness is tachyon "matter." We are not obligated to posit a physical interpretation of tachyons. Therefore, it may not be necessary to retain a three-dimensional superluminal space to piece together the structure of reality.

Parker introduced the idea tachyons may be represented by a one-dimensional coordinate system for spatial location. Other bonuses appear in his interpretation. Using his Extended Principle of Relativity, he postulates there are proper frames of reference in which space, time, and mass have real numbers to represent them relative to subluminal frames. The value of always using real numbers to represent properties is, however, questionable.

For the resulting transformations between subluminal and superluminal frames to not be misleading (according to the Principle of Duality) the assigning of numbers may also have to be arbitrary. Therefore, some tardyon properties that are ordinarily assigned real numbers would be labeled imaginary or complex. When transformations are calculated, some real numbers transform into imaginary ones and conversely. This idea may also extend to energy and momentum.

When a tachyon object experiences a superluminal gravitational force, its energy decreases and its velocity increases. Because it experiences a centrifugal force it accelerates toward the gravitational source just as a tardyon would do. This happens because a tachyon incurs gravitational repulsion, but kinematically experiences a change in motion. This results in the tachyon being attracted to the source. Also, the resulting space-time distribution may be different from that in tardyon mechanics. Generally, tachyons can be viewed from a tardyon perspective as energetically repulsed, but kinematically attracted.

Universe of Imperfect Whatness, and Universe of Perfect Whatness—Universe of Imperfect Whyness Interaction

Although the Universe of Imperfect Whatness exists during Phase I, it is not a dominating universe during any Phase. Without the aid of memory and extrinsic perception, the Universe of Imperfect Whatness would not function. The mechanisms of extrinsic perception are found in the physical body (a Universe of Perfect Whatness entity). The Universe of Imperfect Whyness connection to intrinsic perception produces the phenomenon of memory. The Universe of Imperfect Whatness functions in a deductive mode. Its premises are presented by either extrinsic or intrinsic perceptions. The course of deduction is usually patterned after what the individual remembers. The Universe of Imperfect Whatness arises from the interaction between components of the Universe of Perfect Whatness and Universe of Imperfect Whyness.

Because the Universe of Imperfect Whatness arises from the interaction of Universe of Perfect Whatness and Universe of Imperfect Whyness components, it is unnecessary to postulate a carrying agent. We need only to state a possible explanation of how the Universe of Perfect Whatness and Universe of Imperfect Whyness interact. Further understanding of the "changes" of the Universe of Perfect Whatness and Universe of Imperfect Whyness is discussed in the following chapter. Suffice to say: tardyons (Universe of Perfect Whatness agent) and tachyons (Universe of Imperfect Whyness proposed carrying agent) may interact directly or at least are mediated by electromagnetic phenomena.

Alternating Dominant Universe

A unique event occurs marking the changing of the prevailing dominant universe. The Universe of Perfect Whatness dominates Phase I. The Universe of Imperfect Whyness dominates Phase II.

In a mechanical sense, how can the changing of Phase I to Phase II be described. Here, descriptions direct us toward noncontradictory possibilities that explain the fluctuating structure of the Deltic Cosmos. The reason for a two phase time cycle was explained in the When Aspect. Those explanations set the structure. We now consider descriptions that involve all three universes functioning as a unit.

In describing the change of the dominant universe from the Universe of Perfect Whatness to the Universe of Imperfect Whyness, we are linking tardyons and tachyons with universal expansion and contraction. We are guided by the General Theory of Relativity. We first look at universal expansion.

In Phase I all tardyon matter (Universe of Perfect Whatness) is expanding. The red-shift of distant galactic light shows that the expansion of the Universe of Perfect Whatness overcomes gravitational attraction. However, an observer in another frame of reference, or universe, might perceive this phenomenon as contraction along with gravitational repulsion.

Scientists interpret universal expansion as evidence for the Big Bang theory of Universe of Perfect Whatness origin. According to the Big Bang all tardyon mass was once crushed out of existence and infinitely condensed into a mathematical point—the universal singularity.

Physicists say physical laws break down at the singularity. The singularity is the center of a "cosmological" black hole (or white hole) where space-time is infinitely warped or curved. Maybe only pure gravitational radiation could pass through the singular point that effectively exists apart from spatial extensionality and temporal dimensionality.

Gravitation (or the tidal effect) is so intense near the singularity that it very quickly produces particles and antiparticles by pulling apart virtual pairs (in the vacuum) of matter and antimatter. That is, from virtual pairs, tardyons materialize. The resulting extreme dense mass then creates immense internal pressures causing the primordial explosion. The explosion overcomes gravitational attraction resulting in our expanding tardyon universe (the Universe of Perfect Whatness). An inversion of a black hole is a white hole. The mechanics of a white hole may describe the Cosmic Universe of Perfect Whatness Big Bang.

The Universe of Imperfect Whyness is inactive throughout most of Phase I and has little effect on the Universe of Perfect Whatness. Yet, near the end of Phase I the Universe of Imperfect Whyness tends to have an increasing influence. Both the Universe of Perfect Whatness and the Universe of Imperfect

Whyness diverge (effect of entropy) before the changeover to Phase II. "More bodies; more minds."

What happens before the instant of cosmic changeover (from Phase I to Phase II)? The Universe of Perfect Whatness ceases expanding and the Universe of Imperfect Whyness stops diverging. Phase II commences with the Universe of Imperfect Whyness being activated and dominant. What this could mean for black or white holes is uncertain. Note: The Universe of Imperfect Whyness is present in both phases of the cosmic cycle. At the changeover from Phase II to Phase I the tardyon universe begins anew with another big bang. The Universe of Imperfect Whyness is still present but cosmically dominated by physical matter. The Universe of Imperfect Whyness would have already converged (via negentropy) into one unit before changeover.

Key Points:

- The Deltic Cosmos is a single structure "integrating" three aspects. They are the Where Aspect, How Aspect, and When Aspect. No aspect functions independently from the others. Because the Deltic Cosmos is a three-in-one structure, it is appropriately represented by a triangle.

- An "extended" version of Special Relativity suggests there is no contradiction in the possibility of superluminal existence. These hypothesized "faster than light" particles are called tachyons. Superluminality (dynamics) may represent Phase II of the cosmic cycle.

- Granted the possibility of superluminality, physical matter exists as subluminal particles called tardyons. Tardyons exist during Phase I of the cosmic cycle.

- The Deltic Cosmos structure resolves "what" happens (i.e., where, how, and when). It defines where (the Where Aspect) change occurs. (The Universe of Imperfect Whatness transforms into the Universe of Imperfect Whyness.) It defines how (How Aspect) it changes. (Order in the whatness universe shifts [changes generally] from entropic to negentropic processes [dynamics]—subluminality to superluminality.) It defines when (When Aspect) it happens (there are phase changes—Phase I to Phase II and Phase II to Phase I.)

- Recognizing the significance of entropy means time cannot reverse over what has already transpired. Instead, time reversal is explained by a second phase where time "goes the other way." There is no contradiction in postulating a two phase time cycle where the second phase is superluminal.

- The Deltic Cosmos concept retains the unidirectional nature of time flow yet completes the solution to the problem of time reversal by stipulating it occurs in a different phase of the cosmic cycle. Time flows "forward" in the physical world (Phase I) and goes "backward" in a mental world (Phase II) represented by the monopole side to Maxwell's equations. Problems understanding the beginning and ending of time are completely answered (where and how) by the Deltic Cosmos two-phase cycle. There is no contradiction in postulating the Deltic Cosmos structure.

Our latest survey of the mind expanse was mapped into a construct called the Deltic Cosmos Diagram. It "enveloped" the three Aspects (Where, How, When) representing them as "parts" of a single concept—the Deltic Cosmos. This structure unveiled a "higher level" in reality's structural hierarchy.

The When Aspect answered the question concerning the beginning and end of time flow by posing a two-part-cosmic cycle. The unidirectionality (time flow) of Phase I is "negated" by the "reverse" directionality of Phase II. However, unidirectionality itself is not negated by the Deltic Cosmos structure. The Deltic Cosmos itself must be explained. This is done by postulating an Anti-Deltic Cosmos that accompanies our Deltic Cosmos. The next chapter concerns the combined structures of the Deltic Cosmos and the Anti-Deltic Cosmos yielding what is called the Bi-Deltic Cosmos.

Part C: Bi-Deltic Cosmos

Perspective

- The Bi-Deltic Cosmos is composed of the Deltic and Anti-Deltic Cosmos
- The Anti-Deltic Cosmos is a mirrored inversion of the Deltic Cosmos
- The Bi-Deltic Cosmos negates temporal unidirectionality
- It defines duration without directionality
- The Bi-Deltic Cosmos is one possible course of eventuality development

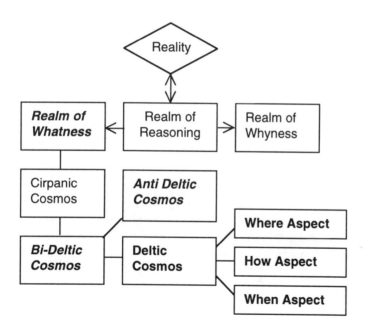

Figure 11:1. Perspective: Bi-Deltic Cosmos

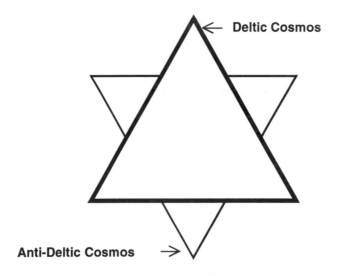

The Bi-Deltic Cosmos is a combined Deltic Cosmos and Anti-Deltic Cosmos. The diagram is the commonly known symbol Star David. The Deltic Cosmos is represented by one triangle. The Anti-Deltic Cosmos is represented by the other triangle. Each triangle is an inversion of the other. The Anticosmos is inverted from the Cosmos. In the Deltic Cosmos, time flows "forward"; while in the Anticosmos, time flows "backward." Spatially extended objects in the Cosmos are "right-handed"; while in the Anticosmos they are "left-handed." The Deltic Cosmos is "charged" negative: It is represented by the electron. Its "counterpart" antiparticle, the positron, is charged positive. It represents the Anticosmos. When the Cosmos is "in" Phase I, the AntiCosmos is "in" Phase II.

Figure 11:2. Bi-Deltic Cosmos Diagram

Chapter 11

Bi-Deltic Cosmos

"It is not for man to assert truth, but for reality to demonstrate it. Reality is the final arbiter of what is true."

The previous part (Part B) concerned the meaning of Relativity equations. What does Special Relativity say about the structure of reality? It was stated the equations have two sides. One side is about subluminal frames of reference and tardyons. The other side implies superluminal reference frames and tachyons. Tardyons were correlated with the Universe of Perfect Whatness. Tachyons were correlated with the Universe of Imperfect Whyness. The Universe of Perfect Whatness and Universe of Imperfect Whyness (described in previous chapter) alternate in dominance in a two-part time cycle. The cycle (the When Aspect) the universes (the Where Aspect) and their order (the How Aspect) comprise the Deltic Cosmos.

The Deltic Cosmos left some questions unanswered. In 1928 P.A.M. Dirac was trying to understand certain properties of elementary particles. He combined Einstein's theory of Special Relativity and Heisenberg-Schrödinger's quantum mechanics and developed the basis for quantum field theory.

Dirac applied his field equations to the electron. He found a theoretical formulation that represents the electron as a particle with spin equal to 1/2 Planck's constant. (Therefore, $\hbar = h/2\pi$.) There were two solutions for spin. Besides the two-spin solution he discovered "another side" to his equation. There were added "mirror" solutions allowing negative energy frequencies of the same particle.

For a short time, it was difficult to comprehend the meaning of Dirac's discovery. In 1932 C. Anderson found a cloud chamber picture showing a cosmic-ray particle having mass practically identical to the electron. The particle's track, however, did not curve in the direction expected of an electron. It curved in the opposite direction. Anderson discovered an oppositely charged counterpart to the (negative) electron. The discovery of this "positron" gave (reality based) meaning to the other side of Dirac's equation.

Dirac's theory naturally led to the idea that a particle could only be created in conjunction with its antiparticle. Therefore, the cosmos should have equaled numbers of particles and antiparticles. Yet, empirically this has not been proven. There should be equal amounts of tardyons and anti-tardyons. Because experimental evidence points to antiparticles, we need to incorporate this knowledge into our understanding of reality's structure. The macrocosmic design of the Bi-Deltic Cosmos structure is heavily premised upon information obtained from the theory and study of subatomic particles.

Bi-Deltic Cosmos Diagram

The Bi-Deltic Cosmos is a combined structure of a Deltic Cosmos and Anti-Deltic Cosmos. One triangle inverted and imposed upon another represents this structure's internal relationships. We use the Deltic Cosmos (our half of the Bi-Deltic Cosmos) as a standard to compare it to properties of the Anti-Deltic Cosmos.

Each point of the Deltic Cosmos Diagram (a triangle) represents one of its three "Aspects" (the Where Aspect, the How Aspect, and the When Aspect). The points of the Anti-Deltic Cosmos (an inverted triangle) represent the "inverted aspects" of the Deltic Cosmos. For example, the inverted Where Aspect of the Anti-Deltic Cosmos is a point opposite the Where Aspect of the Deltic Cosmos.

What do the inverted points of the Anti-Deltic Cosmos mean regarding the fate of the Deltic Cosmos? Why do we suspect there is an Anti-Cosmos?

Anti-Where Aspect

The Where Aspect of the Anti-Deltic Cosmos (denoted the Anti-Where Aspect) has three universes similar to those of the Deltic Cosmos except they are all inverted. What this means is developed by first looking at the Universe of Perfect Whatness of the Deltic Cosmos (our immediate physical world).

The Universe of Perfect Whatness has a principal property called mass. It would be difficult to deal with mass without space because "things" have extension in space. (Rest) mass is an invariant property; extension is a variant property. In this section we treat the mass-object as though it were always in a rest frame. This enables us to consider spatial extensionality.

Let's develop some concepts to better understand spatial extensionality. Spatial extension has three dimensions. Any part of a mass-object can be represented with points on a Cartesian coordinate system.

What is inverted extensionality? Spatial inversion is easily explained by the concept of parity. The word parity is French for "pair" or "even." The initial state has to be duplicated in the final state—even to even and odd to odd—for parity to be conserved. We say symmetrical systems have even parity; asymmetric systems have odd parity.

Parity is conserved, in a spatial-three-dimensional system, by an even number of 180° rotations of any coordinate. For example, the +y and –y coordinates could be switched and the +x and –x coordinates could be interchanged. This manipulation results in an even number of switches or turns. Although positionally changed, the object could still be superimposed upon the original by simple rotation. The resulting object (its condition) will be, point for point, identical to the original.

Parity is not conserved by an odd number of 180° rotations of an object's coordinates in a spatial-three-dimensional system. For example, if one axis is rotated 180° around one of the other two perpendicular axes, the object's position will be inverted compared to the original. (Effectively, we could interchange the +x axis and –x axis.) Such a maneuver reverses one plane of symmetry.

After coordinate rotation the resulting object becomes a reflection of the original. This maneuver cannot be done with a physical object. However, some physical objects can "represent" what happens. A glove can be turned inside out. Borrowing a term from chemistry, we say the "outside" of the glove is enantiomorphic to its "inside" when the glove is turned inside out. A right-hand glove is enantiomorphic to the "representative" state after being turned inside out.

Visualize yourself in front of a mirror to more easily grasp what this means. Hold up your right hand. Your right hand is the original object. In the mirror you see your mirror image. The person reflected in the mirror looks like you, but instead has his left hand up and is structurally inverted along an axis perpendicular to the mirror. The individual in the mirror is your enantiomorph or twin counterpart. The enantiomorph is identical except one of his axes has been inverted compared to your system. Simply, left and right have been reversed. Point for point, your enantiomorph cannot be superimposed upon you. One dimension will always be rotated. Similarly, inverting a right-hand glove turns it into a left-hand glove—an inverted facsimile. A three-dimensional object cannot be superimposed upon its enantiomorph counterpart.

If the individual's system in the mirror is axially inverted a second time, he will again be structurally identical to you because he can be superimposed upon you. Any odd number of rotations will again yield your enantiomorph. Using coordinate terms, in an asymmetric system, any odd number of changes in the sign of the coordinate yields a mirror reflected system. Any number of odd changes is the same as making one change.

When parity is conserved, we say such systems have no preferred spatial orientation. Any disorientation of such a system can be superimposed upon the original without changing the sign of any coordinate. Parity is not conserved when a system is inverted and cannot be superimposed upon the original system. We are ready to see how spatial inversion is used in understanding elementary particles.

Physicists applied parity to elementary particles. A three-coordinate system is sometimes used in describing wave functions. A system is said to have odd parity if it is functionally inverted by the change of one coordinate. This could be represented by the quantum number +1. More explicitly: If parity is conserved, a system has to retain the same parity in its final condition. If a system begins odd, it has to end up odd. If it begins even, it has to end up even.

It was assumed that in any physical system parity would be conserved. If an elementary particle with even parity decayed into two new particles, they would both have to be either of even parity or of odd parity. That is, the product of two evens (+1 and +1) would yield even parity and the product of two odds would yield even parity. Either way, the resulting condition would have a total parity of even (+). However, if an elementary particle with even parity decays into one particle with even parity and one with odd parity the total resultant parity would be odd (–1). The final condition is changed (+1 to –1) from the original condition and parity would not be conserved.

Applying parity to quantum field theory pertains to certain wave functions. Abstract wave functions are easily amenable to spatial interpretations. In this book we continue adhering to a spatial description of parity.

During the early 1950's physicists noticed a paradox concerning the behavior of elementary particles called K-mesons (kaons). Sometimes the K-meson decayed into two pi-mesons (pions) and sometimes into three pions. There was no evidence suggesting the parent K-meson of the two-decay scheme was really two different particles. Assigning parity to the original particle based on each decay mode gave the K-meson contradictory values. The parent K-meson should not have both even and odd parity. If parity were conserved, one of the decay modes would be acceptable, but not both. There was another possible interpretation of those results.

In 1956 two physicists, Tsung Dao and Chen Ning Yang, suggested that parity was not conserved in weak interactions. They proposed experiments to test this hypothesis. Madame Chien-Shiung Wu and coworkers accepted the challenge.

They planned an experiment involving the beta-decay of a highly radioactive isotope. They chose cobalt-60. They cooled the isotope to near absolute zero to stop molecular motion. They applied a powerful electromagnetic field to induce most of the nuclei to line up along a "north-south" orientation. The nuclei continued to emit electrons but in the direction of the two magnetic poles. If parity were conserved there would be an equal number of electrons emitted in both directions.

In January 1957 Madame Wu announced the results of her experiments. More electrons were emitted in one direction than the other direction. Before this experiment the labeling of north and south (or right and left) was considered arbitrary. Now there was a natural standard. Nature does have a spatial preference. Parity was not conserved in weak interactions. (Note neutrino chirality suggests our physical universe is left handed.) Later, it was shown that even strong interactions could violate the conservation of parity.

There is little doubt that spatial objects (extensionality) were asymmetric. The Universe of Perfect Whatness of the Where Aspect in the Deltic Cosmos "went only one way." Because of this evidential fact and for purposes of symmetry, another cosmos is postulated where objects (their extensionality) "go the other way." This "other" extensionality belongs to the Anti-Universe of Perfect Whatness of the Anti-Where Aspect in the Anti-Deltic Cosmos.

What about the Universe of Imperfect Whatness? We need not say anything here about the Universe of Imperfect Whatness since it is considered a byproduct of interaction between the Universe of Perfect Whatness and Universe of Imperfect Whyness.

We do need to quickly look at the Anti-Universe of Imperfect Whyness of the Anti-Where Aspect. If Universe of Perfect Whatness (extensionality) of the Deltic Cosmos is asymmetric, we need to postulate the existence of its enantiomorph. We need to do the same for the Universe of Imperfect Whyness. For symmetry purposes we have to say the Universe of Imperfect Whyness of the Where Aspect of the Deltic Cosmos is asymmetric and goes one way. If we do this, then we also have to say that the Anti-Universe of Imperfect Whatness of the Anti-Where Aspect of the Anti-Deltic Cosmos is also asymmetric and goes the other way. The Anti-Where Aspect is a mirror image of the Where Aspect.

It is useful to note when we postulate an Anti-Where Aspect—we are in effect saying there is an anti-cosmos (the Anti-Deltic Cosmos). It may seem as though we are getting ahead of ourselves and indeed that may be the case. However, we are constructing pieces of the puzzle that we will later fit together. As can easily be guessed, we have two more major pieces to consider for this structure before we can put it together.

Anti-How Aspect

The How Aspect pertains to cosmic order. Order is expressed in laws of science. When we mention the Anti-How Aspect, we are effectively talking about an inverted order compared to the How Aspect. There is a physics law which relates the How Aspect to the Anti-How Aspect. The principle is an extended version of charge conjugation. This symmetry principle relates the order (or laws) of the two aspects.

A symmetry principle of elementary particles, in conjunction with relativity, can represent the relation of our immediate How Aspect to the Anti-How Aspect. Let's develop an understanding of this relationship.

Recall: Dirac found particles behave according to Special Relativity only if there is a mirror reflected system of antiparticles. What does this say about cosmic structure?

Dirac's statement about elementary particles is partially represented by the Principle of Charge Conjugation. The principle states that for every particle there must be an antiparticle (sometimes a particle is indistinguishable from its antiparticle).

Elementary particles and their antiparticles have opposite quantum numbers. One exception is spin angular momentum. Spin angular momentum is an intrinsic property that is measurable. It involves rotational or orbital motion. Motion infers extensionality. Extensionality is a property of the Where Aspect (and Anti-Where Aspect). As noted, spatial extensionality involves parity.

In 1917 Emmy Noether, a German mathematician, found that every symmetry is associated with a conserved quantity. The conservation of spin angular momentum correlates with rotational symmetry (same in every direction) and therefore to the uniformity of space-time. Spatial symmetry infers conservation of momentum; temporal symmetry infers conservation of energy.

Although particle-anti-particle pairs share the same values of spin angular momentum, their spin motions trace mirror images. Their spins can be considered opposites in the sense of parity. Spin angular momentum aside, all particle-antiparticle pairs have opposite quantum numbers. For example, for the Anti-Universe of Perfect Whatness, properties of strangeness and charm take on opposite values from that of the Universe of Perfect Whatness. If a particle has a strangeness of –2, its antiparticle will have a strangeness of +2. If a particle has charm of +1, its antiparticle will exhibit charm of –1 (More about such properties in Chapter 15: "Reality-Unity"). However, in each case the forces between particles are effectively unchanged. Their properties behave the same in their respective systems.

The Principle of Charge Conjugation tells us much about cosmic structure. We understand this principle by saying that if the sign of a charged particle, for instance, is changed from negative to positive (charge in an interaction is itself conserved) we come up with a representation of the particle's antiparticle.

Inversion of electric charge signifies more than a change in charge sign. It implies that most other properties are also inverted. It also means the laws of elementary particle physics are invariant under charge conjugation. A brief note: The concept of charge conjugation also applies to neutral particles. Consider the neutron. It has a net charge of zero, yet is not its own antiparticle because the signs of all its other quantum numbers do not match those of its antiparticle. The difference is easily understood by examining the supposed quark properties of the neutron and antineutron.

Understanding charge conjugation at the micro-level is straight forward. To gain more insight, we look at how Dirac might have realized his ideas. Dirac toiled with quantum theory and Special Relativity. To keep things simple we address a source that undoubtedly influenced Dirac.

In 1862 J. Maxwell developed ideas linking the phenomena of static electricity and magnetism. He had four basic equations representing connections between electricity and magnetic fields. In one equation he characterized the relationship between electric field strength and the density of static electric charge. In another he defined the relation between a moving electric charge (current flow) and the secondary effect of magnetism. We consider these two equations to be one side of his idea.

One side of Maxwell's idea can be used to understand the electron. The electron is considered the basic carrier of electricity. Electric charges are found in whole number multiples of the charge of a single electron. This is M. Planck's contribution. Dirac incorporated these ideas into Relativity Theory. The result is now known as the Principle of Charge Conjugation. This principle envelops matter and antimatter into a single concept. For example, it states that the positron is an inverted version of the electron.

The other side of the modified version of Maxwell's equations concerns a "particle" that has not been experimentally confirmed. Dirac coined this particle the magnetic monopole. The magnetic monopole is the magnetic analogue to the electron. It is the basic carrier of magnetism. The monopole would be quantized just as is the electron. The charge would be some multiple of the smallest unit of magnetic charge—that of the magnetic monopole. Another inserted note: Any magnetized object that we observe in our Universe of Perfect Whatness world has a positive and negative aspect (usually labeled north and south) and is called a dipole. A dipole is not the electron's analogue.

The magnetic monopole was postulated to balance Maxwell's equations because the symmetry of Special Relativity called for it. If the magnetic monopole were found to exist, J. Maxwell's other two equations would represent its effects. One equation would represent the relation of a static magnetic charge to the strength of its magnetic field. The other equation would represent the relation between a moving magnetic charge and its secondary effect of generating an electric field. In the Special Relativity version of Maxwell's equations there are only two formulas: one each for electric and magnetic sources.

In essence, we have two versions of charge conjugation. One would be the Conservation of Electric Charge Conjugation, and the other would be the Conservation of Magnetic Charge Conjugation or pole strength. The latter is the Extended Version of Charge Conjugation (charge conjugation is customarily used to designate electric sources).

Let's apply the extended concept of charge conjugation to cosmic structure. Because of relativity theory, we called our Universe of Perfect Whatness a tardyon universe. We (arbitrarily) also assigned a negative electric charge to represent it (we liken this value to our tardyon electron that carries a negative electric charge of –1). Hence, we say because of the How Aspect (considered, simply, to represent order) that our Universe of Perfect Whatness "tardyon universe" has an arbitrary negative charge of –1. If we use this standard, our anti-tardyon universe, because of its Anti-How Aspect, will be related to the positive electric charge of +1 (this value is connected to the positive charge of a positron +1). Essentially, we are saying, by the Principle of Electric Charge Conjugation, that an anti-tardyon Universe of Perfect Whatness (universe) also has to be created if a tardyon Universe of Perfect Whatness (universe) exists. (This is an oversimplification that will be corrected in the next chapter.)

What about our Universe of Imperfect Whyness? Again, we apply the same ideas. We have already correlated relativity to the Universe of Imperfect Whyness and tachyons. Using our tardyon standard, we say the tardyon Universe of Imperfect Whyness (because of the How Aspect) also carries a charge. However, its charge is not electric, but magnetic. We further append the statement "The Universe of Imperfect Whyness can be represented by a magnetic charge." It may be so represented but maybe only through a physical interpretation. Then it may be, relative to a subluminal frame of reference, that the superluminal tachyon frame only appears magnetic to a subluminal observer. Interpreted within its system, what we call a magnetic monopole may be observed to behave similar to an electron. If this were the case, the same idea would be equally valid referenced from a superluminal interpretation looking at a subluminal frame.

The extended version of charge conjugation allows us to postulate an Anti-Universe of Imperfect Whyness. If we say that the Universe of Imperfect Whyness has a "designated" magnetic charge of –1 (this value represents the tachyon magnetic monopole) by virtue of the How Aspect, we can say that the Anti-Universe of Imperfect Whyness can be assigned a magnetic charge of +1 (this value represents the anti-tachyon magnetic monopole) by virtue of the Anti-How Aspect.

We can also say something about the Universe of Perfect Whatness. It has been stated that the Universe of Imperfect Whatness is a byproduct of the Universe of Perfect Whatness and Universe of Imperfect Whyness interaction. How could this interaction be interpreted by magnetic monopoles (representing the Universe of Imperfect Whatness) and electrons (representing the Universe of Perfect Whatness)? A magnetic monopole in motion could ionize atoms. Remember that a magnetic monopole can create an electric field. Hence, Universe of Perfect Whatness and Universe of Imperfect Whyness interaction could proceed via electromagnetic force. The same idea applies both ways. Electrons in motion create magnetic fields.

Maxwell's equations paved the way to add another cosmic structure. That structure is the Anti-Deltic Cosmos. It should be added that the Anti-Deltic Cosmos is not an immediate part of the world in which we live. However, without it we would not be here. Just as elementary particles must be created in pairs, it should be that cosmic units must be created in pairs. Together, a matched pair of Deltic Cosmos and Anti-Deltic Cosmos units constitutes a Bi-Deltic Cosmos.

Just as cosmic units must be created in pairs, so it is that they can be destroyed in pairs. And so we have another question. Why are complementary cosmic units not destroyed? Perhaps they are properly separated. How are they separated? We have a possible solution in what follows.

Anti-When Aspect

The When Aspect denotes change in order. Change of order happens in either one of two possible directions. Time is one-dimensional. We have already seen in the chapter about the When Aspect that cosmic change (for example in the Deltic Cosmos unit) occurs in one direction but covers two phases. The two phases are enveloped in one cycle. Because each phase represents half the cycle, time or change can be said to reverse when one phase gives way to the other. Time (directionality) "goes" forward in one phase and backward in the other phase. Temporal asymmetry is relative. There is no theory in physics that ascribes absoluteness to temporal directionality (as a flowing of time or motion).

The Anti-When Aspect of the Anti-Deltic Cosmos is an inverted facsimile of the When Aspect of the Deltic Cosmos. Change of order (order is represented by the How Aspect) in the Deltic Cosmos goes one way while change of order in the Anti-Deltic Cosmos goes the other way. Order imposed upon the other major aspects yields the directionality of time. How were these ideas obtained from understanding physics?

We earlier stated that in 1957 it was discovered the conservation of parity is violated. Weak (elementary particle) interactions involving beta-decay are not duplicated in a mirror-reflected system. Thus material extensionality is asymmetric. Charge conjugation also is violated in weak interactions. For example, the cobalt-60 experiment violates the Principle of Charge Invariance. If a mirror image of this experiment could be performed, and cobalt-60 is replaced with its antimatter counterpart, beta emission would consist of ejected positrons instead of electrons. When the rest of the apparatus (this is a "gedanken experiment") is also composed of antimatter, the direction of magnetic spin is reversed. Under these conditions our visual representation shows "positron" beta-emission would be more intense in the direction expected of an electron. This implies violation of charge invariance.

It was thought a form of symmetry could be reinstated if an even number of factors were involved. Extensionality is assigned an odd number of factors represented by the three dimensions. Charge conjugation contributes an added factor. If inverted charge is combined with inverted space, it was originally thought the asymmetric beta-decay of both matter and antimatter would duplicate each other by rotation invariance. Symmetry could be restored. However, the idea was short lived.

In 1964 J.H. Christenson, J.W. Cronin, V.L. Fitch, and R. Turlay performed an experiment that seemingly violated charge-parity (CP) invariance. Besides individual charge and spatial differences, they found that a "CP reflected" system would not invariantly reproduce a certain experiment involving neutral K-meson decay.

Through weak interaction, K-mesons (kaons) can decay into a pion, neutrino and electron (or muon) or into two or three pions. We are interested in the two-pion mode of decay. Christenson and colleagues showed two-pion decay occurred at two different rates. The stage is now set to gain a rudimentary understanding of what happened.

Net-decay behavior, in quantum mechanical terms, is a combined result of constituting decay rates. Rates are determined by decay amplitudes and relative phases of each decay time. Two-coupled decay rates may produce amplitude interference. If CP-invariance were not violated in the experiment performed by Christenson and coworkers, only one of the decay amplitudes would be associated with the two-pion decay mode.

Experimental results indicated coupled interference in the amplitude arising from the two decay times. This suggests both the K and the $\overline{K}$ could decay into the same two-pion mode, but did so at two different rates.

Coherent interference verified that CP-invariance was violated. The reason is the following. When the experiment is CP-reflected (parity and charge together inverted) the interference term changes signs resulting in a changed decay curve. This was interpreted to mean the two rates of kaon decay (into the two-pion mode) were not copied by CP-inversion. In fact, when K is replaced by $\overline{K}$ the two rates reversed. What happened was not difficult to understand.

If the experimental results are CP-transformed, the shorter decay amplitude remained unchanged while the longer decay amplitude changed signs. This difference is significant.

CP-violation has since been verified in other experiments. There was little doubt that overall symmetry was broken again. There seems to be only one reasonable way to restore overall symmetry and that is to involve yet another factor.

Charge-parity invariance violation implies time invariance violation. Time is the only reasonable factor remaining that can salvage overall symmetry. Violation of parity led to extensionality inversion. (Note: Parity, in its original sense, can also be applied to the three aspects taken as a whole.) Violation of charge conjugation induced the concept of charge inversion. Time too has an inverted counterpart. Invariance of these three aspects, individually, is evidently not conserved. However, all three aspects collectively can restore cosmic symmetry. This is the more exacting form of a statement in the last subheading implying charge conjugation alone generates an anti-cosmos.

Knowledge of certain weak interactions implicates temporal invariance violation. Time inversion is required as a part of the Bi-Deltic Cosmos structure. There is just one more step required to place together the two major pieces (the Deltic Cosmos and Anti-Deltic Cosmos units) of this structure.

The basic idea of the Bi-Deltic Cosmos structure is enmeshed in the CPT (meaning charge-parity-time) theorem. In 1957 Gerhart Lüders showed that mathematically transforming a particle into its antiparticle did not alter the laws of physics. The theorem says that, collectively, three (C, P, T) determinants can resolve the symmetry problem presented by CP-violation. Individual violation of the three conservation laws (C, P, T) calls for adding three inverted counterparts. It is the Cosmos and its inverted counterpart (the Anti-Cosmos) which restores symmetry. This is the basic idea embodied in the CPT theorem.

The CPT theorem is the magic mirror needed to understand cosmic structure. However, its meaning has not always been clear. A puzzling problem for scientists has been difficulty in assigning meaning to temporal inversion.

The difficulty is caused by two circumstances. One is confusion of temporal inversion with temporal reversion. Temporal reversion occurs in an inverted universe. The other circumstance is finding the "missing antimatter" exhibiting temporal inversion. Specifically, inversion allows a mirror image cosmos that balances the requirements of symmetry. When one cosmos is in Phase I, the other is in Phase II. Our anti-cosmos is always in an opposite phase from our own. This explains why a balancing of matter by antimatter has not been empirically determined. These problems are resolved by the Bi-Deltic Cosmos structure.

The CPT theorem and related evidence are major premises for the Bi-Deltic Cosmos Diagram. Assigning meaning to this structure is simple. For each cosmos going forward in time, there is an anticosmos going backward in time. The Deltic Cosmos When Aspect is a one way cycle composed of two parts—Phase I, and Phase II. One phase is the temporal reversal of the other. The reasoning is simple. When the Deltic Cosmos is in Phase I, moving forward in time, the Anti-Deltic Cosmos is in Phase II, moving backward in time. When the Deltic Cosmos is in Phase II moving backward in time, the Anti-Deltic Cosmos is in its Phase I moving forward in time. Time flow in one cosmos is always inverted relative to time flow in its anti-cosmos. Although intimately connected, temporal reversion and temporal inversion are different phenomenon.

The temporal aspects of cosmic structure solve the problem of entropy. Entropy increases during Phase I. This means the state of Phase I becomes more disordered as change occurs. The state of Phase II increases in order. As change occurs in Phase II there is a decrease in entropy. We again meet the concepts of entropy and negentropy in the final chapters. Entropy and negentropy need to be better understood and explained.

Key Points:

- The Bi-Deltic Cosmos has (two) complementary "sides." It is a combined structure of "our" Deltic Cosmos and its "mirrored" counterpart—the Anti-Deltic Cosmos. Appropriately, the Bi-Deltic Cosmos is diagrammatically represented by two triangles—one inverted and superimposed upon the other.

- A series of elementary particle interactions show "symmetry" in physics is satisfied if there is an anticosmos. An anticosmos also explains the "other side" to Maxwell's equations (magnetic monopole solutions) and extended Special Relativity (superluminality). (Note: Tardyons are associated with [and can be represented by] electronic monocharged particles called electrons. In that same sense, tachyons may be associated with magnetic monopoles.)

- In the anticosmos, charge is reversed; spatial extensionality is inverted (parity) and time flows in the opposite direction (it is inverted) from our cosmos. These ideas of symmetry are reflected in the consolidating physics principle called the CPT theorem (Charge, Parity, and Time).

- Charge symmetry in elementary particle interactions is stated in the principle of charge conjugation. Its invariance is violated in certain weak interactions. This suggests matter has an antimatter (inverted) counterpart, or symmetry cannot be restored. For example, although identical in every other respect, an electron (representing matter) differs in its electric charge (it is −1) from its antiparticle twin—the positron (+1). Yet, a positron going backward in time is indistinguishable from an electron going forward in time. One would have difficulty determining antimatter running backward in time from matter going forward in time.

- Parity relates symmetry of spatial extensionality. It indicates extensionality can be inverted by a 180° rotation of one spatial axis. Parity suggests antimatter is spatially inverted from that of matter. This means matter is asymmetric. It is characterized by "handedness" (left-right). Antimatter is a mirror image of matter (left hand appears as right hand in mirror). Added note: Spin (angular momentum) arises from the conjunction of temporal directionality and material handedness. We later learn spin infers a deeper understanding of particle creation.

- It was discovered CP invariance was violated. Its violation suggests time (directionality) is also violated or overall symmetry (CPT) is not retained.

- Time symmetry obtainment calls for more than one step. The When Aspect solved the problem of the beginning and ending of time with temporal reversal (occurring in a different phase). The Deltic Cosmos defined what is affected by time (Where and How). Yet, neither of these structures resolved a problem associated with cosmic duration. How is the longest "directed" duration (the complete cycle—combined Deltic Cosmos Phase I and Phase II) "balanced" symmetrically? (Directed dynamic duration defines unidirectionality.)

- The CPT theorem suggests temporal inversion (of directionality) happens in an anticosmos: where time reverses, where spatial extensionality is inverted, and where matter is oppositely charged. Symmetry is "restored" when all three factors (CPT) are simultaneously inverted. Collective inversion (of the Cosmos) defines the anticosmos.

- The CPT theorem indicates a synthesis of Cosmos and Anticosmos negates the following: charge disparity, spatial asymmetry, and temporal unidirectionality.

- The unidirectionality problem (of time) is remedied by asserting an Anticosmos. Deltic Cosmos temporal directionality (of the cycle) is balanced by opposing temporal directionality in the Anti-Deltic Cosmos. Anticosmos temporal directionality "flows" oppositely from cosmic directionality.

- Time (dynamic) directionality symmetry is found in the Bi-Deltic Cosmos synthesis of Deltic and Anti-Deltic opposed directionalities. Bi-Deltic cosmic duration is static. It is nondynamic. (It could retain an arrow showing direction, but the arrow would not signify any dynamic activity or time flow.)

- Symmetry considerations (particularly, the CTP theorem) tell us temporal directionality (time flow) occurs from a complementary splitting of (static) durational time (no time flow). The Bi-Deltic Cosmos complementarily splits into the Deltic Cosmos and the Anti-Deltic Cosmos.

- "Categorically," cosmic (longest nondynamic) duration is defined by the Bi-Deltic Cosmos structure. The Bi-Deltic Cosmos explains the Cosmos and Anticosmos structures because (static) duration splits into two (sub)durations with (dynamic) directionality (time flow). There is no contradiction in this supposition.

Our surveying of the mind expanse is constantly stretching our inductive reasoning ability. Problems need to be solved, or we cannot continue mapping the structure of reality. The Bi-Deltic Cosmos Diagram explains and accounts for the CPT Theorem.

The Bi-Deltic Cosmos structure is conceptually premised on knowledge of subatomic particles. Symmetry is collectively restored with the unity of spatial extensionality, order, and time. From a more general view the Deltic Cosmos and Anti-Deltic Cosmos cancel or negate one another. (Further treatment of this concept is addressed in subsequent chapters.) Any of the three major aspects of one cosmos is effectively negated by its anti-cosmos counterpart. Negation via complement is the final (cosmic) structural purpose of the CPT theorem.

The Bi-Deltic Cosmos structure poses a few more questions. The first question is especially clear. Forward time negates backward time. Upon closer inspection, we see that any "given" instant of time in one cosmos is balanced by a corresponding instant in its anti-cosmos. With each cosmic (or anti-cosmic) instant is associated a direction. It is the directional nature of time that is negated by the Bi-Deltic Cosmos structure. Directionality is not an inherent property of time (the When Aspect). (Time is change.) Time "takes on" the property of direction because of its intrinsic connection with order (the How Aspect).

Special Relativity (SR) resolved the phenomenon of temporal rate and Extended Relativity explained temporal reversal. The CPT theorem provided the solution to the problem of temporal complementation. In simple terms, temporal inversion negates directionality. Yet, neither SR nor the CPT theorem can handle a problem introduced by the Bi-Deltic Cosmos structure. In solving one problem another is created. It is a problem rooted in the Bi-Deltic Cosmos instant (BCI). This instant is a combination of a Deltic and corresponding Anti-Deltic instant. The two combined (diminutive) instants effectively cancel temporal directionality. However, a BCI still exhibits the property of (shortest) duration. This is the second question. To maintain symmetry we must somehow negate the effect of the BCI. We must examine time from another perspective.

A difference between temporal directionality and temporal duration poses the next question. What about the BCI's (so called because they are apparently an infinitesimal duration) which have already been and which are yet to be? How can negation be applied from one BCI to another? In a logical sense, negation by opposing structures is equivalent to maintaining symmetry. Taken as a whole, the Bi-Deltic Cosmos does not have an opposing structure. Here, negation by again using inversion symmetry is not possible. The question is how to negate one BCI against all others. Clearly, the answer does not lie within the Bi-Deltic Cosmos structure. Since the Bi-Deltic Cosmos, as a whole, posed the problem, we have to look beyond its structure to find a solution. To obtain a noncontradictory answer we build yet another structure.

Part D: Cirpanic Cosmos

Perspective

- The Cirpanic Cosmos represents all possible eventualities
- It is composed structurally of all logically possible Bi-Deltic Cosmoses
- The Cirpanic Cosmos constitutes every thing of reality
- It represents reality as a collection of thingness potentiality

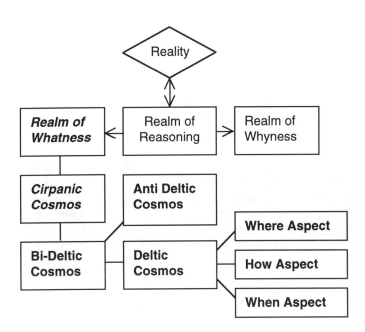

Figure 12:1. Perspective: Cirpanic Cosmos

196

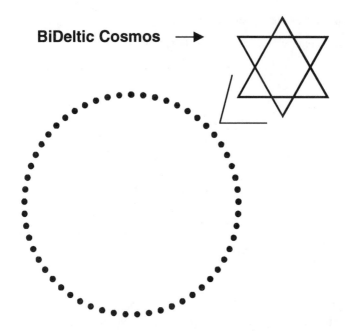

The Cirpanic Cosmos represents every possible event. The Star David symbol represents one Cosmos-Anticosmos pair and every possibility associated with it. The Circle represents innummerable Cosmic-Anticosmic pairs and their possible course of evolvement. Every possible occurrence for every possible Cosmos is represented on the circle. The Cirpanic Cosmos represents potentiality and actuality.

Figure 12:2. Cirpanic Cosmos Diagram

Chapter 12

Cirpanic Cosmos

"Knowledge comes not by faith, but by inductive reasoning. Extraordinary knowledge comes from understanding extraordinary experiences."

The Cirpanic Cosmos is the final structure of the Realm of Whatness. It is an unlimited concatenation of Bi-Deltic Cosmos structures. Its structure is reasoned by inducting knowledge initially premised on intrinsic whatness. Its realization is obtained by intrinsic perception (See How Aspect Diagram, Chapter 4).

Temporal directionality was negated by the Bi-Deltic Cosmos structure. We expect the Cirpanic Cosmos structure to finally negate temporal duration. The key to solving this problem is understanding the Bi-Deltic Cosmos Instant (a BCI). Therefore, how does any given BCI function in the greater scheme (of things)? Before continuing, we examine essential BCI concepts.

Bi-Deltic Cosmic Instant

We evaluate, by classificatory means, whether temporal directionality and temporal duration are separate concepts. If they are, can one of these concepts explain the other?

In Part C, we learned there are two directions of time. Directionality defines temporal asymmetry. Temporal asymmetry implies time flows only one way. A given temporal direction is negated or balanced by an opposing temporal direction in another system. Time flow in the Deltic Cosmos is opposite ("backward") from time flow in the Anti-Deltic Cosmos.

How can time flow be explained? We first showed temporal directionality is categorically distinct from duration. We next described how the concepts of directionality and duration belong to different levels in a hierarchical classificatory system. Directionality is a characteristic of the Deltic Cosmos (and Anti-Deltic Cosmos). Duration is a characteristic of the Bi-Deltic Cosmos. Duration can explain directionality or flow of time.

How did we get from temporal flow to static duration? What is static duration? What do these concepts mean at higher levels of generality?

We found, by correlation with physics, that the Bi-Deltic Cosmos brings the "asymmetric" flow of time into symmetry. In Chapter 11 (Bi-Deltic Cosmos) we first considered each instant of time (of the When Aspect) to have an attached arrow. Each arrow shows direction of time flow. This is not unlike conceptualizing the experience of time. Temporal asymmetry occurs at a lower structural level of the Cosmos. Time flow is experienced in the Deltic Cosmos.

BCI Defined

The BCI is not "embedded" in the flow of time. It represents the inductive essence of time—minimal duration. A BCI is the lowest limit duration considered before it "breaks down" into two "simultaneous subcomponents" exhibiting directionality (i.e., manifested in the Deltic Cosmos [as forward time] and manifested in the Anti-Deltic Cosmos [as backward time]). The Bi-Deltic Cosmos is not a dynamic structure. It is static and categorically supersedes the Deltic Cosmos (which does exhibit "flow of time").

How is a BCI constructed? The BCI is a conjunction of two instants of time. It is a combination of a "forward" instant of time (e.g., from the Deltic Cosmos Phase I) with a "backward" instant of time (from the Anti-Deltic Cosmos Phase II). We conceptually obtain a "combined" instant of time that cancels or negates temporal "directionality." We isolate this "directionless" duration and labeled it the BCI.

How does Planck's constant relate to BCI's? The difference between BCI's is "represented" in physics by Planck's constant—the minimal quantity of energy (and momentum) that exists. Each successive BCI must differ by at least one Planck unit. Planck's constant "measures" the minimal possible shift from one BCI to another.

We have already determined that time flow is "things changing." Change infers some type of shift in space of the thing's extensional aspect. Energy "measures" change. Planck energy is described by the smallest shift allowable in space-time terms. Spatial extensionality and time flow are "collective" aspects of Planck's energy. Simply, cosmic structure arises from the "shifts" or rearranging of energy from one BCI unit to another.

The Bi-Deltic Cosmos is conceptually erected from its BCI's. The Bi-Deltic Cosmos structure includes every BCI composing the cosmic time cycle. It includes all the BCI's of Phase I and Phase II of the Cosmos and Phase I and Phase II of the Anti-Deltic Cosmos.

Before continuing, let's restate the problems of understanding time. Each BCI "self-negates" temporal directionality (inherent in Cosmos and Anti-Cosmos structure). A BCI does not negate itself (i.e., duration). It does not negate any other BCI, or any series of temporal (BCI) durations. However, the "collective" of all BCI's negates temporal directionality.

BCI and Change

Change does not happen during a BCI interval. Change results from the passing of intervals. Let's explain. The temporal instant is the shortest duration. Change happens as the universe "flows" from one BCI to another. Change occurs because the "next" BCI (passing through the present—the "here and now") is different. Each adjoining "frame" (representing a single Planck durational event) though very similar, differs in at least one respect (or change cannot be explained). (This difference is Planck's quantum of activity.)

Consider the "present" to be the lens of a film projector (of motion pictures). "Events" unfold in manifest form—"projected." Shortly, we elaborate this analogy. In the Deltic Cosmos (and its Anti-Deltic Cosmos) the progression of intervals (or cosmic ticks) is (experienced as) a smooth transition. We observe this as change. This is not unlike how we experience movies. Watching the screen we do not observe the stop and go "jerks" of individual film frames because film moves through the projector at 24 frames per second. Each frame is likened to a BCI. There are similar frames of order underlying cosmic dynamics (we negated dynamic temporality [directionality] with the Bi-Deltic Cosmos structure). An "on and off" succession of BCI frames are projected into existence.

It may be argued there is no duration without direction. This is true in the Deltic Cosmos. The question is: Which (direction or duration) can "exist" without the other? From the Bi-Deltic Cosmos perspective, we learn there is no direction without duration. Temporal direction is defined by change in order—successive change in durational instances.

BCI and Order

The term direction also has some merit when considering more than one BCI. This is not temporal directionality. There is a "direction of order" in the interpretation of the Cirpanic Cosmos structure. This order is one we "assign" (reductively) to the Cirpanic Cosmos structure. That assigning is obtained from an understanding of the Deltic Cosmos.

Meaning is ascribed to the BCI order composing the Cirpanic Cosmos Diagram (circle). Here, use of direction is ambiguous since there is no "preferred" direction. Order proceeds in opposite directions. Hence, the order can be "read" in either direction.

Direction has no intrinsic significance when applied to one isolated BCI. Our stated purpose here is to negate one BCI. We use a unit of duration not exhibiting direction because any definition of direction in the Cirpanic Cosmos structure carries little meaning. This is how we qualify defining a BCI unit. It is a static unit of duration. What else does this mean? Simply, a BCI infers order without direction. The Cirpanic Cosmos is a structure exhibiting order, and no inherent direction.

When duration is stripped to a minimum, it is constructively of little value in affixing meaning to this structure. Then what can be said of a BCI in the context of Cirpanic Cosmos structure? The BCI is intimately bound with order. That is all! It is the basic unit of order in the Realm of Whatness. Order in the Deltic Cosmos is traced to the Cirpanic Cosmos. The Cirpanic Cosmos "stages" order for manifestation.

What of the BCI itself? Let's take as closer look. It is the basic unit of duration. Minimal duration is measured against the minimal energy of activity—of change. No smaller energy unit exists in nature. Any given duration (temporal length) is an integral multiple of this basic unit. A given duration acquires direction only when considered from a Deltic Cosmos (or Anti-Deltic Cosmos) perspective.

There is no flow of time in the Cirpanic Cosmos. There is only the analogous "roll of film": the representation of all BCI's that have been and are yet to be. This does not necessarily mean the future is predetermined. It means the future develops from the past. We ask: "But how?" This question is addressed in the next section The Realm of Reasoning. For now, we say the Cirpanic Cosmos limits what and how things happen in the Realm of Whatness.

We say each BCI has its place (order). To explain BCI location we pick the BCI that follows the "change" from Phase I to Phase II in our Deltic Cosmos. (Refer to Chapter 9.) (We could just as well have used a reversed sequence.) This BCI marks the beginning of Phase II. A second BCI follows the first, then a third, a fourth, and so on: one following another in succession. Countless BCI's sequentially come and go every second. Each BCI is successively "counted" by a tick of the Cosmic Clock (the When Aspect). The clock ticks off intervals of order. Here, we are not interested in how BCI's function at a "lower level." We would like to know how these events can be explained. What is the origin of (durational) events?

The cosmic tick does not apply to the Cirpanic Cosmos. It is the Realm of Whatness which "pulsates." Each pulsation, or quantum jump, infers the passing of another BCI "frame" through the present (the here and now) projector lens.

We can serially number each BCI. Each interval (BCI) would have an "order number." This would be like numbering each frame in a motion picture film. The order of successive intervals gives a Deltic (and inversely, its Anti-Deltic) Cosmos "direction" (whether the film is run forward or backward). In reality the film runs both ways at once. Remember that the circle (denoting the Bi-Deltic Cosmos) is two circles in one. When the film runs (set into motion) it does so at a lower level of reality—the Deltic Cosmos and Anti-Deltic Cosmos. This will be more easily grasped when we shortly address this idea by analogy.

BCI Measurement

The BCI is the most diminutive duration. In inexact terms, it is a seemingly infinitesimal "instance" of time. The interval is the most minuscule durational component of the Bi-Deltic Cosmos. The following discussion of the BCI is simplified.

How could we theoretically measure a BCI? What is the BCI? Measurements would probably reflect the diminutive interval as it is noted in a proper (inertial) frame of reference. It is possible the discrete interval (BCI) could be or has been inferred from experiments with elementary particles. At least such experiments suggest a maximum value. The determination would be akin to ascertaining a duration at least as short as the elapsed time of the fastest (elementary particle) interaction.

Each BCI must exist for some minimal duration. If they did not, there would be no difference between "something" held in duration and no duration. Minimum duration also denotes the interval of change from one BCI to another. We standardize the BCI as the minimal durational instance to simplify our concepts. Temporal flow occurs as countless BCI's yield to "future" BCI's. We conclude there is a duration of manifestation, and a duration of nonmanifestation. There is an alternation between these two types.

No two BCI's happen at the same instant. Surely, multiple BCI's do not occur at the same instant. If they did, no discernible change could be observed (even at the greater lengths of time we discern in everyday life). Instead, there would be chaos rather than observance of change (of order). So what is changed? BCI's are changed—one after another. There is also a duration between BCI's or there would be no duration. This is a "blinking on and off" of a succession of BCI's. This is the origin of cosmic pulsation and frequency.

What have we learned from high energy physics? Physicists find that increasingly shorter durations are smoothly continuous to a point. When talking about limiting points and the Planck time we often address the character of virtual particles. Virtual particles exist for brief periods, compliments of Heisenberg's Uncertainty Principle. Virtual particles exist on borrowed energy. The greater the borrowed energy, the shorter the duration of the virtual particle's existence. At the extremely brief duration of 10^{-43} second, borrowed energy becomes so great that gravitational effects intrude upon the immediate structure of space-time. Such effects create enormous distortion in the fabric of space-time.

Events occurring at the Planck level are discontinuous. One frame-event (BCI) passes to the next frame-event by discrete jumps. Quantum physics concerns these types of interactions.

There seems to be a low limit to durational length. Evidently, smooth durational periods only occur for intervals significantly greater than 10^{-44} second. In cosmogony the Big Bang is thought to commence during the Planck era having a duration of about 10^{-44} second. The BCI duration is probably the Planck time of 10^{-44} second.

What does the lower durational limit mean? Measuring a BCI is a Deltic Cosmos endeavor, and it would mean something, but measuring a BCI as it applies to Cirpanic Cosmos structure is over-extending the temporal concept. Similarly, examining a roll of motion picture film does not itself suggest temporal flow. It is only when the roll (Bi-Deltic Cosmos structure) is "set in motion" that time, as it is usually understood (i.e. time flow) becomes apparent. (Refer to Chapter 9: "When Aspect.")

Bi-Deltic Cosmos Duration

At the Cirpanic Cosmos level there is no flow of time—only that which is at the root of time. Therefore, we can only limitedly apply the (temporal) concept to get our foot in the door to discover the real meaning of the BCI. Remember that it is the "apparent" durational aspect of the BCI that needs an explanation. That does not mean the explanation must also be of a durational nature. If it were, the problem of duration would not be solved. We will discover the nature underlying duration (exemplified by the BCI) is order itself.

The BCI is a directionless duration. A single motion picture frame is similarly directionless. The BCI is associated with the Planck duration.

We are left with the question "how are durational events explained?" We are not so much interested in a temporal-like representation (by reductive means) of the Bi-Deltic Cosmos, as we are in connecting the Bi-Deltic Cosmos structure to the

Cirpanic Cosmos Structure. The circular dotted diagram of the Cirpanic Cosmos interrelates all Bi-Deltic Cosmos events. These ideas are developed in what follows.

To balance one BCI calls for a structure capable of (instantaneously) negating every BCI. This is exactly what the Bi-Deltic Cosmos accomplishes. It is a structure "outside" the flow of time. Because any given BCI does not negate itself it must be "balanced" from without. Balanced against what? Balanced against all other (i.e., past and future) BCI's within the Bi-Deltic Cosmos. The total series of BCI's negate any given (individual) BCI.

It could be argued duration is negated in the Bi-Deltic Cosmos structure. That may be but for our definition. For lack of a better word, we still use the term "duration" as a label for the sum of BCI units because the Cirpanic Cosmos structure is heavily (reductively) premised on the BCI unit.

The two phases of each cosmos are combined into a directionless durational time cycle. Let's denote the complete cycle the Bi-Deltic Cosmos Duration (BCD). The BCD is constituted of the totality of BCI's. We shortly enhance our film analogy for this concept.

The BCD itself does not refer to its subcomponents. It does not refer to temporal directionality. It does not refer to phases nor does it refer to BCI's. Rather, it infers them (deductively—reductionistically).

Although the Bi-Deltic Cosmos negates temporal directionality for every instant composing the Bi-Deltic time cycle, it does not negate a higher-level temporal duration. It does not negate itself. The BCD is the cosmic-time cycle as a whole (Bi-Deltic): cosmic duration itself. Our next task is to better understand duration. What does it mean? And later, why is it?

The BCD categorically represents all past, present, and all possible future events (at the minimal limit, viz, the BCI's) within our Bi-Deltic Cosmos. Future events are any event that can happen; therefore, that can become manifest. Such events are potential.

Although the BCD resolves the problem of future and past duration's (minimally represented by BCI's) of our present cosmic cycle, it does not address other possible cosmic cycles, either past or future. To use the terms past and future at this level is to invoke reductionistic terms. There is (now) no past, for it is no more. There is (now) no future, for it is yet to be. That is the idea. There is just the present, here, the present time cycle. The present cycle is all that occurs in this existence (the Deltic Cosmos Phase I and Phase II). So what is the problem? We ask what was there before this time cycle? What follows this time cycle? If a question can be asked, it can be answered.

There is the possibility of yet other cosmic cycles—past and future. We thus, hypothesize to answer the above questions. For example, let's say our Cosmos completes its time cycle only to begin again. This would mean there is more than one cycle. The very idea of "cycle" infers as much.

Let's say that each "time" a cycle begins, events unfold differently from previous cycles. This is the problem exposed. If this happens, no two cycles are exactly alike. If each succeeding Bi-Deltic Cosmos is different, in at least one respect, it cannot negate itself. It is this problem that is finally addressed by the Cirpanic Cosmos.

Cirpanic Cosmos Diagram

The Cirpanic Cosmos is composed of a very large number of Bi-Deltic Cosmos Duration's (BCD). The name "cirpanic" is coined from the words "circle" and "pan" meaning "around all."

Each point around the circle signifies one Bi-Deltic Cosmos Duration (one BCD). Each BCD represents and is reductionistically defined by its combined Deltic Cosmos and Anti-Deltic Cosmos. Each BCD is composed of BCI's. One "given" dot on the circle (the circle representing the Cirpanic Cosmos structure) represents our Bi-Deltic Cosmos.

The Cirpanic Cosmos structure arises from the need to negate all possible cycles (durations) of the Bi-Deltic Cosmos. This means we consider every conceivable BCD for all possible temporal cosmic cycles: past, present, and future (cycles).

The task of the Cirpanic Cosmos structure is to "symmetrize" the totality of BCD's. One BCD is equivalent to summing the BCI's in one temporal cosmic cycle. One BCD negates all BCI's. We asked whether there are differing cycles (of BCD's)? What about possible cycles that are yet to be? What do these questions mean? The Cirpanic Cosmos is a construct that addresses this question. Let's dig a little deeper.

A BCD is duration itself. Every BCI is in this cycle. The BCD represents the complete duration of the cosmic cycle. Now our question is "are there more cycles?" If yes, then there are more BCD's. Note: the BCD is not a dynamic cycle. We use the term "cycle" to define the BCD reductionistically. The BCD is analogous to the complete roll (from start to finish) of motion picture film.

We ask, is there a series of BCD's? If each BCD is different in at least one respect, then we answer yes. Each "time" the cosmic cycle begins again, albeit differently, we find it cannot be structurally represented unless there is a covering construct which includes every BCD. Later, we redefine the Cirpanic Cosmos, not reductionistically, but in a way which better "explains" its role in the scheme of things.

Cirpanic Cosmos Omnipresence

The Cirpanic Cosmos is composed of BCD's. Let's call the series of BCD's, potential or otherwise, the Cirpanic Cosmos Omnipresence (CCO). The CCO is the maximum duration we need address. Admittedly, again, this is a reductionistic approach. (It is reductionistic because it falls back on the definition of time as a flow of events as depicted by the When Aspect.) This approach benefits us in "leaping" to the Cirpanic Cosmos structure. Let's put our concepts together.

The Cirpanic Cosmos structure resolves a problem initially presented by the Bi-Deltic Cosmos Instant (BCI). Stated concisely: although any given BCI is internally time symmetric, because of its two complementary components (the forward time of the Deltic Cosmos and the backward time of the Anti-Deltic Cosmos) it is not negated "externally" against any BCI's of other BCD's (viz, past or future cosmic cycles).

Duration and Motion

Let's elaborate using the film analogy. Again, consider a roll of motion picture film. We experience the dynamism of "motion pictures" when viewing a movie in a theater. We know how each picture frame follows another while running through the projector. This is the normal situation.

Temporal directionality confers dynamism—a succession of "still" durations (i.e., events: each analogous to a motion picture frame). We experience time as a succession of events, but have opted to call each of these events a "still" duration. This we denoted a BCI. These concepts are critical in our understanding of time. You are probably beginning to understand how the flow of time arises. These ideas follow from how quantum physics correlates with the structure delineated in this book.

Let's consider other possibilities. First, the film can "run" through the projector in one of two ways—"forward" and "backward." We normally watch a movie in forward motion—one copying events as we experience them in everyday life. As the film advances, we expect to observe the dominance of entropy. We expect to see a general process of disintegration—things breaking down. Forward time counters backward time—balancing directionality. (Forward time infers an "inverted" backward time.)

Given any frame ("freezing" the movie: "stopping" the film) any other frame precedes it or follows it. Given any BCI (like a frame) any other BCI either precedes or follows it. Each frame or BCI is thus balanced. Inverted time is a consequence of the CPT theorem. (Refer to Chapter 11.)

> To balance BCI's we would run the "projector" in the opposite direction. This was a simplification. Negation of BCI's occurs, not by running the film in an opposite direction, but by using another roll of film. A roll that runs in a way which happens in a different phase (hence Phase II). Entropy balanced by negentropy. It is order, here the effect of entropy, that is balanced by the Bi-Deltic Cosmos. The next digression rectifies this first "simplified" consideration.

Each film frame (an "instance" or minimal duration) represents a "potential act in the flow of time." If we do not run the film through a projector (eliminating dynamism) we would still have the roll of film. A roll of film not "flowing" through a projector is one not yet manifest (viz, experienced: in our analogy—viewing the movie). The roll is "static." It could lay on a table.

> Here is our second consideration. This is a duplex film. Developing occurs by complementary "splitting" of the film. One complement

pertains to the Deltic Cosmos. The other represents the Anti-Deltic Cosmos.

All BCI's are balanced statically (at any given "present" time [viz, no "other" time]). To more easily do this at a higher level we group every BCI (reductionistically, as initially defined by a When Aspect cycle) of any possible BCD together. We construct the Cirpanic Cosmos Omnipresence (CCO). The CCO is the serial archive of all film rolls (BCD's). Some rolls have been exposed and developed (have manifested as experience—viewed film). Other rolls are undeveloped (are potential—not experienced).

Film rolls, in themselves, are static. At this level of cosmic construction all we have to work with, are the rolls themselves. We found each film frame is balanced against all other frames. Each roll represents a different movie. Each movie is analogous to a different cosmic cycle (BCD). The balancing of each roll must also be done "in no time."

Our question now is "how can each duplex film roll (BCD) be balanced?" Each roll is balanced against all other rolls. We must balance (maintain symmetry) without running the film roll through the projector. As the above digression shows, this is a "combined" roll of film. There are no "pictures" on the duplex frames—they are all blank.

All possible film rolls are present at once (the archive is full). Only one duplex roll is projected into manifestation during any given (cycle) time. In the present, from our perspective, there is no flow of time—only the present instant of manifestation. Each film roll represents a potential-cosmic duration (the BCD).

Pick one roll. All other rolls either precede it or follow it. All "preceding" rolls have previously manifested. Past rolls are history and have "developed." All following rolls are potential. Future rolls are undeveloped. The interesting thing here is that there is no order (number) to what follows the present developing roll (being projected into manifestation). The "next" cosmos cyclic manifestation remains undefined (i.e., undeveloped).

Wider Perspective

Let's back off and gain a wider view—regain context. Time is characterized by duration. It was stated in Part C that (temporal) directionality is not an inherent property of time itself. The property of direction is usually assigned to time, not only because of how it is perceived, but because entropy suggests an "arrow of time." Entropy concerns change in order. Order is a characteristic of the How Aspect.

The How Aspect contributes order to time and hence directionality. But what is ordered? The Where Aspect is ordered. The Where Aspect contributes the manifested "events." The How Aspect "orders" these events. The When Aspect imbues these events with motion. Without the When Aspect, potential events (durations) cannot manifest as experience.

Let's use our motion-picture-film analogy. Our Deltic Cosmos is the current "running" of the film—it is dynamic. The Bi-Deltic Cosmos represents the static film roll sans dynamism. The Cirpanic Cosmos represents the archive of all possible film rolls. We can also say that it represents all possible laws of nature. It sets "what" can manifest. Here, we begin to define the BCD in nonreductionistic terms.

We begin to better understand the deeper nature of the Deltic Cosmos. At this stage (the Cirpanic Cosmos) we are explaining what we experience in the Deltic Cosmos. We ask what is the origin of experience? This is a particular angle on the ubiquitous question what is the meaning of life? We answer this question (in Chapter 21) by fully understanding the structure of reality.

Some philosophical scientists thought since the cosmos is evidently "winding down" that entropy implies time itself is inherently directional. The cosmos (duration—series of events) acquires (temporal) direction because of a change in order. Entropy implies a system becomes more disordered. Order and time collectively provide the Where Aspect its dynamic quality.

It is said time stops everything from happening all at once. We can appreciate this statement at this level of structure. Everything—every event, past and future—is simultaneously represented in the structure of the Cirpanic Cosmos. Let's enrich our understanding.

Directionality is defined in the Cosmos and Anti-Cosmos by the How Aspect and Anti-How Aspect. Time proceeds one way in the Cosmos, and it proceeds in an opposite direction in the Anti-Cosmos. Note: At the Deltic Cosmos structural level, the same reasoning applies to extensional asymmetry. That is, extension in space (objects—the Where Aspect) takes on the property of asymmetry because of the How Aspect and Anti-How Aspect. Change from the How Aspect to the Anti-How Aspect inverts order by inverting "manifest" (experiential) space (more appropriately, extensionality).

Time with direction is a more specific instance of time with duration. Directionality is a (subsumed) property of order. A combination of Deltic Cosmos and Anti-Deltic Cosmos temporal intervals, negated "dynamic" duration. That is, temporal directionality. That combination is called the Bi-Deltic Cosmos.

Symmetry balancing of directionality does not balance duration itself (i.e., duration sans directionality). Therefore, combining the Cosmos and Anti-Cosmos as a collective whole (the Bi-Deltic Cosmos) cancels temporal directionality, but not temporal duration. Duration is analogous to the roll of film as it lay on a table: not running through a projector.

Whereas the cosmic time cycle (BCD) is the greatest Bi-Deltic duration, the BCI is the shortest duration. The BCI is analogous to a motion picture frame. The BCD is analogous to the roll of motion picture film. Each Bi-Deltic Cosmos Instant (BCI) defines the minimal property of duration only because of the concatenation of BCI's defines the maximal duration.

It is the manifest series of BCI's that constitutes duration as it is usually understood (viz, time flow). Extended duration is composed of mini-events that denote BCI's. We are warranted in saying this because of the implications of quantum physics. The Cirpanic Cosmos Omnipresence (CCO) is not duration as it is usually understood. The CCO is analogous to the archive of all possible films (i.e., limited by the laws of nature).

Forms

The Cirpanic Cosmos structure concerns the final understanding of duration itself. Because duration is not totally negated in the Bi-Deltic Cosmos structure, we postulate the Cirpanic Cosmos structure to balance the BCI. Particularly, what duration is it that is not negated? The problem stems from the Bi-Deltic Cosmos structure. Its structure is incapable of negating any other possible cycles. Why? If other cycles differ in their potential "form" from one another, they must be balanced by a higher-level structure—the Cirpanic Cosmos.

Every Bi-Deltic Cosmos has one thing in common—form itself. Form has manifest structure. Only in the Universe of Perfect Whatness is form manifested as physical reality. Note, even physical form itself is composed of lower-level structures (molecular, atomic, and so on).

Where does this lead us? Another way of understanding the Cirpanic Cosmos is to say it is constituted of "forms." These are nonmanifested forms. The idea of "formula" is not too far fetched in describing this idea. Equations in physics represent the laws of nature. These laws set limits to what can manifest and how. Most of the laws of physics meet this requirement. We revisit "forms" in later chapters.

History

We can understand this chapter from a physics view in two ways. One way is to understand the structure of the Cirpanic Cosmos from General Relativity. There are higher-level-theoretical aspects of General Relativity that tend to support the Cirpanic Cosmos structure. We can also correlate its structure with a generalized interpretation of Quantum theory. We examine both.

General Relativity

In the 1920's E. Hubble studied light emanating from distant galaxies. It was already known atoms, when properly excited (heated) emit several well-defined-optical spectra. Spectral lines correspond to various atomic energy levels that collectively form a pattern that uniquely identifies the emitting atom.

Hubble examined patterns created by the light he observed. Some patterns were identified, but displaced: appearing in the longer wavelengths in our reference frame.

What would cause such a shift toward the "red" end of the electromagnetic spectrum? One plausible explanation supposes light sources are receding from us. When a source moves away from an observer, the wavelength is "stretched" causing spectral lines to appear shifted toward the infrared region (in our reference system). This is called red-shift. By determining the degree of shift the relative velocity of the source can be calculated. Shifted light patterns are explained by hypothesizing distant galaxies are receding from us. The spectral patterns suggest universal expansion.

Reasoning suggests the velocity of any galaxy is directly related to its distance (from us) by the same proportionality. This is an expression of Hubble's Law. The idea seemingly implies that our planet earth is at the center of the universal expansion. However, there is a more reasonable interpretation. We use a classical" model to explain.

Consider all galaxies as dots on the surface of a spherical balloon. As the balloon expands each dot recedes from other dots at a velocity proportionate to the separating surface distance. In comparing two stellar objects, if one is twice the distance of another, its velocity will be twice as great. This observation led to a formulation called the Cosmological Principle. It states that on an intergalactic scale the universe appears isotropic to all possible observers. So, one can assume any dot or galaxy to be the center of expansion.

Universal expansion can be extrapolated backward in time to that era in which the universe began. Theories of cosmic origin are described in cosmogony. Cosmogony now centers on the Big Bang Theory. The theory states the cosmos (actually, the tardyon universe of our Deltic Cosmos during Phase I) began a long time ago with an explosion of matter-energy from a super dense state.

At the Big Bang initiation all Universes of Perfect Whatness space and time were essentially contained in a dimensionless-mathematical point where temperature and density are infinite. The pressure is so great that matter-energy overcomes the effect of immense gravity (created by such dense mass) and explodes.

The explosion creates a plasma of free particles composed of protons, positrons, electrons, neutrons, photons, and neutrinos. Charged particles in motion give off electromagnetic energy (photons). The particles are bathed in, and thermodynamically interact with, photons. As the Universe of Perfect Whatness expands, it cools.

A few minutes of cooling allowed nucleosynthesis to occur. Helium nuclei formed. Less than a half hour later, helium atoms form. It has been predicted from theory that helium would then constitute 25% of cosmic mass (a one to three ratio by weight of helium to hydrogen). The calculated quantity of helium produced coincides with present day scientific observations of cosmic helium.

Ionized hydrogen and helium interact with photons until thermodynamic equilibrium between matter and light energy is established. After ten thousand years of cooling (expansion) cosmic temperature dropped below ten thousand Kelvin. By that time electrons had coupled with ionized gases forming stable hydrogen and helium atoms.

With formation of stable atoms, photon interaction and scattering markedly diminished thus upsetting the thermodynamic balance between matter and electromagnetic energy. Primordial material continued expanding as a spherical cosmic shell. Decoupling of matter and energy resulted in a general release of electromagnetic radiation.

The released radiation is still with us today. It originally emanated in the near infrared and visible regions (of the electromagnetic spectrum) but has now been greatly red-shifted because of cosmic expansion. The advancing radiation defines the edge of space (extensionality). In our proper frame of reference the background radiation is now observed in wavelengths ranging from thirty centimeters to half a millimeter—peaking at one millimeter. A spectrophotometric curve of the detected frequencies corresponds to microwave radiation emitted by a black body at a temperature of 2.7 Kelvin. (A black body is one in which energy input and output are in a state of equilibrium.)

A. Penzias and R. Wilson discovered cosmic background radiation in 1965 at a research laboratory in New Jersey. All stellar objects lie in front of this background radiation. It is isotropic except for a slight shift in wavelength caused by the motion of earth through space.

On an intergalactic scale isotropy implies uniformity of Universe of Perfect Whatness expansion. These findings further corroborate Hubble's observations in the 1920's that distant galaxies are receding from one another at velocities proportional to their distance. This information constitutes significant evidence supporting the Big Bang theory of cosmogony.

Although the Big Bang theory accounts for the observed state of the Universe of Perfect Whatness, it too must be explained. First, some developments in astrophysics will be described.

Astrophysicists studied what happened to stars that have burnt out. A star's outer layers are usually thought to be supported by energy released from internal thermonuclear reactions. A star begins collapsing when nuclear fuel is exhausted because of its mass.

In quantum theory Pauli's Exclusion Principle states no two identical fermions can occupy the same energy level (quantum state). Stellar collapse induces enormous pressures and densities forcing electrons to fill all the lower-energy levels. Energy is thus not as easily carried away by radiation emission. Subsequently, the "excess" energy fills the higher-energy levels. This contributes to the dying star's kinetic motion that increases internal pressures. This "degenerate electron" pressure hinders further collapse. The star becomes a white dwarf.

In the early 1930's S. Chanrasekhar calculated an upper limit for the mass of a white dwarf. He said degenerate electron pressures would only support a white dwarf if its mass did not exceed 1.25 solar masses (now considered to be about 1.44 solar masses). If a star's mass is greater than the Chanrasekhar limit, contraction of mass overcomes the degenerate electron pressure.

Electrons compress and merge with the protons (of the nuclei) to form neutrons. This creates a degenerate neutron pressure (neutrons also obey Pauli's Exclusion Principle). Calculations reveal degenerate neutron pressures can check further gravitational collapse and form a neutron star. Astronomers have observed stellar objects called pulsars believed to be rotating neutron stars.

Astronomers have observed stars with mass equal to seventy-five suns. What would be their fate if they did not explode off enough mass (as in a supernova) to fall within the limit counteracted by degenerate neutron pressures? No physical force can block stellar collapse of matter surpassing three solar masses.

In 1939, using the General Theory of Relativity, J.R. Oppenheimer and H. Snyder calculated conditions describing a compressed object of more than three solar masses. Much interest was shown in their calculations.

Better understanding of these men's solutions and more astronomical studies renewed interest in the vacuum field equations of General Relativity. The solutions to these equations may explain what happens to space (extensionality) and time (flow) in the presence of extremely dense matter. (We again delve into an understanding of General Relativity in Chapter 14.)

The next step was to use those equations to describe what happens during the collapse of a three-solar mass object. It was found that extremely dense matter causes nearby space and time to become appreciatively curved. It is commonly stated that the laws of physics break down at the center of such a mass. Gravity becomes so overpowering even light (electromagnetic energy) cannot escape to the "outside world." These objects are called black holes because they do not release light.

Black holes are simple objects completely describable by three variables. The simplest black hole is described by one value—mass. In 1916 K. Schwarzschild found a solution to Einstein's field equations later recognized as describing a spherical symmetric nonrotating black hole. The center of this object is called the singularity. Surrounding the singularity is an event horizon most often considered the surface of a black hole. From the singularity the surface lies at a distance proportionate to the quantity of mass. For example, a three-solar-mass black hole has an eleven-mile diameter event horizon. An astronaut would have to travel at the speed of light to remain in place at the event horizon.

Once inside the event horizon even light cannot escape. Any matter falling through the event horizon is "drawn" to the singularity where pressure and density infinitely compress it out of material existence. At the singularity spatial extensionality and time flow are infinitely distorted. Therefore, space loses its extensionality and time ceases flowing.

The Schwarzschild black hole also has another enveloping "surface" called the photon sphere. This outer surface is composed of unstable-circular-light orbits. If light from distant stars arrive near the black hole at certain angles it will orbit above the event horizon. A black hole of three solar masses would have an 18 mile photon sphere.

Black holes have a startling affect on contiguous space and time. Space-time is flat thousands of miles from a three-solar mass black hole. If from afar, in coordinate time, we could somehow view a space ship approaching a black hole, we would see curious things happen. Gravitation causes time to dilate or slow down. We would observe the space ship slowing as it drifted toward the black hole. From our coordinate viewpoint it would take the ship an endless amount of time to reach the event horizon. At the event horizon, coordinate time ceases altogether. However, by the ship's proper time (time experienced by an astronaut aboard the ship) the ship would pass through the event horizon in normal fashion (neglecting the effect of tidal forces). Once the astronaut passes through the event horizon, he is in for another surprise. He is no longer relentlessly embraced in the flow of time. Instead, he is dragged forward in space. The roles of space and time interchange upon crossing the event horizon. The astronaut is drawn forward in space toward the singularity.

Another interesting effect occurs as the astronaut approaches the black hole. Looking out the front window from afar he sees the black hole. When he turns and looks out a rear window he sees where he has been. As he approaches the black hole, he sees out the rear window that the view of where he has been becomes slowly encircled by the black hole. The entire "outside" universe is phenomenally squeezed toward the center of rear view. This results from the closing of an "exit cone" above the event horizon as it is approached. This means oblique light is increasingly gravity attracted the nearer one is to the event horizon. The increasing mass of the approaching space ship also contributes to the phenomenon. At the event horizon even light radiating "straight" out, does not leave the surface.

Physics

There are two sides to the Schwarzschild solution. One side pertains to black holes; the other represents what are now called white holes.

Structurally interpreting the mathematics of black holes has been difficult. Original diagrams depicted each region of space and time to be superimpositions of one on another.

In 1960 Kruskal found mathematical transformations enabling physicists to diagrammatically represent all regions of space and time. In these diagrams the center of a black hole splits into past and future singularities. Between the singularity lay four regions. Two regions denote past and future event horizons. Two areas represent the outside universes. One outside region could represent our Universe of Perfect Whatness. Their diagram represents all space and time by infinitely extended lines demarcating the two outside universes. Their diagram does not display all space and time although it is completely represented.

R. Penrose found a way to portray all universal space and time in one diagram. Regions far from the "here and now" can be shown by a technique called conformal mapping. He partitions the regions of our universe into five infinities (see Figure 12:3). Time is characterized in the usual vertical directions from the here and now. Penrose maps out past and future-like infinities to represent distant past and future events. The "world line" of a fast moving, extremely long-lived object would follow an arc from the distant past to the distant future (see curved vertical line). Inclined at 45 degree angles, are past and future null-like infinities. One represents light source while the other defines light destination. In a Penrose diagram our universe takes on a triangular shape.

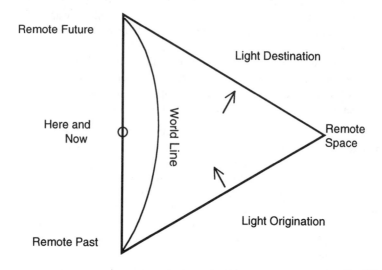

Figure 12:3. Penrose Conformal Mapping of Flat Space-Time

When Penrose's diagram includes a black hole, that "other" universe appears back to back with our universe. This diagram yields interesting information. If our astronaut ventures into a black hole, he observes that other universe before hitting the singularity.

We built our foundation of understanding upon an ideally simple black hole. Such a black hole probably does not exist. We have seen how the understanding of black holes developed from astrophysical curiosity in stellar collapse. We continue erecting a basis for understanding Cirpanic Cosmos structure.

Mass is not lost in a black hole. Mass can be indirectly determined by measuring its effect on surrounding space. Charge is another variable not lost in a black hole. By 1918 H. Reissner and G. Nordstrom presented a solution of Einstein's field equations describing a black hole with charge. The charge can be either electric or magnetic. Realistically, charge would have to be small. A large charge would attract oppositely charged particles that would effectively neutralize the black hole.

A charged black hole exhibits two event horizons, one inner and one outer. If an extremely large charge were possible the two event horizons would merge exposing the singularity. This would be a "naked" singularity. R. Penrose said such singularities should not exist in the universe. But, if they could, an interesting phenomenon would occur. Near such a singularity space would generate gravity forces of repulsion (antigravity) instead of attraction.

Global space-time structure of the Reissner-Nordstrom solution reveals there is also a white hole connected back-to-back to "our" charged black hole. This is expected since the elimination of charge would mathematically reduce the solution to one describing a Schwarzschild black hole.

Although astrophysicists consider charge to play only a minor role in realistic black holes, the same cannot be said of rotation. Nearly all stellar objects rotate. In 1963 R.P. Kerr uncovered another solution to the field equations. It was a unique solution representing rotation.

Solutions incorporating mass and angular momentum reveal more surprises. The singularity becomes a ring around the axis of rotation. It is theoretically possible for an astronaut to pass through the singularity and into negative space! What does this mean? Physicists find it difficult to comprehend such an idea. Some interpret negative space to be a region of antigravity.

The Kerr solution also reveals rotating black holes to have two event horizons: one inner and one outer. With increasing angular momentum the two event horizons draw nearer and eventually merge. Space and time do not change roles upon crossing a conjoined inner and outer event horizon.

Our next task is to connect the Big Bang theory with black and white holes. Instead of using black and white holes to describe astrophysical objects in our universe, we apply the ideas to cosmogony and cosmological structure. To do this we restate some ideas about black and white hole theory.

If we invert a black hole, we effectively describe a white hole. A white hole represents the "other side" of the equation describing black holes. Matter near a black hole is gravity impelled toward its singularity. White holes spew out matter from a singularity.

If we apply these ideas not just to black or white holes in our universe, but to our whole Bi-Deltic Cosmos we obtain possible solutions to the problem of cosmogony.

On a cosmological scale our whole tardyon Universe of Perfect Whatness could have exploded from a single point (the singularity) of space and time. These ideas offer possible explanations for our expanding tardyon universe. In addition, it solves another problem. What was there before the Big Bang? What was there before the cosmological explosion that produced our expanding universe? The answer could be—the collapse of a tachyon universe. We shall see shortly how this later idea can be inferred from a Penrose diagram.

Be mindful: Here we do not have sufficient information to do more than hypothesize about what happens. We begin by pooling information from the Bi-Deltic Cosmos structure and Penrose diagrams. To do this we draw on the Penrose diagram of the Kerr (rotating) black hole. A cosmological-cosmogonical interpretation of this diagram may lead to a better understanding of the Cirpanic Cosmos structure.

Let's look at how Penrose's diagrams are constructed. Simple reasoning dictates certain rules. If charge and angular momentum are excluded, the Penrose diagram reduces to a Schwarzschild black hole. The first rule states there has to be a universe opposite ours that can only be reached through space-like trips. Penrose's technique of conformal mapping yields the second rule: all universes take on a triangular shape with respect to the five infinities. Since all event horizons are light-like the third rule is all event horizons must be inclined at 45°

214

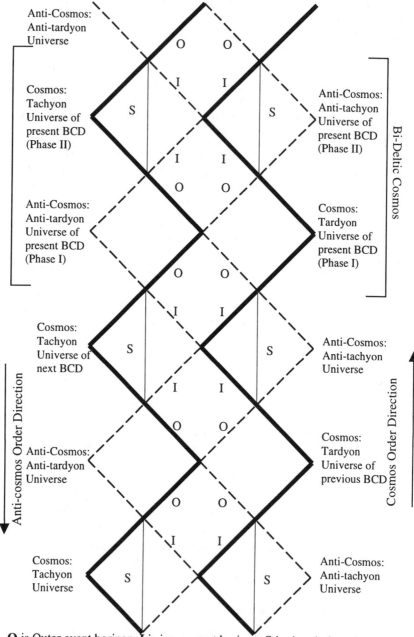

O is Outer event horizon. I is inner event horizon. S is singularity. Thick lines indicate Cosmos. Broken lines indicate Anti-Cosmos.

Figure 12:4. Penrose Diagram for Kerr Black Hole: Enhanced.

angles. The fourth rule states the roles of space and time are interchanged every time an event horizon is crossed.

Two event horizons are crossed when nearing the singularity of a rotating black hole. Therefore, the singularity must be temporally extended. This information leads to construction of a Penrose diagram for the Kerr black hole. The following diagram (Figure 12:4) is for a black hole of moderate rotation.

The diagram shows global structure. Global structure results from mathematically avoiding infinities at the center of a black hole. In this structure, trajectories into the hole need not converge and cease at the singularity. The center can be redefined as a "wormhole" or passage to other "universes."

Duration and Physics

We now locally superimpose Bi-Deltic Cosmos terminology onto Penrose's diagram. We correlate the Cirpanic Cosmos Diagram with a geometric representation of the conformal mapping Penrose used to conceptualize his generalization of the physics of black holes. See Figure 12:4.

The structure represents a series of Bi-Deltic Cosmos Duration's (BCD's). In this structure the "durations" or cosmic cycles are reductively represented. Each BCD unit is composed of four universes. We find a Tardyon Universe (i.e., Universe of Perfect Whatness) a Tachyon Universe (Universe of Imperfect Whyness) and their anti-cosmos counterparts. Each successive BCD is one cosmic cycle advanced from its previous BCD. (In the diagram, corner frames demarcate the BCD.)

The structure is composed of two intertwining components. One substructure (in the diagram, enclosed by thick lines) represents the Deltic Cosmos. The other substructure (depicted by broken lines) represents the Anti-Deltic Cosmos.

We find the Deltic Cosmos is represented as an unending linked series of Tardyon and Tachyon Universes. Any two connected universes on one zigzag substructure represent the Deltic Cosmos. On the right side (of the diagram) we find a unit tagged "Cosmos Tardyon Universe." This is our present physical universe. We also find above and below our "present" physical universe that every other unit is also tagged a Tardyon Universe. These added Tardyon Universe tags represent our Tardyon Universe "during" a different duration (i.e., past and future cycles). Every other unit on the left side of the same substructure, also, represents our Tachyon Universe. The Deltic Cosmos Phase I and Phase II are thus represented.

We see the Anti-Deltic Cosmos is also represented as an unending series of Anti-Tardyon and Anti-Tachyon Universes. Any two connected universes in this substructure constitute the Anti-Deltic Cosmos.

By conjoining these substructures every BCD is represented (past, present, and future). Every cosmic cycle is represented.

Clarification: The above does not mean that our idea of "future," ascribed to the diagram, represents what will happen. It merely suggests that there are future possible cycles. In our motion-picture film analogy, a film roll (representing the "motion" picture) does not and cannot of itself represent a "future showing" of the film. Consider "future" film to be blank and not yet developed. Only past "film" rolls have been fully developed. Only past cycles can have any sense of "memory."

The idea of "beginning and ending" does not apply to the Cirpanic Cosmos structure. We continue using the word "duration" because the "higher level" Cirpanic Cosmos structure explains dynamic duration (which we observe in our Deltic Cosmos).

Duration Synopsis

We define three concepts of duration. The first is the most infinitesimal duration. This is the Bi-Deltic Cosmos Instant (BCI). The second is the greatest duration of our Cosmos. This is the Bi-Deltic Cosmos Duration (BCD). This is a cosmically complete cycle. Inductively extending the BCI idea to an extreme, we obtain the third concept—Cirpanic Cosmos Omnipresence (CCO). Here, duration is completely stripped of its temporal nature. We end up with a series of ordered events: some manifest, and some potential. Let's take a closer look.

A succession of BCI's, each "held" for an instant—an infinitesimal duration, generates the flow of time (in the Deltic and Anti Deltic Cosmos). In our film analogy, static duration—BCI (a timeless event)—is first synonymous with a motion picture frame. Each frame is a "still photograph."

An infinite-like number of BCI "held" durations, or still frames, can be incrementally extrapolated to the everyday notion of dynamic duration. Note if these still events were not held in time for at least some infinitesimal (quantum) duration, there would be no flow of time. If this were so, time flow cannot be explained. This reasoning applies to the instants of time between the "showing" of events. If there were no infinitesimal duration between events, there would not be the "on-off" nature of the quantum. There must be a separation of events for change to occur in the flow of time. Temporal separation is supplied by the When Aspect.

Quantum phenomena are inherent in cosmic structure. For example, When the Deltic Cosmos is in an "on" mode, the Anti-Cosmos is in an "off" mode. When the Deltic Cosmos is in an "off" mode, the Anti-Deltic Cosmos is in an "on" mode. These ideas reflect the meaning of the CPT theorem. We also find these ideas apply to the relation between the Cosmos and yet higher realms.

Duration is also ascribed to the roll of film. Each film frame is a cosmic "event." A film roll is a series of events. Clearly, the film analogy does not itself suggest dynamism. And neither does a BCI or BCD.

Cycles continue. There is continuity between rolls. When the "showing" of one film roll ends, another (movie) begins. When one BCD ends, another BCD begins. Each is just "another" cycle.

A phase change is no different, from a cosmic perspective, than any other change from one BCI to another. This also applies to the "change over" from one cycle to another. Where the "beginning" and "ending" of a cycle is pegged is arbitrary. Cycles themselves do not end. The clock just keeps "going around." There is no beginning and ending.

The extreme concept of duration is attributable to the Cirpanic Cosmos Omnipresent (CCO). Clearly, there is no hint of temporal flow at this level. Yet, this construct offers a conceptual description for "all time." Its explanation awaits later chapters.

Look at Figure 12:4. Locate the Deltic Cosmos (in Phase I) labeled "Cosmos: Tardyon Universe." This is our Universe of Perfect Whatness (physical universe). Further up the page we find a tardyon universe represented for a "succeeding" cosmic cycle. We could also say that this is our Tardyon Universe in a later cycle. For us, now, this could be our universe in a virtual state or one of potentia. In the same substructure, zigzagging to the left, are Tachyon Universes represented for their respective BCD's. One of these represents our Deltic Cosmos Phase II.

The other substructure represents the Anti-Deltic Cosmos (see Anti-Tardyon universe and Anti-Tachyon tags). Notice "our" Anti-Deltic Cosmos is also represented. In one phase (Phase I) it is labeled Anti-Tardyon universe. In its other phase (Phase II) it is labeled Anti-Tachyon universe. In our film analogy, this is a duplex film. The two intertwined substructures relate this idea.

Specific note: This interpretation is not how many physicists would understand this Penrose diagram. Most would say each "universe" is distinct and separate from our universe (Deltic Cosmos). However, at this level of generality it is parsimonious to say each "universe" is our universe (or our anti-universe) at "all possible time." The reason is simple. Either the diagram simultaneously represents our universe at every different (i.e., all) time (cycles) or it represents "many different universes" at any given time. We opt for the former interpretation because there is no "flow of time" at this level. Hence, all time must be simultaneously represented. This is our rationale for eliminating the second possible interpretation. If we considered the second interpretation, we would not have a means to represent all time (i.e., events, viz, every BCI and every BCD) within this structure. There is some ambiguity in defining some of these terms (e.g., All and Every). We address these concepts in following chapters.

Mechanics

Let's further examine the structure. One (zigzag) substructure (in Figure 12:4—enclosed by thick lines) represents the Deltic Cosmos (our tardyon and tachyon universe for all possible BCD's). The other substructure (enclosed by broken lines) represents the Anti-Deltic Cosmos (the Anti-Tardyon and Anti-Tachyon universe for all BCD's). Though structurally intertwined, the two cosmoses function independently.

Our Tardyon Universe (Universe of Perfect Whatness) is temporally connected to its cosmic singularity ("S" in the diagram) through outer event and inner event horizons ("O" and "I" in the diagram). On the other side of the time-like singularity is our Tachyon Universe (Universe of Imperfect Whyness). This is the negative space that some scientists prefer to call an antigravity universe.

Our Tardyon Universe is also connected through more inner and outer event horizons to another Tardyon Universe. This universe is our Deltic Cosmos in either a past cycle or future cycle. (Arbitrarily, we say a tardyon universe tagged toward the top of the page represents a future Phase I of a Deltic Cosmos cycle.) Each of these tardyon universes is one Bi-Deltic Cosmos duration (film roll) advanced or retarded from our own. In our film analogy, again, each successive film ("showing") is consecutively archived. Each BCD can hypothetically be assigned an "order number."

Only time-like trips are permissible in the Tardyon Universe (Universe of Perfect Whatness). In the diagram such trips cannot be inclined more than 45° from the vertical. Only space-like trips are allowed in the Tachyon Universe. Such trips cannot be inclined more than 45° from the horizontal.

We apply these allowable routes to cosmological collapse and expansion. In Phase I our Tardyon Universe is dominant. That is, it is manifested. Immense antigravity forces near the singularity cause it to explode outward much like a white hole. This is how the theoretical mechanics of a white hole concur with the Big Bang Theory.

Eventually, the Tardyon Universe ceases expanding. The obvious question is the following. Does the Tardyon Universe collapse into the singularity? If it did, time would be reversed. That would not mean that collapse would be the exact opposite of expansion. There are problems with this possibility. Although this scenario is possible, it is unlikely. Let's say that this does happen. Would it not take some governing intelligence to make the collapse identical to its expansion? In our analogy this is represented by running a film backward to mimic time reversal. Yet, if a governing super intelligence is needed to assure this reversing duplicity, then an intelligence can just as well reverse time some other way. We apply our understanding of the CPT theorem. This type of time reversal is not required by physics.

What other means for collapse is there? Let's say the Tardyon Universe expands in Phase I. It diverges from the singularity: just as it is theorized from the perspective of white holes—the opposite dynamics of black holes. Black holes and white holes can also apply to the universe as a whole.

Maintaining symmetry does not call for collapse of the Tardyon Universe. Instead, collapse occurs in Phase II where the Tachyon Universe converges or collapses. The universe expands as tardyon matter, and collapses as tachyon matter.

The Tardyon Universe expands from a cosmological singularity. It eventually reaches the point of maximum expansion. It then "dematerializes"—transforming into a Tachyon Universe. The Cosmos begins collapsing as a Tachyon Universe toward the singularity. The two-phase cycle repeats itself over and over. Again, each cycle may differ from each other. This is all speculation. The best we can do is to describe the possibilities in noncontradictory terms.

When we consider cosmological collapse, we include the possible formation of a "naked" singularity. A naked singularity implies there is no "outside" space and time. It is space without extensionality. (This idea is used in the final Section of this book.)

There is another way of preserving the above ideas and having each universe undergo both processes of expansion and collapse. In Phase II our Tachyon Universe is dominant. Much later it converges to a singular state. Consider dominance to be a manifest condition and subjugation to be a virtual condition. Our Tardyon Universe expands manifestly, but collapses virtually. Also, our Tachyon Universe converges, collapses, manifestly, but expands virtually. Again, there is more than one scenario that could just as well depict what happens.

Back-to-back with our Deltic Cosmos is our Anti-Deltic Cosmos. When we "changed over" from Phase I to Phase II, our Anti-Deltic Cosmos changes from Phase II to Phase I. The Anti-Tardyon universe becomes dominant and explodes from a space-time singularity in a manner not unlike a cosmological white hole. The Anti-Tachyon universe becomes subjugate and much later undergoes change toward maximum divergence.

In Figure 12:4 cosmic development proceeds "up the page" as shown by an upward directed arrow. Anti-Cosmic development goes "down the page" as shown by a downward pointing arrow.

Let's look at just one cosmic cycle using our "flat-two-dimensional filmstrip" analogy. If the "film" (BCD—one time cycle) is laid out, double "twisted" into an enhanced Möebius strip, and joined "end to end," we would have a "double inversion" of the film. (This is associated with the "twice around" nature of spin angular momentum.) Ideally, halfway around the circle we would have the "first point of inversion." This is the "first" phase change. The other point is reached where the frames "match" or where they are spliced together. This represents the second phase change. Each of these inversions shows the "beginning" of time for one phase and the "end" of time for the other phase.

Phase I gives way to Phase II for the Deltic Cosmos. Phase II gives way to Phase I for the Anti-Deltic Cosmos. Time flow, spatial extensionality, and charge, invert. The covering concept for these processes is the Bi-Deltic Cosmos.

When the Anti-Deltic Cosmos reaches its maximum condition from the singularity, the Anti-Tardyon universe begins collapsing in virtue. The Anti-Tachyon Universe becomes dominant and much later converges manifestly to a singular state. Keep this in mind. In the view presented here, the singularity forms at phase change from Phase II to Phase I.

There is another interpretation: one that calls for more explanation. When the Deltic Cosmos explodes from a singularity, its Anti-Deltic Cosmos counterpart does also. We might be interpreting that the inverted antigravity universe ceases to expand when that might not be so. Although this idea leaves other questions begging for answers, it does posit one bonus. If both cosmoses (Deltic and Anti-Deltic) simultaneously spring anew from a singular state, another question could be answered.

Scientists often use imaginary or virtual pairs of particles to explain how matter manifests. They know matter is created in pairs of particles having opposite properties. Reflectively, as we have noted, one member particle is an inversion of the other.

Theorists consider space to be filled with virtual particle pairs (this is the most reductive nature of space itself). If enough energy is present to "energize" virtual pairs, manifestation can occur with pair members shooting off in opposite directions. The intense gravitation of a cosmological singularity could impart the needed energy to materialize a whole universe. Matter could explode in one direction as a Deltic Cosmos and in another direction as an Anti-Deltic Cosmos.

The idea that concentrated gravity near a singularity can tear space and time apart to produce particle pairs, should not be ignored as an avenue to cosmological explanation. This book will adhere to the first interpretation because it is less problematic. This allows us to say that when our Deltic Cosmos is in Phase I, our Anti-Cosmos is in its Phase II.

The Penrose diagram for a moderately-rotating black hole extends infinitely. Consider one "diagonal" of the diagram (Figure 12:4). Pick our Deltic Cosmos at a given time. Connected to it are an unlimited number of BCD's representing our cosmos at all other cycles and therefore, every possible instant. Each of these instances is similar to ours except they are of a different order number (recall numbering the film frames). The same idea applies to our Anti-Deltic Cosmos. Here instead of the direction of change progressing forward in time (zigzagging "up" in the diagram) it proceeds in the opposite direction (zigzagging "down" in the diagram).

As each Bi-Deltic Cosmos is checked, we see the procession toward changeover continuing in both directions until a certain point is reached. Of the cycle itself, we observe its "duplex" nature (i.e., BCD). A BCD is composed of two phases. One represents the Deltic Cosmos Phase I; the other represents the Anti-Cosmos Phase II. Each half cycle has two "points of inversion."

The order relation between cosmoses should remain constant once a standard is chosen. We assign other Bi-Deltic cosmos order numbers as we proceed along the series of BCD's. Any designation of a Bi-Deltic Cosmos Duration (BCD) is based on the arbitrary (archive) order number assigned to the BCD that is chosen as our standard. Given a specific cosmic cycle, each order number indicates a placement along the series of BCD's. Every BCD composing the Cirpanic Cosmos is "hypothetically" archived.

Bi-Deltic cosmoses change. Bi-Deltic cosmoses experience cosmic ticks (BCI's) in a sequence of "on-again, off-again" manifestations. Change happens step by step. We have already likened each step to the controlled jerky motion of movie film as it passes in front of a projector lens. The on-again, off-again nature of manifestation correlates with an understanding of quantum theory.

Key Points:

- The Cirpanic Cosmos Diagram is represented by dots arranged in a circle. Each dot represents a Bi-Deltic Cosmos. Every possible (past and future) Bi-Deltic Cosmos is represented by the dots. The Cirpanic Cosmos negates (every possible) "order."

- The Bi-Deltic Cosmos cannot negate itself. The reason is simple. If a cosmic cycle happens in any way other than exactly how our own Deltic Cosmos unfolds, then the Bi-Deltic Cosmos cannot negate itself. How is this problem solved?

- Any given Deltic (forward directed) instant (quantum) of time (about 10^{-43} second) is simultaneously negated by an Anti-Deltic (backward directed) instant of time. The synthesis of these two instants of time yields a nondynamic durational instant. It is called the Bi-Deltic Cosmic Instant (BCI). BCI's negate (time flow) directionality (arrow attached to BCI) but do not negate duration (i.e., the BCI itself).

- One BCI is a "still" frame. One frame is an instantaneous slice of the Bi-Deltic Cosmos. No given (specific) BCI is negated against any other BCI's. One BCI is a (directionless) duration. Duration needs to be negated. How can this durational instant be understood?

- Change (time flow) results from a projected succession of BCI frames. This is analogous to motion picture projection. Although a given frame (by itself) does not indicate which direction the next frame will materialize (it is either forward or backward) we (from a meta-position) can attach an arrow to the frame to show direction of "motion" (which way the frame changes). Note. This is nondynamic directionality. If direction of order is "frame forward" in the Cosmos, then it is "frame backward" in the Anti-Cosmos.

- The problem is to negate one BCI (directionless duration) against all other BCI's. How are BCI's explained?

- All BCI's composing a Bi-Deltic Cosmos are collectively called a Bi-Deltic Cosmos Duration (BCD). It is a directionless, nondynamic series of ("still") durations. It is a "strip of film" representing the whole Bi-Deltic Cosmos. It is a duplex film. It represents (via reduction) both Deltic Cosmos and Anti-Deltic Cosmos. (Nondynamic duration might better be defined as the "different possible ordering" of what can happen [in time]).

- The inductive problem is not "manifest order," but "all possible order." At the least BCD's represent "potential" order. (At best, BCD's represent all possible-manifest events.) And, this idea is akin to what is possible—what is logical. All possible BCI's define the limits to everything—to what can occur. Durational negation simplifies to synthesizing every possible arrangement of cosmic order.

- Rules of order are logical. Logic sets the limits. That was simple. It means there cannot be any contradiction in scientific laws and principles representing everything. This is precisely what symmetry principles do—eliminate disparity (contradictions) by synthesis. Any possible order must be contradiction free.

- The categorical "contradiction-free" synthesis of every possibility is called the Cirpanic Cosmos Omnipresence (CCO). An explicit (meta level) understanding of this structure is not explained until Section Six and Section Seven.

Certain mathematical understandings of Einstein's General Relativity support the structure in this chapter. There is no contradiction is supposing this understanding. Traipsing into the mind expanse has narrowed our questioning to ever simpler problems. We "see" the topographical "lay of the mind expanse." Yet, we are not finished mapping because there are yet higher level questions to address. We are close to visualizing the "big picture"—the completed structure of reality.

Here is a quick review of what we understand. We have just answered the question what is the widest intrinsic whatness mode? The answer is explained step by step in the structuring of the Realm of Whatness. The structure was pieced together by inducting scientific information. Sections of this "puzzle" were premised on solid evidential facts. Other sections (structures) had to be fitted using noncontradictory relative facts. Anyway, the pieces (relative fact) fit together rather well.

The Realm of Whatness is commensurate with the totality of Cirpanic Cosmos structure. The Cirpanic Cosmos structure is made from an unlimited number of Bi-Deltic Cosmos Durations. Each Bi-Deltic cosmos in turn is constituted by a Deltic and Anti-Deltic cosmos. In the system we have used, each Deltic Cosmos (or Anti-Deltic Cosmos) can be represented as an intrinsic whatness mode. In these terms our next goal is to establish the relation between the intrinsic whatness mode and an intrinsic whyness mode. We must establish a causal connection between the Realm of Whatness and another higher-level realm. However, in seeking a connection, there must be an "in-between" realm to tie together the intrinsic whatness mode with its intrinsic whyness mode. The interjacent realm we call the Realm of Reasoning.

Section Three

REALM OF REASONING

Perspective

- The highest structurally complete unit of reality
- Conceptually characterized via two separable natures
- Structurally composed of the Realm of Whatness and the Realm of Whyness

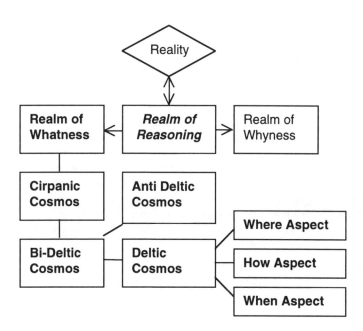

Figure 13:1. Perspective: Realm of Reasoning

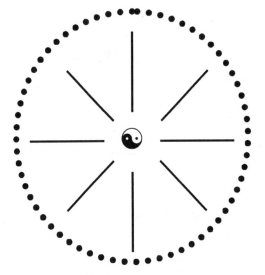

The Realm of Reasoning diagram represents the whole of
Reality and every part, but not as an integrated functioning unit.
This statement becomes more clear in following chapters. The
dotted wheel is the Cirpanic Cosmos. It represents every possible
thing. The inner circle represents All of Reality but no part (no
thing). The "spokes" connect the wheel (everything) to the hub
(All that is). The hub is the singular reality function.

Figure 13:2. Realm of Reasoning Diagram

Chapter 13

Realm of Reasoning

"Memory sustains both the knowing of what can be
understood and the understanding of what is known."

The Realm of Reasoning is the most general conceptual unit of Reality.
Structurally, the Realm of Reasoning has two natures. One is the Realm of
Whatness. It has been delineated and described in the previous section (Section
Two: Realm of Whatness). The other is the Realm of Whyness. It explains the
Realm of Whatness and is described in the following section (Section Four:
Realm of Whyness). The Realm of Reasoning is a combination of the Realm of
Whatness and the Realm of Whyness. Reality, conceptually, is the why-what
paradigm in the grandest sense.

The physics slant on unity (unification of the forces of nature) is
described in Chapter 15. The concept of "unit" is analyzed and defined
in Chapter 18. The meaning of reality's unity is assessed in Section 7.

The Realm of Reasoning is the highest level coupling of "atemporal" cause and
effect. The Realm of Whatness is generated intrinsically by the Realm of
Whyness. We later learn how this occurs.

Early in his work the author recognized the Realm of Reasoning to be the
connecting factor between the Realm of Whatness and the Realm of Whyness.
However, that connection (how it is composed of these two realms) was not
understood until after nearly twenty years of intensive work.

Before the Realm of Reasoning was understood "categorically," it was
considered the next "classificatory explanation" in the hierarchical structure of
reality. The Realm of Reasoning "appeared" to be independent (of the Realm of
Whatness and the Realm of Whyness) but it is not. In this section we treat the
Realm of Reasoning as though it were a separate realm (from the Realm of
Whatness and the Realm of Whyness).

We understand things by categorizing them. We understand reality by seeing
how its different hierarchical levels fit together. We learn before we understand.
Therefore, we defer explaining the complete understanding of this structure until
completing the next four sections. Meanwhile, we will learn again "things are
not as they appear." Here, we describe the apparent nature of the Realm of
Reasoning.

Apparent properties of the Realm of Reasoning mirror the contemporary
scientific approach. We learned from our conceptual scheme of connecting
phenomena that the Realm of Reasoning has sufficient characteristics of both
whatness and whyness that it seems to be "categorically" separate from either.
The Realm of Reasoning is represented in physics by light: electromagnetism.

When the Self conceptualizes the Realm of Reasoning as a basic feature of the
structure of reality, he recognizes the conceptual level denoted "Self-
Essentialization." Attaining this level of understanding draws upon the mental

function called "Reasoning Intrinsic Whyness." (Refer to How Aspect Diagram, Chapter 4.)

Preliminary Consideration

In previous chapters we explained how temporal direction is "reversed" by the cyclical action (two phases) of the When Aspect. The Bi-Deltic Cosmos effectively "negates" (by inversion) temporal directionality by functionally combining the Cosmos with its inverted Anti-Cosmos.

Temporality (as directed duration) and spatiality (as extension) are exclusively intrinsic characteristics of the Realm of Whatness. One purpose of this section is to set up our concepts (and understanding) of the Realm of Reasoning in a way not contradicted by the true nature of the connection between the Realm of Whatness and the Realm of Whyness. The Realm of Reasoning is described in a way not thwarting further explanation.

We hypothesize every manifest property (defining the Realm of Whatness) and indeed manifestation itself (manifest form) is produced by the Realm of Reasoning. We have yet to know how.

If the Realm of Whatness is connected to yet a higher level of reality, how is it connected? Another way of seeking this connection is to pose a question from the Cirpanic Cosmos structure. What basic question is unanswerable by examining the Cirpanic Cosmos structure? Let's reassess the situation that poses this question.

Reexamine the Cirpanic Cosmos structure. The circle is constituted of points (this is understood). Consider the points to be dots. Each dot represents one Bi-Deltic Cosmic Duration (BCD). Not every BCD has the same status. Some BCD's represent a "past" cosmic cycle. These are completed cycles: ones that have manifested. These could be shown by "darkened" dots on the (Cirpanic Cosmos) circle. One dot represents our "present" cosmic cycle. This is an incomplete cycle. It can be represented by a "half filled" dot. The rest of the dots represent "possible" future cycles. These can be symbolized by undarkened dots.

Our basic question now suggests itself. What is the origin of cosmic cycles? If we say the future develops from the past, we are left with the task of explaining how this occurs. A twist on this question is to ask what is the source of BCD's? We also ask what is the origin of Cirpanic Cosmic Omnipresence?

Some cycles have manifested or are manifesting; some have not manifested. Clearly, the Realm of Whatness is not its own source of being (manifestation). We ask, why are there cosmic cycles? We can refine that question by asking why do the cycles take on the "form" that they do? These questions are not answerable from within the structure constituted by even the greatest of all possible "durations"—the Cirpanic Cosmos Omnipresence (CCO). Although the CCO is defined by completeness, it is left to later chapters to "explain" completeness. This is why it is the greatest of all possible durations.

We ascertained the greatest cosmological "Whatness" that is (the Realm of Whatness) and asked why? Why is it? We examine the structural "Whatness"—the CCO—and ask of it "why?"

A question of "Why" is not answered by again pointing to "what is"—for that is of what we asked why! Is there a reason for this Whatness? In short, is there a Reason (Whyness) for the Realm of Whatness? Before answering this question, we must find a way to connect the Realm of Whatness to the reason for it. That reason is the Realm of Whyness. The Realm of Reasoning facilitates the connection. Some clarification follows:

All serious whyness answers have been found by categorizing phenomena into hierarchical levels. In hierarchical systems we obtain answers by either pointing to lower levels or by pointing to a higher level. The first approach is reductive. The second approach is to find a higher-level explanation.

The reductive tactic is suggested when we answer a whyness by ("reductively") pointing to a thing's components. For example, atomic structure is reductively explained by listing its components—protons, neutrons, and electrons. Reductive explanations tend to be confusing. Do subatomic particles sufficiently explain atomic structure? The answer is obviously no. When someone asks for an explanation it is rarely specified whether a reductive or inductive answer is sought. We ask of the "whatness" atom for an inductive explanation. We treat the whyness answer as a higher order whatness and again ask of it why? Therefore, what we are considering—atomic structure—and of it, we ask why? (Meaning: what is the purpose of atomic structure?) We answer, to form molecules. We can also ask of it, why?

Higher-level (inductive) questioning is answered by ascertaining purpose (whyness, reason). Of a whatness (atomic structure) we ask Why (what is the purpose of atomic structure?) and answer, to form molecules. Molecular structure is the (inductive) reason for atoms. It should be clear whether a reductive or an inductive answer is sought when asking whyness questions.

Of a whatness: we ask why. Of a whyness: we ask what. We have already ascertained what (i.e., the Realm of Whatness) by delineating the structure of reality. No purpose is served by answering whyness questions by reductive means. To do so would be redundant and would bring an end to seriously seeking answers to whyness questions.

To continue seeking understanding is to ascertain yet higher-level categories in a hierarchical system. Here, we speak about the hierarchical system representing the structure of reality. We are seeking explanations that "sufficiently" answer the whyness question. Let's look at our diagram (Figure 13:2.)

Realm of Reasoning Diagram

The Realm of Reasoning Diagram is simply described geometrically as a wagon wheel: spokes and a wheel. What does the wheel represent? What do the spokes represent?

Every Bi-Deltic Cosmos Duration (BCD) has one spoke "connected" to it. Each "spoke" (of the Realm of Reasoning Diagram) connects to one "BCD-dot" on the Cirpanic Cosmos "circle." How many spokes are there? There are as many spokes as there are Bi-Deltic Cosmos Durations.

Each spoke represents a BCD, not as a necessarily completed cycle, but as potentia. Where the BCD (as a dot on the Cirpanic Cosmos Diagram) is a cycle that has manifested or has yet to manifest, the spoke represents the BCD "prior" to its manifestation: whether it has earlier manifested or not.

The Cirpanic Cosmos represents "every possible eventuality" which can manifest the Realm of Whatness. The Realm of Whatness is composed of innumerable Bi-Deltic Cosmoses, i.e., Bi-Deltic Cosmos Durations (BCD's). It has been stated some of these durations represent completed cycles. One represents a partially completed cycle—our present Bi-Deltic Cosmos. Other durations have not occurred and are said to be "potential." Therefore, from a "lower level" perspective, every cycle, past, present, and future is represented. The BCD's are represented by dots on the Cirpanic Cosmos Diagram. The Cirpanic Cosmos Diagram represents the greatest possible cycle, one that is, manifestly, forever incomplete. Any and every possibility is represented in this structure. This greatest "cycle" (the outer wagon wheel) has been labeled the Cirpanic Cosmos—Omnipresence (CCO). Minor note. When we say the Cirpanic Cosmos is forever incomplete, we state this from a reductive perspective. This idea is remedied only when we sufficiently understand time. This awaits the final Section.

To acquire some depth in understanding this section, we survey a history of what scientists have learned about certain phenomena. Many experiments and their results are mentioned.

History of Light Investigation

Scientists attempt to group related phenomena by understanding their causes. This section is designed along historical lines to convey how scientists developed methods enabling them to pry information from reality. Various experiments are simply described. Their results are often combined (categorically) into classes (of information). The order of experiments and discoveries will be shuffled to an extent allowing the smooth development of generalities.

During the fourth century BC in ancient Greece, a mathematician named Euclid postulated light propagated in straight lines. From that he got the idea the angle of light incidence was equal to the angle of reflection. One of Euclid's contemporaries, Erasmus, understood light emanates from a source, reflects from objects and produces sensation.

In the middle of the first century AD, Ptolemy, the Alexandrian mathematician-astronomer, understood light was refracted when traveling from one medium to another of different density. We observe this phenomenon when we see that a straight stick appears bent when poked into water.

In the eleventh century the Arabian scientist Alhazen suggested light is an objective constituent separate from our minds. He also said illuminated objects reflect light in all directions.

Interest in light increased substantially in the seventeenth century. In 1604 the German astronomer J. Kepler suggested that vision is a device which "paints" inverted images of objects corresponding point for point to some "original" object. Most philosophers following Kepler considered light to be another aspect of a mechanical universe.

In 1621 the Dutchman W.R. van Snell derived the sine law of refraction stating the angle of incident light to refraction is constant. A contemporary, the French philosopher-mathematician R. Descartes, independently discovered the same idea.

Another Frenchman P. de Fermat soon continued Descartes studies. Fermat demonstrated an idea called the principle of least time. He said light rays follow a path that requires the least amount of time for light travel. His demonstration and interpretation were based on the idea light moves slower in dense transparent media. He stated the speed difference of light between two media is inversely proportional to their indices of refraction.

Nature of Light—Wave or Particle?

The Italian physicist F. Grimaldi experimented with objects placed in narrow light beams. He observed light and dark bands surrounding the edges of their shadows. He observed the phenomenon of diffraction. The English philosopher-experimentalist R. Hooke independently discovered the same phenomenon. He also suggested light exhibited an oscillatory nature and is propagated through wave fronts.

The Danish physicist E. Bartholin discovered the polarization phenomenon. He found certain crystals could split light beams. This phenomenon was inconsistent with the notion light was a longitudinal wave. Experimental results suggested light is transmitted in a transverse wave moving in a plane.

The second half of the 17th century was a significant era for investigations of light. The English mathematician and natural philosopher I. Newton experimented with prisms. He discovered light to be more complex than earlier thought. He resolved a beam of sunlight into a spectrum of colors matching the rainbow. He showed the heterogeneous nature of "white" light. He also used other prisms to reconverge the split light—reproducing white light.

Newton discovered that a one-sided convex lens, when placed down on a glass plate, generates bands of concentric rings appearing around the center of the lens. He attributed this phenomenon to "fits" of reflection and transmission. He opened the door to the idea light exhibited periodic character or phases. Although this experiment supported the wave theory of light, Newton realized such a theory would not then explain the rectilinear propagation of light.

Newton found different colors are closely linked to varying angles of light refraction in a given transparent medium. This and the fact light travels in a straight line were then more easily explained by the particle theory of light. Such a theory asserts light is composed of minute bodies or corpuscles. Proponents of this theory tried to explain all light phenomena by varying the character of the corpuscles.

The latter part of the seventeenth century also produced another brilliant investigator of light phenomenon, C. Huygens of Holland. He thought the particle theory of light could not be varied enough to explain certain phenomena. He wondered how the sun could continually emit particles and not shrink in size. He reasoned if light is corpuscular, there should be an apparent collision between two crossed beams of light. In experiments he observed no obstruction of either beam by the other. Huygens became the leading proponent of the wave nature of light.

Huygens developed the concept of the wave front. It ended the apparent contradiction between the wave theory and rectilinear light propagation. He said from the light source is spread a spherical envelope upon which any point could be considered a new light source that collectively forms secondary fronts or wavelets. If a line is drawn from the light source through the first wave front, it will also intersect the secondary fronts at right angles. This reconciled the wave theory with rectilinear propagation. Citing refraction experiments, Huygens also predicted light velocity was different for all transparent media.

Neither the wave nor particle theory had yet accounted for the recent discoveries of diffraction, periodicity, and polarity. In 1803 an English physician named T. Young endeavored to understand phenomena that Newton and Huygens failed to explain.

Young made a screen with a narrow slit. Beyond the first screen he placed a second screen having two narrow slits. Beyond the second screen was placed a third screen. He carefully positioned a source of light in front of the first screen. Light diffracted through the slit in the first screen and spread outward toward the second screen. Light passed through the two slits in the second screen—diffracted and spread outward toward the third screen where equally spaced fringes of light appeared.

The symmetrically placed slits in the second screen operate as sources of coherent light. Alternating bright and dark fringes appeared on the third screen.

Light emanating from the two slits in the second screen were "in phase" and hence coherent. The two light sources either reinforced or canceled each other. At the equidistant point (from the two slits in the second screen) on the third screen the resultant light was in phase thus reinforcing the fringe. It appeared bright. To either side of the center fringe appeared a comparatively dark band indicating one source was out of phase with the other. These bands were one half a wavelength closer to one slit than the other. Beyond these bands Young again saw bright fringes.

The pattern continued alternating between bright and dark bands. Bright fringes correspond to even multiples of the wavelength coming through the second screen. Dark fringes relate odd multiples of half wavelengths. Where Huygens had used pulse waves, Young's apparatus design allowed continuous periodic waves to illuminate his screen. Young could calculate a single wavelength of light by choosing certain interference bands.

In 1815 a French physicist, A.J. Fresnel, improved upon Young's results. He was able to account for all diffraction phenomena by relating fringes to wave interference caused by wave fronts spreading outward from the edges of diffracting objects.

To explain polarization, Fresnel suggested sunlight propagated in all planes parallel to the direction of motion but vibrated in planes perpendicular to that direction. He said polarization results from one plane being separated from all the others. The accumulated findings of several other investigators firmly established the "transverse" nature of light.

Light Velocity

An interesting avenue of investigation focused on the velocity of light and its means of transmission. Alhazen had thought light was transmitted instantaneously. Later, refraction and other phenomena showed light propagates at a finite speed.

O. Roemer, a Danish astronomer, studied at the Royal Observatory in London. Some of planet Jupiter's moons had just been discovered. Roemer noticed the "observed" period of one Jupiter moon (Io) systematically changed. When the earth is closest to Jupiter, the moon was eclipsed by shadow sooner than when earth was farthest way.

Roemer correctly attributed the apparent period difference to the varying distance of light transmission. He compared the period difference to the variation in distance. In late 1675 he announced his finding that the speed of light was 214, 000 Km/sec. Though noble, his determination was based on inaccurate information. The figure he used for the distance of earth's orbit as well as the moon's periods was inexact.

In 1728 the English astronomer J. Bradley published how to determine light velocity based on the annual shift of 41 seconds of arc in the apparent position of fixed stars. This is the parallax effect. Bradley's findings confirmed the finite velocity of light. Placing today's figures for earth's orbit into his calculations, yield an appreciably close approximation of light speed.

In 1849 the Frenchman A.H.L. Fizeau obtained the first reasonable laboratory measurements of light velocity. He used the teeth of a rotating cogwheel to interrupt a beam of light (passing between the teeth) aimed at a mirror placed 8.63 kilometers away. The intensity of returned light is minimal when stopped by teeth in an intervening position. Light intensity is maximal when it returns through a space between teeth.

At a certain rate of rotation a beam sent out through one space (between teeth) would return through another. Using the number of teeth and rotation speed, Fizeau established light velocity to be 315,300 Km./sec.

In 1850 Fizeau's co-worker J.L. Foucault, using a similar idea, determined the velocity of light using rotating mirrors. The rotating mirror displaced the returning light beam. The mirror rotation rate and displacement angle enabled him to arrive at the figure of 298,000 Km./sec.

Using the same method, Foucault determined the velocity of light in water was considerably less than in air. He thus verified Huygens' prediction of two hundred years before. Variations of Foucault's experiment have since been carried out quite regularly with ever more precision.

Medium of Light Transmission

We briefly reviewed some early experiments designed to determine light velocity. We now describe a controversy concerning the transmission of light. During most of the 19th century scientists thought light moved in a medium. Water waves occur on water. Sound waves are actuated by vibrations of air. It seemed that light waves would need a medium of transmission.

The supposed medium was called the luminiferous ether. If the ether existed, it should extend throughout space. The ether should be detectable because the earth moves through space.

It was reasoned there should be a difference in light speed between the direction of earth's motion and the direction perpendicular to its motion. The difference corresponded to what was called the ether wind. The "wind" supposedly arises from earth's motion through the ether (inertial space).

In 1887 Americans A. Michelson and E. Morley designed an experiment to detect the difference. Using an interferometer (developed by Michelson) the experimenters caused the incoming light beam to split into two mutually perpendicular beams. The two beams are made to come together again and interfere with each other. The apparatus was first aimed in the direction of earth's motion. An interference pattern was observed.

The equipment was then rotated 90°. If the ether wind existed, there should be a time difference in light travel between observations before and after rotation. A time difference would be shown by a shift in the interference pattern. The phase difference, although calculably small, should have been observed in the Michelson-Morley experiment. No phase interference was found. Some scientists suggested the null result could be explained away if the ether wind travels with the apparatus.

In 1902 the Dutchman H.A. Lorentz and G.F. FitzGerald, an Irishman, suggested the null result could be explained if the apparatus shrank in the direction of motion. They said shrinkage could have matched the phase shift caused by the ether wind. If this occurred, double refraction would not be the same in all directions. This experiment was carried out with no observation of change. It seemed light did not need a medium for transmission.

Maxwell's Equations

Another significant development occurred before Michelson and Morley's famous experiment. In 1855 the Scot J. Maxwell decided to attempt mathematically to state M. Faraday's idea of "lines of force."

An electric charge has a free field pointing in all directions from its center. A magnet is surrounded by a force field. Electric current generates a magnetic field along with an electric field. Electric charges and magnetic forces act upon one another.

Maxwell wanted to systematically describe electric and magnetic phenomena. He used vectors to represent the magnitude and direction of fields. In 1864 he succeeded in mathematically generalizing Faraday's ideas. Maxwell's differential linear equations describe certain electric and magnetic properties that are proportional to one another.

His first equation is a restatement of Gauss's Law (which followed Coulomb's law). It states flux lines representing the electric field around a charged body is constant for all distances. Therefore, the field strength (lines per unit area) decreases as the square of the distance.

The second equation, a reformulation of Faraday's law, states the flux line, representing the magnetic field, forms closed loops. Hence, there are no isolated magnetic charges.

The third law generalizes Faraday's law of induction. It states for a given period along a closed loop; the electric field is proportional to the magnetic flux passing through it.

The fourth law generalizes Ampere's law. It is a magnetic analogue to Coulomb's law of electric charges. It states that during a given period the magnetic field in a closed loop is proportional to two quantities. One is determined by the increase in electric flux, and the other is determined by the electric charge passing through the loop. When voltage is applied to the plates of a capacitor, the effective magnetic field can be attributed to either the charge in the electric field between the plates or to the current in the loop.

The first two equations copy Coulomb's law for charges at a fixed distance. The third and fourth equations treat electric and magnetic effects as two separate aspects of one phenomenon—electromagnetism.

Maxwell's fourth field equation implies magnetic effects are produced not only by electrical currents in conducting material, but also by electric induction in nonconducting material. Although his first three equations restate, albeit precisely, what was already known, the fourth equation hypothesized something new. It asserts nonconducting substances can produce magnetic effects. This equation led to the theory electromagnetic waves can propagate through a vacuum (empty space). No luminiferous ether was needed to explain this.

Maxwell's equations predict a phase velocity is propagated in vacuum at a constant velocity (c). (c) is calculable using two nonkinematic parameters: the magnetic susceptibility of the vacuum and the dielectric constant.

Maxwell discovered the value for (c) was practically identical to the value for the velocity of light. Simple reasoning dictated light was an electromagnetic wave. Maxwell predicted visible light was only a part of a wide band of electromagnetic radiations.

Maxwell's ideas were soon tested. They easily explained polarization. In 1845, Faraday demonstrated that there was a connection between magnetism and light. He caused polarized light to rotate as it passed through a transparent substance placed in a strong magnetic field. In 1895 another German physicist, W. Roentgen, discovered high-frequency radiation now called x-rays. Maxwell's predictions were verified.

Einstein's Contribution

Maxwell's equations were not symmetrical for moving objects and did not explain the results of Michelson and Morley's experiment. Einstein questioned Lorentz and FitzGerald's transformation rules concerning object contraction in the direction of motion. He recognized time was closely associated with the velocity of signal propagation. He thought Michelson's null results suggested there is no absolute frame of reference. Coupling that with the idea Maxwell's and Lorentz's equations were also valid in a moving frame of reference led him to postulate the invariance of light velocity. He applied these ideas and derived the Special Theory of Relativity five weeks later.

In 1905 Einstein published his Special Theory of Relativity. It implied a medium at absolute rest does not exist. (Special Relativity has been described in previous chapters.)

Without an absolute frame of reference the idea of ether wind is meaningless. Einstein said physical phenomena only depend on relative motions. He concluded the speed of light in vacuo was independent of the motion of light sources. This explained the null result of Michelson and Morley's experiment. It was determined the speed of light in vacuo is constant for all observers regardless of motion.

History of Quantum Theory

Black Body Radiation

The year 1890 witnessed the auspicious beginning of an idea that revolutionized scientific thought. Scientists had studied what is called black-body cavity radiation. Black body radiation is one in which radiation is in equilibrium with matter (the body) at a particular temperature. The body completely absorbs all incident radiation (all wavelengths) just as would an ideal black object. At thermal equilibrium the energy distribution of such an object matches radiation given off by an ideal black body; hence, the name "black body" radiation. Note the object need not be black. It should only behave as though it were black; that is, it should be a perfect absorber and emitter of radiation.

All objects emit radiation. For example, even at room temperature a piece of coal emits radiation although it is invisible and occurs in the infrared region. If coal is slowly heated to 950 Kelvin it begins (glowing) emitting energy in the visible spectrum. It first takes on a red hue, then orange and yellow. Heating to several thousand degrees causes it finally to appear blue-white. At this temperature coal also emits ultraviolet radiation. It is a task for scientists to examine this phenomenon under controlled conditions.

W. Wien of Germany conceived a clever way to get an object to behave like an ideal-black object (one not reflecting impinging electromagnetic energy). A sealed opaque box with a small orifice (about 1.5 millimeter) in one side can serve as a practical black body. The hole allows electromagnetic energy to enter the container but only lets enough out to be detected. Such a setup does not effectively disturb the energy equilibrium in the cavity.

Radiation (light) enters the hole (in the box) reflects off the inner walls and soon becomes absorbed (into the walls of the box). The radiation eventually settles into thermal equilibrium with the walls at which point radiation absorbed (into the wall) is balanced by radiation emitted. Once this happens, radiation (frequency distribution) in the cavity characterizes the wall temperature. The quantity of cavity energy depends only on the temperature of the walls.

Some radiation comes back through the perforation and can be detected by a spectroscope. The radiation detected is the same as that in the cavity.

The detected radiation (light intensity) for a given temperature, is graphically plotted against the frequency (or wavelength). Numerous readings yield a spectral distribution curve exhibiting a maximum value corresponding to the middle wavelengths. This is the black-body spectrum.

The Black-body spectrum was in need of explanation. If the temperature is raised, there is a substantial increase of energy in equilibrium. This is reflected in the spectral energy distributions for various temperatures.

A law of classical mechanics was first used to understand black-body radiation. Energy was considered exchanged by oscillators in the walls of the cavity. (An oscillator is a charged particle that vibrates thus emitting electromagnetic energy.) Classical theory (statistical mechanics) suggested all frequencies of the oscillators (or modes) had to be of the same energy KT (K = Boltzmann's constant, T = temperature). This was an application of the equipartition theorem that associated a specific energy (KT) with each mode. This is known as Rayleigh—Jeans Law.

Classical thermodynamics led to unobserved conclusions. The Rayleigh—Jeans Law predicts the wavelength decreases to zero as the number of modes approaches infinity. The cavity could thus contain a limitless quantity of radiation energy. Ultraviolet light, in an enclosure, could then be associated with an endless quantity of energy. The Austrian P. Ehrenfest said this was absurd and called it the "violet catastrophe" because classical calculations relating the behavior of highly heated objects show all its energy would explode in a catastrophic release of (ultra) violet radiation.

Although the observed-spectral-energy distribution of black-body radiation agreed classically for longer wavelengths, it did not concur with that for shorter wavelengths. Observed spectral distribution for the shorter wavelengths did, however, coincide with Wien's Displacement law stating that maximum radiation intensity is inversely proportional to the temperature.

Planck's Realization

In 1990 the German, M. Planck, examined Rayleigh's assumption that all frequencies radiate with equal probability. Because experiment disclosed otherwise, Planck began searching for a representative formula. He found a formula that closely matched Rayleigh-Jean's law for longer wavelengths and Wien's Law for shorter wavelengths.

Essentially, his formula is the following:

$$E = nhf$$

E is the average energy

n represents an integer, equivalent to 0, 1, 2,...

h is Planck's universal constant—the quantum of action

f is frequency of oscillation: low frequencies equate with long wavelengths, high frequencies equate with short wavelengths

Either extreme of the (bell shape) distribution curve approaches zero energy. This indicates very short wavelengths (very-high-frequency oscillators) and very long wavelengths (very-low-frequency oscillators) have little chance of emitting or absorbing energy.

A graph for a given temperature could be calculated from his equation. It agreed with observation thus removing the ultraviolet catastrophe imposed on black-body radiation by classical theory.

Planck wondered why the probability of radiation decreased when there was an increase in frequency. Studying his formula he realized a given mode (allowed frequency of vibration) could not take on continuous values of energy. That limited possible values to discontinuous units of energy.

Discontinuous-energy values can be likened to the rungs of a ladder. Varying the spaces between rungs can represent different modes. The distance between rungs is proportional to the frequency of the oscillator. For each wavelength of light in the cavity a certain group of oscillators will respond by vibrating thus taking on energy.

Oscillatory excitation is not induced by just any energy quantity. Excitation occurs only if the oscillator is given its "due amount of energy." The amount varies with the frequency of the mode.

Let's reassess what has been stated. Planck said modes correlate with a quantity of energy (E) that is an integral multiple of an irreducible energy unit called the quantum. The quantum has the value (hf) where (f) is the frequency of the oscillator and (h) is an experimentally determined universal constant (Planck's constant) equal to 6.626 x 10^{-27} erg-second. For angular circumstances, 1/2 π (i.e., 1.054 x 10^{-27} erg-second).

Planck found by inserting the value of (h) into his equation that it would mathematically represent the observed black-body spectrum. He called the constant the "quantum of action."

Planck stated that modes with shorter wavelengths, taken as an average, correspond to such a small quantity of energy that although the number of modes may increase considerably, the total energy would not approach an infinite value. In practical terms, this means there is insufficient thermal energy in the cavity walls at room temperature to excite the higher-frequency modes. This results in a damping effect on the higher-frequency oscillators. Planck's formula represents this by introducing a non-zero maximum value for radiation intensity in the middle frequencies.

Essentially, Planck originated the idea energy is only exchanged in discrete quantities called quanta. His idea suggests energy is packaged in small bundles not unlike that depicted in the corpuscular theory of light. For the time, however, wave theory continued to be the best explanation of light phenomena.

Einstein Extends Quantum Concept to Light

In 1905 Einstein extended the quantum interpretation to cover electromagnetic radiation itself. He studied a phenomenon called the photoelectric effect. Electrons are emitted from polished metal when it is illuminated with a light beam of sufficient energy.

Early experiments used ultraviolet light to irradiate negatively-charged-metal plates. Classical theory suggested light energy would be uniformly distributed over the entire impinging wave. If that were true, electrons near the plate's surface would be ejected regardless of frequency. Observations revealed that below a certain frequency (characteristic of the metal) electrons did not eject despite increasing beam intensity.

Einstein realized if light were quantized with energy (hf) that it would explain certain observations. First, in the photoelectric effect, energy of ejected electrons varied with the frequency of incident light. Second, if the intensity of incident light is increased the energy of ejected electrons is unaffected; however, the number of emitted electrons increases. Third, there is no accumulation of energy on the metal surface. Therefore, an electron may be spontaneously emitted when struck by a quantum of sufficient energy. Any energy above what is needed for electron emission (energy that binds the electron to the metal) is carried off kinetically.

Einstein concluded not only is light energy exchanged in discrete quantities (as proposed by Planck) but light itself is apparently transmitted by little packets of energy. The packet of light energy was given the name "photon."

Einstein's interpretation of the photoelectric effect clearly showed Maxwell's electromagnetic wave theory could not adequately account for all light phenomena.

In 1915 the American R.A. Millikan verified Einstein's formula for the photoelectric effect and calculated from the photoelectric experiments a value for (h). It closely matched Planck's value obtained from black-body radiation measurements.

A couple of years later, Einstein drew upon what he knew about Brownian motion and his understanding of the photoelectric effect and decided light must also exhibit the property of momentum. He said momentum (p) would be equal to Planck's constant divided by the wavelength ($p = h/\lambda$).

In 1923 the American physicist A. Compton observed highly energetic photons (x-rays) often increased in wavelength after passing through matter. Compton attributed the loss of photon energy to collisions with matter.

Compton found that increase in wavelength was related to the angle of scattering. It was assumed the photon collided "elastically" with a stationary electron. Energy and momentum are conserved in an elastic collision. Calculations easily accounted for the phenomenon if the colliding photon, in addition to energy, exhibited the property of momentum.

The Compton effect confirmed Einstein's assertion that photons do have momentum. It was verified momentum was related to the frequency of the discrete energy and momentum. Such conceptions helped explain many phenomena including the fact light exerts pressure.

Here are some added notes: In 1907 Einstein extended the use of quantum theory to explain how the heat capacity of matter varied with absolute temperature. He said atomic or molecular absorption of heat occurred in discrete quantities. It is interesting that the German G. Wentzel explained the photoelectric effect in 1927 without using quantum theory. Quanta began to affect the whole of physics.

Atomic Energy Levels and Bohr

In 1911 E. Rutherford, an English physicist born in New Zealand, from mathematical considerations, conceived the atom to be like a miniature solar system. He said its center was very massive and yielded a positive charge. Negative electrons orbit around the center.

Rutherford's model allowed electrons to travel in any orbit and could thus emit light spectra of any frequency. This was a defect not reconciled with spectroscopic observation. Another problem was posed by Maxwell's theory of electromagnetism. Classically, an orbiting electron would continually radiate energy and quickly spiral into the nucleus.

In 1884 a Swiss schoolteacher named J.J. Balmer became interested in the wavelengths of spectral lines emitted by hydrogen atoms. At the time nine spectral lines were known. Four of the known lines occurred in the visible region.

Balmer conceived a formula depicting the discrete lines as irregularly spaced rungs on a ladder. The distance (frequency difference) of the rungs could be calculated when a certain constant is included in the formula. Inserting different integers into the Balmer formula not only yielded results agreeing with four of the observed spectra, but also predicted other spectral frequencies. By inserting integer values between three and eleven into his formula, Balmer calculated the wavelength of the first nine lines of the hydrogen spectra.

In 1908 the Swiss scientist W. Ritz showed Balmer's idea to be valid for elements other than hydrogen. Clearly, the emission of distinct spectral lines showed the inadequacy of the Rutherford model.

In 1913 a Dane named N. Bohr, a student of Rutherford, realized distinct spectral lines were a clue to atomic structure. Using this clue and substituting electrons for Planck's oscillators he improved upon Rutherford's model.

Bohr used Newtonian mechanics for the atom but placed a restriction on orbits. He said allowed orbits were determined by whole number wavelengths of the electron. Put another way, he hypothesized the angular momentum of atomic hydrogen is quantized. He said the electron could only revolve in certain orbits that are integer multiples of angular momentum. Bohr replaced the constant in Balmer's formula with one having natural units. (Planck's constant h divided by 2π; therefore, $\hbar$).

When an electron jumps to a higher or more distant orbit, it did so only by taking on quantized bits of energy. If an electron dropped to a lower energy level, it released excess energy in discrete bundles of electromagnetic radiation.

Where Planck failed to account for discrete exchanges of energy, Bohr's model offered an explanation. Even the explanation departed from classical prediction. Where classical theory connects energy emission to specific orbits, Bohr's model relates the emission to differences between orbits.

Bohr's theory predicted several spectra series arising from energy transitions from the major-electron orbits (n). Putting integer values into Bohr's formula yields a series of spectral sequences.

The integer $n = 2$ initiates the Balmer series. That is, any electron transition from higher-energy states to the second-energy level gives off spectra represented by the Balmer series. Integer $n = 1$ gives the far-ultraviolet-Lyman series of which the first term was discovered in 1908 by T. Lyman. The integer $n = 3$ gives the Paschen series of which the first term was discovered in 1906 by F. Paschen. Bohr's theory predicted a spectra series for $n = 4$ and $n = 5$ that were discovered in the 1920's by Brackett and Pfund. Integer values greater than $n = 2$ represent spectra occurring in the infrared region.

Using Planck's ideas, Einstein's explanation of the photoelectric effect, Balmer's insight and improving Rutherford's model of the atom, Bohr explained atomic spectra.

Bohr's energy ladder idea of orbits was soon verified in the experiments of J. Franck and G. Hertz in Germany. Franck and Hertz heated a cathode tube filled with mercury vapor. Electrons, ejected from the cathode, accelerated toward a grid. Beyond the grid was a plate having a lower potential. The grid current slowly increased. At a potential of 4.9 electron volts (eV) the electrical current abruptly decreases. At this point electrons lose energy to mercury atoms and cannot complete the trip to the plate. The result is a decrease in current. This implies mercury atoms are excited to 4.9 eV above ground state where they emit light and return to the lowest state. The wavelength emitted is calculable from a formula. When the voltage exceeds 4.9 eV, the wavelength can be seen with a spectroscope. As current increases there is a drop in voltage every 4.9 eV. The Franck-Hertz experiment was the first nonoptical verification of discrete-atomic-energy levels.

Although Bohr's model worked seemingly well for the hydrogen atom, it did not fare well in explaining the spectra of more complex atoms. Bohr's original model was reworked repeatedly to fit experimental findings. This was especially true for information garnered from refinements in spectroscopic technique.

In 1896 a Dutchman named P. Zeeman had already shown spectral lines (like the yellow sodium D line) broadened when (sodium) atoms were placed over a flame in a steady magnetic field. The Zeeman effect also reinforced the idea electrons were negatively charged particles.

Later other scientists using more sensitive equipment showed individual spectral lines could be further split into groups of three or more. In 1913 the German J. Stark used electricity instead of magnets to split individual spectra even more. It is interesting the Stark effect was discovered by observing spectra emitted in the Balmer series of hydrogen. It supported Bohr's model of the hydrogen atom. More quantum numbers were needed to account for these phenomena.

In 1915 the German, A. Sommerfeld, used relativity effects to mathematically explain the so called fine structure splitting of spectra observed with powerful instruments. His theory provided this explanation by taking the angular-momentum quantization condition described by Bohr's circular orbits and changing them into elliptical orbits.

One problem kept asserting itself. The Bohr theory not only explained the observed spectra but also predicted many more lines never observed. In 1918 Bohr heuristically framed what was later termed the "Correspondence Principle" to explain away certain failings in his theory. In simple terms it states that as distance increases from the atomic nucleus the orbits become ever larger with the difference of their energies becoming smaller.

When an electron jumps from one of these "other" orbits to an even higher orbit, the energy change approaches zero. For these "large" jumps energy change should be equivalently calculable from either quantum or classical theory. Their predictions should "correspond." Energy changes quite smoothly in the larger orbits and can therefore be explained classically. Thus the frequency of emitted radiation should equal the orbital frequency of the electron in those orbits. Such reasoning pointed the way to calculate the energy of the electron. By appealing to this throw back to classical theory Bohr was also able to dispose of unwanted spectra predictions in his theory. It also enabled him with some success to predict spectra intensities.

Yet, serious discrepancies between theory and experiment remained. When a fourth quantum number was needed to account for "spin," the Bohr theory had outlived its usefulness (meaning of the quantum numbers will be described in the latter part of this chapter). A new theory was needed.

De Broglie's Matter Waves

Amid interest in Bohr's theory the conflict between the wave and particular nature of light was nearly neglected. There had developed a stand off with major support for the wave theory being interference while the photoelectric effect pointed to the corpuscular theory.

In 1923 the Frenchman L. de Broglie, suggested, because light displayed both wave and particular properties (based on Einstein's equations for energy and momentum of light quanta) that matter too may exhibit wave characteristics. He specifically was thinking of electron waves.

De Broglie had noticed the role integers play in Bohr's theory and recalled how they aided understanding wave interference and other classically treated phenomena. He soon realized electrons in motion around an orbit met the conditions for the classical-standing wave. The condition is that an integral number of half wavelengths can be induced, for example, by plucking a tautly fastened string like those on a violin. The same phenomenon occurs as discrete overtones in the air column of a pipe organ. The waves take on a fixed pattern.

Standing waves can also be treated as a superposition of two (sine) waves of equal wavelength and frequency moving in opposing directions. This means two solutions also yield a solution. (We will shortly take note of a use for wave superposition.) Because standing waves result from integral-resonance frequencies, de Broglie said these frequencies were equivalent to Bohr's requisite for fixed angular momenta (that orbits are quantized and endowed with discrete quantities of energy).

The mass-energy equivalency of Special Relativity states that a given amount of mass is associated with a particular energy. A given amount of energy can be related to a certain frequency. De Broglie said symmetry considerations suggest that a particle be associated with some oscillatory phenomenon.

When periodicity is measured for the rest frame of the particle, it differs from that measured by a moving observer who must include the time-dilation effect in his calculations. De Broglie believed he solved the conflict by saying that according to the observer a wave extends from the particle in its direction of motion. The wave keeps in phase with the particle's oscillation. He called this a phase wave.

The wave must travel at speeds greater than light for the particle's periodicity to stay in phase with its extended wave. Phase wave velocity depends upon the relative motion of the observer. This interpretation was De Broglie's attempt to affix a deterministic meaning to his theory. So what did De Broglie do?

He considered the relativity effects on an oscillating particle of specific frequency traveling in uniform motion. He derived a formula representing its phase (wave) velocity.

Given the same velocity an object of greater mass has a correspondingly shorter phase wavelength. The phase wavelength and its velocity are infinite for an object at rest. Here the phase is the same everywhere since there is no phase difference unless there is spatial displacement.

For an accelerated object, mass increases without bound and the associated phase wavelength decreases without limit.

Since conveyance of information (via some energy vehicle) is not involved it does not present a relativity problem. No energy is transferred, and there is no propagating medium.

Little can be learned from one phase wave. Even phase wave velocity is unobservable and seems to have no simple physical significance. Theoretically, the relativistic increase in an object's mass has been attributed to the increased frequency of the de Broglie phase wave.

Understanding phase waves is a basis for deeper insight. Consider the coherent superposition of two phases differing in frequency and wavelength, and we have a more significant phenomenon. The superposition of the particle's constituent waves often results in a modulating wave or pulse. The pulse is composed of many individual phases of different wavelengths that have collectively interfered.

Individual phases of varying wavelengths travel at different velocities. The pulse generated by the phase group has a velocity of its own. This group of de Broglie wave pulses is called a wave packet. The wave packet is detectable and travels in a group velocity equal to the speed of the particle. The packet represents the action of the matter-wave.

As we have seen, de Broglie adapted his electron-wave theory to atomic orbits by stating the wave should be in phase with itself and fit on the orbit in integral multiples. De Broglie also thought free electrons, like photons, would be guided by waves and suggested this could be detected experimentally. He said a small enough aperture should diffract a beam of electrons and produce interference.

In New York City at a telephone laboratory, C.J. Davisson had conducted certain experiments since 1921. He reflected electrons off metal. In 1925 Davisson and L. Germer, a co-worker, were shooting electrons at a piece of crystal nickel in a vacuated flask. When the metal was very hot, a liquid air bottle exploded wrecking the apparatus. This allowed air to rush in and oxidize the nickel surface. They cleaned the metal surface by heating. Unbeknownst to the experimenters the atoms were "heat treated" in the process and were rearranged into comparatively larger crystals.

The spacing between the atoms in the crystal provided the small aperture that de Broglie had suggested. They expected the electrons to bounce off the metal the same for all angles of incidence. They continued the experiment and discovered the distribution angle had changed. On photographic plates diffraction patterns developed similar to the Laue patterns (after work by the German physicist M. van Laue) produced by x-ray diffraction. They also found the angle of reflection would shift with a change in the velocity of the impinging electrons.

These phenomena eluded explanation until Davisson in 1928 attended a meeting of physicists in London where he heard about matter-waves. The wave behavior of the electrons agreed with de Broglie's formulae. The angles of electron scattering coincided with that calculated from the wavelengths of their matter-waves.

Matter-waves are calculable from an object's mass, velocity (momentum) and Planck's constant. It was previously mentioned that matter-waves travel at a rate equal to an object's velocity. They travel in pulse form in a wave packet. A wave packet in motion results from the superposition of waves (the algebraic addition of two or more waves). Energy associated with the pulse is related to the form of the standing wave. The more massive the object, the faster it moves, the shorter is its matter-wave.

Later experiments confirmed matter-waves for atomic hydrogen, diatomic hydrogen, alpha particles (helium nuclei) and neutrons, as well as whole atoms.

For objects of everyday life no aperture is small enough to cause matter-waves to diffract and interfere. The waves are too short. Theoretically, objects large or small do behave in accord with wave propagation. Motion of large objects, for all practical purposes, is identical with that described by Newton's laws.

Quantum Mechanics

During the 1820's the Irishman W.R. Hamilton revamped Newtonian mechanics. He expressed Newton's laws in a way not calling for description using a specific coordinate system. He pointed out the value of p's (mass x velocity) and q's (indicating position, for example, by displacement from an equilibrium position) in calculating mechanical effects. His mathematical insights influenced Planck, Bohr, and many other scientists. Hamilton's work was again to leave its imprint on scientific thought.

In 1925 the German physicist W. Heisenberg developed the foundation for what later became matrix mechanics. He wanted to build an atomic theory derived from relations between observables. He intended to improve Bohr's Correspondence Principle and offer a better explanation for spectral intensities.

Using matrix algebra Heisenberg tabulated the quantized states of the atom as elements in matrices. The tabulations take a form similar to the square or triangular chart on road maps showing distances between different towns. The matrix itself represents the mechanical properties of the atom.

Heisenberg made matrices for the p's (momentum) and q's (position) of atomic hydrogen. He multiplied $p \times q$ and got his square tabulation. The resulting table conjointly represented the position and momentum of the particle.

It was soon shown the matrix quantity $p \times q$ was not equivalent to $q \times p$. The multiplication was non-commutative. Matrix algebra is operatively different from that used with common numbers. M. Born a contemporary German physicist, informed Heisenberg of a matrix calculus developed in 1858 by the English mathematician A. Cayley. Heisenberg had rediscovered it. M. Born and fellow German P. Jordan undertook Heisenberg's problem and found the difference between $p \times q$ and $q \times p$ was $h/2\pi i$, $(i = -1^{1/2})$.

Dirac found an equation representing the matrices Heisenberg had developed. From Dirac's equation Heisenberg's square tabulations could be generated.

Dirac found a way to adapt classical equations to quantum mechanics. He used Poisson brackets (discovered by the Frenchman S.D. Poisson about 1834). He said we need only write the equation in classical form and apply a different interpretation to the Poisson brackets. Calculate the Poisson brackets using classical theory and multiply it by $h/2\pi i$. The results solved the problem vexing Heisenberg and his collaborators. It gave the difference between $p \times q$ and $q \times p$.

Abstract as Heisenberg's theory was, it soon was shown by W. Pauli, an Austrian physicist, to yield the Balmer frequencies and relative intensities for the hydrogen atom. It was, however, complex and difficult to work with.

Schrödinger's Wave Function

During the same period Heisenberg was trying to improve Bohr's Correspondence Principle, another physicist approached atomic structure using de Broglie's matter-waves. The Austrian E. Schrödinger in 1925 developed a logically consistent theory of matter-waves.

Schrödinger advanced a partial differential equation showing how matter-waves behave through a specified time. He expressed de Broglie's idea that orbit circumference divided by an electron's wavelength must equal a whole number.

Schrödinger calculated wave patterns for the hydrogen atom. Multiplying the frequencies of the wave pattern by Planck's constant again yielded Balmer's formula.

Schrödinger had generalized de Broglie's equation with operators. (In classical mechanics the Hamiltonian function specified the total energy of a system using p and q variables.) An operator is a symbol giving a mathematical command. In quantum mechanics operators represent observables and must be real (have the property of hermiticity).

Schrödinger rediscovered in wave theory what Dirac had discovered with particles. Heisenberg's idea is inherently implied in Schrödinger's theory. Square tabulations can be generated from the Schrödinger equation.

Schrödinger's equation concerns the magnitude (amplitude) of matter-waves. This property is represented by the Greek letter ψ (psi). It is commonly called the wave function (or state function). The wave function describes the state of the system. Anything we can learn about a system is contained within the wave function.

Interpreting the Wave Function

Despite rapid mathematical progress in developing quantum mechanical equations, confusion reigned in interpreting them. Schrödinger had suggested his wave function represented "smeared out electrons."

After a suggestion by Einstein, M. Born in 1926 formally showed the wave function to be a measure of finding an electron in a particular location. Schrödinger's wave function squared (functions as intensity) was soon rendered in terms of the probability (per unit volume) of an electron being in a given region (more about this later). That is, finding an electron within the "packet" is proportional to the square of the amplitude at that point. Let's take a closer look at the meaning of the Schrödinger equation.

Unlike Bohr's distinct orbits Schrödinger's solutions are limited to sequences of discrete wave-patterns. His equation yields a series of patterns each associated with a specific energy.

Close inspection reveals Schrödinger's equation yields many solutions, but a stipulation called the normalization condition restricts the number of possibilities. The wave function must be finite (the particle must be somewhere). The probability must yield single values everywhere, and the wave function must be continuous (the probability of particle location cannot vary discontinuously). These restrictions generate a representation of energy quantization.

Energy quantization can be explained with the standing wave condition used by de Broglie. In simple fashion we will explain what this means for a particle in one dimension.

Discrete patterns emerge from Schrödinger's equation by considering a particle's wave-pattern (in sinusoidal form) to be confined by an imposed barrier. A waveform is calculated for a particle in a "box" (infinite square well). The sides of the box are said to be infinitely thick so that any recoil of the walls will not influence an electron's motion. At the walls of the box the wave amplitude (ψ) must be zero (location of a node). There is no probability of the particle or wave pattern extending beyond the confining walls. (Again, this is done mathematically by "normalizing" the equation.) The "particle" must be somewhere in the box. Only certain wavelengths (of the particle) are allowed for a given length between wall barriers. That is, only an integral number of half-wavelengths can fit between the walls. When the same idea is applied to atomic orbits, the number of half-wavelengths corresponds to the principal quantum number (n).

A resultant wave can be created by the interfering mix of more than one wave traveling back and forth between the wall barriers. This superposition of waves constitutes the standing wave. It can have any whole number of nodes. Its wavelength depends on the number of nodes. The number of nodes can only change by whole integers. Thus, change can only occur discontinuously. Even if a standing wave is a complex mixture, it can still be analyzed into simple standing waves. So what does this have to do with particles such as electrons?

How is information obtained from the Schrödinger equation? First, the equation is linear differential in form. As earlier stated, the solution to the equation is called the wave function. The wave function, as stated by M. Born, represents the distribution of the particle (probability of finding it in a given region) or wave field.

All information about the system is contained within the wave function. Observable properties can be calculated by performing appropriate operations on the function. The equation is unsymmetrical in space and time. The first derivative (slope) of the function represents the particle's momentum. The steeper the slope, the greater the momentum. The second derivative (mean curvature) is proportionately linked to the system's kinetic energy. The magnitude of curvature increases when potential energy is exceeded by the total energy. The swinging of the function back and forth across the axis describes a harmonic wave, hence the representation of matter-waves.

Successes of the Schrödinger equation were immediate. It explained the varying intensities of different atomic spectra. Acceptable solutions describe how electron transitions occur. The theory mathematically explained the Zeeman and Stark effects. It was even shown by Dirac to explain the spectrum of helium.

Nature of Matter Waves

A simple experiment demonstrates the wave nature of particles. A two-slit interference apparatus can be set up using electrons beamed through suitable apertures. The intensity of the beam is reduced until only one electron at a time passes through the chamber between the slits and the photographic plate.

There is no predicting where the electron will "mark" the film. The best we can do is to consider the electron statistically and predict a probable mark (we use the superposition principle to calculate the probability from the waveform coming through each slit). The square of the absolute magnitude of the matter-wave is proportional to detecting an electron in a given region. (This is analogous to light intensity being proportional to the square of its amplitude.) Over a couple of months more electrons impose dots on the film. After sufficient time a pattern shows a distribution of dots in accord with wave theory. That is interesting. Particles aligning in a wave-interference pattern! Probability assures us that the dots will be found most where constructive interference occurs (where solution yields greatest amplitude) and least where destructive interference occurs (where minimal amplitude exists).

The emergent pattern does not depend on beam intensity. Because the interference pattern develops after allowing only one electron at a time to pass through the apparatus we conclude each electron exhibits wave characteristics. The wave nature of electrons is the basis of the electron microscope.

In the preceding experiment there is another interesting effect. A single electron wave (like a photon) must pass through both slits to cause interference. It seems the "particular" (corpuscular) nature of the electron will pass only through one slit, but we cannot detect which one without destroying the interference pattern. So it is meaningless to say the "particle" only passes through one slit.

Heisenberg's Uncertainty Principle

Let's determine by gedanken experiment precisely where an electron will travel. Our experiment is patterned after one proposed by Heisenberg and used by Bohr.

We attempt to determine the electron's trajectory or momentum. We need to locate its position (q) using an ideal microscope. The microscope can use any wavelength of light. (We must use light to see the electron.) We need only to find the electron in the field of view. The electron is traveling in evacuated space. No outside influences initially hamper it while in flight.

We shine the light (Heisenberg suggested using gamma waves) at the electron. Photons exert pressure on the electron, deflecting it from the expected path. We try again by decreasing the intensity of light. To observe the electron at least one photon must hit it. Again, the electron is disturbed in flight if it is observed. Electron velocity will change and so will its momentum.

Whether we visually observe the electron with our extraordinary microscope or capture its image on film we have another problem. Because of diffraction effects the optical image of the electron cannot be smaller than the wavelength of light used to detect it. As we decrease the wavelength to obtain a sharper image; and therefore, better determine the electron's position (q) we simultaneously use photons with greater energy and disturb the electron even more.

Let's go the other route and use low-energy photons to minimize the effect of energy. We perturb the electron less, but its image, because of diffraction effects, correspondingly enlarges. This makes it more difficult to determine location.

Although the behavior of matter-waves is deterministic, the behavior of individual electrons only lends themselves to statistical treatment. This was the assessment of Bohr and Born. Unlike the world we experience, the microcosmic world of individual particles is unpredictable.

In the preceding gedanken experiment, we assumed the electron intrinsically exhibited precise momentum and position. If we measure p we disturb q; if we measure q we disturb p. The precise amount of disturbance for each is conjugately indeterminate. Simultaneously measuring the p and q of the electron introduced the uncertainty. What is the extent of uncertainty introduced in the best possible case? The limit to precise determination is governed by Planck's constant (h). The error in measuring either p or q is uncertain because $pq \neq qp$. The uncertainty is greater than $\hbar$, e.g. $\Delta p\, \Delta q \geq \hbar$. The uncertainty involved in describing the orbit of an electron is nearly equal to the size of the atomic radius. The uncertainty is inherent in the electron.

This problem bothered Heisenberg. He tackled the problem of refining the meaning of the wave function. To define it, he said one must consider the p and q of the electron (the same idea applies to any particle). This is what he thought:

Consider a particle moving in an indefinitely-extended-straight line. Because the particle can be anywhere on the line, its matter-wave also extends over the whole of the line. At any given instant, its associated matter-wave, for all practical purposes, will be localized and centered somewhere along the line within a region where we would expect to find the particle.

The localized matter-wave may be a superposition of many waves constituting a wave packet. Let's reexamine the wave packet. It is described by a wavelength peaked in the region where the "particle" is located. Within the packet, the waves constructively interfere. Outside the region the wavelength virtually decreases to zero where destructive interference occurs. The packet acts as a pulse moving along a line at the velocity of the particle. When the pulse is composed of a narrow range of wavelengths, the packet will be more spread out. The spread out nature of the wave packet already introduces uncertainty. We can only postulate that the particle is somewhere within the packet. If a greater range of wavelengths constitutes the packet it will not be as spread out.

From de Broglie's formula (representing wavelength and momentum) we see the particle's momentum will take on values corresponding to the range of wavelengths in the packet. When the range of wavelengths is large, the wave packet will be small and the particle's position (q) can be shown to reside within a narrow area. Under these conditions the particle's momentum (p) will take on a wider range of possible values. The converse also holds. If the wave packet is comparatively long, the particle's momentum can be more easily determined, but its position cannot be so easily determined.

Two hypothetical extremes can be visualized. The wave packet decreases to zero width when the wavelength takes on composite values from zero to infinity. Under those circumstances although the position is precisely determined, the particle's momentum becomes completely uncertain. The other extreme is also possible. If only one distinct wavelength constitutes the wave packet, its amplitude will range over the whole line of trajectory. The particle's momentum can then be precisely determined from de Broglie's formula. However, since the pulse is not restricted to some part of the trajectory the particle's position inevitably becomes uncertain.

Heisenberg thought uncertainty is built into the very nature of things. Even if we ideally know the initial state of an electron, we still cannot determine its final state by measurement (experiment). When an electron interacts or exchanges energy, the wave packet changes. This happens when one of the photons in a beam of light, used for observatory purposes, interacts with the electron. Conceptually, this creates a new wave function by what is called the reduction (collapse) of the wave packet. This means one of the possibilities, a state vector, represented in the original wave function, is realized through observation or measurement.

Collapse of the wave packet is another way of saying the state vector changes discontinuously. Instead of a continual spreading-out of the wave packet, this idea enables the observer to "reduce" it to a size about equal to the original wave function. After measurement the packet is said to begin spreading outward again.

The use of the wave packet to describe the state of an electron precludes precisely determining subsequent states. In 1927 Heisenberg expressed his findings in the Principle of Uncertainty (Indeterminacy Principle). Although first used to describe p and q, conjugate-uncertainty relations (incompatible-physical variables) also apply to the determination of energy and time. Precise energy is determined only at the expense of not knowing when the system exhibits that energy. This is all a consequence of the wave-particle nature of micro-things.

By including Heisenberg's Uncertainty Principle in quantum theory we arrive at some interesting statements. First, the electron propagates in accord with classical wave theory, but exchanges energy as though it was a discrete particle. (Bohr thought the Uncertainty Relations were the result of trying to describe quantum phenomena in classical terms.) Second, the electron stands at the crossroads where criteria often decide whether a phenomenon behaves more like a particle or a wave.

Bohr's Principle of Complementarity

All phenomena can be described by a solution to a wave equation. For objects more massive than electrons, calculated wave properties become more difficult to detect and are therefore less significant. The opposite is true for particles less massive than electrons. Finally, scientists are led to the inescapable idea that matter, like light, has a dual nature. The nature exhibited depends on the experiment performed. Thus, Bohr was led to say two mutually exclusive experiments are needed to determine both wave and particular properties. He expressed these ideas in the Principle of Complementarity.

The Principle of Complementarity states the knowledge of a particle's position is complementary to its velocity or momentum. This suggests that wave and particle natures are two aspects of one thing. (The particular aspect readily applies to absorption and emission. The wave concept more aptly describes propagation.) In the double-slit experiment, we find determining the particular property (defining which slit includes the trajectory) is complementary to simultaneously determining the wave property (interference).

Heisenberg connected the meaning of the wave function to his Uncertainty Principle. That is, individual particle behavior can only be statistically defined because p and q cannot both be determined under the same exact conditions.

Heisenberg said when dealing with micro-events, we must abandon the classical notion of a trajectory. It cannot be defined by an infinitely thin mathematical line. Again, resolution applies jointly to p and q. One could be exactly determined, but only at the expense of complete indetermination for the other. Recall, Planck's constant governs the inequality in calculating the difference between p x q and q x p in Heisenberg's matrices. It was Heisenberg's trials with this problem that led him to formulate the Uncertainty Principle.

Bohr described the state of quantum theory at conferences during the latter 1920's. He talked about: Heisenberg's Uncertainty relations and his own idea of complementarity (introduced in 1927) based on particle-wave duality, Born's statistical (probability interpretation of quantum theory) concept, and the idea that the very act of observing disturbs what is being observed. Bohr posits a corollary that when we are not observing, we cannot say anything about a system. The conceptual embodiment of these ideas, especially the probability interpretation, is loosely referred as the Copenhagen Interpretation of quantum theory.

When dealing with quantum phenomena, some ideas need to be stated cautiously. A particle is said to not intrinsically exhibit position and momentum. Information about an electron's position or momentum, for example, is quantum mechanically obtained from a particle's matter-wave. This is inherent in quantum theory.

Most of the ideas we use to build constructs are abstracted from familiar everyday experiences. Such notions are just not rigorously applicable to the microworld of physics. That aside, mechanical states generally tell more of particle behavior than they do about the particles themselves.

Scientists ascribe properties to phenomena to establish relations between them. Knowing relations leads to understanding. Philosophers and scientists have often failed to remember this. Many strange ideas have surfaced about the meaning of quantum mechanics (e.g., the wave function). There is a difference between an object itself and its representation.

Neumann Formalizes Quantum Mechanics

In 1929 a significant advance in shaping quantum mechanics took root in the work of the Hungarian mathematician J. von Neumann. He amply showed quantum mechanics to be a complete system within itself.

Von Neumann formalized quantum mechanics as a calculus of Hermitian operators in Hilbert space (after work by the German D. Hilbert during the first decade of the 1900's). Hermitian operators correspond to physical observables. Von Neumann abstractly defined Hilbert space as a "linear strictly positive inner product space, usually involving complex numbers, which is complete via the metric generated by the inner product." Von Neumann explained that Heisenberg's matrices and Schrödinger's wave equation were particular examples of the calculus of Hermitian operators in Hilbert space.

Chemistry is Explained by Physics

The meaning of quantum mechanics extends beyond mere formulation. The physics of quantum mechanics clarifies other phenomena. Here we convey the idea that diverse areas of scientific inquiry are intimately interrelated.

Table of Periodic Elements

As early as 1829 the German J. Dobereiner tried to systematize the understanding of atomic elements. Insufficient information was then available. The properties of many elements had not been discovered.

In 1869 the Russian chemist D. Mendeleev (and independently L. Meyer, a German) began to systematically arrange atomic elements by their atomic weight. The properties of the elements were a periodic function of their atomic weights. He left spaces for undiscovered elements. He rearranged some elements, despite their atomic weights, to align them periodically with their chemical properties. At best his effort was crude and incomplete.

In 1913 the Englishman H. Moseley, using crystal spectrometry, measured the wavelengths of x-ray spectra for many atomic elements. He aimed high energy electrons at target elements. Sometimes an inner electron would be knocked from an atomic orbit leaving a vacancy. Moseley hypothesized an outer electron would replace a knocked out electron. In the process the energy difference would be emitted as a x-ray photon.

Moseley observed the wavelength of x-ray line spectrum varied regularly from element to element. He graphically plotted the square root of the observed x-ray frequency against a "numerical ordering" of the elements. His atomic "number" replaced the role of atomic weight. His periodic plot eliminated problems presented by weight. Some elements are incorrectly aligned when using their atomic weights. Potassium and argon are examples, as well as cobalt and nickel.

The elements aligned into many groups according to their chemical properties. Groups are the vertical columns in the Periodic Table. The elements of Group I (for example, lithium, sodium, and potassium) exhibit a valence of one (a single electron in their outer most shell). Elements in Group II (for example, beryllium, magnesium and calcium) exhibit two electrons in their outer most shell.

Periods are shown by the horizontal rows in the Periodic Table. Each period or series corresponds to a major energy level. Chemists labeled these outer most energy shells by the call letters (K, L, M, N, and so on).

"Moseley plots" resulted in an improved periodic table of the elements. The atomic number previously pertaining to only spaces on the periodic table now took on added meaning. Along with the number of electrons surrounding the nucleus, the plots also represented the number of positively charged units (protons) in the nucleus.

Moseley's findings supported Bohr's energy picture of atomic structure. (Recall that spectroscopic analysis initially contributed to the understanding of atomic-energy levels.) As mentioned earlier, major energy levels of the atom are denoted by the principal quantum number (n).

(n) can have the value of any positive whole number. For any main-energy level there is only one value of (n). It represents both the energy level and wave function of the system. The electrostatic force between the electron and its surroundings, principally the positively charged nucleus, creates the main energy shells.

The ground state or lowest energy level that is closest to the nucleus is represented by $n = 1$. This energy level corresponds to Moseley's K-shell. Successively "filling" this shell relates the elements of hydrogen and helium. The second innermost energy level corresponds to $n = 2$ and is referred by chemists as the L-shell. Successively filling this energy level relates the elements of the second period (lithium through neon). As the principal quantum number increases, the energy difference between successive levels decrease.

More Quantum Numbers

Schrödinger's equation, applied to a particle existing in three-dimensional space, generates more quantum numbers. The second (subsidiary) or azimulthal quantum number (ℓ) represents energy subshells (orbital angular momentum) which are generated by the orbital magnetic moment of the electron. Recall that an electrically charged particle (like the electron) in motion creates a magnetic field. The magnetic moment is a measure of that field. It is a vector quantity.

The value of (ℓ) depends on (n). It can be any positive whole number from zero to a number no greater than $n - 1$. (ℓ) represents several differently oriented subshells (orbital angular momenta) that are generated by the orbital magnetic moment of the electron. Increasing (ℓ) values denote stronger magnetic moment. When an atom undergoes photon emission, (ℓ) must change by ± 1 due to the conservation of angular momentum and the fact a photon has a spin angular momentum of one.

(m_ℓ) is a projection of (ℓ) along a given orbital axis. It represents an orbital refinement largely resulting from the influence of magnetic fields created by electron revolution around the nucleus. It describes the spatial orientation of angular momentum (ℓ). There are $2\ell + 1$ possible values of m_ℓ.

Pauli's Exclusion Principle

In 1925 the Austrian W. Pauli stated his Exclusion Principle. Pauli said because there were only two variations of the wave function for the lowest energy shell, there should only be two electrons associated with it. Further, only two electrons fill any state collectively described by the first three quantum numbers. This solved many problems. If all electrons of the higher-number elements could crowd into the lowest-energy shell, electrostatic attraction (to the nucleus) would increase thus sinking the orbits. Extracting an electron from such an atom would become more difficult. This was not observed. It was this fact which Pauli originally attempted to explain with his (Exclusion) principle.

A fourth quantum number was introduced at the beginning of the 1920's to account for the additional splitting of spectral lines. One case involves what is called fine-structure spectra. For example, consider the first line of the Balmer series that occurs from electron transitions between the third ($n = 3$) and second ($n = 2$) energy levels. One line is expected at a wavelength of 6,563 angstrom units yet there appear two lines 1.4 angstroms apart. Another case involves the effect of a strong magnetic field upon spectra. Besides the expected lines from the normal Zeeman effect, other lines (the anomalous Zeeman effect) appear.

The same year Pauli enunciated his principle, two Dutchmen, S. Goudsmit and G. Uhlenbeck, using the Bohr-Sommerfeld quantum model, proposed that the fourth quantum number was not due to yet another orbital variation, but to the electron itself. They attributed the fourth quantum number to the electron spinning about itself. They thought an electron would possess intrinsic angular momentum independent of its translation and therefore exhibit a certain magnetic moment.

Experimental verification had already been performed, but a few more years passed before proper interpretation was recognized. In 1921, O. Stern and W. Gerlach of Germany designed an experiment to measure the magnetic moment of silver atoms. They varporized neutral silver atoms in an oven: producing a beam of atoms through a line of small holes. The escaping atoms were collimated and directed through a strong nonuniform magnetic field. Classically, the silver atoms were expected to line up uniformly in a vertical band on the photographic plate.

Stern and Gerlach observed that the atoms did align vertically, but in two distinct groups. The inhomogeneous magnetic field acts differentially upon each charge of the "dipole" created by the electric and nuclear charges. It was later recognized the two groups correspond to the two quantized values for electron spin angular momentum. This phenomenon is related by the spin quantum number (m_s). For each (m_ℓ) the value of (m_s) is a measure of angular momentum (regularly $\hbar$) along a given orbital axis in units of either +1/2 or –1/2 $\hbar$.

A reformulated Pauli Exclusion Principle states no two electrons in an atom can share the same quantum numbers. The principle allows two electrons to share the first three quantum numbers. Those two electrons in turn are differentiated by the fourth quantum number stating they must possess opposite spin. Each electron must have a different wave pattern. The magnetic effect of intrinsic spin is now said to change the orbitals of the two electrons. The number of electrons filling an energy level is limited. Electrons fill the energy levels according to Pauli's Principle.

With some rearranging of orbitals, according to actual increasing energy levels, an adjusted number of electrons, comprising the main shells, can be constructed. For example, in the higher elements the innermost subshell of N overlaps with the outer most subshell of M.

As we add one electron at a time to the orbitals, we proceed across the periodic table of the elements. When the table is modified, we observe that the electron configuration, permitted by the reformulated Pauli Principle, explains the valences or vertical groups of the atomic elements.

The modified main shells designated by letters (beginning with K) correspond to the periods of the table. For example, in the first period the maximum number of electrons that can fill the K-shell is two because there are only two distinct sets of quantum numbers assigned to the first period—hydrogen and helium. The number of elements allowed in each period corresponds to the maximum number of electrons (allowed by the Pauli Principle) in each main shell.

The number of possible electrons in each shell (n) is calculated by $2n^2$. Only two electrons are allowed in the first (or K) shell. The maximum number of electrons for either the L-shell or M-shell is 8. The N-shell and O-shell can each contain

18 electrons. The P and Q shells can contain a maximum of 32 electrons. These numbers are identical with the number of elements in their respective periods. The bonus is the number of elements normally in the outermost shell explains how atoms chemically combine. Chemically similar elements are aligned in the same group. Quantum theory explains the Table of Chemical Elements.

Let's look at a conceptual system that combines the quantum numbers in one expression. The possible energy levels of an atom, adjusted to reflect the "initial" electron filling order of each shell, are as follows: $1s^2$, $2s^2$, $2p^6$, $3s^2$, $3p^6$, $4s^2$, $3d^{10}$, $4p^6$, $5s^2$, $4d^{10}$, $5p^6$, $6s^2$, $4f^{14}$, $5d^{10}$, $6p^6$, $7s^2$, $5f^{14}$, $6d^{10}$, $7p^6$, $8s^2$, $8p^6$, $8d^{10}$. Each element of the periodic table is also represented by progressively adding one electron at a time to this series. These letters are a carry-over from early experiments in spectroscopy. Notice that "1s" represents the K shell. It can contain no more than two electrons. 2s, 2p represents the L shell. It can contain no more than 8 electrons (2 + 6). 3s, 3p represents the M shell. 4s, 3d, 4p represents the N shell. The O shell is 5s, 4d, 5p. The P shell is 6s, 4f, 5d, 6s. The Q shell is 7s, 5f, 6d, 7p. Also notice the fourth shell begins to fill before the third shell is completely filled. This also happens with many other elements.

The prefix number is the principal quantum number (n) indicating the major energy shell (values 1–8). The lower case letter is the subshell or orbital represented by the second quantum number (ℓ). The suborbital represented by $\ell = 0$ is s; $\ell = 1$ is p; $\ell = 2$ is d; and $\ell = 3$ is f. The post superscript number represents the maximum number of electrons allowed in the suborbitals. The maximum number of allowable electrons for each orbital divided by two yields the number of orientations (m_ℓ) for each orbital. Only two electrons are permitted per orientation as stated by the Exclusion Principle. Each electron is differentiated by spin. This is represented by (m_s).

The principal quantum number $n = 1$ corresponds to Moseley's K-shell, $n = 2$ is the L-shell, $n = 3$ is the M-shell, $n = 4$ is the N-shell and so on until enough electrons are assigned to represent the heaviest and last recorded element. Atomic element 114 has been detected. There are 94 naturally occurring atomic elements.

We have seen that electrons are bound to a nucleus in quantized orbits. Quantization also applies to free electrons. Instead of hundreds of energy levels available in atomic orbits, free electrons can occupy one of billions of energy levels. The Exclusion Principle applies to these as well. Again, the electrons fill the lower energy levels first.

Particle behavior agreeing with Pauli's Exclusion Principle obeys Fermi-Dirac statistics. The proton and neutron are examples. All baryons and leptons have half-integral spin and are subject to Pauli's Exclusion Principle.

Electron configurations tend to oversimplify the real nature of atomically bound electrons. The idea of "smeared out" electron orbits is no better at describing the functional nature of the electron. The mathematical idea of the wave function offers the best means of describing atomic nature. This poses the next problem. What is the best way to understand the wave function?

Understanding the Wave Function

Experimentally, quantum theory has been very successful. Yet from the beginning, controversy surrounded various interpretations of the wave function.

The Copenhagen Interpretation states the wave function is a mathematical method for making statistical predictions for individual systems. Its supporters say quantum theory is pragmatically complete. Instead of representing a real particle, the Copenhagen view holds that quantum mechanics only represents our knowledge of the particle.

A "popular" version considers the wave function to represent a single system that under observation appears to change discontinuously according to probability. However, unresolvable problems arise if more than one observer is injected into the theory.

Other interpretations followed. A stochastic-process version asserts quantum phenomena are indigenous to observed systems themselves. It posits quantum mechanics describes nothing more than the classical theory of probability. This version denies the possibility of determinism at a very fundamental level.

A statistical-ensemble interpretation asserts the wave function describes a collection of numerous identical systems. It sidesteps the problem of applying the wave function to a single system. If this interpretation is admitted, two possibilities emerge. Either the physical world is ruled by chance (when observed, the state to which a system collapses from the superposition, occurs randomly) or quantum mechanics describes nothing more than the classical theory of probability. We revisit this theory in the next subheading.

EPR Paradox

Many interpretations of the state-function (the wave function was agreed to describe the "state" of the system) were challenged using thought experiments.

One such famous experiment reached published form in 1935. A. Einstein, B. Podolsky (a Russian born physicist) and N. Rosen (an American) in a thought experiment (which M. Born later called the EPR paradox) asked, in a fundamental sense, if the quantum mechanical description (and particularly if the Copenhagen Interpretation) of reality were complete.

EPR said every element of physical reality should have its representation in theory (condition of completeness). An element (quantity) in physical reality corresponds to theory if it can be predicted without being disturbed (criterion of reality). Thus, elements (properties) exist whether they are observed (by instruments or witnessed consciously).

They went on to describe the wave function. They said precise knowledge of one noncommuting operator (representing physical quantities such as position and momentum, i.e., q and p) precludes knowledge of the other (from the uncertainty relations). They continued: under those conditions the two operators cannot both (simultaneously) represent reality if the quantum mechanical description of the state-function is incomplete.

EPR describes an experiment in which the momentum and position determination of one particle, having interacted (both coupled in a singlet state) with a second particle, allows a predicted determination of the second particle's momentum and position without in any way disturbing it.

Even quantum theory permits us to measure the total momentum of two particles and their separating distance. Because momentum is conserved, we can measure the momentum of the first particle and calculate the momentum for the second particle. Here, we measure the position of the first particle (disturbing its momentum) and using the original separating distance, calculate the position of the second particle. We now know the position and momentum of the second particle without in any way disturbing it, contrary to the Copenhagen interpretation (and specifically the uncertainty relations).

Bohr replied that values cannot be ascribed to the second particle unless it is measured. Yet, an apparently reasonable (objective) definition of reality, according to the "criterion of reality," allows us to ascertain the position and momentum of a particle without disturbing it. Thus, a quantity (position and momentum) exists in reality for both particles.

If completeness (of the quantum mechanical description of physical reality) were assumed, these quantities (quantum mechanically represented by noncommuting variables) would be represented simultaneously and would be predictable. Yet the state-function (wave function) cannot predict both. This led EPR to conclude the state-function is incomplete and does not represent reality (because both elements, p and q, are experimentally shown by "thought experiment" to exist simultaneously).

Many debates followed EPR's argument. Some scientists countered by saying EPR used terms that were ambiguously defined. One controversial term was the word "reality." Because they could not agree on what constitutes reality, contradictions were inevitable. For example, what does it mean to determine a measurement "without in any way disturbing the system?"

The following is Bohr's retort. He said we should not impose onto reality what we perceive its "properties" to be. Thus, we cannot infer (by "thought experiment") a property (q and p) of one particle by measuring another particle that had earlier interacted with it. This consideration shows the nature of the paradox. After particle interaction has ceased, how can a measurement on one particle affect what is a property of the other particle?

Hidden Variable Theories

More elaborate theories were developed to rid the contradictions. A "hidden variable's theory" (HVT) was one attempt to back EPR's conclusion that the state-function does not represent reality. Proponents of HVT's suggested the state-function, instead of representing a single system, describes a large ensemble of similar systems.

HVT's begin by stating the state-function is incomplete. The theory then appeals to hidden parameters that if known would lead to a deterministic theory. This would be a theory where individual events could be exactly predicted if we just knew the hidden variables.

A hidden assumption in the "criterion of reality" is that these properties, though existing in "one" reality, can always be considered as separately existing elements. HVT proponents said it is only because the hidden parameters are not known that we must describe the state-function with probability. For instance, using EPR's thought experiment, they argued the two "separated" particles become "correlated" at the time of interaction. (More about this in Chapter 15: "Reality—Unity: Structural Integration.")

Proponents argued if certain "hidden variables" were known the p and q of the second particle could be predicted without assuming that performing the first measurement influenced the second. Thus, HVT's predict more of what would be the case than what is the case. A major downfall of the HVT's is that they could not duplicate the results of the statistical quantum interpretation of the state-function.

Schrodinger's Cat and Wigner's Friend

After studying the EPR experiment E. Schrödinger wrote a paper which included his well known "cat paradox." It is another thought experiment constructed to show the Copenhagen interpretation is incomplete.

Schrödinger describes micro-phenomena using macro objects. In a cage is a cat. The cage also includes a small piece of radioactive material, a Geiger counter, a flask of cyanide, and a hammer connected to the detector.

The cat cannot influence the detector. In one hour there is equal probability the Geiger counter will detect the decay of one atom. If an atom decays, the detector signals the release of the hammer; the flask breaks; cyanide escapes, and the cat dies.

Any time during the hour the cat is either alive or dead. We have no way of knowing the cat's situation until the cage is opened. When the cage is opened the observation causes a "collapse of the wave function" to one of the elements (states) of the superposition. Until the observation is made, according to the strict Copenhagen interpretation, the cat's status is neither one of being alive nor dead, but is a mixture of the two possibilities.

Schrödinger says, in contradistinction to the Copenhagen interpretation, that the cat cannot be both alive and dead at the same time. Some theorists suggested that it is the cat that collapses the wave function.

E. Wigner tackled this problem using what has become known as "Wigner's friend paradox." He says we cannot get an answer from the cat if we ask it to describe the state it was in before the opening of the cage. So he replaces the cat with a human observer—his "friend."

At the end of Wigner's experiment, if found alive, the friend is unlikely to report that he experienced a state describable as one suspended between life and death.

There are other problems. If the state-function does not collapse unless observed (some theorists insist that a mere recording of an event is sufficient for collapse) a bizarre consequence ensues.

Could it be the real world only exists when it is observed? Does the world only exist in the mind of the observer whom, by his acts of observation, actually creates the world? In philosophy this is called solipsism.

In solipsism everything, including other people, exists but in the mind of one individual. The problem obviously is that such an idea cannot apply to everyone. There are many people, and each, it is supposed, has his or her own mind. Therefore, the "preeminent" position is reserved for just one "observer." If each individual exclusively owns his or her own mind or consciousness, this would be an untenable hypothesis.

Another slant on this is the placing of one observer in a large box with another observer ad infinitum. Each observer in a larger box collapses the state-function of the smaller boxes. This results in an endless regression of causes. This type of reasoning solves little.

We continue our quest to understand the EPR paradox and Schrödinger's cat in following chapters.

Quantum Electrodynamics

Quantum theory continued to explain phenomena. It aided physicists in understanding the atomic nucleus and radioactivity. It branched into specialties such as quantum field theory that treats particles as fields. In this respect it now receives the most attention because of its use in elementary-particle physics.

An outgrowth of quantum field theory is quantum electrodynamics (QED) begun with Dirac's work in the late 1920's. It was founded by imposing quantum conditions onto classical electromagnetic theory. It deals mostly with the interaction of fields. An example is the interaction of an electron matter field with electromagnetic fields.

QED achieved full form in the work of R.P. Feynman, J. Schwinger and others during the latter 1940's. Quantum electrodynamics later spun off a branch called quantum chromodynamics (QCD) which is concerned with the deeper nature of elementary particles. (We will see its influence in understanding the function of quarks—in Chapter 15.)

Quantum of Action and Realm of Reasoning Structure

How does the quantum of action explain the structure of the Realm of Reasoning? Many examples (and paradoxes) have been described. Each attempt to explain or contradict quantum theory uses macro-situations that we can easily visualize. We have seen that it is not presently suitable to use the macro-world to describe the micro-world. Maybe it is easier for the micro-world of quantum theory to explain the macro-world.

Bohr said all action is characterized by a certain discontinuity governed by Planck's quantum of action. The quantum of action is a fundamental constant of nature. As stated previously, it is symbolized by the letter ($\hbar$) and has the value 6.626×10^{-27} erg-second. The intent of this subheading is to understand ($\hbar$) using the structure of reality.

One most palpable question concerns the value of ($\hbar$) and its affect on cosmic stability. If ($\hbar$) were a number closer to zero energy, quantization would be nearly undetectable. De Broglie's matter waves would be infinitesimally small. The minutest systems would obey classical laws.

If the numerical value of ($\hbar$) were zero there would be no quantization at all. There would be no regularity or order to what exists, and matter would not exhibit definable properties. Quantization induces physical existence. Although introducing discontinuity in processes of change, the effect of ($\hbar$) is not patently discernible on the scale of everyday experiences. However, if ($\hbar$) had a much greater value, discontinuity would affect everyday life and existence would be very different.

In the microcosmic world of quantum mechanics the concepts of time and space are necessarily more incomprehensible. That compels us to start with the following ideas: Quantum mechanics concerns order. Time is generated as a product of ordered changes in the Realm of Whatness. The big question concerns the cosmic mechanism that is responsible for change occurring in discrete steps.

De Broglie linked several concepts. He said matter is composed of mass. Einstein's equation $E = mc^2$ asserts mass is equivalent to a certain quantity of energy. Planck's formula ($E = hf$) associates energy with a specific frequency.

De Broglie said frequency is a pulsating phenomenon. Mathematically, pulsation can be interpreted as either being confined or spread out. De Broglie used both interpretations.

De Broglie said a particle at rest has a confined pulse that beats in harmony with a spread out universal pulsation. All particles are immersed in—and beat in unison with—that pulsation.

The pulsation is simultaneous everywhere. That sounds as if it disagrees with simultaneity as defined in relativity. However, no problem occurs. At its core, relativity pertains to the transport of energy in space and time. In a tardyon sense, it is energy that cannot move at a speed exceeding (c). Yet, the universal pulsation travels much faster than (c): it is instantaneous.

A single universal pulse is "felt" in every inertial system at the same instant. Physicists have known about extremely fast waves for some time. They are known as phase waves. It was De Broglie who discovered the link between phase waves and a particle in motion.

De Broglie considered the relativity effect of time on an oscillating (pulsating) particle in uniform motion. Time dilation causes a moving particle's frequency to change.

Energy content also affects particle motion. Thus, a moving particle has a certain energy that translates by a quantum relation into a different frequency or rate of pulsation.

When many associated waves for slightly different particle speeds are brought together, a "composite" matter wave is produced. As previously noted: De Broglie said the superposition of involved frequencies results in a constant phase wave moving with the particle's direction and speed. What begins as a pulsating particle, translates into a matter-wave in synchronous motion with the particle. De Broglie found the connecting factor between a moving particle and its associated wave to be Planck's constant.

Let's assess what has just been described. Any given particle is bathed in a universal pulsation. We have already coined a unit for this pulsation—the Bi-Deltic Cosmic Instant (BCI). The pulsation is everywhere simultaneous. It has been described how this unit generates the flow of time.

What is the duration of the pulse or BCI unit? It should not exceed the Planck time that is equal to 10^{-43} seconds.

It was stated in the Chapter 11 that the BCI is an "on-again, off-again" mechanism. This corresponds to the oscillatory nature of the pulsation. Look again, at Figure 12:4 in the previous chapter. When the "charge" is tardyon positive the cosmos is "on" and the anti-cosmos is "off."

We suppose the Realm of Reasoning promulgates the BCI phenomenon experienced by the Realm of Whatness. It should not be difficult to understand why radiating spokes geometrically represent the Realm of Reasoning. Opposing spokes are (like the zigzagging lines in Figure 12:4) in essence, the universes.

Is there any scientific information that supports the structure of the Realm of Reasoning?

Everett's Many World's Interpretation of Quantum Mechanics

In 1957 H. Everett, in his doctoral thesis, proposed a more literal and more general interpretation of quantum mechanics. It offers a reasonable answer to Einstein, Podolsky, and Rosen's argument that quantum theory does not offer a complete description of physical processes.

Everett said the conventional formalism of quantum mechanics is inapplicable to a closed universe because it offers no outside reference from which to observe change. He says the observer, who measures quantum phenomena, should be included within the formalism of quantum theory. Unless the observer is included, the formalism will be incomplete.

Everett postulates a pure-wave theory in which the state-function (wave function) universally (deterministically) obeys a linear-wave equation. His formulation can also be used as a model for an isolated system. Specifically, one that is subject to outside observation when both systems (the observer and that which is observed) are considered a part of an even larger system. He thus considers outside observers to be a special case of normal interaction occurring within the system.

Everett divests quantum mechanics of most remaining classical connotations. His theory is not premised on a probability interpretation of the wave equations. Everett uses a "relative state" interpretation referring to the state of the observing apparatus.

The apparatus takes on a value indicating it has recorded a measurement on a system. Observation implies the observer has become "correlated" (interacted) with the system being observed. This should not be surprising. The Copenhagen interpretation speaks of the idea that the observer disturbs the system being observed. The letters exchanged between Einstein and Bohr wrestled with the idea that the observer disturbs a system by performing a measurement on it.

Everett says the complete state of a single system cannot be represented because it cannot be isolated from the collective of all possible systems. Each "restricted" system is treated as a "subsystem" of the bigger system of which it is a part. This implies an observer cannot determine the total state-function of a specific system because he is limited to observing only his system.

Can an observer obtain information of other possible systems from the perspective of his system? If he can, the external observer of classical interpretations becomes an integral part of the system under investigation.

What can Everett's theory do to resolve questions not satisfactorily answered by conventional quantum theory? For example, quantum physicists have not been at ease with the usual interpretation of the superposition of state vectors constituting the state-function. And, there is the problem of the observer (disturbing the system by measuring it).

It was thought each state vector represented a possible result of measurement (by an observer). In actual practice, of all possible values, only one state-vector is observed. This is the "protective" idea embellished by the "collapse of the wave packet." This is itself a logical problem, not one resolved by conventional-quantum theory.

In Everett's interpretation there is no collapse of the wave packet. What are the implications of his theory? How can the observer eliminate the problem associated with wave-packet collapse? Let's slowly work our way to a reasonable understanding.

Everett says the state-function represents the forming of a wave packet that upon measurement splits into orthogonal packets—one for each possible value (state-vector or component of the superposition). The problem of the observer is not resolved by introducing a second observer to watch over the first—because splitting recurs.

Von Newmann recognized the consequences of such a solution by saying it leads to an infinite regression of observers. Clearly, this is where the conventional theory and Everett's theory part company.

The conventional interpretation asserts the wave packet collapses "in the mind" (of the observer) after observation (after measurement). This is called the "measurement problem." The wave packet reduces to one of the elements (state vectors) represented in the decomposition of the state-function. Yet, a wave equation suggests no such thing. The resulting element is unpredictable. (Schrödinger's Cat Paradox is a visual representation of the measurement problem).

Most physicists have rejected further interpretation of the splitting of the wave packet because they could not find something in reality to which it corresponds. To them the state-function can only represent the probability or possibility of event occurrence.

Is there another way to interpret the splitting of the wave packet? Collaborating on Everett's initiative, American physicists, J. Wheeler, and N. Graham (collectively—EWG) worked out a model that retains the splitting of the state-vector. They said the mathematical formalism could represent the real world if it were a complex structure with more than one universe. B. DeWitt said, under those circumstances the formalism itself could produce an interpretation.

Their immediate task was to address two questions. First, Everett's formulation must be capable of rendering the conventional interpretation. Indeed, the statistical interpretation of quantum mechanics appears to observers at a subjective level as relative phenomena. As in the conventional interpretation, an observer is restricted by the Uncertainty Principle in what he can predict. Second, the idea the state-function does not collapse must correspond to something in reality.

The EWG model suggests there exists a different universe for each state-vector (or component element) represented in the superposition. And just as there are an endless number of distinct possible values in the superposition, there would exist an infinite number of noninteracting worlds. All these worlds exist simultaneously.

Each state-vector (or possible measurement in conventional terms) diverges into numerous elements. Everett and others say this represents the world continually splitting into several branches. The "branched" state-vectors must equal the value of the original superposition.

The splitting of universes results from the interaction of the system and its measuring apparatus (substituting for an observer). Since there is no outside reference from which to determine which branch represents the "real" world, Everett says all branches must be considered equally real.

Everything that could happen does happen when all possible worlds are collectively considered as a whole. Each possible world must be different from any other world in at least one respect. Each world differs from others depending on how many BCI units it is "removed" from the one considered. It is relative. The farther removed a possible world from the "given" one, the less they will seem to have in common.

Everett and his collaborators said a quantum jump anywhere could entail the splitting of our world into many slightly different copies: each of which continues splitting.

Why is the splitting not noticed? There has been no experimental way to discern the superposition from its composition. Everett said it is similar to the situation prevalent during the early sixteenth century. Many people rejected Copernicus's theory of earth rotation because they could not feel the earth move. It was not until the early eighteenth century when Newton revealed his theory of gravity that we began to understand why we do not experience earth's motion. We do not experience the splitting of the universes because of quantum mechanical laws.

That may be one interpretation. However, let's take another look by going back to the quantum jump. When Everett's formulation is applied to the whole universe, one may wonder what the "quantum jump" means. Remember de Broglie's universal pulsation? Each pulsation is a universal quantum jump. The superposition becomes a universal state-function, which by Schrödinger's equation changes linearly.

Although Everett may be on the right track, his interpretation of universal splitting may have to be more satisfactorily worked out. Everett's formulation of quantum theory does provide an interesting approach to cosmological structure. This is especially noteworthy because his basic idea coincides with the structure in this chapter.

Let's see what else can be done with Everett's ideas. What does it mean when a universe splits? A state-vector branches into components that find themselves further branching much as does a tree. Each split component or universe is a slightly different copy of other universes. Could this represent the Cirpanic Cosmos structure and the radiating arms of the Realm of Reasoning structure? Each universe on the "circle" differs by at least one BCI. Although this project does not take a position on actual correlation of Everett's theory with the structure presented, enough questions are posed warranting further research.

Let's correlate Everett's ideas with the structure in this book. Let's consider the splitting to pertain to all the possibilities allowed (by splitting). The source of splitting is not found in the realm of things. The source is not "categorically in" the Realm of Whatness.

Splitting emanates from a nonmanifest realm. This means the conventional idea of wave function collapse can be reinstated. Collapse occurs in the realm of things. It happens in the manifest realm. Its source must be intertwined with the Realm of Reasoning.

The nonmanifest realm infers all possibilities. The laws of nature are associated with all logical possibilities. Logic limits "what" can happen (in manifestation). Can this "realm of law" be conceptually isolated from the Realm of Reasoning? If it can, then we can better understand not only the Realm of Whatness, but also that which is its source.

The realm of law is commensurate with "all possibilities of manifestation." We can simply rename the laws of nature as knowledge in the purest sense. For now, we denote this pure sense of knowledge as "potentiality." Potentiality refers to "a pure form" of knowledge—one that is useable for making something. Potentiality is unrealized possibility.

Let's take a broad view of potentiality. If there is an intelligence (and hence design) behind everything, then its governance over things is at least of a potential nature. The questions are simple. If potentiality (possible design) is not actualized in manifest order, then things are incapable of innate intelligence. And hence there would be no "self design" to things. To the contrary, we easily observe innate intelligence at work. We see design in everything because we find meaningful "constructive" order everywhere.

Yet, if intelligence is actualized as order, then how does there remain a potentiality (of that order)? One point is clear. If intelligence is totally manifested as order with a concomitant surrendering of potentiality, then there is no potentiality for things to be directed as a whole—to fulfill purpose. There would be no governance over manifest order. If that were the case, then there is no direction to the order of events. Scientific observation and understanding suggest otherwise.

There is a direction to things, and therefore, there is a potentiality to which they are directed. Just one example suffices to make the point. An infant grows into an adult. (An infant is a potential adult.) These everyday occurrences, though taken for granted, are amazing examples of partial realizations of purpose fulfillment. Purpose implies potentiality. Directed order infers potentiality. We just have not reached a point in our knowledge to understand it. (We will understand potentiality before the end of this book.) There is every reason to suppose potentiality has a role in reality. The dual nature of everything supports that understanding.

Key Points:

- The Realm of Reasoning "connects" the Realm of Whatness with the Realm of Whyness. Using a "reductionistic" approach to build this structure led to labeling it the Cirpanic Cosmos.

- The Realm of Reasoning is scientifically represented by electromagnetism—light.

- Light is constant. The constancy is often stated as light velocity invariance (in vacuo).

- Light has a dual nature—wave and particulate. Light propagates like a wave, but exchanges energy like a particle.

- Planck's constant defines the smallest quantity of energy exchange. Any transformation of energy from one form to another is governed by an integral (whole number) multiple of Planck's quantum of action. An electron (a manifest photon) exchanges energy in discrete packets of energy.

- An electron, though primarily particulate, can also exhibit a wave nature—matter waves. Matter has a wave counterpart. Also, wave phenomena may have a particulate aspect.

- Heisenberg's Uncertainty Principle shows that precisely defining a particle's velocity (or momentum) precludes knowing its exact position. The same idea applies between energy and time. We are limited in defining the state of an electron by classical physics.

- Bohr stated there is a complementary relation between a particle's position and its velocity and between energy and time. An electron is not a particle in the classical sense. It is a "wave packet" of energy.

- Physics explains chemistry. Atomic-valence electrons explain how atoms combine (chemically) to form molecules.

- Quantum theory describes particle activity (or state) as a wave function—a probability description of "wave packet" energy. A single electron is difficult to define. Because the affect of many electrons is determinable, the activity of a single electron is predictable.

- The wave function is thought to "collapse" into particularity when something is observed. For example, an electron wave packet of energy can collapse into manifestation as a dot on film. This idea was confusing. It was "magnified" and depicted in Schrödinger's Cat Paradox. When we observe a cat, it is either alive or dead. It seemed ludicrous that a cat could be in an "in-between state" when we are not looking at it. How can something be in a state of partial existence? How can an electron be both a wave and a particle?

- Problems associated with understanding the wave function were poignantly addressed by the EPR paradox. Einstein and collaborators thought the wave function could not completely represent reality if everything existed independently. Some scientists theorized "hidden" variables, which if known, would complete an "objective" description of physical events.

- Everett postulated a "many world's interpretation of quantum mechanics," which though an extreme solution, resolves problems associated with wave function collapse. He hypothesized every possibility, represented in the wave function, actually existed somewhere in "other worlds." There was no collapse—we just see the one possibility that is actualized for our world.

- Everett's idea closely relates to our understanding of the Cirpanic Cosmos. Instead of every possibility actually existing "somewhere," every possibility exists potentially. And potentiality is a nonphysical aspect of reality. Bohm wrestled with quantum potential.

Although the Realm of Reasoning should represent Reality in total, there are other questions not yet answered. We do know that the Realm of Reasoning is the "connector" between the Realm of Whatness and the Realm of Whyness. We have yet to develop sufficient knowledge and understanding to determine the nature of this connection. The "connection" is later addressed and explained.

Doing so calls for a multi-disciplinary approach. We develop the multi-disciplinary tactic in Section Six.

We look ahead and wonder: Is not knowledge a property of consciousness? We are not finished until we understand (logically) the connection of consciousness (the "observer") to what it experiences (phenomena—that which is observed).

Can quantum theory provide a way to understand consciousness and experience—observer and that which is observed? Can the "measurement problem" in quantum theory point us in the proper direction? The observer must be part of the system if he affects what he is measuring. We need a greater understanding to assess the meaning of the quantum of action. We must defer grasping deeper meaning until we develop the needed concepts that simplify our methodology. (The deeper meaning of the quantum of action awaits Chapter 15 and Sections: Six and Seven.) Let's continue.

<u>Section Four</u>

REALM OF WHYNESS

Perspective

- Why we are / Why anything is
- The highest level structure of Reality
- Conceptually characterized via wholeness function
- The Realm of Consciousness

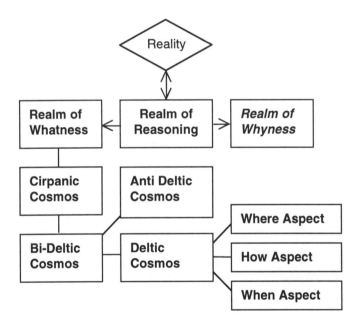

Figure 14:1. Perspective: Realm of Whyness

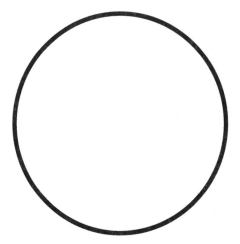

The Realm of Whyness Diagram is a circle. It represents the whole of Reality, but no part. It is All of everything, but not anything, nor even everything. Rather than being a circle, it could just as well be a dot. In itself, it is best represented by a dot. Because the Realm of Whyness accomodates anything and everything, it is also represented by a circle. It is the outer circle of yin yang.

Figure 14:2. Realm of Whyness Diagram

Chapter 14
Realm of Whyness

"The degree an individual's will is limited, is also the
extent his abilities are determined."

The Realm of Whyness is the answer to the widest noncircular why-question
that can be asked. It is the "highest" structure of reality in the cause and effect
hierarchy. It is the ultimate cause in the sense of Type III Whyness-Whatness
Association. It is represented by the intrinsic mode as expressed in the modal
system of classifying knowledge. (Refer to Chapter 2.) The Realm of Whatness
is produced by the Realm of Whyness through the Realm of Reasoning.

The self attains awareness called "self-realization" when conceptualizing this
realm as a basic feature of the structure of reality. This level is obtained by the
mental function called intrinsic perception. (See Chapter 4," How Aspect
Diagram.)

What is this realm in physics? We hypothesize at this stage of inquiry that the
Realm of Reason is gravitational radiation. The Realm of Whyness is devoid of
properties generally associated with the Realm of Whatness (such as: spatial
extensionality and time flow).

Realm of Whyness Diagram

The Realm of Whyness Diagram is as simple as any geometric representation
can be—a "dot" or circle. A dot or circle represents the Realm of Whyness
because it is (in itself) structureless. Yet, it embodies all structure.

Scientists have not studied gravitational radiation. They have studied its effects
by determining how bodies behave in a gravitational field (extended spatiality).
We examine the historical development of gravitational concepts. Real
understanding is achieved only when sufficient knowledge has been acquired: as
the following reveals.

History of Gravitation Investigation

Early cosmologists sought to explain phenomena. During the latter part of the
third century Greek scholar Eudoxes proposed a system to explain the behavior
of celestial bodies. He said the stars were fixed to certain concentric spheres
encircling earth. His theory required twenty-seven spheres to account for various
motions of the then five known planets and the sun and moon. Aristotle
modified Eudoxes's ideas using fifty-five spheres to explain the motions of
celestial bodies. Ptolemy replaced the notion of concentric spheres with a
conceptual facsimile of uniform circular motion. He developed a concept called
the epicycle. Other ideas like the eccentric and equant were added. These
devices enabled Ptolemy to calculate future planetary positions. Early Greek
celestial mechanics survived for centuries.

In the middle of the fifteenth century the German Nicholas of Cusa suggested the earth rotated on its axis. He believed the universe could very well be infinite with no center.

At the turn of the sixteenth century the Polish astronomer N. Copernicus developed ideas initiating the overthrow of Greek celestial mechanics. Copernicus developed a heliocentric theory. He said the planets revolved around the sun. Because he retained the idea of uniform circular motion, he continued to use epicycles and eccentrics. He insisted the stars lay in a fixed sphere beyond the planet Saturn.

Astronomy received a significant boost at the close of the sixteenth century when Danish astronomer T. Brahe made rather precise measurements of planetary motions. He observed the motion of comets and a super nova explosion. The old Aristotelian idea that the stars were fixed on spheres was seriously challenged.

The Italian G. Galilei was an advocate of the Copernican system. In 1609 Galileo built a thirty-power telescope. Within months he made discoveries that forever changed cosmology. He observed the moon was not a perfect sphere but was similar to earth with mountains and valleys. He also saw many stars too faint to be seen with the unaided eye. He observed the motion of sunspots indicating the sun rotated about its axis.

During the beginning of the seventeenth century the German J. Kepler was invited by T. Brahe to analyze his planetary observations. Kepler studied the orbit of Mars. After about seventy attempts to explain an eight-minute-angular discrepancy between observation and theory he found a solution. Mars did not move in a perfect circular orbit, but moved in an elliptical orbit.

Kepler subsequently published three laws of motion. The first law states that planetary orbits are ellipses with the sun at one of the foci. The second law states a straight line from the sun to a planet sweeps out equal areas of space in equal times. Thus the planets in elliptical orbits accelerate when closer to the sun. The third law states the square of the planets' period is directly proportional to the cube of its mean distance from the sun.

The precision of calculating future planetary motion using Kepler's laws eroded the two-thousand-year-old idea that motion of celestial bodies could only follow perfect circular motion.

Although Kepler had described how the planets moved, one question remained. Why did a planet accelerate when closer to the sun (agreeing with Kepler's second law)? Let's examine some earlier notions of force.

The Greeks developed the concepts of force and motion. Aristotle thought a continuous force was transmitted. For example: Throwing a stone imparts a force to the air, which then continually propels the stone. Aristotle thought a thrown object, unencumbered by a medium (air) could attain infinite velocity.

Aristotle's ideas were accepted for a long time. Even in Aristotle's time there was some opposition to his concepts. J. Philoponus said particles slowed in air and in its absence an object would not attain infinite velocity. He thought force was not transmitted to the projectile. He also stated the velocity of falling objects was not determined by their weight.

During the middle of the fourteenth century J. Buridan, a Frenchman, further developed Philoponus' idea of an impressed motive force. Buridan said a body set in motion acquires a certain quantity of impetus (weight x velocity) causing the object's motion. Until the impetus is used up, the object continues in motion. This was an early version of inertia.

Galileo dropped objects of various weights from a tower. He found (neglecting air friction) all objects fall to the ground at the same rate. He realized that constantly applied force was not required to maintain motion. He wanted to measure the rate of free-fall, but found the rate change of velocity was too fast to measure with a water clock.

Galileo reasoned round objects should roll down an incline at a rate between that for an object in unrestrained free-fall and one not moving at all. The incline would slow the ball so its rate of fall could be measured. Lessening the angle of incline proportionately reduces the rate of fall. He marked the distance the ball traveled after two seconds and found it to be four times that after the first second. After the third, fourth, and fifth seconds, the ball had traveled nine, sixteen, and twenty-five times the distance covered during the first second. He repeated the experiment using balls of different composition. He discovered all objects fall to earth at a constant downward acceleration and cover a distance proportionate to the square of the time it falls. Galileo went on to state that projectiles follow a parabolic trajectory.

Kepler thought space could exist in a vacuum and even entertained the idea forces could act in vacuo. He attributed the rise and fall of ocean tides to forces acting from the sun and moon.

Although seventeenth century natural philosophers understood velocity and acceleration, they only vaguely comprehended such properties as force, weight, mass, and inertia. The Frenchman E. Mariotte ascertained the difference between mass and weight.

Many ideas abounded during the preNewtonian era. Many questions were examined by the natural philosophers. Does gravity arise from the action of invisible particles or is it caused by some internal agency?

Newton and Force and Motion

E. Halley, R. Hooke, and C. Wren were members of the Royal Society of England. They pooled their knowledge of physics. By 1684 Halley became more convinced gravity, like light, was a force that diminished as the square of distance. He visited the Englishman I. Newton, a professor of mathematics. He asked what type of path a planet would take if it were attracted to the sun by a force acting according to the inverse square law. Newton replied the planet would orbit in an elliptical pattern. The encounter increased Newton's interest in gravity and motion.

Newton understood a planet is held in orbit by a sideways force (caused by its motion) which counters the force exerted by gravity (drawing the planet to the sun). Newton combined his formula for circular motion with Kepler's third law and showed that the controlling force varied inversely with the square of the separating distance. He mathematically compared this result with the experimentally determined value for the rate of acceleration of an object close to the earth's surface. He discovered the force acting on celestial bodies was the same as that acting on an apple falling to ground. In 1685 Newton said earth's gravity could be considered acting from its center of mass.

In 1687 Newton published his findings stating three laws of motion. The first is the law of inertia. Although the French philosopher R. Descartes first stated it as a universal, it actually is a follow-up to Galileo's work on accelerating bodies. It states, unless acted upon by force, every body will continue in a state of rest or uniform straight motion. It maintains that a reference at rest is indistinguishable from one moving with constant speed. This is Galilean relativity.

The second is the law of force. It states a (instantaneous) change in motion—acceleration (a)—is proportional to the force (f) acting upon a body of mass (m) and occurs in the direction the force is acting ($a = f/m$). (m) refers to inertial mass, which is a measure of a body's resistance to change in motion. The force (f) refers to accelerating forces. A heavier body calls for a proportionately greater force to accelerate it the same as a lighter body.

The downward accelerating force for falling objects near the surface of earth (denoted g) increases 9.8 meters (32 feet) for each second of free-fall. It is the constant of proportionality between mass and weight. Weight is the effect of earth's gravity on mass. This is the usual and older sense of the word gravity.

The third law is the law of action and reaction. It states for every action (force) there is an equal but opposite reaction (force). Or as Newton said, "The mutual actions of two bodies upon each other are always equal and directed to contrary parts." Not only does the earth act upon the moon, but also the moon acts upon the earth. The total gravitational force (F) of attraction depends on both masses. The force (F) acting between two masses (m) and (M) separated by distance (d) is equal to the product of their inertial masses divided by the square of the intervening distance ($F = GmM/d^2$). The two mass terms refer to the quantity of gravity (gravitational mass) inherent in each body. (G) is the universal gravitational constant. It provides the means of calculating the equivalent force from the two masses and their distance of separation.

In 1798 the Englishman H. Cavendish measured (G) using two large one kilogram spheres one meter apart. He obtained a surprisingly accurate answer. Many measurements have since been performed. (G) is generally considered to be the same everywhere (including antimatter) and has the value $6.6726 \pm .0005 \times 10^{-11}$ Newton-meters squared per kilogram squared. The constant is the same whether the attracting bodies are composed of solid lead or cork.

The explanatory power of Newton's laws ended the age-old idea terrestrial and celestial mechanics obeyed different laws. Kepler's laws mathematically describe the fact attracting-forces between bodies diminish as the square of their separating distance. Newton's laws explain Kepler's laws, as well as the ocean tides, the slight bulging of the earth's equator, and the precession of the earth's axis. His laws also explained why planets move faster when closer to the sun.

Halley used Newton's laws to predict a certain comet (Halley's comet) would return in 1758. In 1845 the Englishman J. Adams and in 1846 the Frenchman U. LeVerrier independently predicted the existence of the planet Neptune. They observed a variation in the orbit of Uranus that could not alone be explained by the presence of the then known planets (using perturbation theory). Galle and d'Arrest discovered Neptune in 1846. Its location was found within one degree of that calculated.

The many successes of Newtonian mechanics in describing force and motion were unparalleled in the history of science. However, in 1859 LeVerrier, building upon planetary perturbation theory (departures from Kepler's laws) developed by fellow countrymen P. Laplace, J. Lagrange and others, showed Newton's laws could not fully account for the precession of Mercury's elliptical orbit. The perihelion of Mercury's orbit (nearest point of orbit to sun) advanced 1 degree 33 minutes 20 seconds per century. Forty-three seconds of arc was not explainable by the perturbations of the other planets. LeVerrier suggested there could be an unobserved planet near the sun that perturbs Mercury's orbit. No such planet has ever been observed.

There were some serious questions that were not satisfactorily answered by Newton's laws. His laws were formulated in Euclidean geometry in which time and space are considered absolute. The distance between objects was considered rigid, in not only space, but also time. Time flowed at a uniform rate everywhere. The implication is two observers at different locations would measure the same distance between points and the same separation in time.

Newton's law of gravitation operates through "action-at-a-distance." The effect of one mass on another supposedly occurred instantaneously. Even Newton realized the objections of such an idea. No one could then argue against the successes of his theory.

Another question deserving attention: Was a body's inertial mass (determined by Newton's second law) and gravitational mass (determined by the force it experiences in a gravitational field) equivalent? There was no cogent reason they should be. Newton carried out experiments to detect a ratio difference between inertial and gravitational masses for various objects. Using Galileo's idea of dropping objects of different weights, Newton could not discern a variation in the rate of fall for objects of different composition. His conclusion was based on experiments accurate to one part in a thousand. Gallileo had arrived at the same conclusion by 1610. Maybe a difference existed, however slight.

Beginning in 1889 the Hungarian R. Eötvös, a geophysicist, carried out numerous experiments to detect a difference between inertial and gravitational masses. He wondered if bodies of different composition fall at the same rate. Does gravity act the same upon different materials? This is an updated version of Galileo's dropping various objects from a tower. Eötvös tested brass, copper, cork, glass and many other substances.

Eötvös originally refined the torsion balance to better study the geophysical characteristics (mass) of earth's surface—particularly mountains. He continued modifying the balance. He suspended a horizontal beam from a wire. At each end of the beam he suspended a weight. He lined up the beam in an east-west direction. If inertial and gravitational mass were not equivalent, he expected the beam to rotate due to torque caused by the difference. The torque would be supplied by the centripetal action (an acceleration) of earth's rotation.

Running tests on and off for over thirty years Eötvös found no discrepancy between the behavior of any mass in earth's gravity and a comparable accelerating frame of reference. His conclusions were accurate to five parts in a billion. The equality of inertial and gravitational masses for bodies of different composition has since been determined to better than one part in 10^{11} by the American R. Dicke at the beginning of the 1960's. Others such as the Russian V. Braginski and V. Panov have bettered that figure to one part in 10^{12}.

Einstein and General Relativity

In 1907 A. Einstein began with Galileo's idea. Under the influence of gravity all (uncharged) bodies, regardless of substance, accelerate at the same rate. He proposed a null experiment. An observer in a closed box could not tell by any experiment whether he was at rest on the earth's surface and subject to its pull of gravity or experiencing an equivalent acceleration in space. These ideas are embodied in the weak Equivalence Principle. The principle is only locally valid because of the varying strength of gravity from one place to another. If a small enough volume of space (local condition) is chosen, variations will ideally vanish. This restriction leads to the plausible assertion inertial and gravitational masses are exactly equal. The Equivalence Principle became Einstein's major physical basis for building his theory of gravitation.

Einstein said light rays would bend in a parabolic arc in the presence of an observer experiencing acceleration at right angles to the direction of the rays. Using the Equivalence Principle, he postulated light rays passing near massive bodies would also be deflected or curved.

Carrying his idea further, Einstein said light moving up through or against a gravitational field (or escaping from a massive object) would be red-shifted (losing energy). Light moving down through a gravitational field (or toward a massive object) would be blue-shifted (gaining energy). What does this mean for the Equivalence Principle?

Consider a uniformly accelerating box. Light shines through the top of the box. The light becomes red-shifted when a stationary outsider—at the moment the box passes him—compares the wavelength that the "boxed" individual observes to what he, the outsider, sees. An equivalent effect happens when an observer is in a stationary box near a massive object. Again, light is shined through the top of the box. Here the change in frequency (or clock rate) is induced by gravitational red-shift.

Einstein realized by picking a suitable local frame of reference that the "force" of gravity acting on a body could be effectively eliminated. If a closed box is freely falling under the influence of gravity a man inside will not experience any gravity effects (neglecting possible tidal forces: more about that later).

Einstein's idea of free-fall supplants Newton's law of inertia. Inertia in turn depends on the energy inherent in an object's mass according to Einstein's equation $E = mc^2$. Einstein's reasoning subverts the notion gravity is a force. He began looking for a more suitable explanation.

Because gravitation seemed to be a nonmechanical phenomenon, Einstein thought it related to the geometric structure of space itself. He decided to find a way to generalize Minkowski's work of 1908 describing a four-dimensional space-time characterization of Einstein's Special Theory of Relativity. Einstein suggested space near matter is warped or curved. He was talking not only about the curvature of space but also of time: a curvature of space-time. What does curvature of space mean?

> The idea matter affects space-time curvature is slightly misleading. We later discern that (the rate of) time (flow) is affected only because (extended) space is warped. Time is affected when space is distorted.

As far back as 1827, the German F. Gauss realized a surface has a separate intrinsic geometry. If the amount and type of distortion are known the shape of a surface can be determined. How did Gauss do it? Consider the Pythagorean theorem. The hypotenuse of a right triangle squared is equal to the squares of the other two sides. If two sides are known, the remaining side can be calculated. Can the same idea apply to a triangle on a curved surface? Yes, provided there is some modification. For example, a hypotenuse of a right triangle on a sphere will have a value less than 5 though the other two sides are 3 and 4. The sphere "distorts" the geometry from what we would expect on a plane surface.

What are some features of a curved surface? The surface of a sphere, for example, does not permit parallel lines to exist; instead, they converge and intersect at antipodal points. On a sphere the subtended angles of a triangle are greater than 180 degrees.

Visualize a big triangle marked on the earth's surface by some giant. The giant centers one vertex on the North Pole (antipodal point) with its sides meeting at right angles. He extends the sides as lines of longitude, which intersect the equator perpendicularly.

The little people who live on the surface do not know whether the earth is flat or spherical. They undertake the task of measuring the length of the lines. Near the North Pole vertex they mark off a line connecting the two sides. They do know the surface near the North Pole is apparently flat. They measure the lines of their triangle. Using the trigonometric law of cosines, they determine the polar vertex angle. They use the same procedure on the other angles (on the equator) of the big triangle. They also know the proportionality is governed by the square of the radius of the sphere, which they also calculate. Thus from measuring ground distances they can determine not only that the surface is curved, but by how much.

Gauss found that distance on a curved surface is a function of intersecting coordinates and curvature between points. What does this mean in geometric terms? Again, consider a two-dimensional curved surface. Stretch over it a flexible grid not unlike a fisherman's net: one that would shape up orthogonally on a flat surface.

In any region on the surface there will be two sets of curved lines that intersect. Quite simply, any point on an intersection can be found using one line from each set. Monotonically, number the lines. Having done this how can surface distances be determined? The Pythagorean formula does not work. Corrections must be introduced into the equation for a triangle on a slightly curved surface, but how?

Gauss determined the functional relation between the real-numbered pairs assigned to successive points on the grid. The functional relation is independent of the arbitrarily assigned coordinate values. Gauss found a way to characterize a curved line that is independent of the coordinate system used. Using his new values, he determined the distance to a given point. Gauss then modified the Pythagorean equation to calculate that distance. He formed a triangle off each grid line intersecting at the first point and drew opposing lines such that their intersection occurred at the second point. Geodesics can be defined using g-functions and the coordinatization.

Starting with the basic equation, he added to the two "corrected" squares, a third term, which is the product of the new values for the two side curves. To each of the three terms Gauss also added a coefficient which is a function of the curvature at the given point. The coefficient is proportionately related by the inverse of the radius squared of the circle best duplicating the shape of the surface surrounding the given point. The larger the radius the smaller the curvature. A basketball has a greater curvature than the earth. If the surface has constant curvature the ordered pair of numbers incorporating the value derived from the radius can also represent curvature at other points. The pair can then be translated or rotated and still represent surface shape.

The value of a coordinate point depends upon the distortion surrounding it. Gauss used a single function (the metric tensor or G-function) composed of product values (the g-functions) derived from a matrix based on the values for the two curves representing the coordinates at a given point.

Any finite length on the surface can be calculated from the g-functions when the points on the curve are given in terms of intrinsic coordinatization. Four g-numbers (one is eliminated) are assigned to the point. Gauss used these values for the metric coefficients in the modified Pythagorean formula. The coefficients of connection are functions of position, that is, of the coordinate point and its relation to coordinate lines that are nearby. The coefficient depends on the derivatives of the g-functions. The metric formula or line element fixes the geometric relationship between points on the surface. That relationship remains unchanged by any rigid transformation.

The shortest distance between points on a curved surface is called a geodesic. On a smooth surface, for example, the path of a geodesic can be determined by joining two points with a taut string. Geodesic curves are higher-dimensional counterparts of straight lines in Euclidean space. What is the use of geodesics?

In a plane tangent to a point on a curve a vector can be placed pointing in the direction of the curve. Manipulation of that vector can be used to measure surface curvature. Parallel transport the vector along the geodesic curve. Viewed from a higher-embedding space any change in vector orientation is attributable to the curvature of the surface. If the vector is initially tangent, it will also be tangent after movement.

What can be done if one grid system does not suffice to label all points on a surface? Again, we use the earth as an example. Earth's surface is intrinsically a two-sphere. Two numbers can locate a point on it using lines of latitude and longitude. The intrinsic geometry of a surface can be abstracted as distance distortions using overlapping charts, much as a world atlas is composed of many maps.

A local area on the earth's surface can be represented on a flat-two-dimensional map. Although a small area can be easily represented, a single coordinate grid may not intrinsically relate the larger surface of which it is a part. Two grids are needed to cover both hemispheres of the earth. One grid centers on the North Pole; the other centers on the South Pole. Notice that the grid for the Northern Hemisphere can undergo a rigid transformation (neglecting the inversion) becoming that used for the Southern Hemisphere. More than one grid is needed.

Earth's surface can be illustrated on flat paper. Note what happens in a Mercator projection. Curvature dictates how much the surface must be "stretched" or distorted to obtain a flat representation.

In 1854 B. Riemann, a student of Gauss, developed the concept of the n-dimensional manifold. The dimension refers to the number of variables or coordinates needed to locate a point on a multidimensional surface. Riemann extended Gauss's methods for a two-sphere. The number of components (*g*-functions) for a given dimension (n) is found by squaring (n).

When a single-coordinate grid cannot represent a surface, like the earth, mathematicians use the manifold. The manifold concept refers to the use of many-overlapping spatial-sections to represent a surface. In each section are many points, some of which are copied in other sections. Riemann thought the Pythagorean theorem would be valid if the section is infinitely small so that the surface could be treated as Euclidean. Such a scheme results in a contiguity of points interconnecting each section into a "product" space constituting the manifold. Riemann's idea describes the intrinsic geometry of a manifold as surface curvature.

Many geometers have studied the three-sphere, a three dimensional analog of the two-sphere. Geometrically, space is a three-sphere. Like the two-sphere it is positively curved. Measurements cannot be made on it from an extrinsic position by appealing to a higher dimension as can be done with the two-sphere. (Note: a two-sphere is embedded in three dimensions. For example, we could use a three-axis grid {x, y, z} like the Cartesian coordinate system to find positions on the earth's surface.)

Riemann said an object, having the same dimensions as the surface in which it exists, is free to move around continuously in that surface. He coined the term "continuum" to refer to the intrinsic aspects of any multidimensional surface or manifold. The continuum in modern times is further developed in the study of topology. Riemann's main point is curvature can be determined by carrying out measurements over a sufficiently large area. A set of points can be continuously mapped.

Riemann used a specialized-differential geometry. Einstein was looking for just such a tool. A former undergraduate friend, M. Grossmann, pointed him in the right direction by referring to Riemann's non-Euclidean geometry. Einstein continued developing this mathematical method—renaming it tensor analysis.

Let's take a closer look at some gravitational effects and see how Einstein's use of tensor analysis intimates the curvature of space-time. Einstein thought the geometric properties of space-time itself acted as a gravitational field. What did he mean? Instead of the moon being held to earth by an interacting gravitational force, he replaced Newton's concept of mutual attraction with curved paths on a four-dimensional surface. Einstein thought the moon merely followed the straightest possible path through space. Space around earth is curved.

The structure of space is the gravitational field. Such a field could replace and solve problems associated with Newton's "action at a distance." Einstein began to reason. Gravitation always seems to act in the manner of attraction. Its affect is cumulative. All bodies react the same (whatever the composition) in a gravitational field. Locally, as an ideal, gravitational forces can be eliminated. A free-falling box follows the straightest possible path (geodesic) or time-like worldline. A freely falling frame of reference, however, cannot be extended very far in space or time and is therefore, valid only for short intervals along a geodesic.

A free-fall reference frame is extensible only in the absence of gravitational fields, in which case it approximates the inertial frames of Special Relativity. Let's see why a freely falling frame (a local coordinate system) is only locally valid.

Visualize a box traveling on an earthbound geodesic. As the box falls, one part of it (the bottom?) will be closer to the earth than its other parts. Gravitational fields are invariably inhomogeneous. Picture the free-falling box as a collection of particles. Neglecting other forces, each particle follows its own geodesic. Because the particles closest to earth feel the stronger pull of gravity, the box and its contents begin to stretch and distort. Other effects take their toll. The sides of the box converge on radial lines toward earth's center. What if someone could view two particles on opposite sides of the box (perpendicular to the center of earth)? He would notice in the box frame of reference that the two particles come closer together as the box free-falls on converging lines toward earth. This effect in former times could simply have been labeled a "force of attraction." Such differential effects are named tidal forces because of their inherent similarity to ocean tides. These secondary effects cannot be eliminated through a freely falling frame and are considered real. Einstein thought a gravitational field would more aptly pertain to tidal forces that cannot always be eliminated as in cases involving free-fall.

In electrodynamics a charged test particle will be acted upon by the field in which it is placed. The resulting force can be represented by a single vector that can be decomposed and analyzed as three perpendicular components. (Three coordinates represent Euclidean space.) Einstein applied the same idea to a small test particle placed in a gravitational field.

Back to the falling box. Besides the mentioned effects, the box could also suffer forces tending to rotate, twist and sheer it. A single vector cannot represent all these stress producing forces. This is why Einstein used generalized vectors called tensors. A tensor is a second-rank vector requiring many components (the g-functions). These higher order vectors can represent the direction and magnitude that a given test particle could experience in space-time.

How do tensors describe curvature? Again, parallel transport on a two-sphere is used. Join two points on a two-sphere with the shortest possible connecting line (a geodesic). Place a vector at one point. The idea is to establish a vector of comparable magnitude and direction at the second point. Transport the original vector along the geodesic always keeping its directional arrow at the same angle (or tangent) to the line. When the vector reaches the second point, the arrow though seemingly altered, is directionally equivalent to its original orientation. The vector may be parallel transported along other lines until it arrives at its point of origin. This time the arrow may end up pointing in another direction. If this happens, it signifies a definite curvature of the surface. Curvature can be partially determined by parallel transporting a vector around a small enough closed loop, and comparing the change in arrow orientation with the enclosed area circumnavigated.

Einstein said all physical laws should be the same everywhere. Any arbitrary-coordinate system can be used to formulate those laws. (This contrasts with Newton's concept of absolute space, which calls for a fixed coordinate system.) This statement is identified with the mathematical principle of general covariance that forms the major mathematical basis for Einstein's General Theory of Relativity. It also expresses the principle of General Relativity. The principle pertains to how the laws are expressed. For example, the principle entails the geometric relationship of a collection of points in one coordinate system must be identical when transformed into another. Thus, when a set of vectors represents points on a rigid body, the object must be duplicated point for point in other manifolds.

Vector transport on a two-sphere also applies to four-dimensional space-time, but is more complicated. The idea is mathematically carried out by integrating infinitesimal parallel displacements of tensors. The curvature of a continuum can be characterized by noting tensor reorientation after mathematically transforming the tensor along a geodesic. The tensor portrays, in a physical sense, what happens to a test particle placed in the warped space-time of a gravitational field. The use of Riemannian geometry is used because a particle will accelerate in such a field.

The tensor, often called the Riemann-Christoffel's tensor (named, in part, after the Swiss mathematician E. Christoffel) describes the distortion (or worldlines compared to Euclidean-four dimensions) in space-time geometry. It describes the rate change (the first derivative) of the vectors near a point, but again, is independent of any coordinate system.

The tensor has thirty-six components that fully describe it. The components represent the six separate orientations of a four-coordinate system. Any tensor rotation is specified by these components. Because of certain symmetries, however, only twenty components are needed to analyze all possible motion in four-dimensions. The twenty components are usually divided into two groups of ten (hence the name tensor).

Each set of ten components can be totally transformed as a unit into other manifolds. If the tensor components of one set vanish in one coordinate system, they will similarly transform in any other system. That would mean the curvature is zero and the space would be Euclidean. One set, called the Einstein metric tensor (G) is a slight rearrangement of the Ricci tensor (after the Italian C.G. Ricci) which is built from the Riemann-Christoffel tensor. Part of this tensor is fixed; its divergence is zero and therefore relates the specific-coordinate system used. The Einstein tensor also enables the construction of vector and tensor fields on a manifold akin to vector field divergences occurring in Euclidean space.

By the end of 1915 Einstein had completed General Relativity Theory after deriving the co-variant field equation. It is often written as follows.

$$G_{\mu\upsilon} = 8\pi T_{\mu\upsilon}$$

The G is the Einstein tensor accompanied by its tensor functions or ten g-components. The Einstein tensor $G_{\mu\nu}$ is derived from the Ricci tensor.

The T is the stress energy tensor accompanied by a representation of its ten components.

The equation describes the relationship between the Einstein metric (on the left side) and matter (on the right side). (The metric is the formula used to determine intervals between two neighboring points.) Let's look at the full field equation.

$$R_{\mu\nu} - \frac{1}{2}g_{\mu\nu}R = \frac{8\pi G}{c^4}T_{\mu\nu}$$

$R_{\mu\nu} =$ (after Ricci) the curvature tensor The Ricci tensor $R_{\mu\nu}$ is constructed from the Riemann tensor. It describes gravity effects. If 0, it represents the vacuum equation and reduces to the flat space-time of Minkowski.

$g_{\mu\nu} =$ metric tensor and its coefficients—component functions of the coordinates (position). It describes coordinate differences of two close points by relating them to the space-time interval separating them. It represents space-time geometry.

$G =$ Gravitational constant. It relates to the mean density of matter in universe. This is Mach's contribution.

$T_{\mu\nu} =$ Stress energy tensor. It contains all information about gravity sources.

Einstein solved how to mathematically represent fluctuations in space-time geometry on the left side of his equation. It represents the gravitational field at a given world point by what happens to a vector at that point. Reorienting vectors during parallel transport yields information about geodesic deviation. The rate spread in geodesics, points to gravitational potentials that describe the acceleration of a free particle.

World points are correlated with the intrinsic geometry of space-time as described by the g-functions and the curvature tensor. The world point reflects what happens nearby. Six of the g-functions assure temporal continuity, while four—prescribed by the coordinate basis—impose restrictions on initial conditions.

Einstein realized for particles in free-fall (following geodesics) he needed to connect space-time geometry to gravitational sources (matter and energy). Using the equivalence of mass and energy ($E = mc^2$) from Special Relativity, the right side of his equation describes the mass-energy distribution in the region. It correlates the energy-stress tensor relating mass sources with energy (one vector) and linear momentum (stress, a three-vector).

How does the equation depict the relationship between geometric structure and gravitational source? The equation describes curved lines (geodesics) connecting world points in the field just as lines of force do for electric potentials (in Maxwell's equations). The curve passing through a space-time point, defined by the tensor, represents the direction a point-mass-test particle would accelerate if placed in that tidal field (or curvature). Acceleration is determined by geodesic deviation from initially parallel worldlines. The flux (number of curves per unit cross section) defines the strength (magnitude) of the local gravitational field.

Geometric structure is equated to the field representation of gravitational sources in terms of the energy-stress tensor on the right side of the equation. The number (density) of lines directly corresponds to the quantity of matter and energy in the area. The energy-stress tensor ($T_{\mu\nu}$) has the same formulation (uses the same transformation laws) as Einstein's tensor ($G_{\mu\nu}$—in the first equation).

How did Einstein do this? He combined the stress and mass-energy-momentum four-vector and developed a product tensor having ten independent components like the Einstein's tensor. Einstein said the energy-stress tensor and Einstein tensor are therefore, proportional. This connects the quantity of mass-energy-stress and momentum of the source with space-time curvature. That is the essence of the field equation. This relationship is valid in all manifolds thus adhering to the principle of general covariance. The four laws of continuity are preserved. One is about the conservation of mass (and energy) while the other three represent the conserved nature of the three components of linear momentum. (As already mentioned, the components are graphically delineated by line densities.)

Because Einstein's field equation is differential, it requires the specifying of boundary conditions to obtain solutions. The solutions then reflect the imposing conditions. Besides obeying the law of general covariance the equation must approximate Newton's law for weak fields, low velocities and small sources. To incorporate the inverse square law, curved lines are said to originate at infinity and converge up to the given mass. In a coordinate system, vanishing of the Einstein tensor components (in the left side of the equation) yields what is often called the vacuum (for empty space) field equations of General Relativity. In such a case the space-time metric is said to be flat. It means, in absence of matter, Einstein's equation describes a reference system similar to the Lorentz frames of Special Relativity.

More information about the metric: The metric or interval formula for space-time is not one just representing distance. As in the Minkowskian space-time of Special Relativity, the metric is an invariant interval separating events (or world points). For each space-time point (event) there is an arbitrarily large number of coordinatizations or Lorentz frames of which the one associated with the point is at rest. To each pair of such points are assigned numbers invariant under coordinate transformations. Between two events the tensor g-functions represent the geodesic metric.

The area surrounding any space-time world point is divided into distinct regions. While the square of the three-space metric is positive, the square of the metric of four-dimensional space-time can be positive, zero, or negative. The worldline of any event falls into one of these regions.

Events occurring in the zero classification are light-like and in vacuo follow null geodesics. A null-like event is timeless, and spaceless (lacks extensionality) in its own reference frame. Any interval along a null geodesic is zero. Two events (world points) share the same null geodesic (light cone) when they just become connectable by a common light signal. Any null-like event appears the same to all other observers. An astronomer viewing stars is looking at light that has traveled a past null geodesic. Any observer, on the future light cone, will observe the same effect.

Geometrically, null geodesics divide the other two groups from each other. The positive category refers to events moving forward in time from the world point sometimes called the "here and now." An interval along a time-like event is maximal because the temporal coordinate is opposite in sign to the three spatial coordinates. Time is maximized for a body in free-fall; that is, it most closely follows a time-like world line (geodesic). Time-like events are divided into absolute past and absolute future relative to the "here and now." Tardyon matter falls into this grouping of worldlines.

The negative classification is often called "elsewhere" and is space-like in nature. An interval along a space-like geodesic is minimal. This book has earlier suggested this category is associated with tachyon matter. A free-falling tachyon mass follows space-like geodesics.

H. Weyl, a German, discovered the second set of ten components had the interesting property of conformal invariance. This group of components is called the conformal-curvature tensor or Weyl tensor. Coordinate transformations are used to assign new coordinates point-for-point in another manifold. The underlying geometry is not disturbed. Conformal transformations, however, may alter that geometry. A simple example of conformal mapping is the shadow-like stereographic projection of a three-dimensional object onto a surface. At a given (here and now) point the time-like, space-like, and light-like lines remain unchanged by any conformal transformations. Both Weyl and Einstein tensors conjointly describe the matter induced curvature of space-time.

What are the scientific verifications for General Relativity? There have been five basic tests of which three are considered classic. Most of the tests have been based on the Schwarzschild solution to Einstein's field equation. Let's take a quick peak, without further comment, at the solution as Schwarzschild formulated it

$$ds^2 = (1 - \frac{2M}{r^2})dt^2 - \frac{dr^2}{1 - \frac{2M}{r}} - r^2 d\Omega^2$$

The German astronomer K. Schwarzschild found the Schwarzschild solution in 1916. It is an exact solution representing the field around a spherical-symmetric mass.

Neglecting perturbing effects of other planets, the Schwarzschild solution to Einstein's field equation predicts the perihelion of planet Mercury's orbit will advance the previously unaccountable 43 seconds of arc per century. This agrees with observation. We have already seen LeVerrier had proven that Newton's theory of gravitation could not fully account for this phenomenon.

A second test predicts a phenomenon unknown until after Einstein presented his theory. We have already seen one consequence of the Equivalence Principle is the deflection of a light beam by a gravitational field. The theory predicts light, because of its corpuscular nature, will bend 1.75 seconds of arc as it passes close to the sun.

In 1919 a British astronomical expedition headed by A. Eddington traveled to the island of Principe off the West African coast. (Another team went to Brazil.) Eddington set up apparatus to detect the predicted bending of starlight during a total solar eclipse. His group compared photographs of stars taken during the eclipse with some photographs taken five months earlier when the stars were not in line with the sun. An examination of photographic plates confirmed, within close agreement to Einstein's prediction, the displacement of starlight away from the sun.

During the mid 1970's researchers E. Fomalont and R. Sramek used radio telescopes to detect the bending of radio waves emitted by certain quasars. Quasar sources can be used anytime their radio waves glaze the edge of the sun. Einstein's prediction was verified to an uncertainty of one percent.

A third classical test of General Relativity is gravitational spectral shift. Modern technology presented scientists with new ways of testing theories with greater precision. One such technological advance enabled the Americans R. Pound and G. Rebka to perform an experiment in 1960 using the Mössbauer effect. It involved recoilless emission and absorption of monochromatic gamma-rays. Their setup practically eliminates the Doppler effect.

The experimenters used the iron isotope of atomic mass fifty-seven. Their idea was to use nuclear vibrations as "atomic clocks." The precision of their device is good to one part in a million years.

Rebka and Pound positioned iron nuclei at two locations 74 feet apart in a high tower. The iron is embedded in a crystal-like lattice that virtually prevents recoil from gamma emission. A similar setup is used as a target in which gamma rays are absorbed.

A variation in clock rates occurs between source and target positions because the effect of gravity varies at the two locations. The source at the top of the tower is set in motion toward the target at the bottom to compensate for this difference. The velocity of the source is very slowly increased until a detector records a minimum counting rate. Exact resonance is achieved at this speed. In this experiment the velocity is a couple of millimeters per minute.

Pound, with another colleague J. Snider, improved the apparatus and performed the experiment again in 1964. They found the frequency shift—due to the radiation moving up against earth's gravitational field—correlates with the slowing of clocks near earth's surface. The experimental results correspond to Einstein's predictions. Red-shift predictions can be deduced from the Equivalence Principle without recourse to Schwarzschild's solution.

A similar experiment has also been performed using cesium[133]. In 1975 two airplanes, traveling in opposite directions, were used to detect the influence of velocity and altitude on cesium clock rates. Again, the results fell within the parameters allowed by Einstein's General Relativity Theory.

Mach's Principle and Variable *G*

We now examine some yet unsettled issues concerning inertia and gravitation. Is the property of inertia intrinsic to matter or does it depend on something outside a material body? How can it be determined if a frame of reference is truly inertial? Newton said a body at rest is defined by absolute space that is "immovable and always the same." In his view it was absolute space that resisted acceleration. He also thought it would be possible to determine in principle, if not experimentally, a body's motion relative to this all pervasive medium.

Newton performed an experiment to show the effect of absolute space. He suspended a bucket from a rope, which he twisted. He filled the bucket with water; then waited for its surface to settle. He let the bucket spin. The water first remained still; then it too began to swirl. The surface of the water concaved to the sides. Newton stopped the bucket while it was spinning; the water continued to swirl.

Newton argued his experiment implied the water swirled relative to (absolute) space when its surface concaved. He interpreted the rotation of water as a specific example of absolute motion. That is, absolute with respect to absolute space. Thus, a body was either at rest in—or experienced motion relative to—absolute space. He stated the surface concavity happened whether or not the bucket was in motion with respect to the water. The water surface curved despite anything outside the bucket. He said the water swirling accelerated to the center of the concavity thus showing absolute acceleration.

The idea of absolute space came under heavy criticism. In 1692 G.W. Leibniz, the German philosopher and mathematician, thought Newton was in error about absolute space. He said if motion is change of position then it cannot be absolutely determined which moved—the body or its surroundings. Real motion can only be ascribed to one or the other. If motion is arbitrarily ascribed to one, then the other may be considered at rest. He continues: Space is nothing but arrangement of things. He said nature does not make a distinction between a thing's motion and its surroundings.

About 1710 G. Berkeley, an Irishman Bishop, also said there is no absolute space. He said Newton's bucket experiment would not yield the same results if there were no other matter. He added, the bucket was also in motion because of the rotation and revolution of the earth. He said there could be no motion but what is relative. There would have to be at least two bodies so that one could be used as a standard by which to gauge the other's motion. If there existed but one body, conceivably it would not suffer any motion. He thought even if two bodies circled about a common center that no motion could be discerned without the fixed stars (the galaxies). Berkeley applied the same thinking to Newton's bucket experiment. He thought the water surface concavity was influenced by the fixed stars.

In 1872 an Austrian named E. Mach also said there was no absolute space. He believed inertia arose from a body's interaction with all other masses. He thought the motion of matter is relational. Leibniz previously espoused the same idea.

Mach said there is no geometric difference between the first case of the earth rotating on its axis relative to the fixed stars and the second case where the fixed stars revolve around a nonrotating earth. Mach thought the law of inertia must reflect a formulation derivable from either view. He thought if all except earth disappeared, the idea of a rotating earth would make no sense according to the usual interpretation. That is, there would be no centrifugal force, hence, no acceleration and consequently, earth would not exhibit the property of inertia (other than a diminutive "self-inertia"). Because in the second case the earth at rest would not experience centrifugal forces (analogous to the concavity of water in Newton's bucket experiment) it would not suffer equatorial bulge. To derive the law of inertia in the second case, without invoking the notion of absolute-space, inertia would have to be induced by something else. Mach suggested the fixed stars provided the answer. Stars constitute a reference frame by virtue of their "smeared-out" mass. This contrasts with Newton's idea that if space were devoid of all other matter that earth would still exhibit inertia.

Mach's ideas were not immune to criticism. He never did explain exactly of what the interaction between masses consisted. Anyway, it must depend heavily on the quantity of mass because the multitude of distant bodies had to outweigh local effects to maintain a smoothed-out frame of reference. It must also depend on the body's relative acceleration to other masses and by nature the interaction must be of long-range.

Observationally, there is little difference between the consequences of either Newton's or Mach's ideas. Yet, there should exist a difference, however, small, in a body's inertia when subject to highly distorting local quantities of mass. The idea a body's inertia should increase when placed near a massive object, opened a possible way to test Mach's principle. In addition, the center of a galaxy could serve as a "local" reference frame.

Mach finally thought the interaction between masses that produces inertia was somehow connected to gravitation. However, he could not offer proof. He realized Newton's inverse-square law could not supply an answer. Maybe some other formulation of gravitation would explain the interaction between mass and inertia. Einstein had hoped to provide the answer with his Equivalence Principle.

Einstein hoped to show inertia originated in gravitation. Also, by eliminating the requirement for preferred coordinates (using the principle of general covariance) he sought to eliminate the need for absolute space; thereby, lending credence to Mach's ideas. In this endeavor Einstein was successful. He did show space (space-time) was not a phenomenon which "acts but cannot be acted upon." It was Einstein who coined the phrase "Mach's Principle" to represent the loose collection of ideas mostly associated with Mach.

Despite Einstein's attempt to answer Mach's sought after explanation of inertia, Mach's Principle has not been successfully incorporated into the General Theory of Relativity. The gravitation of General Relativity did not explain inertia as acted upon by other matter as Mach thought. Instead, the inertia of a body is acted upon by the curvature of space-time. Carrying the same reasoning further, space-time structure is not totally described by the mass-energy distribution that functionally replaces the "fixed stars." Other factors such as boundary conditions also play a significant role.

Increased scientific understanding has added to the confusion. For example, the three-degree-cosmic background radiation might serve as an inertia reference.

However, an anisotropy has been detected in the radiation. Elementary particle physics tells us space is filled with virtual particles having net-zero energy. This concept alone could replace the fixed stars as an inertial reference. Scientists, however, do not know what to make of the quantum nature of such a reference frame.

The basis of Mach's idea seems to have as much support as does General Relativity itself. It is just that the two theories do not naturally mesh. Theorists have tried to make one a consequence of the other. If inertia resulted from gravitation, it would depend on the collective action of all masses. Space-time structure would then depend upon mass distribution.

General Relativity does not produce Mach's Principle in all cases. A universe, devoid of all matter, would still satisfy Einstein's field equations but would not include Mach's idea of inertia. In 1917 de Sitter found just such a solution. A universe exhibiting space-time structure but empty of matter and radiation, points out the unnecessary inclusion of Mach's Principle in General Relativity. Einstein later noted the non-Machian character of his theory is attributable to boundary conditions.

Mach's Principle had other consequences that have been incorporated into some nonmetric theories of gravitation. For example, if gravitation is dependent upon the distribution of masses (read galaxies) and yet the galaxies are receding from each other through universal expansion (and there are no compensating factors) it follows the gravitational constant should decrease in value.

The calculated rate of diminishing local gravitational influence varies according to the theory used. One such theory is the Brans-Dicke model that was developed in 1961 by the Americans C. Brans and R. Dicke. It is a scalar-tensor theory in its description of the gravitational field. The scalar quantity determines the strength of gravitational interaction or the gravitational constant (G). The theory predicts lesser values for the bending of a light beam and the advance of Mercury's perihelion than does Relativity Theory. The predicted differences seem to fall within present experimental parameters thus assuring the Brans-Dicke model as an alternative to Einstein's theory.

There are other theories suggesting a weakening G that discount universal expansion as the major factor. General Relativity attributes gravitation to changes in the curvature of space-time. H. Weyl wondered how electromagnetism might have arisen from changes in the local scale (gauge: the standard of temporal duration or spatial length). Weyl thought if equations were developed to describe how the standard differs from place to place that equations could be found, which represent the electromagnetic field.

Dirac followed Weyl's idea of uniting gravitational and electromagnetic phenomena. In 1937 he began developing a theory known as the "large number hypothesis." Dirac examined the ratios of certain physical constants. He first compared the ratio of electric force to the gravity that couples the hydrogen nuclear proton and its orbiting electron. He found the electromagnetic force to be 10^{40} times greater than gravitation. Dirac then states the age of the universe (based on 20 billion years) is 10^{40} times the smallest known unit of time called the chronon (about 10^{-24} second). Dirac believed this, and other ratios, suggested something more than mere coincidence.

If there is a connection between these large numbers what could it all mean? Dirac thought if one factor increased in value, the others must also change. For example, when the age of the universe doubles, it would be twice its present value in chronon terms. To again match that ratio to the electric force/gravity ratio, one of the latter factors had to change. It has been determined by examining the fine-structure splitting of emission lines observed in the spectra of distant sources that electric force does not change. To keep both ratios equal, Dirac finally concluded gravity must decrease in strength as the universe aged.

Dirac's theory led to the idea of continual-matter creation in a closed finite universe. He thought this could happen in one of two ways. The first, additive creation, states matter is randomly created anywhere from nothing. If this were the case, Dirac found when the age of the universe doubles, its mass would quadruple according to another fundamental ratio. However, this idea is also riddled with problems. For example, additive creation via Dirac's theory leads to the conjecture that the solar system is contracting with the earth-sun distance decreasing. If that happened the sun would have burned itself out over two billion years ago because insufficient matter would have been added to the sun.

Dirac formulated a second method known as the multiplicative process. Matter is proportionately created where it already exists only at the expense of a compensating creation of negative mass elsewhere; thus, assuring mass conservation. Dirac's theory, considering multiplicative creation, leads to an expanding solar system where the sun-earth distance increases. Weakening gravity and bodies increasing in mass conflict with General Relativity. Dirac reassessed his theory. He suggests changes in gravity and mass do not actually occur but are illusions. Instead, only mass is added to objects to compensate for that lost by physical processes. Only atoms are affected but not the distances between objects. For example, in one year it will apparently take 2×10^{-10} times longer for sunlight to reach earth.

In Dirac's revision the increased time is not due to increasing distance but to a slowing of "clocks" (atomic processes) in the sun and earth. Carrying the same idea further, we find the red-shift of stellar light could be explained not by universal expansion, but by a change in the gauge or standard of space-time affected during the transmission of light. This is where Weyl's ideas naturally occur in Dirac's large number hypothesis.

The variable gauge theory still does not account for the 2.7 Kelvin background radiation that supports the Big Bang Theory. The variable gauge idea implies an eternal universe that does not expand or contract. Solar system experiments to date have not substantiated a declining G (gravitational constant). In the vein of Dirac's work: theory suggests that G decreases at the rate of 5×10^{11} parts per year. Although testable, experimenters have only obtained negative results. In science, however, no reasonable possibility can be dismissed.

Gravitational Radiation

We are all familiar with the effects of gravity. As with most phenomena, understanding gravity is quite different from the way it is experienced. Einstein's General Relativity Theory does not consider gravitation as a natural effect of something more fundamental. It is not a phenomenon embedded in some more general cause. A certain mass distortion is accommodated by a specific space-time curvature, which then affects the distribution of matter. One phenomenon carries no more import than the other does; in a sense, they are equivalent.

The behavior of matter is restricted to the geometry of space-time. A massive body produces curvature in its vicinity. Another nearby body would also produce curvature. A product curvature develops to which both bodies respond. This common response is described by Newton's theory. Like Newton we perceive the response as two bodies attracted by a certain force. Einstein's theory instead suggests the structure of space-time acts as an intermediating factor that affects both bodies.

Enveloping practically all else, gravitation is an end in itself. There are no symmetry considerations that call for additional-logical structure. This is a brief scientific understanding of gravitation effect. However, there is more to gravitational manifestation, and that is its cause.

In 1918 Einstein published a paper showing his field equations also yield radiative solutions. Such solutions are difficult to obtain because of the nonlinearity of the equations (gravity couples to itself). The equations dictate gravitational radiation travel at a finite speed—the same as that of light.

Gravitational radiation itself is void of spatial-temporal properties. This also applies to electromagnetic radiation. Extended spatial and temporal properties are ascribed to gravitational radiation because it manifests itself in the Realm of Whatness (for example, as a curvature of space-time).

What are gravity waves? P.C.W. Davies said they are "ripples in the curvature of space-time itself." It is a tidal field propagating throughout the fabric of the universe.

How are gravity waves produced? Just as oscillating electric particles emit electromagnetic waves, so it is thought oscillating masses produce gravity waves. Gravity waves are also thought to result from a change in the distribution of matter, which excludes wave production from symmetrically rotating spheres.

Gravity waves carry energy and momentum and affect all matter-energy in its path. As a wave passes through, it sets up oscillations in any matter or energy form it encounters. As far as gravitational radiation is concerned, space-time behaves like an elastic medium.

What type of vibratory phenomenon is gravitational radiation? Because total-linear momentum is conserved, there are no dipole-gravity waves. The first radiative multiple allowed by Einstein's tensor field theory is quadrupole. It is similar to the quadrupole-electromagnetic radiation associated with that emitted by certain FM radio band antennae.

As Einstein noted, the wave might exist in either of two polarized states—both of which act orthogonal to the direction of propagation. When one polarized form of the wave passes through matter, there occurs along one transverse axis a spatial contraction followed by dilation. Along the other perpendicular, there is a corresponding dilation followed by contraction. The second polarized form induces similar effects along a perpendicular rotated 45 degrees from the first state. The net result is a superposition of the two states. The process continues at the frequency of the wave—inducing repeated quadrupole deformations.

Gravitational waves interact with all major forces (interactions). Any form of energy is subject to its influence. However, compared to the other three major means of interaction (strong, weak, and electromagnetic) gravitation is very weak. Gravitational attraction is 10^{37} times weaker than the electromagnetic force binding the electron and the hydrogen nucleus together. Compared to the strength of the short range forces, gravitation is 10^{34} weaker than the weak force and 10^{39} weaker than the strong force.

As gravitational radiation passes through space some of it may be absorbed (imparting energy) and re-emitted (extracting energy from the media). Because it is very weakly interactive, only an extremely small amount of energy would be involved.

Because matter absorbs very little gravity wave energy it is thought by some that gravitational radiation has been accumulating since the Big Bang. The universe is now thought to be bathed in a background-like flux of gravitational radiation not dissimilar to the three-degree-cosmic-background-electromagnetic radiation. The detection of such a background could tell us about events immediately following the Big Bang.

The relative weakness of gravitational radiation also makes it difficult to detect. It is considered implausible that any earth made Hertzian type generator could ever produce measurable gravitational waves. It is equally improbable that gravity waves could be detected on an atomic level. However, unlike the electromagnetic charges that are often negated by their positive and negative properties, gravity waves, because they come in one charge only, can accumulate in large scale phenomena.

From the idea electric charges accelerate in the presence of electromagnetic waves, scientists think masses would also accelerate in the presence of gravitational waves.

A radio pulsar (rotating neutron star, found in a narrow orbit in a binary system) should mass-contract because of loss of energy from emitted gravitational radiation. The loss of energy should be indirectly detectable by measuring a change in the bursts of its radio waves. An increased frequency would signify orbit contraction and therefore, mass loss from radiating gravity waves.

There have been experimental attempts to detect gravitational radiation directly. The idea is to determine the effect of gravity waves on the relative motion of two widely separated particles. One type of experiment uses an aluminum disk, or cylinder kept at cryogenic temperatures. It is configured to resonate under the influence of a strong passing wave. The wave would impart energy to longitudinal vibrational modes of the metal. It is hoped vibrations in the kilohertz range can be detected by carefully placed piezoelectric crystals that could detect mechanical deformations in the metal.

What could supply enough energy to emitted gravitational radiation so it could be detected? A collapsing neutron star, as earlier mentioned, could give off waves, which could be measured. If gravitational radiation were detected and properly measured, it would indicate space-time curvature.

Is gravitational radiation quantized like electromagnetic radiation? In 1958 Dirac found Einstein's field equation could be quantized. He found a quantum of gravity (graviton) is energy equivalent to the product of Planck's constant and frequency of the wave. Like the photon the graviton has zero mass and therefore, infinite range. It is thought to be a boson having integer spin twice that of photons. Gravitons, like all even-spin particles, generate an attractive force.

Quantum theory describes gravitational interaction (force) to be a result of the exchange of gravitons. This is similar to the interaction of charged particles connected with the emission or absorption of photons. The graviton, like the photon, does not decay. It is its own antiparticle.

Associated with the graviton is the 3/5 spin fermion called the gravitino. It has a very short operating range and may acquire mass by spontaneous-symmetry breaking (see chapters, 15, 19, 20, and 21). If gravitons produce gravitinos they do so through pair creation. It is thought most any particle can be transformed into gravitinos and therefore, theoretically back into gravitons.

This book considers that gravitational radiation may be connected to the wholeness aspect of electromagnetic radiation. See Chapters 19, 20, and 21.

Gravity, Particle Creation and Sustentation

Let's examine the concepts of creation and sustentation. Gravitation is so strong near the singularity of a cosmic black-white hole that it pulls apart the very fabric of space-time. The process generates such extreme temperatures and pressure that sufficient energy is imparted to virtual particle-antiparticle pairs.

Because many elementary particles are created in complementary pairs from photons in the presence of extremely strong gravitational fields, another leap of imagination envisages the Realm of Whatness (tardyons, tachyons) to be created from the Realm of Reasoning (photons, electromagnetic radiation).

What about sustentation? Once anything (tardyons and tachyons) is created, it seems reasonable that it must be sustained. Gravitational radiation changes the distribution of matter, and changes in matter can produce gravity waves. As J. Wheeler said: "Matter tells space how to curve, and space tells matter how to move." Because of matter there is gravity and because of gravity there is matter. It seems that one does not exist without the other. Let's look at it differently.

There is no What without Why; no effect without a cause. One is an inventor only because of his invention. The invention exists only because of its inventor. One cannot exist without the other. What causes what? Does matter in motion produce gravity waves or do gravity waves produce motion in matter? There is no creation (Realm of Whatness) without its creator (Realm of Whyness) and no creator without its creation. One only exists because of the other. Like many phenomena the two concepts are complementary and mutually exclusive. Note that there is no problem here with the notion of an endless regression of causes. We ask why is there a creator, and answer because of the creation. There can be no end to asking why, only an end to what that why corresponds (to in reality).

Key Points:

- The Realm of Whyness represents the widest noncircular why question that can be asked (about reality). The Realm of Whyness is the reason for the Realm of Whatness.

- The Realm of Whyness is scientifically represented by gravitational radiation. It functions only via space (itself). It does not function in time. (That is reason enough not to label this "force" a radiation. Its effect, however, does function in extended space and hence functions manifestly in time. We do observe its "effect" on physical bodies. It is called gravitation.)

- Galileo discovered that dropped objects, regardless of composition, fall to earth at the same rate. Newton discovered the force acting on celestial bodies (e.g., planets) and the force acting on falling apples were equivalent. The force diminished as the square of the distance between objects increased. This explained why the speed of planets increased when closer to the sun (in their elliptical orbits). The gravitational force was thought by many to operate via "action at a distance"—instantaneously.

- Newton found no proportional difference between inertial mass and gravitational mass. Eötvös refined Newton's experiment. He also discerned no difference.

- Pushing these ideas further, Einstein stated his Equivalency Principle. It stated one could not detect from inside a closed box whether the force experienced was caused by the presence of a gravitational field or caused by acceleration. He stated they were equivalent. More specifically: One cannot discern a difference between the pull of earth's gravity and the pressing force experienced in an equivalent uniformly accelerating space ship. The Equivalence Principle became the basis of Einstein's General Theory of Relativity.

- General Theory of Relativity correlates space "curvature" with the presence of matter. Einstein's theory predicted light would "bend" in a strong gravitational field. Eddington, during a solar eclipse, verified starlight "curved" near the sun.

- If gravitation is a conceptually isolable aspect of reality, it needs to be represented by the structure of reality. The Realm of Whyness represents gravitation itself (not its effect).

The whole structure of Reality has now been delineated. We just do not fully understand it. We have determined that the Realm of Whyness correlates with gravitation in some grand sense. Perhaps gravitation can "unify" all the other forces and also therefore can explain by "inclusion" everything that exists.

Although we have finished "mapping" the mind expanse, we are still trying to understand the structure we have mapped. We need to ask deeper questions, ones that can be answered either explicitly or in categorical subsumption by that structure. This sounds easy, but it is not. We still do not fully understand the unity of reality.

There have developed nonscientific nomenclatures and concepts that also correspond to the structure of reality. We will look at nonscientific terminology to see what in reality, if anything, they could refer. This will be found in Section Six: Interpreting Structure. However, before that bridge is crossed we have another chapter dealing with the structure in scientific terms.

Reality is structurally composed of many logical units that can be schematically united. This has been illustrated in this book. It would be easy to delineate the structure of Reality and let it go at that. However, we would be irresponsible if we did not show how the individual structures are conceptualized as a unit. To do so is to scientifically piece all the various structures together as a "functioning" unit. The next chapter attempts to answer how the many structures are conceptually integrated in the scientific sense.

Section Five

REALITY–UNITY: STRUCTURAL INTEGRATION

Perspective

- Reality represented as a functional unit
- Structural integration of reality
- Reality reconsidered with conceptual unification via physics
- A deeper understanding of the Realm of Reasoning

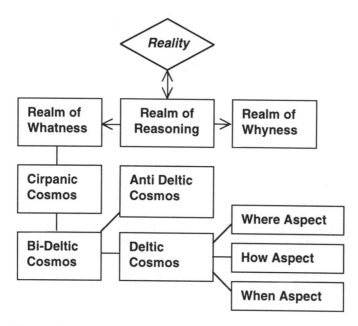

Figure 15:1. Perspective: Reality Unity

The Reality-Unity Diagram is the common yin yang symbol. It
represents Reality as a functioning unit. The diagram represents All
of reality and every part. The outer circle represents All of reality,
but no part. The inner "parts" represent everything, but not
everything as a whole. The two "twists" (yin and yang — female and
male) represent the complementarity of everything: that everything is
created in pairs of opposites. One is the Cosmos; the other is the
Anticosmos. As we "circle" the inside of the diagram, we notice each
"twist" represents the "diverging" (widening) or "converging"
(narrowing) of its respective Cosmos. The two inner small circles
suggest the whole (yang) is in the parts (yin) and the parts are in the
whole. Each Cosmos harbors a little of its opposite.

Figure 15:2. Reality-Unity Diagram

Chapter 15

Reality—Unity: Structural Integration

> "People speak without clearly defining their words.
> People listen without questioning. 'Not to question'
> and 'not to act' are decisions in themselves."

A major goal of science is to show how disparate phenomena can be classified into general concepts. That goal is realized by building structure. Structure is built by interrelating parts. Parts (composing the structure) are explained (holistically—methodologically by induction) by how they fit into the structure as a whole. Also, the whole is explained (reductionistically—methodologically by deduction) by its parts. Here we set up the means to better understand the whole and part.

Restating: Reality, as a whole, is explained by its parts. Reality's parts are explained by "Reality" itself (as a whole). Wholeness and parts (functionally) constitute reality.

"Explaining," as we have seen, is done by relating each part to other parts; then parts to whole and whole to its parts. Structure infers internal relationships. We gain understanding from the effort to connect phenomena. The unity of reality is represented by the why-what paradigm.

This chapter tells of the effort of physicists to unify the major "forces of nature" into encompassing concepts. The ultimate goal of science is to understand All Reality. For physicists this means that relativity and quantum theory need to be understood in one conceptual scheme. That goal is realized by the end of this book.

Reality-Unity Diagram

Reality as a complete structure is easily represented by the Yin Yang symbol. The outer circle represents all of reality (its completeness) but not its parts. The inner "twists" (Yin Yang) reflect the complementary aspects of "everything in" reality (Cosmos and Anti-cosmos). Each twist denotes the converging or diverging nature of the Cosmos through the flow of time.

Overview

Structures in this book are subsumed under three covering concepts called Realms. Each realm categorically correlates with certain phenomena called forces or interactions.

The Realm of Whatness is represented by matter-fields: mass-carrying-fermion particles (tardyons and tachyons). Forces interacting between matter fields operate over a noninfinite range.

Fermions are particles having odd multiples of intrinsic-spin-angular momentum (of value 1/2 Planck's constant). Fermion particles obey Fermi-Dirac statistics.

They have rest mass and are only created in particle-antiparticle pairs. Fermions are subgrouped into hadrons (an oversimplification) and leptons.

Hadrons are subgrouped into baryons and mesons. (Correcting the above oversimplification—mesons, though they are bosons obeying Bose-Einstein statistics [after S. Bose and A. Einstein], are composed of quarks, which are fermions.) All baryons interact by the strong force. This force was thought to be mediated by mesons. A subsystem is now understood to be the underlying mechanism of the strong force. Its exchange-force particle is the gluon (also a vector boson) which acts on the quarks that constitute hadrons.

Certain leptons and quarks are subject to the weak force. No leptons interact by the strong force. Both charged hadrons and leptons are subject to the electromagnetic force.

The Realm of Reasoning correlates with the electromagnetic force (unmanifested electromagnetic energy—radiation). The electromagnetic force functions over an infinite range.

The Realm of Whyness correlates with the gravitational force (unmanifested gravitational energy—radiation). The gravitational force also operates over an infinite range.

Four Forces of Nature

By the early 1930's it was known there existed four fundamental forces. Physicists found that elementary particle interactions could be categorized according to certain uniquely-obeyed-conservation laws. The laws involve various types of charges. The charges often show whether the interaction involves attraction or repulsion. The categories of interaction suggest there are but four basic forces in nature. Other forces can be classified as a derivative manifestation of one of these four. For example, the electric force is but a derivative manifestation of the electromagnetic force.

Quantum physics supplies the necessary information to piece together the interrelationship between the forces. The forces are not directly observable. We learn about the forces of interaction by studying high-energy experiments.

Strong Force

The strong force holds atomic nuclei together. Strong interactions are associated with hadrons. Hundreds of particles have been classified as hadrons. The mesons have intermediate mass between baryons and leptons. Mesons are boson particles and have integral spin and obey Bose-Einstein statistics. Bosons can be created without regard to their antiparticles. Three of the lighter and longer-lived mesons are the pi-meson (the first discovered meson, 1947) the K-meson and the eta-zero meson. An unlimited number of (non-rest mass) bosons can occupy the same energy level and have the same quantum numbers.

Baryons are nucleonic-like fermions. They have half-integral spin thus obeying Pauli's Exclusion Principle. Fermi and Dirac used this principle in formulating concepts of subnuclear states. In any interaction the total number of baryons is constant (excepting the possibility of proton decay, more about that later). Unlike bosons these particles can only be created or destroyed along with antiparticles from their own class. Together these ideas form the basis for the concept of baryon conservation.

Baryons can be classified by spin. Hadrons are divided into bosons and fermions according to integral spin (being half or whole multiples of Planck's constant). It was stated in Chapter 11 that spin is intimately connected with the concept of parity.

Integral spin is a measure of spin-angular momentum (which is the rotational counterpart to the law of conservation of linear momentum). Together angular momentum and rest mass can be used to group hadrons.

A useful chart is developed by graphically plotting spin and rest mass. Certain values of integer spin are paired with particle rest mass. A slanting line can be drawn through dots representing particles sharing most of the same quantum numbers, but exhibiting different spin angular momenta. The connecting lines are called Regge trajectories after T. Regge who first developed the method. The lines connect dots representing allowable integer spins.

Baryons are represented by dots placed at half-odd integers with a separation between different states of two-whole integers on a line. Each dot on a line corresponds to a resonance or higher-energy version of the same particle. The lowest dot on a line corresponds to the most-stable-energy level (which has the lowest value of spin-angular momentum). For example, the nucleon is most stable at an energy level of 939.5 MeV (million-electron volts: a measure of rest mass) where it has an integer spin of 1/2. The nucleon also has a resonance with an effective mass of 1,688 MeV with a spin angular momentum of 5/2.

Summarizing: Dots on a Regge trajectory represent the energies that fermion particles are allowed by spin to appear as physical entities. The trajectory represents a continuous plot representing a family of particles having all the same quantum numbers except spin.

The use of Regge trajectories is one method of classifying baryons that led to the prediction of once unknown particles. By extrapolating a line from known particles to other allowable spins, resonance frequencies (in units of MeV's) are easily charted. These frequencies show, energy-wise, where to look in scattering cross-sections (a method used in high energy physics). These methods also predicted mesons.

Conservation principles lead to another method of classification. For example, the nucleon is a composite particle and is the lowest energy state of the strong interactions. When charged it is called a proton, when uncharged it is called a neutron. In strong interactions the proton and neutron behave alike. Despite this, they are differentiated by the principle of isotopic spin invariance (an idea borrowed from the conservation of spin angular momentum). The idea is patterned after work done in 1918 by H. Weyl. In this respect, other baryons are higher-energy states of the nucleon. These more massive variations include the lambda, sigma, and cascade hyperons.

The hadrons (baryons and mesons) can be further "ordered" by other quantum numbers also based on conservation laws. These properties remain unchanged after interaction. Some conservation laws do not apply to mesons and baryons. This difference orders hadrons into separate classes.

Again, hadrons are first grouped by mass differences. Particles with nearly the same spin can be further grouped according to their electronic-charge states—usually represented by +1, 0, −1. This law approximates what happens because of interference from electromagnetic disturbances on charge. (These disturbances cause a difference in mass between the members of a particle group having the same isotopic spin.)

Isotopic spin usually orders the various particles into "multiplet" groupings of one, two or three: called singlets (for example the lambda particle) doublets (nucleon) and triplets (the sigma baryon and the pi meson).

The number of multiplicity or charges is used in calculating the quantum of isotopic spin. The value of isotopic spin is one more than twice the number of members of the multiplet. (Do not confuse isotopic spin with spin angular momentum.)

Hadrons are further grouped according to the conservation of strangeness (or another form of the same idea—hypercharge) which involves the average charge observed in an isotopic multiplet. The strangeness-quantum number is equal to twice the average of the electronic charge less the baryon number. (Hypercharge is equal to twice the average of the multiplet. Hypercharge is thus equal to strangeness plus the baryon number.)

These laws partially explain why specific combinations of hadrons are necessary for the production of certain reaction products. Strangeness was first used in the 1950's to represent heavy particles (compared to other particles of the same class) that "strangely" took longer to decay. Examples are the lambda baryon and the K-meson. Strangeness conservation was originally introduced to explain why K-mesons were not produced alone in conjunction with hyperons (one K-meson + hyperon = 0 strangeness). The strangeness is now understood by conservation of "color." (More about that later.) The property of strangeness is conserved in strong and electromagnetic interactions.

There is another way of ordering the "elementary particles"; one that led to deeper insights into the subatomic world. In 1961 Y. Ne'eman and independently M. Gell-Mann ordered the hadron family of particles into a system of supermultiplets. They made a classificatory system grouping particles having the same spin (angular momentum) or the same parity.

The particles are arranged in each group by plotting strangeness-quantum numbers with isotopic spins. The mesons can be grouped into supermultiplet families of one or eight members. The baryons can be classified into supermultiplet families of one, eight, or ten members. (A ten-member-baryon family will shortly be profiled.)

The system of ordering particles into supermultiplets has often been called the "eight fold way" because it uses eight-quantum numbers and strikes a parallel with Buddha's path to noble truth. The mathematical basis for arranging the particles is found in algebraic systems developed by S. Lie in the latter 1800's.

The system used in the eight-fold way is called SU(3) symmetry. It refers to a special unitary of matrices of 3 x 3. A "special" condition reduces the elements of the array from nine to eight. Three of the elements represent isotopic spin orientations; one represents strangeness (hypercharge). Two more represent transformations between differing states of strangeness. The last two are associated with transformations between differing states of both strangeness and electric charge.

How is a baryon decimate (or decuplet) assembled? We begin by selecting particles with a baryon number of +1 having spin-angular momentum of 3/2 (each family member—same spin). There are many particles having this spin and strangeness. The chosen particles must also be of the same parity.

The ten particles meeting these criteria are categorized by vertically plotting strangeness and horizontally plotting electronic charges. Values of strangeness range from 0 to –3. Particles of 0 strangeness have the lowest mass, and the particle of –3 strangeness has the highest mass. The electronic charges vary from –1 to +2. Differing values of charge have little effect upon mass. For 0 strangeness, there is a four-member multiplet of charges with an isotopic spin of +3/2. Each member has an approximate mass of about 1236 MeV. These are delta particles having electronic charges of –1, 0, +1, and +2. At strangeness –1 three sigma particles (isotopic spin of +1, and about 1386 MeV) have electronic charges of –1, 0, +1. At strangeness –2, two-cascade particles (isotopic spin + 1/2, and about 1535 MeV) have electronic charges of –1, and 0. At strangeness –3, the tenth particle (isotopic spin 0, and about 1696 MeV) has an electronic charge of –1. This particle was unknown when it was originally postulated using the supermultiplet model. It is appropriately called the omega-minus particle and was discovered in 1964. These methods are used to categorize mesons and other baryons into supermultiplet orders.

It soon was apparent the known hadrons fit quite nicely into supermultiplet families having only one, eight, or ten members. SU(3) symmetries could allow each family to have three, six, or more members, but these have not been observed. The question then is: Why are there only three patterns (groups of one, eight, or ten) of supermultiplet groupings when others are possible?

In 1963 M. Gell-Mann and G. Zweig independently proposed a solution. They said if hadrons were constituted from still more fundamental particles, an answer to the preceding question might be found. Gell-Mann called these supposed fundamental particles "quarks." He originally postulated three quarks calling them u, d, and s for up, down and sideways (strangeness). The idea was that in a strong interaction the net number of each quark type would remain unchanged. There should be the same number of u-quarks represented in the products as in the reactants.

Quarks would be another way of representing baryon number, charge and strangeness. For example, the idea that the net number of s-quarks (representing strangeness) would remain unchanged during a reaction, is just a more fundamental way of stating the conservation of strangeness.

For each quark is added an antiquark. Mesons are made from one quark and one antiquark. Baryons are composed of three quarks (antibaryons are made from three antiquarks). All allowed combinations of quarks soon had a correlate in known particles.

In 1974 a new meson was discovered simultaneously at the Brookhaven National Laboratory and the Stanford Linear Accelerator Center. The new particle (the J or psi) could not be represented by using any combinations of the three originally postulated quarks. Although a fourth quark had been suggested by theoretical considerations (to explain suppression of neutral-weak currents with changing strangeness) and seemingly needed for purposes of symmetry (it would give the s-quark a companion) the idea had no experimental support until this discovery.

A fourth quark helped explain the new discoveries. J. Bjorken and S. Glashow (1964) had previously named the fourth quark (c) for "charm." The property of charm would be conserved in strong and electromagnetic interactions. A fifth and sixth quark were also proposed that were called b (for "bottom") and t (for "top"). This pair was postulated partly to correlate with the tau meson and tau neutrino. (The u and d quarks correlate with the electron and electron neutrino. The s quark correlates with the muon and muon neutrino.)

By 1977 a particle was discovered called the upsilon (or bottomonium). It is explained by supposing it to be constituted of one bottom quark and one anti-bottom quark. In 1983 B-mesons were discovered. They are composed of one u-quark and one b-quark (or b-antiquark). In 1984, evidence indicated the t-quark; it was confirmed in 1995. The addition of quark flavors expanded the families of supermultiplets.

How are hadronic properties dictated by quarks? Usually, quark properties additively correlate with observed hadronic properties. That is, hadron-quantum numbers are equal to the constituting quark-quantum numbers. The six quarks each have a baryon number of +1/3. Hence, three combined quarks would yield a baryon number of +1 for the baryons; a quark (+1/3) and an antiquark (−1/3) would yield a baryon number of 0 for mesons.

All quarks have fractional-electronic charges. The u, c, and t quarks each have charges of +2/3, while the d, s, and b quarks have charges of −1/3 (their antimatter counterparts are oppositely charged). Fractional electric charges easily account for the net charge of observed hadrons. The various three-quark combinations explain the different charges exhibited by the members of the isotopic multiplets.

The s-quark has a strangeness of −1 (its antiparticle has strangeness of +1). A hadron exhibiting the property of strangeness is caused by the presence of a s-quark in its makeup. Strange quarks and quark electronic charge account for the arrangement of particles in the supermultiplet.

The charmed quark has the property of charm +1 (its antiparticle has charm of –1). The addition of the charm quark expands the number of possible-three quark combinations. Each supermultiplet is incorporated into a super-super multiplet. To each of the decimate and octet baryon supermultiplets, is added ten particles having charmed quarks. To the baryon singlet (the lambda particle) is added three charmed particles. To the meson octet is added seven particles having charmed quarks. The original uncharmed particles can still be placed within the super-supermultiplets as they were in the earlier "eight-fold way" arrangements. The groups become even more elaborate by the addition of the b and t quarks.

Here are some specific examples of hadronic constitution. According to quark theory the proton is composed of two u-quarks and one d-quark or (u u d). The baryon number is (1/3 + 1/3 +1/3) or +1. Its electronic charge is found by adding the fractional charges of its constituent quarks (2/3 +2/3 –1/3) or +1. The positive pion is composed of a u-quark and d-antiquark (u, $\bar{d}$). Because it is a meson, its baryon number is (1/3 –1/3) or 0. Its electronic charge is (2/3+ 1/3) +1.

The positive K-meson (kaon) is composed of a u-quark and s-antiquark (u, $\bar{s}$). It has (2/3 + 1/3) or +1 electronic charge. The strange antiquark gives it a strangeness of +1. There have been many particles observed with the property of strangeness.

The J (psi) particle (charmonium) is a meson composed of a charmed quark and a charmed antiquark (c, $\bar{c}$). Here all the quantum numbers yield values of zero except spin-angular momentum, which is +1. Detection of charmed hadrons is difficult because charmed quarks often convert into strange quarks through the weak interaction. This occurs by absorption or emission of the W particle (the carrier of the weak force).

Although the utility of quark-explanatory power is practically unquestioned, what evidence is there that quarks exist? The quark model suggests quark-antiquark pairs materialize from photons. This is indicated by two jets (mostly pi-mesons) diverging in opposite directions from collisions of electron-positron pairs with net momentum of zero. This has been observed. The electron and positron annihilate into a photon. Its energy is quickly transformed into a quark-antiquark pair that forms meson particles. This would not happen if hadrons were not constituted of smaller-elementary particles. More evidence is found in energy spectra of psi and upsilon particles. They have similar atomic spectra and probably represent different levels of excitation. The excited states suggest sublevel constitution.

Quarks were assigned 1/2 integer spin (and hence are fermions) to correctly predict baryon and meson spin-angular momenta. Quarks, like leptons, can only have vertical-spin alignment in only two possible directions. A meson's quark and antiquark either point in the same (in which case the spin would be +1) or opposite (spin would be 0) directions. This accounts for meson-integer spin.

Baryon quarks align in the same direction or one of the quarks aligns in the opposite direction. This accounts for baryon half-integer spins, e.g., 1/2 and 3/2.

Beside spin angular momentum, bound quarks can have orbital-angular momentum; that is, the quarks may spin around each other or around a common center of mass. These phenomena add integer values to the ground state spin-angular momentum. This helps explain the more massive (because of increased energy) baryonic variations of supermultiplets having higher values of half-integer spin-angular momentum. This is observed in the Regge trajectories.

Particles with half-integral spin are subject to Fermi-Dirac statistics and should obey Pauli's Exclusion Principle. No two particles of the same energy in a system can have all the same quantum numbers. This was the problem explaining the omega-minus baryon was composed of three s (strange) quarks having identical quantum numbers.

In 1964 O. Greenberg, Moo-Young Han, and Y. Nambu suggested each "flavor" (u, d, s, c, b, and t) of quark also came in one of three varieties or "colors" (red, green, and blue; anticolors are cyan, magenta, and yellow). The Exclusion Principle could then be satisfied since a baryonic particle with more than one identical quark flavor could still have each quark differentiated by color. (The idea each quark comes in different colors explained how three identical quarks—being fermions, which would be expected to repulse one another—could attract and hold each other.) At any one time all three colors would be equally represented by the constituting quarks so that overall the baryon would be white or colorless. For mesons, the quark and antiquark would be the same color, but at any given time would equally represent all three colors. The forces act to maintain an overall white or colorless condition. The colors represent flavor charges of +1/2, −1/2, and 0 (anticolors are represented by reversing the sign).

Adding the property of color does not predict an increase in the number of particles because the rules state color to be an "internal" quantum number. This raises the number of quarks to 18. Because quarks are not observed in isolation there is no problem for the theory. Thus quarks of the same flavor but different color have the same mass.

The present theory of the strong force is called quantum chromodynamics (QCD) because it is based on quark colors. (QCD is named after QED—quantum electrodynamics.) The color force acts between color charges. Theoretically, the property of color can only be observed in isolated quarks and that is deemed impossible. This "quark confinement," as it is called, is explained by the concept of asymptotic freedom (more about this in the subheading called "Strong and Electroweak"). The QCD coupling constant tends to zero as the distance at which the force is measured decreases to zero.

Unlike flavor, color is based on an exact SU(3) symmetry of color singlets. The SU(3) of QCD represents symmetry associated with color charge. The (3) represents the three colors acted upon. The S implies the sum of the three colors is zero in a baryon particle. An attempt to explain color singlets suggests there is a vector-boson carrier or quantum of the strong force. The supposed carriers are called gluons and are predicted to have a spin angular momentum of +1.

Gluons are said to be the force that binds ("glues") together the colored quarks. Gluons are made from a composite of a color and anticolor charge. Combinations yield eight effective gluons. Some are neutral and some are color charged. The exchange of a colored gluon between quarks changes their color. The fact gluons can be colored automatically assures color-charge quantization.

Just as an electron is surrounded by virtual photons and virtual electron-positron pairs, so it is that a quark is surrounded by a cloud of virtual gluons and virtual quark-antiquark pairs.

Because gluons are massless bosons one might expect the gluon to operate over an infinite range. However, the presence of the color charge resulting in quark confinement eliminates that possibility. Hence we only expect to find hadrons (which are composed of confined quarks) and gluons in a bound state.

Quarks interact by exchanging gluons. Each gluon represents a carrier-force field. It is said the strong force arises from the gluons acting to maintain the colorless condition of hadrons (quark flavor is conserved in strong interactions).

Quarks have been experimentally indicated in interactions showing meson "jets" in the expected direction of quark ejection. In 1979 gluons were indirectly observed in hadron distributions observed in electron-positron annihilations. Experimenters found a three-jet event indicating a quark-antiquark pair moved off in opposing directions and a gluon moved off in a third direction.

The strong force (of interaction between two nucleons) operates over a distance of about 10^{-13} centimeter, which is about the diameter of atomic nucleus. The strong force is used as the standard to compare the forces. Hence it has strength of unity.

Weak Force

H. Becqueral accidentally discovered the weak force in 1896 while conducting experiments in phosphorescence. He placed a photographic plate into a drawer along with a rock happening to contain uranium. Beta emission clouded the plate. The weak force plays an important role in nuclear reactions, which power stellar activity. It is also thought that neutrinos carry away gravitational potential in stellar collapse.

Weak interactions involve quarks and leptons. Some of these are the muons (100 x electron mass) electrons, tau particles, and u and d quarks. Leptons and quarks are fermions and have 1/2 integer spin.

Quarks and leptons exhibit no internal structure and no measurable size. With the exception of differences in mass and charge, all leptons are essentially alike. Each is also associated with a neutral neutrino.

The neutrino interacts solely through the weak force. It was postulated in 1933 by W. Pauli (but named by E. Fermi) to carry away the excess energy in beta-decay. Neutrinos have minimal mass (perhaps less than 1/10 eV.) and no electric charge. There are three varieties. They are the muon neutrino, electron neutrino, and tau neutrino.

The neutrino was finally detected in the 1950's. For many years the neutrino was considered to have zero-rest mass. In that respect it would be like the photon and travel at the speed of light. However, in the early 1980's physicists reported a phenomenon that seemed to demand neutrinos have rest mass. It is called neutrino oscillation. It implies a back and forth shift in its identity (from one type of neutrino to another). In 1998 it was reported that the neutrino has mass.

Weak-charge interaction is dependent upon particle handedness. That is, particles are affected by the weak force according to their spin. In the direction of motion, these particles either spin to the left or to the right. Of the mentioned particles, only the left-handed variety and right-handed antiparticle variety are subject to the weak force. The left-handed (spin is antiparallel to momentum vector) electron and right-handed (spin is parallel to momentum vector) electron have a different weak charge.

Left-handed leptons form gauge symmetry doublets. Right-handed leptons are singlets. Furthermore, only tardyon left-handed neutrinos and right-handed tardyon antineutrinos have been detected. Because the weak force acts on a group of particle doublets (here, left-handed particles and right-handed antiparticles) the neutral and charged weakons are described by SU(2) symmetry (special-unitary matrix with 2 x 2 elements). In the SU(2) scheme the W-particles are massless. (How they acquire mass is described in the subheading titled "Electroweak Interaction.")

Radioactive decay is the most well-known interaction involving the weak force. The principle subject of beta-decay is the neutron. In a free state (not nuclear bound) it is unstable and has a half-life of about 10 minutes. A "free" neutron spontaneously breaks down into a proton, emits an electron and an antineutrino. (The neutron is composed of udd quarks. Quark flavor is not conserved in weak interactions. The d-quark changes into a u-quark by emitting a virtual particle called the negative weakon.)

In the middle 1930's H. Yukawa proposed that in a nuclear interaction there should exist a third particle which mediates and embodies the force of that interaction. He believed the mass of the intermediating particle would be inversely proportional to the range of the force. Theory later suggested three such carrier particles for weak nuclear decay. The carriers of the weak force are called intermediate-vector bosons or weakons and have a spin (+1).

Weakons act on quark flavor (u, d, s, c, b, and t): changing one quark flavor to another. They have no affect on quark color (red, blue, and green). Two weakons are charged particles called the W^+ and the W^- (the W stands for "weak"). A third weakon is the Z^0.

Charged weakons work on transforming doublet particles into each other. For example, left-handed electrons can be transformed into left-handed neutrinos; right-handed positrons can be changed into right-handed neutrinos. Left-handed quarks (u, d) and right-handed antiquarks ($\overline{u}, \overline{d}$) are also subject to the weak charge having a value of either ±1/2. The Z^0 performs identity operations—electron to electron and neutrino to neutrino. The theory describing these interactions is called quantum flavor dynamics (QFD).

Weakons were discovered in 1983. Physicists observed an expected lepton asymmetry caused by the unique spins of decay products arising from the parity-violating effect found in the weak interactions. Weakons decay very rapidly. Usually the W⁻ breaks down into an electron and neutrino; the W⁺ breaks into a positron and neutrino. The W's are also expected to decay into muons and neutrinos, and tau particles and neutrinos.

The weak force operates over a very short range—from about 10^{-15} to 10^{-16} centimeter. So, weakons are very massive. The W's have a mass of about 81 GeV's (billion electron volts). The Z^0 particle has a mass of about 93 GeV's. The weak force is not as strong as the electromagnetic force and is only 10^{-5} the strength of the strong force.

Where does the weak force fit into the structure of reality? Unlike the other forces, there does not seem a clear answer. There is but three realms to correlate with four forces. The odd force out must be the weak one. The weak force does not seem to hold anything together. (It does figure in the burning of stars.) However, by asking questions we do gain some insight about its utility.

Facts, first brought out by understanding K-meson decay, led to the discovery weak charge conservation is violated. This finding overthrew the conservation of parity (left-right symmetry). (See Chapter 11.) So we ask, why have right-handed neutrinos and left-handed antineutrinos not been observed? Based on the parity-violating action of the weak force, it would not be unreasonable to conjecture these unobserved particles are a parcel of the Anti-Deltic Cosmos. It is because of the weak force that we can assert there is an Anti-Deltic cosmos.

We note the weak force affects electrons, which are also operated upon by the electromagnetic force, and quarks, which are predominately operated upon by the strong force. We also observe an interesting relationship between the U(1) charge (which equals the average of the electric charge of the two particles in the weak doublets) and the weak charge. The electric charge of leptons is the sum of the U(1) charge and the weak charge. There is no reason the weak force could not be considered an "adjunctive" force arising from an interaction between the strong force and the electromagnetic force.

Electromagnetic Force

Electromagnetic phenomena are described in Section Three—Chapter 13: the "Realm of Reasoning."

The electromagnetic force acts on particles with electric charge. It is the force that holds electrons in orbit around atomic nuclei. It gives atomic structure stability. Most chemical reactions, including those governing physical life, are directly owed to this force.

Much is known about electromagnetic interactions. The theory describing electromagnetic interactions originated in the work of Dirac during the late 1920's. It details the interaction of charged-electron-matter fields with electromagnetic fields.

In 1948 the theory, called quantum electrodynamics (QED) attained full development in the work of F. Dyson, R. Feynman, J. Schwinger, and S. Tomonaga, and others. QED offers predictions accurate to one part in a billion.

The carrier of the electromagnetic force is the photon. In electromagnetic interactions photons are exchanged by emission and absorption. Photons mediate the exchange of energy and momentum.

The photon is a boson particle of spin (1). Because the photon is massless (has no rest mass) the electromagnetic force extends over an endless range. It shows a strength 10^{-2} times that of the strong force.

Conservation of electric charge (according to QED, the real charge of the electron—the self-energy—is infinite and what is measured is what is not shielded by surrounding virtual particles) and the masslessness of the photon, are two facts related by the mathematical scheme called group symmetry.

QED displays U(1) symmetry (unitary group with one element). U(1) commutative group or phase symmetry indicates a 1 x 1 matrix occupied by a single particle—the photon. U(1) symmetry represents the freedom of altering the phase of an electrically charged field. Although the photon acts on charged particles, it does not alter their charges.

QED is an example of the class of non-Abelian gauge theories. The gauge or standard can be chosen differently for any local space-time without affecting the physics. Introducing a gauge field changes global (or continuous) invariance into local invariance. The single-gauge field of QED is the electromagnetic vector potential which in quantum theory represents the electromagnetic field.

Gravitational Force

> Gravitational phenomena are described in Chapter 14: the "Realm of Whyness."

The gravitational force is the only one that acts on all particles (all fermions and bosons). As far is known it only manifests as an attractive force. Compared to the strength of the other forces, it is very weak. It is 10^{-39} times the strength of the strong force.

Little is known about the actual mechanism of gravitational interaction. The exchanged particle that mediates the gravitational force is the graviton. It has not been observed. Theoretically, it is thought to be a spin-two boson. Attempts to detect its presence experimentally have failed.

History of Unifying Concepts

Conceptual levels (in knowledge) represent structural levels (in reality). Each conceptual level (or its corresponding structure) shows organization. (See Chapter 4: How Aspect Diagram.) Consider the human body: It has many structural components. There are: brain, nerves, heart, lungs, blood vessels, muscles, kidneys, liver, and many other organs. Each has its own tasks to perform.

Body components can also be grouped in various ways. A classifying scheme by function can be used to form systems. There are: the brain-nervous system, skeletal system, muscular system, endocrine system, circulatory system, and the reproductive system, and others. No system functions alone. Each system interacts either directly or indirectly with other systems. The gross interaction results in a functioning-human being. In this sense the human being is the sum (conceptual generalization) of its parts.

Conceptually, each step of generalization is a bringing together: of seemingly unrelated organs, of organs into structural systems, and of systems into the supersystem that is the whole individual.

We can also go the other way. To understand the human body as a whole we must know not only the relationship between organs, but we need to understand how each organ functions. Organs are composed of tissues. Tissues are divided into subsystems. And, there are deeper subsystems.

Eventually, we arrive at the molecular level. It too has subsystems. One system consists of nucleons and electrons. As we have seen, nuclear systems also exhibit substructure. Again these systems are dependent upon other subsystems. Here, abstraction (mathematical) becomes more important in the task of obtaining knowledge.

We now know much about how nature functions. Here, we trace how science supports the interconnection of the three-major forces. Not all connections have been experimentally verified.

Major advances in physical understanding occur when different phenomena are explained by unifying concepts. In 1687 I. Newton published his Principia. It conceptually unified the phenomenon of celestial and terrestrial gravitation. Much later J. Maxwell found equations that formally unite magnetism with electricity. His equations also pointed to the existence of electromagnetic radiative energy.

Unifying concepts need experimental support. Planetary motion was explained using Newton's equations. In 1819 H. Oersted showed that electric current affects a compass needle. In 1888 H. Hertz generated "Hertzian waves" using electric sparks. Today these waves are called radio waves and are understood to be a part of the electromagnetic spectrum.

We find these general concepts, though often abstract, do correspond to actual phenomena. Yet, we only experience these "higher-level realities" as broken manifestations. There is a clear example in Einstein's Special Relativity. Space is experienced as an extended "whereness" and time as an ever changing "whenness." Yet these are but two manifestations of what, at the next-higher level of generality, is called "space-time": which we do not directly experience.

Through generalization we produce a conceptual synthesis of phenomena. By doing so our constructions become more abstracted from that of everyday experiences. The only way to understanding is to learn of the relationship between parts. Unlike the human body, reality cannot be directly experienced as a whole. There is no outside vantage point from which to observe it. Hence, we have built, categorized, and structured the parts into a "super" whole. Let's show how these parts can be unified through physics into the superstructure called reality. To do this is to unify the four major forces.

Electromagnetic and Weak Unification

In the late 1960's a model of the weak force was developed incorporating electromagnetism. It is called the Glashow-Weinberg-Salam model (after S. Glashow, S. Weinberg, and A. Salam). The model is based on the Yang-Mills theory (after C. Yang and R. Mills, 1954).

Yang and Mills extended the gauge concept of electromagnetism. They originally developed their "vector-meson" theory to better understand strong interactions. It was used to describe isotopic-spin invariance. Recall that isotopic-spin invariance is the symmetry principle stating that strong interactions are unaffected by transforming a proton into a neutron and conversely. Arrows (vectors) representing phases (of the wave function) are superimposed onto the particle. Phase shift can be expressed by gauge transformation. This is an internal symmetry where arrow direction, like a dial, identifies the particle.

The Yang-Mills model is based on a local symmetry with respect to isotopic spin (potential field). Local symmetry means certain (up or down) isotopic spin (arrow) transformations can be made in one location without affecting the physics (the equations of motion remain invariant) elsewhere. Each coordinate point is independently transformed. Here, particle (charge) identity anywhere can be chosen at will.

Local-gauge theories are non-Abelian (after N. Abel's work on mathematical group theory). They do not obey the commutative law. For example, sequence of spin (arrow) rotation determines final arrow direction. In contrast, global-gauge symmetry transformation in one location simultaneously affects the physics everywhere else. Under these conditions symmetry is maintained only by performing the same transformation everywhere else—at every coordinate point. Global symmetry obeys the commutative law. Here spin-arrow sequence does not affect final arrow direction. The Yang-Mill's theory was to consider the consequences of changing the symmetry of the model from a global one to a local one.

The electromagnetic vector potential serves as an example of how a change in the phase of a wave function can be understood by gauge transformation. The Aharonov-Bohm effect (1959) in a two-slit electron-diffraction experiment shows that the vector-potential field has physical consequences. The effect shows that the wave function of an electron (matter field) beam could suffer a phase (a measure of displacement from an arbitrary reference) shift solely by the effect of the vector potential (induced by turning on a current in a solenoid placed between the slits). The phase shift is observed in absence of a magnetic field. Because the vector potential alone can cause a phase shift in the wave function, it suggests that any change of the wave function is relative. Its relativeness allows it to be a local variable. Hence the phase shift can be interpreted by a local gauge transformation in quantum mechanics. Therefore, charged particles in interaction can be interpreted through local gauge theory.

The Yang-Mill's model seemed better suited to describe weak interactions. Here their original idea is adapted to the leptonic equivalent of isotopic spin. Their theory used three fields. Because these fields are locally invariant, addition of gauge fields is required.

Gauge fields introduce exchange forces that mediate the interactions. The influence of a field can be represented by virtual particles. The quanta of gauge fields are vector bosons (spin-one particles). In 1957 J. Schwinger suggested a vector-meson theory (incorporating the then hypothetical W boson particle) which might unify electromagnetism with the weak force.

In 1961 Glashow, a student of Schwinger's, proposed a weak interaction theory using four exchange particles: the photon, and three-intermediate-vector bosons. Initially, the four vector-boson exchange particles are massless. One field remains neutral and unbroken. It symbolizes the electromagnetic force. Its gauge particle is a precursor to the photon. Because particles in the massless state operate over endless distances the other three gauge particles (the W^+, W^-, and Z^0) cannot represent the weak force. A way was needed to supply the three Yang-Mills particles with mass.

How can Yang-Mill's fields acquire mass? If the exchange particles acquire mass their range could be shortened in accord with observation. (Effective strength is inversely proportional to the mass of the exchanged particle.) In 1961 J. Goldstone realized an idea called spontaneous-symmetry breaking necessarily involves massless-scalar particles. His ideas embody the Goldstone theorem. Particles associated with scalar-field components exhibit zero mass. The component arrow in direction of the vacuum represents finite mass.

In 1964 P. Higgs and R. Brout discovered a way for exchange particles to gain mass. They added four extra fields (called Higg's fields) having the unusual property that they do not vanish in the vacuum (the lowest energy state of a field). Thus, Higg's fields exhibit lowest energy when having a value greater than zero. These are Higg's bosons and are thought to be very massive. Because the Higg's field is scalar, it has zero spin. A spin-zero particle has only two spin states—parallel and antiparallel. A Yang-Mill's field is a vector; it has spin-one value. A spin-one particle usually has three spin states: parallel, antiparallel, and transverse to the direction of motion. The Yang-Mill's quantum acquires mass, and one additional spin state (transverse) from the Higg's particle. The Higg's particle then vanishes. (We revisit the Higg's boson in Chapter 19.)

312 Section Five: Reality—Unity: Structural Integration

S. Weinberg (1967) and independently A. Salam (1968) combined Higg's contribution with Glashow's theory. They suggested the Yang-Mill's model could unify both electromagnetic interactions and weak interactions if three of the vector-exchange particles acquire mass. They theorized that when vector bosons pass through the vacuum they interact with the charge distribution and pick up mass from the Higg's particles. (Higg's particles also carry weak isotopic spin.) When gauge particles take on mass, gauge symmetry is spontaneously broken (invariance is violated). The three-vector bosons then become the exchange particles (the weakons) of the weak force. This is the Higg's mechanism. It provides the means of associating the massless photon with heavy mesons (the intermediate-vector bosons).

A Higg's field can also be used as a frame of reference to determine the direction of the isotopic spin arrow. (The length of the Higg's arrow is fixed by the state of the vacuum.) This has but relative significance not unlike that imposed on the phase of the wave function by the electromagnetic-vector potential. Thus, particle doublets can be differentiated by measuring their spin arrow against the Higg's arrow. The effect is that gauge symmetry, though hidden, is maintained regardless of arrow rotation (transformation).

Equations describing electroweak theory are symmetric although their solutions are not symmetrical. That is, the W and Z particle states, described by the highly symmetrical theory, need not exhibit symmetry. When the equations are solved, the intermediate-vector bosons take on mass and break the symmetry by (arrow) pointing in a specific direction. The symmetry appears only in the premass state of the exchange particles.

Spontaneous-symmetry breaking is borrowed from solid state physics. An example is the shift in the north-south equilibrium of the magnetized state of a ferro-magnet as it is cooled to low temperatures. Above 1044 Kelvin, iron atoms are very disoriented and the material is nonmagnetic. Below the critical temperature the atoms begin to align and spontaneously exhibit magnetization; that breaks the symmetry. Even better examples are found in the low temperature behavior of superconductors (as in the Meissner effect).

The Glashow-Weinberg-Salam theory was not free from problems. L. Faddeev and others laid down the rules for making the proper Feynman diagrams that could be used for renormalizing infinities appearing in the equations. Although infinities appearing in QED could be renormalized away, the same could not be said of weak-interaction theory. This was primarily because the weakons have mass. But could the theory be renormalized if spontaneous symmetry breaking occurred? Although S. Weinberg thought so he could not prove it. In 1971 G. t'Hooft (with refinements by B. Lee, M. Veltman, and J. Zinn-Justin) proved a way to renormalize the noncommutative theory and maintain exact symmetry. Scientists took the theory more seriously.

Physicists justify renormalization by assuming that virtual particles (from the vacuum) surround the electron, for example, and "neutralize" its supposed infinite mass. This allows the equations to correspond to measurement.

The theory predicts strangeness-conserving-neutral-weak currents similar to neutral-current-electromagnetic interactions. Scientists immediately recognized this idea suggested a connection between the weak and electromagnetic forces (beta-decay involves exchange of charged-vector bosons).

The Glashow-Weinberg-Salam model suggested the existence of interactions involving a neutral intermediate-vector boson—the Z^0 particle. The particle is similar to the photon: except it has mass. The Z particle should be involved in the scattering of neutrinos by electrons, neutrons, and protons. A neutral-current process occurs when two particles interact without exchanging electric charges. It would be observed in neutrino-proton or electron-neutrino collisions where neither particle loses its identity.

In 1973 neutral-weak currents (interactions) were discovered in electron-neutrino collisions. Further, the Weinberg-Salam model specifically implied that neutral-weak-current interactions would violate parity. In 1978 C. Prescott and R. Taylor experimentally confirmed that prediction.

More experimental verification was needed. For quark theory and gauge theory to be both valid, there should exist c-quarks (the isotopic partner to the s-quark) and b and t quarks. As previously stated, in 1974 the J/psi particle was discovered. It is a meson composed of one c-quark and one c-antiquark. The b and t quarks were later detected.

Scientists realized a more important finding would be the detection of the weakons themselves. In 1983 all three weakons were detected. (The W's have a mass over 80 GeV, the Z^0 has a mass of about 90 GeV; by comparison the proton has a mass of about 1 GeV.) The neutral Higg's boson awaits discovery. It may be found in debris from higher-energy collisions of protons and antiprotons. Its discovery would further show that the present electroweak theory is correct.

The weak force is not as strong as the electromagnetic force because of the large mass of its exchange particles. But, if weakons were massless (and hence have limitless range) the weak force would exhibit the same intrinsic strength as does the electromagnetic force.

The Glashow-Weinberg-Salam model suggests the weak and electromagnetic interactions are but different manifestation of a single more fundamental interaction—the electroweak interaction. This is reflected in the gauge transformations of the symmetry group combining the one-dimensional unitary group U(1) representing the gauge transformations of the electromagnetic force with the isotopic spin-rotation group representing the gauge SU(2) symmetry of the weak force. U(1) x SU(2) symmetry is the basis for quantum flavor-dynamics (QFD). QFD contains QED within it. Because these combined forces call for a unified local theory, scientists suspect the strong interactions could also be included in a similar but expanded theory.

Strong and Electroweak Unification

Scientists, encouraged by the success of the electroweak theory, have constructed models uniting the strong and electroweak forces. These models are called "grand unification theories" (GUT). They are usually defined by one gauge-coupling constant representing all three forces.

In 1973 D. Politzer, and independently D. Gross, F. Wilczek, and t'Hooft discovered an interesting property of the strong interactions. They found that because of quark interaction, the coupling constant (or effective-interactive strength) of the strong force tends toward zero at asymptotically high energies (or at correspondingly shorter distances). Bare-color charge does not induce a shield of opposing polarity as in similar electromagnetic phenomena. Instead, the charge is surrounded by virtual gluons (supplied by the vacuum) of the same polarity. The effective charge therefore decreases at closer distances to the bare charge, and the quarks act like free particles. This is called asymptotic freedom because total-particle freedom is not realized but only asymptotically approached.

Within these close distances quarks behave like free particles. Deep inelastic scattering experiments bear this out. Conversely, it is also thought (after Kogut, Wilson, and Susskind) that the strong force remains constant if an attempt is made to pull quarks apart. This may explain "quark confinement" and the fact quarks have not been observed in isolation. If quarks were pulled apart, other processes would intervene. Quark-antiquark (a colorless meson) pairs would be created. A consequence of asymptotic freedom is the force between quarks must also be mediated by non-Abelian gauge theories. Hence, calculations of the strong interaction can be carried out similar to those for the weak and electromagnetic interactions.

In the GUT scheme, the task for theorists is one of erecting a higher symmetry, which includes the U(1) x S(2) flavor quantum numbers of the electroweak interactions and the SU(3) color quantum numbers of the strong interactions. The simplest transformation matrix that accommodates these subgroups is SU(5) symmetry proposed by H. Georgi and S. Glashow in 1974. The matrices include eight gluons, three intermediate-vector bosons, and the photon. The remaining matrices are filled with hypothetical "X-particles." In the minimal model [SU(5)] there would be 12 X-particles. These are vector bosons. They are thought to mediate the interconversion of quarks and leptons. They carry both flavor and color quantum numbers. Like other subgroups the total charge of all the X-particles is zero. This and the fact both weak and electromagnetic charges are represented in bosons may explain charge quantization.

The physical basis allowing an inclusion of gluons in an encompassing theory is the behavior of the coupling constants (a measure of probability of a fermion field absorbing or emitting radiation quanta) and the strong interaction. (One coupling constant describes the intrinsic strength of the gluon field acting on quark color. The other two describe electroweak hypercharge [electromagnetic contributor] and electroweak isospin [weak force contributor]).

Asymptotic freedom suggests coupling of the strong force decreases (because of the effect of surrounding virtual particles) when separations between particles become about 10^{-29} centimeter (or equivalently at energies of about 10^{15} GeV (proton masses). (10^{15} GeV is the grand unified mass at which symmetry breaking occurs.) These conditions negate quark confinement and minimize the effect of X-particle mass. The other two constants (one slightly increases, the other decreases) merge with a decreasing-strong-coupling constant. At this point (of momentum transfer) a single coupling constant (about equal to a little more than the electromagnetic coupling constant) can describe both the electroweak and strong forces. This idea (after Georgi, Glashow, Pati, and A. Salam 1973–1974) serves as the basis for unification.

Under conditions where the three forces exhibit equal strength, it is expected that the particles they act upon (the fermions) will be indistinguishable and would interact alike. Here, quarks and leptons can be interconverted by X-bosons (remember: they carry both strong and weak charges). This means quarks and leptons are but two different members of one family. It is only at greater distances or lesser energies where the differences between leptons and quarks become manifest. The manifestation is a result of spontaneous-symmetry breaking of the underlying gauge group. The symmetry between the forces is broken because the vacuum confers a large mass onto X-particles. High mass suppresses the exchange of X-bosons at lower-energy levels. Its strength only appears greater because our experiments are conducted at relatively large distances. Because of the very high energy needed to create X-particles they are not thought capable of production in any particle accelerator that could ever be built. So, X-particles are unlikely to be directly detectable.

Although W-particles are not observable there are indirect ways of detecting their presence. In 1973 J. Pati and A. Salam suggested that baryon number conservation is violated. Just as W particles are indicated by parity-violating effects, X-particles should be indicated by a violation of baryon number (or net quark number) conservation.

> Baryon number and lepton number have not been associated with any gauge invariance that might explain their conservation. Baryon-number conservation has been but a "bookkeeping device" to support the observation that protons do not decay.

This level of unification suggests instead that there is a collective conservation of baryon-lepton number. (Lepton-number violation could be found in neutrinoless double beta decay—a double nuclear proton decay into two neutrons and two positrons. It would also be indicated if the neutrino has mass.)

If two quarks pass within 10^{-29} centimeter, an X-boson could exchange and convert a quark into a lepton, an antilepton, another quark, or an antiquark. SU(5) and other GUT's, predict (besides possible interaction with their antiparticles) protons (made of uud quarks) and bound neutrons (made of udd quarks) are unstable (u and d quarks exchange an X-boson) and are expected to decay most likely into a positron and neutral-pi meson (or positive pi meson and antineutrino). The positron would quickly interact with an electron and annihilate into gamma rays. The positive-pi meson, formed from the u quark-antiquark pair would soon annihilate also into highly energetic photons. The result is pure-radiational energy. Here, both baryon and lepton number conservation are violated (baryon number changes from +1 to 0). This interaction is thought to be very rare. (Note a free neutron decays in 10 minutes into a proton, electron, and antineutron. A bound neutron does not follow this route because of repulsive electrostatic forces generated by surrounding protons.)

Based on the theoretical mass of the X-boson it is calculated that the proton has an average lifetime of about 10^{+32} years. This is a very weak process. However, experimenters can fill a vast container with enough water to contain 10^{32} hydrogen atoms (each hydrogen atom has one nuclear proton). In this way some proton decay should be observed during an experimental running time of a year. Many such experiments were conducted to search for proton decay. The experiments were conducted underground to shield out muon decay produced from cosmic rays. (The detectors are sensitive to muon decay.) Nucleon decay produces Cherenkov light. Proton decay yields three cones of light; neutron decay yields two cones of light. The energy released would be greater than that observed in common beta decay. Proton decay has not been detected.

Verification of GUT's would help explain the quantization of electric charge (charges are multiples of 1/3) and the opposing charges of electrons and protons (total charge of each submultiplet is zero). They would also explain why quarks have 1/3 integral charges.

How can GUT's be verified? One early success of the minimal SU(5) theory is the accurate prediction of the electroweak mixing angle (or Weinberg angle). The mixing angle, a measure of vector-meson masses, relates the unification (ratio mix) of the electroweak and electromagnetic interactions (coupling constants). The angle (.22°) observed in neutral-current experiments is a natural consequence of GUT's if the proton decays at the expected rate (which is but an arbitrary value in electroweak theory). If the decay is not observed as expected (protons decay into pi-meson once in 5×10^{31} years) the minimal SU(5) theory may have to give way to one of the other more exotic candidates for symmetry such as E(6) [extended group], or SO(10) [special orthogonal transformations in ten dimensions]; there are many such models. Most of the other models call for an increase in particles. [SU(5) calls for 45 two-component fermion fields, 34 Higg's fields, and 24 gauge fields.] Experimental data, as of 1985, indicate the proton's lifetime is greater than that predicted by SU(5).

Most models [except SU(5)] predict massive neutrinos. As already stated "neutrino oscillation" has been reported. It is explained if the neutrino has non-zero rest mass (expected to range from 10^{-5} to 10^{+2} eV). Some models (having more than one Higg's doublet) also predict Higg's-mediated-flavor-changing neutral currents.

GUT's also predict superheavy magnetic monopoles associated with lepto-quarks. They are predicted to have masses of about 10^{+16} eV. These are not the variety predicted by Dirac. G. t'Hooft suggested that three-dimensional solutions (solitons, solitary-wave solutions to nonlinear problems) could be identified as magnetic monopoles. Some of these could exhibit fractional magnetic charges. Superheavy monopoles, if they exist, are thought to reside in iron nuclei. Many could gyrate to the earth's core. If earth's polarity reversed, the interchanging of monopoles and antimonopoles could release annihilation radiation that could be detected.

GUT's have drawbacks. They do not explain fermion representation and offer little clue to why there are second and third fermion families (muon and tau group) which are more massive versions of the first family (electron group—electron and electron neutrino, u-quark and d-quark). The (more) massive second and third families decay into products of the first family through weak interaction. GUT's also do not address the question of why there are particles of different spins. But, the major drawback is that GUT's do not include the force of gravity.

Strong-Electroweak and Gravity Unification

In 1974 J. Wess and B. Zumino, generalizing work of others, proposed a theory connecting bosons to fermions. This was once thought a formidable task because fermions obey the Exclusion Principle. (It states only one fermion can occupy a given point in space and time. Whereas any number of bosons can occupy a given point.) Wiess and Zumino suggested every fermion is associated with a super-partner—a boson of equal mass. They found that mathematically changing a boson into a fermion necessarily involves space-time structure. This and the local invariance of the theory implicate a gauge force for gravitation.

The symmetry is internal. As with other gauge theories the symmetry does not involve the object-particle or its motion. However, it does suggest the invariancy of underlying mathematical laws. The invariancy is demonstrated by reflecting a particle in a "supersymmetric" mirror. Reflecting a boson yields a fermion. Reflecting the fermion yields a boson again, but one that is space-time displaced from its original position. Repeated reflection in the symmetry mirror moves the particles around. This idea is the basis for local-supersymmetry theories.

Local-symmetry transformations associate particles with adjacent spins. A boson particle (X) with whole integer spin is associated with a fermion particle of 1/2 spin. For example, the photon—spin 0—would be accompanied by a fermion particle called the photino—spin 1/2. The W-boson would have a fermion partner called the wino. A spin 3/2 fermion is called the gravitino. Its boson partner is the graviton—thought to be a spin 2 particle. It is considered the quantum of gravity. The graviton is the gauge particle or carrier of the gravitational force.

Fermion-boson pairs are viewed as two manifestations of a single undifferentiated superparticle. Each of the manifested partners is distinguished by opposing isotopic-like spin arrows. Rotating the arrows does not affect the laws of physics. Although rotation angle varies at each space-time point, local invariance is maintained. Arrow rotation and translation represent supersymmetry algebra.

Why is the supersymmetry between boson and fermion particles not observed in nature? Spontaneous symmetry-breaking via the vacuum worked well for the Glashow-Weinberg-Salam model. There is no reason for theorists to think it would not apply to supersymmetry. Broken symmetry is a parcel of the more complex supergravity theories. In these theories bosons acquire mass equal to their fermion partners. It has been difficult to identify real particles with the ones predicted by theory. One reason is the symmetry would only manifest at extremely high temperatures or at extremely short distances. The super-boson partners are thought to acquire a very high mass from symmetry breaking. The mass is probably so high that directly producing them is thought to be beyond the capability of any current particle accelerator. This reasoning is the offered explanation why the symmetry has not been observed. Many scientists believe they must turn to cosmology for answers.

As previously stated, spontaneous-symmetry breaking uses local-gauge theory. Local supersymmetry is called supergravity. General Relativity is a gauge theory. Supergravity includes General Relativity. Supergravity reproduces the predictions of General Relativity at the macro level. However, at the micro level (10^{-32} centimeter or less) quantum effects are expected to intrude.

Supersymmetry combines the geometric description of force used in General Relativity with the exchange of virtual particles characterized in quantum theory. Gravitational waves are replaced by the quantum of gravity. As mentioned, the supposed quantum of gravity is the graviton. It is massless because gravity is long-range. Again, the graviton is thought to be a spin 2 particle. It must have an even integer value because odd integer values represent repulsive forces occurring between similar particles. The graviton acts upon particles in a way characterized by the curved paths predicted by General Relativity.

There are problems with supergravity theory. Adding adjacent particles other than spin 2 and spin 3/2 doublets complicate model construction. Although the quantum gravity interactions are calculable, they incur infinity problems in the one-loop diagrams. The infinities have been immune to cancellation. These theories suffer the added burden of not leading to a unification of the forces.

In 1921 T. Kaluza proposed a theory uniting the then only two known forces—electromagnetism and gravitation. Kaluza suggested there is a fifth dimension along with the four-curved dimensions of space-time in General Relativity. Because a fifth dimension had not been observed, his theory did not kindle much interest. In 1926 O. Klein thought if there were a fifth dimension that it might be microscopically curved and not be noticed. Today these ideas are known as Kaluza-Klein theories. Today we recognize four forces of nature.

Applying Kaluza-Klein theories to four forces extends the number of dimensions needed for proper description. These are called extended-supergravity theories. They envision a higher-dimensional space called superspace.

Superspace describes supergravity with space-time curvature. In the simplest version eight coordinates represent each space-time point. Four are the regular space-time dimensions of Relativity. The other four coordinates represent additional gauge fields in the superspace. The theory describes a reality consisting entirely of gravitons and gravitinos. In these versions theorists face the task of showing how all other particles are derived from gravitons and gravitinos.

Superspace, though more restricting than mere supersymmetry, may allow complete unification. These theories incorporate more anticommuting coordinates. There are eight-extended-supergravity theories. Each represents a specific number of boson-fermion transformations. These theories are represented by the numbers from $n = 1$ through 8. The original supergravity theory is represented by $n = 1$. The number of spin-particles predicted is determined for each value of n. For each additional n-value there are added four more dimensions. The most realistic theory is for $n = 8$. It dictates one spin 2 graviton, eight spin 3/2 gravitinos, 28 spin 1 bosons, 70 spin 0 bosons, and 56 spin 1/2 fermions. These theories allow the graviton to be transformed into a spin 1 boson. In these theories all predicted particles can be traced back to the Higg's boson.

Full unification is achieved by using both supersymmetry transformations with the internal symmetry of particle doublets. Because extended supergravity allows deriving any particle from the Higg's boson, every particle is considered a manifestation of one superparticle. Rotating the internal arrow of the Higg's superparticle appropriately represents all particles. This model posits all particles fall into a natural hierarchy.

In most extended supergravity theories: Gravitational strength is represented by one constant and the other three forces are jointly represented by a single coupling constant. Theorists believe, ideally, to achieve a genuine unification that all forces should be represented by just one coupling constant. There are some theories (after H. Weyl—1923) not based on General Relativity, that do lead to a complete unification of coupling constants.

Although extended supergravity points toward unifying all physical laws, scientists believe other problems remain. Because supergravity is local, it introduces a cosmological term. The term relates the strength of the strong and the electromagnetic force. It dictates a finite value for the size of the universe. The size predicted far exceeds that estimated by observational techniques.

However, the above problem might be solved if considered via the structure involved at this level of generality. There are many universes. The Cirpanic Cosmos structure and Everett's noncontradictory many-world's interpretation of quantum mechanics indicate that. It is thought, however, if spontaneous-symmetry breaking applies to supergravity the cosmological term will decrease in value. Another problem is, there are more particles in reality than are accompanied by even the $n = 8$ theory. However, many particles might be excited states of simpler particles.

The supergravity idea shows much promise. It seems to automatically renormalize infinities that have plagued quantum-field theory at lower levels of generality. For example, infinities appear in QED calculations for electron mass and charge. Those calculated values must be replaced by experimentally determined values. Extended-supersymmetry theories cancel the fermion infinities with a negative counterpart showing up in their bosonic partner.

Infinities arising in photon contributions are canceled by infinities with opposing sign arising from photino contributions. The same problems have arisen in GUT's. (Problems, or contradictions, appearing at one structure of reality only indicate that a noncontradictory solution will probably be found at a higher level of generality.)

Extended-supergravity theories may also lead the way to renormalizing infinities appearing in a true theory of quantum gravity. Extended supergravity theories also seem to place a limit on possible particles. In that respect they would explain those particles that do exist.

Inflationary Universe

It is one thing to ask how particles are produced. It is quite another to ask how particles, or for that matter the universe, originated. Although the Big Bang theory provides a partial answer it does, however, leave some questions unanswered. Some of these questions are addressed in the horizon problem, the flatness problem, and the monopole problem.

The horizon problem arises from the 3 degree microwave background radiation. (The horizon distance is the furthest electromagnetic radiation has traveled since its release.) The radiation is very uniform. Matter itself (as galaxies for example) is evenly distributed compared to the size of the universe.

How does radiation coming to us from opposite directions (from along our past-light cone) settle at the same temperature? How can radiation from one part of the universe match that from another region unless it was causally connected somewhere in the past? According to the Big Bang theory, when electromagnetic radiation was released (about 10,000 years after the Big Bang) the universe had already expanded too far for distant regions (separated by more than 10^7 light years) to have been in causal contact even by light signals. Thus, uniformity could not have resulted from a natural settling of thermal equilibrium. So the Big Bang model assumes the uniformity arose out of the Big Bang itself. Yet, it is difficult to conceive that the Big Bang, like any explosion, would have uniformly spewed out space-time and matter in all directions.

The flatness problem, sometimes called the age problem, involves the energy density of the universe. The idea is similar to the horizon problem but involves the rate of universe expansion; or, why is it as old as it is? Scientists believe the fate of the universe will be one of three possibilities: open, closed, and flat. The most likely fate is the universe either expands forever (if the energy density is less than a certain value called the critical density) and it is open, or it eventually collapses (if the density is greater than the critical value) and it is closed. A slight deviation from the critical value shortly after the Big Bang should increase as the universe expands. The present observed energy density is slightly greater than the critical density. This suggests the physical universe will expand forever.

The third is the magnetic monopole problem. GUT's suggest there should be as many magnetic monopoles as baryons. If there were equal baryons and monopoles, the increased energy density (super-heavy monopoles can be 10^{+16}

times proton mass) would cause the universe to expand at a rate matching today's size in 30,000 years.

There must be a mechanism for eliminating monopoles. One interesting approach is called the inflationary model. It combines information from GUT's with the Big Bang model. The theory allows scientists to reconstruct what happens closer to the instant of the Big Bang than has before been possible.

Physicists now can calculate events up to 10^{-43} seconds of the instant of the Big Bang. 10^{-43} is called the Planck time. It is associated with a length of 10^{-33} centimeter—the Planck length. At these short distances physicists believe space-time loses continuity. Here, quantum fluctuations effect an acausal bubbling of space-time bits, out of and into nothingness. S. Hawking has calculated that mini-black holes of 10^{+19} GeV can evaporate and reform in 10^{-43} seconds. Hence the idea space-time bits can pop into and out of existence much like bubbles in a foam.

A. Guth suggested the universe experiences an early-phase transition in which it, in the form of many bubbles, overcomes the effect of gravity and expands exponentially. The transition, called the inflationary period, affects the course of universe development. It was conceptualized to solve problems plaguing the standard Big Bang model. It predicts super-expansion will smooth-out matter. Hence, it solves the horizon problem. It predicts the uniformity observed today results from all parts of the universe having once been in causal contact. This happens because the universe evolved from a much smaller region than predicted by the standard Big Bang model. Because the universe achieved thermal equilibrium early on, the flatness problem is also eliminated. The universal energy density is predicted by inflation to match the critical density. Inflation also decreases monopoles to an extent that is not inconsistent with the observed rate of universal expansion.

Guth believed our universe inflated at the speed of light from many nucleation sites or bubbles in the quantum space-time foam. Although his theory remedies defects in the Big Bang model, it had problems of its own. Some scientists think our universe emerged from just one space-time bubble. They believe other universes emerge from other bubbles. A multi-universe structure emerges. This is the "new inflationary" model.

In the new-inflationary version, domain walls replace bubbles. And there is no energy difference between the false vacuum and the true vacuum. Regions encapsulated by domain walls have a temperature above 10^{+27} degrees. At these temperatures the four forces, we are familiar with today, are undifferentiated and behave identically. Their gauge particles are symmetrical. This is the superunified force—supergravity.

Up to 10^{-43} seconds after the "beginning" there existed only the superforce. During this era the universe cools to 10^{+22} degrees. A null Higg's field dominates the high pressure-energy density of the universe. From 10^{-43} seconds to 10^{-35} seconds, the effect of the vacuum (in which the Higg's field has positive energy) state, causes the superforce to differentiate into the gravitational force and the grand-unification force (the strong-electroweak force). The "false vacuum" maintains a constant energy density in the universe. And the temperature of the universe becomes metastable by the effects of curved space-time.

As the system rolls over toward a true vacuum state, the equations of General Relativity suggest the universe will expand exponentially. The universe heats back up to 10^{27} degrees during the inflationary period. It expands so fast that at 10^{-35} seconds the supercooling effect spontaneously breaks the symmetry of the grand-unification force. The Higg's field surrenders its energy from the vacuum, forming lepto-quarks and highly energetic radiation.

From this point onward the standard Big Bang theory adequately describes universe expansion. At about 10^{-12} seconds, the electroweak force settles out. Neutrinos propagate freely. Scientists believe a remnant of neutrino release should be around today as a neutrino background with an equivalent temperature of 2 Kelvin. After 10^{-12} seconds, quarks couple to form hadrons and leptons. At one second the Weinberg-Salam symmetry of the electroweak force is spontaneously broken, and the weak force freezes out. At 10^{+5} years the electromagnetic force freezes out and photons propagate freely. Before electromagnetic-freeze out, the universe is optically opaque. We cannot optically see back to any time prior to this phenomenon. The 3 degree electromagnetic-microwave background observed today was released at that time.

Not surprisingly, there has been problems with this model. After all, the reason for inflation was to redefine a critical phase just after the Big Bang. Physicists thought this "adjustment" was necessary for cosmic directionality to evolve into the world we know today. The problem with this approach is that the present might not be completely explained by what has happened in the past. We revisit this problem in the final chapters.

Relation of Realms

The three realms have been structurally integrated through physics. However, the relation between realms entails more than the fact they combine to form the whole of Reality. The question remains how the realms interrelate. What is the relationship between realms? The Realm of Whatness is brought into existence from the Realm of Reasoning by the Realm of Whyness. Let's examine the meaning of this statement.

Realm of Reasoning and Realm of Whatness Structure

After years of controversy, when scientists finally conceded light was a wave phenomenon, Einstein subverted that finding by explaining the photoelectric effect occurred because light exhibited particular (corpuscular) properties. With an equally interesting phenomenon Davisson and Germer performed an experiment using matter in the form of electrons. Electrons beamed at a metal crystal scattered to form diffraction patterns as if they were waves. L. De Broglie's work had already provided an explanation for this effect. Under suitable conditions light and matter both exhibit wave or particular properties. (Refer to Chapter 13.)

It became more difficult to state the elemental differences between light and matter. One most noteworthy difference, however, persists. Light (here the term denotes all electromagnetic radiations) does not exhibit the property of rest mass as does matter. Light only has energy because of motion. Moreover, light velocity is unaffected by energy content whereas nonluminal particles approach light speed as energy increases.

Although light and matter (tardyon or tachyon) share many properties, the fact light exists only in constant motion, is sufficient to categorically separate it from matter. Categorically, the Realm of Reasoning "structurally" represents light. In this book subluminal and superluminal materiality is subsumed in the Realm of Whatness. Let's take a closer look at light.

Light is conveyed through space at a velocity of 299,792.458 kilometers per second. The speed of light (in vacuum) is constant for all inertial references and is denoted by the letter (c). The constancy of light speed establishes a basic symmetry for the Realm of Whatness. At a fundamental level the symmetry applies especially to time flow and spatial extensionality. Other concepts connected with these two properties are also affected. For example, (c) governs the inverse relation between light frequency and wavelength. The lower the frequency the longer the wavelength.

A certain evidential fact presents itself. It is interesting that (c) is an ascribed property of light generated by the relational status of light to inertial systems. For example, it is our superimposition of temporality onto light that endows it with frequency. Frequency is proportional to light energy. Conservation of energy becomes a rumination of temporal irrelevance. The same reasoning applies to wavelength. Imposing spatiality onto light gives it the property of wavelength. Other qualities follow. Wavelength is proportional to momentum. The conservation of momentum reflects the indifference of spatiality.

We observe the relational properties of light through its interaction with matter. When light is emitted or absorbed there is an exchange of energy and momentum. These properties must be conserved in interactions.

Functionally, light mediates electromagnetic interactions. Although the electromagnetic beam is usually described by waves, the beam itself is said to be composed of photons. In QED the use of particle and field concepts is essentially interchangeable. A field is a region within which some quantity is defined at every point. The particular (corpuscular) characterization of light most often is described by quanta. In electromagnetic terms the photon is used to describe a disturbance in the field (a wave pulse carrier of energy). The photon acts on charged particles and electric or magnetic currents.

Properties linked to spatial extensionality and temporal flow, though instructive, do not tell us much about the intrinsic nature of light. Electromagnetic radiation, though the carrier of electromagnetism, is neutral with respect to electric or magnetic charge. And, although temporality and spatiality are complementarily governed by (c) light itself is not intrinsically endowed with temporal or spatial properties. However, we must adopt a simplified time and space framework to gain a logical understanding of light.

Light itself is a wave of force naturally constituted by electric and magnetic properties. The electric and magnetic components oscillate in phase from zero amplitude increasing to a positive maximum value then reverses direction decreasing back to zero. From a nonluminal frame of reference the oscillations appear to be perpendicular sinusoidal waves. Theoretically, the electromagnetic force continues fluctuating to an endless distance.

Using an inertial perspective, we can diagrammatically represent waves of force in Figure 15:3 A. (Although light vibrates in all transverse directions to the line of motion, here we consider but one transverse electric or magnetic vibration.) We can also represent the force in cross section as in Figure 15:3 B.

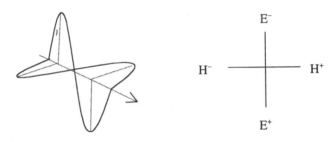

Figure A. Magnetic or Electric Wave   Figure B. Wave Cross Section

Figure B represents the dual properties of light. One line is of an electric nature (E) and the other is of a magnetic nature (H). Each has juxtaposed extremes. The electric component oscillates between two oppositely charged positions. Let the negative position (E–) symbolize the electron. Let the positive position (E+) represent the positron.

Figure 15:3. Electromagnetic Wave

We need to denote the extreme positive and negative maximums of oscillation for each of the electric (E) and magnetic (H) components. Figure A shows an E (or M) wave propagating in space through time. Figure B is a cross section showing amplitude of both electric and magnetic components.

The Realm of Reasoning is the essence of (or extrinsic reason for) the Realm of Whatness. All universes and their defining spatial and temporal phenomena arise from the Realm of Reasoning. Again look at figure B. Let the "point" of intersection of the two lines (zero amplitude) represent light in its frame of reference. The point also represents the absence of spatial extensionality or time flow. It represents a quasi-state seemingly lacking energy or momentum. (Note it is a "point" on the Cirpanic Cosmos structure that, at a high level of generalization, symbolizes one Bi-Deltic Cosmos.)

Imposing a plane perpendicular to light direction endows the point with a simplified time and space character. The operation expands the point into the representation in figure B that can be used to illustrate how fundamental universal properties are derived from light. First, the concept of spatial extensionality generates whereness. The two major universes (mind and matter—tachyon and tardyon) of the Where Aspect (and Anti-Where Aspect) are in essence represented in diagram B. Let's see how.

Generalizing to a cosmological level we arbitrarily say (E^-) represents tardyon matter. (E^+) represents tardyon antimatter. The magnetic component also oscillates between two oppositely charged extremes. Let the negative position (H^-) symbolize a magnetic monopole. Let the positive position (H^+) stand for an antimagnetic monopole. Again, cosmic generalization has (H^-) representing tachyon matter while (H^+) symbolizes antitachyon matter.

In 1931 Dirac suggested magnetic monopoles would help explain the quantization of electric charge. He said angular momentum of electrically charged particles would be balanced if monopoles exist. (The smallest unit of angular momentum is related by Planck's constant. Any other measure of angular momentum would be a multiple of this basic unit.)

R. Adair used a thought experiment to illustrate a violation of time reversal invariance. He imagined a proton moving through a magnetic field generated between two electric-current-carrying-metal loops. If time is reversed, the proton retraces its curved course because both the current and magnetic fields reverse.

Instead of the above, visualize a similar magnetic field generated between an array of north and south magnetic monopoles. What happens if time is reversed? The direction of the field remains the same (polarity does not reverse) and the proton reverses motion but curves in an opposing direction from its original path. This is an example of temporal inversion (in contrast to reversal). Although temporal-reversal invariance is violated in this example, it has been explained in Chapter 11 that an inversion of time, in conjunction with charge and parity (CPT-theorem) restores symmetry. Coupling magnetic monopole behavior with tachyon behavior, seems a reasonable solution to the individual violations of C, P, or T invariance.

Maxwell's equations infer restorative symmetry. The equations are time-reversal invariant. They represent either a charged particle giving off electromagnetic energy to an infinite distance (retarded wave) or represent electromagnetic energy from infinity collapsing onto a charged particle (advanced wave). The former may represent tardyon or antitardyon behavior while the latter may represent tachyon or antitachyon behavior. Corrective interpretation shows tachyon particles absorb and emit E-M energy in a similar manner to tardyons.

One condition must be met for monopoles to exist according to Maxwell's equations. Light velocity must be constrained to travel only at (c) for photons to exhibit zero rest mass. Monopoles are easily allowed once the condition for photon-zero-rest mass is met.

The Where Aspect and Anti-Where Aspect (and therefore, things) are projected by light into manifestation. How is temporality represented in "extended" light?

Therefore, how does light take on the characteristics of extended spatiality and extended temporality? E–M oscillation represents the essence of the When Aspects and Anti-When Aspects. For example, when the oscillations swing toward (E^-) and (H^+) the tardyon universe and antitachyon universe are in their dominant period. The swinging of the oscillations to the other extreme represents the dominance of the antitardyon universe and tachyon universe. The frequency of the wave confers the temporal manifestation of electromagnetic energy in each universe. The How Aspect and Anti-How Aspect are represented by the ordering of positive and negative maximums. Dirac thought magnetic monopoles should have antimatter counterparts.

We easily see how light is tagged the essence of the Realm of Whatness. We also know much about how light interacts with tardyon matter. Yet, it is not so obvious how the Realm of Whatness is generated from the Realm of Reasoning. Much of the rest of this book is focused on unraveling an answer to this question.

Physicists suggest empty space is filled with virtual pairs of particles and antiparticles. Virtual-particles exist under the indeterminate conditions arising from Heisenberg's uncertainty relations. Americans W. Lamb and R. Retherford indirectly detected the presence of these evanescent particles in 1947. They noticed a variance between the fine-structure spectrum of excited hydrogen and that calculated from Dirac's QED. The difference, known experimentally as the Lamb shift, was later attributed to virtual particles.

Virtual particles are created effortlessly from an apparent nothingness and are just as easily destroyed. An electron summons forth numerous virtual photons and other virtual particles. Virtual particles obtain energy from the manifest electron but must return it within a time allotted by the conditions imposed by Heisenberg's principle. The energy borrowed determines the distance the virtual particle can travel. The greater the energy the shorter the distance and the sooner it must be returned. The distance cannot exceed the length of the particle's associated matter-wave.

Virtual particles surrounding an electron are thought to compose its electrostatic field. An "exchange force" is created when two electrically charged particles interact (attract or repulse). Virtual photons dart back and forth between the pair. Each virtual photon creates virtual electrons that produce virtual photons and so on ad infinitum. Field creation is so quick that the process is identical with classical-electrodynamic predictions.

There is no such thing as a bare electron. That applies as well to other manifest particles. The mass of a manifest particle must include the effect of its surrounding virtual particles. This is what Dirac did not consider in an earlier formulation of QED. Calculated corrections for electron mass using the mathematical technique of mass renormalization accurately agree with experimental findings.

Virtual particles explain many phenomena. For instance, when an electron drops to a lower-atomic-energy level, the excess energy is thought to add mass to a virtual photon, which then materializes (becomes a real photon) and radiates from the atom.

Is there a cosmological process that materializes virtual-particle pairs into real particles? We have already learned how the tardyon universe is thought to have originated in a Big Bang. The Big Bang is considered a phenomenon akin to the behavior of a cosmological white hole. The center of such a hole is called the singularity.

At the instant of physical origination a cosmological singularity would include all space and time and hence all virtual particle pairs. Some scientists consider the period of origination to be equal to the "Planck time" (equal to 10^{-43} second). During that instant the whole physical universe would be confined to a diameter no greater than 10^{-33} centimeter. This is less than the matter wavelength of the total mass. Such conditions preclude making sense of material causation so we are not restricted in theorizing. Extreme conditions near a singularity could generate enough gravitational pressure to split virtual-particle pairs apart. From that point the initial dynamics of universal expansion could supply the force field needed to materialize virtual pairs into real (manifest) particles.

Virtual particles are exchanged between interacting particles. Quantum mechanical considerations assume virtual particles mediate the transfer of energy and momentum in elementary interactions. The relation between energy and momentum for virtual particles is different from that of manifest particles. Because virtual particles cannot be directly detected, no problem is introduced by the fact virtual particles flout conservation principles.

Feynman graphs or diagrams (after R. Feynman) are often used to represent particle interactions. These diagrams show how an interaction proceeds in time. For example, in electron-positron annihilations two general cases are worthy of our interest. First, an electron and positron can annihilate by exchanging a virtual electron and by it yield two real photons. Here, two real photons are needed to conserve energy and momentum. In the second case, the electron and positron annihilate into one virtual photon, which in the brief time (about 10^{-25} second) allowed by indeterminacy conditions, decays into two real particles. The particle pair will be matter-antimatter complements.

What is learned from these statements? Matter-antimatter particle pairs can annihilate into one particle—a virtual photon that subsequently decays into one particle-antiparticle pair. A matter-antimatter particle pair can be produced from one virtual photon. It is reasonable to suppose manifest particles are not only mediated in interaction by virtual particles but are created from virtual particles.

Photons come in two varieties, real and virtual. Planck's constant governs which variety we encounter. Electron-positron pairs and magnetic monopole—antimagnetic-monopole pairs are juxtaposed composites of virtual particle analogs. Created (manifest) particles arise from "creating" virtual particles. Virtual particles are the nonmanifest energy "stuff" of the vacuum—space itself.

The Realm of Reasoning correlates with electromagnetic energy and therefore to electromagnetic radiation and photons. Because photons essentially "represent" all virtual and manifest nonluminal particles, it is easy to (extrinsically) attribute the Realm of Whatness to the Realm of Reasoning.

Change in the realm of nonluminal particles is mediated by virtual particles. Let's look at a model that attempts to describe, via composition, how particles are brought into existence.

Many theories are formulated to (fundamentally) explain matter. The quark theory is now well accepted in the scientific community. Yet, quarks have not been directly observed. The quark model does explain and predict many subatomic phenomena.

In 1979 H. Harari suggested a simple and very "economical" subquark model. There is yet no evidence that it represents anything in reality. However, it would explain certain phenomena. For example, it helps understand the connection between color and electric charge; it also seems to explain the exact neutrality of the hydrogen atom.

Harari calls his theoretical subquark entities rishons (after the Hebrew for "first" or "primary"). He suggests there could be just two rishons labeled T and V. Both are thought to have mass and both have 1/2 spin-angular momentum. The T-rishon has an electric charge of +1/3. The V-rishon has zero charge. Thus the positron would be composed of three T-rishons (1/3 + 1/3 + 1/3) or +1 electric charge). The electron would be constituted of T⁻ T⁻ T⁻. VVV and V⁻V⁻V⁻ with zero charge, form the neutrino and antineutrino.

Properties such as quark color and flavor arise from the arrangement of rishons. A TTV arrangement has a charge of +1/3 and represents the u-quark. This combination is said to exhibit the "rishon" colors red, blue, and antiblue. Blue and antiblue cancel yielding a net color of red. TVT and VTT are permutations corresponding to the other color possibilities—blue and green. All fractional combinations are colored. All integer combinations are colorless. This explains the connection between charge and color.

How does the rishon model explain the neutrality of the hydrogen atom? A proton is composed of uud quarks. In rishon terms, this is TTV, TTV, and T⁻V⁻V⁻. And as we have seen the electron is T⁻T⁻T⁻. The T's, and V's exactly cancel with their antirishon complements.

Although the rishon model does well in explaining the first generation of leptons and quarks, it and other similar models have difficulty explaining the second and third generations. However, there may be some unknown factors that may yet account for them. For example, as mentioned above, the more massive generations might be excited states of the first generation.

There are other drawbacks. Leptons already seem to be point-like particles, at least to dimensions of 10^{-16} centimeter. For example, the magnetic moment (related by the "g" factor) of the electron has been measured within ten decimal places. And that determination is based on the idea the electron is point-like, or at least is no bigger than the current acceptable size limit of the electron. If the g-factor can be measured even more precisely, the idea there are subquark entities could have some experimental support.

The main point, as far as this book is concerned, is that there are models of subquark constitution that, at a very fundamental level, shun complicated theories. Some models, like Harari's, build all subluminal matter from just two particles and their antiparticle counterparts. Looking at Figure 15:3 B one cannot help wondering if there is a direct creating connection between electromagnetism and matter. At least such models go a long way in suggesting a "mechanical derivation" of the universes.

We could continue seeking connections (of an extrinsic nature) between light and matter. Anymore such theorizing becomes even more speculative. Instead, we turn our attention in another direction. We next consider the intrinsic nature of reality.

Realm of Whyness is Intrinsic Cause of Realm of Whatness

Purpose: Describe experimental evidence suggesting an immediate intrinsic cause of the Realm of Whatness.

Reasonable arguments have shown how the Realm of Whatness is an extrinsic manifestation of the Realm of Whyness. However, we have yet to establish that the Realm of Whatness experiences sustentation. So how can it be scientifically ascertained that the Realm of Whyness is the intrinsic cause of the Realm of Whatness? The solution to that question comes in two parts. First, it must be shown there is an intrinsic cause. Second, the intrinsic cause must be noncontradictorily identified. We begin by restating the basis of the EPR argument. (Refer to Chapter 13.)

Few scientists, including Einstein, argued with the statistical success of quantum mechanics. However, the meaning of the wave function regarding individual microscopic events remained unclear. Several theorists (especially followers of Bohr and the Copenhagen interpretation of quantum mechanics) continued to claim, for example, that prior to measurement, correlations between incompatible variables at least had no meaning and at most did not exist. This view implies no reality exists apart from observations or consciousness.

Some followers of the EPR argument differ. Using the HVT (hidden variables theory) position they insisted that properties exist although not measured. They concurred with Einstein, in saying the quantum mechanical description of nature (e.g., the Copenhagen interpretation) was incomplete because it did not fully represent reality. They contended if the hidden properties were known that quantum theory would be complete and therefore, deterministic.

Theorists adhering to the Copenhagen interpretation (which partly was Bohr's reply to the EPR argument) of quantum theory, stressed that an electron, for example, has no definite position or momentum (these are incompatible variables governed by the Uncertainty Principle) but covers a large region represented by probability. And that region does not represent anything concrete, but is merely a concept describing what is observed.

Thus, quantum mechanics is but a set of rules for predicting observations. Scientists argued that hidden parameters should not be included in the rules of physics if they cannot be observed. Here the wave function does not entail any underlying reality beyond what is observed. Yet this not only conflicts with the common view held by most people, but also is contrary to three basic assumptions of most sciences.

These assumptions are philosophical cornerstones of modern science. One premise is called realism. There are many philosophical variations. It asserts there is a reality autonomously existing apart from awareness. That does not mean what we experience is always an exact representation (naive realism) of "what" we are conscious of, only that our observations have a source in something separate from our perceptions.

The second basic assumption is induction. It is a practical requirement for most sciences. Induction is the backbone of scientific methodology. For example, based on the color paradigm we inductively infer scientific knowledge represents a "world out there." Each of the first two principles establishes the other.

The third cornerstone is locality or separability. It roughly states events separated in space are independent of each other. Test results in one location should not be affected by tests conducted in a widely separated location. This premise is implied in the EPR argument (as "criterion of reality"). A restrictive version of this premise is Einstein separability, which avers no influence can travel faster than the speed of light. Here its use implies the results of one test cannot be conveyed faster than light speed to another test site. Thus, if a second test can be performed before a light signal from the first test arrives, the second test cannot be causally influenced by the first test.

These three premises constitute what B. d'Espagnat calls the local realistic theory. It can test hidden variable theories as a class. The local realistic theory can be used to make certain experimental predictions. Quantum mechanics conflicts with the statistical predictions of the local realistic theory. Experimental results could demonstrate that quantum theory is an incomplete description of reality, as Einstein's EPR argument suggests, or that it is complete but in some other sense. The results of such an experiment should also inform about the structure of reality.

In 1952 D. Bohm designed a thought experiment elaborating in more concrete terms the basic ideas suggested in the EPR argument. His reformulation is amenable to experiment. He suggested using spin 1/2 particles such as protons.

Two protons could be made to interact—coming together in a singlet state. Spin interaction is conserved; the state is rotationally invariant. Each proton has one of two possible spin values (+ or –). That is, the protons while interacting establish a combined spin state of zero. We saw in the meaning of the quantum numbers that a spinning charged particle produces a magnetic moment. Magnetic moment can be measured in three-space. Spin can be represented as a vector [either spin up (+) or spin down (–)] projected onto one of the components constituting the three dimensional axes (x, y, and z). Here, the vector is a measure of fermion-intrinsic spin 1/2. (Directed spin quantities are shown by a dot with an attached arrow.)

Bohm's thought experiment consists of a series of tests. In the first test an experimenter uses one pair of detectors. Protons are individually ejected to an analyzer (Stern-Gerlach magnet) generating a magnetic field. The field acts on the magnetic moment of the incoming particle deflecting it to one of the two detectors. One detector registers (+) the other (–). The experimenter cannot determine from this test whether the readings arise from the process of detection or actually measure some real property of the particles.

Ideally, the experimenter could split and pass a particle beam through three channels: each leading to an analyzer. One would expect the detectors to register the spin value for each of the three components of the spin. The Uncertainty Principle precludes this possibility (spin values are incompatible variables). The more probable one spin component value is known, the less probable another spin component value can be known. Only one spin component can be measured (in reality) for each particle beam.

The experimenter elaborates his setup. He begins a second experiment. He assembles two sets (A and B). Each set uses one analyzer and two detectors. He places a source between sets A and B that generates particle pairs. Again the properly oriented analyzers allow only one of the three components of the particle's spin through to the detectors. Here the analyzer for each set measures the same spin component.

Our experimenter again uses particles emanating from a singlet state as Bohm suggested. (The composite spin of the two particles is zero: $1/2\ \hbar\ (+) -1/2\ \hbar = 0$.)

With the total angular momentum unchanged the two particles uncouple and travel in opposite directions to their respective analyzer. The experimenter finds whenever set A registers (+) set B always registers (–). (This is simplified. Individual events are not measured: only their relative frequency as in an ensemble of identical systems.)

This correlation (of relative frequencies) is always observed: given enough trials. Whenever set A registers (+) for component Z (Z^+) for one particle of the pair, the experimenter supposes he can infer from the local realistic theory that its sister particle would register (–) for that same component (Z^-) even if it is not measured. The experimenter concludes he is measuring an inherent property of the particle.

The experimenter is confident that for each particle of the pair that he can measure one spin component and infer another. The inferred spin value is a hidden variable. (Hence the variable is a deductive consequence of expected correlations based on the local realistic theory.)

Is there a way in which these hidden variables can be tested? The experimenter begins a third experiment. He changes the orientation of the analyzer used with set B in the second experiment and resumes testing. This experimental arrangement is described in what follows.

Bell's Theorem

In 1964 J. Bell wrote an article showing hidden variable theories (HVT) have measurable consequences in the correlations obtained from tests using the third experimental setup. Bell discovered a mathematical inequality that must be satisfied if local realistic theories represent reality. The correlations can be used to test quantum theory against the local realistic theory.

Building on a generalization of the EPR argument, Bell's theorem (a converse of the separability premise) states that measurements taken in one location cannot be exclusively independent of measurements conducted in another location—even if separated by an arbitrarily large distance. The theorem implies if the predictions of quantum mechanics are experimentally verified then the local realistic theory must be faulty. Particularly: At least one of its three major premises must be in error.

Bell's inequality, and generally—local realistic theories, put a limit on correlation that can be observed in the second experiment. Yet, quantum mechanics predicts that limit will be exceeded. According to quantum mechanics the number of certain component pairs (X^+, Y^+) will exceed the sum of the other two possible pairs $(X^+, Z^+$ and $Y^+, Z^+)$.

The experiment involves three series of double measurements. Each series is like the one arranged in the third experiment. Each tests for one of the component pairs. D'Espagnat has logically proven the inequality in the context of the local realistic theory.

H. Stapp has found a proof of Bell's theorem that does not rely on the hidden variable theory of quantum mechanics. He uses two assumptions. He combines separability with a nonprobabilistic approach to spin measurements and modifies them into the single principle of local causes. It states widely separated detectors function independently. Local causality is entwined in the concept nature can be described because of its uniformity. Stapp shows that though the principle holds statistically for quantum mechanics it fails at the level of individual events.

It should be possible to experimentally determine whether quantum mechanical predictions violate the inequality. Such an experiment could prove there is something erroneous either with the rules of quantum mechanics or with the premises of the local-realistic theory (or local causes).

In 1969 J. Clauser, R. Holt, A. Horne, and A. Shimony began designing an experiment for Bell's theorem. They realized the experimental difficulty of obtaining perfect spin correlations between particle pairs (as suggested by Bell). To develop a practical experiment they had to further generalize Bell's inequality. They modified Bell's inequality and designed an experiment to measure the differences between pair readings.

In 1972 Clauser and S. Freeman conducted the first experiment. Instead of detecting spin components of protons, they measured the linear polarizations of photons. They used photon cascade transitions in the calcium atom. The atoms are excited, and electrons in a singlet state cascade two energy levels emitting photons in the process. (Pauli's Exclusion Principle states the two electrons must have opposite spin.) They succeeded in proving a violation of Bell's inequality. Other experiments followed. Most show that correlation limits are exceeded.

Concerned scientists now concede the confirmation of Bell's theorem because Bell's inequality is experimentally violated in just the way quantum mechanics predicts. Nature behaves in a way consistent with quantum mechanics and not in the way envisioned by Einstein.

What does it mean? Violation of Bell's inequality means there is something amiss with the local realistic theory: at least one of its premises is in error. Scientists have insufficient reason to surrender the premises of realism and induction (one practically needs the other).

That leaves separability. Remember: It states that results of one experiment cannot influence the result of another experiment separated by a large distance. That is, the observed spin component value of one particle should not influence what is measured for the spin component value of another particle separated in space. We find that in using beams of particle pairs from a singlet state, when set A registers (+) set B always registers (–) when both analyzers (or polarizers) are adjusted to measure the same spin component (or polarization). Here, the inequality is satisfied.

This was our second experiment. Our experimenter naturally assumed that each particle pair inherently carries "hidden" instructions, which caused the detectors to correctly register the expected correlation. However, that supposition does not explain some other cases (as in the third test performed by the experimenter where different spin components are measured by each set). In some of those cases, the correlations exceed the inequality. Therefore, the idea particles have inherent properties that influence the two sets, is ruled out as a possible explanation of the correlation phenomenon. The observed properties could not have existed prior to measurement.

So how is the second experiment to be explained? Hidden variables are not a factor (based on the third experiment). Hence, experimental demonstration of the violation of Bell's inequality proves Bell's theorem. The results of one experiment can influence another experiment conducted in a distant location (or at least there is some type of correlation between them). Thus, Stapp decides no theory conforming to the predictions of quantum mechanics involves local causes.

How is this influence conveyed? Are light signals sent from set A to set B to tell it how to register? It is expected that if regular separability is violated, then so is Einstein separability.

Einstein separability has been tested. In 1980 A. Aspect and collaborators repeated the original Clauser and Freedman experiment using two sets of time varying polarizers in which the polarizer is set in less time than needed for a light signal to traverse the distance between sets. The inequality was again violated as expected.

What about faster than light signaling—in contradistinction to Einstein separability? How can an effect precede its cause? Immediate problems with causality arise if faster-than-light signaling cannot explain the violation of Bell's inequality. We learn in the final chapters there is a reasonable explanation how certain "effects" are purposively "directed."

Experimental verification of Bell's theorem proves there are events that, though separated by light years, are correlated. (This is corollary to the fact proved by Aspect that the correlation holds irrespective of possible causal linkage by light signal.) Stapp stated these events could not be explained by common (extrinsic) causes. Bohm suggests the property exhibited by these events must somehow depend on the state of the whole system. But, it means more than that. The whole system (mental and physical) must be simultaneously operated on by a single factor.

There is an underlying cause that links seemingly disconnected phenomena at the quantal level. Otherwise, Bell's theorem becomes difficult to explain. As Bohm suggests, it shows that no theory of reality could be complete unless it incorporates an underlying quantal interconnectedness. And that connectedness calls for a common thread. This means if there is an intrinsic cause for all physical existence that extrinsic causality must break down somewhere. And that "somewhere" appears in quantum phenomena as suggested by the Uncertainty Relations.

Extrinsic causality appears as a one to one correlation of causes and effects for the macro world. It is here where regularity is observed between events. Yet that regularity has its origins in quantum phenomena. Thus, it should not be surprising that Born's statistical interpretation of the wave function is difficult to ignore.

The Copenhagen interpretation continues as a viable theory. It eschews the idea of an independently existing reality. Moreover, unlike classical physics, the Copenhagen interpretation indicates inseparability between what is observed (the "rest of reality") and the observer (the "self"). (See Figure 8.1.) (For example: Observation is said to cause wave-packet collapse.) J. von Neumann realized that quantum mechanical laws could not be complete without including the observer. One event does not cause another at the cosmic-structural level.

We should not confuse the interaction between things (Type I Whyness-Whatness Association) with actions imposed upon things from within (Type III Whyness-Whatness Association). For example, confusion in interpreting the meaning of the wave function arises when mixing the extrinsic causes of macrosystems with the intrinsic cause of microsystems. This confusion appears in gedanken experiments such as "Schrödinger's Cat paradox" and its polished update described in the predicament of "Wigner's Friend." Confusion also arises when we realize intrinsic causality also underlies macrosystems. Intrinsic causality operates upon and sets the limits to extrinsic causality.

Let's look again at the correlation experiment. A simple noncontradictory explanation is both events (the correlation between two experimental results) have a common cause. The correlations do not identify the common cause but merely suggest it. And this common cause is intrinsic to the whole system of physical events. Thus, there is an influence or intrinsic cause (the Realm of Whyness) which lies outside of and is separate in kind from physical reality (a part of the Realm of Whatness) yet is indirectly manifested in it (through the Realm of Reasoning).

An intrinsic cause can be said to propagate infinitely fast. As far as two widely separated systems are concerned this is instantaneous action at a distance. There is no problem with that. No energy or information is exchanged or transferred from one system (experiment) to another. Only, each system is a part of a larger system and is subject to this common influence. Conversely, the whole is implicated in each part or subsystem. Aspect and others reveal this by experimentation.

A hint of this common cause has been implied in the study of other phenomena. Bohm has addressed this idea in his concept called the "quantal potential." He mentions that quantum-many-body systems cannot be analyzed in independent parts. Rather, that each part depends on their arrangement in the whole system. He says, "the quantum potential depends on the state of the whole system." He says it is unmistakable there is an inseparable quantum interconnectiveness of all reality. The proper question is "then what causes the whole (physical and mental) system?" It is more easily stated by ruling out some factors not entangled in the answer.

Temporal causality is not involved at this level—where there is no time flow. Temporality occurs in the physical (and mental) level.

Change arises from a shift in the relationship of the parts composing the whole.

The degenerate change in relations, described by entropy, casts the shadow called the arrow of time. Major parts (like Cosmos and Anti-cosmos, e.g., Bi-Deltic Cosmos) constituting a "whole system" negate temporal directionality. So we find quantum phenomenon affects the whole of the physical world instantaneously. This idea is reminiscent of de Broglie's idea of a universal pulsation.

We also discover at the quantum level that EPR correlations are time reversible. O. Costa de Beauregard has said that just as there is a nonseparability of two measurements issuing from a common preparation, there is also a nonseparability of two preparations converging unto a common cause. Here as expected causality is CPT invariant. This idea is viewed from the manifested (Realm of Whatness) perspective of this pulsating commonality. Invariancy shows these quantum correlations apply as well to other "whole" systems. These systems include other tardyon universe—antitardyon universe pairs and other tachyon universe—antitachyon universe pairs. It is in the higher level of this super system (Cirpanic Cosmos) where temporal duration (arrowless time) cancels out.

We find at this level of generality (all parts of the Realm of Whatness are essentially ONE in the Realm of Whyness) that our consciousness (or "self") is not separate from what we could be conscious of ("the rest of reality").

"Consciousness" and "What we could be conscious of" merge into a quasi-mental-physical oneness: a "wholeness" transcending spatial extensionality and time flow. This is simple to understand. As with many "in-between hierarchical levels" of generality there is always some universality covering in one category what at lower levels would be distinct (separated) categories (or entities). We learn this not only to be the case conceptually, but also in reality.

At the quantal level, reality functions as an inseparable whole. This is why the EPR argument is unverifiable. Particles issued from a singlet state act as though they were one particle. And in the sense of nonseparability they are (a part of one reality). It follows that frequency correlations, although not apparently deterministic in themselves, intimate determinism for individual events. This applies not only for physical events (represented by the proton experiments) but for electromagnetic radiations (represented by the photon experiments) as well.

What does this say about the statistical interpretation of quantum mechanics? We concede individual particles obey the principle of randomness. But, it would be inappropriate to suggest that principle represents a law of nature. It is but a sign of what we do not know. That does not necessarily imply there are hidden properties (of whatness) in the particle, only that whatever the factor, it functions in a common mode to all particles. Similarly, to say some laws of nature (like intrinsic causality—although not well known) are "temporally" suspended (for probability cases) when determining particle group behavior, for example, is also absurd. All extrinsic factors (any factor, which if altered, would affect the particle's course of action) related to a single particle's fate cannot be known. A thing's fate is not solely determined by the activity of surrounding things.

Probability reflects the regularity we observe in nature. We cannot deduce from the Uncertainty Principle that microcosmic events happen by chance. If they did, and there was no intrinsic cause, we would have to suppose all macrocosmic events also happen by chance. That just is not observed. We conclude probability fills the void in understanding created by the overlap between contributing extrinsic factors and intrinsic causes.

What does all this mean in the structure of reality? We know the correlation experiments do not measure an inherent property of the emitted particles. Instead, evidence points to a common cause: one manifestly indicated by the EPR experiments. Just because quantum mechanical predictions point to some underlying cause does not mean that cause will be satisfactorily explained by quantum theory. It seems intrinsic-causal interconnectedness would be difficult to explain by contemporary quantum mechanics alone, if for no other reason than the Uncertainty Principle seems not to permit (quantal phenomenon) it to be its own cause. Although an intrinsic cause is implicated in quantum phenomena, we would expect the cause to be linked to one of the major forces.

It seems reasonable to (noncontradictorily) suppose that the common cause is gravitational radiation. We arrive at this conclusion through two ruminations. First, we consider what the underlying cause could be in the structure of reality. Deductively, it has to be either electromagnetism or gravitation. The previous subheading concerning manifestation of the electromagnetic energy and the causal hierarchy of possibilities (used in unification schemes) point to gravitation.

In its frame of reference, gravitation, like electromagnetism (their energies are manifested forms of their radiation) is timeless and spaceless (lack extensionality). Its quantum pulsation acts instantaneously everywhere. Gravitation manifests in the physical (as tardyon and tachyon universes) via electromagnetic phenomena. Therefore, the manifestation (the Realm of Whatness) also exhibits that pulsation.

We are familiar with that pulsation in quantum phenomena. Using the structure presented in this book we find the Realm of Whatness is an extrinsic manifestation of the Realm of Reasoning. The Realm of Reasoning, unlike the Realm of Whatness, is a temporal processing force. The Realm of Reasoning, although the immediate source, is not the final cause of the pulsation, but is rather the conveyor of the pulsation. It is the "on-off" pulsation of the Realm of Reasoning that directly creates the Realm of Whatness. And it is the Realm of Whyness that intrinsically underlies it all.

Key Points:

- A long sought scientific goal is to conceptually explain and unify all physics in a single equation. At this point, we have the "concept" but not the equation. The Realm of Whyness is the concept. (We derive the equation in Chapter 18.)

- The unity of reality is recognized as the structural integration of the Realm of Whyness and Realm of Whatness. The Realm of Reasoning represents the integrating factor. Everything unifies "under" the Realm of Whyness.

- Physicists recognize four "forces" of (or interactions in) nature: weak, strong, electromagnetic, and gravitation. The weak and electromagnetic forces have been verified to be two aspects of a single encompassing electroweak force. Physicists postulate a Grand Unification Theory (GUT) that unites the electroweak with the strong force. The final task is to unify the GUT with gravitation.

- We have conceptually unified the three basic forces: strong, electromagnetic, and gravitation. (Here, the weak force is not considered separate from the electromagnetic force. The distinction of the weak force is that it differentiates the Cosmos [particularly, the Universe of Perfect Whatness] from the Anticosmos [its antimatter counterpart.])

- All forces "unify" under the umbrella of gravitation. Gravitation is a space function. It acts instantaneously on anything in space. It is not a linear time function.

- Experiments verify that Bell's inequality is violated according to predictions of quantum theory. Experiments indicate certain events are simultaneously correlated: they are affected by an instantaneous function. We suppose this is the space function itself—what we typically call gravitation. Such ideas extend beyond the fringe areas of the common understanding of gravity. To state these extraordinary ideas are associated with gravity means we need to broaden our scientific understanding.

We have considered three major forces: strong, electromagnetic, and gravitational. We have seen they can be conceptually unified through the graces of mathematics. But, we soon learn there are problems inherent in the contemporary logical understanding of mathematics. Resolving these problems should yield a better understanding of the structure of reality described in this book. It should also point us in the proper direction to merge relativity with quantum theory.

In the next section we use another tactic to analyze what we have stated so far in this book. We will use a multi-disciplinary approach to understand the mind expanse. We endeavor to better understand how the realms relate to each other. Pondering philosophical, religious, and logical questions, we find there are yet other more profound ways to understand the structure delineated in this book.

Section Six

INTERPRETING STRUCTURE

Perspective

- Using structure to answer philosophical questions
- Epistemology
- Metaphysical questions
- Using structure to answer religious questions
- Deeper understanding of logic addresses previously unanswered questions
- Ultimate Principle explains Cosmogenesis and unifies Relativity with the quantum

Chapter 16

Using Structure to Interpret Philosophical Concepts

> "If all 'big' questions have been satisfactorily
> answered, it becomes exceedingly difficult to explain
> why those questions are still seriously being asked!"

The word science, from the Latin meaning to know, was coined in 1840. Prior to that time, what is now called scientific experimentation was called "experimental philosophy." Science as knowledge was called "natural philosophy." B. Russell had properly noted science had its origins in philosophical thinking. The goal of philosophy is to understand Reality as a complete system.

Philosophers primarily have pondered big questions. They foremost wonder about the meaning of life. Other philosophical questions beg to answer this one. There has been no shortage of conceivable answers.

Historically, philosophy has been concerned with man and his place in the universe. Reality is only understood by its structure—the interrelation of its parts. To answer the question of man's meaning is to ascertain how he fits into the scheme of things—how he fits into the structure of reality.

The subject of this book has been the conceptualized structure of reality. We also refer to it as "mapping" the mind expanse. Delineation of structure falls within the domain of science. That structure, based on scientific principles, will be used to gain a deeper understanding of philosophical concepts: foremost of which is "the meaning of life." Answering this question is deferred to the final chapter.

Major Philosophical Questions

Philosophy begins with common sense and the asking of questions. By doubting existing dogma, attitudes, and beliefs, philosophers had hoped to acquire a better understanding of man's existence.

Some great questions bearing on the meaning of life follow. Preeminently, what is the nature of reality? This question is answered by the structure of reality presented in this book.

What is the nature of the self? The self is the mind. The mind has been analyzed and defined using mental functions—deductive and inductive. How these functions relate to reality has also been explained.

What is the nature of behavior? Behavior is a natural consequence of the mental functions. People behave in a way to become more secure. Security can be analyzed in terms of wanting and having—especially wanting and having things pertaining to one's security. Wanting and having is applicable also to knowledge; therefore, whether one has or has not knowledge. There is security in knowing and understanding things but there is also a security that comes from understanding reality generally.

What is the nature of knowledge? It has been carefully defined in this project and is again addressed below.

Does the self have free will? This question is analyzed in this chapter and is definitively answered by book's end.

What is the nature of matter? We have defined it as a "whatness" form of experience.

What is the nature of space and time? These two questions are understood scientifically by learning about Special Relativity. Space and time are more deeply explained in the final section.

What is the nature of religion? Is there any merit to religious truth? We turn to this question in the next chapter.

What is the nature of consciousness? Consciousness has defied understanding. It too will be definitively defined in the final chapters.

Most individuals have asked big questions sometime during their lifetime. Many people have adopted answers without seriously questioning their source. Some answers are thought to have been rigorously defined by past philosophers. The question remains; however, do their answers represent reality?

Some philosophers speculated whether truthful answers to the big questions could even be found. It is no wonder. The question itself is fraught with a problem. Serious thinkers since Socrates have not agreed on what is truth. Until truth is noncontradictorily defined, the other questions cannot be satisfactorily answered.

Most philosophical questions have eluded scientific investigation for two intertwined reasons. First, truth had not been properly defined. Second, until now there has not been sufficient-scientific knowledge to understand man's relation to the "rest of reality" (so he could be assured of "what" is truth).

The structure of reality has been delineated in this book. We now know what that truth is: by definition and by what it represents. We have arrived at a point where we can revisit the big philosophical questions and obtain answers based on an understanding of reality itself. The structure of reality can be used to render a better understanding of philosophical questions. We are beginning to fill in the details of what constitutes the Self and the "rest of reality." (Refer to Figure 8:1, Defining the Self.)

Most big philosophical questions can be sorted into one of two categories. How can we be sure of what we know? What is the nature of what we know? The first question is about the theory of knowledge—epistemology. The second question focuses on the structure of reality—metaphysics.

Epistemology

What is the nature of knowledge? What is truth? Delineating the structure of reality enabled us to noncontradictorily define truth. Truth is all knowledge. Knowledge is information that represents reality.

How is truth obtained? If we say knowledge is information that corresponds to reality, then we also need to define how we know that. These questions also have been answered.

Knowledge itself must be coherent and therefore logically consistent. If these conditions are met then knowledge should also represent reality because we suppose, quite properly, that reality exhibits uniformity. Why? We have not found any indication reality is anything except logical. The final test is if reality can be properly delineated. If yes, we can be assured of our premise that reality exhibits uniformity.

All knowledge is inferential. We have extrinsic perceptions. Using B. Russell's terminology, we call this "knowledge by acquaintance." Extrinsic perceptions by themselves are practically useless. We wish for understanding. To understand we must define "what" is experienced. This is "knowledge by description." To understand reality is to systematically arrange the knowledge we have by description. This is knowledge of relationships.

All knowledge is constituted in a hierarchy of universal statements. How much or how little universal statements have in common determines arrangement of knowledge in the hierarchy. This book refers to the relation between hierarchical structures as interphenomenal relevancy. The relation to things within a hierarchical level is denoted intraphenomenal relevancy. (Refer to Chapter 1.)

To classify is to define. Each hierarchical level denotes a classification of phenomena. Near the base of the hierarchy are found the lowest categories. All lower categories are a series of subsumed classes of phenomena. These categories define the characteristics (or properties) that describe particular objects, things, and entities. Intermediate categories consist of the more general classes of phenomena. The hierarchical pinnacle is all-inclusive: causally representing all that is. "All" is inferred by the universal "reality."

The knowledge hierarchy defines the intrinsic nature (constitution) of reality. The hierarchy is representable by structure. Various aspects of structure are represented by propositions.

How can the surety of propositions be determined? Knowledge is relative. Knowledge is only as certain as the tests used to establish truth. Knowledge becomes less tentative the more it is tested.

Science offers the best method for testing propositions as knowledge. The overwhelming increase in scientific knowledge and the concomitant refinement in scientific methodology enables testing of what would otherwise be unyielding propositions.

F. Bacon realized knowledge is derived from experience. Experience helps distinguish what is true from what is false. Propositions about extrinsic perceptions are tested against "external" reality. We know how this applies to low-level generalities—those representing particular objects. It is not so easy to comprehend how testing applies to higher-level generalities that are more abstract. Yet, high level generalities are testable, for example, by the laws of physics and logical consistency.

There is a separation between awareness and the object of awareness. The observed object or sense datum is what is experienced. It represents something in the external world. We label sense "objects" using symbols and words.

Words and knowledge imply inference and hence awareness (of experience). Therefore, there is also a separation between knowledge and what it represents. Our problem is "have we correctly represented the referent in knowledge?"

How can we finally know whether a proposition represents something in reality? Knowledge represents reality. Particular statements to the highest generalizations correspond to something. Although knowledge is reflected in verifiable-experimental results and noncontradictory description, the test is whether it is contradicted by the structure of reality. What if a proposition is too difficult to test? We consider whether the proposition is logically consistent with the body of other well-established generalities. We also ask what level of generality is involved.

Phenomena are categorized according to "levels" of generality. All levels fit into a structure. Although correspondence is the initial test of what is true, it offers little help in the trial stage of arranging different levels of generality. This is where theorizing—hypothesizing plays a role.

How are higher-level structures built? Abstract generalities are potentially pieced together by inductive reasoning. Abstractions must agree to the coherency of knowledge as a whole. Each level must find its place in the hierarchy of knowledge. Coherency cements structure.

Many high levels of generality, as propositions, have been tested experimentally. Some of these experiments have yielded well-defined results. One striking example is the testing of quantum phenomena as a whole with Bell's theorem. Obtaining definitive results from very general propositions inherently adds to the supposition that each universal level infers a specific level of structure.

What is the meaning of coherency? What in reality does coherency represent? Coherency is found in order. Order arises from the uniformity of nature. Knowledge coherency represents order in reality.

What criterion determines coherency? Coherency is ascertained by invoking the idea of noncontradiction. Given nature is uniform, knowledge, because it corresponds to reality, should be immune from contradiction from both noncorrespondence to reality and inconsistency with established knowledge.

For us, knowledge coherency is not absolute. Truth is only approached ideally. Coherency is approached by elimination of contradiction. Some propositions do not fit "into" the puzzle of reality. These propositions are considered "false."

How do we know false propositions from true ones? Testing and experience. In ascertaining truth, errors occur from misinterpreting experience or from not sufficiently considering possible contradictions.

Science has a powerful tool to check coherency. Coherency entails logical representation. In science logic is often expressed in mathematical equations. If equations are found that adequately describe the relation of certain phenomena, the subsequent task becomes one of descriptive interpretation.

Interpreting the meaning of phenomena is easy. Quoting A. Einstein: "As far as the laws of mathematics refer to reality, they are uncertain; and as far as they are certain, they do not refer to reality."

The final task is to test whether hypothesized coherency corresponds to uniformity in nature. The ultimate test becomes one of logic itself.

Although man cannot know all things, he can understand all things in a relative way. The idea is to interrelate enough knowledge to build structure. Before embarking on the search to understand the structure of reality, we assumed the primary premise that nature exhibits uniformity. That the structure of reality was delineated supports the original premise. Correspondence and coherence together test the truth of propositions.

We now understand the meaning of knowledge and how to ascertain what is true. The great metaphysical questions can now be addressed.

Metaphysical Questions

Metaphysics means "beyond physics." Metaphysics pertains to the object of knowledge in the broadest sense—"that which is." The word metaphysics is used here because philosophers have refined and formulated the questions and not because some of their answers may have been later verified by science.

Historically, metaphysicists sought an understanding of the nature of being—of "what is." The structure of reality can scientifically answer questions bearing on "what is." Some often asked questions have been definitively answered in this book; others have yet to be addressed.

The following are representative metaphysical questions. What is the nature of mind and matter? Is there a mind-body dualism? Is there free will? What is the nature of memory? What is consciousness? Is there a God? If so, what is its nature? What is man's relation to God?

Particulars and Universals

A particularity is a specific thing. A universal is usually acknowledged as a mental abstraction (not physical) representing many things.

Plato, the early Greek philosopher thought universals were "forms" or ideas existing independent of any physical thing. This is like a cookie cutter (the mold, or form) stamping out physical representations of itself (the cookies). These ideas much later became known as extreme realism.

For a few hundred years following one thousand AD, scholars debated the issue of particulars and universals. The issue arose when some philosophers believed universals were more than an abstraction, a verbiage string or idea. Are universal concepts merely thoughts in one's mind? Or, do universals represent something existing in reality?

Boethius translated the works of some early Greek thinkers. He wondered if generalities, acknowledged to exist in the mind, also represent something in reality. If they did, he recognized there were immediate problems with this view. One example is that one can imagine unicorns, but they have not been shown to exist in reality. Not every idea represents something in reality.

Boethius believed that universals exist both in the mind and in "outside" reality—our environment. They exist materially in physical things, and immaterially in the mind. The question became "does a universal represent something real or not?"

Guilliume de Champeaux thought form was inherent in every physical representation of that form. Every cookie shows the "universal form" of "cookieness." If this were true, then how could anyone differentiate between one physical form and another? Yet, we clearly can and do observe differences between similar things. He changed his view saying each physical form is alike in some respects and different in others.

Roscellinus believed only particulars existed in reality. Universals are strung-together words that, though representing categories of things in reality, did not itself represent anything more in reality than the things that composed the category. For example, white ducks, categorically, do not represent something specific in reality. Rather, there are specific instances of white ducks. There was not something in reality corresponding to white ducks taken as a collective whole. Whiteness existed only in abstraction. The universality of whiteness existed only as a mental abstraction of physical examples. This idea is known as extreme nominalism—only particulars exist.

Abelard believed universalities are only mental representations of physical things. He believed things provided the basis for universalities.

Why were these ideas significant? Many philosophers of the Middle Ages were theologians who sought to understand God. Some believed universalities existed in the mind of God, but not in the physical things God created.

The issue of particulars and universals rested on a deeper understanding of mind and matter. In Chapter 18, we revisit this issue with the deeper understanding.

Mind and Matter

What is the nature of matter? Philosophers have provided many interesting theories. One common-sense approach is called naive realism. Simple naive realists believe the world is just as it appears: that we see a physical object just as it really is. Everyday-man, without much thought and therefore by default, to some extent is a naive realist.

Philosophers often use the argument from illusion to support the contention sense experience is unreliable for representing what is real. A few overused examples suffice. A person can put part of a broomstick into water. The partially immersed stick appears bent. If naive realism is correct the stick must be bent. By simply reaching into the water and feeling the stick one quickly realizes it is straight. An amputee may feel a sensation in his leg when his sight and memory suggest otherwise.

Reasoning also suggests objects are not as they appear. An astronomer looks through a sophisticated telescope and observes a very distant patch of light. He knows from science that light speed is finite. He sees the object now, as it was billions of years ago. The object may no longer exist. These contradictions imply the world is not exactly as it appears.

Faced with obvious problems presented by naive realism, philosophers, like J. Locke, suggested that what we experience is sense data. Sense data represent physical objects in the "outside world." Locke said physical objects are ascribed two types of qualities: primary and secondary.

Primary qualities are experienced through two or more senses (usually visual and tactile) e.g., solidity, extension, size, shape, texture, velocity, and duration. (Nowadays we probably would include in this category physical properties of science.) In primary qualities, Locke believed, we experience something resembling the actual physical object. These qualities are real and measurable. Seeing a coin, although from an angle it may appear elliptical, means we are in the presence of an object that is round-like.

Secondary qualities are experienced by only one sense, for example, color, sound, odor, and taste. Secondary qualities are not assignable to physical objects themselves, but arise solely because of their affect upon our senses. These are sensible qualities.

Everyday-man, to some extent, is a representative realist. Although representationalism circumvents problems associated with naive realism, it incurs problems of its own.

Philosophers generally concede if sense data merely represent physical objects then the outside world cannot be proven empirically. This would not mean, however, that external reality does not exist. We leave a blazing campfire; upon return the fire is reduced to embers. Something transpired while we were gone. The laws of nature, predictability, and systematic corroboration of evidence, suggest there is an external reality. It would be foolhardy to assert it does not exist.

The idea in science is to verify (our ideas about) what we experience. That is different from saying what we experience exactly duplicates external reality or even that an external reality actually exists (in the usual sense attributed to it). If the external world were exactly copied in our senses, there would be no need for verification.

Can we know the outside world exists by reasoning? This project suggests we can. The color paradigm offers one way out of the difficulty. For example, we find ourselves in the presence of a fruit tree. It is bearing fruit, which we recognize to be lemons. Lemons appear a certain color. By convention we acknowledge the color is yellow. Reasoning tells us the lemon, aside from its appearance, is itself blue. Although the color paradigm is a mere concept, it carries enormous explanatory power.

Except for light, we would not see lemons. White light, a combination of all visible light, strikes the lemon's surface. Some light energy is absorbed into the object's surface; some energy is reflected and chemically transduced (by molecular isomerization) into neural signals by electromagnetic receptors in the retina of our eyes.

Our visual senses react to all the visual wavelengths except what is absorbed by the object's surface. The color of the object-in-itself is represented by the energy absorbed. The color of the object-as-it-appears is associated with the energy reflected to our eyes.

The color of light-energy absorbed is the complementary color of light reflected. A black appearing object signifies all light energy is absorbed, and none reflected. A white appearing object suggests no light is absorbed—all light is reflected from its surface.

Is there empirical evidence supporting the color paradigm? We can photograph the lemon. We get a negative film that, aside from the orange mask, shows the lemon appearing blue. How else can this be explained except that the reasoning of the color paradigm is correct? The lemon appears yellow, but in-itself is blue.

Two colors—one by sense data, the other by reasoning—can represent the same object. If the lemon can be represented these two ways then there is, as I. Kant suggested—the phenomenon, which appears to us, and the noumenon, which exists, but is hidden from our visual senses.

We are sure phenomena, as R. Descartes intimated, infer the mind exists. The color paradigm enables us to use our mind to deduce the existence of a material world. The color paradigm implies the material world is distinct from its appearance.

Was Descartes correct in postulating a "mind-body dualism?" Can we be as certain of a physical world as we are of sense experience itself? Some people would say, "seeing is believing": meaning they foremost trust their senses. A close analytical look at sense experience should settle the issue.

We cannot deny what we experience. Using our example, we do not deny seeing yellow lemons. The initial problem is most people assume sensory experiences themselves are real. The error is one of inference. The color paradigm implies mind and matter are categorically distinct and separate. (Refer to Chapter 3, subheading: "Self Cosmolization.")

Sense experiences arise from an interfacing of mind and matter. We are directly conscious of sense data. By nature people deductively infer, for example (with the help of memory) that yellow lemons exist.

A deduction is only as good as its premises. Although sense data are presented "as is" to consciousness that does not mean sense data in-themselves are real. They merely represent a part of what is real, viz, the material world.

Material objects (in themselves) are not directly observable. In our example, we infer lemons (in themselves) are blue. The unproved premise is that we see things as they really are.

Is external reality understandable? To understand is to interrelate. What is interrelated and how? Phenomena, i.e., sense data, are interrelated by inference. The first inference occurs quite naturally; we identify what is experienced. To identify is to focus our attention and choose certain things from the "stream of experience." In this lies the second problem regarding trust of the senses.

To identify is to perceive; to perceive is to ascribe meaning. Meaning is associated with properties and characteristics. Sense-data labeled "physical objects" are usually perceived to "exist" in the "outside world." Given there is an external reality we are sometimes deceived whether certain experiences represent something real.

Sometimes people report experiences thought to be extrinsic in origin, but are not. Hallucinations, certain drug induced experiences, and vivid dreams are examples. Finally, if we are mistaken, it is because of a perceptual problem (one of inference) and not one of sense experience.

Extrinsic perceptions should be empirically differentiated from intrinsic experiences. Extrinsic perceptions are testable by systematic corroboration of evidence. The question is always "did we correctly infer meaning?"

Even if an experience is established as extrinsic in origin, there can still be a problem. The perception of something is not just a consequence of said sense datum (object). We may perceive a change in a sense object even when there is no change. This was exemplified by E. Boring who in 1930 presented a drawing of a woman. The figure is ambiguous. Looking at the figure we see a young woman. If we continue to look at the figure, something happens. All of a sudden the figure appears to change. Instead of a young woman, we see an "old hag." The figure does not change, yet it perceptively flip-flops back and forth from one identity to the other.

Perspective (affected by saccadic-eye movement) and conditions (of sense organs) also affect how we experience something. Most people have stood on a bridge watching the river flow below. Sometimes for a moment, it seems as though the river is still and the bridge is moving. The same thing may happen when we are in a car and stopped at a red light. The car to the side slowly moves forward. For a moment, it seems as though our car is slowly moving in reverse. We conclude our perceptual apparatus may also deceive us.

Mind-Body Problem

Philosophers and psychologists have not all agreed on the dualism of mind and body. Let's examine this. There is our mind and there is that of which our mind is aware of (e.g., sense experience): that which represents the outside world and even our body.

Many dualists believe matter influences mind; mind influences matter. Sense experience is mental. Psychologists may say sense experience is a response to some stimuli in the (physical environment). An injured leg produces the sensation of pain. Imbibing alcohol affects judgment. We make decisions (mental); we act upon them (in the physical world). We desire food; we try to obtain food.

There is a reciprocal cause and effect relationship between mind and matter. Material events are transduced through the senses and represented mentally. Sensory input registers in "association" areas in the brain.

The brain is composed of matter. If mind is distinct from matter how does it receive sensory impressions? One may be tempted to suggest one way out of the difficulty is to say the mind is manifested in the brain. Although that supposition is a beginning it still harbors the same question. How are sensory experiences produced? This question is a parcel of the mind-body problem—a part of Descartes' legacy.

Philosophers have attempted to resolve the mind-body problem. Monists suggest there is no mind-body dualism. They insist there is just one substance. Their theories suffer from the reductive fallacy. They dispose of the dualism by explaining, either mind in terms of matter or matter in terms of mind.

Materialists believed matter could explain mind. G. Ryle attempted to reduce mental events to behavioral events. He thought the concept of mind, or the "dogma of the ghost in the machine," was a categorical mistake. He believed all allusions to mind could be explained by observed behavior. Behavioral psychologists J. Watson, K. Lashley and others, developed this idea.

Subjective idealists believed only mind exists. G. Berkeley and others believed matter could be explained as a projection of mind. Berkeley suggested Locke's primary qualities, like secondary qualities, only existed in one's mind. If this were the case there could be no way to characterize matter (separate from mind). So, he supposed matter does not exist.

Berkeley believed material objects exist only while perceived. He realized the immediate question—what happens to things when we do not experience them? He says, when things are not perceived that they only exist in the mind of God. This is an interesting idea, as we shall see by the end of this book. His position is defended because the material world could not be empirically proven. Using our campfire for instance, we ask, "how did the fire continue to burn without anyone watching it?" When we walk from the fire, did God take over the watch? Again, because the outside world cannot be proven empirically does not mean it does not exist in some sense. It is that "other" sense we are seeking.

Another monistic idea is the identity (or double aspect) theory. It postulates both mind and matter exist, but are just two ways of describing the same thing, which is itself, neither mind nor matter. Supporters of this theory have used the example that the morning star and evening star, though described differently, are really identical (the planet Venus).

The double aspect theory suggests man can be wholly described either physically or psychologically. Adherents insist these are two ways of perceiving a common underlying reality. Critics maintain this theory does not adequately define the underlying unity behind each aspect. B. Spinoza said the underlying reality is nature or God. Berkeley took the same route to avoid solipsism. D. Hume held a similar idea. He thought mind and matter were different arrangements (or bundles) of the same thing.

These ideas are interesting. This book explains the unity of the Deltic Cosmos by representing it (as knowledge) and breaking it down into three "aspects." We shall see especially how close Spinoza's idea (of defining God as some type of underlying nature to reality) is realized in the structure of reality.

Other philosophers thought maybe mind evolved out of matter. Emergent evolutionists believe both mind and matter exist. They believed there are many levels of creative synthesis. Like electrons forming atoms and atoms forming molecules so matter eventually forms into mind-stuff. Each succeeding level attains properties which previous levels do not exhibit. C. Morgan thought the levels could be represented in a pyramid where each step is a function or process of evolution. This idea looks familiar (see How Aspect Diagram, Chapter 4) but sidesteps the issue.

If neither mind nor matter can be reduced to the other, then a duality of mind and matter exists. Try as some philosophers have, there is no reasonable alternative, but to accept there is both mind and matter.

Without both mind and matter, we would not have sense experience. The question then is—how can one affect the other—how do they interact? Philosophers thought two sets of things, having no common properties, could not affect one another. To avoid this difficulty they proposed theories of occasionalism, parallelism, and epiphenomenalism. We will skip these attempts and examine the heart of the problem. The problem narrows to cause and effect.

This book's introduction lists three classifications of whyness-whatness associations. Philosophers and scientists alike have been puzzled over Type I and Type II whyness-whatness associations.

Type I association represents a series of linked occurrences. Each event is a contributing factor in the series. Each contributes in the sense that if it did not occur the series would be other than what it is. However, that cannot be construed as a series of cause and effects where each subsequent effect becomes the cause for the next effect or event. (If a cause becomes an effect, it is no longer a cause.) That the series continues, albeit in a different direction, stultifies the idea each physical event (or effect) in turn becomes a cause.

Focusing the idea of cause on a succession of events overlooks the question why the series follows one direction instead of another. This is the question that a genuine cause should answer.

During brain operations, surgeons like W. Penfield have stimulated a region of a patient's brain and evoked visual memories. Can such a reaction be induced if the physical and mental are separate and distinct in kind? If our earlier analysis is correct, at the least, mentality is correlated with the physical and conversely. Then the only plausible answer is the physical and mental are correlated by a third factor common to both. The factor would have to operate intrinsically. Functionally, the physical and mental must be two aspects of one thing.

The structure of reality suggests that the Realm of Reasoning is somehow common to both the physical and mental. The universe of matter and the two-sided universe of the mind (deductive and inductive functions) operate together as part (i.e., How Aspect) of the "What Aspect": better known as the Deltic Cosmos.

Cosmic structure is driven by an intrinsic cause. Another way of saying this is that tardyon (physical) matter interacts with tachyon-mind stuff through electromagnetic phenomena. It takes no philosopher to realize an intrinsic cause bespeaks of determinism; or, at least that some things are determined.

Free Will and Determinism

Determinism is the bedrock of science. Scientific prediction is predicated on determinism. Classically, the rigor of the scientific method has been limited to the physical, and very successful it has been. Many philosophers realize this but maintain, however, despite the physical that somehow man is free. If so, how could man translate his will into an environment that is strictly determined? Let's scientifically examine the issue of free will.

In previous chapters we studied the meaning of Bell's theorem and Everett's many-universe interpretation of quantum mechanics. Experimental confirmation of Bell's theorem suggests a fixed common cause.

The common cause correlates with two separated experiments (A, and B) when observed. The observer also cannot escape this determining factor. We realize the experimenter watching set B could decide and change the orientation of one analyzer after the particle has been emitted. The experimenter, watching over set A, does not know what orientation the other experimenter has set his analyzer. Yet their detectors correctly register the correlation.

We cannot suppose the particle registers the correct reading at the whim of the observer. (Neither can we think an isolated particle knows how to align itself to effect the correlation.) Rather, because both the physical event and the observer are parcels of the same system, it is more reasonable to suppose the action (registering) of the particle and the action (decision) of the observer are intrinsically determined by the same cause.

We also find determinism in Everett's theory. If the universal state-vector does not collapse, the world becomes a strictly deterministic system. Either a physical entity (in interaction) or some recording mechanism plays the role of observer. In both cases the question of determinism is brought to our attention.

Is the condition of the Realm of Whatness strictly determined? It is, if physics has anything to say about it. However, we can still ask from a behavioral perspective if that in any sense leaves room for free will.

Philosophers have debated the issue of free will and determinism since the advent of literacy. Fuel was added to the argument after the founding of Christianity. Theologians realized an apparent contradiction arises if God is omniscient and man has free will. If man is truly free how could God know what man would do? Under these circumstances it is difficult to see just what the term omniscience could mean.

Also, if God has no control over man's will (and man is truly free) how can God be omnipotent? If God were in another way omnipotent, one would also wonder why such a God, who is supposedly benevolent, would allow the suffering of humanity. It also seems absurd that a reasonable God would hold present man accountable for the original sin committed by an ancestor.

What happens if man is not free and all his experiences are predetermined? If that were the case, one would wonder what justification there is for God to punish man with eternal damnation for sinfulness, when man is not ultimately responsible for what he does. All such questions are inevitably bound not only to man's conception of himself but to his relationship to God. What is the nature of that relationship? It is this sense in which man has usually analyzed his predicament regarding whether he is or is not truly free.

It is thought free will and determinism, in a strict sense, are incompatible. We know this through the Law of Contradiction. Some philosophers have said either what one does, happens by chance or is strictly determined. However, we must not confuse the idea that the "here and now" limits what can happen next with the idea that the "here and now" determines what happens next.

Some believe if man's actions originate from the self, it can only mean his actions are uncaused. If so, then what man does would seemingly have to happen by chance. At first glance: If man's actions happen by chance then he should not be anymore liable for what he does than if his actions are strictly determined.

By the standards of everyday life, we behave as though we were free, but that alone does not mean we are free. Until proven otherwise, we must premise our decisions on the possibility of free volition. To see why—consider the possible consequences of adopting determinism. For example, what could happen if we later found free volition to be the case, and there is a God and a judgment? Quite simply, we may find ourselves ultimately responsible for our actions. So to play it safe we initially assume responsibility for ourselves. Then if it turns out we are not responsible, what happens to us after death will not be contingent on our behavior before death. The idea is to assure the play-it-safe premise of free will and from it draw conclusions which we can attempt to contradict.

One of our more interesting tasks is to begin a responsible investigation of the free will and determinism dilemma. Let's first define what is generally meant by the word free will. Man is considered free if he is the sole cause of his behavior.

This usually means, aside from one's nature, that there is no factor that causes him to so behave.

Many psychologists believe man is autonomous—meaning that despite contrary influences he is capable of forming and carrying out his goals. That means even his motivations are often self-caused. In everyday life man "feels" free because he believes he has the choice to do what he wants (desires) to do. (Note. The subject of wanting has already been covered in Chapter 4, the "How Aspect.")

A libertarian might say most choices are motivated by the very nature of the self, but that idea carries little import if the term "self" is not realistically defined. Nonetheless, it "appears" as though the self operates upon the environment through its desires. It is easy to understand why man believes in free will. He is aware of choices, which in turn operate upon options offered by the environment.

We must note how the word "will" is used. Is it different from other cognitive activity? For example, what distinguishes wishing from willing? It is usually said of two choices, one can wish either but only will one. This is derived from the idea what one wills is what one translates into action. But, why does one so will?

Free will does not pertain so much to what one does, as it does to why he does it. So our reformulated question centers more on whether man's will (or desire) itself is of his own doing. Before analyzing this intricacy we must refine the definition of determinism.

Let's reexamine the meaning of free will and determinism. We shall look for possible contradictions in a few more prominent concepts and seek out their most noteworthy consequences. The issue of free will and determinism can only be resolved by what reality tells us.

A traditional definition of determinism is "every event has a cause." Everything that happens—happens necessarily. So if "All" is determined then every event is determined. Scientists recognize causality in the chain-linked phenomenon of action and reaction. (Extrinsic) cause appears as an antecedent condition to an event.

Determinism implies every event is compatible with All that is and that causality is inherent in the very nature of reality. That includes our awareness. Determinists assert our will, i.e., desires, is itself caused. Determinism implies no thing can happen or change without reason. Everything happens by design.

Is a person's will reflected in desire? We first probe a little deeper into what free will means in a pragmatic sense. Volition is studied in psychology where it is often analyzed as conscious or subconscious drives. Again, the word "will" pertains not so much to one's physical actions as it does to the immediate causal agent of the affected acts.

The immediate question is why did the person so act? What drove him to do it? We first ask the person. He may say, "because I wanted to." Further questioning may elicit he wanted to fulfill a need. For example, he fetches a drink of water and says he was thirsty. But why is he thirsty? Is it because he needs to replenish bodily fluids?

Here the mind becomes aware of a need. Yet the body is a part of the environment in which one's mind finds itself. Here, we can easily state the supposed cause of the thirst categorically resides in the same environment (the body is also physical) as does the primary effect (obtaining water) which first prompted us to ask why water was obtained.

We easily see in this example that cause can be traced beyond the mind and back into the environment. The question now is did the person really obtain water because of want? Obviously not. The wanting itself cannot be the cause of his obtaining water. He could have the same desire while in a desert far from any water source. In that situation one's wanting alone cannot produce water.

If a person asserts he can do what he wants to do, then why cannot he fly unaided? Why cannot he occasionally grow younger instead of older or even live forever with no change in physical age? One can want to do many things, but he is limited in his actions. We concede the degree to which a person's will or desire is limited is also the degree to which his abilities are determined. We immediately see the contradiction in saying desire or will is a sufficient explanation for what a person does.

It is absurd to say what a person does is what he wills. For example, a person may turnaround only to encounter a projectile. His eyes blink. He has no time to desire to blink, nor even enough time to think about blinking. Here the act of blinking happens without cognitive participation.

There are many acts deemed involuntary. Eye pupils dilate upon entering a darkened room: it is an involuntary reaction. There are many bodily functions not subject to voluntary control.

A not too dissimilar example is a person who is given a post-hypnotic suggestion to do something when a certain phrase is heard. Observers of the act hear the phrase and believe they can explain why the subject acted as he did. The subject, however, unaware of the scheme, may believe he did it of his volition. So we conclude a person's mind does not offer a noncontradictory (all-inclusive) explanation for one's actions.

The two previous examples suggest a nonapparent causal factor is involved in our actions. We later return to this idea. At this stage (of analysis) we can also rule out the idea that free will means that what one does, happens by chance. If chance ruled the will, then we would expect man to act capriciously and not expect to see much order in his activity.

Let's examine the idea man has free will because of the choices he can make; though they be limited. To make a choice is to make a decision. A wise and knowledgeable person has more choices from which to choose. Thus in a sense he is freer than someone less knowledgeable or less intelligent. However, this is not what we are seeking. Free will is not a question of making a decision and then carrying an idea to effected conclusion. Even if this definition is first allowed, contradictions arise because a person is most often unaware of all the consequences of his decisions, viz, all its effects. Therefore, not really knowing in full what one has willed, restricts defining the will as free.

The question is "could a person have chosen to do otherwise, than what he did: given the same exact circumstances?" Now just because a person thinks or believes he could have done otherwise does not mean he could have (done otherwise). Let's examine this. What does "could have" mean? It most probably entails a conditional clause. It might mean he would have succeeded if he had tried (harder). Such a statement asserts effectively that the individual would have done otherwise "if" only things were different, but that meaning alters the circumstances.

To say one could have done otherwise seems to mean he could do otherwise. The problem with that meaning is, it does not address itself to defining free will because saying "one was able but did not," offers no explanation. Yet, it is most often that meaning that is the reason man is held responsible for his actions. Moreover, one cannot get outside his own predicament to ascertain if he could have done otherwise.

The fact is, there is no such thing as "if" in reality. "That which is otherwise" never happens. Things happen only one way. Because man often considers several choices, does not mean he has free will. The contradiction is—choices considered in thought are every bit a part of what happens, as are physical events. What transpires mentally happens only one way just as does that which is physical. Moreover, although one may entertain opposites in thought and even wish to translate both into action, only one thought stands a chance of being physically carried out during a specified time.

To say a person could have done otherwise, than what he did, can only mean he could have been otherwise, than what he was. That type of reasoning leads to a dead end. If a person truly had free will, why did he not pick his parents and why did he not pick when and where to be born? We cannot say a fetus grows and develops into an adult, because it was so willed by the individual it grows into.

Is there any real meaning that could be attached to free will? Yes, one would be that any of one's wishes be realized. That is the crux of the problem. It does not happen. The idea there are such characterizations as wishes tells us something. We may wish to live forever in our present condition, however, we physically exist but for a short time during which we have but limited say in what happens to us.

B. Spinoza realized people are conscious of their own actions. Because they do not know why they do what they do, they assume they are the cause. This idea implies the following question. Is there a deeper-seated will hidden from one's consciousness? Is there a hidden volition behind every act? What would it mean to say there could be a volition of which we are unaware? This twists the meaning of free will too far. To say an act is ultimately caused by an unspecified hidden will is to circumvent self will. It is just one step from this to saying each volition is also an act, which calls for another hidden will ad infinitum. Such an idea uses an undetermined-deterministic account to explain the will's freedom; thus, the absurdity.

To postulate a hidden will as the cause of one's actions is to contradict the very meaning of free will. Furthermore, we most often observe causes are not too similar to their effects. For will to be truly free can only mean our awareness is itself uncaused. It would be strange if such a not-all-knowing-acausal agent, our will, could effect change. The question then would be, how could it?

B.F. Skinner comes close when he says death is the final blow to the idea of freedom. Death is also the final blow to free will. Without proof to the contrary we must assume causality is universally valid. Even events that emerge from activity that is only describable by probability are predictable.

We are left to explore the alternative—determinism. Let's examine some explanations of behavior to seek out one that does not lead to contradictions. Again the question is "why do we behave the way we do?"

Skinner studied behavior for many years. He says most people believe there is an inner man, the autonomous man, a center (the will?) from which behavior emanates. People often allude to this inner being, as the explanation of behavior. As we have seen, it is a Being that is not explained: a causal agent that is itself uncaused!

Skinner realizes the imputations man has hung on the idea of autonomous man. Civil "right and wrong" is built around the notion of free will. People are commended for what they do when it reflects the best wishes of society, and condemned when they fail.

Socially, worth and dignity are characteristics attributed on the bases of man being credited for what he does; credit is given because man is supposedly free and therefore, responsible for his actions. Yet the more one's actions are attributed to forces other than the "inner self," the less credit is given. In that case, even personal sacrifice loses its usual connotations.

In practice, lack of freedom just means a person, against his wishes, is subject to aversive conditions. Therefore, freedom is a condition of unrestraint (in implementing one's wishes—desires). Some societies were founded on this very concept of freedom. This, however, is a different meaning than that required by free will.

Skinner realized that the less that people understand behavior, the more it is attributed to autonomous man. He recognized the traditional causal interpretation of behavior was inadequate and sought a better explanation. He found behavior is shaped and maintained by its consequences. Skinner said the consequences are found in the environment. The environment replaces autonomous man as the cause of behavior.

Skinner uses the action-reaction concept. A person's act produces consequences in the environment, which then influences subsequent behavior. The consequences either positively or negatively affect behavior.

Certain behavior is positively reinforced if it leads to pleasant experiences. A positive consequence is like a reward. An individual naturally behaves to increase the incidence of reward. If a child receives candy for doing an errand, he will probably be more inclined to do it again.

Some behavior is negatively reinforcing in that its consequences are unpleasant. The individual most often behaves in a way that removes the negative conditions that are responsible. If a child suffers a burn from touching a hot stove, he will probably be less inclined to get too near one again. Also, some consequences bear little effect on subsequent behavior and are neutral.

Closely tied to the notion of reinforcement is the concept of contingency. A person's behavior often follows specific circumstances in the environment. It is this sense in which behavior is said to be contingent upon what precedes it. For example, obtaining a can of carbonated beverage (the reward) from a vending machine is contingent on the placement of the necessary coins into the proper slot.

Contingencies of reinforcement refer to the sequence of events required to obtain a certain reward. A person's behavior is closely associated with what he experiences; that cannot be denied.

Environmental conditions predispose certain behaviors. If an individual suddenly finds himself in a deep waterway because of an accident, he will probably swim—if he already has the behavioral pattern to effect the proper response, viz, swimming.

What causes a behavioral response when environmental conditions are unclearly defined? Skinner says much of everyday circumstances fall into this category. He addresses the problem by forming a concept called operant behavior: behavior, which operates on the environment. It is similar to what is most commonly termed "voluntary response."

Operant behavior is a response without a definitive environmental antecedent. Here the question of why certain behaviors occur refers to "something" inside the body that can initiate specific responses. It is in effect—seemingly a reaction not preceded by an action.

Operant behavior is a construct for dealing with responses that cannot be satisfactorily linked to the environment. For example, look at behavior involved in choice making. As a question "why does an individual so choose?"

We know from experience that individuals do not all choose alike, given the same circumstances. Using a Skinner box, we may be prompted to ask why did the chicken peck the lever the very first time? "Trial and error" is no explanation. Saying the lever was there does not fully answer the question either. The lever may have been in the presence of the chicken for a while (from the moment the bird first notices it) before being pecked. So the question becomes, why did the chicken first peck it when it did? That question is not satisfactorily explained by only pointing to the lever in the environment.

Similarly, though more complex, why does an artist so choose to paint an abstract painting in a certain way? And, what about insight? Insight characterizes the sudden realization of solutions to problems. Behaviorism does not even adequately explain W. Kohler's famous-monkey experiment. The monkey in a "flash of insight" connects two poles end to end, enabling him to reach beyond his cage and pull in some bananas.

These types of problem solving are instances where the individual quickly grasps a context or pattern that yields the solution. And these solutions, because they are not presented in the environment, must at least, be products of a mind.

Where many people continually point to autonomous man for an explanation of human endeavor, Skinner thought all behaviors could be explained by environmental conditions. The environment is not sufficient to explain all behaviors.

Perception, cognition, memory, intuition, creativeness, and decision making; all have a part in behavior. It does not follow, however, that such abilities are the sole cause of behavior. Indeed, as we have seen, the causal explanation of behavior must nonantecedently extend beyond man's mental awareness. Such a realization does not mean we refer to indefinite autonomous man nor does it mean we defer to the environment.

Because we cannot get into another person's mind to witness mental processes, Skinner practically factors them out of his theory. For example, he asserts awareness is a by-product of the verbal community. Although these ideas are a result of observation, they do not simultaneously acknowledge what constitutes awareness, i.e., inferential functions of the mind.

Apparently, Skinner neglects these features for fear of reconstituting the concept of autonomous man. Yet, no one can honestly deny that functions of intellect exist (deduction and induction). (Deduction and induction obviously do not occur in the environment.)

Skinner's ideas suffer from two main problems. One is that he does not know how to replace the error-ridden concept of autonomous man. The other and more significant one is linked to his not fully understanding the mind and the cause of awareness.

Skinner's theory goes far in explaining man's behavior. Man's interaction with the environment can be understood in Skinnerian terms. Children's behavior is particularly explainable by reinforcement.

Reinforcement, however, does not adequately explain the abstract intelligence functions. Surely, deductions and inductions are often premised on the environment. For example, we walk into our living room and experience a sense datum that we identify as a lamp (with the aid of memory). Secondary deductions begin to show the problem, for example, that the lamp was a gift. Such information is not presently deducible from sense impressions. Memory is involved.

Secondary inductions are especially troublesome. Memory cannot be sufficiently defined as behavior that has been much reinforced (although there is some truth to that). It would be ludicrous to say such abstract information is a direct response to the environment. Skinner realizes the environment itself is very orderly and determined. He extends those features of regularity to explain behavior. Yet, he does it in a way that practically excludes acknowledgment of the cognitive self. The cognitive self, one's identity, is tightly associated with memory.

We have already seen egocentric and realistic behaviors are subject to the "laws" of wanting (desiring) and having. (Refer to Chapter 5.) Egocentric behavior results from ignorance and is most often derived from the function of intellect called deduction. Realistic behavior results from knowledge and is most often derived from the function of intellect called induction.

Characterizations of wanting usually pertain to something in the environment. "I want this thing or that thing." Skinner would agree with that by saying wanting is connected to contingencies in the environment, but there is more to it. There must be something that experiences wanting. Although the object of want may be of the physical environment that which experiences wanting is not. Information in this book informs us that "wanting and having" deal with the relationship between awareness and the inferred objects of awareness. (Again, refer to Chapter 5.)

Let's look at a broad view of determinism. Quantum mechanics aside, scientists agree the physical world is quite determined. If it were not, we would be living in an almost impossible environment. The environment, though inflicting hardship, is obviously livable. Existence in the physical world is tolerable only because it is predictable.

A hallmark of science is its use of explanation for predictive purposes. That does not imply if All is ultimately explained that every event could be foretold. Neither does it mean foreknowledge itself can be used as a definition of determinism.

If foreknowledge were used as the definition of determinism, it could only mean that events occur because they were foretold. Experience tells us something else. What is called foreknowledge is technically only a prediction until fulfilled because not all predictions are fulfilled. So what about predicted events that become fulfilled? It would be difficult indeed to conceive how physical events could be foretold if their nature were not in some sense determined.

Predictability of physical events shows that changes in the material world are consistent. Consistency is not a characteristic of human behavior.

If some events are "determined" by the individual, then the issue of free will resides in the nature of self-determination. Man must understand himself. It seems all but impossible for man to deductively settle the question of determinism on any less a basis than the total understanding of reality. Surely, the question of free will can only be settled by fully understanding the structure of reality. We endeavor to settle this question.

Is it determined there is limited free will? If there is free will, though limited, there is a reason for it. Evidence overwhelmingly indicates some things are determined. We surmise behavior does not emanate from acausal islands in the midst of cosmic structure. Instead, behavior, like all else, is intrinsically caused. The answer to why a person behaves (determining this and that) the way he does is not found by asking the individual.

What have we accomplished? Causally connecting the Realm of Whatness to the Realm of Reasoning enabled us to ascertain with practical finality that we do not have "unbridled" free will. If individual will is not completely self-willed, then it is not free. Self-will is limited at best. Resolving the problem of will calls for a complete defining of the self through the structure of reality.

We may be pleased to know our ultimate destiny does not depend on our own choices and decisions. We have established by the structure of reality that the Realm of Reasoning is the immediate (intrinsic) cause and sustainer of the Realm of Whatness. How this happens is addressed in subsequent chapters.

Is it not enough we experience life as though we have unbridled free will, though we do not? Surely, it would be far worst to experience life as though everything, including our behavior, were determined though we had free will. If that were so, we would be incapable of responding to what we experience. In that case, we would have no meaning. Because we are able to respond to experience, life must have meaning.

We must assume responsibility for what we do. Because we respond to "things" (we infer ideas about them) we are responsible for our actions. We must live as though we did have free will because that is how we experience everything. We can act upon things.

In examining the question of free will we analyze the relation of inference to what is inferred. Our inferring-awareness and all we experience are "directed." Experience is determined by the following: physical environment, the inferential processes, and memory. We still can ask whether these factors are then governed by something from a "higher" level than physical and mental existence. Another way of stating this question is the following. Do physical and mental factors have a common origin?

If mental and physical experience is also determined from a higher level, then there must be something doing the determining. Some people call this director God. Before tackling the question of God's existence, we should look closer and more specifically define what is being determined.

Experiencing life as though we were free is due in no small way to the intrinsic nature of memory. We are a mind. Mental existence is heavily premised on memory. Let's examine the nature of memory.

Memory and Recollection

There is a positive way to view determinism. A. Schopenhauer once said, "A man can do what he wills to do, but he cannot determine what he wills."

In a way we could say we have free will. One criterion for man to be free is that he is the sole cause of his behavior. This criterion is practically met by the fact that the cause of memory is intrinsic. The problem is that the self must be completely understood via the structure of reality.

To meet the criterion of free will (that memory is intrinsic) a feedback mechanism would have to operate to complete the action-reaction sequences of experience operating from the environment through awareness. Part of that circuit includes memory storage. The immediate problem is memory is categorically distinct from inferential awareness. The cause of memory is nonapparent: it is consciousness itself. That technically invalidates the argument for unbridled free will at least for our level of existence—awareness. The reasoning is simple. Take away consciousness and there would be no awareness. No awareness also means no self memory.

Much of man's behavior is attributable to memory. The fact we remember *prima facie* shows we have some free will. It is best to denote limited free will as self-will. Self-will implies there is some type of memory bank associated with each individual. This is easy to understand. When remembered, memories surface to aid self-awareness.

It seems difficult to understand how memory arises solely from physical experience. Something is doing the remembering. Let's rephrase this as a question. Where are memories when not being remembered? It seems that they would have to be stored in some "personal" filing bank (if each individual has free will).

Are memories kept by some type of molecular arrangement in brain cells? Psychophysiologists have found specific brain areas associated with long-term memory. Do biochemical or neural mechanisms explain memory? If there are localized areas and there is some type of mechanism, another question remains. What is the cause of memory location and what is the mechanism or cause of memory retrieval?

Memory is needed to identify what is seen. The very functions of deduction and induction are wrapped in remembrance. The fact memory has not been adequately explained does not mean what is remembered is only self-caused.

Memory is recollection entwined with awareness. How is awareness defined? Let's review. Awareness is inferring. Inferring is the nature of the mind. If what we are "conscious of" can be called a "stream of experience" then awareness in-itself is a "stream of inferring."

The "flow of perception" constitutes our identity. We infer deductively when we are less than knowledgeable. This typically leads to egocentric behavior. We infer inductively to seek knowledge. This often leads to realistic behavior. These concepts have been explained in previous chapters.

Inferring is perceiving. Perceiving, i.e., cognition and recollection, is something that happens to us. All perceiving originates from the intelligence level called intrinsic perception. Intrinsic perception functions for memory recollection. (Keep in mind that perceiving is different from sense experience.) Now we surpass ourselves. How do we explain intrinsic perception? Where does it originate?

If we had "unbridled" free will we would be faced with insurmountable difficulties in explaining recollection. Scientists have not explained why we remember. The reason is simple. Man is not the cause of recollection. Explaining recollection must be connected to why there is life. Because self-will is limited, we realize if we can explain intrinsic perception we should also be able to explain our very existence.

We have seen the structure of reality and an analysis of behavior allows us to say that all individual behaviors are determined in some sense (it is limited). In short, the range of man's experience is limited.

We are limited from below and from above. Man is limited from below by the physical world. What sets the upper limit to man's experience? The answer will be found when everything is categorically explained. That sense is suggested by "man's meaning to life" or purpose. (Man's purpose is properly explained in the final chapter.) Since "everything," to some degree, is determined, all we need to do is identify (by inductive inference) the source of all experience.

We know intrinsic perception partakes in the stream of experience. Moreover, intrinsic perception is not the final cause of existence. For we can ask "what is the cause of memory?" To address this question is to ask what is memory itself?

Because everything is at least partially determined by an upper limit, it means (in terms of purpose) memory is not confined to each individual. Rather, individuals serve the purpose of memory. Intrinsic perception itself serves as our memory. Intrinsic perception is the bridge between awareness (the mind and its inferential functions of intellect) and consciousness. Just as we do not determine the form of extrinsic perception, we do not determine the form of intrinsic perception. Consciousness is the "observer" that experiences both.

We are mindful that "everything" is what we experience. Experience is "what" we are (or can be) "conscious of." Therefore, we expect consciousness sets the limit of experience.

Key Points:

- Philosophy concerns the nature of knowledge and reality. Speculative philosophy includes "big" questions about reality. Philosophical metaphysics speculates on the structure of reality.

- It is difficult to settle questions about reality unless the nature of knowledge is understood. That is the providence of epistemology. The problem is two-fold. Knowledge cannot be properly defined until reality is understood. Reality cannot be defined until knowledge is understood. That problem echoes in the question what is truth?

- Truth is knowledge. Knowledge is that information which corresponds to and therefore represents reality. We have now defined one (knowledge) by the other (reality). This is a pragmatic definition. It does little to advance understanding. However, it does provide a beginning. As we unravel reality's structure, we should better understand why there is knowledge. Knowledge is surely a curious phenomenon.

- Philosophers have wrangled over "particulars and universals." Universal concepts are easy to state. "All those vehicles are automobiles." The problem has been whether universal concepts represent something in reality. The category "automobile" is not a specific thing. (It represents a classification of things.) We learn that some universal concepts do represent things in reality. We recognize one problem. If generalized statements cannot represent reality, then there is no possibility of finding a meaningful single-all-embracing equation to "categorically" represent All reality. Therefore, if we do find a meaningful equation that can (conceptually) represent "all reality," then it probably does actually represent all reality.

- The problem of mind and body has vexed philosophers for centuries. Is mind separate from matter? They seem to be different. Or, is mind just an epiphenomenon of matter? Each view incurred problems that seemed unresolvable. We have solved these problems to a point. There is no contradiction in supposing they are different. We have defined them "by knowledge" as different. We just have not refined our understanding how they are connected. In Chapter 20, we explicitly define the difference between brain and mind. We also explain how they are connected.

- Philosophers are still perplexed by the problem of objective reality. Scientists act as though it is no problem. (Science is predicated on that premise.) Yet, the problem has not been scientifically settled. This old philosophical problem has resurfaced because of Bell's theorem. We resolve these problems in Section Seven.

- Man behaves as though he has free will. After all, he "acts on his own." Possible solutions, either way, led to other problems that were just as baffling. If man did not have total (free will) control over himself, why should he be fully responsible for his behavior if there is a God and a judgment? If man is limited in his behavior, and he surely is, then he should not be held fully responsible. Some of that responsibility must fall upon God (who bestowed the limitation). We easily see how philosophers can muse about anything and offer possible solutions. Our goal is to understand by elimination of contradiction. Therefore, these problems must await fuller treatment.

Before explicitly addressing consciousness, we must tidy up our concepts. We need a broader basis for understanding consciousness. Specifically, how does consciousness fit into the structure of reality?

Therefore, how does consciousness relate to extrinsic perception (the Realm of Whatness) and to intrinsic perception (the Realm of Reasoning)? In the next chapter, we address some religious concepts. They will help us understand experience and consciousness. We then take everything we have learned and submit it to an in-depth logical and scientific treatment. If we completely understand reality, we must also necessarily understand consciousness. Our quest to completely understand what we surveyed in the mind expanse continues.

Chapter 17

Using Structure to Interpret Religious Concepts

> "The more 'religious' people convince others of the
> validity of their beliefs, the more they believe they are
> doing God's will—contrary to what the Bible clearly
> states. The problem is not what people believe. The
> problem is that they believe."

Can basic religious concepts be understood by science? Yes. The scientific truth of significant religious concepts is established by correlation with the complete logical structure of reality. This project represents the scientific-logical structuring of reality. The structure in this book can be used to gauge the truth of major religious concepts.

Understanding religion through science offers another perspective on the meaning of existence. This chapter also concerns concepts of religion that correlate with our understanding of behavior.

Most religions offer answers to metaphysical questions. By what authority do religious people assert the verity of metaphysical beliefs? For example, Christians state the Bible is their authority. Elaborating, they say the Bible is God's word. The Bible states there is a heaven, hell, Satan, Lucifer, and God. Most religious people unquestionably accept these concepts as true.

We can safely suppose the Bible and other holy books contain some true statements. Some statements are obviously true. However, most significant scriptural statements, even if true, are unexplained. There is no easy way to discern true biblical statements from false ones. The Bible informs of particular "whatnesses," but does not explain "whyness."

The Bible and other holy books are not self-explanatory. They do not describe how to verify (the truth of) given statements. Neither do they offer logical proof. Consequently, the Bible, and other holy books, are not complete: they do not constitute the whole truth. If all the big questions have already been satisfactorily answered, it is difficult to explain why big questions are still being asked.

Are religious ideas relevant to today's scientific inquiry into the ultimate nature of everything? Yes. Modern scientific concepts do correspond to many religious concepts. They are relevant. Consequently, we need to understand the correlation of concepts between religion and science.

Many religions point to present scientific understanding. It is easiest to establish these connections by briefly describing the history of major religions and their basic creeds. First, we show how the major scientific concepts correlate with those creeds. Then we explain how those concepts, when properly understood, fulfill prophecy.

Scientific objectivity is a recent phenomenon. It has existed but a few hundred years. In the early years of science, only a few people understood its methodology and fewer people applied it in everyday life. Even today, serious objectivity is usually found only in the science classroom and experimental laboratory. Practically no one is seriously objective about his or her personal lives. (Refer to Chapter 2 on PEP: the personal explanation principle.)

Scientific objectivity was unknown thousands of years ago. When stories were told and retold, they were embellished. Events become more interesting when exaggerated. Objective reporting of historic events was not a consideration. Truth was what people fervently believed.

It is difficult, if not impossible, to ascertain truth from falsehoods in the holy books. Stories found in the Christian Bible were not recorded until at several decades after the supposed events happened. If biblical events were originally embellished (unobjectively reported) and only much later put in written form, there is every reason to suppose much of written scripture is nonobjective, unscientific, and inaccurate.

Because scriptures are not properly explained, their meaning is open to interpretation. This shortcoming enabled many religious groups or sects to assert their church is the "true church." The "gap" between what is written in scripture and the whole truth is compensated by church doctrine asserting that their leaders are divinely inspired when interpreting (the meaning of) scripture.

The designation of true church assures members that their church's interpretation of scripture is true. Yet these "true" churches often adhere to doctrine opposed by other churches. Enlightened people realize not all claimants can secure the title of true church. Logically, at most only one such title can be granted.

The question of a true religion or true church may be irrelevant. There might not be an existing religious organization entitled to the claim of "true church." Why? To date all religious organizations share one commonality. None have scientifically verified their claim (of being the true church). Church members do not grasp the real meaning of "true church." If this is the case, what is the significance of their activity? What is the true object of their worship? Probing questions quickly challenge claims of any group asserting true churchhood. These questions are definitively answered and explained in this book.

There is a crucial distinction not recognized by most religious people. There is a difference between the written word and the reality it represents. Knowing words and concepts does not mean the individual knows the reality represented. Many religious people believe they know God, merely by reading and studying the Bible and becoming emotional. (Refer to Chapter 7: heading, "A Crucial Difference.")

Is there an afterlife? Does God exist? If there is a God, what is its nature? What is man's relation to God? To answer these questions (the word) God must be noncontradictorily defined. This is done by connecting the word God to the structure of reality. There is no real alternative. We finally address this correlation in Chapter 21.

If God exists (that God is) it means God is real and hence at least a part of reality. The question then is—what part (or aspect)? Before analyzing this, some other questions of religion should be addressed. That is the purpose of this chapter.

Can we have knowledge of God before being cognizant of what constitutes knowledge of reality? No. We have inductively reasoned enough cosmic structure to correlate many religious concepts to the structure in this book. These "lesser" questions need to be understood (via the structure of reality) before the word God is correlated with some aspect of reality.

Some minor notes: The Bible and other "holy" books can only be true if every indicative statement in them is true. One can guess there are many statements in "holy" books that are verifiable and therefore, true. It is also probable the Bible, for example, is not the whole truth, otherwise, it would have (by categorically subsumption) answered (through scientific verification and logical proof) any big question one could ask. It does not.

The best one can hope is for the holy books to contain some true statements. Again, our problem is whether these statements and concepts are true according to reality. (This is backwards from proving using one's uncritical personal method of belief. In religion, proselytizers, when asked for proof say "you have to prove it to yourself.")

Before assessing religious concepts we analyzed the import of metaphysical questions. In the previous chapter many philosophical questions were understood using the structure in this book. Answering metaphysical questions using reality's structure helped establish the meaning of religious concepts.

In 1962 Bruce Mazlish wrote an updated preface for the book written by Andrew White (published in 1896) called *A History of the Warfare of Science with Theology in Christendom*. In the new abridged version (it is still over 500 pages) Mazlish states that White's purpose in writing his History was really to free science and religion from theology. This is the idea: Because theologians (and anyone who insists on interpreting what they really do not understand) do not know to what in reality various biblical statements correspond, they are not warranted in stating the meaning of those statements! This clears the way for an idea.

If there is any meaning to these religious books, it is because reality supports that meaning. This is where science enters the picture. After all, scientists define truth as that information which corresponds to and therefore represents reality. We are also cognizant most of the religious books define God as (represented by) truth. If these definitions hold, the word "God" and the word "reality" correspondingly are the same, i.e., they refer to the same structure—the structure of reality. (Note. We learn in the final section that this correlation is not so simple.)

Science can provide the basic understanding of (true) statements found in the holy books. If there is any real meaning to these statements, they must correlate with and be testable by reality. Terms must be defined—definable in a way precluding contradiction. If a term is explicitly definable then it represents some situation in reality. And, if it represents reality, it is explicitly definable—a paraphrase from Hegel's "the rational is real and the real is rational."

People who say reality is the way they interpret—are fools. That rings a bell—that "the wise of this world are not wise; they are fools." (Bible Romans 1:22). If it can be proven that man is foolish, then maybe a stronger case can be made for at least some statements we find in holy books.

An ignorant man asserts the truth of statements he cannot prove. A belief is a statement one asserts as true but cannot prove. It is a lie that a belief represents what one knows to be true. ("Man exchanges the truth of God for a lie." Romans 1:25.)

Man reifies words and concepts. People in religion use words like spirit, hell, heaven, God, and love, as though they know to what in reality these words or concepts correspond. They assume the listener, or reader also knows the real meaning of these words.

People do not seriously question authority figures. Religious leaders speak without clearly defining their words. People listen without seriously questioning them. Religious authorities "act" as though they know what they are saying. One should question these "dispensers of wisdom." Ask them to define their words. Look for contradictions in their definitions. If a contradiction (or lie) can be found, then clearly the speaker does not know (in reality) of what they speak. If so, why do religious leaders speak at all? Answer, they are ignorant of the truth.

Is man that foolish? Surely, we understand man is not wise unless he can prove (demonstrate) the truth of a statement via reality. That being the case—man's beliefs represent his fantasies—that is foolishness. ("Man's thoughts become useless" [paraphrased] Romans 1:21.)

We look back a mere hundred years and say how ignorant those people were. Yet a hundred, nay, even fifty years from now, what will be said about how ignorant man is today?

Let's examine some concepts from the world's major religions and ascertain if they correlate with some of the structure in this book. We focus on some religious holy books, which state something about the structure of reality, or state something about the nature of behavior. We examine only those religions that existed up through the founding of Christianity. They are addressed in chronological order.

Religions Briefly Described

Early Religions

Appealing to nonmaterial entities helped man "explain" events beyond his control. What man feared—what he did not understand—was attributed to spirits or "acts" of the gods.

Gods could act capriciously. Curious events were assigned an unseen cause. Later, rather than attributed to whims of the gods, certain events were interpreted as rewards and punishment. A bountiful crop was reparation. A poor crop harvest was a godly admonition.

Death is seen everywhere. Plants and animals succumb to death. Early people asked what it means to die. Supposing death was not the end of life led to belief in some kind of afterlife. Such concepts were often used to coerce people into socially acceptable behavior. A person is rewarded with a good afterlife (heaven) if he behaved well. If he behaved "badly," he would be condemned to a bad afterlife (hell).

As villages became incorporated into kingdoms, local superstitions and rituals developed into socially sanctioned religions.

Early religions developed along the Tigris and Euphrates, and Nile rivers. Each settlement typically had its own gods. Assyrians, Babylonians and Sumerians developed elaborate religious traditions. The idea of the great flood and ark is traceable to Sumerian beliefs. Babylonians believed: in a burning hell, resurrection, and a "Satan" that was cast out of heaven. Mesopotamians believed in an afterlife.

Egyptians believed Ra, the sun god, was the chief god in a divine hierarchy. They also denoted a god of order, rta.

In many early cultures people also recognized the importance of a mother goddess. She was often considered a protector. She was also associated with the cycles of nature (for example, summer-winter.)

Greeks and Romans developed an extensive system of myth and religion. Homer's epic tales Iliad and Odyssey (middle of ninth century BC) elaborated many myths told in the city-states of his time. The Homeric world view included Hades (Hell) the underworld beneath the earth. There was a hierarchy of gods. Many gods were human like. Greeks believed the gods influenced much of what man experienced. Homeric notions had an impact on later Greek philosophers.

Plato, a student of Socrates, believed only universals (ideas or form) were real, that they existed independently of things. Aristotle, Plato's student, believed forms only existed as an aspect of material things. Therefore, universals did not exist apart from matter. Other Greek philosophers, like Pythagoras, believed the psyche or spirit was an independently functioning entity.

When the Roman armies of Caesar invaded the Celts of Gaul, they learned of the Celtic belief that everything happened by the will of all the gods. Stonehenge (nineteenth century BC) was a Celtic religious structure. It was built by a sect called the Druids. They used it for ritualistic purposes. A sacrificial stone is found at its center. The layout of Stonehenge could be used to determine the summer solstice by angle of sun rays. Scandinavian and Germanic cultures believed in an underworld called Hel.

Greek and Roman oracles, soothsayers, like the famous one at the Temple of Apollo at Delphi, told of the wishes of the gods. Socrates often quoted an inscription at the temple stating: "Know thy Self." Many cults developed around the oracle tradition.

Many religious concepts are traceable to beliefs held by early cultures. During the seventh (?) century BC in Medea, a child named Zarathustra (in Greek, Zoroaster) was said to have been born of an immaculate conception.

Zarathustra had many visions. He met an angel in his first revelation. The angel "lifted" up Zarathustra from his physical body. He was raised to heaven where he met the god Mazda who spoke with him in a flame. Mazda said Mainyu, the embodied spirit of evil, opposed him. Zarathustra was instructed to speak for Mazda.

Zarathustra learned everything is a struggle between the forces of good and evil. That struggle happens in each individual. He said the god Mazda represented the good. Each person would be judged at the end of his life. At that time, each person steps upon the Bridge of Separation. Some "go" to paradise; others are condemned to darkness. Good people enter the kingdom of the eternal light. Evil people fall dead into the realm of darkness. They remain in darkness until the final conflict between good and evil is resolved at the end of this existence—when Mainyu is destroyed.

Zarathustra mentions a final battle. He believed the dead would be resurrected at the end of every cosmic cycle. He said during the "end days" a descendent of his seed would return as a savior to govern the resurrection. A new life—a different existence—would begin.

Zarathustra's ideas were rejected by his people, but accepted by neighboring Persians. The *Avesta* (Book of Laws) is the source of Zoroastrian scripture. "Adam and Eve," and a great flood are other concepts once entertained by Zoroastrianism. Zarathustra said only Mazda was worthy of worship, hence its monotheism. Mazda was helped by Mithras, the Light.

Many cults flourished during the Roman Empire. Each cult borrowed from earlier beliefs. One secret cult was Mithraism. It became widely spread among the Roman soldiers during the first few centuries AD. Romans adapted Persian and Egyptian ideas. They believed Mithras was born of a virgin: a birth witnessed by some shepherds. They thought believers would be raised on Judgment Day; nonbelievers would perish. Mithras was a sun-god and was worshiped on Sundays. Mithras was said to be "the Light, the Son of God, the way, the life, and the truth." In Mithraism was a scared rock, called the petra.

Ancient China developed a different type of religious tradition. Lao tzu, possibly not a historical figure, lived at the turn of the 7th century BC. He wrote the *Tao Te Ching*. The Tao is the Way, the Path. It also pertains to the underlying actuality of everything: of existence. The Tao is not activity, but underlies all activity.

The main symbol of the Tao is Yang-Yin. It represents the principle of complementarity. In the same are opposites. Yang is masculine, positive, active, odd, and light. Yin is feminine, negative, passive, even, and darkness. Everything, every activity, is inherently dualistic. There is a bit of oppositeness in everything. It explained the seasons, sexuality, and many other things. There is some male in a female and some female in a male.

Everything inherently was composed of both yang and yin. The interaction of yang and yin underlies all change. Lao tzu taught one should show kindness even to those who were unkind. He set such high standards of conduct that his ideals were only practiced through ritual.

A book of Chinese divination called *Yi Jing* (The Book of Changes) used Yang-Yin to foretell events using sixty-four hexagrams. The hexagrams are represented by a series of solid and broken lines. Too much Yang or too much Yin was not beneficial.

Confucius was a younger contemporary of Lao tzu. Confucius wrote that remembering lessons of the past guided decision making of the present. Confucius's writings were woven with morality and ethics. His ideas embodied the principle of *Li*. *Li* is the standard of everything sacred, which was almost everything. It set a tone of harmony: that one should behave in harmony with custom and nature. He stated the idea of the Golden Rule. Do not do to others what you would not want them to do to you.

The proper order of things was called *Dao*. Applying *Li* was to follow the *Dao*. Confucius was not an innovator of ideas. He was trying to get society to return to the ways of the elders. He influenced a social reformation.

Hinduism

Hinduism evolved from loosely developed beliefs associated with the Indus valley civilizations and Aryan settlers who, in the second millennia BC, moved into what is now called India.

Early Hindus believed there are three realms. The lowest is the earthly realm. Above it is the atmospheric realm. The highest realm is that of eternal light. The eternal light was believed to be the dwelling of the gods. It was also thought to be the highest level abode for those who lived a virtuous life. Order in the realms is governed by the balancing principle of change, rta.

About the eighth century BC Aryan priests wrote down many stories of the natural gods. Of the collective work, the *Rig-veda* (knowledge) is significant. The latter pages of the Rig-veda mention the creator. The creator was thought to be the whole of reality.

Vedic prayers were originally denoted by the neutral word Brahman. It later represented the collective of "all" Vedic scripture. Eventually, Brahma (male genderization of Brahman) represented "All" of nature. Brahma became the whole of reality.

Brahma is omnipresent. By its own contemplation, Brahma ("It, That One") creates everything. Brahma is the Supreme Being—"creator." It is consciousness—Beingness itself. Vedic gods are considered partial manifestation of Brahma.

Hindus later believed the realms are sustained by the god Vishnu—the preserver. Though much revered, Vishnu has its source in Brahma. Another god is Shiva—the "destroyer." Brahma creates existence; Vishnu sustains it; and

Shiva destroys it. Because destruction is also a creative process, Shiva is also associated with creation.

Hinduism espouses the (real) inner nature of the self is atman. The inner self (atman) although an aspect of the physical person, is functionally independent. It is identical with Brahma. To the early Hindus the real Self is synonymous with the transcendent whole of reality. The atman is the aspect of the whole in the part (the individual self). The highest realization of the self is to connect to the absolute nature of Brahma.

Rituals were associated with seeking the inner self. Hindu priests wrote down their rituals. The collective writings of Brahmanic priests became the Upanishads (Equivalencies).

Theistic concepts evolved and are told in the *Bhagavad Gita*. It relates the story of Arjuna, a warrior preparing for battle. He is helped by Krishna, a messenger of Vishnu. The Gita explains the caste system: where everyone has "given" responsibilities in the social order. Yet, it stresses that one finds salvation only by detaching from engagement. For many, asceticism became a way of life.

Vishnu can manifest as individuals called avatars. Krishna was one such incarnation. At least nine avatars have appeared. Some latter incarnations can be linked to historical figures. Hindus believe Buddha was an avatar. Supposedly, Vishnu will incarnate again. The tenth incarnation is called Kalki. He appears toward the end of the cosmic cycle.

The Gita states that nature is a projection of Brahma. There is no separation between things. There are no "separate" individuals. The immediate question is why do we experience things as though they are distinct? Why do we perceive ourselves as though we are separate from other people?

A dilemma is how can reality exhibit a plurality of things and nonseparability at the same time? The Hindu answer is found in Maya. Maya is the "appearance" of things—phenomena. The plurality of things (e.g., people, indeed everything) is illusion attributed to Maya. There is no real plurality of things—there only appears to be a plurality. Appearance of things is neither real nor unreal. Maya is the source of self deception. To obtain real truth is to understand that which is beyond Maya—beyond the phenomenal world.

Two interesting ideas of Hinduism are karma (an action) and reincarnation (rebirth). Karma is nature's system of reward and punishment. Hindus believe there is an inner aspect of the individual that survives death. The inner self can inhabit future physical bodies. When between physical incarnations, the individual inner self does not desire.

A baby might be "incarnated" by a previously existing "inner self" individual. If a (physical) person lived a virtuous life, he is rewarded by rebirth in a better life situation. If an individual behaved unacceptably, karma could later bestow its consequences upon the doer. He may return as a lower animal life form.

Karma and reincarnation developed in step with the caste system. Karma and rebirth easily explained a person's life situation. It made it easier for people to accept their station in life.

Hindus believed an exceptionally lived life might allow an individual to escape the cycle of rebirths. There are three paths to salvation. They are, the way of works, knowledge, and detachment. The highest level of afterlife is commensurate with the abode of Brahma. Here the individual would be "released" from earthly cares: released from the flow of time, from illusion of separability, from change itself. Brahma represents the essence of everything, but itself is not anything. It is the single principle upon which everything is derived. Brahma is the basis of Hindu monotheism.

In Hindu cosmology the universe began in a state of void and formlessness. First, there was just the "I am," Selfness itself. Before cosmogenesis, the Universal Self (Brahma) was deeply "asleep." Desire itself led to the genesis of life. Brahma is associated with the form of things but is not a part of anything. The universe is thought to "pulsate." There is a cycle of cosmic destruction and recreation. "Existence" is absorbed into Brahma and then recreated by emanation. As the god Shiva dances, destruction of the cosmos occurs. The universe is created and destroyed countless times. Energy (sakti) is associated with a goddess named Sakti (wife of Shiva).

Buddhism

In the sixth century BC Siddhartha Gautama was born near the Himalayas Mountains. He was born into a "wealthy" family and was surrounded by affluence. Before Siddhartha's birth his father was told his son would either replace himself as chieftain of his clan or become a religious leader. His father, favoring the former possibility, sheltered Siddhartha from the harshness of everyday life.

Siddhartha married and fathered a child. When 29 years of age, he left his family to seek spiritual fulfillment. For six years he wandered. He practiced many rituals of Brahmanist Hinduism and Jainism. He did not bathe. He mortified his flesh. He fasted. While near death from starvation, he experienced an epiphany. While sitting under a Bo tree, he realized such rituals were not the way to enlightenment. He questioned Vedic scripture.

He said desire is the cause of suffering and dissatisfaction. Absence of desire led to enlightenment. He believed there was a middle way between detachment and self-indulgence. He adapted the idea of karma and rebirth. He thought if craving was eliminated and along with it the unhappiness it brings that one could escape the ceaseless wheel of rebirth. Reincarnation is transcended, and one obtains nirvana—Buddhahood—a state of blissfulness. Siddhartha became The Buddha—the Enlightened One.

How does one aspire to a state of nirvana? Buddha summed up his dharma with Four Noble Truths. (Dharma represents "elements" [things] or doctrine to understand the plurality of nature, and eventually, to understand the ultimate order of reality.)

Four Noble Truths:

1. Existence is permeated by suffering. Life has little apparent purpose.

2. Desire, craving, and indulgence cause suffering.

3. Suffering ceases by eliminating desire.

4. Desire could be eliminated in stages by following the "Eightfold Path." Applying wisdom to life mitigated suffering caused by desiring.

The "Eightfold Path" described Buddha's "middle way." It enabled an orderly approach to eliminating desire.

Eightfold Path:

1. Perception: fool oneself not by superstition and delusion.

2. Purpose: pursue only worthy activities.

3. Speech: be truthful.

4. Livelihood: bring no harm to any individual.

5. Behavior: be honest and unprovocative.

6. Diligence: practice self control; strive toward self knowledge.

7. Thoughts: use the mind; question things.

8. Contemplation: reason to understand.

The first two stages develop one's proper "mental" direction. The next three concern ethical practice. The last three are meditative practices that culminate in nirvana.

Buddhism developed as a reaction to Hinduism. Much of what Buddha taught was not written down for four hundred years. After the third century BC, Buddhism began spreading across Asia. Many of its followers believed nirvana was synonymous with nondistinction of identity. In most areas Buddhism absorbed much of the local customs.

Some Buddhists, following in the Hinduism tradition, believed that the inner nature of the Buddha would come again. He will be completely awakened, full of knowledge, and with total understanding. He will explain everything to people who will listen. He will speak of the real nature of truth and reality. He will be the Maitreya, the future Buddha, the one who explains the "higher life"—the one who explains ultimate reality.

Later aspects of Buddhism incorporated cosmological tenets. It was believed everything originated at one time. Buddhism does not admit of a single underlying Self. Ultimate reality entails contradictions. Nirvana or Buddhahood, instead of being a state of mind, became associated with the void—the Absolute undivided whole.

Judaism

During the twentieth century BC a nomadic Hebrew tribe left (the city of) Ur in Babylonia (along the lower Euphrates River). Hebrews learned about their god through revelation and vivid dreams. A few "selected" people, called prophets, experienced revelations. Among them was Abram (Torah, Genesis Chapters 11 and 12). Abram met a god so overwhelming; he unquestionably obeyed him. His god was very powerful and demanded strict obedience.

Through Abram, God made a covenant with the Hebrew people. The covenant originally applied to the head of the family. They were specially chosen to live according to God's will. God would help the Hebrews; they would serve his purposes. He was to be their only God. They were to pray, live a righteous life, follow dietary laws, and obey the moral code. In return, God would protect them. He would assure them a fertile harvest and a prosperous life. God told Abram that he was hence forth to be known as Abraham (Genesis 17:5). As a sign of the covenant, God dictated that Abraham (at 99 years of age) and his male descendants were to be circumcised (Genesis 17:12).

God sometimes tested certain worshipers by requesting an extraordinary sacrifice. Abraham was willing to sacrifice what was most precious to him. He was asked to sacrifice his son. Before he could carry it out, God stopped him. A ram had caught its horns on nearby brush. God allowed Abraham to sacrifice the ram instead of his son.

God promised Hebrews the land of Canaan (Genesis 12:31, named after a tribe living in that region). Canaan is the present land of Palestinians and Israelis.

Abraham was to pass the covenant to future generations through his children and their offspring (Genesis 17:7–11). One of his grandsons, Jacob, also had an encounter with God. His name changed to Israel. Jacob had twelve sons. They became the ancestors of the twelve tribes of Israel. Keepers of the covenant became known as Israelites.

A prominent son of Jacob was Joseph. Joseph led a group of Israelites who eventually settled (probably during the middle of the eighteenth century BC) in Egypt. A pharaoh later enslaved them to work on civil construction projects.

In the fifteenth century BC, a pharaoh condemned to death all first-born-Israelite males. He feared a divinely inspired individual would emerge and lead the Israelites in rebellion against Egyptian rulers. First-born males were traditionally destined for leadership. When mature, they often took over clan duties.

After the beginning of the thirteenth century BC a woman from the Levi tribe gave birth to a child and secretly cared for him. After three months she could no longer hide the infant. She put the infant in a basket (made of reeds and patched with pitch) and placed it among the bulrushes along a river. The pharaoh's daughter discovered the infant. She named him Moses.

Moses was saved by God to deliver the Israelites from bondage. (This story parallels an Egyptian story of an infant named Horus who was placed among papyrus reeds.)

Moses, a Hebrew, was raised in the pharaoh's family. One day Moses observed a foreman whipping a slave. Moses slew the foreman and fled the area. He journeyed to the land of the Midian Arab tribe. He married the daughter of the man (Jethro) who befriended him and settled in the area.

Moses was tending sheep for his father-in-law while on Mount Sinai. He heard a voice speaking to him from a burning bush. He asked the voice to identify itself. The voice answered that it was God. He revealed his name was Yahweh ("I am"). He gave assurance he was the God of Abraham. Yahweh said if the Israelites kept the covenant, he would free them from bondage. Moses would deliver his people to the Promised Land. The covenant was sealed. (In the King James Version of the Bible, Yahweh becomes Jehovah.)

Moses attempted a negotiated release of his people. Moses implored: "Let my people go, so that they can worship God." God told Moses to perform some uncommon tricks to convince the pharaoh that Moses was doing God's bidding. Moses threw down a walking stick. It turned into a snake. He waved his stick, and river water turned into blood. The pharaoh's magicians duplicated his tricks.

Moses took some ashes from a furnace and threw it in the air. Animals and Egyptians became inflicted with boils. The magicians, overcome with boils, did not duplicate this feat. Moses foretold of a very destructive hail storm. After the pharaoh witnessed the storm, he realized his situation. But, he would not let everyone leave.

The Israelites were allowed to leave after plagues weakened the pharaoh's soldiers. Led by Moses, they began their trip out of Egypt. But the Egyptian soldiers gave chase when the pharaoh realized he was losing all his workers.

Although the Israelites had been in Egypt for over 430 years, some were apprehensive about leaving. To show the Israelites that God ordained their departure, they were led by a cloud during the day and by fire at night. They arrived at the edge of the Reed Sea (then considered the northern part of the Red Sea).

The Egyptian soldiers were closing in on the Israelites. They feared for their lives. Moses raised his stick and the waters of the Red Sea "parted." As many as a million Israelites marched across the opening. After they reached the other side in safety, the opening closed, swallowing up the pursuing Egyptian soldiers in its wake. The Exodus ("departure") became a powerful symbol for the Israelites.

The Israelites journeyed to the base of Mount Sinai. Moses ascended the mountain. God spoke with him face to face. He was given a broader covenant to include any people who submitted to teachings of the Torah ("law and teachings" prescribing righteous living). He received the Decalogue (Ten Commandments) inscribed on stone tablets. It summed the essence of the covenant. The Hebrews were specially chosen to worship and serve the one and only God. They were to have no other gods before them. The stone tablets were placed in a sacred chest called the Ark of the Covenant.

Moses led his people to the land of the Canaanite people. He looked out from a hill top and could see the land promised Abraham. Moses was old (120 years of age) and unable to lead his people. Before dying he appointed Joshua to take his place.

Joshua sent two spies to check on the Canaanites. They returned and reported the land was rich in resources.

The Israelites advanced toward Canaan. As they marched, they followed a short distance behind the Ark of the Covenant, carried by priests. They overtook the land.

Because the word Hebrew pertained more to wanderers, after settling, the Hebrew designation was dropped and they all became known as Israelites. Their land became the kingdom of the Israelites. Its northern region was called Israel. The southern area around Jerusalem was called Judah.

When the Israelites were disloyal to God, they experienced undue hardship and disaster. When they repented and obeyed God's commands, God helped them. If the people experienced difficulties, it was because they did not meet the terms of the covenant.

Although Israelites staved off polytheism, they did absorb ideas from other cultures. They adopted some items used in Canaanite worship. Unleavened bread was used in the Feast of Passover. Spring lambs were sacrificed. Burnt offerings were used in some of their rituals. Some local Canaanite fertility rituals became a part of Israelite practices. Among other attributes, Yahweh also became a fertility god. Celebrations of harvest became an attribute of Yahweh's compassion for his people.

The Israelites enacted restrictions to limit "outside" influences. They were not allowed to marry Canaanites, probably because of some of their more unacceptable rituals. The Canaanites sometimes sacrificed the first born to their tribal gods. Yahweh did not require such sacrifices.

In the eleventh century BC seafaring people called the Philistines invaded the Fertile Crescent. To fight these marauders, Israelites brought the Ark of the Covenant to the coastal plain to indulge Yahweh's presence. The Israelites were overwhelmed, and the ark was taken from them.

A person named Saul became the first king of Israel. The story of David and Goliath, a giant Philistine, emerged from this period. David slew Goliath in battle. However, the Philistines still controlled the area.

After Saul committed suicide, David (beginning of tenth century BC) became king of the Jews. He successfully led his people against the Philistines. The Israelites recaptured the Ark of the Covenant. David restored Israelites to power. He extended his rule to the northern regions of Israel.

David ushered in a golden age of Hebrew history. He was so well respected that in later times when Israelis experienced national distress, they hoped for a scion or "son of David" to become king. It would signal a "return" to the predominance of the Hebrew nation and their God. This belief permeates all subsequent Israelite thought.

David's son, Solomon, became the ruler and built the first temple in Jerusalem (962–922 BC). The ark was placed in the temple. Solomon had a long and successful rein. Prosperity ushered in a golden age. After Solomon died, his son Rehoboam ruled, but the northern area, Israel, seceded.

Assyrians in 721 BC overran Israel. Assimilation and lack of religious focus engendered the tale of the "ten lost tribes of Israel."

During this time, prophets continued to reinforce the Mosaic Law. Amos, a shepherd, was informed by Yahweh to tell the people to renew the covenant. He preached in Israel. He spoke of social wrong doing. He understood the injustice of the wealthy prospering while the poor were oppressed.

Hosea warned the people their kingdom was in danger of collapse. His prophecies were imbued with his personal experiences. He had married an unfaithful wife, actually a prostitute. He said Yahweh was the god of steadfast love.

The book of Isaiah is named after a wise prophet of the eighth century BC who had lived in Jerusalem. He foresaw the kingdom disintegrating. He envisions the time of world peace and the coming of a "son of David" who would be the ideal king. He said Yahweh would appear and save the kingdom.

Jeremiah reiterated Yahweh's way of righteousness and stated the covenant was still valid. He also predicted the kingdom would fall apart. Local priests denounced him for speaking for their lord. Jeremiah fled to Egypt.

The term Jew is derived from the Latin Judaeus, meaning "one who lives in Judah" (Judea is a later name for practically the same region).

Babylonians (under Nebuchadnezzar, about 586 BC) overtook Judah. Judaic temples were razed including the one in Jerusalem (built by Solomon during the tenth century BC). Affluent Jewish families were taken into captivity. Some captives were assimilated into the populace of the conquerors. Others believed the time would come when God would restore them to their rightful land. The Babylonian captivity happened during Jeremiah's lifetime.

Quickly taken into captivity was Ezekiel. Israelites in exile believed their lot resulted from the sins of their fathers. Ezekiel said each person was responsible for his or her own life. He said all lives equally belong to Yahweh. When the land of the Israelites is returned to them, they will know Yahweh guides their destiny.

A Second Isaiah (appended to the first Isaiah) said Israel was the child of a divine mission. Israel would bear witness to a world that will later acknowledge the sovereignty of Yahweh.

After Persian conquests, Cyrus (Persian emperor 538 BC) permitted Jewish resettlement. He authorized rebuilding of the temple in Jerusalem. But, many Jews would not leave Babylonia. Later, many did return and the temple was finally rebuilt under the direction of Zerubbabel (515 BC). Jerusalem again became the center of Jewish culture.

The temple was an important centerpiece and symbol of worship. In it sacrifices were offered. Goats, pigeons, and sheep were often sacrificed to gain God's favor. The burning of flesh on a large stone altar was a common ritual. Rituals were believed to compensate for one's sins.

By the fourth century BC the Torah reached its final form. The book of Daniel mentions a time would come when the world order, as it was known, would end. In its place would be a New World order of the righteous. It mentions for the first time the idea of an afterlife. Those asleep, who were righteous, would awake to everlasting life. Until then there would be a succession of empires and their destruction.

In the book of Daniel, Satan, borrowed from Zorastrianism, became an independent power. Yahweh was seen to have his own problems. Evil resulted from supernatural influences. This also explained why evil people sometimes prospered. These ideas were derived from apocalyptic (Greek: "disclosures") visions. Daniel suggested there would be a confrontation between the forces of good and evil. This led to the idea the righteous would be spared and restored by Yahweh to their rightful place.

Alexander the Great of Macedonia (north of Greece) with his conquering army swept through Judah (332 BC). Some intellectual Jews embraced Hellenistic influences; particularly those based on Platonism. The idea of body and soul, immortality, and wisdom, are examples. These concepts found their way into biblical literature (Proverbs).

For the next few hundred years, priests argued over many issues. One of the major conflicts was who had the right to define and interpret the Torah. A group of upperclass landowners called the Sadducees, believed the Torah did not include the oral dialogues—some of which were finally written on scrolls. Sadducees did not believe in an afterlife. They said no scripture supported that idea. They believed when a person died that their soul perished. They did not believe in fate. They denied the existence of angels and spirits.

An upstart group called the Pharisees said only those who accepted both the Torah and oral traditions were warranted to interpret scripture. Pharisees believed there to be an afterlife and a resurrection. They believed in reward for the righteous and punishment for the wrong doers. They believed man, though having free will, was also an instrument of Yahweh, and in that sense had a special destiny.

In 63 BC Roman general Pompey took over Judea and it became a Roman province. In 37 BC, Herod, in return for Roman loyalty, became the procurator of the region. In 34 BC Herod authorized rebuilding of the temple. Building the temple began in 19 BC. The third temple was completed in 64 AD.

In the latter seventh decade AD, the Roman emperor Nero ordered his armies, commanded by Vespasian and led by his son Titus, to storm Jerusalem—the last city of resistance. In seventy AD the temple was again destroyed.

Pharisaic scholars began the "rabbinical" (rabbei—teacher) tradition that survives to this day. They decided which books were to be in the Torah. Because the temple in Jerusalem lay in ruins, critical rituals once performed there tended to lose significance. The Jerusalem temple was the only significant center of worship of the Israelite people. Yet, because of temple destruction and exile of many Israelites, the Pharisees said worship in any synagogue (originally a "meeting place") was an acceptable alternative. This pronouncement sustained the covenant when the Israelites dispersed to other lands. Rabbis reasoned how to apply the Torah to the myriad problems of everyday living.

One of the last resistant groups, were the fanatical Zealots. They not only fought against Roman tyranny, but also opposed other Israelites who did not actively resist Roman occupation. They used the fortress of Massada as an enclave and staging area for attacks on Roman troops. The fortress was built by King Herod (ruled from 37 BC until 4 BC) on an uplifted rock formation near the southwestern coast of the Dead Sea. In 73 AD Flavius Silva, a Roman governor, marched thousands of troops against Massada. Rather than subject themselves to foreign ideology, nearly a thousand Zealots committed suicide. Other than their fervent belief in the God given homeland, their core beliefs were not dissimilar to those of the Pharisees.

The Torah is a compilation of how the Hebrews fared in keeping the covenant. To reinforce Jewish tradition, later prophets restated and reinterpreted the ideals of Judaism (named after the land of Judah).

The Bible (Old Testament) is composed of three parts: The Law (Torah) the Prophets, and the Writings. The Torah includes the books of the Pentateuch, Genesis (story of creation) Exodus (Hebrew liberation from Egyptian slavery) Leviticus, Numbers, and Deuteronomy. These are attributed to Moses, but were probably written between the ninth and seventh century BC. The word Torah also refers to all Jewish biblical literature.

Originally the Mosaic faith was difficult to live. It prescribed uncompromising justice for violators of the law. Later prophets muted some of its unforgiveness. Judaism changed from one of ritual to a way of life. Judaic faith became more tolerable. God took on a merciful quality. The covenant originally applied to Hebrew people as a whole. Later it was extended to individuals. Each person was to live the covenant in everything he or she did.

How were people to abide by the covenant? The Torah prescribed how to live and how to worship God. Living the Law meant God would be closer to the people of Israel in everyday life. By worshiping and living a righteous life, God would sustain them. Everything was to be sanctified. Blessings are offered for everyday tasks. Even food and drink are blessed and prepared according to rabbinical tradition.

Judaism contends man cannot escape sin. Because of free will, man makes choices between good and evil. Man is to love as God loves and should emulate God's will and do good. Love was synonymous with practicing the teachings of the Law (Torah).

Jewish people, as God's chosen, were to bring salvation to the Gentiles (non-Jewish peoples). Legal banishment of Torah study and the Diaspora (Jews in exile from their homeland) did not stop Jewish people from maintaining the covenant.

Every significant word and idea found in the Bible was scrutinized and commented upon. For centuries the arguments were not put in writing for fear they would rival biblical scripture in importance. They were passed on orally. By the middle of the third century AD they were committed to writing. The written codification of the oral tradition is called the Mishnah ("teaching"). Commentary on the Mishnah (known as the Gemara [completion]) was completed in the sixth century AD. It is known as the "Babylonian" Talmud. It records the thoughts of thousands of wise men covering an eight-hundred year period. The Talmud helped sustain the covenant.

Many prophets amended Jewish-religious tradition. We have already seen that Isaiah was one of these. He believed Yahweh directed the course of history for his own purposes. Isaiah mentions a "mashiach" who would come and rule, and consolidate the land of the chosen people. (Mashiach is Hebrew meaning "anointed [one]." It refers to someone occupying religious office whose head was anointed with oil.) Saul, David, and Solomon were anointed into kingship. Saul was called the Mashiach Yahweh—anointed of the Lord (1 Samuel 24:6).

People who were appointed by Yahweh for a special mission were said to have been anointed. Prophets are said to have been anointed.

The Jewish people as a whole were thought by some to be "anointed." (Psalms 28:8.) The word messiah (from the Greek) became associated with a future king who was to preside over restoration of the Israelite nation into a new kingdom (2 Samuel 7:16). Nathan reported to David of his vision (2 Samuel 7:12) that a descendant of David would succeed him as a great leader of the Israelites.

The original idea of messiah pertained to someone in high religious office—high priest or king. The Old Testament did not ascribe the later meaning of messiahship. The closest it comes to today's understanding is found in Daniel. Daniel mentions a "son of man" (Daniel 7:13) in his vision. He is one who comes with clouds. He is destined to rein with sovereign power over those of the coming kingdom: a kingdom that cannot be destroyed. It was to be God's kingdom on earth.

After the destruction of the second temple in Jerusalem the phrase "son of man" became associated with the more grandeur notion of "messiahship." It later permeated all Judean thinking. Some had thought Zerubbabel, who reconstructed the temple during the close of the sixth century BC, fulfilled criterion of messiahship.

The Book of Daniel mentions an "afterlife" for the righteous. Renewed interest in apocalyptic eschatology (end times) is shown by Judaic literature dated from 200 BC through 100 AD. At the end of time, as we know it, the righteous will awaken to an afterlife. Jewish people believe the afterlife to be a physical resurrection in this existence. The kingdom of God would be on earth.

The story of the Jewish people is rich in history. In 1948 a Jewish state was carved out of Palestine. Many children of Israel returned to the land promised Abraham by their God. That the Jewish people have returned to their homeland is said to signal the nearness of the coming of the messiah.

Christianity

Beginning in sixty-three BC and through much of the first century AD Judea was still under Roman control. Many Israelites believed the Messiah would return and command an army against the Romans. His appearance would also signal nearing of the end of time.

In 1947 the Dead Sea Scrolls were found in a cliff-cave near Qumran on the northwestern shores of the Dead Sea. The scrolls support the idea the Messiah was soon expected by the Essenes. The Essenes were an ascetic sect. They believed the Messiah would punish the unfaithful and reward the righteous.

Israelites had many preconceptions about the Messiah. They did not agree on how to recognize the Messiah if he did appear. The Jewish people have not accepted any of the many people who either claimed or were said to be the Messiah. There were many such claimants in the first century.

Many Jewish religious cults developed during the period of Roman occupation. Some cults arose because certain individuals believed the Messiah did appear. From one such cult a major religion developed. It is called Christianity. (The word Christos is Greek for Messiah.)

In the fourth-year BC a child named Jesus was born in the region called Galilee, probably in a settlement called Nazareth. (Later, legend states Jesus was born in Bethlehem. This is probably because of the foretelling in the Old Testament that the Messiah would be born in Bethlehem.)

No distinguishing events occurred during most of Jesus' life. He began to have some interesting experiences when he turned thirty years of age. A person named John (the Baptist) baptized Jesus. That ritual triggered a transformation in Jesus.

Jesus soon began preaching. He told about the coming of God's kingdom. He said some of those who heard him would not die before the end of time would come. He said he would soon return and deliver the believers from oppression and inaugurate them into a new life. His immediate followers did not write down what Jesus was saying because they believed the end of the present era was imminent. His ideas were passed on orally for at least two generations.

Early followers of Jesus did not consider their religion to be separate from Judaism. They believed their Messiah, Jesus, created a new covenant for everyone. In their eyes, Jesus fulfilled the Jewish prophecy of a Davidian Messiah.

Jewish traditionalists naturally rejected Jesus as their Messiah. They consider Christianity to be incompatible with Judaism.

Destruction of the Jerusalem Temple (70 AD) forced active Jews and Christians to move to other cities. Many Pharisees escaped to Jamnia, a city in western Judea, where they continued the Judaic tradition. Many Christians fled to Pella, a village east of Jordan. The result was a physical separation of Christianity and Judaism.

Having not observed a Second Coming of Jesus, his later followers, who did not know him, began writing down what they had heard. Mindful that stories are modified with each retelling, we suppose that biblical events took on the characteristics of the teller's perceptions and biases.

Original meaning of events and pronouncements were altered. How much the changes represent what originally happened is practically impossible to ascertain. The only reasonable approach is to gauge past events by what is presently known and understood. Regardless, these writings became the New Testament and the basis of Christianity.

There are very few new ideas presented in the New Testament. Many concepts were "borrowed" from other religions and philosophies of that time. Foremost among nonchristian sources are the Jewish Torah (Old Testament) Mithraism, and Greco-Roman ideology.

Let's just mention a few borrowed ideas. From the Torah were obtained the ideas of the kingdom of God, the Davidian Messiah, and the Lamb of God. From Mithraism came the idea of Sunday worship, Christmas, Easter, Son born of a virgin, Jesus being both divine and human, and judgment day. Greco-Roman culture contributed the Logos as the Word, the Word represented by a Son of God, that God was the All-Father, and resurrection of the dead.

Let's examine the books of the New Testament and note if some of its scripture has merit in the structure of reality. Before beginning, we also note Jesus never publicly announced he was the Messiah expected by the Israelites. Yet, Jesus did tell his disciples some secrets that were not disclosed. Though not explicitly stating he was the expected messiah, he never said he was not. Nor did he do anything to discourage anyone from believing he was the Messiah. Moreover, he did everything he could to "fulfill" Judaic prophecy concerning the coming of their Messiah. Deuteronomy 18:18 informs that Moses was told a prophet like himself would come from his people. One could surmise one of Jesus' secrets was that he believed himself to be the Jewish Messiah.

Another interesting observation about Jesus: His teachings never indicate faith is a requirement to enter the kingdom of God. Yet this idea is later added by letter writers who did not know Jesus. (He does mention faith in his miracle healings.)

The Christian Bible (Greek, biblios meaning book) is not a biographical account of Jesus' life. Surprisingly, his physical appearance, stature, and eye color are not even mentioned. His education is not cited. His friends, prior to preaching, are not listed. It does not say whether he was married.

Neither is the Bible a historically accurate accounting of the life and times of Jesus. There is no record Jesus wrote anything. Nor is the New Testament a personal reminiscence of Jesus' life. There is no definite chronological listing of events. Historical documentation is severely lacking in the Bible.

Biblical authors are unknown. The exceptions are some of Paul's letters. Undisputed Pauline letters are Romans, Corinthians, Galatians, Philippians, and Philemon. Disputed Pauline letters are Ephesians, Colossians, and 2 Timothy. Paul's other letters are thought to have been written by someone else. It was not uncommon for writers to "put words into someone else's mouth." Most biblical scholars agree the titles of the books of the Bible were not added until the second century.

The Bible is a theological rendering of stories orally passed from one person to another. Originally, the oral message of Jesus was called the gospels. The stories were put in writing between thirty-five and seventy years after Jesus' death. The first four books of the Christian bible are called gospels (after the Anglo-Saxon word godspel meaning "good news").

The gospels are not first hand accounts of events surrounding Jesus life. They are a reconstructed accounting of Jesus teachings. The gospels offer, at best, a bias and propagandized version of what actually happened. The writings were redacted: edited by the authors from earlier sources.

The gospel consists of the books of Matthew, Mark, Luke, and John. Again, these biblical chapters were not written by their namesakes. The names of the gospel books probably were not affixed until the second century. The fifth book, Acts, is considered written by the author of Luke. The next series of chapters are letters written by a person named Paul. The authorship of some of Paul's letters is disputed. The rest of the chapters, titled as though they were written by the titled name, are letters by unknown authors.

The first three gospels are very similar in content. Most scholars believe the book of Mark was written first. These books could have originated from a common source that did not survive the first century.

There have been many attempts to understand the gospels. There is little total agreement among scholars. Many scholars believe Matthew and Luke were copied from Mark. Aside from other clues, one particular reason suggests this. Mark misquotes a prophecy stated in the Old Testament. He attributes to Isaiah 40:3 a statement found in Malachi 3:1. It prophesies the coming of a messenger whom presages the coming of the Messiah. Mark uses it to support John the Baptist as the messenger. The authors of Matthew and Luke avoid mentioning this misquote. Ninety percent of the content of Mark is duplicated in Matthew. Some scholars believe Luke was copied from Mark and Matthew. Some say Luke was the earliest written book.

Throughout the gospels the unknown authors diligently portray Jesus as fulfilling Old Testament prophecy. This is especially noted in Matthew. In Psalm 78:2 is foretold of the coming Lord speaking in parables. In

Matthew 13:34–35 the author says Jesus used parables when speaking to a crowd to fulfill what was spoken through the prophet of the Old Testament.

Very little in the Bible is defined or explained. The apocalyptic eschatological theme, commonly written of during that time, permeates much of the New Testament. Coming of God's kingdom, judgment of dead and the resurrection and redemption of mankind, are all connected to the acts of a savior. Yet even these all-important concepts are never explicitly defined nor explained. Maybe the reality of these concepts is what Jesus spoke of in parable.

Either biblical concepts were too difficult for the crowd to understand or Jesus himself did not really understand them. Jesus asked people to believe him. Later, the Bible asks the reader to accept on faith the truth of what is stated in it.

Mindful of biblical shortcomings (it is nonbiographical and historically inaccurate) let's try to make some sense of it. We begin examining some books of the Bible, looking for statements that correlate with what we now understand (in this book) about reality. Scripture is stated and appended with comment. The predominant source is the King James Version of the Holy Bible. Verses are rewritten into modern language and rephrased to more accurately represent reality.

The author of Matthew stresses the teachings of Jesus. He refers to Jesus as the "Son of David" (Matthew 1:1). By so stating, Matthew is trying to position Jesus as the Messiah prophesied by the Old Testament. Matthew implies Jesus presents a new covenant. It is stricter: calling for greater righteousness. His messiahship is said to be verified by his miraculous healings.

- Matthew

  Matthew 1:23. A virgin will give birth to a son who will be called Immanuel (meaning, "God is with us").

  > This passage copies the Old Testament Isaiah 7:14. The Lord himself will give you a sign: a young woman will bear a son and he will be named Immanuel. The phrase "young woman" was translated into the Greek as "virgin" in Matthew 1:23.

  3:2. John the Baptist states the kingdom of heaven is approaching.

  > At the time of Jesus many Jews believed their future kingdom, headed by Christ, was near.

  3:11. John states there is one who comes after him, who is greater than he is.

  > Some people thought John was the messiah. (See Luke 3:15)

  11:14. Accept that John is Elijah whose coming was foretold.

  > In the Old Testament we find in Malachi 4:5 that the prophet Elijah would return prior to the Lord's coming.

  > The mind of John once inhabited the body of Elijah.

17:12. No one recognized John as Elijah.

> Being a mind, it is understandable why Elijah was not recognized as the mental occupant of John's body.

3:16–17. John baptizes Jesus. A voice from Heaven said, "You are my son."

10:7. Jesus states: The kingdom of heaven is approaching.

> Jesus "prophesies" the kingdom of God will shortly begin.

10:23. Jesus says, people will not travel far before the Son of Man arrives.

> Jesus states that the Son of Man is other than he was at his First Coming. But, he implied the Son of Man would be (an aspect of) him at his Second Coming.

11:3. While in prison, John the Baptist asked what Jesus had done. He asks if Jesus is the expected one.

> Is Jesus the Christ? The idea of a savior returning to earth to save humankind from evil was also an idea that was expressed by Roman writers (e.g., Horace and Virgil) in the first century BC.

11:6. Jesus sends a reply. Bless those who have no doubts about him.

> Jesus clearly implies he is the messiah.

11:27. Jesus says that he obtains special knowledge from his Father. The Father can be known by listening to the Son.

> There is an intimate relationship between the Father and the Son. The Bible does not define nor explain the relationship. (This book does explain the relationship.)

16:13. Jesus asks, Who do people suppose is the Son of Man? Simon Peter answers by saying Jesus is the Son of Man. Jesus says, the Father told you so.

> Jesus continues to push the idea he is the Messiah. Yet, the phrase "Son of Man" should only pertain to the mortal aspect of Jesus. After all, this phrase states that this individual is but a "son" of man. As such, it refers either to Jesus' physical body or to his mind functioning deductively. (In the physical body, the mind is asleep: it functions deductively.) The problem of referring to the mind as the Son of Man is twofold. First, A person either acquires his mind through his immediate physical parents, or he acquires it through "reincarnation." If reincarnation is the source, it means it is an "intact" mind that previously had functioned in a former physical body. Either way, if the mind is the referent, the term can also refer to Jesus (that is, his mind) in a future incarnation. The mind does not function by itself in a physical individual: it functions through the brain. Other authors of Biblical books refer to Jesus as the Son of God. This phrase cannot refer to an individual's physical aspect. Therefore, the phrase "Son of Man" should only pertain to the physical body. The phrase "Son of

God" should only represent Jesus' mind, not his body. To be more precise: "Son of God" should only pertain to an active (self-functioning—inductively "awakened") mind (spirit) Adam-Christ—the Second Coming.

16:27. Jesus says the Son of Man will come before some of those present will die.

> Yet here, Jesus indicates that he is not the Son of Man in any commonly held sense. Here, Son of Man seems to refer to his (mind) coming in another body. It refers to another physical individual. The above explanation should resolve any confusion arising from this nomenclature.

22:30. Resurrected individuals do not marry.

> This implies a different society; one that is quite different from what is known in contemporary culture. Raising the dead was a common Greek idea. Asclepius, believed to be a son of Apollo, was said to have raised the dead. Zeus did not like the idea humans could escape death and he killed Asclepius with a thunderbolt. Asclepius is said to have often come to earth to visit and aid humans.

22:41–46. Jesus asks of whom Christ is a descendent. Some Pharisees said he is David's descendant. How can that be? No one answered.

> The explanation is that Jesus was once the spirit (mind) who, in a former existence, occupied David's body. And he who was David's mind was once the mind of Adam.

24:23–25. Do not believe anyone who says he is the Christ. False Christs and false prophets will come and perform wonders to deceive even the very elect. You are forewarned.

> Here, Jesus provides the criterion to determine who is not the Christ. If people say there is the Christ—because someone implies he is the Christ by demonstrating great wonders and performing miracles—do not accept that he is the Christ. At the time of Jesus, prophets were expected to perform wonders. During Greek and Roman rule, miracles were thought to be quite common, especially healings. No less was expected from someone who suggests he is the Christ. Even people of today call an event a miracle when they do not understand it or cannot explain it. That someone does miracles only proves he does things people do not understand.

24:27. Just as there is lightning, the Son of Man will come (here this phrase seems to represent the Christ).

> This could mean there would be a strong lightning storm when the Son of Man is indicated. This may not signify any more than the individual who is the Messiah learns about the nature of the Son of Man on a day of much lightning.

24:44. Jesus says the Son of Man will come when not expected.

> People of Jesus time were expecting his quick return as the Son of Man.

25:34. Those who are connected to the Father will inherit the kingdom that has been prepared since the beginning of creation.

> This scripture suggests the kingdom of God is of a long term plan. Part of that plan calls for the existence of the physical world.

Scholars believe the book of Mark was written in Rome in the eighth decade. Mark stresses Jesus uniqueness in his death and resurrection. Where Matthew emphasizes Jesus' teachings, Mark focuses on his miracles. Mark tries to show how Jesus' death and resurrection fulfill Old Testament prophecy. Mark presents the gospel from the perspective of Peter, a disciple of Jesus.

- Mark

    Mark 1:14–15. Jesus came to Galilee to preach about the Kingdom. He said the kingdom of God would soon come.

    > We now know Jesus' prophecy about the imminent coming of the kingdom of God was not very well prophesied.

    3:21–22, 31. When Jesus' family heard that people believed he was crazy or possessed by the devil, they went to get him.

    > There is no indication members of his family accepted him as the promised Messiah. Their actions show they did not accept him as the Messiah. They were concerned about Jesus' mental health.

    3:33. He responds: Who is my mother or my brothers?

    > Jesus mocks his concerned family.

    6:4. Jesus says, A prophet is honored everywhere except where he lives.

    > This strongly suggests his family even rejected his status as a prophet.

    4:34. Jesus explained everything to his disciples.

    > There is no indication whatever Jesus understood everything, let alone, able to explain it to his disciples. There is much indication he did not really understand what he espoused.

    4:11. Jesus tells his disciples the secret nature of the kingdom of God.

    > The only secret is Jesus, and his disciples, did not understand the kingdom of God.

    1:15. The kingdom of God is imminent.

    8:29–30. Peter says that Jesus is the Christ. Jesus tells his disciples not to tell anyone.

Jesus asks his disciples to keep a secret, viz, his messiahship. This is the messianic secret.

9:1. Jesus said some people who saw him would not die until after they have seen the kingdom of God come.

Obviously, Jesus did not do well as a prophet. He states the kingdom comes in a way to be seen.

11:28. Jesus is asked by what authority he cleans out the temple.

11:33. He does not answer.

14:61–62. A high priest asks Jesus whether he is the Messiah, the Son of God. Jesus answers: "Yes I am." The "I am" is the definition of God given in the Old Testament when God revealed his name (Yahweh—I am) to Moses.

Whether Jesus actually said this, or the author of Mark put those words (Son of God) in Jesus' mouth, is indiscernible. If he is the promised Messiah, in what way is his messiahship defined? Again, by saying this ("I am") Jesus is trying to fulfill Old Testament prophecy.

The Book of Luke was written between the eighth decade and the turn of the first century. Some interesting characteristics of Luke are its emphasis on salvation for Gentiles and Jesus as their savior. Luke uses Paul as his authoritative source.

- Luke

Luke 1:15. It was predetermined that Jesus would be filled with the Holy Spirit.

2:11. The Savior, Christ the Lord, is born in the city of David.

Luke is the first writer to use the word "savior" in conjunction with Christ. This is also the only site of the biblical phrase "Christ the Lord."

3:6. Everyone will see God's salvation.

This clearly states there is no individual who will not be saved. This again echoes the Old Testament scripture that all humanity will witness the revealing of the Lord (Isaiah 40:5).

3:38. Jesus' genealogy is traced to David and eventually to Adam who is the Son of God.

Such a genealogical tracing is to state Jesus is the Davidian Messiah expected by the Israelites. Yet, there is no indication Jesus ever said he was a direct descendant of David. The writers of the Gospels used every possible means to indicate Jesus was the expected Jewish Messiah. Note, in Matthew is listed a different variation of Jesus' ancestral lineage. Obviously, this is another one of those stories that developed around the myth of Jesus's messiahship.

9:51. The time was coming when Jesus would ascend to heaven; he headed to Jerusalem. The disciples believed the end time was near.

In Jesus' time, people believed heaven was "out there" among the stars. Hell was believed to be beneath the earth. This was typical of Greek and other regional religions.

19:11. As Jesus and the disciples neared Jerusalem they believed the kingdom of God was ready to appear.

Luke mutes the common idea the kingdom was near. In what follows, he asserts the kingdom is not physical. This is contrary to not only the Judaic thought of the time, but also to a statement in Mark stating some people, who were living at the time of Jesus, would see the coming of the kingdom.

9:22. Jesus says the Son of Man will endure much suffering. He will be rejected by the best of people and put to death. On the third day he will come back to life.

Here again, Jesus talks as though he is and is not the Son of God.

11:20. Jesus says because he can cast out demons that the kingdom of God has already arrived.

Jesus says, because he drove out demons proves the kingdom of God has already come to those he speaks. Such a statement is a poor rendition of proof. It does not prove anything.

Jesus did not prophesy very well because the kingdom did not come in a way to be seen as he had stated. That does not bode well for a Messiah.

13:18. The kingdom of God is like a mustard seed: it grows and branches like a tree.

This is a possible cover-up explanation why no one noticed the coming of the kingdom. These statements do not mean there is no "coming of the kingdom." Rather, Jesus did not properly understand its nature. Jesus did not know when the kingdom would come. It is also evident he did not properly understand the kingdom of heaven.

13:30. The last people will be first, and the first will be last.

This pertains to the coming of the kingdom. The last individuals (minds) of this existence (Cosmic Phase I) will be the first ones to awaken (function on their own) in the next existence (Phase II). The first individuals (minds) of the next existence, will be the last of that existence. Adam (the first mind) returns at the end times (as the Messiah). He reacquires the knowledge of everything to help him govern the next existence.

Epictetus, originally a Roman slave, wrote (in the latter first century) about cosmic cycles. He believed God's will was revealed in universal order. He believed the universe operates in cycles. Periodically, the world ends in a conflagration, and then it is recreated just as it once was.

17:20. The kingdom of God cannot be seen.

17:21. People cannot observe the kingdom of God because it is within them.

This further explains why no one can see the coming of the kingdom. This also covers up why no one has seen the coming of the kingdom of God as prophesied by Jesus.

17:22. Jesus said there will be a time when people will want to see the Son of Man, but they will not see him.

Jesus says the people of his time will not see the days of the Son of Man. Yet, it has been said Jesus was the Son of Man. Maybe the nature of the Son of Man has not been properly understood. The Son of Man is different from the Son of God. They are, however, intimately related as previously explained. More will be explained later.

17:24. As lightning streaks across the whole sky, the Son of Man will be in his day.

The Son of Man is linked to the phenomenon of lightning. We surmise the Son of Man will be revealed on a day of much lightning.

17:28–30. It will be a typical day when the Son of Man is revealed.

This does not mean the Son of Man will necessarily be revealed to the populace on that day of lightning. As noted above, people will not observe when the Son of Man is identified. It may only mean that on the "day of lightning," it is revealed to a certain individual that he is the Son of Man.

17:33. Those who try to save their life will lose it; those who die will live again in another existence.

Man cannot save his life. That a man is saved is due first to his death (where he "loses" his life). One is not saved by "faith" or believing (or believing not) any given thing or any given statement. That man is saved via an extended existence in "heaven" is not of his doing. That individuals live after death is part of the design of everything. How this occurs will be fully explained by the end of this book.

17:34. Jesus says when the kingdom of God comes, there will be two men in one bed: one will be taken, the other left.

This is not about two men sleeping in one bed. It is about the two natures (of man) in each individual. One is human nature (the physical, which is left behind) and the other is spirit nature (which is taken away).

17:35. Two women will be grinding grain: one will be taken; the other left.

17:36. Two men will be in the field: one will be taken; the other left.

19:10. The Son of Man came to seek and save what was lost.

What was lost? Functionality of spirit and its knowledge. The Son of Man loses his life so that the spirit nature in him develops (like a seed) into a mind that seeks deeper truth (knowledge) about reality. The Messiah is God's instrument to reacquire knowledge lost in the creation process (More of this problem later). Scientific methodology is the means to obtain the knowledge of reality.

20:41. How can the Messiah be David's son?

This was prophesied in the Old Testament. The spirit is the mind doing its own work (functioning via inductive reasoning). The mind, which occupied David's body, is the same mind that later becomes the mind of Jesus (First Coming) and finally functions on its own as the Messiah (Parousia, the Second Coming).

John is the fourth gospel. It was probably not written before 90 AD and not after 120 AD. It is decidedly different from the first three gospels. It stresses faith: that believing in Jesus assures eternal life.

▪ John

John 1:1. In the beginning the word was synonymous with God. Where God was, so was the Word.

The "Word" represents knowledge.

1:2–3. Knowledge was with God from the beginning. Using knowledge God created everything through Adam. (Adam later returns as Christ—first and second coming.)

The "word" is a Hellenistic idea. The word is logos. Greek Stoics used the term as "reason." Philo wrote of the logos and associated it with God. God is characterized as omniscient.

1:4–9. God sent John as a messenger to tell about the real Light—light that shines in the world.

1:10. Knowledge became manifest in the world. The world was made through Christ, but that is something people did not understand.

The Logos (Reason) permeates the whole world. This book insists Whyness (Creator) permeates all Whatness (Creation).

1:14. The knowledge used to create everything becomes manifest in Christ.

The Messiah, by reasoning, develops an awareness of Reality that is acquired only by truthfully seeking the complete understanding of everything.

1:29. Jesus is the Lamb of God who takes away the sin of the world.

John the Baptist makes this reference as a parallel to the Israelite tradition of sacrificing lambs during Passover. The implication is Jesus is sacrificed to save humanity.

3:3. Jesus said one does not enter the kingdom of God unless he is born again.

The first birth is physical (leading to human nature). The second birth is mental (developing the spirit).

3:5. No one enters the kingdom of God unless he is born of the Spirit (the mind).

The spirit is the inner self. Only it "becomes" of heaven. It already is a part of heaven in the individual. The spirit just does not function on its own when occupying a physical body. After the mind is freed from the body, and at the inauguration of the kingdom of God, the spirit (mind) becomes active: it begins functioning by itself (awakens by functioning inductively). These statements will be further explained in the final two chapters.

3:6. Flesh is born from flesh and spirit is born from spirit.

Physical bodies give birth to physical bodies; minds give birth to minds.

3:21. One seeks truth by questioning what he learns from interacting with the light. The light shows us what is true. Following the truth will lead us to God.

To get to God, one must finally understand light. Light itself, is the final defining of truth.

3:31. He who comes from heaven above is very great.

This refers to heaven as being "above in the sky." Other passages also seem to say the same. (Isaiah 55:10) Rain and snow falls from heaven above.

5:31. Jesus says to not accept anything he says as true merely because he says so.

What anyone says on their own, as in "I believe, I think," is not to be taken seriously. Assertions verified by reality can be taken seriously.

8:44. Your father is the devil and it is your nature to be like him. The devil does not tell the truth; it is against his nature to be truthful. When he lies, he is just being his natural self. He does not realize he is a liar.

Jesus is referring to Jews in this passage. However, Jewish people are no different from anyone else. People in the physical existence, by their human nature, are opposed to truth. Moreover, a devil is only doing what is natural for it to do. A devil does not know he is a liar. To know that is to know the truth, and that is not a characteristic of a devil. The mind functioning deductively (defines devil) is naturally opposed to seeking greater truth.

8:58. Jesus says he existed before Abraham was born.

This indicates his pre-existence. It also refers to the Old Testament definition of God that was spoken to Moses ("I am"—I exist). Again, this suggests Jesus' messiahship is a physical body harboring a unique mind, one that preexisted.

10:38. Jesus says the Father (Adam) is in him, and he is in the Father.

Adam presages the one to come. He is in Jesus and Jesus is in him. The Father is not God. Adam is the Father of all who followed. Adam, the mind, later incarnates into the body known as Jesus. An individual's spirit is the mind. If one's mind does not function at all, then he has no awareness—not even awareness by deduction. A body without a mind means there is no self-identity.

11:25. Jesus says he is the resurrection and the life. Those who partake of his spirit will live, even after they physically die. The spirit never dies.

Jesus implies those who believe in him already have eternal life on earth and will not be judged (3:15). Those who do not believe in Jesus are already judged (3:18). In John we find the idea of eternal life and judgment.

Belief has nothing to do with whether an individual becomes of the next existence. The plan of destiny has little to do with what people believe or believe not.

14:6. Jesus says he is the way, the truth, and the life. No one can go to heaven and be with the Father except through him.

Because the mind of Jesus is the original mind (Adam) everyone is a descendant of Adam and therefore has the same original mind as Jesus. All have come from Adam and all return to Adam. One's mind gives life and awareness.

14:10. Jesus says again that he is in the Father and the Father is in him. The Father in him performs his own tasks. When Jesus speaks about the Father in him, he is not speaking about himself in the sense most people understand the self-identity (which is usually associated with the physical self).

Jesus's "inner mind" functions inductively (when active, awakened). It is a separate function from his mind functioning deductively. In his "First Coming" his mind was not inductively active. The Father in him was functionally inactive. The First Coming was an extreme deductive functioning of the mind. The Second Coming is an extreme inductive functioning of the same mind.

Yes, the Father-mind was the mind of Jesus (First Coming) but it was not functioning on its own. It was inactive. It functioned only with respect to the physical world. Only when the mind is active (in itself—reasoning inductively) does it "function" as the Father. Yet the Father-mind, though inactive, did incarnate the body of Jesus. His mind when inactive (in itself) is the mind of Jesus (First Coming). When inactive, the mind functions deductively. Jesus' mind (deductive) is in the Father. The mind functioning deductively is just an unawakened mind—one not yet born again.

14:16. Jesus says people obtain their spirit from the Father.

This conveys the idea that people's minds will one day begin to function (on their own).

14:17. The world cannot see a spirit because it is not physical. Light does not shine on it. But you know there is spirit because it lives in you.

The spirit is the mind functioning on its own. When activated, the mind seeks truth. It seeks the logos—the reason for everything.

14:18–20. Jesus says he will not abandon the people. He says the world will no longer see him, but people will know he is here because their spirit came from the Father through him. He lives as spirit and so they will live as spirit. When Jesus returns, people will know that he is in the Father, and that people are in him.

When Jesus comes back (as the Messiah—the Second Coming) he will explain things left unexplained during his First Coming. This includes nearly everything.

Jesus was once Adam, the Father of everyone. In that sense Jesus is in everyone. Every mind is a part of the original mind, although it has become (functionally) split into a plurality.

14:28. Jesus says the Father-Spirit that is in him is greater than he is.

It is the Father (originally Adam) who is the Messiah.

16:8. At the Second Coming the Father-Spirit will prove that people do not understand the nature of sin, righteousness, and the judgment.

The nature of sin, righteousness and the judgment was not explicitly defined during the time of Jesus.

16:13. At the Second Coming the Father-Spirit will tell the whole truth and people will understand.

18:36. Jesus says to Pontius Pilate, procurator of Judea, that his kingdom is not of the physical world.

The kingdom of heaven is within you: it is your mind.

18:38. Pilate asks: "What is truth?"

Socrates was asked this question. Truth is the collective of all information that represents (and therefore corresponds to) Reality. Seeking truth brings one closer to understanding God. The Bible suggests God is (represented by) truth.

- Acts. The fifth book is Acts. It was written as a continuation of Luke.

Acts 2:21. The time will come when those who call on the name of the Lord will be saved.

Yet we already read in Luke 17:33 that those who will try to save their life will lose it and those who die, will live again in another existence.

17:26. All spirits (minds) are created from one individual—Adam—the Father-Spirit who returns as the First Coming and later as the Second Coming—Christ. God determines when and where people will live.

The original man was Adam—the Father-Spirit. Everyone is created from the Father-Spirit—the Holy Spirit. God has fixed certain events to control the course of history. This is a common Judaic belief.

17:27–28. God is everywhere: we live and exist in him.

God is all there is. That which is All includes anything and everything. More about this in the final chapters.

17:31. God has set a day when Christ judges everyone.

Each person has two natures. One stays (the physical body and its human nature); the other goes (spirit nature—one's mind).

Paul wrote a series of letters to support the idea of Jesus death and resurrection. There is no indication Paul knew Jesus. Paul, formerly called Saul, a Jew and a member of the Pharisees, had once persecuted Christians. He had a profound experience while traveling the road to Damascus. He had a vision that transformed his life. He never totally left Judaism nor totally embraced Christianity. Paul believed his "calling" was similar to those of the Old Testament prophets. Paul says a person follows Jesus through faith. He says Jesus' death results in forgiveness of sin for those who have faith in him as the savior. Paul's letters are considered the earliest writings of the New Testament. Each letter concerns specific problems of the addressed community or individual. Let's look at some of Paul's letters. The first letter is to a Christian church in Rome.

- Romans

Romans 1:4. Paul says Jesus is the Son of God because he was raised from death.

If people saw Jesus lifted up, then the "raising" was of a physical nature. Yet, we know the physical body stays behind. One does not see the kingdom of heaven—because it is within you. If heaven is "in you," it is not something one can see. Physical resurrection is one of those miracles of which we were warned. (A great pretender doing physical miracles may indicate the antichrist.)

1:17. The Bible says righteousness is achieved only by faith.

Paul's writings change the tenets of Christianity. Jesus had said he would return. He did not return as he prophesied. Because his expected return did not occur within the lifetime of some of those who saw him, Paul adds another "condition" (faith) for the individual to be received into the kingdom of heaven.

Gnosticism, espoused by a religious group of the first couple of centuries AD, stated salvation was obtained by "knowledge" and not by faith.

1:18. Man is naturally evil according to God's standards. Sin hinders man from seeking truth.

Man's human nature (deductive mental function) leads to reductionism: away from knowing All. Deduction is explaining the whole by its parts. Induction is explaining the parts by the whole (of which they are a part).

Gnosticism espoused the duality of good and evil. Matter is evil; spirit is good.

1:21–22. People who believe they worship God are fools. Their own ideas and imagination fool them.

"Intelligent" men deduce ideas and believe they are wise. This is foolishness. Men only deduce because their inductive-mental function is inactive (asleep). A mind not functioning inductively, by default, functions deductively.

1:25. Man is drawn to lies instead of the truth about God. He worships the creation instead of the Creator.

Belief versus truth: Man deduces beliefs. Beliefs arise because of the insecure nature of one who does not understand God (the whyness for everything). Reductionism is the mental "reducing" of information to its components, the "whatness" stuff of creation.

3:4. Man lies about the true nature of God. God is represented by truth.

The "God" of the Bible is defined as (is represented by) truth. We will learn such a characterization is not so simple.

3:9–12. Everyone is affected by sin. No one can do what is right by God's standard. No one seeks to understand God. Everyone has turned away from God. Without exception no one can do what is good.

This reminds us of Proverbs 14:12, 16:25. The way that seems right to man ends in death.

This means that no one, because of sin resulting from his or her human nature, can worship God. Deduction versus induction. One cannot assess an understanding of everything and then know God by deduction (reductionism).

5:12. Sin entered the world through one man—Adam. Sin brought death. No physical being will escape death.

Adam is the name given to the first being. Everyone is a descendant from the first being. When Adam (defined by "his" mind) became a physical being, everyone who descended from him took on a physical nature.

5:14. Adam is the figure of the one who will come.

Adam comes again as Jesus—the First Coming. Later, he returns as the Messiah—the Second Coming. Adam is the Father. The same Father who incarnated the body of Jesus. The same Father who is the Messiah of the Second Coming.

6:7. People are freed from sin when they die.

The mind is free from deducing things about the physical world when the physical body dies. The mind is set free after death. But, that does not mean it immediately begins functioning inductively. That awaits the beginning of the next existence (Phase II). Looking forward to 2 Peter 3:13, we read: God will give us heaven and a new home where righteousness reigns.

7:14–25. The law is spiritual, but I am a mortal being, sold as a slave to sin. I do not understand why I do things. I cannot do what I want to do; instead, I do what I hate. That I cannot do what I want to do demonstrates the law of sin. What I do is controlled by my sinful nature. Goodness is not a part of my human nature. My wanting to do good prevents me from being good. I cannot do the good I want; instead I keep doing evil. What I do is contrary to what I want to do. This means I am not really the one who does evil. Rather, my behavior results from the wanting that arises from my physical and emotional nature. Because I cannot do good, evil is the only choice I have. My inner being cares about truth and God. I can seek the truth about God only by using my mind (spirit), but, in the physical world, my mind is overwhelmed by my human nature, which perpetuates wanting (sin).

Sin is desire. Buddhism concerns itself with the problem of desire. Wanting and not having are similar. One wants what he does not have. Human nature results from a mind not functioning on its own. Not self-functioning, the mind depends upon the physical world for its premises. The premises are images (sensory data) representing whatness things. In the physical world, the mind only needs to function deductively (deducing ideas about the physical world). The very act of wanting (desiring) to do good prevents one from doing good. It seems ironic, but the opposite of what one wants by human nature is typically what happens. An inductive-functioning mind is of little help in the physical world.

8:13. When one's mind is controlled by what human nature wants, one is against God, and death results. When the spirit controls human nature, life and peace are the consequences.

> One is not controlled by the spirit (mind) in the physical world. Material things are what act upon one's desires. The mind is controlled by the physical world. The physical world overwhelms the mind (spirit). In the next existence, the mind regains control of itself.

12:2. Do not conform to the whatness patterns of this world, but allow your mind to develop by seeking the truth about God. Ask why and you will be seeking after God. (God is the reason (whyness) for everything.)

> This is practically irrelevant. Moreover, that people find themselves in the physical existence is not of their doing, and is not their responsibility. The "doing" is the responsibility of Adam—Father-Spirit—Messiah.

> Paul implies "another existence." God will not change people until the time is right, when the next existence, the kingdom of heaven, begins.

14:10. We will all stand before the judgment seat of God.

> Paul says at the end of (forward) time everyone will be judged. He believed judgment day would happen within his lifetime.

▪ 1 Corinthians. Paul's first letter to Corinthians.

1 Corinthians 1:20–21. Man cannot know God by his own wisdom. The wisdom of this world is foolishness.

> Man's wisdom, obtained by deduction, is foolishness. It leads him in the opposite direction of seeking greater truth and understanding God.

2:10–16. The spirit seeks to understand everything. Only man's spirit really knows him. The spirit is the mind. Only it can discern the nature of everything. Our mind originated from Adam-Father-Christ.

> The spirit (mind) can search for deeper truth, but only if activated. By nature, it is typically not very active in the physical world.

6:19. The body houses the Holy Spirit.

> Bodies are occupied by minds.

15:20. Just as Jesus was raised from death; those who sleep in death will also be raised.

> A person's mind is the real self. It is nonactive in physical existence: it sleeps. Because the Messiah's mind becomes active, so too will the minds of others. The raising up of Jesus in the First Coming was a physical resurrection. It represents symbolically (by physical demonstration) what happens to the mind. The body is a vehicle that

carries the mind around in the physical world. Paul combines the Greek idea of immortality with Jesus' idea of the spirit resulting in the afterlife heaven. When the mind is "raised," it awakens and begins to function on its own (reasoning inductively).

15:42. The physical self is buried as a mortal being. When the self is raised, it will be immortal—a spiritual being.

When a person dies, his mind leaves the "housing" of the body. One's mind is immortal.

15:45–53. The first Adam was created a physical being, but the last Adam is the life giving spiritual being. The physical comes first, then the spiritual. The first Adam was physical; the last Adam is spiritual. The physical cannot enter God's kingdom; what is mortal cannot have immortality. Physical death is not the end. A time will come when we will be given immortality. Mortality is wrapped in immortality.

The time will come when heaven (the collective of all minds) becomes manifest—begins to function (Phase II of cosmic cycle). Paul was a Pharisee and believed in immortality and resurrection. Physical death is part of an extraordinary process to create many beings from the first being (Adam).

- 2 Corinthians. Paul wrote a second letter to Corinth.

2 Corinthians 4:16. The spirit can continue developing even as the body functionally deteriorates with age.

The spirit naturally seeks truth (inducts knowledge). However, one's human nature opposes truth seeking (because it naturally deduces information about the physical world).

- Galatians. In a letter to Galatians, Paul states:

Galatians 3:3. We are all descended from the original Father-Spirit called Adam.

5:16–18. One can overcome the desires of human nature if the spirit guides him. Human nature (emotion and deduction) tends to hinder spiritual development.

This restates in scripture what was mentioned above. Reason can overcome desire.

- Ephesians. Paul's letter to Ephesians.

Ephesians 1:4. Before God created the world, he planned to create individuals by using Adam (later to become Christ) to initiate the process.

We have the spirit (mind) which is traceable back to Adam. Adam-spirit (mind) later incarnated in Jesus.

1:10. At the right time, God will bring everything together under Christ-Father.

This refers to the next existence. The CPT theorem in physics decidedly calls for a universe, which inverts temporal flow. Physical bodies "split" the spirit-mind (originally and functionally "nonsplit" in Adam) into a plurality (resulting in many minds).

5:6. Do not let anyone fool you with ideas when they cannot verify what they mean in reality.

People only deceive themselves.

5:14. Light shows us what is real.

We learn from the world because of light. Electromagnetism (light) supports the things of the physical world. Chemistry, the binding of atomic elements, results from electromagnetic activity.

- Colossians. Paul's letter to Colossians.

Colossians 1:15–20. Christ is God in manifestation. God created Adam-Christ as his first born. Everything was created by him and for him. God exists in creation as Adam-Christ who predates all things. All things exist because of him. God's selected his Son, Adam-Christ, to represent him. God controls the whole universe through his Son. And through his Son he will pull the whole universe back to himself in the next existence (heaven).

This is trying to say the Messiah represents God in knowledge. God uses the Messiah to "get back" the knowledge lost in creating the world. God has awareness of himself through the Messiah. More about this later. Here again, Christ is called the first born of God. The Son of God is Adam, the "Father." The Father is not God. The Father is God incarnate (manifesting himself in his creation). The Messiah brings the universe back to God. When the kingdom of heaven (of God) begins, the universe is brought back to God.

2:3. In Christ is hidden all God's wisdom and knowledge.

The Messiah (or Christ) understands reality. Again, the Messiah is the Father in Jesus, not the bodily Jesus. This pertains more to the Second Coming. The Father was aware of all knowledge prior to his role in physical existence. His mind ceases functioning inductively in the physical world, and he could not sustain the knowledge "of everything." Knowledge became "hidden" from his awareness.

- 1 Thessalonians. Paul's first letter to Thessalonians.

1 Thessalonians 5:21. Test all things.

Things need not be tested for truth. Things are what they are. Testing concerns statements about things. Test any given statement to see whether it is true: that it corresponds to reality—that it correctly represents some "thing" or event.

- 2 Thessalonians. Paul's second letter to Thessalonians.

2 Thessalonians 2:3–4. Let no one deceive you by any means. The time will come when the Great Deceiver will be revealed.

> This refers to the antichrist. He is against truth. How shall we recognize him? He plants the methodology that encourages people to continue deducing (reducing an understanding of things to their components) so that they are further distanced from the "whole" truth. He asks people to "believe" what he states. The very act of believing hinders truth seeking.

2:9. The Great Deceiver (or antichrist) is allied with Satan. He performs miracles that seem to imply that he is God's servant.

> Jesus has already warned us about such a person. That person turns out to be Jesus himself. The First Coming is the antichrist. Jesus does all kinds of miracles and gave false signs and wonders. Though miracles were expected of prophets, in biblical times, Jesus figured he had to outdo the others: walk on water, multiply bread and fish, and raise the dead. When one has experienced little else except the physical, it is not difficult to imagine how easy it would be to get people to deceive themselves. Miracles, wonders performed, false signs, do not come close to replacing verification and proof of a statement's truth.

2:11. People are naturally deluded so they will believe a lie. This results from inheriting sin.

> God has no choice but to create a side of people that is against himself (God). Creating something out of nothing calls for the creation of things in opposition: matter-antimatter, and forward time and reverse time. Christ-antichrist. Heaven-hell. We will discover God also creates its opposite, a Goddess. Everything is created in complements.

- 1 Timothy. Paul's first letter to Timothy.

1 Timothy 2:3–4. God the savior wants all men to be saved and to know the truth.

> God is not the savior. That God is found wanting is absurd. About truth: If God is (represented by) truth, then we cannot know God until we first know and understand not only the nature of truth but be able to define truth and falseness. This awaits the next chapter.

2:5. Christ is the intermediary between man and God. He brings man together with God.

> This suggests man cannot get to God without going through the Messiah.

- 2 Timothy. Paul's second letter to Timothy is considered by some scholars to be written by a different author than First Timothy.

2 Timothy 4:4. Human nature turns people against truth. They direct their attention to legends.

There has not been a time in this existence when man did not oppose truth. And sure enough, the legend of Jesus has captivated those of certain religious persuasions. Asserting one is a truth seeker is not the same as being a truth seeker.

- Hebrews.

Hebrews 2:9–10. Jesus was made lower than angels so that he could die for all men. God made Jesus suffer to make him righteous.

This indicates God sustains his creation.

9:27–28. Everyone dies at least once. Likewise, Jesus died to symbolize that he is responsible for the sin that entered the world via one of his previous incarnations named Adam. He will come again, as the Second Coming, as a being without sin (desire) and he will complete the unfinished task of truth seeking.

This refers to the Second Coming of Jesus (Parousia). The Father did inhabit (the body denoted) the First Coming. But, the Father in the First Coming was primarily inactive. In the Second Coming he becomes mentally active (functioning inductively).

11:1. Faith concerns things hoped for, the evidence of things not seen.

This is a reasonable definition of faith. This definition is different from saying faith is a requirement for "going to heaven."

- James. Letter from James.

James 1:6. Believe and do not doubt.

This is a trick. Not to doubt is not to question and seek truth. One cannot doubt reality (it is what it is). One doubts by questioning his own ideas. When one can verify statements by reality, he comes closer to understanding everything. The final understanding is to know God. But, God is not understandable until one has sufficiently understood everything. Everything is understandable in God. This will become more clear.

2:24. Man is not reconciled with God by faith alone, but by what he does.

Incorrect. The reason everyone's mind does not function is traceable to Adam. He comes to earth more than once (in different bodies) to "atone" for his past doings. But this is just God's plan—to create many individuals from one given individual, namely, Adam (Genesis 1:28). In a later incarnation, Adam must reacquire the knowledge of reality (which, according to the Bible, represents God) that he once had. When his mind fell asleep (Genesis 2:21) no longer could he sustain that knowledge, and it was forgotten. Yet, we know the knowledge used to create the world is manifest in its structure. Scientists are working to understand the nature of reality. In doing so, they are retrieving the knowledge inherent in reality's structure—the very knowledge used in its creation.

2:26. A body with no spirit is dead.

> A body without a mind does not function with awareness. Faith alone is insufficient to gain entrance to heaven. Only active mental pursuit of truth pushes one along the path to understanding God.

- 1 Peter. First letter from Peter.

1 Peter 2:22. Jesus did not sin; he did not lie.

> Jesus lied aplenty. He made contradictory statements. He did not understand nor could he explain what he said. Jesus, the First Coming, was not (functioning as) the Christ.

- 2 Peter. Second letter from Peter.

2 Peter 3:4. Jesus promised to come again. Jesus said he would return before many of those he knew would die. Where is he? There has been no change since the world was created.

> The followers of Jesus were confused when he did not return as he had prophesied.

3:10. Christ will come. The physical world will disappear: it will be annihilated. The earth and everything on it will incinerate.

3:13. God will create a new world for spirit beings where individuals are inherently honest with reality.

> A time will come when the next existence (Phase II) begins. The next existence is different from the physical world. It is a world of the spirit (minds)—heaven (all minds).

- 1 John. First letter of John.

1 John 1:5. God is Light.

> If this is true, then the Biblical God is the God of Light, Lucifer. This is an issue that will be addressed more fully before this book is finished.

2:18. The end is near.

> The author of John is telling followers of Jesus that they should not become discouraged because heaven was soon to come.

2:21. No lie comes from the truth.

> True and false must be understood and explicitly defined. This will be done.

2:22. Who lies? Answer: those who say Jesus is not the Christ.

> Making a claim proves little. No, a liar is one who asserts the truth of statements he cannot verify or prove. If someone says Jesus is the Christ, but cannot prove he was, then the asserter is lying. He, who proves what he states, allows reality to assert itself. Let reality (or God) verify and prove who is the Messiah. Jesus did not verify or prove what

he said, let alone that he was the Messiah. If Jesus (First Coming) had verified he was the Messiah, no faith would be needed to accept him as the Messiah.

2:23. Those who deny the Son also deny the Father. Those who accept the Son, also accept the Father.

This is more about the intertwined nature of the Son and the Father. In the Bible this relationship is too ill-defined to help anyone discern what it means. This book defines these concepts.

4:1. Do not blindly accept statements as true even if it comes from someone claiming to be spiritual. But, test what they say to learn if what they state represents reality.

Again, test every statement to ascertain whether it is true and therefore represents reality. If a statement is shown to correspond to reality, it is a true statement. Ask people to verify or prove their claims.

Discussion

Understanding reality helps make sense of the Bible. Let's listen in on a dialogue between a scientist answering questions posed by an individual who is skeptical about biblical scripture. We examine some biblical contradictions. We then correlate some religious concepts with our understanding of science.

Dialogue Between Explainer and Skeptic

It is not surprising (so called) "religious" people do not study the Bible in an unbiased manner. Let's establish some correlations between what we have learned and certain biblical statements. Surely, serious Christians cannot cry "unfair" if they are indicted by their own standard of truth—the Bible.

Consider a dialogue between an inquiring Skeptic and an Explainer. The inquiring skeptic poses questions. The Explainer answers using biblical verse. The purpose is twofold: One, to show the Bible does contain statements worth considering. Two, to counter the idea science is contrary to certain biblical statements.

Whether a statement is found in the Bible is beside the point. We are concerned with a statement's truth. If true, it corresponds to what reality tells us about itself. Someone could very well pose a statement and be unaware it is represented in scripture.

Maybe the Bible can support the seeking of truth as practiced in science. Science is the serious discipline that uses an exacting methodology to understand reality. This is contrary to many "religious" peoples' preconceptions. (Bible verses are again rephrased to more accurately represent reality.)

Bible: Beware of the teachers of the law who walk in long robes and like to attract public attention. They take advantage of people; then do long prayers. Luke 20:46–47.

Do not believe every spirit, but test their pronouncements to learn whether they represent reality. 1 John 4:1.

Let no one deceive you by any means. 2 Thessalonians 2:3.

Explainer: "Lesson: we should be skeptical of all teachers. This follows from being skeptical of asserting the truth of every statement we hear. The problem is not the teachers. The problem is the methodology employed to ascertain statement truth."

Skeptic: "Is the mind the same as (correlates with) the spirit?"

Bible: God's spirit is one's mind. Romans 8:9.

Explainer: "We know the mind can seek truth; it is not much useful for anything else. People who claim to be God's servants often say, 'We feel the spirit today, yea!' Do feelings and emotions search for truth? Has anyone ever seen a feeling or emotion do this? Or, are feelings and emotions obstacles to understanding Reality?"

Bible: The spirit seeks to understand everything. Only man's spirit really knows him. Only it can discern the nature of everything. Our mind came from (Adam-Father-Spirit) Christ. 1 Corinthians 2:10–11.

Ask (questions) and you will given (answers); seek and you will find. Matthew 7:7.

Skeptic: "Well, apparently the spirit functions like a mind. It could be that 'spirit' is an archaic word for mind. Why do people have trouble seeking truth?"

Explainer: "It is difficult to seek anything if you cannot properly define what you are seeking. The mind functions to cope with physical reality (in which the mind finds itself interacting through the brain and physical body). Reasoning itself does not put bread on the table. Consider: Perhaps one's mind does not function in itself, but is asleep. It only responds to the particulate (whatness) world of matter via deduction. Premises (for deduction) are given by experiencing the physical world as sense data. Let's examine this."

Explainer continues: "Perhaps sleep is only a mystery because it is not sufficiently understood. The body rests; the mind sleeps. Notice what happens. Go to bed and shut your eyes. You soon 'fall' asleep. What happened? When you are tired and sleepy, stimuli from the physical world are minimized. Your mind ceases responding to the distractions of the physical world—it barely functions at all. Could it be the reason you sleep is your mind (in itself) is not even functioning on its own? Yes, it appears to you (i.e., to your consciousness) that your mind is awake. But, if your mind were to do some simple reasoning, maybe it would realize the difference. Perhaps, people are not cognizant of the difference (between a mind functioning and not functioning on its own) because their minds do not function by themselves in this existence. It is difficult to know a difference unless one has experienced a difference."

Bible: Adam-Christ is the first born. He existed before all things. Colossians 1:15–17.

God put the first man into a deep sleep. Genesis 2:21.

Men are by nature opposed to the truth—their minds do not function to seek truth. 2 Timothy 3:8.

Skeptic: "You mean science, which tests everything to ascertain if statements (about them) are true, is the true religion?"

Explainer: "Science offers the methodology for critical testing."

Bible: Test and prove all things. 1 Thessalonians 5:21.

Skeptic: "Is the mind and body categorically distinct? If so, how can this be demonstrated?"

Bible: There is one body, and one spirit-mind. Ephesians 4:4.

Explainer: "Consider an individual who is brain-dead. His body continues to function with the aid of machinery, but there is no brain activity, and no hint of a mind. The same idea applies to someone in a coma. However, someone in a coma still has an EEG (electroencephalograph) reading and does not need special machinery to sustain his life functions."

Bible: A body without spirit is dead. James 2:26.

Explainer: "If a body can function without a mind, is there any reason a mind cannot function without a body? In this existence mental activity occurs in conjunction with the physical world. Does this mean the mind can function on its own, but in another time and place? Many things suggest this. For example, in diagrams conceptualizing General Relativity equations, there is a region labeled 'elsewhere.' We know we receive a body and a mind from our parents (this confers the two natures of man). Each individual is really two people in one: two aspects—the physical (human nature) and the mental (spiritual nature). Most people received both natures from their parents. This also how we explain immortality. The individual's mind pre-existed potentially in his or her parents."

Bible: Flesh is born from flesh and spirit is born from spirit. John 3:6.

Skeptic: "But did you not say the mind does not function on its own (in the physical world)?"

Bible: The kingdom of God is within you. Luke 17:21.

Explainer: "In this world you have a physical birth. You also receive a mind, but it does not function on its own. You will also have another birth, the birth of your mind awakening. If it awakens, therefore, functions on its own, it will function with respect to Reason itself (the whyness world). The whatness world is a lower level creation. The whyness world is associated with the creator (or reason for everything.)"

Bible: Unless one is born again, he cannot enter the kingdom of God. John 3:3.

Man worships and serves created things instead of the creator God. Romans 1:25.

Do not love the (physical) world or anything in it. If you do, the love of the Father is not in you. 1 John 2:15.

Skeptic: "Many people try to save themselves through religion. Must I save myself to be born again?"

Explainer: "No, you cannot save yourself."

Bible: Those who try to save their life will lose it; those who die will live again in another existence. Luke 17:33.

Explainer: "Did you, or anyone else, ask to be born into this world?"

Skeptic: "No."

Explainer: "That you find yourself in this existence is not your responsibility. Did you not receive your mind from your parents and they from their parents and so on? Do not these minds have an ancestral lineage to the first mind?"

Skeptic: "Well, yes."

Explainer: "Ah, then does not the responsibility for you being in this (physical) existence, rest upon who was originally responsible for you being here?"

Skeptic: "Well, yes. Is that the responsibility of Adam? Why him?"

Bible: The first Adam was created a physical being, but the last Adam is the life giving spiritual being. The physical comes first, then the spiritual. The physical cannot enter God's kingdom; what is mortal cannot have immortality. Physical death is not the end. A time will come when we will be given immortality. Mortality is wrapped in immortality. 1 Corinthians 15:45–47.

Sin entered the world through one man. Romans 5:12.

Adam was a figure of him that was to come. Romans 5:14.

Skeptic: "You mean, Adam, the name given to the first mind, is the Father of everyone who has lived? How can that be?"

Explainer: "Again, did not you receive your mind from your immediate physical parents? And did not they receive their minds from their physical parents? Someone (a mind-spirit) was the first being, it is difficult to speculate if the first mind-spirit resided in a one-celled animal or not."

Bible: The body houses the Holy Spirit. 1 Corinthians 6:19.

Skeptic: "Does that mean the difference between plants and animals, is that animals have a mind?"

Explainer: "Yes, Just as you obtained your physical body through your parents, and they suffered no physical difference, so it is most people receive their mind from parents, who suffer no difference of (identity) mind. Many minds develop from the one original mind. You, as a mind, have your origin in this first mind. And that mind has always existed—it is immortal. And if your mind is immortal—it must also have existed before you were physically born. This understanding explains immortality. Your mind existed in your parents prior to your physical birth."

Bible: We all began in the spirit. Galatians 3:3.

Christ is in the Father, and the Father is in him, just as he is in you. John 14:20.

Just as Jesus was resurrected from death, those who sleep in death will also be resurrected. 1 Corinthians 15:20.

The physical self is buried as a mortal being. The resurrected self will be immortal. 1 Corinthians 15:42.

Explainer: "Since Adam is the one responsible for us being here, then our eventual fate is also his responsibility."

Skeptic: "So you mean he is also the Messiah, who really did come to redeem for this situation in which we all find ourselves? So the Bible is about Adam (the mind) returning more than once. Then the First Coming was Adam, the mind—the father of all who exist—returning in another physical body?"

Explainer: "Yes, but it is not that simple."

Skeptic: "Why?"

Explainer: "Did he not say—"

Bible: Jesus said people knew neither him nor his Father. If they knew him, they would also know the Father. John 8:19.

Jesus said the Father is in him, and he is in the Father. John 10:38

The Father in Jesus does his own work. John 14:10.

Explainer: "You see, the Father, who was Adam, was incarnated again (reincarnated) as the First coming, perhaps also as the Buddha, and maybe others. An entire religion is built around the possibility of reincarnation—Hinduism."

Skeptic: "But why does the Bible seem to turn people against truth?"

Explainer: "The Bible does not turn anyone against truth. It is human nature and ignorance that turns man from seeking truth."

Bible: People are fooled by their own ideas and imagination. Romans 1:21.

Explainer: "The Bible, tells us about human nature."

Bible: Men's foolishness prevents them from knowing the truth. Romans 1:18.

Man changes the truth about God for a lie. Romans 1:25.

Explainer: "Question. Which comes first, the First Coming or the Second Coming?"

Skeptic: "Ah, the First coming?"

Explainer: "O.K. Are not the signs (patterns) practically complete for the advent of the Second Coming?"

Skeptic: "Looks like it."

Explainer: "There is no reason a greater Being could not direct a series of events we would recognize as a sign. Fulfillment of those events indicates the imminence of the Second Coming."

Skeptic: "But is not the antichrist supposed to come before the Second Coming?"

Explainer: "Well, consider this. If the signs are fulfilled for the coming of the real Messiah, and the antichrist was to come after the appearance of the signs (but before advent of the real Messiah) then would that not make a liar of God (Reality)?"

Skeptic: "Well, yes."

Explainer: "For Reality not to be a liar (it is not because it is the final arbiter of truth) the antichrist must have already come, right?"

Skeptic: "That makes sense. Well then who was it? Was it Hitler?"

Explainer: "No, the antichrist was the First Coming. And you understand, as the First Coming he comes as Prince (of darkness) and as the Second Coming he comes as King."

Skeptic: "How can that be? Does the Bible tell of the signs of the antichrist?"

Explainer: "Again, look at the Bible. It says—"

Bible: Let no one deceive you by any means. The time will come when the Great Deceiver will be revealed. The Great Deceiver is allied with Satan. He performs miracles to imply that he is God's servant.
2 Thessalonians 2:3–9.

Skeptic: "Does this mean, because the First Coming did not have sufficient knowledge and understanding, that he was the Messiah, but could not prove it and that is why he says to believe, until he comes a second time to make his case? Like most people in this existence, was his mind subfunctional (asleep) during his First Coming?"

Explainer: "Yes, the antichrist (because he could not yet prove his own predicament) used miracles instead of proof. If he was to prove his case (and he could not) no one of that time had sufficient knowledge to understand it. There is an extraordinary difference between yesteryear and what is known today.

What does the Bible say the First Coming did? He multiplied fish, multiplied bread, walked on water, raised the dead and cured people of physical ailments. Are not these miracles?"

Skeptic: "Why, yes he did perform some interesting things. These could be called miracles."

Explainer: "But Jesus' performances were 'physical' occurrences. We were warned not to place importance on the physical."

Bible: Do not believe anyone who says he is the Christ. You are forewarned. False Christs and false prophets will come and perform wonders to deceive even the very elect. Matthew 24:23–24.

Do not love the physical world. If you love the physical world, you will not love the Father in you. 1 John 2:15.

Explainer: "That someone multiplies fish, walks on water, only proves he can multiply fish and walk on water—nothing more. Construing 'physical meaning' in any way other than what is proven is foolish. The word 'miracle' typically denotes something not well understood. These events were called miracles because no one understands how they were done. Like anything else, once a miracle is explained, it is no longer considered miraculous. Miracles only prove people make unwarranted deductions based on what they do not understand. Anyone who performs miracles and makes claims, but cannot prove them—is not behaving like the Messiah. Then, does not this answer the question about who is or was the antichrist?"

Skeptic: "Well, yes, I guess it does. But, why did it happen this way? If these religions are not what they appear, why do they exist?"

Explainer: "Aside from cosmic structure, especially in terms of the Principle of Complementarity (which dictates things are created in pairs of opposites) consider the following. Let's say, hypothetically, there was no First Coming, but there really is a Messiah. In the extreme—what would happen if there never were any religious concepts?"

Skeptic: "O.K."

Explainer: "Let's say the signs of the Second Coming are fulfilled (but obviously not recognized) and the Messiah again reincarnates. He shows up in modern times in another physical body and announces he is the Messiah—the first son (remember, he was once Adam) of God. What would people say? They would ask, 'What is a Messiah, and what is this called God?' The Bible and the other religious books contain many unverified statements that give significance for the proof (that God is) which is offered by the Messiah at his Second Coming. There was no serious science during the time of the First Coming to test verification and proofs if there were any. This is very simple to understand. People of Jesus' time had no serious conception of verification and proof. Using what we know today, we can look back and say that people can make all kinds of claims on their own, but proof is predicated on Reality (and ultimately affirmed by God)."

Bible: Jesus says to not accept anything as proof merely because he says so. John 5:31.

Skeptic: "But the people of religion say, 'Surely we do good work,' If they did not, it becomes painfully difficult to explain why they would do what they knew to be bad. What does the Bible say about people doing good?"

Bible: Everyone is affected by sin. No one can do what is right by God's standard. No one seeks to understand God. Everyone has turned away from God. Without exception no one can do what is good.   Romans 3:9–12.

My inner being (the mind-spirit) rejoices in the law (not the Torah, but the laws of nature) of God. But, there is different law that applies to my body—a law that is against the law of my mind (that it functions via wanting—desire). I only can serve God's law by using my mind (spirit)). The law of sin (desire) dominates human nature (feelings and emotions). Romans 7:14–25.

One can overcome the desires of human nature if the spirit guides him. Human nature is opposed to spirit nature. They are enemies. Galatians 5:16.

It is impossible for man to know God by means of his own wisdom. 1 Corinthians 1:21.

People are naturally deluded so they will believe a lie. This results from inheriting sin. 2 Thessalonians 2:11.

Skeptic: "If these statements are true, it sounds as though man is by nature anti-God; it makes men look like devils."

Bible: Your father is the devil and it is your nature to be like him. The devil does not tell the truth; it is against his nature to be truthful. When he lies, he is just being his natural self. He does not realize he is a liar. John 8:44.

Explainer: "Should we be surprised man has two natures? Scientists know that practically everything comes in complementary pairs: matter-antimatter, day and night, male and female, and heaven and hell. Even the question about color reveals two answers for the same phenomenon, e.g., yellow-blue. Behavior also has two natures: egocentric and realistic. Either one's attention is to things in the 'rest of reality' or it is focused on one's self. One's emotions tend to keep his attention fixed on himself. Why? To answer this question calls for an in-depth understanding of behavior. Simply, a mind not functioning on it own functions deductively. Well that is appropriate; we live here in this 'whatness' physical world. However, the mind, if functioning on its own, infers inductively—toward higher levels of reason—toward, perhaps, the reason for everything. Have you ever noticed when someone asks you a 'why' question that you usually have trouble answering. And, if you are asked to further explain (why)—you quickly learn that you cannot do it. And, by now you are familiar with the 'split-brain experiments' dating from the early 1950's in the work of Roger Sperry et al. Notice that the left brain functions in a logical-deductive manner. It works well in dealing with the physical whatness world. The right brain functions in recognizing faces—patterns. It functions intuitively; that is, it does not function very well. If it were to awaken, it would function inductively—placing evidential facts into patterns denoting higher generalities. Potentially, the right brain has a nature, which seeks higher-level understanding. The right brain is the seat of the 'Father in you'—your inner being. Compare this with memorizing and noncritical acceptance of statements characteristic of the left brain."

Skeptic: "If people in the physical world are devils, does that mean they are in hell?"

Bible: Those who try to save their life will lose it; those who die will live again in another existence. Jesus says during that night, there will be two men in one bed: one will be taken, the other left. Two women will be grinding grain: one will be taken; the other left. Two men will be in the field: one will be taken; the other left. Luke 17:34–37.

Explainer: "This is about the coming of the kingdom. Remember that each person has two natures—the physical (human nature and its deductive mental function) and the mental (spiritual nature and its inductive function). The two men sleeping together represent the two aspects of one's nature. These statements are not about two men (women) with one nature—but it is about one man (woman) with two natures. Examine the quote. One is taken (the spirit); one is left (the physical body). Notice. There are only two places mentioned. There is that place where the spirit goes, and there is where the physical body stays. Use a little logic here. We know this place is not the kingdom of God (heaven) therefore, this (physical) place is Hell."

Skeptic: "Those are interesting points, but let's get back to a question bothering me. Do not these people of religion, for example, Christians, worship God?

Bible: All men are under the power of sin. No one is righteous; no one seeks for God. All people have turned away from God. Romans 3:9–12.

Explainer: "A belief masquerades as a fact. When someone believes a statement to be true, it is an act that they really know it to be true. But, look at this world, it is a big act. People in religion act as though they know God. This is a serious sickness—stemming from sin (desire, here—the desire to do right). Surely, you understand one only desires to do right because he cannot do what is right. If one could do right, there would be no reason for him to want (desire) to do it. And because one cannot do it, he merely acts as though he can do it. He becomes self-righteous. A premium is placed on how well one acts (desiring to behave righteously) in this world. The best actors earn much money for acting in movies. The best actors involved in religion collect much money from fools. This is observed every day. Religious leaders have deceived themselves. They are very insecure and need approval of their fellow man. They hear their 'followers' say things like, 'Oh, the minister is such a wonderful person, and he is surely a man of God.' And ministers, the more they convince their 'flock' of what they say, the more they become convinced they are doing God's work—contrary to what the Bible clearly states. The Bible states no man does what is right. All have turned from God."

Skeptic: "Although by nature, people (in this existence) are incapable of worshiping God, are they not capable of worshiping Christ—after all, they call themselves Christians? And what about these individuals who claim to be enlightened? Are they really not enlightened?"

Explainer: "If they were Christ-like (the real Christ) they would have to prove what (statements) they assert to be true, just as Adam's descendant would have to do for him to be the Messiah. Only to the extent one verifies (or explains without contradiction) is he Christ-like. That someone can worship God through one's own beliefs is foolishness. One is Christ-like only when he tells the truth—and therefore that he can prove what he states is true. Christ represents Reality as (in the form of) knowledge—truth. Those who say they know not (for example, whether there is a God or whether Jesus is the Son of God) tell the truth and in that respect are like the real Christ (if such a person exists). For so called 'self-professed' enlightened people, if they cannot prove what they claim, they are not enlightened. Like most insecure people, they crave attention by making such claims. They fancy themselves as leaders of men."

Bible: Jesus said he was the way, the truth, and the life. No one will meet the Father except through him. John 14:6.

Men's thoughts are foolishness. Romans 1:21.

Do not be deceived by foolish words. Ephesians 5:6.

At the Second Coming the Father-Spirit will prove that people do not understand the nature of sin, righteousness, and the judgment. John 16:8.

Skeptic: "Let's see if I understand this correctly. Even devils and hell are a part of what is God (Reality). Is that correct?"

Explainer: "For how could it be otherwise, if God is All (of Reality)?"

Bible: God is close to us, within him we live and move and have our beingness. Acts 17:28.

We are not independent functioning beings; we are a functional part of God. 1 Corinthians 6:19.

Skeptic: "It seems as though there is a big plan, by which God (reality) 'resets' the world."

Explainer: "Yes, you can see entropy predominates negentropy in this existence. Societies disintegrate, relationships fall apart, individuals exhibit self-destructive tendencies, and physical bodies break down and eventually die. The Principle of Complementarity:

That there are pairs of opposites, e.g., matter-antimatter, night-day, male-female, a color and its complementary color, plants breathe in carbon dioxide and breathe out oxygen while animals do the opposite. Here, people are told to "get their act together." There may be another place and time where one "gets it together by getting rid of the act." There is nothing inherently "bad" about any of this—it is the way of nature. Opposites also include the complementarity of Christ and antichrist.

(Complementarity)—suggests since we observe (according to Bible scripture) that man is foolish in this mortal existence, there just could be another complementary existence of the awakened (wise) mind—an immortal being. There could be a plan where God (having the characteristic of negentropy) pulls all these things in creation into another phase—another half of a time cycle as mentioned in Hinduism. And we know a proper understanding of the CPT theorem (charge-parity-time) in physics practically calls for a temporal phase where time is inverted. The CP aspect of this theorem has been verified and understood. The T (temporal aspect) of this symmetry principle has not been properly understood."

Skeptic: "This sounds as if God, eventually, counters the 'breaking up' effects of entropy (which we observe in creation) by 'bringing together' (negentropy) the separated parts."

Bible: At the right time, God will bring everything together under Christ. Ephesians 1:10.

A time to break down, a time to build up. Ecclesiastes 3:3.

Skeptic: "I am now beginning to understand a little."

Explainer: "You should understand. You have been skeptical enough about your preconceptions, and you have asked questions. You have sought truth by asking questions. The Bible supports skepticism. After all, if one is not skeptical, he erases the only real reason to seek truth. The Bible says to seek truth, and it will set you free." John 8:32.

Skeptic: "But, free from what?"

Explainer: "Free from ignorance."

A glass is as empty as it is not full. A person is as ignorant (about reality) as he is not knowledgeable. If a person sets aside his preconceptions, maybe he can attain an open enough mind to ask questions about everything. That does not mean he will acquire an in-depth understanding of reality. But it betters his chances.

Statements must be tested to ascertain whether they do represent reality. Statements representing reality may be considered suppositionally true. This idea shows the whole purpose of science. Let's apply basic logic to biblical verse to test their actual truth.

Biblical Contradictions

The Bible states God is true (John 3:33 and Romans 3:4). Yet, when Pilate asked: "What is truth?" He did not get an answer. Neither truth nor God is defined in the Bible.

Reading, praying, and reciting verses from the Bible do not lead to knowing or understanding God. It cannot be said one knows anything unless he has had contact with it (with what a statement represents in reality). Moreover, the difference between knowing and understanding suggests knowledge without understanding is practically useless and meaningless.

A person can memorize information and not know the reality it represents. One can recite statements and not logically understand. One knows his automobile. Yet, if he does not understand how it works (functions) when there is a problem, he is helpless. Merely knowing something (by acquaintance) is not sufficiently beneficial. Understanding is needed to solve problems.

Is the Bible the word of God? To "truthfully" (and therefore ultimately) answer this question, is to understand God. We cannot understand God until we understand the structure of Reality.

The Bible does not explain God; it does not prove "God is." Understanding God is developed in the next four chapters. Meanwhile, we do have a reasonable working definition of truth. Applying this definition to biblical verses is to test their truth. If there is one lie in the Bible, then not every verse is true. If so, one must be skeptical of any statement found in the Bible.

Our "working" definition of truth is "that information which corresponds to and therefore represents Reality." We are mindful such a definition should eventually lead us to an understanding of God (if there is such a Being). We seek truth just as the Bible recommends. Test all things. (1 Thessalonians 5:21).

We use the Law of Contradiction to discern biblical truth. Opposing statements (about anything) cannot both be true at any given time. If we find contradictory statements in the Bible, it means one of two things.

> First, there is a possibility there are no actual contradictions. The paired statements only appear contradictory. Apparent contradictions are attributed to misconstruing meaning, viz, that the statements are not really in opposition. This happens when one of the statements is not an explicit negation of the other.

If a different meaning is offered for one of these statements, it too must be questioned and explained. The context of many such conjectures is open to challenge. We ask if these statements mean the same? Then we must also ask "why, or why not?"

To assert a statement does not mean what it states is perhaps to say it is a metaphor. Saying some biblical statements are metaphors presents its own set of problems. To say a meaning is different from what is stated can only mean that some biblical verses themselves are truthfully unreliable. Metaphorical meaning falls in a "gray" area between truth and falseness. Metaphors are no substitute for factual statements.

Most statements about reality pertain to a specific level of structure. To understand, we "categorize" statements by "gauging" their generality. Occasionally, we miscategorize a statement and it becomes associated with the wrong hierarchical level. This is rare. When this does happen, errors in reasoning do not go undiscovered.

Categorical misplacement is usually limited to "some" and "all" statements. A statement's truth or falsity, considered at one hierarchical level, may be otherwise when considered at another level. An "all-statement" either pertains to the whole structure or to everything composing the whole. "Some-statements" represent parts.

A more common problem arises when someone "believes" the context (level of structure) represented is otherwise, than it actually is. Hypothetical concepts are used to test not only a statement's viability, but also its level of generality. There is little room for "believing" in seeking truth and understanding. If a statement means something other than what it explicitly states, a satisfactory reason must explain why.

> The second possibility is that contradictory statements found in the Bible are actual contradictions. If there are contradictions, the Bible contains lies. If God is truth, and the Bible is wholly inspired of God, there should not be any lies found in the Bible. There should be no contradictions.

The Bible states (2 Timothy 3:16) All scripture is inspired by God and can be used to understand truth. We do not expect to find lies in a book that is inerrantly inspired by God. If we do, then we must readjust our "preconceptions" about the meaning or purpose of the Bible. The very truth of 2 Timothy 3:16 will be tested. 2 Timothy 3:16 is true for the whole Bible if "every" scriptural verse is true.

Defenders of "biblical truth" say those who question its scriptural integrity change the meaning of what is stated. They do not realize that they construe (interpret) scripture to support their particular beliefs. They twist actual meaning of biblical verse to suit their own needs. Such needs, through belief, contribute to emotional stability. Intellectual security is only obtained through knowing and understanding. The latter statement reflects the methodology used in this book.

If sufficient contradictory statements are found in the Bible, one must question the source of biblical inspiration. Did God wholly inspire it, or did some other intelligence infiltrate what man had considered solely God's word? We are cautiously mindful (Hebrews 6:18) God cannot lie. Therefore, if we discern a lie in the Bible, it is not wholly of God. We have often heard it said Satan throws in some truth to deceive people with some very critically placed lies.

If God is the source of biblical inspiration, why did not "he" reveal himself in the Bible? The Bible does not prove "God is." That God did not explain itself is unexplainable unless God is not truth or some other intelligence (other than

God) is responsible for inspiring men to write the Bible. That other intelligence is called Lucifer and its transformation is called Satan.

Our treatment of contradictories includes some "apparent" contradictions. There are endless ways of explaining away contradictions. Some are even devious. We do not say "apparent" contradictions cannot be explained away. Some may be valid; others may not be valid. Yet, if a biblical verse means otherwise, than what it explicitly states, the idea the Bible is not wholly of God looms as a likely possibility.

The Bible cannot be understood literally if some biblical verses are more metaphorical than factual. If some biblical verses are merely symbolic, they can mean almost anything. Saying a verse does not mean what it states, evades the issue. Saying a verse means something different from what it explicitly states is to say it is a lie. Which is it? Is a given statement (about anything) true or is it false? It cannot be both (at the same time and at any given time).

Let's use our definition of truth to gauge the significance of some specific biblical verses. Do the following chosen passages represent something in reality? We use our understanding of what has been presented in this book to examine some biblical contradictions. Let's begin sifting some biblical statements through the Law of Contradiction.

- (2 Timothy 3:16) God inspires all scripture. ~ This means there is no scripture that is not inspired by God. This contradicts:

    (2 Corinthians 11:17) Paul said, he was then not speaking as inspired by God. ~ This self-professed scripture states God did not inspire it.

    Therefore, not all scripture is inspired by God. Or, God does not inspire some scripture.

- (Acts 2:21) The time will come when those who call on the name of the Lord will be saved and (John 3:16) everyone who believes in the Lord is rewarded with eternal life and (Luke 13:3, 5) Jesus said unless you repent, you will die: contradicts:

    (Romans 2:6) God judges every person according to his deeds and (2 Corinthians 5:10) We must all appear before the judgment of Christ and every one will be rewarded according to his deeds and (James 2:24) Man is not reconciled with God by faith alone, but also by his deeds.

    Merely calling out and saying "God, I want to be saved," and it will be done, is vastly different from being judged according to one's deeds. Talk is easy; people can say anything, including "I repent." A lifetime of unselfish behavior is not so easily dismissed.

    Consider a "bad" individual who, during his life, among lesser offenses (lying, cheating, and stealing) raped, tortured, and murdered people. In old age and on his death bed, he calls out to the Lord in repentance. Is he saved? Acts 2:21 states he is saved merely by speaking a few words such as "I repent." There is no biblical requirement of sincerity. If there were, it would be defined without serious contradiction. Unjustly or not, the "bad" person is saved despite his lifelong disregard of others. Most people would say such an individual is depraved and the very embodiment of evil.

Consider a "good" citizen. He lives his whole life giving of himself: working hard for the good of everyone. Though aware of the historic individual called Jesus, he never could honestly overcome doubting the verity of a statement such as: "Those who call on the name of the Lord will be saved." He considers such a statement could be a trick. He asks "how could honestly not accepting the truth of a statement on faith, and therefore without verification and proof, lead to the unjust reward of eternal damnation?" He studied the Bible, and could not discern how such a statement could be true if God is just. He considers the fate of bad people. How could the asking of forgiveness have precedence over bad deeds?

Now, who is honest with reality here? Who is honest with the truth? The truth is that neither of these two individuals could verify or prove the truth of Acts 2:21. Those who believe they worship the Lord will be saved. Only one of these individuals is realistically honest.

The "bad" individual is not honest with reality. He does not know whether such an unverified, unproved statement (as Acts 2:21) is true or not. But by merely accepting and deciding "on his own" the truth of this statement, he is rewarded with salvation and entrance to heaven! His "self" decided what is truth or right—this is "self-righteousness." Does he gain entrance to heaven by being self-righteous?

The "good" individual says he could not, in all honesty and truth, say whether a statement such as Acts 2:21 is true or not. How could he know? He did not have sufficient wisdom or knowledge to determine its truth. He did not really know the truth of that statement.

Further, if Acts 2:21 were true it would mean the "inspired" source of the statement, and perhaps the entire Bible, is not just! How could such injustice be explained?

Similarly, (Galatians 2:16) Man is only reconciled with God through faith in Christ: contradicts:

(Matthew 16:27) Jesus said everyone would be rewarded according to his deeds.

- (John 1:18) God has never been seen and (1 Timothy 6:16) God will never be seen and (Colossians 1:15) God is invisible. These contradict:

(Exodus 33:11) The Lord spoke to Moses face to face, just as a man speaks with friends, and (Exodus 33:23) the Lord said if he passes by that he would take his hands away and his backside might be seen, but his face would not be seen, and (Genesis 32:30) Jacob said he saw God face to face, and (Isaiah 6:1) Isaiah says he saw the Lord in a vision.

Either God is ever invisible, or it is possible at some time that he is visible. He cannot be both ever invisible and also visible at selected times. One of these statements is true; the other is clearly false.

1 Timothy 6:16 clearly states there is no time when God can be seen. If this is true, then anyone who claims to have seen God is lying (not telling truth). If the latter statement is the true, the Bible contains lies.

- (Matthew 16:28, Mark 9:1, Luke 9:27) Jesus said some of those who saw him would not die until after they have seen the Son come with the Kingdom of God. ~ When the kingdom comes, people will see it. Later, Jesus visits Jerusalem. He reads some verses from Isaiah (Isaiah 61:1–2) stating that the time would come when it would be announced that: the Lord has come to save the people. After reading the passages from Isaiah, Jesus says (Luke 4:21) that: these scriptural verses had now been realized, just as it was prophesied. ~ This states the kingdom of God had already come. (Luke 11:20) Jesus says because he has cast out devils that it proves the kingdom of God had already come. These statements contradict:

  (Luke 17:21–20) The kingdom of God cannot be seen. People cannot observe the kingdom of God because it is within them. (Romans 8:23) People will wait for God to redeem them and free their spirit. ~ Paul writes this a few generations after Jesus' death. These statements indicate the Kingdom of God had not yet begun even after Jesus's death. To state the Bible is inerrant seems to be an errant assertion!

  If the Bible is inerrant, and therefore totally inspired by God, it means it contains no lies. Clearly, if the kingdom of God had not occurred after Jesus's death, then Jesus lied about it when he said it had already come. Do people see the kingdom of God come, or do they not see it come?

- (John 3:17) Jesus said God did not send his Son to judge the world, but to save it, contradicts:

  (John 9:39) Jesus said he came to judge this world and (John 5:22) he states again that he is to judge this world.

  If Jesus judges, then it is a lie he does not judge! And, if he does not judge, it is a lie he judges. One of these statements is not true; one is a lie. Which one?

- (1 Corinthians 15:52) The dead will be become immortal and (Luke 20:37) The dead will be resurrected, contradicts:

  (Ecclesiastes 9:5) The dead are deceased. They will not receive a reward; their memory is forgotten and (Isaiah 26:14) The dead are deceased and will not live again.

- (Matthew 23:2–3) Jesus spoke to a crowd and said they must obey the Law. Here the Law refers to the Torah, the Old Testament. (Romans 2:13) Those who obey the law will be just before God. And (Luke 16:17) Jesus says it is easier for heaven and earth to disappear than for even one Law to fail.

  Did Jesus himself obey the Law? If not, then was he not a hypocrite?

We learn (Matthew 12:1–2) that on one Sabbath Jesus was walking with his disciples through the grain fields. They picked some wheat and ate some grain. Some Pharisees witnessed this and told them that it was unlawful for them to do this on the Sabbath.

There are other examples of Jesus and his disciples breaking the Law. Paul tries to cover up such inconsistencies. He says (Romans 3:20) No man is reconciled with God by obeying the Law. Also (Galatians 2:19–20) Obeying the Law is not sufficient to gain eternal life. Rather, one is reconciled with God only through faith in Christ.

- (Proverbs 4:7) Obtaining wisdom and understanding is foremost important, is contrary to:

   (1 Corinthians 1:19–21) The wisdom of the wise will be destroyed. The wisdom of this world is foolishness. It is impossible for man to know God by his own wisdom.

   Proverbs 4:7 encourages wisdom while 1 Corinthians 1:19–21 discourages it.

- (James 5:11) The Lord is merciful and compassionate, contradicts:

   (Jeremiah 13:14) Compassion and mercy will not stop God from destroying them.

- (James 1:13) God tests no man, contradicts:

   (Genesis 22:1) God tested Abraham, and Moses said (Exodus 20:20) God came to test them.

- (John 3:13) No one has ever ascended to heaven except the Son of Man, contradicts:

   (2 Kings 2:11) Elijah ascended to heaven.

- (1 Corinthians 4:5) Do not judge anything until the Lord comes, contradicts:

   (1 Corinthians 2:15) A spiritual man is able to judge everything.

- (1 Samuel 15:29) The Lord does not change. (Malachi 3:6) The Lord says he does not change. (James 1:17) The Father does not change, contradicts:

   (Genesis 6:6) The Lord was sorry he ever made man on earth and (Jonah 3:10) When God saw the works of the Ninevehians, he changed his mind

   Also: God said to David (Psalms 89:3, 20, 33–37) that he (God) would not break his covenant with the chosen people nor alter any promise he had made, contradicts:

   (Luke 22:20) Jesus inaugurated a new covenant and (2 Corinthians 3:6) gave them the ability to obey it by using the spirit.

   Does God change his mind or does he not change his mind?

- (Acts 10:36) God sent a message through Jesus to the Israelites proclaiming there would be peace, and (John 14:27) Jesus says he gives them peace, contradicts:

  (Matthew 10:34) Jesus says he did not come to bring peace. He brought a sword. ~ What Jesus says will lead to divisiveness and conflict.

- (John 5:31) Jesus says to not accept what he states as real proof, contradicts:

  (John 8:13–14) The Pharisees said to Jesus that his testifying on his own behalf proves nothing. Jesus replies that even if he does testify on his own behalf that what he says is true.

- (Matthew 11:14) John (the Baptist) is Elijah, contradicts:

  (John 1:21) John is not Elijah.

- (1 Peter 2:13–15) Submit yourself to every human authority, this is God's will, contradicts:

  (Acts 5:29) Obey God, not man.

  Obey man over God is opposite to obeying God over man.

- (Acts 2:22) Jesus (the supposed Christ) was a man directed by God to perform miracles, wonders, and signs to demonstrate he represents God and (John 4:48) he adds that no one would believe what he says unless he does wondrous things.

  (If these acts characterize the Christ, they should not also characterize the antichrist. We quickly learn these acts do characterize the antichrist. These characteristics describe Jesus. These criteria suggest Jesus could be the antichrist.

  (2 Thessalonians 2:3–4, 6–7, 9–12) Do not be deceived by any means. The antichrist will come. He is against everything holy. He will place himself above everything that is worshiped. He will sit in God's temple and claim to be God. (Jesus did go to the temple; he did claim to be God.) The antichrist will have the power of Satan and perform extraordinary miracles and give false signs and perform deceptive wonders. He will use every possible trick to deceive those who will perish (all physical bodies and their deductive functioning of mind). The time will come when the antichrist and his charade will be explicitly revealed. People will then learn about the nature of deception.

  The above criteria cannot be used to distinguish the Christ from the antichrist unless the real Christ does not perform "physical" miracles and wonders. In that case, Jesus was not acting as the Christ. This does not mean the real Christ (Father-Spirit) was not in Jesus—he was in Jesus, but that the Father was not functional while in Jesus. (John 14:10, that the Father functions independently.) Our only recourse is to understand the difference between what is true and what is false: to understand the difference between trueness and falseness.

- (Ecclesiastes 1:4) Generations come and go, but the earth does not change and (Psalms 104:5) The earth was made and will not be changed, contradicts:

    (Matthew 5:18, 2 Peter 3:10) The Lord will come and the material world will vanish. Celestial bodies will be consumed and the earth with everything on it will incinerate.

- (Matthew 21:22) Believers will receive whatever they want through prayer. ~ This means "believing" is a sufficient condition to receive whatever is asked, contradicts:

    (2 Thessalonians 2:10–11) The antichrist will use every possible trick to deceive those who hate the truth. God uses the power of delusion so that people will believe what is false. (This means believing is not sufficient to receive whatever is asked.)

    Similar verses. (Matthew 7:7–8) Jesus said those who ask, will receive; those who seek, will find; knock, and the door will open. Everyone who asks will receive, and they who seek will find, contradicts:

    (John 7:34) Jesus said, said those who seek him will not find him.

    We also read (Romans 5:12) Everyone sins. (John 9:31) God does not hear sinners. (Romans 3:10–12) No one is righteous, not even one. No one understands. No one seeks God. Everyone has turned from God. No one does what is good.

- (Luke 3:6) Everyone will see God's salvation, contradicts:

    (Matthew 7:21) Not everyone will enter the kingdom of heaven. Also (Matthew 19:24, 26) It is easier for a camel to go through the eye of the needle than for a rich man to enter the kingdom of God. This is impossible for man. (A camel cannot pass through the eye of the typical needle.) And (Romans 2:6) God will reward everyone according to his deeds and (Psalms 100:5) The Lord rewards everyone according to his deeds. ~ These verses mean some people will not be rewarded with salvation: not everyone will see God's salvation.

There are many passages, which by today's scientific knowledge are just plain erroneous. Consider the following:

    (Psalm 75:3 and Job 9:6 and 38:6) The earth has pillars, which support it (like a table supported by legs). ~ Pillars are not a metaphor for gravitation.

    (Chronicles 16:30, Psalms 93:1 and 96:10) The earth is stable; it cannot be moved. ~ Today we know it is in constant motion through space.

    (Matthew 4:8) The devil took Jesus to a high mountain from where he could instantly see all the kingdoms of the world. ~ There is no mountain high enough from which he could see the opposing side of the earth.

There are many Biblical verses stating different quantities or different lengths of time for the same situation or event. (1 Kings 4:26) states Solomon had forty thousand stalls for his chariot horses, contradicts (2 Chronicles 9:25) Solomon had four thousand stalls for his chariots and horses. Jesus said (Matthew 12:40) he would be in the tomb for three days and three nights (hence, more than three days). Yet, we also find (Mark 10:34) stating Jesus would be raised to life on the third day. Is it three days, or more than three days?

Assume all scripture is inspired of God. If we say one of the writers made a mistake, then he was "inspired by God" to not tell the truth. Yet, supposedly God does not inspire that which is not truth (lies).

Accounts of Jesus' genealogy differ. (See Matthew 1:1–17 and Luke 3:23–38.) How can anyone have an inconsistent ancestral lineage? There is no end to covering up a lie. If the whole Bible is not inspired by God, then how are we to know which biblical verses are inspired by God from those inspired by man's own thoughts? As we are well aware that (Matthew 16:23) men's thoughts are Satan's thoughts. ~ Perhaps God inspires some scripture, but Satan inspires others.

There are so many contradictory statements in the Bible it is a wonder every reasoning person who studies the Bible does not question its truthfulness. Scriptural discrepancy is practically endless. Contradictions are easily explained if most biblical verses are merely inspired by man's own foolishness.

The Old Testament commands people to adhere to the Law (the rules of living set down in the Torah). Yet the New Testament sets aside that very important command. Although the New Testament states everyone must obey the Law of the Old Testament, it then turns around and states faith is more important than obeying the Law. How can these discrepancies be explained?

The answer is simple. The Bible is not the whole truth. Otherwise, there would be no big questions it does not answer. The big questions are not answered. They are not explained; not verified; not proven. This deficiency has left the door open for numerous interpretations of biblical verse. Differing sects developed because the Bible does not prove itself.

Faith is the backbone of most religions. Faith is only countered by truth. Having faith a given statement is true does not confirm its truth. Truth is only obtained by looking to Reality. Does the reality (of the situation) correspond to the given statement? If not, then the given statement is not true.

When we know the truth, biblical statements will not want for faith. A verified statement is much better than accepting its truth on faith. It is not for man to decide by his own mind what is true. Only Reality demonstrates truth. We can surmise that the real Christ must be defined as one who knows the truth about Reality and God. Therefore, to be "Christ-like" is to be a serious truth seeker. To follow Christ is to seek truth. One seeks truth by first denying himself (Matthew 16:24, Mark 8:30, Luke 9:23). The arduous task is cajoling Reality to surrender its secrets. An honest God will not deny an honest man.

The Bible does not explain itself. An honest person does not know whether biblical statements are true or not. Until we fully understand the nature of reality, we cannot discern the truth of significant biblical concepts.

The source of the Bible must be identified. That will be a clue to its true purpose. What role does the Bible play in the scheme of things? Because the Bible does not seriously explain anything, it must be explained.

Religion—Science Correlations

The Bible does not explain science. Maybe science can explain the Bible. Many statements found in the books of the major religions have a correlate in science.

If there is any validity to religious concepts it is because some aspect of reality supports that contention. Science-religion correlates are easily recognized. Lay out two highway maps: each in a different language. Where two maps correlate, the same area is represented. Our one and only common reality is representable in different languages. The same reality is also representable by different conceptual schemes. Religious and scientific concepts should correspond when they both refer to the same thing in reality.

Establishing religious-scientific correlations should validate some religious concepts. All statements are testable. Testing is the providence of science. Let's review some religious-scientific correlations.

Mind—Spirit—Heaven; Body—Materiality—Hell

The kingdom has been prepared since the beginning of creation (Matthew 25:34). It was predetermined that Jesus would be filled with the Holy Spirit (Luke 1:15).

Matthew uses the term kingdom of heaven instead of kingdom of God. Jesus says the kingdom of God had already come (Luke 11:20). The kingdom of God is within you (Luke 17:21). The kingdom is not of this world (John 18:36). Jesus says he lives in the Father, and the Father lives in him (John 14:10). Likewise, the Father is in you.

Heaven is within you. It is the mind. Your mind came from your parents and their parents: unless it came via reincarnation. The original mind was called Adam. Adam comes again in the body of Jesus. Everyone has the mind of Jesus because every mind is traceable to the first mind (Adam). The mind is the spirit.

Jesus states that there will be two men in one bed; one will be taken, the other left. Two women will be grinding grain; one will be taken, the other left. Two men will be in the field; one will be taken, the other left (Luke 17:35, 36).

The mind (spirit) is taken away. It is of heaven (which is in you). The (physical) body is left behind. There are not two "places" (heaven and hell) from the physical world where the individual goes. There is just one—heaven. Hell is where the body is "left behind." The physical world is Hell.

All physical materiality, including the human body, is Hell. Tell anyone who contests this truth to ask someone who has been: burned by fire, maimed in military action, injured in a serious accident. All materiality is of Hell. All minds are of heaven.

Birth, Rebirth and Reincarnation

Consider physical birth. You are first, physically conceived. You are physically born when you are conceived and living in the womb. But, to be fully born into the world you must be delivered from (detached from) the security of the womb. After delivery it may be a while before you realize you are physically an independently functioning being. But this is your first birth—physical. The mind functions deductively while in the physical world. It functions at the bequest of the physical world. It "deduces" inferences from sensory-data input from the physical world. Without physical "input," the mind sleeps.

> Note: when you retire from the chores of the physical world and begin to rest (the body rests) stimuli from the outside world are minimized and your mind (deductively) ceases to function (it sleeps). Hence the mind only functions at the bequest of the physical world.

The second birth is the awakening of the mind. It begins functioning independently of the physical world.

Like physical conception, when the mind begins reasoning inductively—it is born. But that does not mean it is fully born (or functioning independently) in reality. The mind must become secure in itself to be fully born. It must somehow free itself security-wise from its attachments, viz, its security blankets (just like physical birth is finally one of physical detachment from the security of the womb).

All present minds, and all that have existed, are derived from the original mind (called Adam). Just as physical bodies retain their identity (from their parents) so it is that minds retain a separate identity.

There are over six billion people physically living on earth at the beginning of the second millennium. The first billion was reached during the nineteenth century. At the time of the First Coming it is estimated there were about 225 million people. Let's extrapolate further back in time. Eventually, we come to a time when there are only two individuals (minds) existing in some kind of animal life form.

Minds are immortal and survive physical death. It is that simple. Flesh is born from flesh and mind is born from mind. Ordinarily, each person acquires his mind from parents just as each person acquires his body. However, if minds are immortal, instead of obtaining one's mind from immediate parents, an individual might have acquired it via reincarnation.

There are two ways to understand "born again." First, that the mind is the next birth after the physical. Therefore, that the mind at least existed in one's parents. Second, that one's mind might have existed prior to the birth of one's parents. That it "reincarnated" (from an earlier existence). Whatever its immediate origin, when it awakens to function on its own, we say the individual has a second birth: he is "born again."

If one's mind is reincarnated, the individual may have a "pre-existent" identity—one of which he may not be aware. One's identity is related to memories. An individual may take on an identity and not remember specific incidences that contribute to his general sense of identity. Therefore, someone can acquire a vague sense of identity and not recall those "acts" in a pre-existent life that are usually associated with gender. Gender roles usually define identity.

Interesting possibilities arise. What might happen if a mind that previously existed has a primarily feminine identity but reincarnates into a male physical body? Let's consider this possibility from the physical male's perspective. Our male physical being, having a "pre-existent" feminine-mental identity, is "naturally" attracted to "her" opposite sex. Being caught in the physical world, she does not distinguish the opposite sex by mental identity. She associates the sexuality of others with their physical identity. A female mental identity in a male physical body is naturally attracted to a male physical body.

A strong male mental identity finding himself in a female physical body naturally seeks her sexual opposite. Because a person does not distinguish the difference between a partner's physical and mental identity, a male mental identity in a female physical body seeks a physical female for a sexual partner. That their mental and physical identities do not correspond, is not their fault.

Let's describe a symbolic narrative taken from Genesis (in the Old Testament). It is about the first beings—Adam and Eve. Scientists suppose the first animal life forms evolved in water. In Genesis we read that a great mind (spirit) blows over the sea. The first mind is Adam. Adam (his mind) is fully awake and has the knowledge of God (Reality). Being a mind, he is an immortal being. There is no one else; he alone exists.

The mind, which is Adam, is manifested into a physical-life form. Adam acquires a physical identity.

God places him into the Garden of Eden and tells him not to eat the fruit of the tree in the center of the garden. That tree represents knowledge of good and evil. He is told that if he eats its fruit his eyes will open (his mind will only function by way of the physical world) and he will experience physical death. Adam is still alone so God puts him into a deep sleep (he loses his [inductive] reasoning ability).

> Again, this is why people sleep at night—they do not mentally function on their own: they only function at the bequest of the physical world.

From Adam is born Eve. Adam and Eve were originally one-being. They pre-existed in the "oneness state." She is detached from him, and each is born both mentally and physically as individuals.

Eve picks some fruit from the tree in the center of the garden. The serpent—representing the darker (deductive) side of herself tricks her—into believing the fruit (representing the knowledge of right and wrong and good and evil) was acceptable to eat.

Adam is tempted by Eve, and he eats some fruit. He knew better (he remembered he should not do this). But, having lost his reasoning ability (and hence could not maintain his knowledge of reality) he ate some anyway. They both eat. Their eyes open. They realize they have physical bodies. God said as punishment Eve would reproduce and bear children. So Adam and Eve became involved with each other on a physical basis. They sexually reproduce. Because Adam and Eve eat from the wrong tree, they lose the knowledge of God and face physical death.

Adam and Eve both lose the knowledge of God (or Reality) and fall into a deep sleep. Their minds only function at the bequest of the physical world. Reproduction and death enter the world through their physical identities. Desire for each other's physical identity is maintained. Offspring are produced and propagate. This results in bodies and minds continually splitting into many. Adam and Eve become special instruments of reality to begin propagating physical beings.

The trial of Adam and Eve is the reason we live in this short form of existence—to produce many minds. But why many minds? Answer, to produce people to inhabit the next life when it begins. Remember that in the beginning there was only Adam. (Note: Original sin [desire] is not the real reason that death entered the world.)

Reality uses Adam as a special instrument to produce a plurality of minds. Those minds, when the time is right, will populate the next existence. The first will be last and the last will be first (Luke 13:30). Adam's work, as the Second Coming, is to reacquire the knowledge he once had. He needs to obtain that knowledge before the advent of the next phase of the cosmic cycle.

Light—Lucifer; Deductive Minds—Satan

What color is the print on this page?

red, white, blue, yellow, orange, pink, none of the choices

Answer: white. Many people pick "none of the above"—believing the answer is black. Black is not one of the choices. The question is "What color is the print?" The question is not "What color does it appear to be?" If there is a difference between appearance of things and their underlying reality, there should be two answers indicating this difference.

Immanuel Kant suggested there are phenomena—that which appears to us, and noumena—the underlying reality. If so, he thought, it would explain an external reality distinct from our perception of it.

Noumena explain phenomena. One answer (phenomenon) indicates appearance; the other (noumenon) indicates the underlying reality. The print appears black. The reality is the print is white.

Appearance of things is explained by a "hidden" reality discovered by simple reasoning. Minds seek truth about Reality. If we experienced reality exactly as it is in itself—we could almost dismiss the need to have a mind to infer what is true from what is not true.

The physical explanation is the following. White light (all colors) impinges on an object's surface. The color that appears is the color reflected from the surface. Reflected light is not the color of (light energy absorbed into) the surface.

The color of any surface is "All color but the color seen." A black appearing car becomes very hot in summer—it absorbs most of the impinging electromagnetic energy (in simple terms—light). An object that appears blue is (in itself) yellow (the complementary color of blue). We cannot trust "apparent" truth. We can trust reality.

We experience reality in a twisted—backward (complementary) way. By saying the print is black, we get the "true" answer backwards. It is a lie that the print is black. It only appears black because its underlying nature is white. The true color of this print is white. Photographic negatives (apart from the "orange" mask) support the idea there is a difference between color appearance and reality. Digitally inverting color using a computer also demonstrates the complementary nature of color.

The typical response (to an object's color) is backwards because of the nature of Light and how it interacts with and controls how we experience matter. By deducing (based on what we have been told—from our memories) we get an answer backwards (not true, e.g., that this print is black) from the one obtained had we inductively reasoned (that print on this page is white).

People deceive themselves. Most people naively assume we experience reality as it is in itself. We do not. Some people are aware of this, but make no further inferences.

One's mind develops only if used. One only knows something if he can explain it. Trying to explain things develops the mind. To understand the truth about this (and anything else) one needs to question everything. That means using the mind to derive inductive inferences. Inductive inferences are associated with the explanation for things. There are different conceptual ways to explain things. Those different ways can be correlated.

There is a correlation between understanding Light and some Biblical statements.

> Those who seek the truth come to the Light, and the Light will eventually show them that they sought the truth via God's will (John 3:21).

> God sent John as a messenger to tell about the Light. This is the real Light, the Light that shines on the world (John 1:6–9).

> God is Light (1 John 1:5).

> Satan can change himself to appear as an angel of Light. It is easy therefore for false servants of God to appear as servants of God (2 Corinthians 11:14–15).

It is also easy to understand how the nature of Light fools people. If we use our mind (to ask, "why is this the color it appears?") we would understand the reality behind "apparent color."

Lucifer, as mentioned in the Bible, is the "Light-Bearer." That is how dictionaries define it. In physics, Light is electromagnetic radiation. Lucifer is electromagnetic radiation.

Satan is translated (from the Hebrew *mal'ak*, meaning messenger) into Greek as *angelos*. Saying Lucifer becomes Satan is easy to explain.

Light shines upon an object's surface. The object becomes "illuminated." Illumination does not mean we experience the object as it is in itself. Something happens to the light when it "reflects" from an object's surface.

The significance of "Light" is also found in its role in the creation of everything. We are created from Light (1 Thessalonians 5:5). You are born of the Light. Physical matter is a manifestation (born out) of Light. Let's explain.

The Principle of Complementarity shows itself in the nature of light. (Light itself is described in Chapters 13 and 19–21.) The Principle of Complementarity describes how particle-pairs are created out of Light. For example, let an electron represent matter, and let its complementary counterpart, the positron, represent antimatter. If these two particles collide head-on with a combined energy of about 1 MeV (million electron volts) they annihilate into gamma rays (very high frequency light). Existence (in our example—matter and antimatter) is created in complementary pairs: Cosmos and Anti-cosmos. The Cosmos and Anti-cosmos are born out of the Light.

The significance of Light is also found in its constancy (of velocity—c) as stated in Special Relativity. Besides light constancy, c indicates, at the speed of light there is no flow of time and no matter. There just is pure (electromagnetic) energy.

How is Lucifer related to Satan? The color paradigm typifies the Lucifer-Satan relation. We have heard that Lucifer becomes a "fallen" angel. As the fallen angel, Lucifer becomes Satan. (Explaining how Lucifer came into being in the first place is described in Chapter 21.)

> Lucifer was thrown from heaven and struck the ground (Isaiah 14:12). Jesus said he saw Satan (actually, Lucifer) fall from heaven (Luke 10:18). ~ Lucifer refers to pre-fall; Satan refers to post-fall.

How does Lucifer become Satan? Lucifer is Light itself—electromagnetic radiation. Light strikes a physical surface and is reflected. Lower level beings (like physical human beings) "see" physical things because of "reflected" light. The reflected light is complementary to light absorbed into the surface of seen object. The object itself is "all color but the color seen." The color seen is the complementary color of the object's true color. A grapefruit appears yellow. We understand through reasoning the grapefruit is itself blue.

A person who says grapefruit is yellow is not speaking the truth. The grapefruit is blue. It is a lie grapefruit is yellow. These statements symbolize how Lucifer becomes Satan. Satan is the "deductive mind" Adam became when he lost his (inductive) reasoning ability. (He [his mind] was put into a deep sleep, and he became a physical being.)

Because practically all physical descendants of Adam still have a nonfunctioning mind (does not function in itself) they too share the characteristics of the "Fallen Adam." The collective of all deductive minds is what religions denote as Satan.

Satan is Lucifer's counterpart in the physical world. People lie because they do not know the truth. Their minds only function with respect to the physical world (Hell). "People (their minds) exist in the physical world." Saying the same thing using religious concepts is "Satan exists in Hell."

Although Light (Lucifer) is transformed into people (viz, deductive minds, or devils) Light (Lucifer) maintains its own identity. Lucifer does not lose its identity (anymore than a mother loses her identity when giving birth). Light is still Light; Lucifer is still Lucifer. Satan (or individual devils) misinterprets the true nature of things because "reflected" light is opposite (and complementary) to that "absorbed" into an object's surface.

The transforming of Lucifer into Satan is an ongoing process in the physical world. Again, reflected light yields the premises (sense data form) upon which people in the physical world make deductions. It is natural for the mind to function deductively in the physical world.

Light gives individuals "messages" about the "hidden" (noumenal) physical world. Lucifer is an angel. An angel is a messenger. Because light reflects off things, an object's true color is experienced "backwardly." The true message of an object's color is not directly experienced. One must use his mind to understand the real nature of things to acquire truth.

If we experienced reality just as it "really" is, there would be no necessity for the mind to develop. The mind develops by seeking truth.

Deductive is opposite to inductive. To induct knowledge in the physical world is to "go against" one's own nature. To seek truth using one's mind is to develop spiritually (mentally).

Cultural tradition, as unwritten rules, sets the norm for acceptable citizenship. People, "wanting" to be socially accepted, willingly comply. The effect of a group (of people) on an individual is overwhelming.

If Satan is the collective of everyone, each individual is a devil. People (everyone) try to make-over everyone else in their own image. This is particularly typical of "religious" people. Many "believe" their religious duty is to proselytize others. The effect is very powerful. Treading an unsocially acceptable path might enable an individual to free himself from peer constraint and develop his mind's reasoning ability.

Desire—Sin

By sinning man is said to oppose God. In modern language, sin is desire (or wanting). Sin pushes people away from God. Substitute desire for sin, and truth for God. Man desires truth, but worships what is not true (i.e., lies).

Lies are represented by beliefs. Beliefs represent self-fantasies. Beliefs are a futile and foolish attempt to seek truth.

Beliefs are anti-truth. Supposing God is (represented by) truth, sin turns people against God.

In believing, people are against truth. Beliefs represent the Self—not reality. Believing is accepting a statement's truth without regarding evidence: without regarding reality.

> People seek truth to know. We have sufficiently explained how "wanting" is defined by "not having." Therefore—not having knowledge. (See Chapter 5: subheading "Wanting and Having.")

Do not live like a fool; instead, be wise (Ephesians 5:15). The more we learn about reality; the more we understand. Sin is a natural expression of egocentrism or self-centeredness (or in biblical terms, self-righteousness). (See Chapter 5: subheading "Basic Natures of Man.")

A mind manifesting a physical body is (in itself) nonfunctional. The mind functions only with respect to the physical world. (Without a physical presence, the mind does not function during Phase I of the cosmic cycle.)

Sin (desire) is not in itself bad (or good). It is the inherent nature of a mind that inhabits a physical body. It is just a phenomenon precipitating out of the design of things. When people make a judgment (deductions) on things that they do not understand, they avoid seeking the truth about their real nature.

Redressing Biblical Transgression

There is no problem in reality—it is the way that it is—no problem here. The problem is in the methodology one uses to seek truth. If one believes (on faith) that he has already found truth—he eliminates the only real reason to seek truth on reality's terms. Faith is accepting a statement (as true) without verification. No verification—no truth. Not knowing is the reason to seek truth.

How did people easily get so carried away by faith? The Bible says (Romans 3:28) "WE conclude man is put right with God only through faith." Who is the "WE" concluding this? Not every indicative statement in the Bible is true. We know this from studying biblical contradictions. When the Bible says man is foolish (by God's standards) it is because of his "believing." Believing hinges on faith for its methodology.

Let's redress a serious problem presented by the Bible—faith. Faith is the greatest trick imposed on unsuspecting followers. Serious biblical contradictions explicitly implicate the source of the Bible is not strictly of God (if God is [represented by] truth). If the Bible contains lies, it cannot be wholly of God. Let's explain how to honestly approach biblical verse. The following applies to any statement or idea.

Separation Between Words—Concepts and Reality

The purpose of scientific methodology is to verify statements, concepts, and hypotheses by allowing reality to speak for itself. Reality informs whether a statement is true or false. Once a term (e.g., spirituality) is defined, and we can find no contradictions (lies) in the definition, we test its truth via reality.

There is a "separation" between words and things. There is a separation between concepts and the reality they represent. (Refer to Chapter 7: subheading "A Crucial Difference.")

Knowing words and concepts does not mean one has connected them to reality. There is the word car; there is the car-object—parked out on the driveway. There is the word spirit; there is something in reality to which that word corresponds.

Having studied numerous books (including religious holy books) many people believe they know the reality that words and concepts represent. Knowing words and concepts does not necessarily mean one knows their reality counterpart. Knowing the word "car" does not mean one knows the car in reality.

There are two ways to learn. Indirectly from books (and other secondary sources) and directly from reality.

To make a point we restate a hypothetical situation (described in Chapter 2). A person can study books and become an expert, for instance, on auto mechanics. He becomes a teacher, yet never "worked on" cars. Every time a student asks a question, the teacher gives an answer. His students are unaware their teacher did not directly learn by working on cars.

The book-learning teacher did not use his mind at all. He just memorized information about cars.

Consider the other extreme. There are people who have worked on cars for much of their adult life, but they never learned to read. Such a person can also be a teacher. He can teach by showing how the car works. This person reasoned to understand automotive mechanics. (Especially if he learns without guidance.)

The "learn-from-reality" teacher is not separated from what he teaches. People learning directly from reality need to reason. They use their mind to understand.

The above mentioned examples are extremes. It is easier to make the point using extremes. A combination of the two methods of learning is the best approach. The idea is to not become too reliant on just one of these methods. Both book learning and reality-based learning are practically necessary to acquire in-depth knowledge and understanding.

Learn (in statement form) from secondary sources, then verify their truth by reality. Simple statements representing physical things are verifiable by direct correspondence to reality. Complex concepts are verified by elimination of contradiction. The second method is very difficult and calls for highly developed-inductive-reasoning ability. It is not an ability that is natural for individuals during Phase I of cosmic cycle (when minds "occupy" physical bodies).

The separation between knowledge and things also applies to the separation between man and God. Reading about anything, including God, does not mean the reader knows what he is reading. Memorizing the Bible does not bring anyone closer to God. Praying to "one knows not what," is also of no help. Only reasoning to understand reality as a whole brings one closer to knowing God. An actual car is different from the knowledge (word, car) representing it. Knowledge is different from the reality it represents.

Skepticism

If we say someone "knows" when he cannot prove (a statement true) then we must also say when someone can prove a statement by reality, that he does not know. This is obviously a foolish statement. It is incorrectly defined.

Therefore, we must say an individual is lying (not telling the truth) when he claims to know yet cannot prove a given statement by reality. He can say he knows in his heart, but that is just foolish talk. Those who trust themselves are fools (Proverbs 28:26). It cannot be both ways. The Law of Contradiction does not permit it. Then we remember God is true and every man is a liar (Romans 3:4). If this biblical statement is true then we should observe it in the world we live.

Does this mean every man is a liar? Look around, what do we observe? If people speak for themselves (as heard in such phrases as "I believe" and "I think"); then they are self-centered, self-righteous. Is this what it means? Beliefs represent and denote individual's fantasies: beliefs do not represent reality. Are we surprised? No man can do what is right (Romans 3:10). If an individual merely believes a statement (to be true) then he is not speaking for reality. He speaks only for himself. He is effectively saying his self is right—this is self-righteousness.

If a person speaks for reality, he does not speak for himself. He is not self-centered, but reality-centered.

Ask the question, which is greater, a single person or reality? If people say reality is greater than they are, why are they trying to project their fantasies (beliefs) onto reality? Instead, they could develop their mind to understand reality.

To not be skeptical, is to categorically and absolutely to end doubt (or skepticism).

Familiarity with a statement does not necessarily mean one knows what it represents in reality. Religious leaders try to get people to accept "their" personal interpretation of what they really do not know. They say, have no doubt; do not be skeptical! Many imply what they say comes from God. The Bible says so. It must be true because the Bible is the word of God.

Religious leaders mislead (though usually unintentionally). They do not understand what they read in the Bible. They fill gaps (in their understanding by interpreting) with beliefs. It is no wonder, the Bible sanctions belief.

Some followers ask excellent questions only to be disarmingly misled. For example, a religious person questions his faith. He speaks to his church leader. The follower says everyone in the church believes he a good Christian. Deep down he has many doubts. He sometimes wonders if any of the Bible is true. Maybe long ago it was just something people made up. He wonders why he doubts so much? The religious leader answers:

> Doubt can destroy faith. Yet more faith enables one to overcome doubt. Why do you doubt? Perhaps you have not had some of your questions answered. Seek from those who have sought answers. Godly men have answered every question you could ask. Doubt might arise from a subversive spirit. Deep down it is man's nature to be free from God. Doubt and faith are opposites.

The person questioning his religion is told by a church leader to have even more faith to overcome skepticism. Where does it end? Here is a person who is beginning to be skeptical about religion—and the church leader is doing his best to stop the questioner from being skeptical. The church leader is trying to stop the skeptical person from seeking truth. In a vague way the skeptic knows the Bible has not been properly understood. He also knows it has not been proven God is.

The religious leader commits the very act he speaks against. He "believes" he is "Turning people to God." Instead, he foolishly turns people from truth and away from God. (We cannot know God until we first understand all truth.)

The leader says all are sinners, but he behaves (in advising his followers) as though he were (telling the truth and is) exempt from this law (of sin-desire—which causes him to lie). He is not exempt from the law of sin. So why do people accept unverified statements as true? We understand people behave in a way to become more secure. The problem is not what people believe—the problem is that they believe.

People take the easy route because they are ignorant (about reality). It is easier to accept unverified statements as true rather than expend the effort to verify them by reality.

There is another way, but it affects one's insecurities. Just be realistically honest and say you "know not" when you cannot verify a given statement by reality. One cannot be anymore honest than to say, "He does not know" (a statement to be true) when he lacks proof. That admission is a powerful reason to learn. One must begin somewhere to travel the road to greater knowledge and understanding. Skepticism is a beginning.

Beliefs are incapable of distinguishing true statements from false ones. Have you ever believed something to be true and were deceived (about it)? Examples abound. Beliefs are not a surety of what is true. Consider another example.

You cannot find your car keys. Where are they? You believe they are in the car. You go to the car and see them in the ignition switch. You say, I was right; they were in the car. The reality is the following: The keys were not in the car because you believed they were. You can believe this and that, but will never begin ascertaining what is the truth until you seek out in reality the meaning (reality correlation) of a given statement.

No matter what you believe, either the keys are in the car or they are not in the car. (It is one or the other, but not both.) You found the answer to the question by seeking in reality the object (the keys) corresponding to the word "keys" mentioned in the question.

Lesson: If beliefs are not sufficient for determining the truth of simple denotative statements (representing everyday experiences) where does one get the idea such foolishness (beliefs) can verify the more difficult (more general) concepts such as define God?

A belief is a statement accepted as true based on faith. If someone says he "knows" something (viz, that a certain statement is true) but cannot demonstrate it by reality—he is very foolish and a liar (just as the Bible states).

If an individual is not skeptical and does not question things, not only will he surrender his individuality; his mind will not develop to answer the very questions he now asks.

There is another problem noted in believers. They use adjectival words to describe reality. They say, This is good; that is bad (evil). These are personal judgments and do not reflect any "real" discernment about reality itself.

Science is different. A botanist does not say "there is a good tree; over there is a bad tree." Instead, he describes how the tree functions: he uses adverbial words (describing activity). If adjectival words are used to describe something in reality, they must be standardized according to some concept that gauges relative likenesses or differences.

Beyond Skepticism

The first step in seeking truth is to be skeptical—doubt yourself. Doubt the methods you use; doubt what (information) you have been taught. Students do plenty of memorizing in school. Memorizing does not develop the mind. Constantly question your ideas. If one travels down the road to greater knowledge and understanding—he must use a methodology that drives him as a seeker of truth.

Skepticism itself is not enough to push one beyond skepticism! One must use a critical scientific-logical methodology to seek truth. To go beyond skepticism, is to not only question, but to test everything until reasonable answers to all the big questions are found.

Question (the truth of) all statements until everything is explainable without contradiction. Until that level of understanding is reached—the journey is not over.

The first question to ask is am I really skeptical or am I only skeptical about things, which do not bear on my own insecurities? Remember, the very thing (reality) of which one fears, is the very thing, which if understood, would bring real security!

If you finally understand All Reality (on its terms)—you will have found the meaning of life, and consequently even learn why you ask questions. You might even learn why you are skeptical. Do not be afraid of what you do not know and understand, be afraid because you do not know and understand.

How do you begin? It is a long journey. Here are some hints. Do not commit the "Fool's Fatuity." Instead, use the "Wise Man's Wisdom." (Refer to Chapter 2: "Methodology.")

Before going beyond skepticism—you must first be a real skeptic. Are you really a skeptic? Let's take a test. To the extent you assert the truth of statements you cannot prove by reality, you are not a skeptic. To be skeptical is to doubt. To doubt gives one a reason to go beyond skepticism and to search for truth by allowing reality to show itself to you. Again, reality is the final arbiter of what is true.

So what is an easy way to ascertain if you are not a skeptic? (It is easier to prove a negative.) If you speak saying "I believe" such and such, and to you it means you know (but you cannot verify) you are not (this is proving a negative) a skeptic (about your own ideas—one's in which you have invested personal security). If you are not skeptical about your own ideas, then you (in reality) are not a skeptic, no matter what you believe. Merely being skeptical about things outside your (personal) ideas does not make you a skeptic.

When you learn the secret to a magical trick, you wonder why you did not already grasp how it was done. Is the concept of "belief" the trick that stops people from better understanding reality? Are miracles no miracle when properly understood?

If there is a God, it must be real. If real, God must be at least some aspect or function of reality. If not, then the notion of God is but a fantasy. Even the Bible says all men have turned from God. Look to Reality to determine if a given statement is true. If this is how to seek truth, then should not people be skeptical about their own wisdom? There is no doubting the wisdom of reality—it is what it is.

Deception?

Listen to people who have deceived themselves. Listen to religious leaders. They lie with every breath. They constantly make grandiose statements that they cannot prove. The practicing idea of those self-interpreters of the religious books is to get people to end doubt (and hence skepticism) altogether! They do this by supplying their own (self) answers. To answer one question about biblical scripture, they cite another.

Many people say they have been deceived. Not true. No one (or thing) can deceive someone. Did these people (who supposedly deceived a person) torture them into submission? Or did that old propaganda technique (the big lie) work again? It is amazing in the present age the propaganda ploy still works. Tell a lie many times. After awhile, people say they know it is true although either they cannot prove it, or it has not been proven to them.

You have never been deceived by anyone except yourself. Do you not selectively reject information (such as certain beliefs of others) as not being true? If so, then what information you accept (as true) is solely of your own doing (responsibility).

Are you skeptical about what you learn? Do you ask for verification? You cannot know the truth about any given statement until it is proven. If you demand verity from others, why do you not at least demand it of yourself?

> When you say someone has deceived you, by pointing to the "supposed deceiver," you attempt to avoid the blame (responsibility) that you deceived yourself. Security-wise, it is easier to project the sad truth about yourself (that you deceived yourself) onto something "out there"—the supposed deceiver.

What are the consequences of not verifying statements as true? (Except for defining yourself through fantasy.) If statements are not verified as true—you may be gambling with your life and others—needlessly. Example: A major airline flight departs from an airport. A pilot (evidently) does not verify the wing "slats" are in the down position for take-off. Why do pilots have a check list to verify (by reality) if all they need to do is "have enough belief" (that the slats are down) and it would be so? Beliefs are not sufficient to deal with reality on its terms. The pilots deceived themselves (made an assumption). That is not reality's fault. They ignored Reality. They, and everyone else on board, paid the consequences. No, reality did not deceive them: they deceived themselves. No one has ever deceived you. No one made you believe this and that—you needlessly accepted statements as true because of your own foolish nature.

People die everyday because someone believes something. Infants and children die because someone believes they are somewhere they are not. Children fall into pools and drown. They are forgotten and die in overheated cars. Every day, people die because someone did not verify and check whether a given thought is true.

Have you been brainwashed? If so, then you allowed it. How can you test to see whether you have been brainwashed (i.e., brainwashed yourself)? Here is the test. If you say a statement is true and cannot verify it (by reality) or cannot explain it in such a way in which no contradictions can be found—you have deceived yourself (about what statements you know to be true). If you can point to reality and verify (empirical) statements then what you say is as true as the evidence to which you point (corresponds to what you stated).

You know to the extent you can explain (why) something is. If you understood the meaning of Newton's work with the prism and light, you could have capitalized on his work and not have been fooled about an object's color. Scientific knowledge serves as the basis for knowing. Scientific methodology paves the way to understand Reality.

All religions have one thing in common. No religions have proven the truth of their dogma. Yet most followers of those religions claim to be members of the true church—their church.

There is a difference between belief and knowledge. Logic tells us there can only be one true church. None of the churches, claiming to be the true-church, are worthy of that title. If one were—why have church leaders not proven it?

There are nearly as many churches as there are interpretations about what is true. Church leaders do not define their terms in accord with reality. Ask them to define their terms and then look for contradictions in the definitions. If contradictions are found—then it is a lie they know of what they speak (state to be true).

Notice science is the same everywhere. Reality is the same everywhere. Science is the only discipline developing a serious methodology to understand this world (the reality) in which we live.

Ascertaining truth: A statement is only true if Reality says so. What applies to one statement, applies to every statement regardless of source. Biblical verses are not immune to questioning of their truth.

Definition Testing: An Example

"Spirit" is one of the most misunderstood words. What is "spiritual development?" The word "spirit" must be defined, otherwise, how could we know what it is to seek spiritual development? This book correlates spirit with mind.

A "teacher" (or religious leader) is incapable of helping a person's "spiritual development" unless they know what the spirit is. Teachers who cannot properly define spirit are not true teachers (of the spirit). If there are false teachers, we ask what is happening here? Let's backtrack.

The first purpose of cognition is to define what we experience. The second is to ascribe meaning to particular experiences. Meaning pertains to connecting experience to other closely associated phenomena. This is how we seek understanding. Did we correctly ascribe meaning, or did we misinterpret meaning? It is the second step (ascribing meaning) where errors most likely occur.

How do we ascertain if meaning has been misconstrued? Let's examine this question by determining whether an instructor is a true teacher.

Is the "teacher" a teacher of truth or not? Recall there is a difference between knowledge and what it represents. How do we determine whether a teacher knows the truth of what he teaches? Ask the teacher to define critical words that he uses. If he cannot, will not, or does not clearly define his words; then, his status as a true teacher is immediately suspect. If the teacher does offer reasonable definitions, a third step asks for explanation. Ask why are these definitions true instead of false? This question tests meaning.

Ask whether there are any other possible explanations that could account for the given definition. What does it represent in experience? For example, the teacher might suggest that doing a certain "exercise" develops the spirit. An astute student would ask how else can this exercise be explained?

Let's apply a simple test. We ask the teacher to define the term spirit. When the teacher offers a definition, we examine it. It takes a "greater teacher" to do it this way. Did the teacher clarify any meaning not explicitly clear?

We must do something to limit errors in our defining and understanding of things. How do we limit errors in seeking truth? Look for possible contradictions. Ask for clarification when a discrepancy is found in any given definition. Strive to define words and concepts in noncontradictory terms.

> Noncontradictory definitions are so close to the truth that they are practically immune to academic challenge. Welcome the wisest of scholars to try to find problems in the hypothetical definitions. Always stay open to criticism. Amend the definition until extraordinary effort finds no fault (with it).

Answers from false teachers become more absurd as we probe deeper into their "supposed" knowledge. False teachers cannot get past the second question (pertaining to meaning). If a teacher does not give a definition or offers excuses like "You must find that out for yourself" then the teacher is probably an impostor. A false teacher cannot deceive wise people, although they might have deceived themselves and others.

When the teacher does clarify meaning, ask why is that? Why that particular meaning? If they cannot satisfactorily answer, we establish they do not really understand. If they do not understand, it is a safe bet they also do not know of what they speak. Why? Because: Being able to explain without contradiction is the test of understanding. Understanding (reality) is the basis for properly defining words, concepts, and knowledge.

Reasoning ability is needed to understand. What does this mean? The seeker of truth needs to inductively reason to higher levels of generality. This is done by eliminating contradictions in higher-level concepts. A person becomes "spiritually" (mentally) inclined when he develops his truth seeking ability. When that happens, his right brain begins working on its own.

People will not deceive themselves if they seriously question their teachers. This is the ideal in "courtroom cross-examination." The interrogator tries to find a lie in testimony (definitions).

To test definitions for truth we need to know much. A bonafide teacher or guide is useful. The student must test the teacher. If the student does not openly question the teacher, he or she should do so in his or her mind.

There is also the other side of the true teacher problem. If the student does not know enough to test definitions or does not consult a knowledgeable person, then their only recourse is to test results.

Testing Results

A true teacher helps students develop abilities to better seek truth. If someone implies they can help increase one's spirituality then there are ways to test whether they can.

To increase spirituality is to increase one's mental ability. Specifically, a person is "awakened spiritually" if his or her mental "inductive" abilities begin to function (on his or her own, apart from the physical world). The right brain begins to function on its own.

> The only way a teacher can help someone gain spirituality is to describe the methodology that helps develop the mind. (Refer to Chapter 2: subheading "The Path.")

How can "increased" spirituality be tested? It is easy to test a mind that functions on its own. Continue questioning what you have learned (from the guidance given by the teacher). What do you know (about reality) that you did not know before?

We can determine if an individual is very spiritual by asking him about reality. Does he know about things that have not yet been verified by science? Has he solved previously unsolved problems (about understanding reality)? Have we seen his work in scientific literature? If not, why? An individual only knows when he can verify (a given statement) by reality. Someone knows, in this manner, if his statements can be independently tested by others and corroborated.

A Quick Test

The spirit is the mind. Any other definition leads to intractable contradictions. There is no need to continue using the word spirit. Let's just use the label "mind." To become "more spiritual" is to become "more mental."

We now have a third way to test if someone knows what he or she says about the word spirit. It is a quick check of definition. Do they still use the word "spirit?" If they do, they do not know that spirit is the same as the mind.

It is interesting to read articles and books, and even listen to people who do not have a clue about the real nature of things. The word spirit is widely used. It is used as though everyone knows what it is. Evidently, these speakers and writers have not been seriously questioned about using vague words having nonphysical or abstract denotations. They do not question themselves (about what they believe). Not questioning themselves and not doubting a statement's verity shows they are not serious seekers of truth.

Mentality—Spirituality

Developing the Mind

How does one develop mentally? One must "exercise" to develop. It is not easy "getting started." Most people do not know how. Let's use an analogy.

It takes much time and effort to achieve great physical strength. Physical exercise develops physical strength. One does not develop great physical strength in one day, one week, one month, or even one year. It takes time and step by step development.

A person could consult a physical trainer. A trainer could say he will just sit in front of you and 'magically' give you increased physical strength. It does not happen that way. The same applies to someone trying to give you mental strength. Developing mental strength is something people must do for themselves.

Getting started is the most difficult. Learn "technique" (methodology). First, efforts will not show much "mental strength." It takes dedication to become mentally proficient.

Do not be deceived about what people claim. Ask for verification. Not asking for explanations and their verification means one is not a serious seeker of truth.

One way to test an individual's mental strength (ability) is to ask him to explain questions concerning meaning (definitions) and reasons (why) a given definition is true.

The first clue someone is a fraud (intentionally or not) regarding mental development, is his or her continued use of the word "spirituality." If they do not use the word MIND, we already have a clue they do not know what they say. Let Reality tell about what is true and mistakes are avoided. What does this mean?

Advancing What is Known: Test of a Developed Mind

A person can study by reading and memorizing and not develop mentally. So what is the advantage of studying books?

No one can reduplicate every past scientific experiment. Book learning is extremely beneficial in "filling the gaps" between book learning and our direct experience with reality. The mind develops by directly interacting with reality.

We cannot directly interact with reality to learn every little thing. Our life is too short to repeat past experiments. Again, there is no reason to experiment with prisms, as did Isaac Newton. It has been done. We do need the knowledge obtained from experimentation. It is easier to learn by questioning.

Questioning strengthens the mind just as exercise strengthens the body. The more one questions—the better seeker of truth he becomes. Learning helps solve (previously unsolved) problems. Were we given a mind—only not to use it?

Significant experiments are a clue to understanding something greater. Parts add up to wholes. Each part of the puzzle (of reality) contributes to an understanding of the whole puzzle.

Our task, as serious seekers of truth, is to push back the frontier of scientific understanding. To do so means we must acquire an in-depth understanding of present day knowledge. We use that knowledge and apply it to the reality we directly experience.

> Note: If we only study books, and not strike out on our own, then we will not begin directly interacting with reality. Take contemporary scientific understanding and use it as a basis to advance the understanding of reality.

People go to school, college and some study medicine. After years of studying, students are ready to apply what they have learned from books. Medical students really begin to learn when they directly interact with real patients. Practicing physicians make a connection of the words and concepts learned in school to the health conditions they find in real patients. Reality is the "real" teacher.

Back to our problem. How does one exercise the mind? The mind is not sufficiently exercised by merely reading books. Yet, if one tries to understand what he reads, his mind begins to question things. For example, is a given statement true or false? To ascertain the difference (between true and false statements) one needs to look to Reality. This is the purpose of scientific experimentation.

Pragmatically, the mind develops by applying "book learning" to reality. The mind develops by testing propositional truth against reality. Maybe it sounds too simple. The truth is simple. A practicing physician learns fast.

In Chapter 2 we encountered the "Fool's Fatuity" and the "Wise Man's Wisdom." Only a fool would use unverified and unproved concepts to understand the deeper aspects of reality. A wise man uses only concepts that continue to have scientific merit (tests continue to indicate their viability). Viable propositions (defined) are ones not yet contradicted by established scientific knowledge.

Those who believe (the theists) and those who believe not (the atheists) are in the same (sinking) boat—neither can prove their claims. One who is skeptical and says he does not know (the agnostic) because he cannot cite the proof, is honest with Reality. Further, he who says he knows (a real scientist) can be tested by his ability to produce proof upon demand. Let's use these "viable" definitions and apply them to what we have learned in this chapter.

JUDGMENT DAY?: A Parody

Let's use what we learned to show in the extreme what would happen if each person were solely responsible for himself.

Many so called "religious people" often push the point each individual is somehow responsible for their salvation. They say each individual must come to

Christ to be saved. Yet, if Christ is the one who knows the truth (...I am the truth. John 14:6) about God; then, an interesting situation develops.

What if there really is a Judgment Day and each individual is solely responsible for themselves? Let's consider this possibility.

There is a day on which the world will be judged. Bible, Acts 17:31. This statement does not necessarily mean each individual is responsible for themselves.

> If each person were solely responsible for himself or herself, there would be no need of a Savior. Yet most "religious" people behave as though they had some significant responsibility in saving themselves. This is evident when they ask others if they have been saved? The implication is that these religious people have saved themselves and are now out to help save others. Many even assert they have been saved. The Bible states people cannot save themselves. Those who try to save their life will lose it. Luke 17:33.

We can classify and place people into four groups by what they say about the word God. Let's see what would happen to these individuals on Judgment Day. Each group is represented by one individual. Let's call them by the following names:

Theists	Those who "believe" that God exists (or that there is a God). They often say they "know" but cannot prove that "God is" and hence only believe they know. Christians are an example.
Atheists	Those who "believe not" that God exists. They are also classified as believers. They cannot prove that God is not.
Agnostics	Those who say they "know not" whether there is such a Being that can be labeled by the word God
Scientists	Those who say they know that God is, and can also prove it by reality using scientific and logical means.

These definitions are formed from what the individuals say about the word God and whether they can prove via reality what they assert. An individual may be a scientist regarding other statements or words, and yet not be a scientist according to what he says about the word God. If he asserts that God is, but cannot prove it, then regarding the word God, he is a theist rather than a scientist.

Let's hypothesize there is a Judgment Day. How would each of the above fare in Judgment Day Court? Picture the defendants seated in the court of last resort. They stand before the judge and hear some instructions.

Court in Session

Judge: "You are all here for a reason. Either you have been truthful, or you have lied. It is the task of this court to determine whether you have told the truth or have lied. Do you understand the purpose of this court?"

All defendants reply: "Yes."

Judge: "You are in court to answer some questions. You have all made some statements concerning the word God. Today is your day in Judgment Day Court."

Case One—the Theist

Judge: "Will the first defendant please step forward, and state your name."

First defendant, the Theist: "My name is Theist. I want a lawyer."

Judge: "Please, just stick to the questions asked and this proceeding will be kept short. Sir, we once allowed lawyers to represent clients, but they were never an asset to the client. They had trouble unlearning the lawyer-ways they learned on earth. They were more hell-bent on winning the case at practically any cost, instead of seeking the truth concerning their clients. Here, truth is what we seek. Since truth is the criterion by which an individual is judged, it matters not whether the defendant has a lawyer or not. Either you speak the truth, or you do not. Theist, I understand you were a minister of Christianity, is that correct?"

Theist: "Yes, I even had my own church!"

Judge to himself: "It surely was not one of ours."

Judge: "You have said that you know God. Is that correct?"

Theist: "Yes, I do know God."

Judge: "You have made an assertion of truth. You have claimed to know God. Please, now make your case."

Theist: "I know in my heart—"

Judge: "Wait a minute, you do not have a heart now; you do not have a physical body. You are a mind. Please continue."

Theist: "I am a Christian and proud of it."

Judge to himself: "What is this pride foolishness?"

Judge: "Theist, you are evading the questions, please just answer the questions. Why do you suppose you were given a mind?"

Theist: "I made up my mind."

Judge: "If you continue this foolishness, your trial will be over before you can make your case. Sir, you have said you know God. Make your case—prove that you know God."

Theist: "I know God."

Judge: "Then prove that you know God. If you know God, it can only mean you can prove that God is."

Theist: "How do you prove God is?"

Judge: "Sir, to save time, I ask the questions here. Sir, if you cannot prove 'God is,' then you are not warranted in saying that you know God."

Theist: "But I really do know God!"

Judge: "Theist, then why are you stalling?"

Theist: "God is good, God is kind, God is omnipotent—"

Judge: "Sir, if God were omnipotent, you would not be such a fool. Please desist in making judgments in lieu of verification and proof. If you cannot prove you know God, then it is a lie you know God. If this is the case, then you are a liar. No liars are permitted here. You are defective, and are impossible. On earth, people may listen to you preach the gospel because many of them are as foolish as you are. If you ever read the Bible, you would have found it states, that man's wisdom is foolishness by God's standards. (1 Corinthians 3:19.) Everyone is under the power of sin. No one understands or seeks the truth about God. (Romans 3:9–11.)"

Judge, continuing: "You would not be welcomed here. People who live here will not listen to you. They do not like listening to lies. And if you cannot prove what you assert is true, you are a liar."

Theist: "But—"

Judge: "For the record, this individual's mind (spirit) did not develop, he sought no truth. Instead, he convinced himself he already knew the truth about God. He said he knew God. Let it be noted that another individual mind was wasted and did not develop. No mental (spiritual) development here. Not much can be done with this individual. He does not honestly respond to questions of the court. Send him back, and put him into another body. He will be given another chance."

"Next case please. Step forward and state your name."

Case Two—the Atheist

Second defendant: "My name is Atheist, Where am I and what am I doing here?"

Judge: "You are in Judgment Day court. We understand you have said God does not exist (or that it is not the case that God is). Have you asserted these types of statements?"

Atheist: "Yes. There is no God."

Judge: "Go ahead, make your case."

Atheist: "There is no God, otherwise, why is there all this suffering?"

Judge: "You are answering a question with a question, and that is not an answer. There are problems in creating something out of nothing. To create something out of nothing means everything is created in complementary pairs of opposites. Hence the principles of symmetry discovered by physicists. Please continue, prove 'God is not.'"

Atheist: "No one can prove that God is."

Judge: "You are weakening your case, however, please continue, you must now prove that every man who ever lived cannot prove God is."

Atheist: "Wait a minute. That is impossible."

Judge: "Then why did you assert, 'No one can prove that God is?'"

Atheist: "Ah—"

Judge: "Atheist, all I have to do is to prove the negation of your assertion to show that yes there is at least one individual who can prove God is. He who is God's first manifestation as an individual being—known to you as the Father—and later as the Messiah—can prove God is. And I—the Judge—am that individual. That is why I do the judging."

Atheist: "I do not believe it (I believe it not)."

Judge: "It is not a question of belief, either I can do it or I cannot. Your Believing I can or cannot prove 'God is" has no bearing on whether I can or cannot prove 'God is.'"

Atheist: "I still do not believe it."

Judge: "Atheist, you, like the theist, have insulted this court and the truth for which it stands. You have a closed mind. Like the Theist you often fought against, your mind did not develop because you did not seek knowledge—truth."

Judge: "For the record, we find this Atheist in the same situation as the Theist. He ascertains, in his fantasies, that he is the arbiter of what is true instead of God. He will not fit in here with people who are honest with reality—God. We do not need any foolish people running around here trying to convince people there is no God, anymore than we need theists trying to tell people that God is and yet cannot prove it. It is a lie there is no God. For God is represented by knowledge or truth. The Atheist said there is no God. He also said no one can prove 'God is.' He is guilty of lying because he could not prove either statement. Liars are not welcomed here."

"Next case please. Step forward and state your name."

Case Three—the Agnostic

Third defendant, the Agnostic: "I am Agnostic."

Judge: "What do you mean you are Agnostic concerning the word God."

Agnostic: "I do not know whether there is a God or not. Surely, there is evidence of an interesting design in the world. But, I do not have sufficient knowledge to understand how design can prove 'God is.' Frankly, I am very skeptical about any claim that 'God is.' Further, I have not seen any proof that verifies God is."

Judge: "Interesting answer. Surely, if you cannot prove God is, then you are only warranted in saying you do not know (know not). And surely, the proof you cannot prove 'God is' is that you have no proof—and that is your case. You are honest with reality and hence God. You have made your case. You have an open mind. Your mind can still learn. Welcome."

Agnostic: "I do not understand."

Judge: "That is just the point. You are honest with the truth (the truth you did not understand enough to know whether there is a God or not). In being honest with the truth (or knowledge) you were being Christ-like, for he has the knowledge and understanding needed to prove God is. He has to prove he knows God, or he is not the Christ. As long as you prove what you assert is true—you are like Christ. Understand?"

Agnostic: "No, I do not understand."

Judge: "Excellent, you are ready to learn. We will send you to school."

Judge: again: "Next case."

Case Four—the Scientist

This case differs from the others. Here, we have a different Judge. The Judge here is God. The defendant is the real Christ, who was in charge of judging everyone else.

God-Judge: "Defendant, please come forward. I understand, you are a scientist concerning the word God."

Christ-Scientist: "Yes, I am."

God-Judge: "You have said that God is. Correct? Go ahead, make your case."

Christ-Scientist: "This will take a while, it is not a simple subject. It involves an understanding of everything that exists."

God-Judge: "Then begin by giving a simple and very general definition."

Christ-Scientist: "Very well. God is that which is. It just is. Reality just is. God is Reality as a whole."

God-Judge: "O.K. That is a beginning. But, you must describe how reality is constituted, before you are warranted in saying that all of reality is (can be correlated with) God. That means you must be able to prove everything necessarily and logically follows (is deduced) from God as a given premise. Moreover, you must also establish the premise (inductively) that God is. Are you prepared to do that?"

Christ-Scientist: "I am as prepared as you have made me. I am your first manifestation. You have manifested yourself as an individual-cognitive being to represent you. And am I not this individual?"

God-Judge: "Only if you can prove that 'I am' (Exodus 3:14) and do so without any contradiction. For you to do so, is to prove God is."

Christ-Scientist: "First, I will present a coherent understanding of the hierarchical nature of everything in creation. This describes Reality's structure. Secondly, I shall noncontradictorily explain everything. This is the proof that 'You are.'"

God-Judge: "Interesting! Please continue."

Key Points:

- Religious concepts are archaic. If they represent something in reality, they should have a present-day scientific corollary. Religions can be understood scientifically.

- Many stories found in "holy" books were embellished. Inevitably, information becomes exaggerated with each oral retelling. Statements in the New Testament were not written down until several decades after the events on which they were based. Moreover, objective reporting was uncommon to people of that time. At best: biblical scripture is inspired, but inaccurate. At worst: truth in scripture was compromised. Confusion is the result. Most holy books exhibit the same problem. That problem has expected consequences. It explains proliferation of religious denominations. Each group construes their holy book differently.

- There are many contradictions in the Bible. Many statements are outright false (based on present day scientific understanding). If some of those very statements were posed today, people would say they are obviously false. Yet, when it is mentioned the same statement is found in the Bible, religious people stretch the circumstances to make the statement feasibly true. Or, they walk away saying, "God works in mysterious ways!" Regardless, it is safe to suppose some biblical statements do represent reality.

- There is a crucial difference between a word and what it represents (in reality). A commonly committed error (called reification) occurs when people do not differentiate between words and what they represent. Many religious people assume they know the reality just by knowing the word. Knowing a word does not mean one knows the reality. The problem worsens. Knowing the word and the reality does not mean one understands. If someone is asked: do you know what the word 'actinon' represents? The word, by itself, does not tell what it represents! Actinon is a "word symbol" for something. A symbol is not reality, and a word is not the thing it represents. The word God does not tell what it represents. Vague ideas help not at all.

- Not understanding biblical statements via reality, increases the tendency for people to interpret scripture, or to espouse some group's interpretation.

- Belief and faith are cloaked in "personal" interpretation. Belief is contrary to knowing and faith is contrary to verification. People believe when they do not know. People develop faith in a statement's truth when they cannot verify it. The problem of believing and faith is overlooked because people do not differentiate between words and what they represent. Believing hinders serious truth seeking. Why would anyone seek verification (of a statement's truth) when he already believes (it is true accepted on faith). He "knows" the word (or statement) and unwittingly supposes it represents reality. This normally happens when someone hears an idea repeatedly. Familiarity is not necessarily understanding! Belief and faith sustain ignorance. Truth seeking via verification leads to knowledge.

- Despite the shortcomings of belief and faith, if there is a God, he could be partially responsible for some of the holy books. (He could be responsible for statements in many books.) That possibility forced us to undertake a critical assessment of holy book statements and concepts.

- Let's grant the holy books do inform about some phenomena of reality. Each book emphasizes different things. Some interesting concepts are: reincarnation, redemption, resurrection, life after death, sin, spirit, devil, Satan, Karma, Maya, heaven and hell, love and hate, Goddess and God.

- Most words and concepts, found in holy books, are not explicitly defined and are definitely not explained or verified. If they were, there would be no compulsion (because of one's insecurity) to accept (the reality of) these concepts on faith (as true). Statements need to be tested against reality.

- No "holy" book seriously defines God; no book has verified there is a God; no book has explained God. Problems in understanding result. The same problem haunts other words and concepts. Why are not these words or concepts: explicitly defined, explained, and verified, if the holy books are truly inspired by God? Theologians posed possible answers. "We have free will and must be 'tested.'" But why? Defining the Bible reduces to a simple question. Is it the inerrant word of God? If it is, how are its contradictions and false statements (lies) explained? Why does it not define God? Obviously, if the Bible is the word of God, it is incomplete. That would partially explain its errors and contradictions. Yet, if it does contain errors, then it is ultimately unreliable for knowledge about the human condition. Why are we here? If God is represented by truth, then why is not truth properly defined? Why is not God explicitly defined and explained? These questions deserve an answer.

- Perhaps the Bible is incomplete and indefinite because it informs only of the possibility of things. Meantime, lacking a truth-seeking methodology, an honest person has no choice but to be skeptical. Question and doubt (the truth of) undefined and unverified statements. Just say, I do not know. One cannot be any more honest with the truth (than to say he does not know, when he cannot verify statement truth). If God is (represented by) truth, then we are not straying from God (by being realistically honest) by admitting we do not know. The Bible tells of things (and possible events); it does not define or explain them.

- There was no science to speak of during biblical times. Bible statements are open to question until they are verified (as true). The Bible says to "seek the truth." In that vein, it is possible there is a Second Coming of God's representative. Maybe he will explicitly define and explain everything. Until then, we just do not know.

Is the Bible the word of God? The Bible contains many contradictory statements. This seems confusing. Why would God inspire errors in his own book? Perhaps things are not as they seem? Throughout this project we have overcome "delusion by appearance." We will not deceive ourselves.

There are some true statements in the Bible. That is a given. However, some biblical statements are contradicted (by other biblical statements). Even Jesus spoke contradictory statements. Scientific knowledge also undermines many biblical statements. There are plenty of errors. The Bible is not wholly inspired of God. If God is truth, it is unthinkable he inspires lies. Maybe God's message was altered "in transit."

Considering the many writers who contributed to the Bible, it would be a wonder if there were no errors. If God does not lie, then man, though "inspired" by truth, misconstrued God's message. This is a reasonable explanation. After all, if the messengers cannot be trusted, the message cannot be trusted.

Is the Bible partially inspired by God? For any biblical verse to be inspired by God, means there must be a God. In short, is there a God?

We have linked many religious concepts to scientific concepts. We have correlated God with Reality mostly because God is (defined by knowledge) as truth. It is easy to do correlations. It is more difficult to establish truth by verification and proof.

Before determining whether the Bible is the word of God in any manner, we must fully understand the nature of God. This task is deferred to the final two chapters. In the next few chapters we build the foundation to prove God is.

Chapter 18
Mathematics, Sets, and Logic

"While seeking a logical conclusion, we stumbled
upon the conclusion of logic. A proposition proving
anything and everything—proves nothing at all."

There is ONE Reality. We are defined by how we "fit in" with the design of things, viz, the structure of this one reality. The Structure of Reality is the supporting theme of this book. Once we elucidate the structure of reality we may come to understand the meaning to life: the meaning of existence.

How did consciousness come into being? What is the nature of consciousness? Is there a creator? Is there a correlation between consciousness and the creator? What is the relationship between consciousness and energy? These are lingering questions. This chapter provides the means to answer these outstanding questions. Their answers are keys to the "meaning of life."

Before addressing these final questions, we must understand the basic nature of mathematics. The word mathematics is traced to the Greek *mathematika* (or *mathema*) meaning "to learn" or "understand."

To conceptualize this ONE reality is to know and finally understand the mathematical, set theoretic and logical meaning of "one." To attain this realization, we must be guided by some simple, but obvious rules. They are:

Rule #1. Assume nothing: begin with nothing.

Rule #2. Conservation: we cannot end with more than we began.

This chapter follows a personal approach in developing answers to our basic questions. This chapter is presented in a way to represent the progression that led to understanding the Ultimate Principle.

I had reached an understanding of how everything relates (by delineating the structure of reality in this book). Cosmic structure denotes everything. I also understood consciousness and the mind are different. The mind is included as a something amid everything (it is represented in the Where Aspect Diagram). Consciousness did not correlate with any "something in" Reality. Neither did it correlate with everything. Interesting problem. If the problem were solvable, I needed to broaden my scope of understanding.

The initial tactic was to establish a conceptual link between mind and consciousness, then try to ascertain how it (consciousness) relates to everything else. How does mind connect to consciousness? This line of questioning was unproductive. The next idea was to see whether consciousness could be connected to everything (which includes mind). The question became "how does consciousness relate to the structure in this book?"

It is easy to hypothesize that everything was created from consciousness if it (consciousness) refers to what religion designates God. Even considering that, I did not understand how everything (Cosmic structure) was derived from consciousness. Determining how everything is derived should lead to a basic understanding of consciousness.

The Primary Equation: Mathematical Understanding

The above questions are answered by understanding an equation. Although I had delineated the structure of reality, I still did not fully understand it (reality). For example, I had yet to correlate the concept of entropy with anything in reality's structure. Trying to understand the basic nature of entropy provided the reason to explore the meaning of the primary equation.

The equation is a variation of $1/0 = \infty$. In this form the idea is zero (0) should divide into one (1) an infinite (∞) number of times. It has been a controversial equation. The following is our variation of that equation.

$$1 = 0 \times \infty$$

This equation looks impossible at first glance. Do we not assert $0 \times n = 0$? If this is granted, would not $1 = 0 \times \infty$ reduce to $1 = 0$? This does not "appear" acceptable. This conclusion is contradictory; hence, its apparent unacceptability. Regardless, let's proceed and see what we can learn.

There are three ways to begin an understanding of the equation. They are the following.

First consideration: the simple and conventional mathematical understanding of the individual terms of the equation.

Second consideration: the factors ("0") and ("∞") are understandable in terms of each other.

Third consideration: each factor ("0", and "∞") can be understood in terms of the ("1").

Let's examine each of these interpretations. We use physical examples to better understand.

First consideration. Convention posits the following:

One ("1") denotes the least of any number. It can also represent the least of any something. Using "water" for instance, ("1") would denote the least of that which can be denoted water, namely, one water molecule: two chemically combined hydrogen atoms with one atom of oxygen.

Zero ("0") denotes lack of a number: a "placeholder." It also can represent nothing: "no thing"—not something. In our example: No water, and no molecules of water.

Infinity ("∞") entails the natural numbers without end, viz, 1, 2, 3,.... It can represent an unending number of things. The earth (or the universe) could be said to harbor a quantity of water molecules approaching an infinite number.

We concede the definition of infinity "representing" any limitation to the number of things is less than mathematically satisfying. For example, to say there is an infinite number of things, like water molecules, is foolhardy when we also acknowledge there also exist things other than water molecules.

Even if infinity represented everything, we still find ourselves in the same uncomfortable situation. What if one of these "things," after our "defining" of infinity, gives "birth" to another "thing?" Doing this adds one more thing to "everything."

A more serious problem is defining "thing." For example, how do we consider the atoms (oxygen, hydrogen) composing a water molecule? To say "adding numbers to every number does not increase the value of infinity" does little to understand such a concept.

Let's restate our progress in understanding. We notice an inherent problem in defining infinity in any countable or uncountable manner. If uncountable, adding one more to "infinity" changes it. The definition of uncountability appears unsatisfactory because it is "indefinite." Infinity must be defined in terms other than mere endless sequentiality.

We learn one thing in the first consideration. Using things to "represent" numbers may benefit understanding the fundamentals of mathematics. Existents might help recognize problems that were artificially created in the effort to solve certain problems (viz, paradoxes and contradictions).

Using conventional definitions we conclude the following: Zero infers no value (of any number or anything). Zero represents no number or no thing. Infinity represents the greatest value (of any number or anything). This definition is "countably undefinable." But, is this not a definition? Yes, and no.

There is an apparent contradiction. A term should not be definable and undefinable at the same time. (We defined infinity as endlessly countable and therefore undefinable.) That is a problem. A contradiction exists by these definitions. Maybe the problem is not what is being defined. Perhaps the problem is the definition itself.

Question: Should the factors ("0" and "∞") defined individually as in the first consideration, dictate the understanding of the equation? If they do, what are the consequences?

Let's use an approximation of the first consideration to better understand the equation. Here, we consider defining an approximation of both zero and infinity in terms of each other. In the second consideration we examine the factors as approximations to their ideal.

The second consideration. Pragmatism suggests the following:

Zero ("0") is approximated by 1/1, 1/2, 1/3,....

Infinity ("∞") is approximated by 1, 2, 3,....

One ("1") is a constant of proportionality.

Pragmatism contends via the equation the two factors (the ["0"] and ["∞"]) vary inversely to each other because of the constant one ("1"). This is a relation of complement.

To understand this "pragmatic" definition (of the factors zero and infinity) let's do the following. Instead of the zero ("0") substitute a series of ever smaller fractions. 1/10, 1/100, 1/1000, and so on. (We could use the sequence of the reciprocals of the natural numbers 1/2, 1/3, 1/4,....) Let's represent this graphically (using Cartesian coordinates). As the fraction decreases in value, it more closely approaches zero ("0") (as an asymptote). The zero is approached infinitesimally.

Similarly, proportionately substitute the infinity factor (∞) with 10, 100, 1000, (or $10^{n\cdots}$) and so on. As the value increases, it approaches ever closer to ideal infinity.

The second consideration posits zero is approached as the fraction (e.g., $1/10^{x\cdots}$) decreases in value. And, infinity is approached as the natural numbers (here, in exponential multiples of ten) increase without limit.

The second consideration approximates the meaning of the individual definitions offered in the first consideration. Therefore, approaching zero is said to come ever closer to zero itself. Approaching infinity is thought to come ever closer to infinity itself. We just do not know what "zero in-itself" or "infinity in-itself" means. They still are undefined. If we cannot specifically pinpoint our terms, can they be explicitly defined? The catch, zero is nothing, and infinity is "endless" by definition. What can be accomplished with definitions like these?

In the second consideration, we conclude zero is less than any enumerable value while infinity is greater than any enumerable value. The difference between the first and second consideration is that zero in the second consideration is only approached (approximated) as "nothingness." This may not be the same as nothingness in-itself.

However close we approach either zero or infinity that (itself) is not necessarily the same as either zero or infinity in themselves (we differentiate between an "apparent," or approximate, and "actual" zero and infinity). And that leaves an unsettling problem. If there is a difference (between apparent and actual zero and infinity) however minute or undefinable, there may be a difference in understanding.

Question: Should the factors ("0" and "∞") define the understanding of the equation; therefore, define the "1" (as previously asked in the second consideration) or should the equation ("1") dictate the understanding of the individual factors ("0" and "∞")? Maybe this consideration will lead to a difference in understanding (between "apparent" and "actual" definitions).

Specifically, can we incorporate the individual definitions of the factors, as in the first consideration, into an understanding of the equation? For example, can we define zero as nothing and yet maintain that definition within the equation? Does the equation permit defining zero as nothingness? Here, we suggest the proper meaning of zero and infinity depends on their role (or meaning) in the equation. A vague way of stating this is the following. "Do the terms define the meaning of the equation as is customary, or in this case, does the equation (uncustomarily) define the terms?" Let's examine this possibility.

The third consideration. Each factor can be defined by how it functions in the equation.

Zero ("0") represents a "zero divisibility" aspect of the one ("1").

Infinity ("∞") represents an "infinite divisibility" aspect of the ("1").

One ("1") is unity. Unity equivalently represents a complementary combination of zero and infinity.

The equation's left side (i.e., "1" or unity) is equivalent to its right side (0 x ∞). In the second consideration: We found the "approximating" factors (respectively 1/n..., n...) for zero and for infinity are complementary. Zero and infinity are complementary aspects governed by the constancy of unity.

There is a leap from an "apparent" (approximating) definition to an "actual" definition when defining zero and infinity in themselves. Our task is to follow this leap and maintain complementarity.

One ("1") is said to be unity. What is the meaning of unity? Specifically, what defines unity? The equation ($1 = 0 x ∞$) suggests that "1" is definable by two factors—is equivalent to a relation between zero ("0") and infinity ("∞"). How do the terms zero ("0") and infinity ("∞") explain one ("1")? Can we understand zero and infinity by how they relate to unity?

The third consideration suggests zero is an aspect (of "1") indicating "no divisibility" (of the "1"); i.e., ["1/0"]. Infinity represents an "endless divisibility" aspect of the "1"; i.e., ["1/∞"].

We surmise unity is definable by two aspects—the factors zero and infinity. Unity is analyzable in terms of "no-divisibility" (0) and "infinite divisibility" (1/∞). Unity is also analyzable in terms of any division: it need not be a limitless division. A single unit of water is analyzable through its components, viz, oxygen and hydrogen. Unity implies we are concerned with something that is a completed composition. Let's pursue this line of inquiry.

Most people would say one ("1") signifies a unit. Offered for examples are: one foot, one yard, one meter, and one kilometer. The same idea applies to any individual thing: One car, one person, and one country. The concept of unity abstractly delimits and demarcates identity. One water (identity) molecule meets this criterion. A water molecule is identifiable by its composition (H-O-H).

Using the above understanding: We say the concept of unity (categorically) denotes an individual object. It does not explicitly pertain to an object in terms of its composition. Although composition is suggested, it has not been traditionally considered an explicit aspect of the concept of unity.

Though implied, unity has not "conventionally" inferred divisibility of itself. Unity typically describes the individual (or object): with or without regard to its composition (or divisibility). Be mindful, we are associating numbers with (physical) things to better understand the concepts entailed by the Primary Equation.

Conceptualizing the Equation

One of the tasks of theoretical scientists is to add meaning to equations. The idea is to extend our understanding of reality. We ask: What is the possible meaning of the equation $1 = 0 \times \infty$?

We correlate the (1) with the "unity" of Reality.

Let's say one ("1") is space-time (i.e., as depicted in Special Relativity) itself. It remains constant.

Let's correlate the zero ("0") with spatiality.

This would be spatiality in-itself (as noumenon).

We correlate infinity ("∞") with temporality.

This is time itself, aside from how it appears.

("x") is the interactive factor. It means the zero ("0") (spatiality) and infinity ("∞") are entwined. They are complements: one does not be without the other. ("x") also means each factor ("0", "∞") is an aspect of the one ("1").

Mathematically, the two factors are extreme and diametrically opposed values. To maintain the constancy of "unity" (i.e., of "1") the two elements, here represented by ("0") and ("∞") are inversely related.

Defining Space—the Problem

Let's examine the problem in understanding the equation in terms of reality itself. Certain conceptual contradictions appear.

There was no problem in understanding the equation on a purely mathematical basis. However, when the equation was translated into physical theory, we encountered problems in obtaining a noncontradictory conceptualization.

Specifically, a contradictory understanding of space ensued from opposing possibilities. When a contradiction appears, it usually indicates an error in understanding. Let's trace the source of the problem.

Let's attempt to "fit" the concepts of space-time, space and time to our Primary Equation. Let's postulate that the unity ("1") of reality represents "space-time." Space-time is composed of two parts, viz, space and time.

Using the equation, can we explain how unity ("1") as space-time, is realized as space and time? If we correlate space with the infinity factor, it means space is an aspect of Reality (e.g., space is infinite). However, if space is associated with infinity, and therefore, an aspect of reality, it (space) loses it status as non-reality. Space is nothing (i.e., not reality) by definition and therefore seemingly should not be considered a constituent aspect of Reality. Space is not a thing.

How can space be real (and therefore, a constituent aspect of Reality) when it is by definition not something—not real? How can space be nothing and still be a constituent aspect of Reality? This is our first major problem.

In our attempt to correlate space as an aspect of Reality, we find there are deeper considerations. If space is an aspect of Reality; then, surely would it not behave as though it were something? Does not space exhibit definable characteristics? Special Relativity and the Big Bang Theory suggest space is dynamic. It expands or contracts. Highly accelerated objects also exhibit these characteristics according to General Relativity.

Space must be represented in the equation to make conceptual sense. Yet, if the concept of space is kept in the equation, it seems to lose its status as nothingness! Either space is essentially nothingness, or it is not. The logical Law of Contradiction does not permit both.

If we say space is not real, then it is not any constituent aspect of Reality. If space is outside of what is real, it easily retains its definition as nothingness. (It is easy to state reality exists in space, if space is nothing.) If this is the case, space seems to be unrepresentable in the equation. This is the essential problem.

> The problem is more serious. What is space expanding into, if it is not space? If space is expanding into space, then there is a difference between expanding space and the space into which it is expanding. Otherwise, stating space expands makes little sense. This added problem is one we will revisit.

At this point, without violating the Law of Contradiction, there was one possibility. Let's consider the hypothesis that the energy of the vacuum and the vacuum itself (space) are different. We could seemingly have it both ways. Infinity had to at least stand for the energy of the vacuum. (There is said to be an infinite amount of energy in the vacuum.) That would take care of the infinity factor in the equation. We are left to suggest that zero represents space in-itself: apart from its energy. However, that presupposes via relativity that universal expansion is of energy (galaxies) expanding in space instead of an expansion of space itself. Our problem remains.

Let's examine these possibilities using a balloon. (Or, visualize the following.) This is a common depiction. Blow some air into the balloon.

> First case: Place black ink marks on the balloon. Continue slowly blowing up the balloon. The marks, like galaxies, continue to separate through expansion of the space between them. The marks (galaxies) are merely caught in the expansion of space. Here space is dynamic.

Many people still have difficulty in understanding how galaxies recede because of an expansion of space. Some people picture the galaxies as expanding in space. Let's look at this possibility.

> Second case: Blow up the balloon. Do not allow its air volume to change. Tie the balloon's opening. Rub the balloon against some wool cloth. It becomes electrostatically charged. Place small bits of paper on the surface

of the balloon. Now move some pieces of paper further apart (from each other). Here the marks (pieces of paper) representing galaxies, separate from each other by moving further apart in space. In this case, space is static.

Experiments like Michelson-Morley's were conducted to ascertain if space is static or dynamic. They obtained null results. Hubble's investigations of spectral red-shift indicate galaxies are receding from each other in a universal expansion of space. There has been little to indicate space is anything but dynamic.

The problem seems intractable. If space is definable as static as implied by the notion of nothingness, it can be represented by zero in the equation. However, that evades explaining spatial dynamism. Moreover, if space is nothing, how can it be an aspect of something, i.e., reality? If this were the case, we would not expect to find a representation of space on the right side of our equation.

If space is dynamic, how can it be said to be nothingness in itself? If space is dynamic, apparently, it would have to function as an aspect of reality itself. If so, we could include the concept of space in the equation. We suppose by stipulating space is dynamic, that it is only representable in the equation as the infinity factor.

The idea is to maintain the constancy of space-time (represented by [1] in the equation). If we say space is representable as the zero factor; then, we could maintain the idea of space being nothingness. Therefore, it shows no characteristics or properties such as dynamism. Does that leave us saying time is infinite? If so, are we saying the energy of the vacuum somehow characterizes time?

We are seeking a conceptualization of the inverse relationship between space and time. (Our mathematical treatment, in understanding the equation, suggests zero and infinity are inversely related. But, how is this to be conceptualized?)

Surmising our problem:

If we say space is a part of reality; then, it assures utility of the equation, but the cost is inability to define space as NOTHINGNESS as it is customarily understood (i.e., nonreality).

If we say space is not an aspect of reality then space retains its definition as NOTHINGNESS (or nonreality) but then reality (of space-time) cannot be explained using the equation (because the term space, itself, would be excluded from definition). (Space would not be associated with a factor in the equation.) If this is the case, the equation cannot be conceptually linked to the physics of Relativity.

The problem is quite simply one of contradiction. Either the equation completely explains reality, or it does not explain reality. In either case, the problem focuses on the definition of space. Either space is a part of reality, or it is not a part of reality. The Law of the Excluded Middle states space must be one or the other, and the Law of Contradiction states that it (space) cannot be both without contradiction. Namely: "Something cannot both be and not be."

This is an interesting dilemma! If we cannot rely on the Law of Contradiction, then what can we trust? The Law of Contradiction has been the very test of (logical) consistency. It applies to anything and everything. There has not been found anything in experience that defies this law. Maybe we need an understanding that supersedes both the immediate mathematical understanding of the equation and its physics' conceptualization. But what could it be?

Conceptualization led to a contradiction regardless which way we attempted to define space. There is just one noncontradictory solution. Allow (and define) space to both be a part of Reality (in which case it would be dynamic and hence explain the universal expansion of space) and allow it to represent nonreality (in which case it would be static and space could unqualifyingly retain its definition of nothingness) at the same time. But this idea violates the very understanding of logical contradiction.

If this tactic is allowed, it means there is something amiss in the present understanding of the nature of negation itself. Has the Law of Contradiction ever failed before? If the Law of Contradiction is not universally valid in defining consistency, it suggests there is a flaw in logic itself! Logic contradicts itself!

If there is not an inherent contradiction in certain uses of logic, there must be a flaw in the present understanding of contradiction.

If there is a flaw in logic itself, how could everything represented in this book fall so easily into place? We have not encountered a serious logical problem in building our hierarchical system until now. Where do we go from here?

Set Theory and Russell's Paradox

In trying to understand the dilemma, I had exhausted not only most physics' perspectives, but also any philosophical conceptions that might help solve the problem. Because I could not conceptualize the physics using the equation, I turned to the fundamentals of math and logic. I asked, what fundamental understanding lies between math and logic? I decided to try set theory. It is the last opportunity to resolve the contradiction. (Namely, that space exhibits contradictory properties.)

I learned that in May 1901 B. Russell, the English mathematician and logician, found an apparent flaw in deductive logic. Russell was concerned with sets.

A set is a class of elements. For example, "man" is bodily classified into various levels. Each level represents a "set" of general characteristics. The human body includes subsets. There is the circulatory system, muscular system, nervous system, and other systems. Each of these classificatory systems is composed of subsystems: tissue, cells, and lesser systems. These are typical sets. Notice the class or "set of elements" is not itself duplicated in the set. The notion of tissue is not included within the notion of cells. (Yet cells are included within the notion of tissue.)

Consider certain letters of the alphabet. Let's pick out the vowels: a, e, i, o, u, (forget y). These "elements" form a special set having defining characteristics (determined by how their sound is formed in the mouth). Let's call this set "V" (for "vowels").

Let's also erect another set and label it set "D." Let Set D represent the digits (fingers and thumb) of the typical human hand. We assign a number to indicate the digits. (For example, let the number one (1) stand for the first digit, the thumb; let 2 represent the first finger, until there is a pairing of numbers to digits.) To do this numerically we use the first-five positive integers.

We now have two sets. Let's display them.

$$V = \{a, e, i, o, u\} \qquad D = \{1, 2, 3, 4, 5\}$$

The first one is read: "V is equivalent to a set of elements a, e, i, o, u." The second set is read: "D is equivalent to the set that includes as members the 1st digit (thumb) 2nd digit, 3rd digit, and so on. These are very typical sets much as we find every day. Each set itself is not included as an element in the set. We do not see any "V" as an element of the first set. And, we do not see a "D" as a member of the second set. These sets are not members of themselves.

Russell was concerned with the very large "class of All classes that are not members of themselves." He was talking about the class of natural numbers (1, 2, 3,... ∞). Simply, all we need to do is explain the general idea. For our purposes, we "collect" the aforementioned sets of letters and numbers.

These sets do not list themselves as an element of the set. That is the requirement for an element to be included as a member of the set. Again, these are very simple sets. Nothing tricky here. Now we shall classify these sets at a "higher level of generality." Let's erect a set which includes both (above mentioned) sets. Let's call it Set "S". Therefore:

$$S = \{V, D\}$$

Now we should have a set (S) of "All the sets (V, D) which do not list themselves as members." Oops, we now have a higher-class set (set S) which is not included as an element of the set (S). (Only V and D are elements, which are defined by "sets that do not list themselves.") Therefore, this set is incomplete because, in the listing of sets, which are not members of themselves, we obtain the one additional set (viz, S) which is not included as a member of the set of "all sets which do not include themselves as members."

The very nature of listing sets, which do not include themselves as members, inevitably leads to a contradiction. Namely, that the set cannot be the "set of All sets which do not list themselves," if "S" is not also represented as a member of its own set. The problem is that these are the common "garden variety" sets.

Now let's try to rid the contradiction. We saw S is a set that did not include itself as a member. Let's include S as a member. It now meets the criterion for being a member of its own set. We can therefore complete the set so that it meets the requirement that "All sets are included." Let's list the set of "S" as one of the elements and see what happens. Therefore:

$$S = \{V, D, S\}$$

We now have a "complete" set. All sets are now classified. But upon closer examination, we find that Oops: we now have another problem. Our set no longer meets the definition of requirement, which states that only sets, which are not members of themselves, are allowed to be a member of the set (S). We clearly see "S" is now a member of itself! Hmm, another contradiction.

By making the set complete (i.e., that it be a "set of All sets") it becomes a set which is no longer a "set which does not list itself as a member!" Instead, it becomes a set, which does list itself as a member. This is contradictory to the requirement (that this be a set of "All sets which do not list themselves as members").

When we include the set as a member of itself (to complete the "set of All sets, which do not include themselves") it violates the requirement defining the set (that the set does not include itself as a member).

Each time we attempt to solve the problem, we end up in the opposite predicament. Because of this, one would expect and maybe conclude "there is no set in which the 'set of All sets which do not include themselves as members' will be complete." Each alternative solution leads to a contradiction! It is an unending cycle. This, in brief, is Russell's Paradox.

Gödel's Theorems

In 1931 Kurt Gödel, a mathematician, supposedly proved there will forever be mathematical "truths" that will not be proved with logic. We will not detail Gödel's Incompleteness Theorem. Briefly, it states any consistent mathematical system which can handle the necessary logic for all mathematics of which we may ask, is also not complete in that it cannot prove its own consistency. Or, any system strong enough to include elementary arithmetic cannot prove its own consistency. We shortly address Gödel's problem in a very simple way.

We learn there is a problem in logic. Logic was believed incapable of solving all "logical" problems. Many eminent mathematicians and logicians accepted there would forever be truths that are unsolvable within sufficiently complex systems.

The supposed limitation of logic is perhaps the same obstacle we encountered while attempting to correlate the concept of space to our equation. Maybe we missed something? Perhaps there is an erroneous assumption.

I began to ask, what permits the Law of Contradiction? How is it justified? I have not seen anything to explain this. Let's state the problem in familiar terms. How can the Law of Contradiction be explained? This problem should be solved before returning to our problem of defining space. There should be an explanation for the Law of Contradiction. We have yet to come across an instance not amenable to explanation. If it is unanswerable in that form (i.e., explanatory) there must be a reason. Either way, we would have an explanation. (Either there is an explanation for the Law of Contradiction or there is a reason ["explanation"] why there is no explanation.)

Solutions are usually quite simple. That is easy to say once we understand! Let's reexamine the problem. The problem is the following: A contradiction is found in "the set of all sets, which does not include itself as an element." This is what we learned. If the set is complete, it is inconsistent. If it is consistent (hence tautological) it is incomplete.

We will use a simplified (nontautological) version of Russell's Paradox to illustrate the solution. Before examining Gödel's problem, let's glimpse the concept behind Cantor's construction of transfinite sets.

Consider the set of All numbers. We might say there are an infinite number of them. Well that does not help much. There is a difference between ordinal numbers and cardinal numbers.

Ordinal numbers are indicated by the following, 1st, 2nd, 3rd,... nth. When we count "one," "two," "three," and so on, we are using ordinals. These numbers indicate "placement" in a series.

All elements of ordinal numbers are the same. There is no (quantitative) "abstract" difference in an object (element) sitting in the 2nd place and one sitting in the 25th place. Picture pennies laid in a row. A penny, in kind, is no different in one place than one found in another place. An ordinal number only represents placement in an ordered series.

> We could find in some of those places a cardinal number. We could look in the tenth place and find the number "six." This number "in itself" would be a cardinal number. But in this example, there is only "one" set in tenth place. There could be sets of sets in that place, but there is only one set of the included sets.

We ask whether there is an infinite ordinal number. There is no last natural number. For any ordinal number we pick, regardless of position, there is always another "numbered" position that is greater. To say there is an infinity of ordinal numbers is to say there is no last number. Therefore the term infinity cannot refer to the "last" ordinal number. Translation: "Actual" infinity cannot be included in the set of ordinal numbers. (We labeled the positive integers "potential" infinity.)

Can the word infinity refer to the set of All ordinal numbers? And if so, how? Before answering, let's look at cardinality.

A cardinal number indicates amount. It is reflected in the question "how much?" For example, 2 apples, 150 dollars, and ten billion grains of sand. We have already seen that the word infinity is not one of the elements of the ordinal numbers. So what is infinity? It seems the word "infinity" can be used to denote the set of ordinal numbers in the sense of cardinality (i.e., how many?). It answers by saying "an endless number."

We see a reasonable definition of infinity by connecting ordinality and cardinality. "Infinity" can be used in the cardinal sense to indicate the "set of" ordinal numbers. The infinite cannot be reached, even ideally, by counting (the elements of) the ordinal numbers, because of the following. Like most typical sets of things: the set of things is different from the elements that compose it. Actual infinity (cardinal) is approached by potential infinity (ordinal).

We readily see the parallel between this problem (defining infinity) and Russell's Paradox. Gödel examined mathematical truths against Russell's Paradox. Actual infinity is not included within the set of infinite elements (numbers). Rather, actual infinity is the set that includes every element. Let's briefly look at what Cantor was doing.

Cantor's Confusion

Cantor attempted to understand infinity. He used the method of diagonalization. Diagonalization combines and alters "selected" numbers from an array of numbers to obtain a number not previously represented. We cannot combine things into something else without those things being yanked from their former position. The Law of Conservation (typically used in physics) does not permit addition without something else changing. This idea applies to numbers. If we yank a number from an array, thus altering it, its original setting (array) is changed. The whole setup becomes other than what it was. The linear series of numbers or things is changed.

Be mindful of the following: Let's tie down the symbolic notation to what it represents. Unintentional errors easily arise when symbols take on a "life of their own." A symbol should not take on a meaning disallowed by the reality it represents.

If we line up the elements of a series (1, 2, 3,...) we cannot yank any of them out of place to combine into a functionally different "other" number or thing because that changes the original lineup because of missing numbers. What we can do on paper is not necessarily duplicable in reality. Reality sets the standard of what is allowable. Principles we use to understand and explain must be reality-compatible. Using diagonalization to create numbers not previously represented is an illegitimate maneuver.

We ask "how can we say there is no difference between infinity before and after adding something to it?" To say there is no difference is to ignore the issue. Moreover, that Cantor uses the power set to add infinity upon infinity evades properly (categorically) defining the term (infinity).

Cantor's work led to a definition of an infinite set as one that can be put in bijection (exhibiting a one to one correspondence) with a subset of itself. The idea is that two sets are of the same size when there is a bijection between them. The power set is another reality-incompatible concept.

To add something to a concept that was to be all-inclusive does little to aid understanding. Cantor's Paradox results (the set of all sets, i.e., that a power set of a given set is also a member of the set). The very idea of a set of all sets is said to engender a contradiction, hence the paradox. More about this problem later.

Again, we cannot duplicate the diagonalization method using things to represent numbers. A power set is not applicable to things without their duplication. A duplication of something is not the same as the original. A duplicate photocopy is not the original.

Mechanistically, we cannot allow a duplicating of numbers (even though added to other numbers). We cannot add "1" to "1" and get "2," nor can we add "1" to "2" to get "3." The very idea of numerical iteration has been poorly conceived.

Let's see what happens when duplicating numbers or let's see what happens when an array of numbers is carried to extreme.

What have mathematicians said about generating numbers? Consider the series of natural numbers "1, 2, 3,...." Given the unending series of natural numbers, we can only obtain a greater value by including the set of those numbers in the series. It is said we end up with one more element (being the set of all sets) which is not included within the set of All ordinal numbers. Consider the following:

Omega (the set of potential) infinity (Ω) = { 1st, 2nd, 3rd,...}

Read "infinity is equivalent to the set of the ordinal numbers."

The infinity of the set of natural numbers is often denoted "omega infinity."

Mathematicians have said "when representing the set of all natural numbers, we end up with one more element, viz, omega infinity itself." Their justification for saying this is the following. Because we can assign yet another ordinal number to this infinity (on paper, it appears to be another element) we end up with the following:

Omega infinity $\Diamond$ 1 = { 1st, 2nd, 3rd,... Ω (omega infinity)}

We could repeatedly do this, by always, in the next move, including the last element (which includes the set) within the set. There is no end to this maneuver. We continually obtain an endless buildup of ever higher-level infinities. It is open-ended. Yet this maneuver is contrary to our equation ($I = 0 \, x \, \infty$). This equation suggests infinity is of a "closed system" (indicated by unity). If there is an error here, where is it?

Infinity: Error in Understanding and Correction

The error is easily recognized. Mathematicians and logicians have committed the error of using quasi-reductionism to extend a generality. To extend a generality is to (inductively) infer a higher-level universality. (This statement becomes more understandable as we continue.)

Once an element is inclusively classified (by a set) it (in meaning) is no longer available (to be used) for further explanation at higher levels! For example, if we list three elements, 1st, 2nd, 3rd, we have done so ordinally. But, if we ask "how many" elements are there—we are asking how many ordinals are there in the given (cardinal) set. Adding these elements yields three items represented, viz, the 1st item, the second item, and the third item. Remember, each item, like one penny, has a value of one ("1"). Three items have a cardinal value of three ($1 + 1 + 1 = 3$).

Meaning changes from ordinality to cardinality from listing the elements to naming how many (there are). Saying, this is the third and final element is different from saying there are three elements.

Listing cardinal numbers denotes sets within sets. In a cardinal sense, 1, 2, 3, means a set implied by each element (i.e., a one item set, a two item set, a three item set). Adding these up (1 + 2 + 3) yields six items represented. By ordination (like match sticks) this is {1st}, {1st, 2nd}, {1st, 2nd, 3rd}. Note, each cardinal number (e.g., 1, 2, 3) is a set in its own right, because each listed cardinal element represents a series of ordinals. (Again, the cardinal number 3 is represented ordinally by three ones.)

Any natural number, used at a higher level as a cardinal, cannot exhibit the same meaning as it did when listed as an ordinal. Consequently, the set of natural numbers (denoted by actual infinity) must be defined and explained by other means. Let's restate the definitions of infinity.

> Potential infinity ("∞") signifies the endless natural numbers. "Ordinally" speaking, infinity can only mean there is no last number. There is no last (discrete and definable) number that corresponds to infinity (within the set of ordinals).

> Actual infinity "defines" the set of the natural numbers. It encapsulates "ideal" infinity. Infinity itself (actual) infers a leap from (hypothetical or) "potential" countability to "actual" uncountability. Actual infinity "explains" potential infinity by "inclusion."

It is not legitimate to allow (actual) infinity to be a member of the series of elements (denoted by potential infinity) composing the set. Actual infinity is the "cardinal" limit and is never reached by any sequential numbering of elements. The elements (ordinal numbers) of the set (only approach actual infinity but) can never be juxtaposed (in a one to one correspondence to [the set of ordinals defined by] actual infinity). "Many" juxtaposed (in the diagonal sense) to a "one" makes no sense. Reality-wise, Cantor's constructions are illegitimate.

Infinity (in itself, i.e., actual) cannot be treated as though it were an ordinal occurring at the same level of the elements (i.e., the natural numbers) comprising the set of natural numbers. A set of anything, and the set of everything as well as the set of natural numbers, is still a cardinal notion. Numerically, set and cardinality are practically interchangeable.

What does all this mean? Let's simplify. All natural numbers are "classified" by actual infinity. The natural numbers are not legitimately usable to continually assign yet another ordinal to yet another (higher) infinity. Because ordinals are already grouped by the set denoted by (actual) infinity, they (the ordinals) are not available for use in any other "one to one" correspondence to any higher-level generalities. Any other higher level generality must be considered in a different way.

If a number, specifically one (1) occurs at higher levels in a hierarchy—it infers a different (level of) meaning—and is not the same one (1) included within the natural (ordinal) numbers. If we find a one "1" at a higher hierarchical level, it is not used in the ordinal sense. If one "1" is used at a higher level, it signifies "inclusion." There is a difference in meaning between the cardinal notion of "inclusion" and the ordinal meaning of "placement."

Refreshing: The number "25" cardinally (how many?) means we are talking about 25 "individual" elements (or ordinals). That is, ordinals are represented individually by listing (counting) i.e., {1st, 2nd, 3rd,..., 25th}.

Does Cantor leave a legitimate problem? There is no ordinal infinity. Thus the label "potential infinity." Actual infinity can only represent the ordinal numbers in a cardinal way. If we say (actual) infinity cannot be represented as some type of ordinal, then there is no problem! Therefore, the problem (of higher level infinities) is created by erroneously considering infinity to be some type of ordinal. The set of all ordinals is not (itself) an ordinal (at least regarding the ordinality of the natural numbers).

That is, if we say actual infinity can only act cardinally in representing the set of All ordinal numbers, there is no problem. The set (denoted by the term actual infinity) of all (ordinal) numbers is not only not a member of itself (of the set) it stands alone and is not an aspect of any other set. We will shortly clear up this understanding.

(Actual) infinity should not be considered in any way to be a part of the ordinal numbers, which compose the set when the set is (defined as) infinite. It is a categorical mistake to do so.

Let's simplify. Ordinally, there is an approximation of (actual) infinity. It is (defined as) an uncountable sequential series (but hypothetically countable in an "endless sense") of numbers—each representing one item. This is "potential infinity."

"Actual infinity" can only be defined as the set of ordinal numbers. Actual infinity is a cardinal definition. Therefore, the complete set (cardinally—actual infinity) is only approached by the unending series of ordinal numbers (potential infinity). Actual infinity is the (cardinal) limit of potential (ordinal) infinity.

What have we learned? Quite simply, there is a categorical difference between a set and the elements that compose it. The difference, by abstraction, is between the ordinal and cardinal understanding of number. The typical error is supposing an immediately-higher-level generality (of somethingness) can fit into a lower level generality. How can it? The same idea applies to the numerical (representation or) abstraction of anything. Actual infinity cannot be included as an ordinal number. This understanding precludes extending potential infinity into ever-higher-level (transfinite) infinities (through power sets).

A given set is a higher generality of the elements that compose it. One might argue we could say this of higher-level sets, which are abstract, but what about specific things? Using familiar physical objects helps elucidate understanding. Let's consider three cases.

First, let's say, we place three match sticks into a box. We now have a set (box) of three matches. The box now represents, by inclusion as members, three matches. We cannot place our box into itself! Nether can we redefine the box as an included element (match sticks). The box cannot legitimately be counted as the fourth item following (in the same way as) the three matches.

We can fill other boxes and begin counting them. Numerical labeling of boxes is counting at a different level of inclusion. We may have a room full of boxes containing matches. We could ask again, how many (other) rooms (in the building) full of boxes, containing matches, are there? We cannot put a room into one of its boxes! A box is a higher level concept (in the sense of inclusion) than are matches. A room is a higher level concept than a box. A building is a higher level concept than its rooms. There is a hierarchical order to the concept of inclusion and set.

Instead of matches, we could use three small boxes. We can place the smaller boxes into a bigger box. We could count four boxes. When we place the three small boxes into the larger box, the relative use or function of each item or box must be considered. Not to do so is to commit the categorical error (mentioned above).

The error is not citing the role (of defined inclusivity) for each box. Bigger boxes are used (the role) to include smaller boxes. The categorical error results from not recognizing different levels of inclusion. Inventory numbering of boxes without regard for their function (size) is to "group" boxes of different sizes (and hence functional level) as though there were no differences among them. Trying to establish a coherent-inventory system for box types (to categorize and number) but not differentiating between box sizes, helps us not at all. Saying there is a categorical difference, then overlooking the difference when convenient, is to evade the role different levels (of box sizes) play in building the "inventory" system. There is a box count for each box size. Listing only the total number of boxes regardless of size, is no inventory at all. Cantor was trying to inventory sets of sets. He erected reality-impossible sets.

This idea also applies to the cardinal numbers. Because a cardinal number is characterized as a single entity it treats its "included" ordinals as though they occur at a lower-hierarchical level: a big box of smaller boxes

Second, consider one's automobile? An automobile is a more specific example (element) of the more general term (the set of all) "automobiles." Yet even here, upon closer inspection, we observe, for example, that the automobile is composed of lesser (hierarchical) level sets.

Consider the cooling system, the engine, the transmission, and the electrical system. And are not these systems broken down into even "lower level" sets (of specific components)? For example, the cooling system includes the radiator. The engine includes pistons, and other components. Specific automotive components, such as the alternator, are composed of even lower level components (brushes, armature, and other parts). These lower levels too are composed of yet even lower level components (molecules and atoms). Atoms are composed of electrons and nuclei. Nuclei are considered composed of quarks. Quarks have been hypothesized to be composed of lesser components (for example, rishons, named by H. Harari) having a charge of "0", or "1".

> Third, consider a number, such as the number 10. Is this number a higher-level abstraction representing a group of elements? Yes. Cardinally, the number "10" means we are considering a set of 10 ordinal elements. These elements are identifiable and countable, i.e., {1st, 2nd, 3rd,..., 10th}. Denary numerals are also representable in simpler terms. Breaking down the number "10" into a more basic representation gives us the following: 1010. This is binary for the number 10 (in the denary system). It cannot be represented by any simpler means other than the one to one representation of the unary system (represented by our match sticks).

Notice in these examples: A generality (denoting a collection of elements) ultimately can be broken down into a set of "0", and "1's". One ("1") indicates a match, a box, or a room. Zero ("0") is a place holder. Is not it interesting so much is explainable using the concept of "0", and "1"? We later return to the binary notation.

Mathematicians concluded (from Gödel's incompleteness theorem) that there is no complete set of mathematical truths. They state mathematics is inconsistent (because of the problem of logic akin to Russell's Paradox) when they consider infinity in an ordinal way. Why? Because infinity supposedly represents the set of all ordinal numbers. We have established by simple reasoning and redefinition that we cannot in any way consider "actual" infinity in the ordinal sense (of the series of natural numbers). A set is different from the elements composing the set. We may have three items that do not immediately compose a set. For example, three matches may not be "grouped" into a collection as happens when we place them into a box.

The big question here is: Is mathematical truth incomplete? Essentially, mathematicians believe it is incomplete if infinity cannot be categorically included as an element composing the set of ordinals. Every instance of including one more element (or infinity as another set) into a given set generates one more set not previously encountered. Each additional infinite set is a power set of an earlier established set. This is Cantor's legacy.

Instead of solving the problem of proper inclusion, Cantor constructed an endless piling of power set upon power set. He calls the inclusion of elements within each power set "transfinite" numbers. Transfinite numbers and their power sets defy final categorization. Cantor's transfinite constructions suggest there is no completed system. Like potential infinity, there is no end to it. Cantor's construction correlates with Russell's Paradox. If this is granted, there is no set of all sets. Accepting this we would concede the transfinite set building of numbers will forever be incomplete.

Instead of getting swept up in Cantor's wake, let's reexamine some fundamentals. When we see a concept more complex than what it is supposed to explain, we are already forewarned something is amiss. A higher-level concept should be simpler than what it includes. (Ockham's Razor: the Law of Parsimony.)

Infinity should be defined in a way consistency is maintained! And consistency is easily maintained if we do not consider infinity in an ordinal way. This book does not use the word truth to pertain to self consistency without (testing) correspondence (to reality). In short, there is no ordinal number (or even an unending series of ordinal numbers) that will ever correspond to actual infinity. Any number "higher" than the natural numbers, is not an ordinal. It must be a cardinal.

The set of all numbers (a cardinal notion) cannot be included as a member of the set of ordinals. "Everything" is not identical with the "set of everything." A box of three matches is not the same as three matches. One room is not the same as one box; one box is not a match. Yet, "one" ("1") in an abstract sense, looks the same. Without noting that one (item, or one set) may pertain to different levels of inclusion, we could easily suppose that any given "one" ("1") is "one and the same." The idea of sets infers a hierarchy of different levels of inclusion.

Let's reexamine our Primary Equation ($1 = 0 \times \infty$) using what we now know. We have solved the problem of the infinity of the natural (ordinal) numbers by redefining infinity in a simple way. Infinity is included as a factor in our equation. If this equation is correct then it is (mathematically) complete.

How can the Primary Equation be correct if logic is flawed? Admittedly, the infinity factor (in the equation) represents the unending series of natural numbers. Each ordinal is discrete and is therefore noncontinuous. Each ordinal is an explicitly definable number. This includes every (positive) integer. In short, the term infinity does not include nonordinal numbers. It does not denote the cardinal numbers. Cardinal numbers are sets of ordinals.

Infinity does not include the following:

It does not include irrational numbers. For example:

Pi, the ratio of the circumference of a circle to its diameter (3.141592653589...)

Square root of 2 (1.414213562...)

"i" the imaginary number is also not included. It is the square root of minus 1. This number is neither positive nor negative. It too is ordinally undefinable.

These numbers are nondiscrete. They are continuous. "i" behaves as a constant. If these so called "noninclusive" numbers are to be categorized, where can we place them? They are members of the ("0") factor (in the equation). How can this be justified?

Recall mathematician Georg Cantor showed that just as we cannot understand "continuous infinity" by anything less, so it is that a (continuous) line of zero length has as many points as an infinitely dimensional universe. Consider this also when we later ponder the meaning of the mathematical singularity.

Let's take a quick look at what we have accomplished. We have defined infinity in a way that removes the objectionable addition of enumerable (infinite) varieties of infinity. A reductionistic approach to the problem of infinity (which leads to an infinite regression of infinities by generating ever-higher-dimensional power sets) is not only not the answer, it leads to unsolvable problems. The reductionistic approach led to the idea that the ultimate foundation of mathematical truths was believed to be "forever" inconsistent (as suggested by Gödel's Incompleteness Theorem).

Pointing out mathematical difficulties, which we have circumvented, what problems have we solved? We redefined infinity in a way that ends its infinite-Cantorian regression. Recall the problem prompting this investigation. It was our attempt to define space. Overcoming the mathematical difficulty was easy. But, we still need to face the problem behind the mathematical difficulty, for it is the same problem behind our problem of space. It is a problem in logic itself.

Defining Space: Revisiting the Problem

In the sixth century BC the Greek philosopher and mathematician Pythagoras hypothesized nature was built according to mathematical principles. In the early seventeenth century Galileo said mathematics is the language of reality. Knowing this, we had trouble understanding the equation conceptually by how it relates to Reality. The equation can abstractly explain practically everything. The lone exception, to this point in understanding, was Nothingness itself. In physics, when a contradiction appears, it indicates something is not sufficiently understood.

Is space an aspect of reality or is it not an aspect of reality? Specifically:

Either Reality exists in space, in which case space is "outside" or "beyond" Reality, or
space is a component of Reality (in which case it [space] is an aspect of Reality).

Does the definition of Reality include the concept of space or not? This is our current problem.

The only way we could solve our problem is to have it both ways. The Law of the Excluded Middle and the Law of Contradiction are at issue. Space had to both be a part of Reality and non-Reality. Space had to act simultaneously as a dynamic somethingness and a static (non-dynamic) "nothingness." The question became "how could it be both ways?"

We saw how our Primary Equation can easily be understood mathematically. Reality, as a unit, is defined by the one ("1") in the equation. (We redefined infinity ["∞"] to rid the traditional problem of it [infinity] being defined in such a way, which becomes meaningless; therefore that it is open ended. We did this by "categorizing" the ordinal numbers [viz, potential infinity] into a cardinal set [and denoted it actual infinity]).

To solve our remaining problem, we turned to logic and set theory to learn if there had been, historically, any problem in understanding logic. We learned from Russell's assessment of Gottlob Frege's arduous work. Frege attempted to derive all the theorems of arithmetic from simple axioms. We found there was a serious problem. In tracing the problem to Gödel, we come to understand the problem of infinity and how the present understanding of it led to contradictions.

After Russell, many logicians, when confronting this obstacle, concluded there is a "flaw in logic itself." Mathematicians, after Gödel, also believed mathematical truth would forever be incomplete. Both logicians and mathematicians faced the same obstacle. That obstacle narrows to the nature of negation.

In trying to solve the problem of space, I faced the same problem logicians and mathematicians had encountered. After studying their efforts, I decided to look for a problem in their reasoning or lack of it. Because reality seemed so consistent in its order (up through understanding the equation mathematically) I decided there just could be some nuance logicians and mathematicians may have overlooked.

Overcoming Russell's Paradox

We know the problem of defining space converges on negation. What is the essential nature of negation? What permits it to operate (by default) on something? What is "this" that can negate somethingness? If I could solve Russell's Paradox, I could probably translate the solution to understand the problem of defining space in a way not leading to contradiction.

Let's assess our predicament. The only noncontradictory solution is to allow "space" to occupy a unique position—that space both be an aspect of Reality or somethingness, and be nonreality or nothingness. However, fulfilling this requirement, meant an apparent violation of the Law of Contradiction and its immediate consequence—the Law of the Excluded Middle.

Attempts to end the contradiction cause another contradiction! If this is viewed using the Law of Contradiction, we are finished before we begin. Can we look deeper?

Back to Russell's Paradox. Russell's problem arose from considering the ordinary set—a set that is not a member of itself.

We will not use the set called omega infinity that denotes All ordinals (natural numbers). However, this is the more suitable set to use since it relates directly to our equation and is the set Russell had in mind. Instead, we enlist our "higher-level set" (S) the one we used to introduce Russell's Paradox. It is a set of the subsets V, and D. "V" is the set of vowels and "D" is the set representing the digits of the typical human hand.

$$S = \{V, D\}$$

"S" is the set of All sets (in our simple example) which do not contain themselves as elements. The paradox arose in considering "the set of All sets that do not list themselves." If we cursorily examine this set, we would conclude, as have most before us, that this is indeed a set that does not list itself as a member.

If we can somehow show ALL sets include themselves as members, we will solve our problem defining space. How can it be done?

Surely, we do not see the "S" as a member (of itself). Let's examine this set a little closer and ascertain its meaning. Let's pick the subset "V" and take a closer look. Let's replace "V" by reintroducing V's equivalency back into set (S). Set "S" is the set of All sets that do not include themselves as members. "V" is equivalent to {a, e, i, o, u}. We insert it into the equation in place of "V." Therefore:

$$S = \{\{a, e, i, o, u\}, D\}$$

Perhaps, the specific members of the set are confusing the issue.

Let's strike out (or negate) every member of set "V" except element "a". Therefore:

$$S = \{\{a, \text{e, i, o, u}\}, D\}$$

We now have the following:

$$S = \{\{a, , , \}, D\}$$

Simplifying, we have:

$$S = \{\{a\}, D\}$$

Here, we are mindful there is a difference between a set and its members. Therefore, we can have elements, e.g., "a", "e", "i", "o", "u", which are not yet collectively denoted as a group or set. Only when these particular elements are considered as a "group" of vowels, do they form a set. Also, we can have a set with no members. This is denoted by the "null set" or { }. By convention, the word "set" in itself (devoid of members, and hence by itself) means an "empty set" or null set. Back to our set (S). Let's strike the remaining member "a" of subset "V". We have:

$$S = \{\{ \ \}, D\}$$

We have already established that a set with no members is still a set (null though it be). We can also establish that a set with specific elements (for example, "a") is nothing more than an "empty set" { }, which includes members! We remember from high school that a null set is, by fiat, also a member of itself. Are not All empty sets themselves, the same? Yes. Therefore, a null set is a member of All sets.

Now let's strike out subset "D" and remove it from the set. We have:

$$S = \{\{ \ \}, \ \}$$

Here, we clearly see in pure set language that a set is a member of itself. We could have originally included a null set as a member of our set. Thus:

$$S = \{V, D, \{ \ \}\}$$

We conclude from considering null sets that "there is no set, which is not a member of itself."

Now, we can clearly see Russell's Paradox arose because the "hidden" empty set was not considered. There is no paradox. An empty set is a member of All sets! There is no set that is not a member of itself! This solves the logical problem. What else does this mean?

We now understand the very nature of negation itself. Back to our previous sets. When we strike the member "a" from set "V", we effectively remove all distinguishability from the set. Therefore, the "V" designation no longer applies when we maneuver to erase the last member of a subset (here "a"). Removing the last somethingness "a" from the set, leaves us with its negation—nothingness.

Does not nothingness imply somethingness and somethingness suggest nothingness? Contradictories imply each other. It may seem ironic nothingness by its very nature implies somethingness. The set remains, although there is no differentiation of the set from an empty set. It is because the null set is a member of every set that generates the principle of contradiction.

The very idea of any set is foremost that of "an empty set" that can "take on" members (or subsets). It is simple. The very idea of negation is built right into the concept of set!

Since all setness, whether alone or including members, are by their very nature (in themselves) empty, the concept of "emptiness" enmeshes every set. There is no set that does not include this property. Every set and all sets are members of themselves. Therefore, the problem enmeshed in Russell's Paradox arose because it was believed no set could be a member of itself without violating the Law of Contradiction, viz, that something cannot both be and not be.

Defining Space—the Solution

How does our present understanding of sets affect solving the problem of space? Our problem was to define space in way "that seemed" contradictory. We needed space to be defined as "nothingness" while also treated as though it were a "somethingness."

First, we did not see how contradictories could be meaningful. But, when negation itself is the issue, we found the concept "negation" is built into the concept (set) of "somethingness." A set of nothingness (the negation of somethingness) can have (elemental) somethingness as members. Nothingness, as exemplified by the empty set, permeates all somethingness. All somethingnesses are members of the empty set, viz, nothingness.

We conclude not only can space be defined as nothingness, but also by its very nature, space is implied in every "set" of somethingness. The empty set is a *de facto* member of itself. It is a set that is always a member of itself. Our problem is solved. Space can be defined as nothingness and still be an aspect of reality. And, there is no contradiction.

Another differentiation: There is a categorical difference between an element and that element being a member of a set. This also applies to the concept of an empty set. There is the set of all sets: complete emptiness, or space in-itself. There is also a hierarchy of sets within sets. This suggests an "apparent" difference between space in-itself and space as an included aspect of Reality. Space, as generally understood (i.e., spatial extensionality) is seemingly different from space in-itself. Space "itself" is dimensionless (has no extensionality). Space in-itself infers (spatial) extensionality. Yet technically, there is no difference between space itself and its extensionality. Space appears extensional because of "dimensional" things in space. The difference is one of function. This becomes clearer as we continue.

What else can we learn? We can use our present understanding of set theory as a basis to gain a deeper understanding of reality.

Form—Content; Arrangement—Structure

Let's redirect our attention to some concepts that may help us better understand and define space. We begin with the concepts of "form and content." The concept of form (or pattern) and content goes back to Plato. His notion of form more closely refers to the notion of idea. Idea or form pertains to the visual aspect of an object. Yet, it more appropriately suggests an idealistic portrayal of the form. (Plato called his idea "the Good.") Therefore, form exists independently of the various things, which exhibit that form. Some philosophers later connected the concept of form to universals. This is the track we pursue in this book.

Form is identity specific. It bestows distinguishability by differentiation of shape. On an ocean beach we observe a sand castle. Its FORM is (identified as) the castle; its substance or content is sand. The substance (grains of sand) is all sameness—(practically) undifferentiated. For future reference: The grains of sand, concerning their role, act like the (quantum-virtual particles) "energy-stuff" out of which everything is constructed. Recall the author's original consideration of form arose from his need to understand entropy and how it fits into understanding the evolving structure of Reality.

A typical "thing" in physical manifestation has form. Consider the domesticated cat. It exhibits a certain expected form. A thing's form or shape is the usual means of identity. We see a certain form and identify it as a cat. "Meow," it is alive. What constitutes form? A cat, for example, is composed of substance or content—physical matter. Form is an expression of the "arrangement" of content. Everything shows two aspects—form and content.

There is a difference between form and content. We observe representations of a given form composed of different substances. We have seen the cat-form copied in plastic, aluminum, iron, and wood, for example. These copied forms, however, do not exhibit life. No meow here. (Understanding life itself is more fully explored in subsequent chapters.)

We also observe a given substance in different forms. We see wood carvings in different forms. We see wood statues of different animals, for example, bird forms and reptile forms. We also observe clay and bronze "art" statues of many different forms.

We observe "same form having different content." We also find "same content in different forms."

Form and content produce structure. Consider one molecular unit of water. A water molecule has structural components, viz, an oxygen atom and two hydrogen atoms. Any less than what constitutes one water molecule is not "chemically" definable as water.

If a water molecule is divided, it is no longer identifiable as water. Merely isolating an oxygen atom and two hydrogen atoms does not represent a water molecule. These atoms must be physically (chemically) bound together to function as a water molecule.

We note that an item's form is inadequately defined by its components. The reason is simple. We also must ask how content is arranged into form.

Identification of something corresponds to the item not merely in terms of its individual components (content) but by how its content is structurally arranged into form. Here we address the issue of different arrangement of content.

There are many examples of "same content, but different arrangement in form." A clay model of a cat can be reshaped into a model of a dog. Let's use a specific example from organic chemistry. Look at Figure 18:1.

We find the same number of atoms of hydrogen, carbon and oxygen composing (normal) n-propyl alcohol and isopropyl alcohol. There is a structural difference in the arrangement of their constituent atoms. Not only do we find the same atomic content arranged in a different form, but we find them (propanol and isopropanol) exhibiting different properties (such as, specific gravity and boiling points).

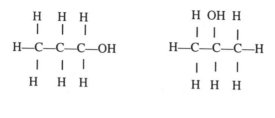

n-propyl alcohol   isopropyl alcohol

Figure 18:1. Alcohol Molecules

A question for later: What is the relationship between form and content? Is form a function of content or is content a function of form? Or is each a function of the other? Whatever the answer is: the critical question is "how is arrangement of content into form achieved?" Better yet, why is content arranged in one form rather than another? We later pick up on this question.

Whole and Parts: Reconsidering the Equation 1 = 0 x ∞

Any given object, as a unit, has two aspects—form and content. We correlate the notion of an object's form with the idea of "setness." We connect an object's content with the "elements" of a set. Instead of using form and content, we can use the concept of whole and part.

The notion of set is represented by the zero in our equation ($1 = 0 \times \infty$). The elements of a set are represented by potential infinity ("∞"). The combination of set and its elements, is represented by unity ("1").

In the context of unity, we recall zero meant "zero divisibility of unity." Infinity meant "endless divisibility of unity." How else can the equation be understood from the perspective of form and content? Let's look at yet another example of a unit.

A practical example of the two aspects of a unit is illustrated by a jigsaw puzzle. Dump every puzzle piece onto a table. Although every piece is present—they do not function as a "whole" until properly placed together. Only when puzzle pieces are sufficiently "collected" into their "proper arrangement" do we see the whole picture (of what it represents). When the puzzle is completed, there are two functions: one of wholeness (the picture) and one of parts (the pieces). The wholeness function transcends its parts. Let's go back to the numbers.

Zero ("0") conceptually, denotes "zero divisibility of the unit (one)" and therefore is a representation of the unit as a WHOLE. (If the unit is undivided, then it exists as a whole.) Zero ("0") represents the whole-aspect of the ONE as a FUNCTION (whether there are any parts or elements). In practical terms many people would argue unity is the same as the whole. But there is a difference. A whole ("0") is just that—no parts, whereas a unit ("1") (besides being a whole) almost certainly contains parts. A unit ("1") at least carries with it the partner of wholeness (or in the context of the third consideration of our Primary Equation—zero divisibility of the unit—oneness: viz, the whole). And, if we translate our abstraction (the equation) to represent something in Reality, a unit always infers a divisibility of itself. And if it does not, it does so at least potentially.

Although a unit infers divisibility, it is often not acknowledged. We note this issue focuses on identity. Recall a molecule of water is no longer identified as water when it is electrolytically decomposed into hydrogen and oxygen. When an item is deconstructed, its identity changes. This is probably why people do not venture further than the form, which defines identity, to consider that a unit also inherently has content.

The point is: Unity infers divisibility as well as wholeness. A yard stick infers three feet or thirty-six inches. A unit infers divisibility, but not necessarily sameness in identity. Our castle (on the beach) is different from the sand of which it is composed. The castle is identifiable because of the arrangement of sand (content) into a castle (a specific form).

The infinity ("∞") factor means there is also a FUNCTION of the ONE that is found in its INDIVIDUAL parts. The infinity factor means "infinite divisibility of the ONE or unit." (If the unit is endlessly divided; then, it also has a limitless number of parts.) In short, the infinity factor denotes the PARTS (or elements) which compose the ONE.

Let's reiterate:

The ONE or unit shows (is defined by) two FUNCTIONS.

The first factor zero ("0") depends on the ONE "unit" ("1") acting as a whole. The other factor, infinity ("∞") depends on the ONE "unit" acting through its parts (or elements).

Let's take another look at our Primary Equation. We have said there is little benefit in using "lower level" elements to ("reductionistically") understand higher-level generalities. The ("1") ("0") and ("∞") of the equation infer more than their "everyday" ideation. Let's briefly restate their meaning under ordination (as it occurs in everyday usage).

A zero appearing as an element of ordination shows placement. We see this in the denary system and the binary system. For example, zero is a place holder. In denary language, for instance, we find it in the number 10. In the binary system we see place holding in the number 1010, which corresponds to number 10 in the denary system.

Zero holds a potential position and gives added meaning to the natural digits (1, 2, 3, 4, 5, 6, 7, 8, 9) or in the binary system to the lone digit ("1"). Here, place-holding of value is "relevant" to the system used. But, meanings of relativeness is not conveyed by the one ("1") of the equation.

The one ("1") of the equation ($1 = 0 \, x \, \infty$) signifies completeness. One "1" means "unity." Unity means completeness. Unity is not necessarily pertinent to anything outside itself. In everyday usage a unit pertains to some part of a greater whole. For example, an automobile engine is a unit-part of greater a whole—the automobile. Regardless, we find:

Unity, as a completion, is inherently equivalent to two aspects or functions:

A wholeness function, represented by zero "0", indicates "zero divisibility of the unit" ("1").

A parts-function, represented by infinity "∞", signifies a "potential infinite divisibility of the unit" ("1").

The unity factor (itself) in the equation functionally transcends the rational numbers. If we insist on calling it a number, then it is a super-natural number. It stands alone. This is not (in meaning) the same "1" we find as a member of the positive integers. This super unit is only immediately interpretable in terms of its two aspects—zero and infinity.

In terms of each other, i.e., ("∪") and (infinity) we have:

> The zero ("0") factor is absolute. It signifies All (but not every or even any). It may "imply" an infinite number of parts, which are represented as a "collective" whole. Use the jigsaw puzzle as an analogy. Zero represents the whole puzzle, but not its parts. The whole is a function of parts. Zero is a function of infinity.

> The infinity ("∞") factor indicates plurality. It (infinity) is unity represented in terms of its "individual" parts. Infinity represents jigsaw puzzle pieces, but not the puzzle itself. Parts are a function of the whole. Infinity is a function of zero.

Numbers, the Continuum, Numerical Iteration, and Negation

We stated that the one and zero of the Primary Equation do not carry the usual meaning of the one and zero we encounter in everyday usage. What is the difference? How are the natural numbers created?

Pythagoras, the 6th century BC Greek mathematician, discovered the numerical relationship between the length of a musical instrument's string and the sound vibration it produces. Harmonious combinations of the relative length of vibrating strings are representable as a ratio of whole numbers. Pythagoras believed numerical integers were essential to understand All knowledge.

Near the turn of the twentieth century, G. Frege worked on deriving all the numbers from "nothing." Frege was motivated in trying to see how a creator could create something from nothing. In 1973 John H. Conway proceeded with his version of deriving the numbers from nothing. His work was a brief script entitled *All Numbers Great and Small*. He represents a "number to be derived" by the colon mark (":").

Conway begins with a null set { }, as did Frege. He uses two rules to derive the numbers. One gives the logical definition of number; the other shows how to order the numbers. He says: "Taking two of Frege's null sets { } : { } we can obtain "0". If we take a set of this zero and a null set {0} : { } we obtain "1". He said that a null set { } is nothingness which is not potential. He said a set of nothingness {0} is "Nothingness realized."

Using Conway's rules, one can successively generate all the natural numbers. The number to be derived is flanked on the left by the numbers already derived, and flanked on the right by a null set. He then derives the irrational numbers by using a set with element "0" flanking on the left, and a set of all natural numbers flanking on the right {0} : {1, 2, 3,...n}. Conway shows conceptually, there is no limit to how much an entity can be divided.

The idea of creating the numbers from nothing is a noble one. The problem with Frege and Conway's methodology is not what they did, but how they began. Specifically, what is the justification for "leaping" from zero to one, i.e., leaping from nothing to something?

We can accept the premise of "nothing." Merely leaping on paper from "0" to "1" is not a justification. We need a context or equation that allows us to derive the natural numbers. The best possible scenario is that zero ("0") itself generates unity ("1"). Once we obtain the positive integers, Conway's work succeeds in creating all (the other) numbers, including the irrationals and complex numbers.

All numbers can be generated from our equation ($1 = 0\ x\ \infty$). The equation is a clue to the context we are seeking. Let's examine this idea in terms of numbers.

Unity (graphically "0, 1", otherwise known as the real number line) is representable as the "continuum." The continuum itself is the wholeness function or zero. The continuum itself has no discreteness (or it would not be a continuum!). Denoting (nondiscrete) numbers (like the nonrepeating decimals—irrationals) is an attempt by mathematicians to discretely define the continuum itself. That which is continuous is not definable within itself as a PART without introducing an artifice in the understanding.

Yes, we can divide the continuum and derive the positive integers. When we do this to the continuum, it is not a characteristic of the continuum itself. It is a derivation from the continuum. This derivation is that of the parts-function—infinite divisibility of the whole function. Let's examine these ideas in more detail.

When we partition the continuum, its continuity is "broken" (it is no longer continuous). Discrete parts are the result. How is this done? We can perform a "cut" on a ("continuous") line. If we cut the line, it is no longer a whole unit.

Let's cut the line in half. We now have two discrete "sub" units. We could number them "one and two." As we will see, it is not quite this simple to justify. The cutting is acceptable; the labeling is not. Cutting is acceptable because we end with no more than we began, viz, the "WHOLE" which is the continuum itself. The whole (zero) is no thing: no number. The labeling is a minor problem. Who is doing the labeling? The cutting yields two items. But, to say there are two items, suggests cognizance is needed to evaluate the "reality of the situation" (here that there are two items). Then we also ask how is the "cutting" done (in the beginning)?

If a function is continuous, then does it not represent the ONE as a WHOLE? The whole is the continuum. What is the connection between discreteness and continuity (the continuum)?

The zero factor ("0") itself, in the absolute sense, is not discrete, but is CONTINUOUS. In terms of SETS, the zero ("0") is the set, but not its members—the positive integers (which are represented by the infinity factor). Zero ("0") represents the "set" of natural numbers, but not the natural numbers themselves.

We conclude "set" is more than the elements that compose it. Therefore, zero is greater than infinity. This conclusion is contrary to those we found in the first and second considerations in assessing our Primary Equation ($1 = 0\ x\ \infty$). Zero, being absolute, is the "set" of the natural numbers, i.e., potential infinity (but not the natural numbers or potential infinity itself). (Elements of a set are different from the set itself.) Because zero itself is the limit to which potential infinity aspires, it (zero) can also be known as absolute (or actual) infinity. Absolute infinity is the set of potential infinity.

The infinity factor ("∞") itself represents the traditional potential infinity denoted by the natural numbers (1, 2, 3, 4,...). It is the natural numbers we are trying to derive from the continuum. However, in our equation the infinity factor becomes inverted. Its role (in the equation) defines it as an infinite partitioning of the whole (i.e., of zero). It becomes "$1/\infty$".

If the value represented by the infinity factor is enumerable, its value will be represented as a fraction of the whole. The infinity factor will then represent fractions of the whole (i.e., 1/1, 1/2, 1/3, 1/4,...). This is the unending series of the natural numbers inverted. If we were confined only to this series of numbers, we would not see them as fractions of some whole, but as "individuals" in their own right. Instead of seeing each member of this series as fractions of a "greater whole," we would label them ordinally. We would count off or number them. One, two, three, and so on. This is our derivation of the numbers. We still must justify the leap from the whole (zero) to unity (one). We do this at the end of the chapter.

> Recall that in the beginning of this chapter our second consideration of the equation ($1 = 0\ x\ \infty$) found the infinity factor representing potential infinity. Our third consideration of the equation represented infinity as fractions of the whole. If the value of the infinity factor is less than infinity, the zero factor will represent fractions of the WHOLE (therefore, 1/1, 1/2, 1/3, 1/4,...). Here, the infinity factor represents a value less than potential infinity. The infinity factor was represented in our third consideration of the equation as fractions (of the whole). Notice that 1/1 represents unity as both whole and part without partitioning of the whole.

Let's look again at the meaning of "one" ("1"). ("1") is a unit. A unit is a SET with elements—a whole with parts. SET and the wholeness function are synonymous. A set is a group (of members) acting (functioning) as a whole. The set's elements are its parts. We easily see how the numbers are derived from the equation. The difference between our method of derivation and Frege's and Conway's is that we explain the origin of the numbers. Each number is a part of a greater whole. The natural numbers are categorically distinct from the "0" from which they are derived.

In deriving the numbers from the continuum, we note the following. We represent the numbers as "line lengths" rather than as points on a line. This also suggests that we cannot have a unit exhibiting both the wholeness function and parts-function at the same time, or at any given time.

Let's reiterate our understanding by considering a few examples. First, let's reconsider our "yard" measuring stick. A yard stick is usually considered a unit. But, it has parts—36 inches. A unit is not just the whole. A unit has two aspects—the whole and its parts. The whole and its parts have different functions. That is, their functioning occurs at different levels in a hierarchy. Functionally, we use the yard stick as a whole, or we use it to gauge inches.

The whole (the yard measure) is the "set" (the null set itself) having elements or members (the measure of inches). The yard is defined by 36 inches.

A water molecule is a functioning whole. Its elements are two hydrogen, and one oxygen. Atoms are functioning wholes in their own right. Oxygen and hydrogen atoms, independently, exhibit different defining characteristics than those of the water molecule. In this example, we could say that the atoms were derived from the water molecule. We end with no more than we began. We did not "materialize" atoms from nothing, as it is usually understood. We isolated the atoms from a greater whole of which they were a part—the water molecule.

The molecular function (in our example, the water molecule) results from the "collective activity" of the (chemically combined) atoms composing the water molecule. Molecular function transcends atomic function.

If we say that a part is not the wholeness function, and that the wholeness function is not the parts-function, then we are saying that each is the negation of the other. If it is not a whole, then it must be a part. If it is not a part, then it must be (functioning as) a whole. Think of the jigsaw puzzle. If it is no (not any) piece, and is not even every piece; then, it is the whole puzzle. If it is not the whole puzzle, then it must be at least one of the puzzle pieces. Each is the negation of the other. Negation of the whole is its parts. Negation of the parts is the whole.

However, the question remains: How can we derive the numbers and overall not add to what we started. Adding to nothing, without properly defining nothing is the problem that confronted Frege and Conway. They did not properly understand nothingness. Stating rules about how to accrue numbers is distinct from "explaining" the derivation of the numbers. We cannot "materialize" numbers or things without explaining how they "originated."

Negation is built right into (is intrinsic to) the concept of setness. Starting with "0", we can derive all the positive and negative integers by simply creating "pairs" of positive and negative numbers. Over all, we still have nothing. Positive and negative numbers are opposites and cancel, or negate, such that overall there just is nothing. For example, the negation of the number 5 is minus 5. Together they combine to give zero, or nothing. We learn this from the CPT theorem in physics. We still must discern a justification for this "separating out" from nothing—the something we denote "numbers."

Back to the real number line—the continuum. When we "cut" the line representing zero-unity {0, 1} into two pieces, we need to label one of the halved lines a negative value, and the other a positive value. Overall, they cancel or negate. We now have two lines. Each is a unit in its own right. Let's call the first cut line "plus one," and label the other cut line "negative one." Each line represents a 1/2 of the whole. But, if we were to count them, we would say we have two lines. If we happened upon the scene after the lines were cut, we would say there are two lines and not consider that these two lines are just parts of a greater "whole" line. That is, each line has a value of 1/2 (because each line is half of the whole).

We can continue cutting each successive line and obtain more lines. After halving a line, and then halving each again, we obtain more lines. We continue to halve each line. After a while we obtain a succession of broken lines. We generate a geometric progression of lines. We first generate two lines, then four, then eight, and so on. Or, in terms of the whole: 1/2, 1/4, 1/8, and so on. 1/8 means we have cut the original line into eight pieces. Each piece is an eighth of the original line. Ceaseless counting of this progression defines potential infinity. We note for future reference that infinity is somehow linked to the nature of time. Counting "takes time." This looks like a reductive understanding of infinity!

Let's continue our analysis now that we have "some" mathematical understand of the terms of the Primary Equation.

Zero is different from any other number. Zero transcends all (other) numbers. Note that an empty set necessarily implies a set with at least ONE element in it. (Nothing implies Something.) Whole implies parts. Each is the negation of the other. Zero implies the natural numbers. Specifically, zero can generate the positive and negative integers. When a picture (e.g., used for a jigsaw puzzle) is fractionated, it is no longer whole. Similarly, when we generate the integers, zero loses its wholeness function. Each negates the function of the other. If we translate this abstraction to Reality, we must add a caveat. We cannot have a wholeness function and parts-function "acting" on the same unity at any given time. (The time constraint will later be explained.)

Summarizing: If a zero includes another zero as an element then we have a set of a set. And that is the null set or empty set having an empty set as a member. According to Conway (1973) a SET of an empty set is a set with ONE element (another empty set) in it. Let's call this set by the mathematical notation (1). Therefore, we can see how the number ONE can be generated out of an empty set. Here, unity has two functions—the wholeness function and the part function.

Complex Numbers

There is more to numbers than those represented by the "reals." Certain equations do not have real solutions. Consider the power "$a^{1/n}$". If "a" is negative and "n" is even: no solution in the reals is possible. For instance:

$$x^{1/2} + 1 = 0 \qquad \text{or} \quad x = \sqrt{-1} \quad \text{or} \quad x = -1^{1/2}$$

This polynomial equation has no solution in the real numbers. Yet, if we allow $-1^{1/2}$ as a solution, we do obtain an answer. "$-1^{1/2}$" is not a real number. It has been given the name "imaginary" and is denoted "i". "i" allows solutions to equations that would otherwise be unsolvable. "i" is not like zero, or one. There is no "self order" to imaginaries. The term imaginary was coined by Descartes; the symbol "i" was first used by Euler.

Complex numbers are graphically representable in an Argand diagram (after J.R. Argand, a Swiss bookkeeper, who published his findings in 1806). Argand and Gauss realized complex numbers are representable in a complex plane. In simple (two dimensional) complex geometry the vertical axis is "imaginary." The vertical axis above the vertex is assigned "positive" imaginary values (+1i, +2i, +3i,...). Below the vertex on the vertical axis are negative imaginary numbers (−1i, −2i, −3i,...). To the right of the vertex are the traditional positive integers (+1, +2, +3,...). To the left are the negative integers (−1, −2, −3,...).

Multiplying a complex number by "i" corresponds to a 90° counter clockwise rotation in a complex plane. This was Argand's discovery.

$i^1 = -1^{1/2}$ symbolically representing a two-dimensional vector in the first quadrant.

i^2 or $-1^{1/2} \times -1^{1/2} = -1$ vector in second quadrant.

$i^3 = -i$ vector in third quadrant.

$i^4 = 1$ vector in fourth quadrant.

$i^5 = i$ back to the first quadrant.

This process repeats itself.

Complex numbers take the form "a + bi," where "a" and "b" are real, and "i" is the imaginary value.

Every number can be represented as a complex number "pair." If "b" equals zero, the pair "reverts" to the real number "a." Thus, a + bi reduces to "a" when b = 0. When "a" equals zero, the number is purely imaginary. Consider the following:

$$\sqrt{-16} = \sqrt{16x - 1} = 4x\sqrt{-1} = 4i$$

Complex numbers are included in our universal set of numbers. The one (1) in our equation ($1 = 0 \times \infty$) includes the complex numbers. Therefore, unity is a "super-universal set" (of the reals). The "reals" (rationals and irrationals) are subsets of complex numbers. The reason is that the imaginary number "i" just does not fit in anywhere else. This is a "bookkeeping" chore. We do note "i" cannot be derived without the concepts of negation, integer, fractions and exponents. And, these concepts are obtained after the concept of unity is a given. Our interest here is that −1 can "decompose" into two i's. Once −1 is given, the imaginaries can be generated. (−1 is derivable from +1; therefore, +1 = −1 × −1.)

We have mentioned that Conway derived the numbers from zero ("0") using two concocted rules. We are still left with the chore of deriving unity ("1") from zero ("0") without recourse to an "outside" rule. Zero ("0") itself should generate one ("1"). This is paramount to asking how we can derive something from nothing. Something cannot be derived from nothing unless something is inherent in nothing. We noted that the numbers had not been legitimately derived from zero (nothing).

Logic

Logic concerns inference and deductive proof. Logic has been studied for thousands of years. Early Greeks discovered the axiomatic method and applied it to geometry. From given principles (i.e., axioms) they could systematically deduce other geometric propositions as theorems.

Euclid in 300 BC, in his work *The Elements*, stated all the known postulates of geometry. He listed five postulates as the foundation of geometry. For example, between any two points only one straight line will connect them. A straight line can be extended indefinitely. Euclid also stated many other axioms and theorems. Euclid's postulates were found inappropriate for non-Euclidean geometries—curved spatiality. They do not represent curvilinear geometry (i.e., Riemannian and Lobachevskian geometries [after N. Lobachevski and G. Riemann]).

For centuries it was thought geometric principles could be derived from arithmetic principles. In the early twentieth century D. Hilbert demonstrated the consistency of Euclidean geometry hinged on the consistency of arithmetic. The basic issue capitulates to establishing the consistency of arithmetic itself.

Peano's postulates (after Giuseppe Peano 1894) were an attempt to establish basic principles for the arithmetic of natural numbers. Peano listed five axioms defining the generation of the natural numbers. (Peano used some ideas already developed by R. Dedekind.)

Peano begins by assuming the "natural" number one ("1"). He derives the natural numbers by stating every number (except zero) has a successor. He says one "1" is not a successor to any natural number. No two numbers have the same successor. Peano's last postulate states by continually adding the number one ("1") to each successor number the natural numbers are implicitly derived. Therefore, there is a "natural" unending sequence of numbers. The terms "One" ("1") and "successor" are not explicitly defined. Peano's last postulate infers the recursive idea of numerical induction.

Listing axioms to justify generating the natural numbers is not sufficient to explain their derivation. Why? The rules (axioms) themselves are in need of explanation. As inferred by structural levels represented in this book: to explain something calls for something else as a basis of explanation. For example, it has already been stated geometry is "explained" by arithmetic.

Because we have not found any "something" explained in terms of itself, we expect explanations to appeal to something other than what is being explained. Mathematics is used to explain "some things" in reality. Eventually, "somethingness" itself will be found in need of explanation.

Properly "deriving" or "explaining" the natural numbers has been a goal of many philosophers of mathematics. It would be better to derive some basic operations that could be used to generate the rules of arithmetic. Frege, Russell and others attempted to explain arithmetic (and hence all mathematics) using logic.

Logic pertains to allowable order or possible arrangement of existents (some things, every thing). We ask, how can somethingness (categorically) be explained? This is our present task. Here, "somethingness" pertains to the natural numbers (which can "numerically" represent things in pure abstraction). Our first "directional" guess is that an explanation for somethingness will be found in an appeal to that which is other than somethingness. This "otherness" must be identified, described and defined.

The essence of logic is that argument can be cast in "form" (or structure). The result is certain statement combinations are valid regardless of their content. Theorems are valid because of their form (structural combination) alone. Valid forms are called tautologies. Our present interest is to understand the most basic forms of logic.

True and False

A logical proposition is either true or false. A "true" proposition is (logically) valid. A false proposition is invalid. The nature of true and false is not explicitly defined in logic, except that true propositions are acceptable as premises for further deductions.

Though considered invalid, do false propositions have "real" meaning? Though the question is fraught with philosophical overtones, it implies falseness itself infers more than merely "not true." For example, is not it true that false propositions are not true? This circularity arouses our curiosity. To understand this is to answer why it happens. To do that requires we ascertain the "inherent" nature of true and false. We must examine meaning other than that offered by the "operative" definition that a true statement corresponds to something in reality. It also means that merely saying a proposition is "valid" must be further qualified.

We know from every day life that a true statement represents "something in reality." The words true and false carry a different meaning in logic than they do in daily usage. Though a logically true statement may also represent something in reality, we must first clarify its use in logic. We ask two basic questions:

What is the logical meaning of 'T and F' (true and false)? How are T and F used in logic?

What is the inherent nature of true and false? How do T and F function in their own right? Is there a "self-function" of T and F?

Obtaining understanding of the first question should point us in the proper direction to gain a foothold in grasping an answer to the second question. True and false "themselves" cannot be defined until certain problems in logic are better understood. Let's begin a basic understanding of symbolic and propositional logic.

Propositional Connectives

Basic logical "rules" show how propositions can be combined by various connectives. The usual connectives (or copulae) are the logical constants; "not," "and," "or," "if and only if," "if—then."

What are the basic logical forms? Consider two propositional statements abstractly represented by **"p," "q"**. The following are basic ways of combining propositions.

Not" (symbol "¬")	indicates negation or "contradictory of 'p" (p, ¬p). Other symbols used are the following: "~" (tilde) "–" (minus sign) "‾" (over-bar) p´ (accent).
"And" (symbol "∧")	indicates conjunction (p & q, we use p ∧ q). Conjunction is a joint (simultaneous) assertion of two enunciations (propositions in this case, called conjuncts). Other symbols used: "x", "." (dot indicating logical product or logical multiplication) "∩" (set intersection) "pq" (in this form the conjunctive copula between p and q is understood).
"Or" (symbol "∨")	indicates disjunction or "alternation" (p ∨ q). At least one of the two propositions (disjuncts) is asserted. Other symbols used are "+" (logical sum or logical addition) "∪" (set union).
"If and only if" (symbol "↔")	indicates equivalence, or "double implication," or "biconditional" (p ↔ q). We also use "=". Another symbol used is "≡".
"If—then" (symbol "→")	indicates conditional or "implication" (p → q). Is asserted when ¬p ∨ q. Not asserted only when p ∧ ¬q. Other symbols used are: "⊃" after Russell (meaning "includes" or "contains") "⊂" (set theory notation for "is subset of". "p" is called the implicans (antecedent) q is called the implicate (consequent).

Table 18:1. Propositional Connectives

Logical Consistency: the Problem

G. Frege and B. Russell thought logic was the basis for all discourse. They believed logic could serve as "universal" principles of inference concerning experience. In short, they thought nature, reality, was logical: meaning consistent.

Why did Frege and Russell believe language could be formalized in logic? Logic concerns "form" or arrangement of content. (We have addressed the concepts of "form and content.") In logic, content is an abstraction. Logical truth depends upon the arrangement of content and not upon the nature of content itself. Whether a compound statement is true (T) or false (F) depends upon the logical connective and the arrangement of the truth values of the clauses composing the statement.

We can check the truth value (either F or T) of various logical arrangements (of propositions) by looking at a truth table. They could very well be indicative statements (declarative statements representing "something" in reality). However, their truth only depends upon accepted logical construction. To not be distracted by content, we use the traditional "p, q" symbols.

Let's examine the truth table (a device invented by C.S. Pierce) for the arrangement of content denoted conjunction (i.e., p ∧ q). p and q represent propositions in abstraction.

1	p	q	p & q
2	T	T	T
3	T	F	*F
4	F	T	*F
5	F	F	F

18:2. True and False Conjunction.

We note in the second row, only in the case where p is true and q is true will **p ∧ q** be logically true. We also note from the last row that **p ∧ q** is false when both p and q are false. Our interest, however, is in the third and fourth rows. **p ∧ q** is false when only one conjunct is false (when either p is false, or when q is false—marked by asterisk).

F = T ∧ F: A compound statement (of conjunction) is false when at least one of its component propositions is false.

The above formulation is often mistakenly interpreted as the Law of Contradiction. It can, however, be "reformulated" as the Law of Contradiction as represented by the following:

$$\neg\, [p \wedge \neg p] \quad or \quad \neg\, [T \wedge F]$$

Verbally: "Not both p and not p."

$$Or \quad T = \neg\, [p \wedge \neg p]$$

"A truth cannot both (simultaneously) be p and not be p."

This most important of logical rules was recognized by Aristotle. He wrote of contradiction in his work *Organon.*

> Aristotle was primarily concerned with metaphysical truth instead of mere "correct thinking." We are interested in both. Aristotle argued against Heraclitus who believed it possible for "the same to be and not be." Heraclitus was defending the concept of "Becoming" that, if viable, seems to violate the Law of Contradiction.

Simply: A proposition is true if its contradictory is false.

Why has the nature of contradiction been misunderstood? The Law of Contradiction (T = ¬ [p ∧ ¬p]) has also mistakenly been called the Law of Noncontradiction because applying it to "everything" (every indicative statement) disallows the truth of its contradictory. Many logicians have not distinguished the difference between the formulation of the Law of Contradiction and its negation. The Law of Contradiction and the Law of Noncontradiction are not synonymous. This crucial difference is the focus of what follows.

Notice how "time" intrudes itself in understanding the formulation of the Law of Contradiction (T = ¬ [p ∧ ¬p]). If time is omitted, the formula loses its meaning. Loss of meaning is particularly noticeable when the propositions denote something in reality. A coin may be "tails" up (p) at one time and "heads" up (¬p) at another time. Even without specifically representing something in reality, temporality is "built-in" to understanding the basic rules of logic. In the next chapter we address the issue of time and its connection to this formula.

The Law of Contradiction has been the test of consistency. It applies to "everything." There has not been found any something, i.e., "anything," or (even) every something that poses a serious violation of this law (allowing for the temporal conditional). Let's state the Law of Contradiction in the following form. ("T" represents "something.")

T = ¬ [T ∧ ¬T] or in statement form:

"Something (1st T) can (=) not (¬) both ([ ]) be "something" (2nd T) and (∧) not (¬) be "something" (3rd T)."

If the formula pertains to some things in reality; then:

"T" is that which exists. A "something" exists.

"¬T" is that which does not exist.

The Law of Contradiction is a symbolic conceptualization of everyday experience. We do not expect to (simultaneously) find on a table a single coin both laying tails up (T) and heads up (¬T, or F).

"Something cannot both be and not be." The Law of Contradiction is typically applied to existents. Thus, "something cannot both exist and not exist (at any given time)." Therefore, the "logical form" of the Law of Contradiction tells us about the nature of content (any given something) without specification (of the content). After all, it was observing the nature of things that led to this "Law of Thought." The Law of Contradiction not only defines, but it also becomes a general law of "somethingness."

Using language-logic relevance, we simplify:

A true statement represents something (or its condition) existing (an event, e.g., a "tails up" coin) in reality.

A false statement does not represent something (or its condition) IN reality. If the coin is "heads up," then a statement indicating the coin is "tails up" is false. Note: The preposition "IN" signifies the idea of inclusion.

It was thought the Law of Contradiction was suitable to test consistency of any given system. If a tested system is logical, then it is representable by T valued propositions. T valued propositions are reducible to the basic laws of logic. That is to say: T valued propositions constitute logic itself.

A logical system means its formulation is representable within the basic rules of logic. Here, we can confine our questioning to a basic understanding of logic.

Can consistency "of" a logical system be determined from withIN the system, or is the question of a system's consistency a meta-question "about" the system? (A meta-question is one answered from "outside" the system.)

> First, do elements (e.g., p, q) of a system themselves indicate arrangement? Answer, no. Individually examining every element of a system does not show their arrangement. It is arrangement (of elements) which constitutes the system.

> Second, the arrangement of parts is dictated by the rules of logic. Logic is the system of arrangement.

What are allowable arrangements by the rules of logic? Compound statements exhibiting a truth value of T are allowed. If we say that T (true) values represent possible "things" in reality; then, the rules of logic pertain to "every possible arrangement of things." The rules of logic (themselves) dictate consistency withIN any system arrangement permitted by logic.

Next question is: "How can T values, themselves, be explained?" Briefly, how can something (represented by T values) indeed every something, be explained? Does this question take us "beyond" logic? Surely, it takes us beyond the basic laws of logic. This question can be reformulated by asking how logic itself is justified? What possible avenues of explanation are left? Our only recourse is to consider that which is otherwise, than the accepted rules of logic. Can we gain understanding by looking outside the logical system for answers? Can logic itself be explained?

We find two questions of consistency:

> One concerns the nature of the basic rules of logic. Therefore, a system is "logically consistent" only when T valued propositions represent its laws or is deduced as theorems in the system.

> Parts (elements) give rise to systems (arrangement of parts). A system, in itself, is an arrangement of parts acting as a whole. This was Cantor's original idea of set. He thought a set acted as a single entity.

Let's restate: Within a system (logical) consistency, pertains to the arrangement of the elements constituting the system. If any one element is "incompatible" with the "rest of the elements," the system will be inconsistent. "Arrangement" of the elements is dictated by and reducible to the basic laws of logic.

Let's again use our jigsaw puzzle analogy. If even one puzzle piece (part) is improperly placed or missing, the puzzle is inconsistent or incomplete. We could count and examine every puzzle piece and declare every piece is present and undamaged. We could place together (arrange) every puzzle piece to ascertain that they "logically" fit together.

Does the same test for consistency (of parts arrangement) apply to (the) parts (p, q) themselves? Yes, the same rule is used, but there is a catch. That is what we will soon determine. Logic dictates allowable arrangements. Compound propositions relate arrangement. The proper use of logic is to test arrangement of parts.

What test or rule is applied to system arrangement? It is the Law of Contradiction. It states a proposition cannot both be and not be. Applied to things or "parts" it states "a part cannot both be and not be (at any given time)." Puzzle arrangement easily passes this test. Puzzle pieces, however, pass this test only if each piece is composed of lesser parts. We address this caveat in the next subheading. Simply, if a thing exists, it passes this test. Applied to reality: the Law of Contradiction means there is that which is "something."

Other basic rules of logic are derivable from the Law of Contradiction. Acceptable arrangements depend upon what logic allows (through "T" truth values). The test of consistency (for elements and their arrangement) is logic itself, i.e., its tautologies. Any acceptable system is logical. In a single (irreducible T valued) definition, the Law of Contradiction defines acceptable logical arrangements of elements of any given system. It also decrees that there are things. (There are things [elements] or there would be no "things in arrangement.")

We immediately recognize if logical consistency is, at best, about the arrangement of (the elements of) a system, then it (logic itself) may not be testable by appealing to its own rules.

The lingering question is: Can logic be the test of its own consistency? Is logic itself, besides its applications, consistent? Is logic as a whole testable from within the system (of logic)? Is the Law of Contradiction the test for the consistency of logic (itself)?

The problem simplifies when we realize the question of consistency is answerable from two directions. One is reductive; the other is inductive. The reductive direction pertains to "what is" the nature of logical consistency. The inductive direction is "about" logic using a "meta-language." It concerns the consistency of logic.

The reductive question is answerable by applying logic itself. The reductive answer is obtained by applying the Law of Contradiction as the test of logical consistency. Why? It insists on T valued propositions for acceptance of logical consistency. In reality, it means (the given) parts are arranged.

The inductive question can be formulated as the following. Is logic as a whole consistent? Here, we cannot (reductively) examine logical arrangement for an answer. Recall the admonition. No one has witnessed a case where something explains itself. Using that track record, we do not expect logic, using tautologies or even its basic T valued "laws," to "reductionistically" explain itself.

In 1931 K. Gödel used the Law of Contradiction as the definition of consistency when testing his theory about the completeness of axiomatic systems. In short, the following summarizes his work.:

If a system is complete, it is not (logically) consistent.

If a system is consistent, it is not (logically) complete.

According to Gödel, by using the definition of logical consistency, i.e., the Law of Contradiction, a system cannot both be complete and consistent.

Keeping it simple: If we accept logic as self-referentially (reductionistically) consistent, then it is incomplete. This means the following: Even if we properly place every puzzle piece together, there is still a sense in which it (the system) is not complete.

Many scholars have since accepted that any (well-formed) system could not both be complete and consistent. Yet, it still seems reasonable that a system cannot be complete unless it is also consistent. Here is why: Saying a system is complete, only when it is inconsistent (and hence contradictory) makes no sense at all. The next question is obvious. "Has there been a misunderstanding about consistency?" We have set the basic ground work to gain an understanding of consistency that was overlooked.

If mathematics, as a (logical) formal system, cannot be complete and consistent, then using mathematics to elucidate a completed axiomatic understanding of physics is impossible.

Surmising: The problem of consistency extends in two directions (in a hierarchy). One direction is found within the system of logic. The other is found extending beyond the domain of logic.

The Law of Contradiction suffices to test the consistency of a system thought to be logical. The Law of Contradiction also is the test for any acceptable T valued rules of logic, as well as any tautologies and deduced theorems. The Law of Contradiction is the test of consistency for the domain of logic.

Is "logical consistency," defined by the Law of Contradiction, the appropriate rule for testing the "consistency of logic" as a completed system? Gödel applied the Law of Contradiction to a completed system and found it to be inconsistent. A problem crops up when applying the definition of logical consistency to the system as a whole. The problem is that the completed system is, well, contradictory. Either this is not acceptable or there has been a misunderstanding.

Consistency of Logic: the Solution

Gödel's methodology is insufficient to resolve the problem arising when applying the Law of Contradiction, the definition of consistency of logical systems, to (logic) itself. To resolve the problem of logical consistency is to examine logic from a meta-level (a higher-level inductive generality).

Accepting that the criterion for logical consistency is the Law of Contradiction, we ask, what permits the Law of Contradiction? Is there a higher-level law that generates the Law of Contradiction? If there is, it might shed some understanding on the nature of consistency as it pertains to All (and not just to elements and their allowable arrangements—Every).

Is the Law of Contradiction derivable from a more "primitive" law? Is there a "final" law that is the ultimate test of consistency? If there were a final law, we would expect it to be "outside" of logic proper, i.e., we would expect it to not be a tautology (if it were a tautology, it would be "included" within the domain of logic). If consistency is about logic as a whole, we do not expect consistency to be found within logic as a tautology or even a basic T valued-basic law of logic. Translation: We do not expect the Law of Contradiction to be the bellwether of the consistency of logic itself.

We suspect there is a patented difference between logical consistency and the consistency of logic. Let's develop means to further analyze this problem. Our immediate question focuses on whether "Every" is legitimately synonymous with "All?" We will find Every and All are not synonymous.

"Every" represents "every possible logical arrangement." "Every" is a parcel of logical consistency.

"All" represents "every allowable arrangement as a whole." "All" relates consistency of logic.

Let's probe more deeply into the nature of consistency. Essentially, Gödel asserts consistency is tested by applying the Law of Contradiction. There has been a problem with this approach. Applying the Law of Contradiction to a completed (whole) system leads to a contradiction. Why? The Law of Contradiction is just that, a law that is about contradictories! If there is that which can both be and not be, then we have a contradiction as expected when applying the Law of Contradiction (as the test of consistency).

The Law of Contradiction is the appropriate test for consistency among members of a system. A coin cannot both be heads-up and tails-up at the same time. The Law of Contradiction also tests possible arrangement. It most appropriately applies at once to every possible T valued logical arrangement. Two coins cannot occupy the same space at the same time. We can also interject there is only one arrangement of "Every" thing (at any given time). (There is only one sequential arrangement of the natural numbers.)

Is the Law of Contradiction a suitable test for the "set" which includes the rules of logic and any deduced T valued propositions? More pointedly, is the Law of Contradiction the proper test for logic itself?

Consider an example. A system is not completely defined by (numerically) totaling members. Confusion reins when not specifying whether a number refers to ordinality or cardinality. Ordinality, at best, by enumeration, refers to "every" (single individual member). Cardinality refers to a "collective Every" where the collection is represented as a single entity. A cardinal collection of members is denoted "All." We have distinguished between a set and its members. In our terminology we have discerned "every" is not equivalent to "All."

> Note the following. "(x)" means "for All x...." (It is also translated as "for any x", "for all values of x" exhibiting the defined form.) (This is not the same "x" denoted the "logical product.") ("x") is the Universal Quantifier. It represents a quality predicated of every member. In this use, saying "every member exhibits quality x" is considered equivalent to "all members exhibit quality x." In predicate logic there is no difference between "All" and "Every." Why is this mentioned? Logicians did not know an easy way to state "All" using sentential connectives (viz, "and", "or"). ("Listing" using disjunction could be practically impossible.) Our treatment of the word "All" carries a different meaning than the word "Every." "All" represents the system (set-class) containing members (and perhaps their arrangement). The term "Every" represents the members contained within the set. It may represent "every tautology." A potential problem arises when not distinguishing (a difference between) "Every" and "All." If the Universal quantifier represents only every "individual" member (exhibiting some property) then surely it cannot also represent the set of those members. Although the word "Every" can represent every T valued proposition, it cannot simultaneously represent the "set of" every T valued proposition. The "set of" is denoted by "All."

What is "the same which can both be and not be" that by the test of the Law of Contradiction is said to be a contradiction? It is that which can be represented by an empty (null) set. It is "set" itself without regard to its elements or their possible arrangement. It is no-thing, nothing, "formness" itself, the wholeness function, space, and the mathematical continuum.

> That which can both be and not be is the F (falseness itself) on the left side of the formula F = F x T.

What is the problem? When asking about consistency, we must be mindful of what we are addressing. The answer we find depends upon our direction of questioning. Are we seeking a "reductive" answer or an "inductive" answer? Are we (reductively-deductively) looking within the system for an answer or looking beyond the system (inductively) to explain it (the given system) from a higher level of generality?

To test something (anything, even every "individual" thing) for consistency, we expect to use the Law of Contradiction. A thing can also be an arrangement or system of things. The Law of Contradiction applies to analyzing the logical validity of the arrangement of parts.

Let's hypothesize testing parts (or a unit—arrangement of parts) for consistency. Consistency (for anything and everything) is defined by the following:

> "Something cannot both be and not be." Or symbolically, $T = \neg[T \wedge \neg T]$

This is the formula for the Law of Contradiction. Perhaps the usual negation of this equation hides its real meaning. How? Consider the following:

Negating each side we have: ¬T = [T ∧ ¬T]   This is verbalized as the following.

"No (true) statement can both be true and false."

"No thing (no something) can both be (a something) and not be (a something)."

The initial problem is readily apparent. Merely examining the first example (i.e., "no statement") does not yield much of a clue about any deeper meaning. Moreover, there is a common misdirection in understanding. "No statement" is typically a "cued" conclusion from "this statement." We look at a particular statement ("this"...) and derive a negative conclusion ("no"...).

The second example does give a hint of a deeper way of understanding the formula. Yet, even this verbal translation is unyieldly.

"No thing" is often construed to mean "not anything." The problem was that many discerned scholars thought no further than "no thing" (¬T) meaning "not this thing." This lack of reasoning tends to conceal the idea any given "no-thing" can be (and is) synonymous with (the whole of) "nothing." This will be explained.

There is a difference between the two understandings (no-thing versus nothing). "Nothing" asserts more than mere "no-thing." Saying that "it" (i.e., nothing) is not a part, or even not every part, is to say it is a whole. The whole does not simply mean every part. No-thing pertains to the "All" consisting of the collective "whole" of "every" no-thing. Every and All are opposites.

Refer to the analogy of the jigsaw puzzle in terms of the whole puzzle and its pieces—parts. For instance, we could take away one jigsaw puzzle part and ask what is left? Answer: the rest of the puzzle parts. A part and the "rest of the parts" are complementarily related. We continue taking away parts until there is, but one part left. We again ask what is left? Answer one part.

Let's take away the last puzzle piece. We ask what is left? Answer there is nothing left. The "whole" is equivalent to "not every" (part). "Every" (part) is synonymous with "not whole." Rather than meaning the same: "All" and "Every" are complementary. Here, we are applying what we learned in this chapter under the heading "Overcoming Russell's Paradox."

No part (representing a something, a unit) can exist and not exist (at any given time). This is a typical understanding of the Law of Contradiction when each side of its logical formulation (i.e., T = ¬ [T ∧¬T]) is negated (i.e., ¬ T = [T ∧¬T]). Merely examining the Law of Contradiction in these formulations usually does not arouse any suspicion about the problem of applying that understanding to a completed system. The reason is probably that a completed system has been customarily defined as "Every." "All" was considered synonymous with "Every." This is suggested when we read "All parts exhibit property 'x'."

Why has EVERY been confused with ALL? Again, let's personify a jigsaw puzzle piece—a part. It (the personified puzzle piece) looks around and sees that only parts exist. He surmises this because he too is a part. He reasons: Take away every part and there would be nothing. In this "nothing," he does not recognize a wholeness function. To him, absence of "Any" functional thing (part) implies no function at All. Yet, before there was a puzzle of pieces, there was the whole of which it was made (prior to the puzzle picture being cut into parts).

After the puzzle picture is made into parts, it is less evident there was first a wholeness function. Maybe this is why the consistency test for parts (the Law of Contradiction) has been (erroneously) applied to wholes.

The wholeness function should not be mistaken for the unit function. Not distinguishing the difference created the problem (of applying the Law of Contradiction to the wholeness function). This lack of understanding greatly contributed to certain problems in the fundamental understanding of logic (Russell's Paradox and Gödel's Incompleteness Theorems).

These errors in reasoning led to misunderstanding the nature of consistency. This culminated in logicians applying the wrong test of consistency to completed (whole) systems. Remember that a set is different from its members and their arrangement. Setness in itself is also functionally different from a set with members. Setness is absolute—relationaless. A set of elements automatically invokes the notion of relationship—whole to part(s) and part(s) to part(s).

Arrangement of members is a matter of logic. And logic dictates what arrangements are acceptable (i.e., T valued compound propositions).

Applying the test of parts' (even "every" part) consistency to the (completed) whole ("All") as expected, leads to contradictory (and therefore, inconsistent) results. Let's elucidate:

Any part is representable by and is defined by $T = \neg [T \wedge F]$ (the Law of Contradiction). Simply: PARTS (as unit-thingnesses) are not allowed both T and F values (in their conjunctive arrangement).

Any whole is representable by and is defined by $F = T \wedge F$ (the Law of Noncontradiction). Simply: a WHOLE, conjunctively, allows both T and F values.

Each law is the opposite of the other. What is whole—is not every part. The whole is opposite to every. A whole is not a part; a whole is not "every" part. A part is not a whole; "every" (constituted of parts) is not a whole.

Applying the test of parts consistency (which allows either T or F but not both) to the whole (which allows both: $F = T \wedge F$) creates an unnecessary problem. Applying the test for parts consistency to the whole results in undecidability (between T and F) because it (the whole) permits both T and F values whereas the conjunctive test for parts states that the conjuncts cannot both be T and F.

Concerning the whole: There is "nothing" (not "anything") to be decided. "Undecidability" is seemingly inherent in the formulation of the Law of Noncontradiction ($F = T \wedge F$). (It allows both T and F.) Let's look closer. The word "undecidability" is a poor one to describe the Law of Noncontradiction. Both T and F are "decidedly" allowed. The Law of Noncontradiction declares it.

Decidability is not really a problem until "something" is considered. Either it (something) is or it is not; a decision must be made. Decidability is inherent in the formulation of the Law of Contradiction ($T = \neg [T \wedge F]$). The Law of Contradiction, although the consistency test of thingness, only seemed to be a test of a thing's wholeness aspect.

Applying de Morgan's Law to the Law of Contradiction results in the formulation called the Law of the Excluded Middle ($T = F \vee T$). It must be either T or F, but cannot be both simultaneously. "Something either is or is not." The Law of the Excluded Middle forces the issue of decidability.

Applying the test of wholeness consistency (formulation of Noncontradiction) to parts engenders a contradiction. Where the Law of Noncontradiction yields both T and F, the Law of Contradiction does not allow both T and F.

Consequently, if a system were deemed consistent (by the Law of Contradiction) it was incomplete (because the formulation "every" was misrepresented as "All.") Although "All" is representable "within" a sufficiently-well-formed system (lower level in the hierarchy—as a subset) in that capacity it is not "apparently" representable of the whole system. Instead, it "represents" the "wholeness function" of a subunit "in" the system. Regardless, how it (the wholeness function) is "apparently" represented, there is no differentiation of the wholeness function. If there were a difference between the wholeness function in itself and its representation in lower-hierarchical levels (subsets) then it would no longer be the wholeness function. There is no real difference between the wholeness aspect of one thing and another. The difference is only apparent—not real.

When a wholeness function is "characterized" at a lower level—it may be described differently because of the "role" it plays at that level. The difference between the wholeness aspect of one thing and another is one of function. This will become clearer.

The Law of Noncontradiction is (conventionally) not truth-functionally true and is not considered a law in the (logical) system. It is derived from negating the unit-system itself, viz, by negating the Law of Contradiction. It is derived by negating the test of logical consistency, which applies to every logical rule—every T valued proposition, and any derived theorems in the system.

The Law of Noncontradiction is found in the meta-language (relative to members and their arrangement "in" the "logical" system). Although the Law of Noncontradiction can be "codified" within a system, it still will not yield a decidable truth value (i.e., either T or F, it yields both T and F values). Note, within a reasonably complex system we expect subunits (in a hierarchy) to exist. Each subunit is also "partially" (poor choice of words—better—aspect) defined as a wholeness function in its own right. Reminder: A unit has two functions: a wholeness function, and a parts-function. The Law of Noncontradiction lays outside the domain of logic. The bonus is such a characterization leaves open the possibility that it (Law of Noncontradiction) may yet "explain" logic. Can it do that, and if so, how?

Completeness and the Ultimate Principle

If a system is complete (is whole, i.e., includes conjunctive derivation of both T and F values) it is inconsistent (using the Law of Contradiction). The usual definition of a Law of Logic or theorem thereof is that it be "T" valued only. The "F" on the left side of the formula $F = T \wedge F$ suggests it not be a theorem. Only "T" valued compound propositions (truth-functionally true) are conventionally considered theorems.

If we negate the (usual) test of logical consistency (the Law of Contradiction) we expect to "inductively" derive the test of the consistency of logic (defined by the Law of Noncontradiction). The Law of Noncontradiction is itself only "inconsistent" when applied to "T" valued compound propositions. This is what we expect when applying the Law of Noncontradiction to that (any somethingness) which cannot both be and not be.

The Law of Contradiction defines "T" (trueness). To maintain consistency we do not expect to apply the Law of Noncontradiction to "T" valued propositions. This is obvious. Then, why would anyone apply the Law of Contradiction to "F" valued propositions? Doing so yields a contradiction as expected. Remember that there are no discrete F values. Every F value is synonymous with the whole of nothingness. There is just a single "background" F value which "covers" for any "apparent" F valued proposition.

The consistency of logic (as a whole) cannot be proven from (within the system of) logical consistency. To test consistency of parts arrangement as a whole system is to appeal beyond the system (of relationships) itself apart from its members and their "logical" arrangement. Systems are typically defined by the arrangement of parts. Every allowable arrangement of parts is dictated by the rules of logic. In this sense, "system" means "Every" logically possible arrangement. Further, the word "system" can also pertain to the parts arrangement as a collective whole (i.e., set). Here "system" means "All." "System" can mean both—All and Every. Its use must be specified.

Let's use our jigsaw puzzle analogy: We gain understanding of the (meaning of) puzzle pieces by viewing the whole puzzle. We understand the whole puzzle by viewing its parts when "pieced together." We understand the whole by the arrangement of its parts. We understand parts in terms of the whole.

Let's state some of these views in more familiar terms. We process information in a hierarchy by induction—appealing to ever higher levels to "explain" the arrangement of subsumed levels. Conversely, beginning at high levels of generality (universality) we reductively (by deduction) analyze (by "breaking down") the "given" (whatever) into "lower" (classificatory) levels. The Law of Noncontradiction explains the Law of Contradiction. The Law of Contradiction is explained by appealing to the Law of Noncontradiction. The Law of Noncontradiction generates the Law of Contradiction by negation of itself.

There is no question, which is the Ultimate Principle—the one that cannot be without the other. In jigsaw puzzle terms: parts are generated from a "fractionating" of the whole from which they are derived. Negating the whole yields parts. The whole can "Be" regardless whether there are any parts. Parts cannot Be unless there is a whole from which they are derived. Parts cannot exist unless there is a greater whole of which they are a part.

Can "something" (represented by "T", and defined by the Law of Contradiction) both be and not be? No. Only "nothing" (represented by Not T or "F", and defined by the Law of Noncontradiction) can both be and not be. One is the test of the other: yet they are opposites. For example, "it is true (T) that nothing can both be and not be." But, if "it" (the Law of Noncontradiction) is true, then it is indeed false because it is falseness that the Law of Noncontradiction defines. When we ask "Is this false statement true?", we answer, yes it is true; to do otherwise is to deny that it is false. If it is true then it is what it asserts, therefore, that it is false. Once we "get into this loop," we meet the notion of circularity.

Circularity is not an intractable problem; rather, it is an inevitable product of unceasing questioning. It occurs because of "jumping" out of "what" is being questioned. The test of trueness is the Law of Contradiction. The test of falseness is the Law of Noncontradiction. There is an inescapable dichotomy between trueness and falseness.

Another way of stating this problem is the following. The parts (obviously, of a system) and their arrangement cannot prove the consistency of the system as a whole. The system "includes" the T-valued laws of logic and its tautologies and theorems. The crux of the problem is that the system "in itself" is different from what results from its inclusion.

To prove consistency of logical arrangement requires "assessing" the arrangement of inclusions from a higher "meta" level. Meta language is also the rationale of invoking the "Theory of Types" used by B. Russell and others in an attempt to rid logic of the problem posed by the then poorly understood "set of all sets."

> Meta language infers a hierarchy. The historical problem has been applying the meta-language in a reductionistic sense. For example, one could say that a "power set" is of a higher level than its initial members are. Continually erecting sets of power sets, though endless, is not a proper inductive (categorical) grouping of original members. Instead, this maneuver is a synthetic grouping that is endless. It does not solve any problem. It only evades categorizing members and sets into higher level generalities. Saying such a maneuver (power sets) is synthetic is to assert that it can only be done "on paper." It is not something realized in reality.

The Law of Contradiction applies only to parts and their arrangement. It should not be used as a test of consistency for wholes. That which functions as a whole is complete. When the Law of Contradiction is applied to wholes, a contradiction results. The test of wholeness can prove "nothing" about parts.

Gödel would say consistency is undecidable or unprovable within a completed system. The problem is apparently caused by applying the test of completeness (set of "all" sets) to "every" possible logical arrangement of its members. This has been the reason for the problem in defining consistency. The wrong definition was used.

Completeness of a system does not concern tautologies, theorems, or even the basic "T" valued laws of logic. To erect a statement that "we know to be true" (as embodied by the Law of Noncontradiction in the sense mentioned above) but not provable within the system is to confuse the proper use of such an "outside" statement. Yes, the Law of Noncontradiction is easily derivable from "within" the system, but it is not a "part of the system." Rather, it is the "whole" acting upon a part. Just as a part acts within a whole (of which it functions as a part) so it is that the whole can act upon a part.

When we say "Not Every" it engenders the notion of "All." Any time we use the concept of "negation," "Every" and "All" interchange roles. When we erect an inclusive-categorical ("All") statement (about "Every") we can easily find ourselves automatically "kicked outside the (tautological) system." The same happens when we negate a part and obtain the "rest of the parts," only again to negate and include every part in the negation.

The Law of Contradiction defines the nature of every T valued proposition. In short, T defines "acceptable" logic. Encoding every T valued statement as an All-embracing whole, is to "jump" outside the logical system itself. "Outside" the system we find the single F valued statement represented by the Law of Noncontradiction. "Not Every is All."

Can we "push" the solution of the consistency problem and either find an error in understanding or learn more about consistency? It is a simple problem, and we should expect there to be a simpler understanding of its resolution. Let's reassess what we have learned.

The negation of the Law of Contradiction yields the Law of Noncontradiction. If what is being tested is not a part (or an arrangement of every part) then it is a whole. If it is not a whole, it is a part or every part. There are no other choices. The Law of Noncontradiction is the only choice for testing consistency of wholes.

If it is not false (F) it must be true (T). When we say a "false" statement (e.g., the Law of Noncontradiction) is true, it does not mean the statement itself is true, but that we are saying "from outside" that it is true that the cited statement is false. The problem of circularity is only encountered when we do not understand our "position of questioning."

We question falsity using trueness; we question trueness using falseness. What-why; why-what. We continually tumble in questioning if we do not embrace a standard. Choose the highest level reason (whyness) and cease the circularity of inquiry.

The Law of Noncontradiction is the test of that which is: no-thing, not something, not anything. The Law of Noncontradiction pertains to wholes (that which is functioning as wholes—wholeness function). It does not apply to parts.

Let's say we shall test the wholeness function for consistency. Back to our jigsaw puzzle analogy. Consistency (for that which is "no part" or even "not every part") is defined as the following.

"A whole can be and not be."

This is the Law of Noncontradiction. There is a wholeness function, the subject—the whole picture, before its cutting into puzzle parts. What is predicated of the whole? The whole has two aspects. Let's explain.

First aspect: "it (the whole just) is." Absoluteness—Nonbeing (zero).

Second aspect: "it (the whole when also entertaining a part) can be." Beingness (unity).

After the (whole) puzzle is cut into parts and thrown onto a table top, there is no "apparent" wholeness function. Neither is there a "collective arrangement" of the puzzle parts. (Yes, there is disarray of parts—chaos.) We usually reserve the term arrangement to mean "orderly pattern" (i.e., logical). Until the jigsaw puzzle parts are assembled back into a "collective whole" there is no "apparent" wholeness function. Does the wholeness function persist no matter what the order?

When it (the whole) "is not," it is a parts-function. (It may be a "part" functioning as an arrangement of sub-parts.) Yet, wholeness function is always (time-wise) there in a potential sense. It is always there whether time (flow) is or is not considered. The wholeness function persists independently of any consideration.

Parts cannot exist but that there is a greater whole of which they are, well, a part. Yet, wholes can be with no parts. We discern a difference here in meaning not only between whole and parts, but between false and true, between (null) set and a set containing members.

The wholeness function seems paradoxical. Take away every part and there is not anything left. When applying the test of consistency to the whole—what are we testing? It seems that we are testing no thing—nothing. This is correct. From the perspective of parts there seems to be no wholeness function. Taking away "Every" leaves us with "not any" or "All." "All" is the wholeness function. We learned no-thing—nothing (itself) is the wholeness function.

The equation representing the Law of Contradiction is the negation of the truth values listed in the third or fourth row in the truth table for conjunction.

Therefore:

$T = \neg [T \wedge F]$ is derived from the negation of $F = T \wedge F$

We obtain: $T = \neg [T \wedge F]$ by negating $F = T \wedge F$; therefore, $\neg F = \neg [T \wedge F]$

Negating the Law of Contradiction yields the Law of Noncontradiction.

Negating the following statement:

"Something cannot both be and not be."   yields

"Nothing can both be and not be."

Some suppose that "something cannot both be and not be" is another way of saying, "nothing can both be and not be." Yes, and no. Yes, each is another way of stating the other; no, they are not synonymous. They seem to say the same thing. But do they? In one case we are talking about "something," in the other, we are talking about "nothing."

"Something" pertains to some-thing, or any given thing. Categorically, the concept of "something" can even represent "everything." Any "something" is a unit "1."

"Nothing" pertains to no-thing, not anything. Categorically, "nothing" very appropriately represents "not (even) everything." Nothing is zero "0."

Clearly, "everything" and "nothing" are antonyms. Generally, negating one yields the other. It is easy to see this when juxtaposing whole against part. It is not so easy to see this when a unit (composed of whole and part) is negated and we are left with only the whole. Each is the complement of the other. "The whole is": regardless whether we consider the "existence of unity." We note that "existence" of "somethingness" is only possible because of partitioning of wholeness itself into parts.

Nothing-something is a complementary relation. If we take away "every" puzzle piece, the whole still is. It seems "puzzling" at first, but "the whole just is," regardless whether puzzle pieces (parts) exist or not. The word part implies each (part) is of a greater whole. Nothing is All. Everything and All are complementary.

We have found that F and T are derived from F alone. False generates true. We learned F is also a function (wholeness function). What else does this understanding entail?

We have found the higher level principle from which we can derive the Law of Contradiction. It is called the Law of Noncontradiction. The Law of Noncontradiction is the most "primitive" principle. It assumes Nothing. It defines nothingness.

Negating the Law of Noncontradiction generates and defines the Law of Contradiction. The Law of Contradiction is defined by somethingness. Nothingness generates somethingness.

The logical conditional ($p \rightarrow q$) also generates true propositions from false ones ($F \rightarrow T$). A false proposition implies any proposition what so ever. The Law of Noncontradiction logically implies the Law of Contradiction.

Traditionally, a true proposition cannot imply a false proposition. Let's examine this convention.

Frege pondered material implication. B. Russell embraced it, and discerned a difference between formal implication and material implication. Russell said material implication means it is not a matter of fact that p can be true and q false. (A true proposition cannot generate a false proposition.) Therefore, it must be the case that either p is false, or q is true. Russell believed that only in the sense of material implication is deduction or proper inference possible.

The idea a true consequent is derivable from a false antecedent seems illogical. In a way it is. But, it is far from meaningless.

A true proposition is derivable from any proposition. Specifically, logicians did not like the idea that a true proposition is derivable from a false proposition. The idea (a true proposition is derivable from a false premise) has been accepted because "it is the best that can be done with the conditional."

The everyday use of implication is defined within the notion of material implication. Everyday use of implication is represented if the consequent (q) is true and the antecedent (p) is true. The idea is that the implicans (p) must be known true or at least asserted. If it is not asserted; then, an implicative relation cannot be established between the implicans (p) and the implicate (q). Logical implication is important because it enables the derivation of theorems.

However, the idea a false proposition can generate any proposition, instead of being meaningless, actually represents something profound, viz, that anything, a given something, can be generated from nothing. Parts are derived from wholes. Easy to understand when couched in terms of whole and parts. Now we can understand the conditional.

When thrusting the same idea upon "form and content," the reverse seems correct. Parts build into form. That is how we typically observe things in the physical world. However, asking why a specific form arose from a disarray of content, sets us back on the proper course to understand reality. Why this form, rather than another?

Let's review. A unit function (T) is derived from a wholeness function (F). Analogically, the parts of a jigsaw puzzle are derived from the puzzle as a whole.

The Law of Contradiction defines "T." The Law of Noncontradiction defines "F." The Law of Contradiction is derivable from the Law of Noncontradiction. The Law of Contradiction is derivable not only by negating the Law of Noncontradiction, but T is a component of the Law of Noncontradiction.

The Law of Noncontradiction is the first principle we have been seeking. Now that we have elucidated this understanding, what can it explain? Explanation is also understood by the words—reason and why. Let's apply our present understanding and probe a little deeper into Reality.

When we set out to explore the mind expanse at the beginning of this book, we mentioned three tests of truth. The Correspondence Theory is the typical everyday use of truth testing. Scientists use correspondence testing everyday. Does a statement (hypothesis) correspond to something in reality? If it does, we say it is a true statement. We have used the correspondence idea in our treatment of logic. By assigning "meaning" to the terms F and T we expanded our understanding (and our mental ability). The Pragmatic theory of truth can be construed as applying the appropriate rule (for example, for testing consistency).

We have obtained a better understanding of consistency. Consistency follows from the Coherence Theory of truth. When we reach a threshold so abstract that consistency itself becomes the guiding principle of truth seeking, correspondence (of information to what it represents in reality) must take a second seat (to coherency). We can later apply what we learned about (logical) consistency to the real world and see whether it can tell us more about reality. We do this in the next chapter.

Explaining the Primary Equation *1 = 0 x ∞*

Using what we have learned, let's reexamine and better explain our original equation (*1 = 0 x ∞*). How can logic explain this equation?

Again, to explain something is to appeal to that which is otherwise. We examined Frege and Russell's ideas. To explain arithmetic we solicited logic. To explain logic meant, "jumping" outside logic proper. This meant "explaining" any and every "T" valued proposition. We learned to explain every somethingness is to appeal to nothingness. "T" is correlated with the unit function. A unit is a whole and part. "F" is correlated with the wholeness function. A whole (itself) has no parts.

To understand parts, we appeal to the whole system of which the parts are members. Also, to understand the system as a whole is to appeal to its parts. The context is categorical and best exemplified by a hierarchy of parts.

Each level "in" a hierarchy is usually a unit "in itself." (A unit is a whole, and part.) The exceptions are the top most level and bottom most level. The top level of the hierarchy (the "set of all sets") is a whole and no parts. The bottom level, i.e., the most elemental of parts, is an infinity of parts but no whole. Obviously, the bottom level is constituted of parts of the greatest whole (which is the set of all sets).

The bottom level is not testable "in itself" for consistency. It is not a unit or a whole. The bottom most level is not testable "apart" from its inclusion in the "set of all sets." Therefore, the bottom level is only testable (for consistency) when it is considered along with the "set of all sets."

The Law of Contradiction is the test for consistency of a set and its members. When a set is treated along with its members (or subsets) it functions as a subunit. In a set hierarchy there may be many levels of subunits. The Law of Contradiction tests the consistency of any subunit.

The Law of Noncontradiction tests the consistency of "setness" (the notion of set itself) apart from any members. There are numerous subsets included within the "set of all (every) sets." Any subset "in itself" is still a null set. A null set is a member of every set. (Earlier in this chapter we determined there is no set that is not a member of itself.) The Law of Noncontradiction also tests the "setness" of any subset. As stated, there are no actual "individual" subsetnesses in themselves. It only appears there are such sets. Every subset in itself is just the ubiquitous "All" showing itself in Every. Parts are in the whole; the whole is (represented) in the parts.

Mindful of the difference between a set and a unit, there should not be any problem of what consistency test to apply.

If we do not understand the proper nature of the Law of Noncontradiction, we might say there can be no "final principle." The other problem alluded to above, about the illegitimate erection of power sets and transfinite sets, has also muddled the water. Philosophers of mathematics and logic could not clearly see what they had wrought (categorizing [e.g., power sets] extreme reductionistic processing of accepted elements [e.g., the natural numbers]).

The empty set is represented by zero ("0"). When a set is included within itself, it becomes a member of itself. This is a set denoted unity ("1"). Zero is F (false). Unity is T (true).

A unit consists of two aspects: a wholeness function and a parts-function. This is a unit or one ("1"). One ("1") is denoted "T" in logic. We now better understand the meaning of the "T" in the equation $F = F \times T$ or $\varnothing = \varnothing \times 1$. ($\varnothing$ is the null set).

A unit functions as a whole and as a part. The simplest and first functioning of a part that a unit can have is its functioning as one part. The unit functions as a whole and as a part, but with the whole not divided. How can this be?

It is easy to see the functioning of parts when the whole is cut into half or into quarters (each part a 1/2 of the whole, or each part a 1/4 of the whole). More discernment is needed to recognize a unit can function as a part and not be divided (1/1). A unit can function as the whole and as a single part. The reason is each function occurs at a different (hierarchical) level. The difference is between a set, and a set containing one member. It is the difference between a set and that set containing an empty set. Unity is minimally a whole function and "single" parts-function.

Given the same unity, either we are testing the consistency of the whole, or testing the consistency of its parts (we could also test the unit as a collective, i.e., whole and parts).

If we couch the Primary Equation ($1 = 0 \times \infty$) in terms of logic, we gain a better understanding of the equation and eventually learn what it says of reality. How?: By defining form as arrangement of content. We have connected form to wholeness function. We have connected content to parts-function. In our present treatment, we are attempting to connect the wholeness function to the parts-function. How do we derive the one ("1") from zero ("0") how do we derive something from nothing? Better yet, how can zero ("0") generate one ("1")?

How are the natural numbers produced? Specifically, given zero ("0") how do we obtain one ("1"). We can assume zero. It is nothing. Zero can be accepted as the ultimate premise from which every number is derivable.

In 1879 Frege published his work (*Begriffsscrift*, translated as "Concept-Writing"). G. Frege wanted to "codify" mathematical reasoning. He distinguished between a proposition and whether the proposition is asserted to be true. Asserting a proposition is true is to view it from outside itself. Our above example was saying it is true that the Law of Noncontradiction defines falseness. Frege's work heralds the beginning of modern logic (predicate logic). Frege's logic could also represent set theory.

When Frege's second volume of his work entitled *The Foundations of Arithmetic* was at the printers, he received a letter From B. Russell (dated June 16, 1902) stating that the "set of all classes" leads to a contradiction. This is Russell's Paradox.

We have already addressed Russell's Paradox. Briefly, if there is a grouping of objects exhibiting a certain characteristic, then there also exists a set of those objects with those characteristics. The idea was that a set was a collection acting as a whole. Yet, as Russell stated, this conceptualization leads to a contradiction. We discerned All is different from Every: they are opposites.

Because of Frege and Russell's problem, the German mathematician E. Zermelo by 1908 developed an axiom system for set theory that avoids the paradox noted by Russell. Zermelo's axioms were slightly modified in 1922 by A. Frankel and now are known as Zermelo-Frankel set theory. (Zermelo did not distinguish between set itself and the property of set.) Although Zermelo-Frankel's axioms appear to circumvent Russell's Paradox, they have not been shown to escape Gödel's Incompleteness Theorems even though one of the axioms state that "x cannot belong to itself." We note these axioms are a set of rules. The following are some of Zermelo-Frankel's axioms.

Extensionality: sets exhibiting the same members are identical. This defines set(ness).

Null set: there is an empty set: one having no members.

Infinity: infinite sets exist.

Power sets: power sets exist.

Regularity: x cannot belong to itself. This implies there is no set of all sets.

Changes have been made to Zermelo's original work. Changes or additions were made to counter possible problems (e.g., paradoxes).

Frege also contributed the idea of embedding the concept of "true and false" in logic. S.C. Pierce added understanding by correlating logical addition ($\oplus$) with the "union" of classes ($A \cup B$) and logical product ($\otimes$) with the intersection of classes ($A \cap B$). He also incorporated de Morgan's Law used in transforming logical equations into different but equivalent forms (for example, converting a conjunction into a disjunction). In 1847 de Morgan (in his *Formal Logic*) had suggested logic concerns relations in the abstract.

Boolean Algebra

In 1854 G. Boole's book, called *An Investigation of the Laws of Thought on Which are Founded the Mathematical Theories of Logic and Probabilities,* was published. It is often designated the "Laws of Thought." Boole expanded upon his earlier work with a book titled *The Mathematical Analysis of Logic.*

Aristotelian syllogistic logic could be represented in his system. For example: "all men are mortal" means that the class of men is a subset or part of the class entailed by the notion "mortal."

Boolean algebra relates classes. Boole interpreted "classes" (sets) as an algebra of logic: an algebra of calculus. (It is interesting the word "algebra" is derived from the Arabic *al-jabr*. It means "reunion of broken parts.") Boole's system became the basis for propositional calculus.

Logical propositions relate predicates or properties to a subject. In propositional logic the subject and predicate are abstract and devoid of content. Boolean algebra is pure logic.

Let's get on with our task.

"Nothing can both be and not be" is the Law of Noncontradiction (or, $F = T \wedge F$).

Let's translate our logical truth function for falseness ($F = T \wedge F$) into Boolean algebra.

$$F = 0, \qquad T = 1 \qquad \wedge = x$$

We have:

$$0 = 1 \text{ x } 0$$

Zero ("0") is equivalent to (the logical product of) one ("1") and zero ("0").

In set theory "x" signifies intersection—what is common to both. Intersection (in set theory) is comparable to conjunction (in logic).

We ask: What is common to zero ("0") and one ("1")? Rephrasing: What is common to nothing and something? Answer, nothing. Yet, we have learned "nothingness," far from being free from characterization, is a function (i.e., the wholeness function). Does one ("1") have an aspect that denotes wholeness? Yes. Unity, one (1) infers two aspects: a wholeness function and a parts-function. Zero ("0") is common to both zero ("0") and one ("1"). A whole pie (not cut into pieces) is equivalent in one sense to the same pie cut into pieces. In both cases there is no more than the whole—and that (the whole) is what they have in common. Let's elucidate.

Venn diagrams (after J. Venn: described in his 1881 book *Symbolic Logic*) representing set definitions, have been useful for visualizing set ideas. A Venn diagram of an empty set can be represented as a circle. An empty set of an empty set is a circle within another circle. (We do not depict Venn diagrams in this project.) It is common to find the empty set (representing zero "0") stated to be a subset of every set. No problem here. It is often stated the universal set (denoted by one "1") is a superset of every set. This is not so. Rather, the empty set is the "super set" of every set. The empty set should properly be called the "set of discourse." The universal set is a composite. It is the ultimate "set containing members." The idea of discourse is implied by and contained within the empty set. It is the empty set that delimits discourse. In set theory it is recognized that a null set is equivalent to the negation of a universal set and conversely ($\varnothing$ = U´, and $\varnothing$´ = U). This is the correct portrayal of the extreme of complementarity. Not everything (U´) is nothing ($\varnothing$). Not nothing ($\varnothing$´) is everything (U). What is lacking is the proper understanding of the empty set as the wholeness function. The empty set is common to (both) the universal set as well as the empty set.

Let's represent the Law of Noncontradiction (F = T ∧ F) in set theory using braces.

F, or zero "0", nothing, is represented by an empty set { }

T is unity "1", an "empty" set which includes an empty set as a member { { } }

"x" is intersection: linking what is common to both components:

F	=	T	∧	F	Logic
0	=	1	x	0	Boolean Algebra
{ }*	=	{{ }}*	x	{ }*	Set Notation

Table 18:3. Notation Comparison

An asterisk "*" marks an empty set (outside braces). The left side of the "braces" formula (an empty set) is shown to be common to both members of the right side of the formula. Again, unity is an "empty set" containing an empty set as a member. Here, we clearly see that "nothing" (represented by the empty set on the left side) is common to both sets on the right side.

The empty set is common to itself (is identical with itself) and is also common to a unit.

In a deductive manner, how are the "elements" of somethingness derived? We start with pure emptiness. We cannot (using existents) represent emptiness except by stating that "it is not even every somethingness." Let's see how to generate somethingness from nothingness. We begin with nothingness—an empty set. Let's use curly-bracket-set notation (braces) to understand our equation *1 = 0 x ∞*.

This is our equation (*1 = 0 x ∞*) in set theory (braces notation). This equation shows that unity (1) breaks down into a wholeness function (0) and a parts-function (∞): an infinity of parts. From this equation we originally learned that unity has two aspects.

{ }	The notation for set (nothingness) represents that which is empty in itself. An empty set is equivalent to 0. This is the zero in our equation.
{{ }}	This is the "minimal" set of a set. It is equivalent to 1 (1/1). This is the 1 in our equation. An empty set is a member of every set (which has members).
{{{{...}}}}	This is a "self-duplicating set." Notice there is an infinite number of sets included. (n = 1, 2, 3,....) An empty set is a "default" member of every set.

Table 18:4. Braces Notation

Let's relate these sets:

$$\{\{\ \}\} \quad = \quad \{\ \} \quad x \quad \{\{\{\{...\}\}\}\}$$

$$1 \quad = \quad 0 \quad x \quad \infty$$

Table 18:5. Braces and Numeral Comparison

Axiomatization of Arithmetic

The word axiom is derived from the Latin *axioma* meaning principle. The word arithmetic is from the Greek *arithmos* meaning number.

Axiomatization of arithmetic is an attempt to "formalize" the fundamental assumptions of arithmetic in logical notation. Nicolas of Cusa in the fourteenth century believed God's plans are revealed in mathematics. Many mathematicians, after the seventeenth century, believed by studying mathematics that they could discern God's design of nature.

Most mathematicians are Platonists. Platonism (after the Greek philosopher Plato) posits that mathematical objects, patterns, and relationships, "exist" beyond space and time. The idea is mathematics, in itself, is independent of man's conceptualizations (of it). Plato believed he proved this idea by eliciting correct answers from students. For Plato, that a student could reason out correct answers in geometry showed he already had that knowledge before birth.

Plato thought there exists a "realm" of mathematics, which he called the "Good." Today, we probably would say the Good is constituted of pure Form. We easily recognize this ideal form to be the essence of logic itself. If Plato's idea is not mirrored by reality, then it means mathematics is another arbitrary doing of man.

Is mathematics discovered or is it an invention of man's reasoning? Maybe the answer is a little of both. If so that might explain some obstacles faced while searching for mathematical truth and certainty.

Plato thought the ultimate nature of reality could be comprehended by "abstract" reasoning. According to Platonists, when a mathematician realizes a mathematical truth, he discovers it. The idea mathematics preexists before discovery (by man) also suggests that any inconsistency in mathematics might be attributable to man's lack of understanding rather than logic being inherently inconsistent.

Leibniz thought mathematical truths were derivable from logic. He believed all necessary truths were logical in nature. Frege went further and thought mathematical truth cannot say any more than what is inherent (implicit) in logic itself. Leibniz and Frege thought the laws of logic were *a priori* truths.

The movement called "logicism" began with Frege and was developed by B. Russell. Its core premise is summed up by saying mathematics is a branch of logic. Pragmatically, logicism is the codifying of arithmetic in terms of logic. The only remaining problem centered on the question of logical consistency. B. Russell believed that consistency would be established. Logicism became the effort to reformulate set theory in a way to avoid Russell's Paradox.

How is mathematical consistency established? Let's take a broad view. The consistency of curvilinear geometries reduces to the consistency of the arithmetic of real numbers. Descartes proved geometry (using Cartesian coordinates) could reduce to algebra. Real number consistency reduces to the rational numbers. Rational number consistency depends on the consistency of the integers, which depend on the natural numbers. The consistency of all mathematics depends on the consistency of the natural numbers "1, 2, 3,...", which represent an enumeration of "member" elements. The natural numbers are representable as classes or sets. The consistency problem is simply depicted as Russell's Paradox concerning the set of all sets (e.g., of natural numbers).

Without logical consistency mathematics could not be grounded in certainty. It is "certainty" that serious seekers of truth are searching. Descartes once said, "I shall seek until I find something that is certain, or until I find that nothing is certain."

L. Kronecker, former teacher of Cantor, wrote in 1889 that he believed all mathematics could be "arithmetized" on a single number concept—in its narrowest sense. Perhaps borrowing from Augustine's idea that "God was beyond the natural numbers," Kronecker thought: "God made the integers, all the rest is the work of man." No wonder Kronecker's criticism contributed to Cantor's mental problems. Cantor's transfinite sets (an infinity of sets) were directly opposed to Kronecker's work of attempting to unify mathematics.

To gain a deeper understanding, we went beyond the mathematical equation ($I = 0 \times \infty$) and defined one ("1") and zero ("0") using: set theory, logic, and Boolean algebra.

We found how every number (or part) is generated from ("0"). Zero is defined as the wholeness function. The whole fractionates into parts. There are no parts but that there is a whole from which they are derived.

Zero ("0") is nothing. One is unity ("1"). Let's again translate the formula $F = T \wedge F$ into a sentence.

Nothing (0) can both be (1) and not be (0)

We, again, recognize this as the Law of Noncontradiction. "Unity and zero have 'nothing' in common." Let's look at it in a different way.

From a wholeness function ("0") we can derive two aspects: a unit function ("1") and another wholeness function ("0"). Using the Boolean algebra equivalent of the Law of Noncontradiction ($0 = 1 \times 0$) we have derived one ("1") from zero ("0"). Inherent in the one ("1") is a wholeness function and a parts-function. The parts-function is represented in our original equation ($1 = 0 \times \infty$) by infinity ("∞").

In set theory terminology: We used a null (empty) set (F) to derive a set of elements, a unit (T) and another null set (F). The terms on the right side of the equation ($0 = 1 \times 0$) are components (aspects) of the left side of the equation. Is zero ("0") allowed on the right side of the equation? Answer, a null set is a member of every set and is allowed to be a component.

So, we find there is that which can both be and not be—zero or "nothing."

But, what about the issue of consistency?

In common arithmetic we can negate an equation and not violate rules of consistency. For example, we may exhibit a credit by the following equation.

$+\$5 + (-\$3) = +\$2$ representing a credit of two dollars

We can also multiply each term in this equation by "-1" (minus one). This is negating the equation.

$-\$5 + \$3 = -\$2$ representing a debit of two dollars

The meaning of the second equation, although in itself consistent, is opposite that of the first equation. Where a credit was exhibited in the first equation, a debit is shown in the second equation. They are opposites.

In logic, like arithmetic, we can negate a formula and maintain consistency.

First, some definitions:

Via negation equivalence, $T = \neg F$, and $\neg T = F$.

Or in terms of existence, $T = BE$, $F = $ not BE.

That which can BE is "Something."

That which cannot BE is "no-thing" or "Nothing."

Let's negate our logical equation:

1. Given: $F = T \wedge F$ or $F = T \wedge \neg T$ (nothing can both BE "something" and NOT BE "something"). This is Law of Noncontradiction.

2. Negating #1, $\neg [F = T \wedge \neg T]$ or $\neg F = \neg [T \wedge F]$ or $T = \neg [T \wedge F]$ Here is derived the Law of Contradiction.

3. Using de Morgan's Law $T = \neg T \vee \neg F$, or $T = F \vee T$

De Morgan's Law yields the following. For a disjunction to be valid requires only one of the disjuncts be true. Something either is or is not. This is the Law of the Excluded Middle. There is no middle ground. This Law limits the Law of Contradiction to "contradictories."

In existence, something either exists, or does not exist. If it does not exist, it is no-thing or nothing.

Existence defines something; something defines existents. Any something is synonymous with an aspect of existence.

Nonexistence defines no-thing or nothing. Nothing defines that which does not exist. At least that would be one way of understanding these equations. However, we are not finished in our pursuit of understanding.

We were left with the task of deriving one ("1") from zero ("0").

How are the natural numbers generated? The Law of Noncontradiction is capable of generating the number one (1). Below is the Boolean algebra for the above logical equation #1. In Boolean algebra the Law of Noncontradiction is the following.

$$0 = 1 \; x \; 0$$

Zero ("0") can generate one ("1") and another zero ("0"). We understand that the left side of the equation could be construed as "going outside" of the two aspects (viz, "1" and "0" elements on the right side) of the system. We could also say the right side is encoded in the left side of the equation. This is a better way to understand this formula. Our understanding infers more than this.

A system itself (left side of the formula $0 = 1 \; x \; 0$) can generate members within itself. Unity (1) is a set with members. We still have no more than that of which we began (we began with nothing—zero "0" on left side of equation). Therefore, there is no more than the "whole" of which we began. It is just that the whole has fractionated into parts—perhaps an infinity of parts.

Gödel would say a false proposition is not a tautology and is therefore not provable. We have seen that the Law of Contradiction is derivable from the Law of Noncontradiction. The Law of Noncontradiction, though not typically considered a tautology, actually is true, but we must assess its truth from "inside" the formula itself. ("T" appears on the right side of [within] the equation.) Subsequent application of the concept "T" to the equation (representing the Law of Noncontradiction) yields the following:

It is true that $F = T \wedge F$.

Negating the Law of Noncontradiction yields the Law of Contradiction. The Law of Contradiction defines:

"T"   ($T = \neg [T \wedge F]$).

These are the definitions for F and T. And, we have learned that negating a formula, though changing its meaning, does not make it inconsistent.

Discerning our position of questioning is paramount to maintaining proper understanding. We do not examine a ledger to learn that it shows a debit; then, announce we made a profit. Yet, there is not anything inconsistent about debits. Inconsistency arises only when we confuse what we are asking and not knowing what criteria are applicable to ascertain consistency. We would not place a credit into the debit column.

The left side (F) of the Law of Noncontradiction is defined (on the right side) by the operations of logical negation and logical conjunction. We note because a zero ("0") can generate a one ("1") that the concept of negation is built right into the fractionating (of the whole) process. T is the negation of F. One ("1") is the negation of zero ("0"). Logical conjunction and negation become the first two rules derived from zero (nothingness).

Applying the Law of Negation to the left side of the equation, we can derive the Law of Contradiction. From the Law of Contradiction we can derive the Law of the Excluded Middle. From these laws the other laws of logic can be deduced.

The principle of identity is implied. T is T. That F is equivalent to F x T does not mean that each side of the equivalence is identical in kind. Parts are not identical to the whole of which they are a component, nor is a whole identical to its parts. The difference resides in their role or function. There is a difference between an apple, and an apple cut into parts. One thing we note: no matter how often a whole is partitioned, we never have more than with what we began. Each cutting continually reduces the size of the pieces.

Key Points:

- We "conceptualized" an interesting equation: the Primary Equation. The first task was "fitting" the concept of space to the equation. The problem was space behaved as though it possessed at least one characteristic, viz, that space expands according to the Big Bang Theory of the universe. (Tests of Special Relativity indicate space can expand and contract.)

- That space can expand and contract suggests it is characterizable as though it is a "something." An immediate question was "how can space be defined as nothing and yet be characterized as though it were a something?" It seemed that space had to be defined as though it were both nothing (otherwise, how can we call it nothingness) and (also) something (in that it can expand and contract). How can space both be nothing and not nothing (i.e., something)? The issue of temporality also sneaks into any possible resolution.

- We began our search for a deeper understanding of the basic nature of everything. We settled on correlating space with the zero ("0") in the equation. There is no other choice. We acknowledge this concession despite not seeing how we could reconcile this complementary definition (of nothing—something) with the Law of Contradiction. How can the same (space) both be and not be? Permitting such a realization violates the logical Law of Contradiction.

- If the Law of Contradiction is violated, the very understanding of consistency becomes the issue. This realization forced us to examine logic.

- We said one (1) represented the unity of Reality. We could say unity is "space-time." The next question is how can the other two factors of the equation "explain" unity? After saying zero represents space, we are left to say the infinity factor represents time. (We continue our quest to understand the nature of time in the next chapter.)

- Unity is the constant of proportionality. Zero represents "actual" infinity. The infinity factor represents "potential" infinity (1, 2, 3,...). Actual infinity includes potential infinity. How else can the equation be interpreted? Unity signifies completed Reality (space-time). Reality is constituted of two aspects. Space (zero aspect) includes the infinite (infinity aspect) energy of the vacuum.

- There is a complementary relation between zero and infinity. As the infinity factor increases toward infinity, the zero factor decreases to zero: toward nothing at All. Eventually, we take a (conceptual) "leap" from an infinitesimal fraction (which "approaches" zero as a limit) to nothingness (zero itself). Nothingness includes infinity. This is a role (functional) reversal of the "apparent" idea infinity (potential) includes the natural numbers.

- We reinterpreted our understanding of the equation using set theory. We noticed the problem of defining space without contradiction (such that it could remain "nothing" and yet behave, as though it were something) was akin to Russell's Paradox in set theory.

- We reasoned the following. A unit is a set and its members. Zero is setness itself. Infinity represents every possible member of the set. Setness itself can be understood as a "form" or a "wholeness" function. Elements or members of a set are correlated with "contents" or "parts" function.

- Unity is the "set of a set." Parts are generated from a wholeness function by its (the whole) "fractionation." Relevancy: The "infinity" of natural numbers are generated by a continual partitioning of zero ("0"). "Zero to one" (unity) is the continuum. The empty set is a de facto member of every set.

- Every possible discrete number is found within the partitioning of the continuum. Nondiscrete numbers result from the "synthetic" appending of irrational numbers to the continuum. Complex numbers are not derivable without a basis in the natural numbers.

- The primary problem of understanding space (represented by zero) reduced to justifying the generating of the natural numbers. Specifically, how is the number one ("1") originated? Merely erecting a series of "operational" axioms, telling us how to carry out numerical inductions, does not in itself "explain" how unity ("1") is originally derived. One ("1") is obtained by functionally fractionating zero ("0").

- Why are we interested in numbers? Arithmetic is manipulation of numbers. Physics is "formulated" in mathematical terms. Physics concerns "things." Mathematics reduces to the arithmetic of the natural numbers. Numbers can represent every possible thing in abstraction. Understanding how zero can generate unity is to possibly explain how something (the focus of physics) can be generated from nothing. Abstractly, it (zero) also enables the axiomatization of arithmetic.

- The problem narrowed to explaining arithmetic. Peano axioms are an example of the attempt to "explain" arithmetic. Russell's Paradox seemed to make it difficult to erect a complete axiomatic treatment of arithmetic. Other attempts were made to "patch up" an axiomatic basis for arithmetic using logic. Zermelo-Frankel's axioms and later attempts were made to readdress the logical basis of arithmetic. Gödel's Incompleteness Theorems suggested there cannot be any complete and consistent axiom for B. Russell's logical treatment of arithmetic.

- The search for understanding of things reduced to the consistency of logic itself. Is logical consistency, as defined by the Law of Contradiction, also the test of the consistency of logic? We found the answer to be no. The Law of Noncontradiction defines the consistency of logic.

- Law of Noncontradiction is representable in Boolean algebra as $0 = 1 \times 0$. We found an equation that allows the derivation of one (1) from zero (0). The negation of the Law of Contradiction yields the Law of Noncontradiction, which defines the consistency of logic. It is inescapable, $0 = 1$. Why? Because zero is the whole, one (1) "has" the same whole as (which defines) zero (0) but one (1) has functioning parts. No one will deny that a whole pie is equivalent to the same (whole) pie cut into pieces. Although the whole pie is still present (equivalent) the functionality has changed (from wholeness to unity). There cannot be more than the whole of what is. These are not definitions that typically characterize zero and one in the usual sense of numeration (where 0 does not equal 1). These definitions reflect the understanding of zero and one in their own right (wholeness and unity).

- (It is true that) zero is "nonfunctioningly" equivalent to ONE ("1"). It is false that zero is "functioningly" equivalent to ONE ("1").

- A proposition proving anything and everything—proves nothing at all. All is no-thing: nothing. Nothing proves anything and everything. The Law of Noncontradiction defines the criterion nothingness must meet. The Law of Contradiction defines the criterion for anything and everything.

The next chapter concerns applying what we learned in this chapter. Can what we learned about logic (through Boolean algebra) and our equation ($1 = 0 \times \infty$) better explain physics? Particularly, besides explaining space and time, what else in physics can logic and our equation explain?

520

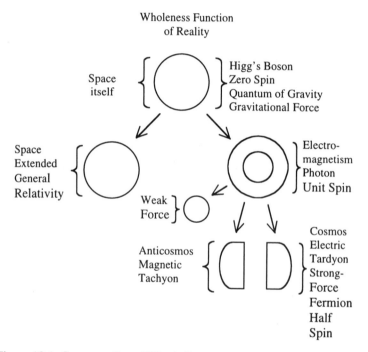

Figure 19:1. Commonality of Higg's Boson and Space

Chapter 19
Physics Interpretation of Ultimate Principle

> "If things happen only by chance, there is no chance
> of explaining anything. Man seeks truth to explain
> experience."

A mathematical equation (by itself) says little about reality. In physics, concepts are "attached" to equations, giving them meaning. If a concept is verified to correspond to something in Reality, it is said to be "true." Presupposing Reality is uniform or coherent, we also expect concepts depicting reality to be self-consistent. The previous chapter resolved the problem of defining consistency. We will be applying our new found understanding of consistency to the physics of cosmogenesis.

There are two basic means of obtaining (scientific) equations. One is to develop equations from experimentally determined values. The other is to develop equations on theoretical grounds.

From experimental data, graphs are plotted. An equation is written to represent the plotted values. In the early nineteenth century, Ampère, Faraday, and Oersted sought to understand electric phenomenon. Oersted showed electric and magnetic phenomena are intimately connected. He observed electric current deflects a compass needle. Later, Maxwell, in a brilliant leap of understanding, showed mathematically that electricity and magnetism were two aspects of one phenomenon—electromagnetism. His equation predicted that electromagnetic "radiation" could exist independent of either electric or magnetic phenomenon. Hertz soon demonstrated the effect of radio waves: verifying Maxwell's prediction.

Another example is the plotting of black-body-radiation spectra. By graphing data an equation was developed resulting in Planck's constant. The quantum of action was born.

Theoretical equations serve as hypotheses. This method is often used to better explain phenomena. Einstein's theory of Special Relativity is such an example. It explains away the need for the luminiferous ether. Radio waves need not propagate in a medium—the Maxwellian ether. Relativity suggests there "exists" the phenomenon of space-time: a higher level phenomenon than the notion of a separate space and time, which it "explains." Experiments are designed and conducted to determine whether a "theorized" equation represents something in Reality.

Either way, the scientific idea is to assign conceptual meaning to equations and establish their correspondence to Reality using empirical data. Equations serve to extend the range of understanding (what we experience). Our questioning now centers on whether there is a reasonable physics interpretation of the Ultimate Principle. If the principle has a cogent connection to physics, it should be a high-level concept that unifies the basic interactions of physics. The loftiest goal of science is to explain experience itself.

The focus of this chapter is the physics understanding of the logical formula $F = F \times T$. This is the logical Law of Noncontradiction. It is the Ultimate Principle. To what, if anything, does it correspond in a physics understanding of Reality? In the last chapter we used this principle to "explain" mathematics. Mathematics explains physics. Can the Ultimate Principle also yield a direct understanding of physics?

Infinity and the Vacuum

Let's backtrack to the previous chapter. We ask, how are things created? Things, abstractly, are representable as numbers. To answer our question (in the abstract) we first recast the question using numbers. "How are numbers generated?"

Consider the following. If we place an apple in front of a child and ask, here is an apple: how can we get two from this one apple? The child can only get two by cutting the "whole" apple into two parts. An elementary-school student was asked this question. She answered correctly. The cutting of the apple results in the lost of its original wholeness function. Once sliced, an apple does not exist ("functionally") as a whole. If a water molecule is partitioned into its elements, it no longer functions as a water molecule. What does this mean?

Some brief statements: Unity of Reality has two aspects: the wholeness function (numerically represented by zero) and a parts-function (an infinity of parts).

To obtain something from nothing is for the whole (viz, "no-thing") to split into parts (things). Continual splitting eventually results in an "infinity" of parts. Whether beginning the cutting process at the whole of Reality or at a lesser level, typified by the apple, the result is the same. For example: Theoretically, the continual "cutting" of an apple or a water molecule eventually breaks them down into lepto-quarks. The same reasoning, in the end, applies to the whole of Reality. Reality could be analyzed by a continual dissection—a peeling away of "functional" layers.

Do lepto-quarks correlate with the infinity of parts? Almost. Lepto-quarks probably fall into the category of manifested particles. Are there lower level nonmanifested particles from which lepto-quarks emerge? If there is a nonmanifested parts-source for manifested particles, what would it be? The infinity of parts is represented in physics by the limitless energy of the vacuum. The vacuum is space (continuum). Space is "filled" with infinite (potential) energy.

Physics suggests everything is built from the energy of the vacuum. All manifested elementary particles have parallel representatives in the vacuum called virtual particles. Electrons have virtual counterparts called virtual electrons. Force particles are also associated with virtual counterparts.

Virtual bosons, the particles of force, mediate elementary particle interaction. The photon is the exchange particle for interactions between (electrically) charged particles.

Bosons are based conceptually on manifested particles. Note: Manifested particles also exist in the vacuum, but in a different sense than that of virtual particles. Think of the vacuum as space and this statement immediately becomes clear. So what is the difference between virtual-nonmanifested particles and those denoted manifest?

All virtual particles are potential particles. Given sufficient energy and proper conditions, i.e., splitting of virtual matter-antimatter pairs, a virtual (nonmanifested) particle materializes into a manifest particle. Virtual particles in themselves are not distinguishable from space in itself. Yet, they act (on things—Lamb shift) as though they are themselves "things in space."

The infinity factor represents the potential ("functionally" discrete) energy of the vacuum. Being nonmanifested, it is easy to say it is infinite. Note an empty set proliferates infinitely in a hierarchy. It is a set of every set. A nonmanifested particle of the vacuum is space (an empty set) functioning minimally. The vacuum has an infinite quantity of empty sets. Again, though this reasoning is acceptable, there is no difference between infinite-empty sets and the whole, which is THE empty set. The only difference is one of function. More about this in the next chapter.

The Primary Equation: Infinity and Time

Before addressing the utility of the Law of Noncontradiction let's take another look at the equation that led to this Ultimate Principle. That equation is $1 = 0 \, x \, \infty$. This is our Primary Equation. In the previous chapter we found the Primary Equation defines two aspects: continuity and discreteness. It is the relation between these two factors on the right side of the equation ($1 = 0 \, x \, \infty$) which defines the unity of Reality on the left side of the equation. Zero defines continuity; infinity defines unlimited discreteness.

We learned about the mathematical nature of the equation and how it could be understood through set theory and logic.

Understanding the equation $1 = 0 \, x \, \infty$ mathematically is one thing, correlating it to Reality is quite another. The "0" and "1" factors were correlated to space and space-time. This understanding suggests that the infinity factor somehow represents time. Can infinity represent time, and if so, how?

Here (in understanding) a pattern began to emerge. Physicists assert there is an INFINITE quantity of ENERGY in the vacuum. The vacuum is space. Space is represented by zero ("0"). From the previous chapter, we learned the concept of infinity is included within the concept of zero. Instead of presenting an intractable problem, this is exactly what we surmised by understanding the equation using set theory. Zero is "setness"; infinity represents the members of the set.

What does the Primary Equation ($1 = 0 \, x \, \infty$) represent in Reality? Let's quickly review what we have learned and set the basis for further understanding. In our Primary Equation:

One ("1") represents the unity that is Reality. A unit has two aspects or functions: a whole and part. One ("1") is the "set of elements." In physics terminology: "All that can be" is Reality. There is not anything or any nothing outside Reality. The unity of Reality correlates with space-time in physics.

Zero ("0") represents the whole of Reality (i.e., zero divisibility of the unit). Zero ("0") represents the set, but not its elements. Zero is the whole, but no parts. In physics, zero ("0") represents space itself, i.e., the "universal" singularity. Space is no thing, no somethings. It is also not everything. Everything is contained within space, and in a twisted sense, space is also an aspect of everything. Besides expansion, and perhaps curvature (distortion) via physics: space has no defining characteristics (and in itself, contains no elements: this is its definition). Space is emptiness itself, nonmanifestation. The whole of Reality is space.

Infinity ("∞") represents discreteness or partitioning of the unit Reality (i.e., infinite division of the unit). Infinity ("∞") represents the individual elements, but not the set of the elements. Infinity represents the parts, but not the whole. Infinity ("∞") represents things (every something) contained within space. Everything (every somethingness) exists in space. Everything is constituted of manifested particles. Everything is synergistically derived from the infinite energy of the vacuum.

Infinity represents the unlimited member elements of zero. Time exists in space. Without space there would be no flow of time. Back to our question. How does infinity represent time? Energy is associated with time. Without motion, there would be no flow of time and no use of energy. Motion of things, of anything and everything, is interpreted as the flow of time. If everything ceased, no motion—no awareness, there would be not only no perception of time, but also no temporal flow at all.

Rate (or cycles) of motion, in every day use, is a measure of the flow of time. An hour is the time earth rotates 15 degrees. A day is the "time" needed for earth to rotate on its own axis (of rotation) once (360°). A month is the time for the moon to revolve around earth one time. It takes a year for earth to revolve around the sun one time. No motion, no flow of time.

Material things are unit-parts that arise from the infinite energy of the vacuum. Arranging of manifest parts results in higher-level structures. Units within the structure, experience motion because of the arranging of its parts. Without motion there would be no time. Flow of time corresponds to motion of parts. Manifested particles (unit-parts) are derived from the limitless energy of the vacuum. Without the infinite energy of the vacuum there would be no flow of time. The infinity factor represents the basis of time. Without "functional" discreteness—there is no temporality.

Functioning Things—Zero's and One's

What simple concepts describe energy use? Let's describe the basic operational design of the computer. At the core of a computer is a central-processing unit usually abbreviated CPU. Its operation is very simple. It processes instructions in bits.

A bit is a binary digit. A bit represents the elementary unit of information. It is information in terms of yes and no, true and false, one and zero.

Boolean algebra is the language of the binary system.

Boolean algebra is used in designing computers. In the binary system there are only two possible values, viz, 0, and 1.

Applying these concepts to electrical circuits, we find the following:

Zero ("0") represents an "off" electrical switch. In logic, zero ("0") is typically represented by F (false). A CPU represents zero ("0") with "low" voltage.

One ("1") represents an "on" electrical switch. In logic, one ("1") is usually represented by T (true). A CPU represents a transistor switch in the "on" position with "high" voltage.

A conjunction (A ∧ B) can represent an electrical circuit connected in series. Current can only flow when both switches "A" and "B" are closed (turned "on").

A disjunction (A ∨ B) can represent an electrical circuit connected in parallel. Current still flows when only one pathway is closed.

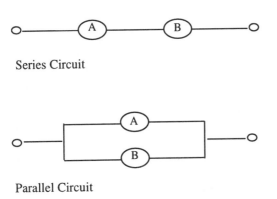

Series Circuit

Parallel Circuit

Figure 19:2. Electrical Circuits

Throwing an electrical light switch to "on" (i.e., "1") turns on a lamp. Turning the lamp switch "off" ("0") turns "off" current flow, and the light goes out. Notice something interesting. There are many different things that are "activated" by throwing a switch to "on." An "on" switch uses electrical energy—the pulse flow of oscillating electrons. There are many different functioning things (when "turned on"). Yet, when every possible "functioning" thing is deactivated, they all have one thing in common—they are all "off." They are all "off" in the same way—nonfunctioning.

When a thing is "activated" (switch is "on") it "functions" for its designed purpose. Things are designed for different purposes. Whatever a thing's purpose—when things are in an inactive mode—they exhibit no function. (Not functioning for their purpose). Inactivity is, well, inactivity. All nonfunctioning things have the same commonality of inactivity.

Simple information can be encoded in bits. A bit represents the lowest level of (what can be called) information. "On" or "off" possibilities represent the smallest bits of information. To a question, "on" means "yes"; "off" means "no." This correlates with the idea a statement or concept is either true or false.

Any number or alphabet character is representable in the binary system. Only a limited number of bits are needed to represent language characters and single digits. In modern-computer architecture usually eight bits are used. Eight bits form a byte (of information).

A binary byte can represent a "character," e.g., a letter of the language alphabet. Each character is correlated with a given byte: for purposes of standardization. For example, in ASCII (American Standard Code for Information Interchange) characters, the capital letter "A" is assigned to the number 65. In byte terminology 65 is "01000001." Counting from the right, the bit (a one "1") in the 7th position represents the number 64 (from 2^6). The bit in the right most position is the number 1 (from 2^0). We have $64 + 1 = 65$.

Sequences of letters can represent words. Arrangement of words can represent statements. A series of statements might represent a concept. Arranged in a hierarchy, concepts can represent the structure of Reality. Within any of these hierarchical levels, things (actually, their function) are still representable as ones ("1") and zero ("0"). Yes and no, true and false, existence and nonexistence.

Although any given hierarchical level is representable as zeroes and ones, such representation does not tell us the import of the concept. This is still logic. Logic is abstract. It does little to inform except to state abstractly that a given logical concept is either true (and therefore logically consistent) or false (and therefore not logically consistent). The import is complexity in reality can be built from an arrangement of basic zeros and ones. Structures can be represented as a hierarchical abstraction using zeros and ones.

Minor notes: The lowest level of "charged" particles has been theorized as representable using zero's ("0") and one's ("1"). H. Harari's theory of quark constitution is one example. It was mentioned in Chapter 15. We also learned in the previous chapter that logical inconsistency (by default) necessarily pertains to the consistency of logic.

How does zero ("0") and one ("1") correlate with the primary forces of nature? These forces are the gravitational, electromagnetic, strong and weak-nuclear interactions. In short, how does the Ultimate Principle correlate with the structure of Reality described in this book?

Correlating Ultimate Principle with Reality's Structure

Stephen Hawking and many other scientists suppose there is a single governing law from which all other laws can be derived. Scientists are searching for a principle (or 'EXPLANATION') which encompasses ALL REALITY. This has also been a goal of this project.

For thirty years Einstein tried to unify electromagnetism with gravitation. Knowledge of the strong and weak force was not properly recognized when he attempted this task. Unifying concepts are discussed in Chapter 15.

The previous chapter concerned logic. In our treatment, logic depicts "acceptable arrangement of parts" in abstraction—devoid of meaningful content. Here, we supply material meaning to those abstractions. We "conceptualize" certain fundamental formulas of logic using knowledge of physics.

Physicists seek the fundamental laws of nature. Conservation laws, in physics, represent invariance principles. What is conserved is not changed. Physicists have correlated invariance principles with symmetry. Isotopic spin invariance (concerning hadrons) is an example. It suggests there is no functional difference (hence the symmetry in an "abstract isotopic spin space") between protons and neutrons regarding the strong nuclear force. In this sense, protons and neutrons (a doublet) are called nucleons. This functional "sameness" is reflected in the "nucleonic" symmetry between neutron and proton. (See Chapter 15.)

The physics' idea of gauge symmetry is encapsulated in the concept of group theory. Each member of a group is representable by a change in the direction of a "gauge" arrow (like a single hand of a "phase" clock). Each position identifies a different particle. Rotate the arrow and another particle is represented. Symmetry is conserved through each rotation. Let's use a simple analogy for an example. The neutron is representably "transformed" (gauge transformation) into a proton by a suitable "gauging" of the nucleon.

In supersymmetry, invariance suggests the laws of nature remain the same whether determined in a laboratory at rest or rotating. Spatial-rotation indifference (not violating the invariance) naturally incorporates Einstein's General Relativity into symmetry, hence, supersymmetry.

The significance of supersymmetry is found in its utility. It brings together two broad classes of elementary particles called fermions and bosons. Recall fermions are the particles acted upon; bosons do the acting.

Fermions and bosons are thought to be two aspects of a single bosonic "superparticle." Not unlike the nucleon, supersymmetry "connects" fermions and bosons via the superparticle.

Change in particle identity (e.g., from fermion to boson) happens in an abstract "geometric superspace." Changing a fermion into its boson superpartner and back displaces the fermion from its original position. This implicates gravitation. Remembering Einstein's General Relativity concerns the effect of manifest objects on spatial extensionality (and conversely) supersymmetry is thought to unite gravitation with the quantum (of action). We show how by chapter's end.

In global symmetry when one point is changed, all points change along with it. In local-gauge symmetry each point can be changed independent of other points. Local symmetries place constraints on the theory, but divulge more about the nature of the symmetry.

Each local space-time point has its own phase clock. A change from one point to another corresponds to a local-coordinate transformation (altering the frame of reference).

General Relativity is "naturally" incorporated into quantum theory by using local-gauge theory. Local gauge invariance invokes a gravitational field. That field, quantized, implies the quantum of gravity. The hypothesized quantum gravitational "exchange" particle is called the graviton. It is thought to be a spin 2 boson. (The space-time metric transforms under spin 2—a symmetric rank-two tensor.)

Supersymmetry unites elementary particles via an "internal" symmetry. An internal-phase arrow conceptually identifies the particle. Rotating the arrow or "rotating the particle in space," (against its superparticle counterpart) changes the particle's identity. Mathematically, the change of a boson into its fermion partner is one of gauge transformation.

The electroweak force has been explained by gauge transformations. Maxwell's theory of electromagnetism is also a gauge theory. Gauge theory is thought significant in grouping everything together. However, we have learned every time something is explained the explanation is of a higher order. We expect this trend to continue up to the ultimate-explanatory principle. The grouping of everything, though necessary, is different from the understanding (interrelating of) everything. Everything constitutes manifestation. How can "everything" (here represented by fermions) be explained?

Ultimate Principle is Theory of Everything

Fermion particles are the proper stuff of manifestation. Fermions exhibit one-half-integral spin (spin-angular momentum times Planck's constant). Fermions obey Pauli's Exclusion Principle. Only one fermion can occupy a given point (of space and time).

Quarks (for example u [up], d [down]) and leptons (most well known are electrons) are fermions. They are considered the basic stuff out of which "higher level" particles are made. Protons and neutrons (i.e., nucleons) are composed of quarks. Atoms are made of nucleons and electrons. Atoms "chemically" combine into molecules. Higher level physical structures are built from atoms and molecules. Fundamentally, "everything," every something, is of a fermion nature.

Boson particles (generally function to) bind fermions together. For example, bosonic exchange particles combine basic fermions into stable nucleonic structures. Bosonic gluons (glue balls) hold quarks together. Virtual photons "hold" electrons to nucleons through attraction of "opposing" electrical charges. The negatively charged electron is attracted to the positive electric charge of the nucleus. Electromagnetism is responsible for atomic and molecular structures.

Photons, $W^\pm$ and Z^0 particles (mediators of the nuclear weak interaction) and the hypothetical graviton, are examples of bosons.

Bosons have integral spin. They do not obey Pauli's Exclusion Principle. There is no limit to how many bosons can occupy any given point (energy state).

Because bosons, under suitable conditions, are thought capable of changing into fermions, supersymmetry conceptually links particles of integral spin to particles with one-half-integral spin. For example, the bosonic photon with a spin of one ("1") is linked to the hypothesized fermion called the photino with a spin of 1/2. Other than a difference in spin (and perhaps rest mass) the photino has similar characteristics to the photon (zero rest mass). The hypothesized graviton with a spin of two (2) is linked to a supposed fermion called the gravitino with a spin of 3/2. Integer spin particles take on values of 0, 1, 2,…. Half-integer spin particles have values 1/2, 3/2, 5/2,…. (Refer to Chapter 15.)

Physicists have postulated a particle called the Higg's boson. It has zero spin. A zero spin particle (represented by a scalar field) has no spin axis: all its orientations are equivalent. By a rotation or translation of the Higg's particle (by its identifying "internal" arrow) all particles are conceptually representable.

The Higg's particle has been uniquely identified with the vacuum. It may be the "ground" particle from which all other particles are produced. As mentioned in Chapter 15, fermion-boson pairs (e.g., photon and photino) are two aspects of the Higg's superparticle. The neutral Higg's particle may be the superparticle which "explains" all particles. How?

Fermion particles gain mass from the energy of the Higg's particle. This occurs by the Higg's mechanism. Particle symmetry is (functionally) broken and the fermion acquires mass from a non-zero-expectation value (arising from an energy shift in the field) of the Higg's particle. This is calculated from performing a gauge transformation. The field (gauge) changes from a scalar field to a tensor field. The Higg's particle provides a way to explain fermion mass. (Note again, manifest particles can be defined as those exhibiting rest mass.)

We saw in the last chapter that "nothing" explains "every something" (reductionistically). In physics "nothing is space." Nothing explains itself. Space explains itself: space is self-explanatory. Further, space (in itself) is dimensionless, yet accommodates (includes) infinite dimensionality. (Dimensionality is conceptually superimposed on space.) Dimensionality is explained by non-dimensionality. Spatial extension is explained by space-itself.

Because the empty set or zero (0) represents space, it is easy to see how the Higg's particle is identifiable with space (the vacuum). Because of its supposed role, it has spin zero (and hence equivalent in any rotation) and is directly related to the vacuum. The Higg's particle is correlated with the first zero (in the numerical equivalency) of our logic formula representing the Ultimate Principle, namely, $0 = 1 \times 0$ (from $F = T \wedge F$).

"Everything" (every something) is derived from the wholeness function. In physics, we suggest the wholeness function is represented by the Higg's superparticle. Every something is an unfolding of the wholeness function, which is no-something, nothing.

> We also recognize the wholeness idea provides a way to reconcile extreme reductionism with extreme synthesis. (It does not necessarily yield the same understanding.) The Higg's "particle" (it is not a "part") supersedes everything because everything is derived from it by (its) self fractionation. All particles "unify" via the Higg's particle. It conceptually encompasses all particles. This enables a merging of General Relativity with quantum theory. (We shortly explain how.) The resultant quantum gravity is not, however, understood as spatial curvature. Rather, it defines space and time (themselves) aside from spatial extensionality and temporal flow. (More about this in the final chapters.)

Nothing explains every something. Parts (things) are explained by the whole (no-thing: nothing) from which they are derived. The Law of Noncontradiction, the Ultimate Principle, defines "nothing." The Law of Noncontradiction is the Theory of Everything (physicists are seeking). Let's explain the reasoning behind this statement.

On the bases of our understanding of the set-theoretic meaning of the Law of Noncontradiction, we suspect the zero Higg's particle represents the wholeness aspect (function) of the unitary nature of electromagnetism. Let's elaborate:

Electromagnetism is usually defined through a melding of space and time. (Its wave nature is a function of—depends on—both space and temporal flow.) Electromagnetic energy is portrayed by frequency (oscillations per second, for example). Cessation of temporal flow disallows characterizing electromagnetic energy by frequency.

Take away temporal flow; there is no past, no future—no dynamism. There just is the present. "Fixed" in the present, electromagnetism is "frozen" in a given "here and now." In the temporal present, there just is the wholeness function. The wholeness function is space. Time enters the equation only when motion of parts is brought into the fray. Only when electromagnetism is "viewed" from a manifest perspective, does it take on temporal characteristics.

Conceptually focusing only on the wholeness function of electromagnetism is to eliminate a consideration of dynamism. It is perhaps easiest to define the physics' wholeness function by what it is not. It is a wavelessness (not a wave) function: a function "outside" the flow of time.

> Be mindful: Electromagnetism is entangled in space-time. Without space, there is no manifest dimensionality; without motion of manifest parts, there is no flow of time.

What is the physical evidence of the wholeness function when applied to electromagnetism? The wholeness function (represented by zero) is associated with the "undivided" wholeness, faced in quantum physics. This contention, as previously mentioned, is supported by A. Aspect's (et al.) experimental confirmation of the violation of Bell's inequality. (Refer to Chapter 15.) Local realism must obey the inequality. Because the inequality is contraindicated by definitive evidence, the only reasonable recourse is to understand the evidential implication of Bell's theorem by the wholeness function.

How do we get from an electromagnetic wholeness function (represented by the Higg's superparticle) to a parts-function? Any whole is potentially divisible. The Higg's boson "decomposes" (fractionates) into parts. "Parts" (in themselves) are fermions. How any whole becomes functionally divided is a clue to how fermion particles are produced.

Before explaining how (in our conceptual scheme) the zero spin (or spinless) Higg's boson transforms into 1/2 spin fermions, let's take a closer look at the photon. In our conceptual treatment: the photon provides the intermediating step between the wholeness function and the parts-function (considered separate from the wholeness function). The physics' unfolding of the wholeness function is patterned after the derivation of the natural numbers outlined in the previous chapter.

The photon's spin value is one ("1"). In the previous chapter we explained how one ("1") is derived from zero ("0"). It seems reasonable the Higg's particle yields a spin one particle (photon) before producing fermion particles. Here, we speak of the photon categorically as the collective of every photon—electromagnetic unity. Let's elucidate these concepts.

> Special note: Nomenclature is often deceptive. It is thought light is a particle and a wave. Not having the proper conceptual scheme to understand how, and at what level these aspects (particle and wave) interrelate, causes confusion. We are presently "conceptualizing" logic principles to better understand the nature of light. We should allow our logic formulas to dictate how to interpret the nature of the photon; hence, to reveal a deeper understanding of electromagnetism.

The particle of light, or photon, exhibits a spin of one ("1"). Light (or electromagnetism) has a dual nature. Light is both particle (as indicated by the photoelectric effect) and wave (as exhibited in the double-slit diffraction experiment). Let's decipher what this means.

Particularly troublesome is understanding the wave nature of light. To say electromagnetism has a wave nature is misleading. Yes, in a manifest system, light (electromagnetism) exhibits a wave nature. But, that does not tell us about the nature of light in its own frame of reference. In its own frame of reference, light is better characterized as waveless. Why do we say this? This is an understanding, not only from Special Relativity, but one obtained by applying our logic principles to interpret the nature of light. Let's explain.

In our "logic" treatment, there are three considerations. There is the unity function represented by one ("1"). There is the wholeness function represented by zero ("0"). There is the parts-function represented by infinity ("∞"). Let's apply these definitions to electromagnetism.

First, a quick review: Unity functions as two aspects. One is the wholeness function; the other is the parts-function. However, unity is a function in its own right. A unit function is a (collective) whole and part. Zero indicates the wholeness function: no parts. Infinity indicates parts-function: sans wholeness function.

> Unit function—unity. The photon is the unit (whole and part) function of electromagnetism. In this sense (i.e., unity) it (the photon as a part, i.e., "particle") is addressed in conjunction with its wholeness function. The notion of the "wavicle" captures defining electromagnetic unity from the perspective of a tardyonic inertial system.

> Wholeness function—zero. The wavelessness nature (wholeness function) of electromagnetism (in itself, without consideration of its particulate aspect suggested by the unity function) is captured in the significance of Bell's theorem and Aspect's experiment. Aspect's experiment demonstrated a correlation of "parts" by an "instantaneous" nonseparability of those parts. That communicative instantaneity is the wholeness function (detected by observing a nonseparable connection [singlet state] between parts).

> Parts function—infinity. Granted there is a wholeness function (besides its role in the unit function) we expect there to be a parts-function separate from the unit function. As a simplistic question: What is the nature of the parts-function in itself? Or, in electromagnetic terms, what is the nature of the "wavicle": solely through its parts? The obvious answer is the photon. But this is the "nonmanifested" particle of electromagnetism.

Let's ask rhetorically, what is the manifested nature of electromagnetism? Simply, light manifests as tardyonic electrons and positrons (antielectrons) and tachyonic-magnetic monopoles and their anti-cosmos counterparts. These are electromagnetic parts considered separately from the unity in which they function.

> Note infinity is implied in unity. We learned this in the previous chapter. Although there are no parts except there be a whole from which parts are derived, our Ultimate Principle, the Law of Noncontradiction does not directly address parts separate from the whole. Therefore, from $0 = 1 \times 0$ we have stated the "one" (collectively) represents both whole and parts. Parts (in themselves) are represented in our Primary Equation $1 = 0 \times \infty$ where infinity represents parts (disentangled from the whole).

The unity function of electromagnetism is reflected in the constancy of the velocity of light. In its reference frame, electromagnetism is timeless and unchanging. The photon in itself (in its frame of reference) does not possess a wave nature. Instead of asserting the photon is a wavicle (wave-particle: defined from an inertial perspective) it is (in itself) better characterized as a "waveless-particle" unit. This conceptualization is derived from applying (to physics) our understanding of the set-theoretic interpretation of the logical Law of Contradiction. More about this shortly.

An endless quantity of photons can occupy a given state. In this sense there is no difference between one photon and another. Differences in frequency or energy of a photon are attributable to either velocity differences between inertial systems, or to the photon source.

Numerically, the Ultimate Principle ($F = T \wedge F$) is $0 = 1 \times 0$. Zero is defined by the Law of Noncontradiction. It states zero (or an empty set) can produce unity—a set with members (and another empty set). Nothingness can generate somethingness. Zero ("0") generates one ("1"). Moreover, one ("1")—in our set-theoretic treatment—is "explained" by zero ("0").

In principle, the logical Law of Noncontradiction permits the derivation of one ("1") from zero ("0"). Zero ("0") and one ("1") are also equivalent in the sense there cannot be more than the whole with which we began. Our physics correlation, the zero-spin-electrically-neutral-Higg's boson—the possible superparticle, can generate a spin one ("1") particle—the photon. In our understanding, the Higg's superparticle itself produces the photon.

From the Law of Noncontradiction we derive the Law of Contradiction. Negating $F = T \wedge F$ yields $T = \neg[F \wedge T]$. Negating "zero" yields the definition of "one." Negating zero (the wholeness function) yields unity (a whole with parts). The Law of Contradiction defines unity.

What does the Law of Contradiction say about the photon as a unit? First, electromagnetic radiation-photon is a unit—a "collective" whole and part. Applying the Law of Contradiction to electromagnetism, we find that a "unit" photon cannot both be and not be. It cannot be no-thing (wavelessness) and something (wavicle) at the same time (or, at any given time). The idea of somethingness also applies to tardyonic objects—things in manifestation. Remember that fermion particles also exhibit a wave nature.

We suggest, for now, that electromagnetism is of three functions. It is both a wave and a particle—a wavicle—a unit function. As a unit, "light" cannot both be and not be, at any given time. When it is not, it is wavelessness—not something (i.e., nothing). When it is, it behaves as a unit-something—electromagnetism. The particulate aspect of the unit photon, considered separately from its unit function, is what we denote as its manifestations, e.g., electrons, positrons, magnetic and antimagnetic monopoles. Fundamentally and categorically, these represent all things, i.e., everything.

In this chapter, we intend to simplify understanding. Here, we deal with basic concepts. Details are not needed to explain the reasoning. For the sake of completeness, we here list some other "parts." Besides electrons, there are other leptons and their antiparticle counterparts. Magnetic monopoles and their antiparticle counterparts are represented mathematically by "complex" numbers (the imaginary number "i"). These and other particles are implied for a complete understanding of the structure represented in this book. These "parts" are more fully addressed in Section Two on cosmic structure.

Applying de Morgan's Law to the Law of Contradiction yields the Law of the Excluded Middle $T = T \lor \neg T$. It states the following about the photon. At any given time, a photon either is or is not. We can "read" (by the Law of Excluded Middle) the Law of Contradiction as saying (depending on our experiment) we can test for either the particulate (as in the photoelectric effect) or wave aspect of light (as in diffraction experiments). We cannot test for both (in the same experiment) at the same time, or at any given time.

Another confusion in physics is the idea the photon represents only the particulate nature of light. Yes and no. The photon is the "part" considered as an aspect of the whole. We suggest the photon is the unity function as mentioned above (defined by the Law of Contradiction). The photon is not the part considered separately from the whole. That notion is reserved for manifest particles (fermions) e.g., the electron and positron.

What else can we learn? Because unity is a whole and its parts, it can generate parts (which are separate from the whole) by a "functional" partitioning of itself. Let's look at another way to describe partitioning. From unity (1/1) a continual fractionation can be generated. The next fraction is 1/2. This signifies a halving of the whole. This produces two halves. Each is a function in its own right. Fractionation was described in the previous chapter. What does this partitioning mean in physics?

Spin-angular momentum is conserved. Again, we cannot end with more than we began. Not forgetting this, we ask: What does spin 1/2 angular momentum tell us about cosmic structure? It means we are dealing with a particle that functions at a level, which is half of the unit. (We suggest this based on an understanding in the last chapter concerning how the natural numbers are generated.) Let's clarify:

A one-half-spin particle implies its place in the structural scheme. It is separated from the unit function. It functions at a "lower" level (than the unit from which it is derived—the photon) in the elementary-particle hierarchy. Simplistically, a different function means we are considering a separate "somethingness." That "something" is manifest particles with 1/2 spin. Fermion particles exhibit 1/2 spin (or multiples thereof).

Fermion particles are derived from a partitioning of the unit function. From nothing (zero-Higg's boson) something (unity—electromagnetic photon energy) is generated. Fermionic parts (e.g., electrons that exhibit 1/2 spin) are produced from electromagnetic energy (photon: our wavicle).

We note only left-handed-spin (fermion) particles and right-handed antiparticles are subject to the weak force. Their antimatter counterparts are not subject to the weak force. The weak force is a benchmark for differentiating cosmos from anti-cosmos.

Lesson: The first fermion partitioning of the unit-photon creates the particle-antiparticle dichotomy. Chirality ("left-right handedness") is represented as a partitioning of Reality into Cosmos and Anti-Cosmos. The CPT theorem informs about the charge, parity (chirality—things affected by placement in spatiality) and temporal aspects of the Cosmos—Anti-cosmos. (Chapter 11 dealt with cosmic-anti-cosmic structure.)

Manifest particles (fermions) are derived from the phenomenon of electromagnetism (radiation-energy). Again, there is no part but there is a whole of which it is a component. The wave nature of the quantum is always there: always in the flow of time. Admittedly, even fermions can express a wave nature. Electrons exhibited wave diffraction in Davisson and Germer's experiment in 1925. Understanding their results supported De Broglie's 1923 thesis that matter is endowed with a wave nature. (Matter waves were discussed in Chapter 13.) The electron is a point-like particle. Anything physically greater than an electron does not have a wave nature of much consequence. It (the wave nature of inertial mass) is, in a practical sense, more of a theoretical consideration. That even fermion particles have a wave nature implies (from the perspective of temporal flow) every part is intimately connected to the whole (of which they are a part). The part (functionally) is in the whole; the whole is (functionally) represented in the part.

Our next question concerns the "unfolding" nature of symmetry. If there is symmetry between bosons and fermions, why is it not recognized in everyday life? Let's approach this topic from the perspective of Cosmogony.

Cosmogony

Physicists suppose (from extrapolating backwards from the present) that the physical universe exploded from a Big Bang. The Big Bang originates from a cosmological singularity (mathematically represented by zero). At the moment of Big Bang, under conditions of extreme high temperature and pressure, there is symmetry between all particles. At the very instant of creation there is no (particle) interaction, just the whole "super" force. (This book suggests the Big Bang instant marks time flow inversion—the change between time phases [Phase II and Phase I: see Chapter 9].)

At the moment of the Big Bang, energy is greater than an equivalency of 10^{19} GeV (giga–electron volts). In this supersymmetric condition, particles are massless and nonphysical. At that temperature there just is the superforce.

Under the umbrella of the superforce all particles are indistinguishable. They are invariant: meaning, under supersymmetric conditions, their manifest nature is only potential. For all practical purposes, they are of one identity—conceptually represented by the Higg's boson superparticle. All manifest particles (potentially) "reside" in the Higg's particle. (See previous pages in this chapter and refer to Chapter 15.)

All physical particles (fermions) have their origin in the Higg's boson. Conservation of superparticle identity is broken by a "gauge shift." (Again, refer to Chapter 15.) What is the source of this idea? Symmetry applies to the equations describing particle invariance. Particle differences are thought to arise because of asymmetry in the solution to those equations.

Manifest-particle states do not exhibit the super-symmetry (associated with their origin in the superforce-Higg's particle). Particles become distinguishable because asymmetric conditions evolve. Particles take on (rest) mass and assume a fermionic nature—becoming manifest matter. Let's review how this occurs as described in this book.

The Higg's boson represents space (in-itself). Space is the whole of reality, but no part. Partitioning of the whole precipitates a change from static nothingness (the superforce) to a dynamic somethingness. The wholeness function becomes broken, producing a unit function. Zero (spin "0" Higg's boson) becomes "one" (represented by the spin "1" boson). The unit function (spin "1" boson) is the electromagnetic quantum (the photon).

The photon functionally splits into pairs of particle-antiparticle fermions—the thingness stuff of manifestation. Therefore, on a grand scale, electromagnetic energy functionally splits into Cosmos and Anti-Cosmos. Things in motion generate the flow of time.

The cosmic interpretation of the Standard Model (of bosonic-elementary particles) explains how the other forces are derived from the superforce. As the physical universe expands and cools, the grand unification force (strong-electroweak) uncouples from the superforce. Temperature and pressure drop. The strong force disentangles from the grand unification force. Later, electroweak symmetry breaks resulting in the disuniting of its components. The separation process yields the four forces of nature we recognize today. These are the electromagnetic, strong and weak nuclear forces, and the gravitational "force." (Refer to Chapter 15.)

The idea of a commonality of particle origin suggests all things, including the (temporal) forces of nature, are derived from some kind of "sameness" (in identity) source. In physics, that also means that Relativity and Quantum Theory must somehow merge at this sameness level. If we have done our work correctly, this merger should be evident in the physics interpretation of the Ultimate Principle.

Merging Relativity with Quantum Theory

General Relativity is Einstein's theory of gravitation. Gravity is now understood by how space curvature affects the motion of physical bodies. (Refer to Chapters 14 and 15.) Newton's "attraction between masses" is replaced by how matter affects the intrinsic nature of extended space. Dimensional space is only properly understood by defining space itself. What is its function? We have answered that question. Space is the wholeness function of Reality. Let's backtrack a little.

Physicists believe General Relativity dictates the graviton (representing gravity's force) is a spin two ("2") boson "particle." This represents its supposed quadrupole nature. There is another way to understand the spin "2" boson.

The 2-spin graviton may pertain to the electromagnetic force considered reductively. Electromagnetism has two natures: electric and magnetic. Each is a dipole. Together they exhibit a quadrupole nature. Therefore, the hypothesized 2-spin graviton could just be the Higg's superparticle (the wholeness function) when it "becomes" the electric and magnetic aspects of electromagnetism. That is: electromagnetism reduced to its separate electric and magnetic components—each considered in its own right as a dipole (plus and minus electric charge; north and south magnetic pole charge) but together acting as a quadrupole. Their individual dipole nature (either electric or magnetic) occurs in the flow of time.

To suggest the other forces (electromagnetic, weak, and strong) unify with gravity at the highest level is to say that the quantum of gravitational energy (the graviton boson—quantum of gravity) should have zero spin. This is based on our understanding of the Ultimate Principle applied to physics. It has been explained in this chapter how that principle is the Theory of Everything. The Ultimate Principle can be used to understand the connection between Relativity and Quantum Theory. (Refer to Figure 19:1.)

There is a functional difference between space itself (represented by the zero-spin Higg's boson) and dimensional space. The processing of space into extensionality corresponds to the Higg's superparticle transforming into the electromagnetic quantum or photon.

The superforce occupies the highest level in the (hierarchical) structure of reality. We do not suppose the superparticle (itself the source of all other particles) exhibits a dual nature. (Yes, it is an [wholeness] aspect of the dual nature phenomenon—electromagnetism.)

We expect the superparticle, as a unifying force, to be a wholeness function. The Higg's zero-spin-super particle is described by the Ultimate Principle as the wholeness aspect of the unit-photon. The super-force "particle" is the wholeness aspect of electromagnetism. (Better: the super-force is the holistic aspect of electromagnetism.)

If gravity, in its pure form, is the ultimate source of the other forces, then it seems reasonable no graviton exchange particle is needed. Instead, the zero-spin superparticle merely suggests all processes (interactions) are by nature "precipitated" from the superforce (embodied in the super "particle"). The superforce reductively behaves in a way that we call gravity (from a physical perspective). This is a carry-over of Newton's idea. Einstein's relativity states that space is distorted or curved by massive physical objects. Space curvature in turn affects the motion of matter. No exchange particle is needed to explain this.

Gravity, once thought to act by "attraction" of masses, is now better explained by space-time curvature. Massive objects appearing to attract each other are merely following merging paths in extended space. If there is no real attraction between masses (and hence no requirement of exchange particles between them) then is the concept of a force (gravitational) needed to explain why masses appear to "attract?" Answer: No. That is the significance of Einstein's General Relativity. Celestial bodies travel the intrinsic geometry of space-time. This book's understanding of his ideas is that the graviton is not an exchange particle. It does not function in the flow of time.

Simplifying: The "unifying" superforce is not expected to function in time flow. Rather, it affects parts in a way not recognized as a cause and (then) effect force. We suggest the spin-zero-Higg's superparticle is the real graviton—the quantum of gravity. All forces and manifest particles unify under the auspices of the Higg's boson. The superforce is not expected to behave as an exchange particle. The Higg's superparticle probably functions independent of time flow. The superforce "acts on" parts in a timeless manner. (We explain how in the next chapter.)

For now, let's look at this problem through Einstein's eyes. It is granted that spatial curvature occurs in the flow of time. Things exist in time. ("Attracted" bodies move closer through time.) That is how we understand gravity. What is gravity itself? It is not inconceivable gravity, aside from how it affects things (in time) functions outside or beyond temporal flow. If this is so, then applying the physics' conceptualization of our Ultimate Principle differs from the idea there is an exchange particle that accounts for gravity.

Our "logic" understanding posits that the superforce acts instantaneously on parts. Treating the Higg's superparticle as the "embodiment" of that force is not to say it is an exchange particle. Rather, the Higg's particle represents the wholeness function—the whole of reality. We are reminded of Aspect's test of Bell's theorem.

If the superforce (as a wholeness function) does not act on parts through a force, then it acts holistically (by the greatest whole). Therefore, if the "action" of the superforce is instantaneous, no exchange "part-icle" is needed.

If gravity has no source in some higher sameness, and it only applies to physical things in space, then it would not be identical with the superforce-Higg's boson. It would only be a lower-level "force," like the other forces. Because of Einstein's work, we do not suppose this is the case. Plus, there are other ways of understanding gravity. We examine a more generalized way to explain gravity in Chapter 21: subheading "Entropy and Negentropy."

What does all this mean? The "unifying" role of the graviton (itself) is assumed by the Higg's boson. The Higg's quantum boson is the wholeness function (of reality).

The wholeness function (of reality) is synonymous with space itself. Space (itself) envelopes extended spatiality (and its possible curvature). General Relativity concerns space curvature.

The wholeness function (of reality) is also represented by the Higg's boson. The Higg's superforce "particle" conceptually encompasses all other forces of nature and all manifest particles. All particles and forces are "reductively" represented by the Higg's boson, except gravity. The Higg's superforce boson is the real graviton—the quantum embodiment of gravity.

Relativity and Quantum Theory naturally merge in the "pure" zero spin quantum. The Higg's boson quantum is the superforce (the "unifying" graviton)—space itself. Relativity and Quantum Theory merge at the pinnacle of reality's structure—represented by the Higg's superforce boson.

Key Points:

- The Ultimate Principle easily correlates with certain theoretical and evidential concepts in physics. Understanding zero is definable as a wholeness function enabled a direct correlation with certain aspects of the waveless (timeless) nature of the quantum. Again, this justification is supported by the implication of Bell's theorem (see Chapter 15). It strongly suggests, at a certain "level," there is a nonseparability between all things. That level is the wholeness function. At this level, there is no flow of time; everything is intimately connected. Additional meaning is found in the next chapter.

- It has been stated that "physics breaks down at the (cosmological Big Bang) singularity." Little sense (in physics) is made of a point (represented by zero) exhibiting infinite pressure. Zero (in itself) cannot be easily understood "within" mathematics proper. Physics breaks down because mathematics breaks down. Zero (itself) is only correctly defined outside (or beyond) mathematics proper. We have done that by (categorically) understanding mathematics in terms of set theory and logic. Let's more closely tie the Ultimate Principle ($0 = 1 \times 0$) to the structure delineated in this book.

- The zero ("0") on the left side of the Ultimate Principle (the logical Law of Noncontradiction) is represented in physics by the neutral Higg's boson. It represents the superforce (exhibited by the singularity): the unifying force of nature. (The physics interpretation of zero is implicated in Section Four—Realm of Whyness structure. See Chapter 14.)

- The one ("1") on the left of the logical Law of Contradiction (and the Primary Equation) corresponds in physics to electromagnetic energy. Unity speaks of a "collective" whole and part: the whole of reality and every part. One (1) is the unified function. (It was discussed in Section Five—Reality-Unity structure. See Chapter 15. It also is discussed independently [of other forces] in Section Three—Realm of Reasoning structure. See Chapter 13.)

- 1/2 spin-fermion particles represent a functional division of the whole into two roles: each diametrically opposed to the other. Things composed of fermions constitute (manifest) cosmic structure. Structures composed of fermion antiparticles constitute the Anti-Cosmos. Tachyons and anti-tachyons also are incorporated into the structure via temporal "inversion" cycles for each cosmic structure. (Cosmic structure is described in Chapter 10—"Deltic Cosmos." The anti-cosmos is discussed in Chapter 11—"Bi-Deltic Cosmos.")

- Parts, without regard for the whole, are represented by the infinity on the right side of the Primary Equation ($I = 0 \times \infty$). Here, the infinity symbol does not differentiate between layers of hierarchical functions. Infinity also represents the potential energy of the vacuum (space). (Hierarchical functions are described and explained in the next chapter.)

- Zero is represented by space (or vacuum) without regard for anything that occupies it. Infinity represents a limitless functional partitioning of space: considered separate from space itself. Hence, there is no actual difference between space and its infinite partitioning. The difference is functional only.

- Quantum theory and Relativity should merge at the cosmogonical level. Cosmogenesis is conceptually represented by the Higg's boson—the cosmological singularity. All manifest particles have their origin in the Higg's boson superparticle. The Higg's boson is the real graviton (the quantum of gravity). The graviton is space itself. Space envelops spatial extensionality. That includes spatial curvature and hence gravitation. The Higg's boson envelops all spatial extensionality and all quantum particles. The Higg's boson (quantum) conceptually represents both all space and all (quantum) particles. It merges Relativity with Quantum Theory.

We are left with lingering questions. The Big Bang has been difficult to explain. How can the Big Bang be explained as an expansion of space? What is space expanding into—if not itself? If space is expanding into itself, then there is a difference between expanding space and the space into which it is expanding. We need to learn more to answer this question. It will be clearly answered.

Time has not been satisfactorily explained. We now understand the "flow" of time. But, what is the nature of time itself? How are various levels in the hierarchical structure (functionally) differentiated? The wholeness function (zero) pulls parts (infinity) into a unity (1). What is the nature of that which the wholeness function is pulling against? Is there a better way to understand gravity? If gravity is just space pulling things together (on converging paths) then space (the wholeness function) seems to act for some purpose. What is the purpose? Purpose implies intelligence. Intelligence ultimately implicates consciousness. And finally, what is the basic nature of consciousness? We can now confidently make a bold statement. Experiments can not tell us what consciousness is—only what it isn't.

How does what we have learned thus far, enable us to tackle these remaining problems? We turn to Section Seven and the Conclusion for those answers.

<u>Section Seven</u>

CONCLUSION

Perspective

- Entropy and negentropy_
- Design and purpose
- Significant experiences_
- Intelligence levels
- Higher functions
- Consciousness Defined
- Memory
- Deeper explanation
- Life origin
- Evolution of life forms
- Explaining biological life-form complexity
- Life forms designed for purpose
- Remaining religious questions
- Meaning to life
- The Hierarchy
- Final Correlations

Chapter 20

Consciousness, Experience, and Memory

"Nothing explains itself.
Nothingness is consciousness.
Consciousness is self-explanatory."

Much of this project concerns exploration of the mind expanse (Section Two through Section Five). As we explored, we continually refined our methodology to push deeper into the unknown. Concepts were constructed and interphenomenally linked. We continually asked, how can these concepts be further explained? We produced "enveloping" concepts that eliminated contradiction. We completed our exploration and delineated the structure of reality. We now know how everything "fits together."

At the end of Section Five, we realized the structure of reality had not answered all our questions. We studied our structural mappings and reexamined our methodology. Some questions in philosophy, religion, and physics were unanswered by the structure of reality. Although every puzzle piece was "in place," none of the pieces or their "placement" (in the structure) revealed anything about consciousness and God. We also had not discovered the principle that ultimately "unifies" everything in physics.

The ubiquitous dilemma of definition was again the obstacle to deeper understanding. What has not been verified cannot be properly defined. What has not been explicitly defined, cannot be tested and verified.

We focused on the bedrock of our methodology—the Law of Contradiction. We studied set theory and logic. We scrutinized attempts to "axiomatize" arithmetic. Major issues were resolved. We enhanced the effectiveness of our methodology by completely and nonreductionistically understanding the nature of consistency. We applied what we learned to physics. The principle that defined the consistency of logic, categorically (and therefore, inclusively) "explained" the structure of reality. We uncovered the Theory of Everything (T.O.E) equation.

We now seriously contemplate the nature of consciousness. Fully understanding consciousness should enable us to answer other unsettled questions. What are the major cosmic processes? We have yet to answer the question that led to this project. "Does life have purpose?" (Refer to "Letter" in Chapter 3.)

Self-realization (see Chapter 3, Where Aspect Diagram) is complete when consciousness is fully understood. That is unachievable until everything is conceptually interrelated. Consciousness was not definable by anything in reality.

Remaining problems converge on a simple question: "Can everything be explained?" Is everything explainable by consciousness?

If consciousness is the explanation for everything, how does it explain everything? Understanding consciousness should enable us to determine if life has meaning. If it does, life-form evolvement should also be nonreductionistically explainable. These questions are answered in the next and final chapter. We begin this chapter by reviewing some of the problems associated with understanding consciousness.

Defining Consciousness: the Problem

Consciousness has eluded understanding. If the nature of consciousness were obvious, it would have been understood long ago. Philosophers and scientists alike have not formulated an explicit conceptualization of consciousness to examine and test. Let's set the parameters for understanding consciousness.

If awareness and consciousness are not synonymous, they surely are entangled. This is indicated by the puzzling phrase, "conscious awareness." If they are distinct, but functionally intermingle, that would account for the puzzlement. Awareness and consciousness need to be "conceptually" untangled.

The problem worsens. Awareness is closely intertwined with experience. Our task is to conceptually isolate each and determine their characteristics. Fortunately, we already understand much about awareness. (It is described in detail in Chapter 4: the mental functions.)

We can make one broad distinction. There is consciousness, and there is that which we are "conscious of," namely, experience. The initial problem is to define where experience ends, and consciousness begins. Awareness falls somewhere between physical experience and consciousness.

Here, we briefly describe the relationship between awareness and physical experience, then explain why awareness is not consciousness.

Consider the experience side of awareness. One type of experience is presented to us through the senses. We "observe" phenomena—things and events. Sensory experience "represents" the (external) physical world (the Universe of Perfect Whatness). From the field of sense data we select and identify visual images. Here is a table; there is an automobile. Identified sensory images constitute extrinsic perception.

(Past) experiences are "formulated" as visual constructs. Things are recognized by referring to abstract constructs stored in memory. Memory harbors a catalog of "forms." The automobile has a very recognizable form. Physical things are abstractly characterized as whatnesses. "What is this (form); what is that (form)?"

Identifying phenomena is the most basic of inferences. Inferring is a parcel of experience. Inferring defines the nature of awareness. We understand awareness by defining inferential functions. (Refer to Chapter 4.)

There are two modes of inferring—deductive and inductive. Identifying sense data is accomplished by deduction. A "whatness" datum is compared to referential forms and the item is identified. "There is an automobile." Sense experience is analyzed (mentally broken into parts) and inferences deduced. "That automobile has a flat tire." We deduce the nature of the physical world.

How do referential concepts (in memory) develop? Statements are "generalized" from extrinsic perceptions. Universal concepts are built from inductive inferences. "The automobile is a vehicle." Properties and relationships are "appended" to these referential constructs. "The automobile is usually made of metal." Induction "classifies" extrinsic perceptions at higher levels of abstraction. Referential concepts are of a visual form.

Refined inductive methodology enables us to know "why" something is. The course of inductive inference is not determined by the physical world. We ask, what is the source of induction? Put another way, to what are we inducting toward? A simple answer is to say we induce toward reason itself (whyness). Unless "reason" is explicitly defined and explained, that idea is of little help. We need to know why anything and everything is.

Corresponding to the two inferential processes, are two "structural" characterizations of the mind. One is associated with the deductive function. It is called the Universe of Imperfect Whatness. It arises from awareness of "whatness" sensory data—producing extrinsic perceptions. Extrinsic perception serves as an intermediary between awareness and the physical world we experience. The other inferential process is associated with the inductive function. It is called the Universe of Imperfect Whyness. It utilizes intrinsic perception (i.e., "intrinsic-referential generalizations or visual forms [constructs]) from memory. Intrinsic perception serves as an intermediary between awareness and consciousness.

The physical world exists in-itself (meaning it is separate from mind—we know this because of the color paradigm). It has been explained that sense experience (in-itself) is not real. (The mind acts upon sense data "as it is perceived by the individual.") Deduction does not solely arise from the mind: it depends on the whatness world for premises. Though the mind is real (meaning it is "functionally" separate from the physical) it contributes only the deductive function.

> Sense experience does not exist except for the physical world, and it does not exist apart from one's "perceptive reaction" (to the physical). No two people observe a given event the same. One person may suffer from poor visual acuity or be "color-blind."

We discern from split-brain experiments that the deductive process is usually associated with the left hemisphere and the inductive process is associated with the right hemisphere. A split-brain patient's right brain can do things of which the left brain is unaware and vise versa. It seems they work together but function independently. Some psychologists thought the split brain experiments indicate each person has two minds.

Sensory experiences are "presented" to the individual (viz, his mind) perceptively. Sensory "input" results from experiencing the physical world. One asks, what is this? He can also ask, why is this? The left brain asks what; the right brain asks why. Although there are two modes of awareness, there are not two minds. Each mode (asking—what? why?) is functionally opposed to the other. Each person has one mind functioning two ways—deductively and

inductively. One deduces what something is, and inductively determines why it
is.

We only have extrinsic-sense experiences while "awake." When sensory activity
falls below a certain threshold, the physical world is no longer experienced. The
deductive function ceases with minimal sensory input. This happens whether
one is asleep or knocked unconscious. This suggests the (hypothesized)
deductive mind (an aspect of the Universe of Imperfect Whatness) if considered
real while awake, does not exist when one is asleep. What is its status when not
functioning? If it exists, "in-itself," it would mean the deductive mind must be
somehow resurrected every morning before awakening. This is a weak argument
for actual existence of a deductive mind. If it "disappears" while one sleeps,
perhaps it does not exist (in the same sense as does the physical world). We
grant the deductive function does not persist, as does the physical world.

Most people are awake (aware of extrinsic perceptions) while making deductive
inferences about sense data. However, the sensory image is not the object it
represents. The deductive function extracts information from sensory
"representations" of the physical world. "The thing 'I see' is an automobile."
We conclude the sensory image is not the real thing (i.e., the actual automobile)
and the deductive mind (itself) does not exist. We therefore ask, how can the
deductive function (an aspect of the Universe of Imperfect Whatness—which
does not exist in itself) originate from something (extrinsic sensory image)
which does not (itself) exist? We wonder how do extrinsic perceptions arise?
Simply, how are extrinsic perceptions explained?

If the deductive function is not generated by a "deductive" mind, there is only
one agency left—the inductive mind (an aspect of the Universe of Imperfect
Whyness). The inductive mind is as real as the physical world. The deductive
mind itself is an illusion. Each individual has but one mind.

How can the same mind (an aspect of the Universe of Imperfect Whyness)
produce opposing inferential processes? How does the mind, which by nature is
inductive, produce deductive inferences? The mind naturally functions
deductively while "inhabiting" a body in the physical world. How can the mind
both be aware (more precisely, fully functional) in one ability (deduction) and
only subaware (subfunctional) in another ability (induction)?

There is only one mind (for each individual). Is it aware or subaware? How can
it be both? It is easily answered, though it is paradoxical. A glass half-empty is
the same as a glass half-full. A glass is as empty as it is not full. Although
opposites are described, they are but two ways of representing the same
situation. In a given circumstance, a mind not functioning inductively functions
deductively. Deductive ability results from the lack of inductive ability.

This is not surprising. The color paradigm suggests there is a
complementary nature to the physical world. There is the "apparent"
(phenomenal) physical word we experience through sensory data. Then
there is the real world we can only appreciate through reasoning. The
same idea seems to apply to the mental world. The deductive function

is an "apparent" mind. It is the backside to the real mind (the inductive function).

Considering intelligence functions: An individual can only infer to the extent he is aware. The more one can realistically infer (and therefore, inductively) the more aware or "mentally awaken" he becomes. A person analyzes things only by whatness when he lacks the ability (i.e., is subaware) to assess the meaning of things by whyness (which he can do when he is fully aware). Both functions are products of the same mind.

The mind is characterized by awareness: it makes inferences. Inferences inform us about "things" in the physical world. Sensory experiences provide premises for inferences. The mind experiences ("imperfect" whatness) representations of the (more "perfectly" formed whatness) physical world. Mentality and physicality exhibit different functions. Awareness (inferences) and what one is aware of (sensory experiences of the physical world) are categorically distinct.

Is awareness consciousness? (This question is not answered by untangling awareness and sense data.) Is the mind (itself) conscious in any form? Or, does this question unintentionally merge two phenomena into a single category? The answer depends on whether consciousness can be "isolated" from awareness.

Let's examine some interesting experiences. If consciousness is the same as the mind, the mind would be fully functional, i.e., inducting when we are fully "conscious" (viz, awake). This does not happen because the mind seldom inducts knowledge to higher-level generalities although we are awake (to the physical world). Furthermore, even when we are inducting high levels of generality (our) consciousness is not increased at all—only awareness is increased. Consciousness stays the same whether our intellect is functioning deductively or inductively.

When two individual's minds "go into the oneness state" (this rarely happens in the physical world) each person is not conscious of the other's thoughts and each is not conscious of the other's memory. Each individual is conscious only of the "mental-togetherness" sensation experienced in that state. If the mind (awareness itself) were conscious, we would expect each individual to know the other's thoughts. (This would imply a temporary merging of each person's consciousness.) It does not happen. This suggests the mind (in itself) is not conscious. This shows awareness and consciousness are distinct and functionally separate.

There is no paradox why two people, in the "oneness state," do not experience each other's inferential thoughts. For us, consciousness is a passive observer. The problem is awareness has not been explicitly understood.

Awareness can be gauged by one's level of understanding. (See How Aspect Diagram, Chapter 4.) Even when the mind is awake to the physical world, it is (itself) asleep because it is not inducting knowledge. People are fooled about this because they are not more aware and because their minds typically do not function inductively (while inhabiting a physical body).

One is unaware of what he has not experienced, does not know, or does not understand—this is experiential ignorance.

We conclude mind (awareness) and consciousness are different. Mind does not exhibit the property of consciousness. Furthermore, there is no reason to suppose the mind is the source of consciousness.

We still have two small hurdles. Let's answer a previous question. It seems most people are not conscious when asleep or knocked "unconscious" after suffering a blow to the head. How is this explained? Quite simply (since the mind and consciousness are separate and distinct) if there are no (extrinsic) perceptions for consciousness to be "conscious of" (something) then we are only conscious of very little (only intrinsic perceptions—dreaming). That people become knocked unconscious also shows mentality is closely associated with brain neuro-functionality. (We will revisit this connection.)

> The deductive process occurs in the presence of extrinsic perceptions. When extrinsic perceptions cease, because of a lack of incoming stimuli (from the physical world) the deductive function also ceases. The idea we are not conscious when asleep or knocked unconscious is easily explained. If the mind is nonfunctional (not processing information from the physical world) there is no "input" for consciousness to experience. It seems as though the deductive function depends upon the physical world. And, in a way, it does. The physical world is the source of whatness premises. However, the mind is functionally associated with the brain. The mind only appears to be active (in itself—working on its own) when it functions (deductively) with respect to the physical world. The mind really functions by itself only when it is processing data inductively. The mind does not function by itself while in the physical world. (It does, but not much, hence its subfunctionality.) This also explains why the mind is asleep when one rests from the weariness of the day. The mind itself is asleep even during daily activity. One is only aware (while inhabiting a body) because his mind is functioning with respect to the physical world—not functioning in itself (i.e., inductively). These ideas led philosophers to consider consciousness might be an emergent phenomenon of the mind (i.e., of awareness). That possibility could be considered an extension of epiphenomenalism, viz, that consciousness emerges from the physical world. That is, consciousness emerges from awareness arising from brain activity. The mind is not an epiphenomenon in the usual sense. It does not arise solely by action of the physical world. If mind is in any sense an "emergent" phenomenon, it is "guided." (This will later be explained.)

The second hurdle is the following. Why do we remain "conscious" when the mind functions inductively, although we are "asleep" (to the physical world)? When functioning inductively the mind does not need help from the physical world (contributing sensory data—everyday activity). It can function on its own "outputting" (inductive inferential) data above the threshold allowing input to consciousness. Mental activity (inductive inferring) keeps us "aware" and hence awake even when our body is resting (and the deductive function is "asleep" to the physical world).

After anesthesia, some medical patients report a "visual" experience "so real" they "cannot believe it 'did not' happen." Imagination "mimics" reality. The source of this experience is intrinsic perception. We note in passing: These experiences are difficult to remember because the mind (awareness) is not actively processing (identifying and trying to understand) these intrinsic visualizations. Inductive inferring (activity) passes the threshold for input to consciousness.

While in the physical world, minds function below their potential. They are subfunctional. This not only explains sleep, but it is easy to understand. In this (physical) phase (Phase I of the universal cycle) we live in the material world. On an everyday basis, a whatness-oriented mind (deductive function) is suited to cope with the whatness world. A whyness-oriented mind (one that is fully functional and inducting information) is not required. Asking why is there bread, does not put bread on the table.

Stating the mind ("individual" aspect of Universe of Imperfect Whyness) is a subfunctional part of consciousness (the Realm of Whyness) does not explain something anymore than saying one's house is a part of the physical universe. However, it (classification) is a step in the right direction (to understand mind). Asserting the mind is a subfunctional aspect of consciousness is not sufficient to understand either. It does not tell what something is: only what it is not.

We learned inferential functions define mental activity. We discerned mind is not consciousness. Though we can conceptually isolate mind and consciousness, we still do not understand the extent or nature of their entanglement. What is the relationship between mind and consciousness? We cannot know how mind relates to consciousness until we understand how consciousness connects to (the wider category of) experience (in general).

Experience is found in the structure of reality. Then it is easy to see why the nature of consciousness is not revealed until after reality's structure is completely understood. The Universe of Imperfect Whyness (mind) is a part of the What Aspect of the Deltic Cosmos. The Deltic Cosmos represents one structural level of the Realm of Whatness. The Realm of Reasoning generates the Realm of Whatness. The Realm of Reasoning connects the Realm of Whatness and the Realm of Whyness. Understanding this connection should explain the relation between experience and consciousness. We just need to find other ways to understand it.

Key Points:

- The brain is not the mind. The mind is not consciousness. Consciousness is distinct from the brain and (mental) awareness.

- Consciousness is not something we experience. We are only conscious of experience.

- Sense experience is "something" that happens to us. It is not something we (our awareness) directly cause. Sense data and awareness are "things" we are conscious of.

- We also experience the ability to infer. Sensation, perception, and inferring are all classifiable as experience. These phenomena are parcels of "what" our higher self (the "observer") is "conscious of." We are (conscious of being) aware of (inferring there is) consciousness. Consciousness is inferred.

- The function of mind is inferring (either deductive or inductive). We determined a mind not functioning inductively, functions deductively (by default). Each individual has one mind; one not contemporarily shared by anyone else.

- Our individual self is defined by awareness—the ability to infer. We are conscious of awareness, but can we be aware of consciousness? Specifically, we are aware of the mind making inferences (we infer that we infer) but from experience, can we inductively infer the nature of consciousness (itself)?

This project delineated the structure of reality by defining hierarchical levels. Most structural levels are definable by their function (in the scheme of things). Each function serves a purpose. Our task is to define consciousness by how it relates to the structure of reality (as described in Sections Two through Five).

Multi-level functions provide the stepping stones to understand consciousness. Each human individual is defined by awareness—by mind. Is there a greater Beingness of which we (our minds) are a (functional) part? There seems to be a "super passive observer" of our awareness (inferential functioning). Because we have not found any phenomenon "in" reality corresponding to this Beingness, we speculate it is "beyond" existence—transcending experience.

Experience is what consciousness is conscious of. Is consciousness our higher-level self? If it is, we seek to understand it. If we cannot ascertain the nature of consciousness by a deeper understanding of reality, then either there is no consciousness or we are left wondering how and why we can observe.

Inductive awareness implies consciousness (itself). Inductive reasoning suggests a reason for beingness. It asks why? It seeks noncontradictory answers. The critical question is why is there existence? Answering this question should reveal the nature of consciousness.

Consolidating What We Learned

Probing deeper into reality would be imponderable except for already understanding the structure of reality. One cannot directly determine the nature of consciousness. That approach fails. "Getting ahead of one's self" does not get the job done. Consciousness is knowable only by completely understanding why reality is structured the way it is.

A solid foundation is needed for any structure. That is no less true of reality. To understand reality calls for an in-depth understanding of fundamental physics. We mapped reality's structure by understanding the meaning of pertinent physics principles. We stepped outside science to broaden the capabilities of our methodology. We finally reckoned with the logical basis of the methodology. Logic is the foundation of structure. Not surprisingly, logic offered the most generalized way to understand experience. Can logic point us to an understanding of consciousness?

Let's pull together concepts that deposited us at the threshold of understanding consciousness. We further explain ideas culled from the previous two chapters. Let's describe the basis for obtaining a contradictory-free understanding of consciousness.

Experience is "everything." Consciousness is not a part of experience—it is "that" which is doing the experiencing. To comprehend consciousness, we must completely understand experience. We know "things" are experienced. Categorically, experience includes everything. How did "categorizing" everything lead to understanding consciousness?

We began Chapter 18 by attempting to understand the nature of space. How is space linked to everything? We defined space as "nothing." We examined the Primary Equation ($1 = 0 x \infty$). We let numerical zero represent space.

To understand space, we needed to fully understand the meaning of zero. We turned from a mathematical treatment (our Primary Equation) to set theory. In set theory, zero is the empty set. We let the "empty" set represent space.

We discovered Russell's Paradox ensnared set theory. Russell thought the "set of all sets" (which are not members of themselves) could not be explicitly defined unless it includes itself as a member. However, including the "set of all sets" as a member was believed to make the set contradictory because it becomes a set, which is a member of itself.

If we say space is "All" (pervasive) we encounter a problem similar to Russell's Paradox. Set theory suggests space (the set of all sets) must include itself as a member (to be complete) but how can it (be complete without being contradictory)? We concede everything is in space and space is in everything. We needed to explicitly define (formulate) the relationship between space and everything.

Understanding space should not be difficult, but it is not as simple as one might suppose. Spatial extensionality must be considered because space includes "extended" things. There is a difference between space (itself) and its extensionality. Not realizing this distinction was an early problem in understanding space. Recognizing there is a functional difference between space, and its extensionality was the beginning of understanding space.

Space appears contradictory in another sense. Physicists affirm space expands from the universal explosion of the Big Bang. How can space be both dynamic and nondynamic (at the same time)? If space is nothing, how can it also act (expand) as though it were something? How can that which is nondynamic also be characterized as dynamic? Nothing and something are opposites. Nondynamic and dynamic are opposites. Space seemed contradictory.

We wrestled with two intertwined issues. Space is in everything and space is dynamic. At the least, things (e.g., galaxies) exhibit extension because space expands. Space seems problematic. However, our puzzlement probably occurs because the methodology used to solve this problem was not sufficiently refined for the task.

Either space is both nondynamic (and hence nonextensional) and dynamic (and therefore, extensional) or the rule (or law) used in resolving this conflict was inadequate. We turned from set theory to logic. The Law of Contradiction was useless in settling the issue (whether space could be both dynamic and nondynamic). (It says it cannot be both.)

Before correlating logic to reality, we needed to "generalize" the understanding of "thing," i.e., "thingness." Thingness must be highly conceptualized, or space (no-thing) cannot be understood (by defining it in relation to its opposite [i.e., thingness]). Somethingness needed to be conceptualized in a way easily represented in logic.

Things are inherently of a dual nature. Physical things have "form (shape) and content." Physical form depends upon arrangement of content. The automobile has a recognizable form.

(In the physical world) form has structure. Structuring of content produces manifest form. How is content structured into form? We are developing the means to answer this critical question.

We identify things by their form. We see a certain object and call it an automobile. Solid things have form and content. Nonsolid things (liquid and gas) are not so easily identified.

What is form and what is content depends upon which is addressed. Contents have (their own) form. Automobile components are an example: tires, engine, transmission, and seats. A thing's content participates in the greater form of which it is a part. Many forms have a role in some greater form.

A thing's form and content needed a "formulation" easily amenable to treatment by logic. Logic concerns the values true and false.

A true statement is predicable: it affirms something about a subject. Collectively, verified statements (truth) represent phenomena. This is trueness in the epistemological sense. By this use, we would say, false does not represent anything IN reality. Though correct, it is deceptive. A false statement could be reality-meaningful. When we assess the deepest meaning of falseness and trueness, we do so in the ontological sense.

> Knowledge is different from what it represents. A statement is not its referent. The word "automobile" is not (the same as) the object automobile. Saying "the word automobile represents something in reality" is to assert there is a direct correspondence between the word

and the object it symbolizes. The everyday use of true is found in statements representing things "in" reality.

Logical form usually pertains to abstract compound statements. True-compound statements are tautologies. Tautologies dictate acceptable forms of combining (by logical connectives) true and false propositions.

Here, our interest in form is limited to its ontological sense. The ontological sense conveys the singular nature of falseness and trueness. It denotes the intrinsic connection between falseness and trueness. What does logical negation tell us about structure? More pointedly, how is falseness "structurally" negated to produce trueness?

We are developing means to answer some fundamental questions. First, can the basic laws of logic be understood using form and content? If they can, there is a second question. Do those laws suggest how content is structured into form?

To establish an ontological correlation between reality and logic, we ask to what in reality do the logical values of true and false correspond?

Form and content together constitute a unit. Things are units. The automobile is a unit. A thing's form is its wholeness aspect. A thing's content is its parts aspect. A unit is a whole (form) with parts (content). The automobile is a whole with parts.

If true and false have a singular meaning, they should tell us something about thingness-structure. Let's briefly follow how we connected true and false to wholeness, unit, and parts-functionality.

False (F) and true (T) are translated (through Boolean algebra) into set theory. False (zero in Boolean terms) becomes the empty set. True (one in Boolean) becomes unity. The next task was to associate the empty set and unity with reality.

If falseness represents an aspect of some thing—it (the aspect) is somehow associated with space. We already conjectured space (no-thing) corresponds to the empty set (setness itself) and therefore corresponds to the false value.

There is a distinction between setness and members of a set. An empty set (in itself) has no members. Setness just is (regardless whether it has elements). A set (which is otherwise empty) may contain elements. Space (the empty set) may contain (include) things (elements—members of the set). A unit is space and something (in space). Reality is the greatest unit function.

These ideas are more helpful when stated by function. The empty set ("0") is the wholeness function (either potential or realized). A unit ("1") is a function in its own right (the unit function). A unit function has two aspects: a wholeness function (setness—which "in itself" is empty) and parts-function (its members). The wholeness function is realized when it is an aspect of a unit function. Parts are a necessary aspect of the unit function because there is no unit functionality without parts. Unity implies (a whole with) parts.

Let's state these ideas by generalized structure. Thingness (any given thing) is a unit having form and content. Setness is (represented by) form. Elements are (denoted by) content. Content is contained within form. Contents also function as subunits.

Things are (ontological) units and are represented in logic by true values. Things have a wholeness function which is directly associated with the nature of space. A unit-thing contains "lower level" components (subunits). It seems puzzling to assert a thing's form is intimately associated with space. Yet, a little thought (about spatial extensionality) reveals it could not be otherwise. A thing's dimensionality confers extensionality on space.

To better understand (interrelate) the singular nature of falseness (the wholeness function) and trueness (the unit function) we examined closely the Law of Contradiction.

The Law of Contradiction defines logical consistency. It asserts (it is true that for any given) something (a unit-thingness that it) cannot both be and not be. If a thingness does not meet this criterion, it is not consistent with experience. Experience (of things) after all, is the source of the definition of consistency. However, we quickly learned this simple definition of consistency was incapable of covering that which is beyond things and their experience. It only applies to that which is "extended"—the extensionality of space and time. The consistency issue teeters on the difference between space (itself) and its extensionality.

Logic has not been properly understood. It has been encumbered by Gödel's Incompleteness Theorems. If a system (a set) is complete (includes itself and every set and member of it) it was inconsistent (self-contradictory). Logicians assert when a system is consistent (not include itself as a member and tested by the Law of Contradiction) it is incomplete. Gödel's work suggested logic itself is ultimately inconsistent. This idea is easily dismissed because a system can only be complete when it is also consistent.

We needed a system that is both complete and noncontradictory (consistent). Our attempt to understand space "structurally" led us to suppose there has been a misunderstanding about logic. We found everything in reality to be uniform—without contradiction. We surmised the problem of completeness and consistency arose because the singular notions of false and true were left undefined.

Hypothesizing reality is complete and uniform, and accepting logic is also complete: the problem reduces to the definition of consistency. Again, the Law of Contradiction has traditionally defined consistency. This test (of consistency) is valid "within" logic. That does not mean it (the Law of Contradiction) can be successfully applied to logic from without. The test of logical consistency cannot be extended beyond its domain of validity, viz, its tautologies. This is why encapsulating and self-referential statements fall outside tautological boundaries: becoming "undecidable" (though "true"). If this statement is false, then it is true; if true, then it is false.

The Law of Contradiction and the Law of the Excluded Middle were steadfast guides for piecing together the structure of reality. All tautologies meet these tests. These laws fundamentally define (the "purest" epistemological requirement of) every true compound statement. True compound statements inform us how things interrelate. However, we are now concerned with the ontological meanings of trueness and falseness.

We have been concerned throughout this book with what exists. The Law of Contradiction in the singular sense provides the definition of trueness: it characterizes existents. The problem is, an existent has not been defined in the pure abstract sense.

Trueness has not been understood. True and false need to be connected intrinsically. Merely stating each is the negation of the other, is not enough to understand them in the singular sense—what they abstractly say about structure.

What is the ontological difference between true and false? We are actually asking for a clear abstract understanding about the "categorical" nature of existence and nonexistence.

Things exist. Nonexistence just is. The Law of the Excluded Middle tells us that "something" either exists or does not exist. Things exist in the flow of time. Nothingness is not characterized by existence. Nothingness itself is not characterized in the flow of time. Nothingness is (conceptually) "outside" the flow of time. This was stated by reductionistic analysis. Inductively we would say temporality is enveloped by space itself. (Later, we explain how.)

> Let's dispose of an issue associated with the Law of the Excluded Middle. It has been controversial. It is easy to understand why. It asserts we must choose either true or false: there is no middle possibility. That works fine for actual existents.
>
> The Excluded Middle fails when we are unsure about a phenomenon's existence. Probability has been one of its stumbling blocks. Probability statements take a value between 1 (true) and 0 (false). Probability has its place. Quantum probability tells us much about reality. Chaos theory suggests recognizable (meta-level) events (represented by true values) can develop from random events (described by probability).
>
> Probability statements indicate our ignorance of random (or potential) occurrences. True ("1") is considered certainty. False ("0") is considered uncertainty. Certainty operates on (or is derived from) uncertainty. Between the extremes of knowing (certainty) and not knowing (uncertainty) is probable knowing. Not knowing, just means we do not yet know the result of a possibility. Either it will occur ("1") or it will not occur ("0")—according to the Law of the Excluded Middle. We soon learn certainty and uncertainty, on closer inspection, switch roles.

Most problems can be formulated as contradictions. When applying the Law of Contradiction did not lead to an understanding of space, the law itself became the problem. The actual problem was our "habit" of using the Law of the

Excluded Middle to test definitions instead of understanding why it failed when applied to space. Either the Law of Excluded Middle is appropriate for "testing" concepts of space, or it is not. The domain of the Excluded Middle does not extend beyond the Law of Contradiction. This is why "(logically) undecidable but true statements" arise. We know the Law of the Excluded Middle is a reductive rendering of the Law of Contradiction. The Law of Contradiction should not be turned upon itself. Let's translate this problem into reality.

Thingness is defined as an existent. Either something exists, or it does not exist. The same cannot be said of no-thingness (space). That is the problem. It is paramount to asserting we cannot say that it is true this statement is false. Once we realize this (statement usage) represents an epistemological assessment and not an ontological correlation, the problem begins to unravel.

Temporarily suspend the idea of (true and false) statements; consider only what trueness and falseness represent (in their ontological referent). Let's restate the above problem. It is not meaningful to say "space exists." Space (itself) is not existence: space just is. Simplifying. It is not beneficial to say, "false implies true" unless properly qualified.

> Undecidability appears when we say "it is true that this statement is false." The Law of the Excluded Middle suggests it must be true or false but not both. But by saying it is false, it means it is true (it is true that it is false). There is a way to understand that false is true. It is true (reality-meaningful) in the ontological sense. We learned from Chapter 18 that in a nonfunctional sense, $F = T$, (i.e., $0 = 1$). Space is the whole of what is. In that sense, space is no different from (the same) space containing things (existents). In both cases, space is equivalent to itself. There is no more than the whole of what is. We shortly elaborate that the Law of Noncontradiction defines this understanding.

No-thingness pertains to nonexistence (space). No-thingness is represented in logic as falseness. However, falseness (itself) is practically meaningless "within" logic. (Within logic false merely means, "not true.") Is falseness itself definable from "outside" logic? Knowing there is no trueness without falseness suggests falseness somehow is more intimately connected to trueness. Can falseness be understood in a way other than the traditional estranged idea of (trueness) negation?

How does nothingness connect intrinsically to somethingness? Specifically, what aspect of somethingness is nothingness? Nothingness (nonexistence) is space. What aspect of things is space? (We treat nothingness as though it had some type of function.) Space becomes the dimensional aspect of thingness. Formness (0—setness, the wholeness function) functionally "projects" spatial extensionality when manifest content is produced. Extensionality occurs by reductive fractionation of the wholeness function. This is how things (1—unity) are created.

The Law of Contradiction defines "thingness" conditionality in the singular ontological sense. It tells us of what is "not" (logically) acceptable of "things" (existents). They cannot both exist and not exist at any given time. This also can be interpreted (by the Law of the Excluded Middle) that a thing might not exist. Any given time is minimally an instant—no time (flow). In an instant, there is no temporal existence. (The) persistence (of things) depends on time flow—"highlighting" successive "frames" in a concatenation of instances. Existence persists because of change (activity—changing of frames). Activity (functionally) shifts back and forth from wholeness to unit. (Refer to Chapter 12, on the subject of the BCI.)

We find there is a vague understanding of the Law of Contradiction that does not obey the Law of the Excluded Middle. It seems to say the same as the explicit definition of the Law of Contradiction ("something cannot both be and not be"). Its vague phrasing is "no thing can both be and not be" (emphasis—thing). Its indefiniteness can be understood in a way opposite to the explicit meaning; hence its vagueness.

The other way to understand the phrase "no-thing" is to say it represents nothing—nothingness (emphasis—nothing). Nothing is opposite to thing. Nothingness and thingness are complements.

In Chapter 18, we negated the Law of Contradiction and "inductively" obtained the Law of Noncontradiction. Negation of thingness is no-thing (the opposite of thingness). No-thing is synonymous with nothing. It represents the value of falseness. The Law of Noncontradiction defines this "other" understanding (of no-thing).

If trueness tells us about things, then falseness tells us about no-things. The Law of Contradiction defines the nature of things; it does not tell us much about "no-things." The Law of Noncontradiction defines falseness. It informs us about no-things—nothingness. Each law complements the other.

For clarity, there are three ways to understand "no-thing."

The first and explicit reading of the Law of Contradiction is "something cannot both be and not be." This does not refer to no-thing. It definitely refers to and focuses on something. The Law of Contradiction definitely refers to somethingness. The Law of the Excluded Middle interprets this as meaning "something either exists or does not exist."

The second and "indefinite" reading states "no thing can both be and not be." Its referent is vague and uncertain. "No thing" could mean the same as the first and explicit reading (emphasis on thing—as defined by the Law of Contradiction). It could mean the same as the Law of the Excluded Middle. No-thing could also mean the Law of Noncontradiction—which defines nothingness. These laws are not synonymous. The second (vague) reading is indefinite because we are not sure of its referent. Does it (no-thing) pertain to something (not this thing) or nothing (nothingness itself—no thing in a super-categorical sense)?

The third and opposite reading (from the first reading) states "nothing can both be and not be." Not any and not even every thing—is nothing. The third reading focuses on nothingness (itself). Categorically, any no-thing and every no-thing are the same as nothing. The third reading definitely refers to nothing. The Law of Contradiction does not define nothingness. The Law of Noncontradiction defines it. It definitely refers to nothingness. The Law of Noncontradiction and the Law of Contradiction are opposites. Each law can be derived from the other law by negation.

We also asked whether the Law of Noncontradiction is in some way "interpretable" as a true statement. Yes, in the following two senses.

It is true that there are false statements. The Law of Noncontradiction is meaningful. This tactic applies an epistemological treatment to ontology. Be mindful of reification. There is a difference between knowledge and what it represents.

The Law of Noncontradiction is also true in the sense mentioned in the previous digression. F = T: meaning there is no more than the whole (of what is). There is no more to an uncut pie than there is to the same pie cut into pieces. In that sense, they are equivalent. The only difference is functional: whole pie versus sliced pie.

Applying the above yields the following. "It is true (epistemologically—detached from the referent) that falseness can both be and not be (ontologically)." This translates into accepting the possibility "nothing can both be and not be" has real meaning (makes a statement about reality's structure).

In a twisted way (i.e., epistemologically) we accept the Law of Noncontradiction as a true statement. Although this tactic is acceptable in searching for truth, such "outside" (of structure) shenanigans are unusable in understanding the intrinsic relation between falseness and trueness. But, it does imply there is a system that is both complete (the set of all sets) and is also a member of itself and is consistent.

Review: Logic, as a whole (completeness) is consistent. We now recognize the error in the "accepted" definition of consistency of a completed system. It is not defined by the Law of Contradiction. The Law of Noncontradiction defines consistency of a completed system. The Law of Noncontradiction defines completeness of logic. To explain, we look to our "structural" (or ontological) interpretation of true and false.

The test of consistency for wholes (falseness) is opposite of the test of consistency for units (trueness) and parts (subunits).

The test of "completeness and the consistency of logic" (not logical consistency) is the Law of Noncontradiction. Completeness is represented by the wholeness function (0—falseness).

The test of logical consistency (not the consistency of logic) is the Law of Contradiction. It pertains to the unit function (1—trueness). It relates also to subunit functions (parts-function).

The Law of Contradiction states what cannot be (the case for something). From the Law of Contradiction ("Something cannot both be and not be") is deduced the Law of the Excluded Middle.

The Law of the Excluded Middle makes a statement about structure. It states that something can be the case. (Something either is or is not.) Formally, it says of opposites (complements) only one is admissible (as true) at any given time. As we have seen, this has relevance to cosmic structure. Either Cosmos or Anticosmos—exists during a given temporal phase. The Law of the Excluded Middle is also usable as a test about structure. Epistemologically, it forces a decision that a statement must be either true (certain) or false (uncertain). This is different from its structural interpretation. Ontologically, it asserts the duality of existence (unit function—in the flow of time) and nonexistence (wholeness function—outside the flow of time). This is also interesting. It provides a deeper clue about the nature of change. It helps define the overall process that gives directionality to change. (See the next heading Entropy and Negentropy.)

Form (itself—i.e., apart from its content) is the wholeness function. There can be a wholeness function without content. A unit is both form and content. There is no content but that it is a part of some greater whole. There can be no trueness without falseness, but there can be falseness without trueness. Zero can be, but there is no one ("1") without zero ("0"). (These crucial insights begin to answer our remaining questions.)

Let's lineup our present understanding of logic. The Law of Noncontradiction and the Law of Contradiction are stated in various ways.

The Law of Noncontradiction:

False can both be false and true.

False is equivalent to (the conjunction of) true and false (in logic).

Zero ("0") conjunctively equals unity ("1") and zero ("0"). (In Boolean algebra.)

The continuum is (functionally) divisible.

Setness (an empty set) can both be a set with members and retain its setness.

Falseness is inherent in trueness. False is an aspect of true.

There is no trueness in falseness.

The Law of Contradiction:

Trueness cannot both be a trueness and falseness.

One ("1") cannot both be one ("1") and zero ("0").

That which is true cannot be false.

A set of elements cannot both be a set of elements and an empty set.

There is falseness in trueness.

What do these laws of logic tell us about reality (in abstraction)? Some of the following are epistemological interpretations; others pertain to the ontological nature of true and false.

The Law of Noncontradiction can be interpreted via reality as follows:

> Nothing can both be and not be. Nothing can give rise to something. Nonexistence can give rise to existence. Nothing is inherent in something.

> The wholeness function can be a whole and a whole with parts (a unit). A whole can both be itself (i.e., whole) and divided into parts. A whole can generate parts. A whole can fractionate into parts. A whole can produce a unit (a whole with parts).

> A wholeness function itself is not a part function. A part is different from the whole of which it is a part.

> Whyness can be a whatness. Whyness is inherent in whatness. Reason is inherent in logic.

> Form (nothing) can both be itself (i.e., form) and have content. Form can produce content.

> Space can both be and not be. Space can give rise to spatial extensionality (by functional division of itself). Space can both be space and thingness (i.e., not just space).

Assuming nothing, we derived something. Somethingness is defined by the Law of Contradiction.

The Law of Contradiction can be interpreted via reality as follows:

> Something cannot both be and not be.

> Something cannot both have (form and) content and not have form.

> A unit has a wholeness function. A unit cannot both be a unit (a whole and parts) and not be a whole. A whole with parts cannot both be a whole and parts unless there is first a whole.

> There is no existence without nonexistence (space).

> Whatness does not exist without whyness.

> A unit (form and content) cannot both be a unit (form with content) and form without content.

> A whole and parts cannot both be a whole and parts and merely be a whole.

> Things cannot both be a thing and only be space.

Applying what we learned from these ideas: How can nothingness and somethingness be connected?

Nothingness:

Nothingness is defined by the Law of Noncontradiction. False can generate true. A whole can produce parts: becoming a unit. Space can "fractionate" into dimensional things (space and its apparent extensional fractionation).

The Law of Noncontradiction defines falseness, form, wholeness, and spaceness. It is "binding" for no given instant—in the flow of time. Therefore, it applies to all time—but no given instant of time. This is later more fully explained.

Form (in itself: the value of logical falseness—zero, the empty set) is represented by the wholeness function. The Law of Noncontradiction defines the wholeness function.

Somethingness:

Somethingness is defined by the Law of Contradiction. There cannot be any something without nothing. There is no unit without a whole that includes parts. There are no parts (subunits) unless there be a greater whole of which they are a part. There are no things but that they "exist" in space.

Unity (logical value of trueness) is "setness" (an empty set) with members. A unit is form and content. Parts are subunits. The Law of Contradiction defines the unit function.

The Law of Contradiction defines false-true (unit) form-content, whole-part, and space-thingness. (It is "thingness" valid for any given time [but not all time.])

The unit function is represented by the value of trueness. A unit has a wholeness function with a parts-function. The wholeness aspect is represented by the value of falseness. The "whole with its parts" is represented by the true value. Unity "structurally" conjoins the logical notion of true and false: for any given time.

What do the laws of logic ultimately tell us about the structural nature of reality? How do true and false (ontologically) define reality?

There is no existence (unit) without nonexistence (whole). There are no things, but they exist in space. Further, space is also in everything. (Everything has a wholeness function.) There are no things, but there is space for those things. Unity intertwines existence and nonexistence. Reality (true—unity) connects things (true—subunits) and space (false—whole).

Nothingness is space. Nothingness is the wholeness function. A unit's wholeness function, its form, is the spatial aspect of (unit) thingness. Setness is an "automatic" member of itself. Space gives form (and hence dimensionality) to things; things give extensionality to space. A unit incorporates spatial extensionality into its form (or wholeness function). A thing's form (its wholeness function) is sustained by the ("collective") action of its parts. Every thing's wholeness aspect is sustained (in time) by a reductive function of space.

A simple analogy: There can be a pie (not cut into pieces). An uncut pie retains its wholeness function. In this case, parts are potential. There are no pieces of pie unless there is first a pie. If the pie is sliced, but no pieces removed, there is still a whole pie, sans wholeness function. A piece is also a unit (as a subunit) by itself. A part (piece of pie) is a subunit. However, sliced (whole) pie is only potentially (a functional) whole.

The Law of the Excluded Middle reductively interprets the Law of Contradiction. It states that a unit-something either has form (falseness) or (form and) content, but not both at the same time. (It must either be one or the other.) Content is of the parts-function. The pie is either whole or cut into pieces: at any given time. In the latter case (sliced) the pie is a whole and parts. That is, the whole pie is still there, but the pie has lost its wholeness function (because of it being cut into pieces). There is a functional shift between whole and part. The pie either functions as a whole or as pieces (but not both at the same time).

If we discover a phenomenon that seems to violate understanding of the Law of the Excluded Middle, "either" the understanding (of the phenomenon) is incorrect or the Excluded Middle does not apply (e.g., probability statements). A form is only a possibility (potentiality) until it manifests as a unit. (The possibility of form manifestation [as a unit] is sometimes measurable by probability.) In this respect, we would say a form "materializes" (by functional fractionation) as a manifest unit. The Law of the Excluded Middle (epistemologically) applies to the Law of Contradiction and logic. It does not (ontologically) apply to the Law of Noncontradiction because this law does not define something (a unity).

No wonder understanding somethingness, eventually, focused on the consistency of logic. Understanding logic points to the (inherent) meaning of form itself. Our treatment of form is more like "formness" (from the context of content, viz, form without content). In this chapter, we primarily use "form" in this restricted singular (ontological) sense. (As opposed to the compound [epistemological] nature [form] of logical statements.)

How does logic fit in with the development of this book? Defining "somethingness" caused epistemological confusion. Structure is composed of "things." We had reached a point in this project, where "every" puzzle piece (thing) "was on the table." Things are interrelated by (logical) structure. Moreover, we had properly aligned every puzzle piece (to represent the structure of reality). Yet, the structure did not render a "complete" understanding of "somethingness" (everything).

If everything was "categorically" understood—what was missing? There was a clue. We realized the ability to interrelate everything is not synonymous with grasping everything as a whole. It is the difference between everything (as a collection) and the singular (categorical) notion of everything (i.e., "All").

Understanding is the ability to interrelate things. There had to be a higher level meaning of categoricity. Here our concern is not "what" is categorized, but (the notion of) categoricity (itself).

To understand somethingness is to ascertain how a whole (thing) relates to (its) parts. Recall our water example: Minimally, a single molecule—H_2O—defines water. If these atoms are not "chemically" combined into a whole, they do not constitute a water molecule. Its wholeness function is molecular; its parts-function is atomic.

Being too close to trees makes it difficult to view the forest. Focusing on "parts" (things) tends to obfuscate the "whole." Overly tempered by the mind-set of "placing pieces" together (to understand) thwarted immediate recognition of the puzzle picture (reality) as a whole. The trick was to back off (intellectually detach) from being among "every" tree (thingness and structure).

(Accustom to) the "separability" of everything hindered "recognizing" the "nonseparability" of everything. There is a difference between a unit and its wholeness aspect. The difference was defined by invoking the concept of function. "Categoricity" (in the abstract sense) can also be a function.

Any given unit-something also has a parts (subunit) function. If the parts-function pertains to the separability of things, then the wholeness function ultimately pertains to the nonseparability of everything. Space is a continuum and (in itself) nonseparable. Space is formness (form without content). That is a telling clue about the ultimate nature of everything. Separation is a temporal concept any way it can be construed. Nonseparability pertains to no (given) time—and therefore, encapsulates all time. Space envelops time flow.

How were we led to the Ultimate Principle? (Refer to Chapter 18.) During most of the development of this book, we deferred dealing with the "exhaustive sense" of "not-something" (i.e., "nothing"). "Not A" at least means "not this A" and can mean "not any A." Exhaustively, it means "not even every A." No pie pieces typically means there is no pie (none). Yet, we discern that no pie pieces (parts) does not necessarily preclude an uncut (whole) pie. "No pieces now," does not necessarily exclude the possibility there was once a whole pie. Observing parts now does not preclude the possibility there was once just the whole (and no parts).

The "exhaustive sense of 'not A'" did not become an issue until "every-thing" was "conceptually" pieced together. With respect to "A" (every-thing) the exhaustive sense of "not A" is "not every-thing." What does this mean? Initially, it only meant "all but (except) everything." This is the best an epistemological assessment provides. Placing every pie piece back onto the pie plate "reproduces" (a representative facsimile of) the whole pie.

Contradictions are resolved by inductive reasoning to higher-level concepts. As we "categorically" included more of "everything else" as a part of "everything," "not A" was left undefined. We "roughly" knew the following. "When everything (every piece) is taken away—nothing is left." "Nothing" seemingly defied definition. The lone exception was that it could be defined by "negating" everything. But, that is not "all" there is to it. The problem was the following: We just did not completely understand the meaning of "not everything" until we used the idea of function.

Some details: Though not easily recognized, higher-level concepts represent both "A" and "not A." "Nothingness" (itself) is a function in its own right (the wholeness function). We did not properly realize in an exhaustive sense that "not A" included "A." "Nothing included everything." "No-thing" universally (inclusively) and reductively meant (inferred) every system and part thereof. A whole pie infers its pieces. The whole includes parts. The nonseparability of things includes the separability of things. Ultimately, space "encloses" everything. We now understand logical implication: False implies true.

Though we did not "formally" understand the inclusive sense of "not A," we used that very tactic, intuitively and inductively, to delineate the structure of reality (refer to Chapter 2, subheading "Problems, Solutions, and Knowledge").

The Law of Contradiction works well in deductive analysis. It formalizes "what parts" are acceptable within a system (intraphenomenal relevancy, see Chapter 1, Standards—Phenomenal Relevancy: Intra-, Inter-). The Law of Contradiction (generally) defines the legitimate state of anything and everything. Traditionally, the Law of Contradiction defines the final appeal (as a generality) to what is logical.

Two methods were used to assess possible inductive explanations. Either a concept is dropped from consideration, or it is included within a "covering" generality or part thereof.

The first method uses the Law of Contradiction. In practice, it offers insight (when it is "interpreted" by the Law of the Excluded Middle) in eliminating unviable constructs (of thingness possibility). If a considered concept does not "fit in," we throw it out. By dropping unacceptable concepts (not: A, B, C, and so on) the sought after explanation must be otherwise. Dropped concepts often point to more viable concepts.

In the second method, concepts were "eliminated" by inclusion. A concept "fits in" when it can be "classified" as a subunit (or lesser wholeness function). A pie piece itself is a whole. Classification is obtained by ("reductively") placing the considered concept into hierarchical structure. The point is: We can use this same tactic in an inductive sense. Of what greater whole is the considered thing (or construct) a part? (Our "operational" inductive methodology reflects this. Therefore, how can a given thing be explained?) A pie piece is a part of the pie. Pie can be classified (at higher levels) as dessert; dessert is classifiable as food, and so on. Each higher-level category pushes us closer to THE explanation for everything. We just did not realize the Law of Noncontradiction generalizes this inductive tactic.

Throughout this book, we were concerned with understanding things in a step by step manner. This is how we connect one thing to other things. We did not recognize the following. If by inclusion we exhaustively eliminate "any and everything possibility," eventually, only "no-thing"—nothing is left. There is, each step, every step, and all steps, until no steps are left. This is the result of using the inductive method of exhaustion (it ends any and every possibility by deferring to all possibilities—categorically). Eventually, we were deposited at the point where no possibilities were left—only certainty. This inevitable finality is embedded in the Law of Noncontradiction. We proceeded intuitively until the Law of Noncontradiction was stripped bare and stood alone. We also note if certainty actually applies to falseness, then trueness is uncertain. This is a role switch from the customary assessment. This is also reasonable in another sense. The future course of events (constituted of things and hence of "true" value) is unknown (and therefore, uncertain).

Applying the Law of Contradiction and the Law of Noncontradiction in the exhaustive sense impelled us along the path to greater knowledge and understanding. Unviable concepts were eliminated. Some concepts were classified by inclusion. Eventually, however, we must deal with everything in the highest categorical sense. This meant everything had to be categorized from "outside" (of everything).

Inevitably, the Law of Contradiction and its testing sibling, the Law of the Excluded Middle, are useless for eliminating the ultimate contradiction. The Law of Contradiction defines "any and every something." The ultimate contradiction was the Law of Contradiction. The test of anything and everything was incapable of dealing with everything by appeal to the next higher-level category (that included everything). That higher-level category is no-thing (it—that). Nothingness includes everything. Our task narrowed to fully understanding nothingness.

What is the nature of NOTHINGNESS? We learn (in Chapter 18) the Law of Contradiction is inductively "explained" by the Law of Noncontradiction. Nothing (functionally and indeed categorically) includes anything and everything. It (nothingness) even includes itself as the wholeness function of subunits. Nothingness (by categorical inclusion) "explains" any and every somethingness. It also explains itself in a thing's form. Space embeds itself in a thing's dimensionality. Nothingness is self-explanatory.

When nothing includes itself (as a member) it becomes a somethingness. Yet, though there be somethingness, the ubiquity of nothingness is still there; it is always there. An empty set is a "regressive" member of itself. Regressively, it becomes the (functional) form (aspect) of any given unit and every subunit.

Moreover, we now understand the Law of Noncontradiction guides the inductive process (reasoning to higher-level generalities). Employing the Law of Contradiction in the exhaustive sense is but the first step to seriously determine the ultimate meaning of the Law of Noncontradiction.

Before the Law of Noncontradiction can be cited as the *a priori* basis for everything, it must be understood in a way everything (every something) can be "mechanically" derived from it. To ask how everything can be explained is comparable to asking about the origin of parts. Simplistically, where do parts come from? How are parts produced? We described this in Chapter 18 by deriving the natural numbers. This is cutting the pie into pieces. Parts are derived from the whole by (its) fractionation.

The problem remains how nothingness can cut itself into functional parts. Technically, because a set can be a member of itself, we ask for further consideration. "Under what conditions, does a set become a member of itself?" Knowing that it can, and knowing how it can, does not answer why it does (become a set of itself). This question goes deeper than asking how "content is structured into form." It asks about origin of content. Why are there "parts?"

Does a set become a member of itself automatically (which probably means randomly) or by design—deliberately? Moreover, that a set can become a "set of (a member of) itself," does not mean that it will. These questions, though uncharacteristic of mathematical inquiry, are considerations for philosophical discourse. These questions are answered and fully explained in the next chapter.

Space and things (in space) form a unit function. Reality is the ultimate unit function. Space is the ultimate wholeness function. The energy of the vacuum represents the extreme fractionation of the wholeness function into an infinity of parts. Physicists recognize the minimal activity of parts as the quantum of action (signified by Planck's constant). Recall the infinite value in our Primary Equation ($1 = 0\,x\,\infty$).

The difference between space as a whole and space in its extreme fractionation is functional. Everything that exists falls into a gradient between the whole and its extreme fractionation.

Things exist in space. Compositionally, everything (and any given thing) has a wholeness function (whyness) and parts-function (whatness). (Again, there is no part but there be a whole of which it is a part.) In an exhaustive inductive sense, a thing's wholeness function is attributable to space. A thing's parts-function, in extreme, is reductionistically attributable to the infinite energy of the vacuum—space functionally fractionated.

The functional "splitting" of reality occurs binarily: Wholeness functionality ("0"); unit functionality ("1"). It is a process of digital bifurcation (tree branching). Cosmos—Anti-cosmos, and so on. Eventually, quark—antiquark pairs spit into substructures, and quickly become undifferentiated from the vacuum's infinite energy—virtual particles. All manifest particles have virtual counterparts.

How can space be nothing and act as though it were something (viz, expand)? Understanding that the Law of Noncontradiction defines space resolved the problem of explaining universal (space) expansion (Big Bang).

> Zero multiplied by any number is still zero. (Recall multiplication is really division of the whole.) There cannot be more than the whole. Space can expand and there still is no more than the whole. Space accommodates its expansion. It is dynamic yet not dynamic because there is no change in its status—it remains whole.

Key Points:

- There is consciousness and there is what it is conscious of—experience.

- Experience is everything.

- Functionally, space is consciousness.

- The functional fractionation of space "creates" thingness—the stuff of experience.

- The Law of Noncontradiction defines falseness, nothingness, space, wholeness function, formness and consciousness.

- The Law of Contradiction defines trueness, somethingness, and unit (extended) thingness (form and content).

- Space can be a subset of itself. When it is a subset of itself, a unit somethingness emerges. Space gives form to things (in whole and part). But form is of space. Hence, things (through their wholeness aspect) expand along with space. Because space expands, it accommodates any dimensional extensionality. There is no difference between nondimensional space and multi-dimensional space. (Any difference is because of function.)

- Space can both be and not be. It is both dynamic and nondynamic. Space is "all" embracing. All includes everything. Space includes everything. Space is the whole of reality, but no parts. "Parts" (every part is implicitly of a greater whole) define "every something."

- The difference between things and space is one of function. Functionality suggests everything has utility or purpose. (To ask why there is purpose rather than no purpose takes us into the next chapter.)

- We finally ask why is there a wholeness function? The answer is that it cannot be otherwise. There necessarily has to be the ultimate whole (of what is).

The previous chapter correlated our logical understanding of space with today's understanding of physics. Realizing zero or space is definable as a wholeness function helped us understand electromagnetism. Electromagnetism is a unit. We learned a unit has two aspects—a wholeness function and a parts-function. Space is the wholeness function (of unit-reality—of electromagnetism). Every some-thing is of "either" the electric or magnetic nature (parts-function) of electromagnetism. The "separation" of thingness into electromagnetic aspects is described by the CPT theorem. (See Chapter 11.)

The Higg's "particle" is thought by some to "encapsulate" the four forces of nature. At least "unifying" the forces of nature has a theoretical end. Physics is also unifiable under our above conceptualization. Space includes everything. We might say space itself is the Higg's particle. In this respect the Higg's particle is not a unit particle. It is a wholeness aspect of (unit) reality. It is the wholeness aspect (i.e., the space [in itself] aspect) of electromagnetism. Reality's unit is the "particle" (whole and part) denoted the (electromagnetic) photon.

Everything (including the other forces) responds to gravity. Gravity is definable as the curvature of space. Gravity is space. We only notice a difference (the "pull of gravity" or tidal effects) when spatial extensionality is distorted ("deformed") "telling things how to move (in space)." Things tend to "pull together" under the influence of spatial curvature.

The forces of nature affect the form of unit-things (parts-function). The electromagnetic "force" is responsible for things as we experience them in everyday life. Physical things result from the chemical binding of atoms and molecules. Understanding the nature of chemistry hinges on knowing the functioning of (an atom's) valence electrons. Electrons are the physical manifestations (parts-function) of electromagnetic energy (unit function).

The forces of nature (gravitation [refer to Chapters 14 and 15], electroweak, and strong [Chapter 15]) distort space in a way producing content (parts—subunit functions). Space is its own subset. Each subset (realized as protons, neutrons, for example) differs from space (in itself) by a difference in functionality. Each functionality is the hallmark of the forces of nature acting back upon space itself. These particles (of reality) "functionally separate" themselves from spatiality. (Space is nonseparable.) The final-functional fractionation of space is the limitless energy of the vacuum. (Vacuum energy is space functionally divided.)

We have reached the pinnacle of structural understanding and logical necessity. We correlated our structure of reality with an understanding of physics. We learned the wholeness aspect of reality (space) is fractionated into the plethora of things we experience.

How else can the relation between space and things be explained by physics? There is another way of understanding the "pulling together" of things. It is another conceptual way to understand gravity. Gravitation describes change in a cosmological sense.

Let's consider how things change. Our particular interest is to determine what processes alter structure. The same processes affect everything at all levels.

Entropy and Negentropy

The First Law of thermodynamics is the conservation of energy. Total energy of the universe is constant. Energy cannot be created (within the physical universe) but only transformed from one state to another. (Our initial consideration concerns the physical universe.) What happens to the "given" energy of the universe? Matter is a form of energy.

In the 1820's a Frenchman named S. Carnot tried to understand the mechanics of steam engines. He understood the energy process results from a state of high-heat concentration—"flowing" to a state of low-heat concentration.

In 1854 the German physicist R. Clausius generalized the idea heat cannot pass from a cold area to a hot one. The process of heat distribution ends when both areas achieve equilibrium, viz, "settle" at the same temperature.

In 1868, Clausius coined the word entropy. Clausius believed entropy (of any closed system) would evolve toward a maximum value. That is, as heat energy dissipates it becomes less available for work (activity).

Entropy defines the Second Law of thermodynamics. Entropy is sometimes likened to an energy "fluid" or carrier. Entropy is the thermo-process that is expressed as a function of a system's energy. The energy content of a system is measured before and after an interval of time. Energy (difference) is a measure of activity. Temperature is a measure of heat—typically correlated with molecular-kinetic energy. Entropy is associated with the disordering of parts.

Englishman B. Thompson believed entropy has cosmological consequences. The German H. Helmholtz realized when all physical energy is expended—the universe ends in a "heat death." When all activity ceases, there is maximal disorder (and minimal order). No available energy means no transformation of material arrangement is possible.

Let's use what we learned under the previous heading and connect it to extreme cosmological settings. The extremes of reality are the wholeness of space and its infinite energy (i.e., of the vacuum).

Space is the ultimate wholeness function. It is formness itself. It is complete order: pure form—form without content. There is no disorder in formness because (considered alone) there is no "activity" of parts (which can "deform" the form). At first thought, it may not be easily grasped how there could be a connection between "that which has no parts" and the form of every possible thing. Once it is understood that things have no form but for the wholeness function, the difficulty of grasping this connection lessens.

The energy of the vacuum is the final-functional breakdown of space itself. There is no order (form) in the random activity ("foaming") of the vacuum's energy. It is total chaos—formlessness. Chaos describes the final parts-function. This is easily understood. Order is contrasted with disorder. If there is that (space as whole) which is characterized by complete order, then there must be that which is characterized by total disorder (viz, vacuum energy).

Between the extremes of reality are found everyday things. We have seen that ordinary things are "structurally" constituted in the extreme by (the wholeness of) space (itself) and its opposing-infinite energy (molecular to atomic on down to vacuum energy). Considering the cosmological setting, what is the relationship between energy and change of structure?

At the beginning of the physical universe (inauguration of Big Bang and extremely high-energy density—high temperatures) order is "potentially" maximized. As the universe expands and matter cools, some energy is freed for use.

Energy can be transformed. Electrical or heat energy is usable for mechanical activity. Activity implies motion and hence (potential if not actual) work. Energy makes possible the building of structural forms.

Form changes as energy is consumed. Burning coal heats water—transforming it into steam. Steam turns an engine turbine generating electrical energy. Electricity can be harnessed for work. After coal burns to ashes, it has practically no form and no usable energy.

Thermodynamic processes have a "direction." Entropy has been linked to the flow of time. Eddington said entropy is the "arrow of time." The idea entropy defines temporal directionality is deduced from Helmholtz's idea the universe is winding down (to a heat death). At that point, motion ceases and energy is unavailable. Until then, things change inexorably. Direction of change defines time flow.

Physical motion is attributable to the universe being in a state of nonequilibrium. Without motion, there would be no determination of temporality. Time flow is associated with motion of parts and temporal directionality is determined by entropy. The natural process of energy use in a given universe specifies temporal directionality.

Temporal directionality is defined by increasing entropy in Phase I of the cosmic cycle. The physical universe becomes more disordered as its energy is consumed. Entropy defines the primary process of change in the physical universe.

Negative entropy, also called negentropy, is the opposite of entropy. Negentropy defines a system that becomes more ordered. Order is associated with structure building (and work).

Negentropy is not the dominant energy process in the physical universe. Even where life forms are thought to develop contrary to entropy, there is no problem in understanding that the source of free energy (used to increase order found in structure building) defers to outside (of living) systems. The main energy source for physical life is sunlight.

Sun energy (electromagnetic radiation) "drives" life processes. Plants obtain nutrition through the photosynthetic process. Sun energy drives photosynthesis. Plants are food for animals.

Negentropy occurs in the physical universe only because it is enabled by the free energy by-product of entropic processes (system degradation). If dynamic symmetry applies to cosmic structure, we expect there to be a universe where negentropy dominates (system building). Phase II of the cosmic cycle is probably dominated by the negentropic process.

> Saying development of physical life forms, which defy entropy, is balanced by greater entropy elsewhere, avoids addressing the "bigger picture" (actually balancing entropic energy processes).

Entropy has also been applied to the loss of information about a system. The more disordered a system becomes; the less information represents the "order" of the system. We presume the more ordered a system is, the more it is associated with knowledge. Complete order (no disorder) is then associated with total knowledge—formness itself.

Key Points:

- Entropy is shown by a high-level-ordered system changing to a lower-level-disordered system. Let's state it differently. Entropy is the "breaking down of wholes into parts." Bodily life functions entropically degrade with age—system breakdown. Physical life forms die and decompose.

- Negentropy is system building. Negentropy is the "coming together of parts" into higher level functionality. People "grow" by the negentropic process.

In Phase I of the Cosmic cycle, physical (tardyonic) systems "naturally break down into (their component) parts." (See Chapter 9.) We presume negentropy is the primary energy process in Phase II of the cosmic cycle. To maintain symmetry, we suppose in Phase II that mental (tachyonic) parts (minds) are eventually "brought together" into a single functionality. (See Chapter 9.) Energy processes suggest there is a design to everything.

Pivotal Events, Design, and Purpose

It happens some things (events) occur for a "higher purpose." A higher purpose is one not of our own design. These "pivotal events" are not always recognized when they happen. Pivotal events are recognized when they have significant consequences. The passage of time is usually needed to ascertain whether any given incident is significant in the course of events. Pivotal events are sometimes called "turning points."

Although critical "turning points" are often not easily recognized, we observe everyday events that forever change an individual's life. For example, an automobile driver's attention is temporarily diverted; his car crashes; he becomes injured. Whiplash breaks his neck—paralyzing him from the neck down. A fraction of a second either way could have resulted in a different outcome. The driver's injury is a significant event. The consequences of such tragedies show they are turning points.

A significant event can be attributed to many turning points. A series of connected events is called seriality. Taking away any one turning point alters the outcome.

What is the nature of turning points? Do they happen by chance or by design, or a combination of both? Some turning points occur by happenstance; others seem to occur by design. We concede the exact course of one's life is beyond his control.

Civilization is "pulled" toward an uncertain future. Yet, many significant contributing events of civilization are enmeshed in the lives of certain individuals. Some people have a wider affect on the course of civilization than others. Usually, a person's contribution to society, if any, has no lasting affect until after his or her death.

It is not easy to determine before the fact, which personal events are pivotal for civilization. Some events are quickly recognized as potentially pivotal. Others take time to develop before their effect is realized.

Is the general course of events determined? Granted, the past affects the future. The past partially determines the course of the future because the future is limited only by what the past allows.

Is the future solely a development, or "push," of what has transpired? Or, does the future also "pull" the present (time) toward an unseen destiny (by some teleological agent)? If this occurs, to what degree are events determined by the future? Is the present racing toward a destined future? If it is, then there is an implicit design to the course of present events.

If there is an implicit design in (the course of) human events, why has it not been recognized? We might have difficulty recognizing a teleological agent. Do some events suggest a teleological agent (operating on the course of events)? Stated otherwise. If the course of events is directed toward a certain future, how could we discern if there is any effect of a teleological agent?

How can we discern mere accidental events from those playing a determined pivotal role in the destiny of humankind? Does an agent outside the flow of time determine some events? These lofty questions deserve an answer. Before answering these questions, let's tackle the issue in a less grandiose way.

Maybe a better question is to ascertain whether any events result from some design (in the course of events). What would indicate some purposeful design in the flow of time? Let's take a simple idea and ask of its significance. Is a series of apparently entangled occurrences more than mere "coincidence?"

Some events appear "coincidental." These are sometimes called synchronicities. C. Jung was fascinated by these phenomena. Einstein thought these occurrences were worthy of study.

Austrian biologist P. Kammerer began collecting and studying examples of coincidences during the early twentieth century. Here is an example of the type of coincidences that particularly intrigued him. In Orleans France, a Monsieur de Fortgibu gave a boy named Monsieur Deschamps Christmas plum pudding. The passing of ten years finds M. Deschamps in a restaurant where he sees a serving of plum pudding. He requests the last pudding piece only to be told a certain M. de Fortgibu had just ordered it. After many more years, M. Deschamps found himself in the rare situation where he was again offered plum pudding. He told friends about his former experiences with M. de Fortgibu and plum pudding. He then remarks: "If only M. de Fortgibu could now join them." The door opens and an elderly man enters by mistake because he had the wrong address. It was M. de Fortgibu.

Many people have experienced similar "not so coincidental events." How can these events be explained? How can something occurring now be "aligned" with something (an event) in the future, which is also just another "coincidental" incident in the course of events? M. Deschamps' experience with M. de Fortgibu is an example of strangely isolated events linked solely (in the physical world) by similarity of experience. Is this seriality of linked events solely a coincidence?

The story of M. de Fortgibu does not appear to have much bearing on the course of pivotal events that change the course of civilization. Yet, it signifies the possibility of a teleological agent (outside the physical world) capable of "directing" the course of future events. If this is not surmised, we are left with the notion of "well that was just a coincidence; it happened because of the chance outcome of random events." It is foolhardy to say all coincidences are by nature accidental if only because we have not developed the proper concepts to explain them.

Mere coincidence infers the course of events is devoid of any overriding design. If "coincidence" is granted as the only explanation for seriality, linked events become exceedingly improbable. If seriality is only coincidental, their occurrence is impossible to explain by design.

Which is easier to explain? That there is some design to the course of events, or that every event, even those outside our control, is by nature accidental. Accepting a design in the nature of things is not only more aesthetically satisfying, but by reasoning is more acceptable. For example, if we surmise every event is fortuitous then we must also accept the extremely improbable "chance" that a seriality of linked events can occur. And yet, they do occur.

Too many "coincidences" overwhelm any idea of an "accidental" linked series of events. However, if we accept there is a possible design to events, how can it be explained? Appealing to "God" (or some other teleological agent) without properly defining it (God) and without explaining how it "creates" such events, does little to promote the idea there is a teleological side to reality.

> Scientists accept Darwinism to explain evolution because there has not been a logical model upon which to posit a teleological explanation. They fall back upon a reductive explanation. They point to "parts" to explain the "whole" course of life-form development. For some scientists, "survival of the fittest" has been supplanted by DNA as the (reductive) explanation for "structural development." Accepting DNA, or any reductive agent to explain the course of evolution, is to say the "parts" control the overall (whole) direction of evolution. Even "turning points" in evolution would have to be considered no more than coincidental. Appealing to chance events (evolutionary adaptation, and survival of fittest) is almost too much to ask of reasoned thought. Yet, if these events are not coincidental, they must be explainable. If life has purpose, it will be explainable. We fully address the evolutionary problem in the final chapter.

If we were to embrace the notion of randomness as the sole cause of events (outside the arena of manmade causes) it becomes exceedingly difficult to "explain" anything. Particularly troublesome is understanding how any higher-level functions could arise from lower-level functions. For example, how do atoms know how to group themselves into molecules? How does DNA know how to map out the "design" of a life form? Saying such grouping of parts occurs because of the laws of nature, explains little because of the next question. What determines those laws? Are general laws of nature determined by the parts or somehow determined by the greater wholes of which they partake? Confusing what is "allowed" in building parts into greater wholes is different from asking why a given form arose rather than another.

Surely, we do not expect atoms to meet and say, this is the plan of how we will assemble into molecules, and it be done. And when molecules are assembled into living cells, we do not suppose a certain group of cells decree in advance they will evolve into living things. These are reductionistic explanations. Yes, without atoms, there would be no molecules, and without molecules, there would be no living cells. But being a contributing factor—of a "design" (parts and atoms, for example) is not (the same as) the completed design. Wood and blocks, used in home construction, are not the house.

Constructed arrangements constitute design. Are arrangements designed? A designer manipulates things (i.e., contributing elements) into arrangement. A man designs and from "parts" builds a house. To make our point and keep things simple, we omit designs by man's handiwork. But, we can ask about the arrangement of parts, which defines man as a physical being. Is man part of a design? What evidence suggests we (and everything else) are part of a design?

Parts (as atoms are to molecules, and molecules are to living cells) are contributing factors, but not necessarily causative factors. It is more convincing and reasonable to consider design in the course of events (in the flow of time). Yet, granting this possibility leaves us to explain how order is directed (in changing arrangement of parts in the flow of time).

We do observe some series of events are reasonably explained by appeal to a higher (level) intellect. People involved in religions often point to spurious events and say, this is God's will. Asserting statements is not sufficient to establish truth. We need some structure or step by step reasoning to support a statement's truth.

Asserting (a statement) and proving (it) are not synonymous. Our task is to develop the means to verify and prove there is a higher-level intellect. This calls for a structure (or function thereof) which logically "explains" the overall course of events. The explanation must also describe how the directing occurs.

Pivotal events suggest the possibility of a "higher purpose" intertwined with the events of everyday life. It is difficult to gauge the meaning of any single event. The usual means of ascertaining if an event is "pivotal" (in that it signifies a higher purpose) is to observe a "pattern" over a long time. History is useful for reviewing the course of events. Our interest is man and his experiences.

Our birth is a pivotal event. Did we ask to be born? No. Did we ask when and where to be born? No. And did we ask to look as we do? No. This is kept simple to make a point. Where and when we are born, how we look, and how others treat us, shape our views. Lesson, we do not have total say in how, when, and where we live. Yet, these local factors affect what we experience and learn.

Key Points:

- Pivotal events are recognized by their consequences.

- Turning points shape the course of history.

- A series of "coincidences" is explainable by design.

- Design implies purpose.

- Purpose implies a telic agent—a designer.

Not all conditions or events affecting us are by nature fortuitous. There is an evident design to some experiences. Design indicates a higher purpose. How can we know for sure that experience has an overriding purpose?

Significant-Consequential Experiences

We need to scientifically assess whether there is a design or purpose to existence. The author of this book uses his extraordinary experiences to answer this question.

The "Letter" (see Chapter 3) represents a stage of mental development called Self-Actualization. This stage occurs when the mind begins to reason (inductively) by accepting the responsibility for one's meaning in life. The How Aspect Diagram includes the major ideas presented in the Letter. Previous "pivotal events," here the Letter resulting from the author's then understanding of life (posed by the death of his brother) sometimes have significant consequences—in this case, the Where Aspect Diagram. The (Where Aspect) diagram was made over two years after the death of the author's brother. This death spurred the author to seriously seek the meaning to life. The following is presented from a personal perspective.

A "pivotal event" is one that is necessary in the sense it must occur for other even-more significant events to develop and emerge. For example, the death of the author's brother—the letter to another brother—the diagram, are all "pivotal events." Traditional "cause and effect" reasoning suggests the following. A removal of any of these events would halt the development of even-lesser-necessary events contributing to the subsequent realization of significant consequences (for example, the diagrams in this book). Seeing significant experiences in context of (previous) pivotal events and telic consequences suggests a "design" in what we experience.

The traditional concept of "effect following cause" is generally inadequate for analyzing a series of linked events—seriality. If there is a design (or pattern) in our experiences, then there must be a "higher purpose" (or reason) to explain them. The next chapter addresses the meaning of these experiences.

For now the question is "how can pivotal events be explained?" The "meaning" of (pivotal) events is better determined by the significance of their consequences. It is as though significant consequences explain (previous) pivotal events. This is "backward" from the traditional "cause effect" concept. This "analysis" suggests a cause that follows effect. Pivotal events are more easily explained by subsequent consequences. Let's look again at our above example.

The Where Aspect Diagram (Refer to Chapter 3) is based on the answer to a question posed by a student in a science class (taught by the author). The student fumbled ahead in his textbook and saw something that prompted him to ask: "What is color?" The author's response became the basis for answering a significant philosophical question. The question concerned the following: Is "what" we experience in the physical world actually "out there" independent of our consciousness, or is "what is out there" no more than a projection of our consciousness? The eventual answer to this question is not as simple as it originally seemed. (The difficulty occurred from not understanding awareness and consciousness.)

> Science is predicated on the idea there is a "world out there." Yet, philosophers know there are inherent problems (contradictions) in this view. This is the problem of "objectivism versus idealism." The answer to this question is obviously linked to the philosophical problem of "free will and determinism."

The meaning of significant events is evident. Let's refocus on the author's example. The answer to the color question explains many things. It provides the basis for the 1970 diagram (the Where Aspect Diagram). More will be stated about how this happened. Without hesitation, I addressed the student's question: "What is color?" I answered: "The color of what a thing is in-itself (i.e., apart from how it appears) is all color but the color that you see." For example, if you are wearing blue jeans, the actual color of the jeans in themselves is yellow (yellow and dark blue are complementary colors). So, I realized immediately the significance of the question and its answer. I pondered the meaning of the question for seven days. My mind was ready to infer more knowledge inductively derived from the immediate answer to the color question. I asked a student to find a large sheet of paper. With brown construction paper in front of me, I just began to fill it in. The writing flowed effortlessly—automatically. It was astonishing. I just wrote it out. There was no planning. I wrote as fast as I could write. My mind was racing—and cognitively, I was just trying to stay up with what was happening. It was quite an experience: to say the least!

What was the meaning of the diagram? The Where Aspect Diagram later became a part of an even larger-encompassing diagram—called the Deltic Cosmos structure (see Chapter 10). And that later becomes a part of yet another more encompassing diagram (representing a greater structure of reality). The point is, what seemed to be an isolated event, later contributes to a greater—more enveloping—understanding of reality. The Letter became a part of the Where Aspect Diagram, which later contributes to understanding "higher level" structures (of reality). Obviously, I did not write the Letter for the intention of later including it into a representation of "some" higher-level construct (conceptual structure)!

I could see there was not only a "design" to what was occurring, but also what was occurring was a revelation of design (constituting cosmic structure). Note that man is defined by the structure of reality. The ultimate purpose of science is to totally understand reality. If I am "to find (define) myself," it can be done only by how I (and mankind in general) fit into the scheme of things—the structure of reality. Structure implies design.

After many years, I finally delineated the structure of reality. Now, obviously, I did not do this on my own. I was guided in this search. That is the point.

Remember what was stated about where and when we were born? We did not have a choice.

We do not have complete say in what we become. What else could this mean? Some of us may have an identity, which supercedes contemporary acceptance of what constitutes an individual's identity. Is a "higher" identity possible?

Have not people (at any given time) espoused wrong ideas about what is real? For example, that mysterious (usually silent) "right brain" that recognizes "patterns" such as faces—what is its real nature? What is the nature of the left-right brain phenomenon? What is being stated?

Are some individuals more than they appear to be? Man is more than merely a physical body—he has a mind. Is it possible that the spirit (therefore the mind) can incarnate different bodies at different times? Some people may have an additional identity. A person's brain may be a manifest structure harboring a mind with an "extended" identity.

We learn from two main sources—physical experience and mental experience. Learning extraordinary things occurs because unusual things happen. "Extraordinary knowledge requires extraordinary experiences." The first diagram of 1970 was a revelation. It did not occur by individual intent. It just happened.

The Letter (see Chapter 3) showed the author was taking full responsibility for his life (and hence its meaning). From the time the Letter was written until the revelation of the diagram—he was not security dependent on anything outside himself. His security was dependent only upon himself and that meant dependent upon how well his mind functioned. Self-responsibility pressures the mind to seek answers. Recognizing one's ignorance spurs the mind to seek truth by questioning. The Letter expresses the author's commitment to help people by learning more about reality. After all, people have problems when they face something in reality they do not understand.

So, what happened to the author in 1970? His mind began to function on its own. What does this mean? His mind awoke and began reasoning inductively. In Biblical terms—He was born again.

> People, who claim to be "born again," like those involved in religion, are lying—just as we would expect—based on the Bible. People are not born again merely because they make that claim.

The author's mind was resurrected. What is the basis for that statement? The mind once functioned on its own apart from the physical world. The mind (or spirit) is immortal and survives physical death. And surely, if the mind is immortal—it must have existed before inhabiting a physical body. Everyone's mind has this condition of immortality. Even if one's mind is not reincarnated, it already existed through the minds of his parents.

Just as physical exercise leads to a stronger functioning body—one which can do physical things requiring greater strength, so it is that the mind, with the "exercise" of seeking deeper truth, can function with greater reasoning ability to answer "higher-level" questions.

Let's take a closer look at change and consequences. Significant discoveries often occur by what appears happenstance—accidentally. One cannot study the history of science and not recognize this. Einstein tried to find a way out of the (apparent) contradiction posed by Maxwell's theory of electromagnetic phenomena and the ether drift experiments of Michelson and Morley. The null results of those experiments indicated light does not travel in a medium like waves on water. So, was it a "coincidence" when Einstein rode the trolley and wondered what it would be like to ride a beam of light? Is this what a real paranormal experience is like? If he did not have the trolley ride, would he have discovered the laws underlying Special Relativity? Einstein had a series of interesting experiences leading him to formulate the Special Theory of Relativity. What about the supposed apple falling on Newton's head? Look at the consequences. Does a telic agent cause these pivotal events? Surely, they did not happen by intention of the individual experiencing them. Are these real revelations (paranormal activity) from a higher-level Being?

When ideas "just pop into the head," what is their source? Many scientific advances occur just this way. Are these illuminations paranormal revelations? As created beings, is part of what we experience (e.g., 'insight') for a higher purpose? Many "insights" have broad consequences. Does experiencing life in the way we do, seem too normal to be sometimes, paranormal? Surely, while a child, we did not say, "I shall 'will' myself to grow up in a specific way." Regardless, we still grow-up without willful intention on our part. So, if we did not create ourselves and grow by our own intention, who or what did the willing? These ideas suggest there is some type of creation process.

Things in long-range development strain our capacity to understand how unforeseen sequences of events culminate in a finished product. Consider the electrostatic cathode ray tube. It was developed in the late 1800's. In 1895, Roentgen, a German university scientist, was studying "cathode rays." He experimented and discovered the rays could penetrate solid objects like playing cards. He held up a lead disk to see whether the rays could penetrate it. His hand was incidentally "x-rayed." He later x-rayed the hand of his wife, Bertha. Roentgen's achievements were rewarded with the first Nobel Prize for physics in 1901. J. Perrin discovered the rays carried a charge. Two years later J.J. Thomson discovers the rays have a charge to mass ratio—the beginning of understanding this phenomenon. He determines the rays cannot be light (they are high-energy electromagnetic radiations generated by electrons). Using this information, Einstein soon explained the photoelectric effect. Medical science found a valuable tool. These scientists did not experiment with the intention of advancing medical science.

Maxwell studied what was known about electricity and magnetism. He wanted to understand the relationship between electricity and magnetism. How does one induce the other? He summarized his findings with four equations (the first three equations resulted from work of earlier experimenters). Scientific understanding, numerous discoveries, and many inventions (like the cathode ray tube) were later brought together resulting in the manufactured product called television. None of these scientists envisioned radio or television as a consequence of their work. The cathode ray tube, used in computer monitors, helps people communicate through the Internet.

If we say the extraordinary experiences of the above scientists are not paranormal—then how do we explain their extraordinary consequences? When viewed in a wider context, these consequences defy the "normal" concept of explanation. It is almost like television, computer monitors, and the Internet were waiting to be invented. After the fact, this supposition seems inevitable.

The idea of deliberate-intentional-sequential development does not include an unforeseen-unintentional-end product. A possible test one has had a paranormal experience could be that the knowledge acquired solves previously unsolved problems. Perhaps we are "given" problems so we will come to understand greater things. Surely, if we had no problems—there would be no pressure to change. (Inertial also seems to apply to people. They will not change by themselves—pressure must be applied.)

That we read—learn and seek truth, must be a clue to the "meaning of life." Consider the consequences of the invention of the printing press. There was a problem in duplicating enough reading material. More people learned to read as time passed. Learning by written instruction accelerated technological development. Television and the Internet have also greatly enhanced dissemination of information.

Is widely available information a main clue to understanding the purpose of the printing press and television? We do know the mind develops by trying to understand things. Mental development is a clue to man's purpose in life.

Key Points:

- The trail of pivotal events is explainable by significant consequences.

- Some individual experiences shape the course of history.

- Purposive instrument design suggests a telic agent guides the development of scientific knowledge and understanding.

- Scientific knowledge and instrument development enhance the understanding of reality.

- Understanding things suggests there is a reason for everything.

- Reason implies purpose.

- Purpose suggests there is meaning to life.

Just as increasing physical ability is a step by step process, so it is that increasing-mental ability is a step by step process. The mind takes time to develop. Increase mental ability is synonymous with "spiritual growth." Pushing beyond one's current mental ability increases his powers of reasoning. Reasoning ability is tested by solving unsolved scientific (mathematical, logical, physical, and behavioral) philosophical, and theological problems. The immediate purpose of many "pivotal experiences" is just to acquire this knowledge.

Implication of Design and Purpose

Many events defy the label of coincidence. Mathematicians are awestruck by an equation discovered in the nineteenth century. The Swiss L. Euler polished its final form. It is $0 = e^{i\pi}$. It interrelates the major constants of mathematics. (Base of natural logarithms: e = 2.718..., imaginary number: $i = \sqrt{-1}$, ratio of circle's circumference to its diameter: π = 3.1415....) Biologists are dumbfounded by life-form evolution. There is order everywhere we look. Order infers design. Discovery and explanation create wonder about everything.

Discovery and explanation not only advance understanding, but also have consequences in extraordinary inventions. We mentioned the developmental process culminating in television. The American, P. Farnsworth's work in the 1920's greatly contributed to its actual development. Invention of the vacuum tube and later refinements such as the transistor (W. Shockley, et al.: 1947 Bell labs) and the integrated circuit (J. Kilby, 1958 Texas Instruments) and other parts that were independently developed, contributed to development of reliable television. Most scientists have little idea of the long-range consequences of their (?) scientific discoveries and inventions. What is happening here?

The "course" of instrument development, like television, in the wider sense, cannot be dismissed as one composed of a series of "unconnected" minor-scientific advances culminating in an unintentional end product—television.

Is there a better way to explain complex instrument development? The historical process of instrument development implies "design." Design implies "purpose." Inventors of the cathode ray tube did not say once this is perfected we will have television! Surely, if there is a "greater Being" of which we are a part—could it not "arrange" some of its "parts" (us) according to its own purposes? Surely, we are not the whole of "what is." Is the whole of "what is," a greater Being? That possibility cannot be denied.

It is reasonable to suppose there is a greater Being. If there is demonstrably a higher purpose then there must be some kind of "greater Being" purposively orchestrating what we experience as pivotal events leading to significant consequences.

Physicists are nearing a complete mathematical (logical) understanding of reality. Many discoveries and significant events can be understood as a part of a grand design. The author's experiences have become a part of this process.

We do not happen to life—life happens to us. Many people have a special purpose though they do not realize it. The death of the author's brother was a "pivotal event." The brother had a special purpose of which he was unaware.

The author experienced many "pivotal events" demonstrating there is a "design" to some of what we experience—thus indicating a "higher purpose." The immediate purpose is to obtain "more knowledge" about reality. A critic could rightfully dismiss such a statement (extraordinary knowledge following extraordinary experiences) as "coincidental." Extraordinary knowledge thus obtained tends to refute such criticism. A series of "coincidences" (pivotal events) cannot so easily be dismissed as occurring by chance. These events are not "coincidental": they only appear that way in a "narrower" time frame.

The author has been guided by reality itself. His mind began to function at higher levels of awareness to understand reality at deeper levels. (See the How Aspect Diagram, Chapter 4.) The author realizes there is an inordinate design intertwined in the experiences begetting the knowledge explained in this book. The design is undeniable.

Key Points:

- Purpose presupposes a greater Being that somehow orchestrates the overall course of history.

- Extraordinary experiences in time infer that any active participation of a greater Being is directed from outside time flow.

"To get to the truth lies must be gotten out of the way." By eliminating contradictions in higher-level concepts (enveloping generalities) we uncover deeper truth.

Design implies purpose. We ask how are the many significant events (seriality) which contribute to the development of anything, connected by purposive design? Let's develop some enveloping concepts that help explain event interconnectivity.

Hierarchical Intelligence Levels

Are there other ways of testing what we have learned about the structure of reality? Can it explain things not presently understood? Philosophers have grappled with the problem of first cause. Religions hang labels on it. People supposedly worship it. If there is a creating Being, and the structure of reality is complete, then we should be able to define this creator using the structure of reality. And, we should be able to explain much about religious concepts.

If there is something divine about religious books, there must be something (a God or some kind of greater Being) behind their production. Let's examine this possibility from a scientific perspective.

Concerning higher-level generalizations: "Picture someone's car." Then classify that car at ever-higher levels—until it (the car) is eventually subsumed as a more specific (deductive) instance (of the higher-level generality). For example, the category "automobile" is a supercategory of the specific car. Yet, it (category "automobile") is a subcategory of the more encompassing classification "modes of transportation"—which includes bicycles, boats and planes. Another example: The plant and animal phyla are excellent representations of the encompassing category "life forms." Eventually, everything gets classified as a "part of reality." It is not this simple. To categorize an "understanding" of reality requires the proper notion of "function," viz, differentiating functions in a hierarchy. This idea is shortly addressed.

At a low level of specificity, we find lepto-quarks. Lepto-quarks are composite parts of a higher organization—atomic structure. Atoms are organized into patterns of higher complexity—molecular structure. Complexity suggests there is a guiding principle at work.

As we "ascend" to higher-level complexities we find functions not found at the lower (composite) levels. Some molecules compose cellular structure. The life function appears for the first time in cellular structures. One-celled plant and animal life forms have negentropic (evolutionary) consequences in higher level structures. These ideas strongly suggest there is a guiding intelligence "beyond" manifest things. Order and complexity imply an underlying intelligence.

Why do some molecules take on a life function while others do not? Again, further (reductively) studying molecules and atoms does not answer this type of question. By seeking to understand higher-level functions in the (hierarchical) complexities, we tread in a more promising direction for an answer.

Cells are organized: into tissue, tissues into organs, organs into organ systems (circulatory, nervous, and muscular systems, for example). There are many such systems. When these systems function as a whole, we find another characteristic—awareness—a mind that can comprehend (make inferences—deductive and inductive).

Reasoning toward higher-level functions shows promise, but for what purpose? We need not understand much about nature to ask this question. And sure enough, it has been asked long before a serious study of nature ensued (using the scientific method).

Before going further, let's digress and personify two (human) muscle cells. They are Matt and Mark the muscle cells.

Matt, in a moment of relaxation asks: "Mark, Do you suppose there is a greater Being of which we are a (functioning) part?"

Mark: "I don't believe there is. After all, do we not have free will? If we have free will, we have no need of a greater Being."

Matt: "Do we not receive sustenance from our environment—food, water, oxygen? That we are not self-sustaining shows we are not an independently functioning organism. That evidence proves we are at least limited in free will. The extent we are limited in free will, is the extent we may function as a part of a greater Being. We understand all these lower levels below us (such as molecular). Just because we don't observe any higher-type Being does not preclude the possibility one exists."

Mark: "I don't believe we respond to any higher Being, and therefore have no reason to suppose one exists, right? Matt please excuse me, I have to contract now."

Matt, wondering to himself: "It is easier to respond to this stimulus than to ignore it. What is the source of this stimulus? Why are we stimulated (to contract)? If we could identify and verify the source of our existence by delineating a higher-level structure, we should be able to ascertain our purpose (or meaning in life). If we ask 'why' we do what we do (in this case, contract) we should be going in the right direction to seek our meaning in life. We definitely have a natural proclivity to contract."

Matt, the muscle cell said, "To the extent we are limited in free will is the extent to which we may function as a part of a greater Being."

In Matt's simple statement, we see a hint how "coincidental events" occur. This idea also explains the nature of real "paranormal" events. Perhaps the muscle cells, Matt and Mark, marveled at other muscle cells—all "coincidentally" contracting simultaneously.

Key Points:
- A greater Being should be definable.

- Pushing the concept of multi-level functionality to its limits should yield an understanding of the greater Being of which we are a part.

- The greater Being should be definable by "categorical" inclusion of "every" hierarchical complexity.

Though necessary, parts are not a sufficient explanation of higher functionality (e.g., life forms). Life forms are part of experience. Experience needs to be explained. We suppose consciousness is the sufficient and necessary explanation for experience and life forms. It should also explain awareness.

A "greater Being" not only explains phenomena like "a series of improbable coincidences," but also explains the high-level intelligence functions necessary to explain life. How do these higher-level functions interrelate?

Transcendent Functions

Higher-level functions connect to (their) components by a concept called transcendence. Consider a hierarchy in terms of interphenomenal relationships. If different hierarchical levels exhibit functions, they might be "linked" by transcendent association. This will become clearer as we proceed. Let's use two analogies for examples: government bureaucracy and the automobile.

Government bureaucracy: Picture a map of homes on a city street, let's say, New York City. Picture another more encompassing map of the city's streets. Each map represents some part of reality. The maps are not reality—they only represent reality "on paper" in the form of mapping. Various structures are represented. In a detailed street map of New York City, we see symbols or markings for buildings, parks, tunnels, bridges, railways, airports, streets and highways.

Consider a map of the State of New York. It marks location of many cities and towns. Also included are townships and counties. Examine a map of the United States. It shows state boundaries. Notice what happens as we move from maps of city streets, to maps of counties, states and country. Each successive map encompasses more (of reality's) structure. Location of countries within continents can also be mapped. We also note: as higher-level structures are viewed, we lose focus of lower-level structures.

Notice something else. We not only mapped physical land—we demarcated different levels of governmental functions, viz, city government, county government, state government, country (federal) government. Each level is governed by a higher-level in the bureaucracy. Each level is responsible for resolving conflict between lower-level functionalities. The United Nations functions to solve problems between countries.

We see two forces at work at every level. There are forces of construction (negentropy) and forces of destruction (entropy). Every government level strives to balance construction and destruction of its function. Each level of bureaucracy tends to maintain itself. Each person tries to maintain his or her position within the system. Parts tend to sustain their own functionality.

What is the point? Each governmental level transcends (in function) the level below it, and is transcended by the level above it.

Consider the automobile. Proper assembly of automotive parts constitutes (functions as) an automobile. Though a wholeness aspect (e.g., of the automobile) is "complete" does not necessarily mean it is operational.

Automotive systems (component groups) serve a purpose. The alternator, battery, and voltage regulator are components of an automobile's electrical system. Each contributes to maintaining electrical activity. Electricity helps sustain the purpose of the automobile—transportation. An operational automobile depends on the systematic activity of its parts. Most things function for a purpose.

A thing's form is its wholeness function. We look out a window toward the driveway. We say, all those parts assembled into that thing's form, functions as the automobile. A thing's "parts-functions" are lesser-wholeness functions (in their own right). Parts are a function of the whole.

There are nonfunctional automobiles. Perhaps an automobile, though complete, is nonoperational and resigned to the wrecking yard. A wrecked or parked automobile is not fulfilling its purpose. A nonoperational or "nonactivated" automobile is only "potentially" functional. Someone says, it (the whole automobile) isn't working.

An automobile cannot function unless its parts and systems are properly assembled and operational. A thing's purpose (utility) is not realized until there is activity of its form (of its wholeness function). A thing (the unit function) may not exhibit the "super functioning" of its wholeness function. (An automobile not in self-motion, does not exhibit its super functionality.) Some things are only potentially functional.

Let's limit our example to an "operational" automobile. The purpose of the automobile is transportation. Transportation is neither the whole automobile nor its parts. Transportation is a "transcendent" function—it transcends the whole automobile and its internal workings (lesser functions). Transportation is realized by an automobile's self-translation. (Auto—self, mobile—capable of moving.)

There is a distinction between a thing's form (wholeness function) and its utility (use of its super functionality). Consider water. Two hydrogen atoms and one oxygen atom chemically combine to "form" a functioning water molecule. One functioning-water molecule, however, is not synonymous with water, as we know it. Water in everyday usage is fluid (defines its form). Water is a product of the gross activity (super functionality) of uncountable H_2O molecules. (Liquid) water results from the loose binding of its molecules.

> There is usually some type of energy or "attractive" force involved in "bringing together" a collection of parts. Water molecules are "held together" by Van der Waals forces (after J. van der Waals, Dutch physicist). "Hydrogen bonding" between water molecules "holds" water molecules into a (liquid) collective. (The force originates from dipole magnetic fields produced from shifting of nuclear electromagnetic charges.) The super function characteristic of water molecules (solid, liquid, and gas) depends on its level of activity (as governed by temperature and pressure).

There are many levels of (parts) functionality. The gross properties of water exhibit functionality that supersedes and transcends its constituent molecules. Molecules arise from chemical activity of atoms. Form (or thingness) is exhibited by the concerted (interactive) functioning of many levels (of parts). There is a hierarchy of systematic functionality.

"Ascending" the hierarchy, we find higher level functions. What these functions are called, depends upon our vantagepoint. Given that understanding, there can be a "super function" of a given thing's form. Water is a super function of its molecules. Utility usually pertains to a thing's super functionality. Water sustains life forms. Transportation is use of automobile motion (i.e., its super functionality). A wholeness function is realized by the "concerted activity" of a thing's parts. A given form can contribute to some greater form, super functional activity, or transcendent functionality (utility).

Parts functions tell us about the subservient (and therefore, reductive) role of content. Automotive parts sustain automotive super functionality (self-motion).

Tidying up: Inherent in a unit (thingness) is a wholeness function (its form) and parts-function (its content—supporting the wholeness function). A wholeness super functionality suggests a unit's purpose—its transcendent functionality. Regardless whether a thing's transcendent function is realized, its purpose (as indicated by its transcendent function) is often suggested by the (potential) activity (i.e., its super functionality) of its form. Super functionality and transcendent functionality are often synonymous. For that reason, transcendent functionality, if different from super functionality, is at least implied.

Transcendent functionality is generated by unit activity (action of whole and parts). Transcendent functionality, in our treatment, is not realized unless there is a collective activity of parts (producing it). There is no automotive transportation unless the components (subunits of the unit-automobile) function collectively to put it into motion. There is no governmental function unless people do the work.

A properly "activated" wholeness function produces an emergent transcendent function. The transcendent-wholeness function conveys the reason or purpose for the form (the wholeness function). The purpose is usually the role a thing has in a greater form or higher level function. An unrealized transcendent function remains potential. We later delve into the deeper nature of potentiality.

Asking about the "whatness" of something is to name the unit by its form. For example, what is that? It is an automobile. Asking the "whyness" of something is to ask about its purpose (or use of its super functionality—viz, its transcendent function). "Why an automobile?" What is its purpose? Asking about a thing's parts is to ask about the supporting role of subunits. Consider an example of structure.

Structures of reality tend to maintain their function. This is called homeostasis in human physiology. Structural stability depends upon maintaining the balance between construction and destruction of function. Specifically, construction is the building-up of systems (from parts). Destruction is the breaking-down of systems into their (component) parts.

How is functional stability sustained? There is constant activity between system building and system breaking down. Why is this activity unceasing? The universe is in a perpetually unbalanced state. Answering such a question is to look at dynamism from the "big picture" perspective.

We observe there is a process that system builds-up lower-level structure into higher-level functions. Physicists and scientists generally have a difficult time with system building. In physics, we recognize the building process as negentropy. In the physical universe, entropy always wins over negentropy. Physicists have little problem with the notion of entropy—systems breaking down. It is the dominant process of the whatness world during Phase I (Universe of Perfect Whatness). Entropy is the predominant process underlying physical dynamism.

Enough is stated here to describe what we learned from the Primary Equation ($1 = 0 \times \infty$). The infinity factor ("∞") functions (conceptually) to represent (end product of) entropy—that wholes are broken into parts—toward an infinity of parts. The zero represents negentropy—the building up of parts into higher-level structures—eventually to the whole of everything.

The following are functional levels of structure: energy of vacuum, lepto-quarks, atoms, molecules, macromolecules, cells, tissue, organs, organ systems (circulatory system, muscular system—remember Matt and Mark the muscle cells) physical organisms, mental awareness (mind) and finally consciousness.

How are structural levels fashioned into a hierarchy of transcendent functions? By resolving Russell's Paradox, zero ("0") and infinity ("∞") in the Primary Equation can be "extrapolated" into the logical Law of Contradiction. There was just one more equation to go that explained the problem with the Law of Contradiction. The problem was that it would not allow space ("0") to be defined as nothing and therefore, nondynamic and dynamic (at the same time). (Space expands according to the Big Bang theory of cosmology. Hubble's astronomical observations and Wilson and Penzias' discovery of background-microwave radiation support this observation.)

By understanding set theory, we recognized the shortcomings in Gödel's theorems. This insight enabled us to push ahead and define the "set of all sets." This all-embracing set, the empty set, is represented in physics by space. We first analyzed generalities using "form and content." We later used "wholes and parts." Using these concepts, we finally traced the Law of Contradiction back to the logical Law of Noncontradiction—the Ultimate Principle.

The Ultimate Principle made it easier to explain things. It carefully became a matter of mapping out the structures of reality according to what scientist presently understand. Recall the Wise Man's Wisdom (refer to Chapter 2). If lower-level structures have not been properly mapped, delineating higher-level structures would be next to impossible.

Mapping different orders of structure is amply represented in our analogical mapping of city streets—county—state—country. Demarcating governmental-bureaucratic structure exemplifies encompassing functions. Yet, there are minor problems.

What appear as individual cities, are not so easily separated when one city merges into another. There are no marked boundaries. Streets lead from one city to others. We can drive from one city or state into another and not necessarily encounter natural boundaries. The separation of one city or state from another is usually unmarked. There is not always a separating landmark or river.

There is no necessary separation or barrier between cities, states and countries. Cities and states are defined by function—governmental jurisdiction. They are separated by function. In physics, this is the nature of Bell's theorem and Aspect's experiment. As far as space (itself) is concerned, there is only a functional separation between things. Yes, there appears to be a separation between cities and states. What appears as separation, is one of function (concentration of activity in a structural hierarchy). A city's boundary limits (sets) its jurisdiction.

Summarizing: Every city is a part of a state; every state is a part of the country. These separations are described by function. There is also the functioning of the country as a whole: in our example, the United States. Each lower-level function is just a part of a greater functioning whole.

Is it possible there is a greater Being of which we are a functioning part? Surely, we, as individual beings, are not the whole of what there is. Do we not receive sustenance from our environment, one teeming with activity and life? Is there a level of structure corresponding to the collective activity of all our minds—a higher-level intelligence—which directs the course of civilization? Matt and

Mark, the muscle cells, noted the curious "concerted" action of their contractions. They wondered what it meant.

We have a natural proclivity to seek truth. Why? Are we (by reasoning) responding to a higher Being by seeking truth (although it is against our emotional nature to do so)?

If there is not something, e.g., a higher Being, the fact entropy pulls all systems into greater disorder—entails it would not be long and there would be no order at all, only chaos. Obviously, a case can be made for postulating a greater Being that counteracts the general tendency of entropy. After all, life forms develop in an anti-entropic manner. And, we quickly appreciate if there was no entropy there would be no change. So, is entropy a part of a bigger plan—a cosmic one in which we find ourselves as practically unwitting participants? Is there evidence indicating work of a higher Being? Yes. We commonly encounter improbable events, which defy coincidence.

It seems we are caught between the extreme effects of entropy and negentropy. Negentropy is the building up of higher-level systems out of disparate parts.

> Is there a greater Being directing the consequences of our activity (and development)—just as we direct an activity to which our muscle cells respond? Maybe muscle cells do not see how they fit into a plan to break the record for the 100 yard dash, or how, for some people, they (the muscle cells) fit into other plans: plans as simple as someone running to the local store to get more beer. All the muscle cells know is they sometimes experience extreme exercise and get very tired (when participating in a 100 yard dash) or become a little uncoordinated (too much beer). Like the muscle cells, we just do not easily see the plan for our own existence. Whatever the plan, it must entail consequences enmeshed in "life's big picture"—consequences that defy entropy.

There is no change unless there is a way for change to occur. Is that the purpose of entropy? This has another side. Obviously, something cannot be continually created unless there is some way to counter entropy. Besides things "breaking down" (effect of entropy) we observe things "coming together" (effect of negentropy). Things are built up; otherwise, there would not be anything to break down.

If things were continually built-up without a way to break them down, it would not be long and everything would be "frozen," one big whole without any functioning parts. And this would be no different from the result of complete breakdown through entropy—no differentiation—all sameness. Is this sameness the vacuum we learn about in physics—the ground state from which all differentiation arises?

> Hinduism mentions a creator (Brahma) and a destroyer (Shiva) as well as a preserver (Vishnu). Negentropy effects a creating process; entropy effects a destructing process. Both are required to sustain change. Chinese philosophy talks about changes.

Can the "bringing together" (negentropy) be cited in a shorter time span? There are many examples, which point to some higher intelligence. Let's pick a simple one that precludes man's mind (and intentions) as a factor.

Occasionally, we learn about lost pets (usually cats or dogs) that, though separated from their family unit, find their way home. If a higher intelligence is not guiding them—how can finding their way home be explained? If we do not hypothesize a greater Being, we are left with dismal explanations. The notion of "instinct" is one such attempt to fill this "explanatory gap."

> Instinct is often cited for unexplained behavior. "The dog instinctively found his way home." However, instinct itself is unexplained. It is not understood how birds find their migratory destination. How can the Monarch butterfly find its way from northern climes to southern climes? Instinct is cited when intelligent planning by these life forms seems incapable of providing an explanation. Instinct can only be scientifically accepted as an explanation if the nature of instinct itself is understood.

Consider another absurd explanation. Suppose the pet (without any paranormal help) reasoned how to get home. We could suppose, the pet sneaks into a gasoline station and "borrows" a map. Yet, if the pet can reason directions from a map, why is it not smart enough to get a transfer of funds and obtain a plane ticket and fly home?

It is not difficult to prove that the pet, after being separated (effect of entropy—breaking down a higher-level system—the family unit) from its family, did not find its way home solely by its own reasoning. It is more reasonable to postulate there is a higher Being countering the "breaking down"—separating effects of entropy—by meaningfully "bringing together" (negentropy) disparate things (like the pet) into a higher-level system (the family unit—structure).

Other examples abound. How do ants and bees function as a whole unit, yet individual members are functionally divided into labor groups? Are we to suppose each ant is cognitively aware of how its role contributes to the welfare of the whole nest? The collective activity of the nest behaves functionally as though it were one whole organism—a collective intelligence. Is this paranormal behavior? Ants do not have reasoning abilities to do things with intention. The "word" instinct (alone) does not suffice as an explanation. Instinct needs to be explained. (It is explained in the next chapter.)

Is there a "higher level" orchestration to what we experience? Are we bit players in a cosmic play? Are we a part of a greater functioning being? (Telic purposiveness suggests there is a greater Being, one implied by functionality—activity.) If there is a greater Being, how could it convey higher-level knowledge to us (if it so intends)? In what form is this knowledge given?

On a day by day basis, it would be difficult to see the overall plan of such a Being. Perhaps everyday life is unique—fitting into a bigger pattern—one of which we are unaware. Again, that we seek truth must be a clue to how we fit into a bigger plan. Events happen only one way. There is no "but what if" in reality. Though someone contemplates that things could have been otherwise, than what they are, his very thoughts too only happen one way. Perhaps our shortcomings result from the poor methodology we use to understand. We find ourselves in a world, due to no fault of are own. If our existence is not our direct doing—whose is it?

It is one thing to experience the world and know the experience. It is another matter to understand it. We have a short (physical) life span. Is knowledge fed to us by piecemeal over generations? The development of complex instruments, like television, needed decades of scientific achievement. Are we an instrument of a higher Being?

We speak a different language than the muscle cells. How would we convey to Matt and Mark (the muscle cells) a message about the purpose of their existence? We would show them a pattern: one they could comprehend. The pattern would be conveyed by macro-molecular structures. The task would take much effort on our part. The muscle cells would constantly misinterpret our intent. They would have to set aside their preconceptions, beliefs, and deny their foolish notions about the way things are. Yet, if they sufficiently discipline themselves, perhaps with a concerted scientific effort (on the part) of many (realistically honest) "reasoning" muscle cells, they could delineate the structure of this greater being—us.

Only when muscle cells understand the functioning of this greater structure, can they begin to ascertain the purpose of their existence. Are we not in the same possible predicament as the muscle cells? Scientists struggle to understand our world. Many of them do not understand why they seek truth.

Scientists are attempting to understand all reality. Is this reality a greater functioning Being? Since most of us are not very wise, it would probably take a while for this greater Being to reveal itself. And, those who project their preconceptions and beliefs, will not be the ones to participate in this effort. Only serious seekers of truth will contribute to its understanding.

Left-right split-brain research shows each hemisphere exhibits different behaviors. Under controlled conditions, one brain hemisphere does not know what the other is doing. What does this mean? The right brain may exhibit a higher-level (transcendent) identity: one that already existed (and incarnated) in other bodies. We muse: How can a transcendent identity be sustained? How can there be transcendent functions?

> We suppose the left brain, or deductive function, is a super function of the brain. If so, then the right brain, or inductive function, is a transcendent function of the brain. For simplifying purposes, we will not "push" this distinction because the inductive function is "silent" while the mind inhabits a physical body. Yet, if the individual has a transcendent identity, it is associated with the right brain function. The mind is only potentially functioning inductively at this stage.

The automobile was "designed for a purpose." A properly functioning automobile offers transportation. Transportation is a transcendent function. A transcendent function is one that arises from the collective activity of parts. It is a function not found in its parts, or in its wholeness function—hence its transcendent nature.

Though necessary, parts are not a sufficient explanation of transcendent functions (e.g., life forms). Life forms are part of experience. Experience needs to be explained. Transcendent functions do not arise by accident.

Key Points:

- A thing's form is its wholeness function. A thing's content is its part function. Whole and part constitutes a thing's unit functionality.

- Each hierarchical level is a wholeness function.

- A wholeness function is either potential or realized.

- A potential wholeness function is one that has not manifested.

- A manifest wholeness function is realized by the "concerted" activity of parts.

- Hierarchical levels are linked by super functions (or transcendent functions) above and linked by reductive functions below.

- A thing's purpose is associated with its super functionality.

- The use of a thing's super functionality shows its purpose "resides" as a part in some greater (wholeness) functionality.

- Entropy is functionality dissolution.

- Negentropy suggests there is an ultimate wholeness function.

We have yet to fully understand the wholeness function. By understanding why content is structured into form, we should be able to grasp an understanding of the wholeness function (on its terms). "Whyness-questions" suggest purpose. Purpose suggests a designer. We acknowledge the reality dichotomy: experience and consciousness. Let's recast the question in the most general sense. Is experience the result of design?

Experience is defined by logical consistency (the Law of Contradiction). Its opposite pertains to the consistency of logic. And that is defined by the Law of Noncontradiction. It is the Ultimate Principle. The Law of Noncontradiction should help us understand not only consciousness, but also explain experience. If it does explain experience, then it also should explain the sustentation of transcendent functionality—including the nature of transcendent identity.

Defining Consciousness: the Solution

Consciousness is conceptually "isolated" by dividing reality into two categories. There is "consciousness" and there is "what we are conscious of," viz, experience. Though this duality is easily stated, it is more difficult to scientifically fill in details. We now realize that consciousness refers to All Reality—but no part. (Refer to Figure 8:1.)

Fortunately for us, details are found in the structure of reality. We delineated the structure of reality and are grappling with its understanding. That structure led us to the nature of consciousness. How does it function?

There is no experience without consciousness. All categories in reality's hierarchy are subsumed by consciousness. G. Berkeley came close when he thought all is a projection of a mind. (The "Self" in Figure 8:1 refers to awareness—mind.)

We eventually applied the Law of Noncontradiction (see Chapter 18) to our why-what paradigm and hence to our construct representing the structure of reality. The Realm of Whatness is "what" consciousness is "conscious of," i.e., experience. The Realm of Whyness is consciousness. Consciousness is "why" there is experience.

Before explaining how consciousness functions, we again cite our analogy of government hierarchy. We understand how different levels in a bureaucracy function. Higher levels govern lower levels. We are cognizant of two curious facts. Each bureaucratic level is functionally separate from the others. Each higher level covers a wider area of responsibility, but is less "attuned" to lower level events. Before examining the connection between consciousness and experience, let's briefly restate what consciousness is not.

We are conscious of awareness. Awareness defines the functions of the (individual) self—the mind. Awareness refers to ability to infer. Our minds commonly infer that we (independently) conceive what we infer. This is why many people suppose mind and consciousness are the same. Most people deduce each mind has its own (exclusive) consciousness. It does appear this way. Most people construe this to mean other individuals do not share the same consciousness. If this were so, there would be as many consciousnesses as there are minds. This is not so. Typically, appearance is "backwards" from the reality. There are as many minds as there are individuals, but there is only "one" consciousness. Awareness is not consciousness, and consciousness is not awareness. The mind is not consciousness.

We easily grant experience is constituted of whatness (physical) things. With some thought, we also recognize awareness is a type of experience. Our (individual functioning) "selves" (defined by awareness, i.e., the ability to infer) are parcels of experience. When we assert there is a table, we infer (in statement form) its reality (the actual table) in an imperfect way. This idea is reflected in the nomenclature of the "intelligence levels" (see Chapter 4). (Statements are labels for thingness visualizations.) Experience is not consciousness.

What have we learned? Our individual being or identity has a single source in consciousness-whyness. All that is manifested in the Realm of Whatness is a deduction of consciousness. Every "whatness" thing is produced from the reductive partitioning of consciousness-whyness. Existence is created by the "functional" fractionation of consciousness into a hierarchical plethora of whatness things.

Awareness partakes of the nature of consciousness and experience. Things take on a life of their own; minds take on a life of their own. Awareness has two functions. Deduction directs one's attention to the physical world. Induction directs us to an understanding of consciousness. We later learn the inductive function is associated with the purpose of mentality.

We are reductive products of consciousness. Consciousness is our "shared" higher-level-functioning Beingness. Just as the personified muscle cells (Matt and Mark) contemplate whether they are a part of a greater being, so it is we (specifically, our minds) are a part of a greater-functioning Being. Matt and Mark are functional parts of a single higher-level being of which they are "members." The higher-level beingness for muscle cells could be a human being. The mentality of each individual is a functional part of the higher-level Being of which we are all members. That higher-level Being is consciousness. Functionality is the key to understanding consciousness.

There is a functional separation between a muscle cell and the animal life form of which it is a part. There is a functional separation between mind and consciousness. There are no muscle cells unless there also exists the greater being of which they are a part. Without consciousness, there would be no minds, and no inferring. For every part, there is a greater whole. There can be consciousness without awareness, but no awareness without consciousness. Regardless of parts, there is always the whole that is consciousness.

Minds are a lower-level function of consciousness. Consciousness is the single greater Being of which every mind is a part. We are to consciousness as Matt and Mark, the muscle cells, are to us. There is "consciousness" and there is that of which it is "conscious of"—what it experiences. We are a part of experience.

Let's examine what is understood about some things experienced. Again, consider the automobile. It is composed of tardyonic matter. Matter is formed and machined (designed) into parts. For example, the alternator is a system within itself. It has many components including the armature, field coil, and brushes.

The alternator, as a unit, is also a component of the ("higher level") electrical system. The electrical system includes other components, e.g., battery, and wiring.

Automotive systems are arranged into functional units. There is the cooling system, power system—the engine, for example. Each system functions for a purpose. All systems collectively constitute the complete unit—the automobile.

Purpose is found by asking about use. All subsystems of a unit function for a purpose. The automobile alternator generates electrical current. Current provides the necessary energy to operate the automobile. To determine use or purpose, one looks beyond the system or unit in question.

The purpose of the automobile is not the automobile (itself). The reason for the automobile is not found by "reducing" it to component parts. Reductionism is an insufficient explanation for the design of the automobile. Asking about (whatness) parts is going in the opposite direction needed to ascertain purpose. Parts serve purpose.

The reason for the automobile is found by its use. If someone never before saw an automobile, he could ascertain its purpose by observing what it can do. A thing's activity (super functionality—here represented by automotion) suggests its purpose (transcendent functionality—here represented by transportation).

"Why" does the automobile function the way it does? It is designed for a specific purpose. Function is enmeshed in a thing's design (its form).

Although automotive parts are necessary for the "whole" automobile to function, its purpose is not sufficiently explained by its parts and systems. Purpose is "inductively" explained by appealing to the reason for a thing's design. We could ask why was this machine designed? The answer is found by stating its purpose. The purpose of the automobile is transportation.

The reason (for something) is found by looking for a more encompassing system (or unit). Encompassing systems are noted by citing a wider context or a higher-level generality. Muscle cells have a functional role in the higher-level life form (body) of which they are a part. The "reason" (for something) is the explanation "why" something is. Life-form mobility is the immediate reason (or explanation) for muscle cells.

The word explanation is often used in a narrower sense. Common practice is to choose a "lesser factor" as the explanation. The human body is composed of systems within systems. A medical physician would probably say (as an operational context) each system is explained by its parts. (Interaction of parts is usually analyzed using Type I Whyness-Whatness Association—described by extrinsic modes. Refer to Chapter 2.) For example, an individual feels chest pain and falls to the floor. Symptoms and cursory examination suggest a possible "heart attack." Later tests, using catheterization, reveal occlusion of coronary arteries. The heart surgeon concludes the immediate cause of the "attack" is an internal occlusion of coronary arteries.

Proper blood flow (supplying oxygen and energy) is needed to sustain physical life. By reducing the "reason for life" to necessary functions which support (as contributing factors) physical life, higher-level functions (reasons) are overlooked (in explaining life itself).

Why do we suffer physical death? If we are doomed to die—why do we have life? These questions are not sufficiently answered by "reducing" the explanation (for life) to internal-interacting systems of the body.

Let's categorize some systems of the body. Various systems can be arranged in a hierarchical order using Type II Whyness-Whatness Associations. Let's begin with very low level systems. Leptons and quarks combine yielding atoms. Atoms combine forming molecules. Special molecular arrangements produce cells. Cells become organized into tissue. Tissues are "molded" into organs and body systems.

Each structural level (unit) shows different functionality. Each (wholeness) function is not characteristic of the components comprising the unit-system. Rather, each (wholeness) function (of a unit) arises from the collective activity of its parts (functionality). This poses an interesting question. What sustains the necessary activity of parts producing the function for which the parts are designed? This is the next step beyond asking how content is structured into form.

Each functional level is adapted to higher systems. Each system has a participatory role in higher-level units. The physical body is itself a functional system. It exhibits life—it functions as a life form.

What explains what? At first thought, like the automobile, our body does seem to be explainable by (its) parts. Body locomotion (physical activity) is a product of the synergistic action of muscles and nerves. Surely, the subsystems are "necessary" factors in the sense that without them the higher unit, in which they functionally participate, cannot sustain its purpose.

A heart attack might contribute to an individual's death. The same reasoning applies to the automobile. The automobile will not function properly unless its subsystems operate as a cohesive system. Cessation of a part function can "cause" an end to some higher-level functionality (in which it participates). Yet, in both cases we realize, subsystems do not "sufficiently" explain the super functioning of the unit to which they are adapted.

Life is properly explainable only by higher-level systems. (The whole draws parts into higher-level functionality. How this occurs is explained in the next chapter.) A malfunctioning of a subsystem or part can terminate (the function of) life. We are not surprised of the supposition "parts explain life" because part malfunction "causes" death. But, this is different from saying parts "explain" life itself.

Each higher unit is "functionally" separated from its components. Each higher level exhibits properties not sufficiently, and therefore, causally, explained by its subsystems. Plants exhibit life. Life is not a property satisfactorily (and therefore sufficiently) explained by plant structure. This may at first be difficult to accept from examining plant structure. However, this reasoning becomes more apparent as we peel away layers of structural subsystems. Clearly, at the subatomic level, there is no difference in the elementary particles of our plant specimen and those found in inorganic material, e.g., in our automobile. Therefore, no adequate explanation is found by (analyzing and) continually reducing a life form to its substructures. Rather, like the super function of the automobile (viz, its motion that provides transportation) the property of physical life transcends the subsystems that support it. A whole and its super functionality are not explainable by components.

Activity of the whole is more than the sum of its parts. Each higher system reveals a function not found in its parts. Life is detected by motion—by activity.

Body parts function as a collective unit (of activity) from which physical life emerges. Physical life is a transcendent function of body parts and systems.

Besides (physical) life, animals have awareness. Awareness, a higher level function, is not sufficiently explained by physical life. It is also not explained by body structure although the body is physiologically adapted to carry out mental directives.

(Mental) awareness transcends physical life and the systems that compose it. Mind is defined by its functions, namely, the deductive and inductive inferential functions. (Extrinsic) perception bridges the physical world and awareness.

The brain is not the mind just as the automobile is not transportation. Minds would not exist in the physical world but for the brain. (Mental) awareness is a transcendent function of neural-brain activity. The collective activity of brain parts (and its systems) necessarily sustains, if not produces, mental activity. Mind is a transcendent function of the brain. Regardless, a transcendent function does not emerge unless the system is designed for that purpose.

Though each function necessarily depends upon its subsystems, each (functional) level is not sufficiently explained by the subsystems that compose it. Rather, the higher units, to which they are adapted, categorically and functionally explain lower-level systems. The automobile is (inductively) explained by transportation. Mind (inductively) explains the activity of the brain. Reason (why) is the proper inductive use of explanation.

We are at the point where classification analysis using Type II Whyness-Whatness Association (refer to Chapter 2) is insufficient to explain the properties of (physical) life and awareness. Let's reexamine our topics.

A (structurally) simple acorn develops into an oak tree. The simple ovum and sperm unite forming a zygote. The zygote changes into a blastula; a blastula becomes a gastrula. The gastrula continues developing and finally evolves into a structurally complex-human being.

The purpose of the acorn and the human zygote is to develop into their mature counterpart. Purpose is inherent in the very conception of the acorn and the human zygote. The acorn and the human zygote strive (in the sense of Type I Whyness-Whatness Association) toward attaining purpose. The "potential" for achieving purpose is ever present in function. Emergence of higher-level functions is inherently associated with the unfolding of purpose.

If purpose is implicit in design, then asking why "things" are structured should eventually lead us to the "reason for everything." Can we apply transcendent functionality to explain cosmic structure? If we can, then it should also suggest the reason for reality's structure. A thing's form (and structure) suggests its purpose.

Delineating cosmic structure is different from understanding cosmic processes. Are entropic and negentropic processes inherent in cosmic structure explainable by transcendent functions? (We addressed entropy and negentropy earlier in this chapter.)

As matter "breaks down," it becomes less organized. (Whole) systems automatically deteriorate into parts. When the purpose of the wholeness function (its transcendent function) ceases, the wholeness function soon breaks into lesser wholenesses (therefore, its parts). This is just the process that ensues when the functioning of the whole ceases. This is entropy.

In the midst of entropy, there is life. Life forms emerge from complex structures. Structure building is clearly a negentropic process.

Are there cosmic examples that defy entropy and therefore, fault reductive explanation? In the 1930's Dirac noticed something unusual about some dimensionless numbers associated with certain physical constants. First, the gravitational coupling constant in atomic units is equivalent to the reciprocal of the age of the universe. Second, the number of massive particles in atomic units is equivalent to the square of the age of the universe. Third, the gravitational coupling constant is equivalent to the reciprocal of the square root of the quantity of massive particles.

Dirac wondered if these extraordinary correlations were coincidental. If we suppose they are merely coincidental, we are left wondering about their improbability. The more improbable their occurrence the more difficult they are to explain by happenstance and therefore, by reductive explanation.

By linking these numbers (using differences in orders of magnitude) Dirac believed there was an unknown causal connection among them. Why are these numbers related in such a way (for example, correct temperature range for biochemical activity) which permits our presence in the universe?

To explain cosmic processes meant man's existence would first need to be explained. In 1961 R. Dicke reversed the idea and suggested, by what he called the anthropic principle, that if Dirac's numbers were correlated in some other way, it would be difficult to explain the conditions which opt for the development of life forms—particularly mankind. Dicke thought the explanation was because of man's existence. He thought the Hubble age of the universe could not be otherwise and allow development of mankind.

Dicke fills in the unknown cause to which Dirac alluded. He explains the age (or past events) of the universe by the present existence of man. This reasoning disagrees with traditional-scientific thought, which tends to explain present conditions by past ones. This seems to be an adaptive explanation for man's development and is teleological.

Other scientists have also invoked the anthropic principle. B. Carter thought our existence limits what we can learn about existence. He wonders why our world, of all possible ones (even considering Everett's "many-worlds interpretation" of quantum mechanics) just happens to be the one actualized.

Carter answers by saying our world is the one in which prior and existent conditions enabled us to develop and exist as observers. Carter also noticed if the coupling constant (that unifies elementary particle interactions) were otherwise that the strength of the strong nuclear force would not permit the development of atoms higher than hydrogen. If carbon, along with other atoms, did not evolve, we would not be here as observers. S. Hawking and others concluded galaxies could not form and there would not be large-scale isotropy, if the galactic recessional velocity were not equal to the escape velocity. They too conclude isotropy exists because of our existence. These assessments suggest existence has purpose.

J. Wheeler reexamined the history of science. He discovered that most scientific concepts explained things by reduction: by alluding to a thing's parts. For example, heat (in a given thing) is explained as molecular motion. He realized the inadequacy of reductive explanations.

Wheeler believed we are not here by accident. He realized the improbability life arose by chance. He pushes the anthropic principle into metaphysics and suggests we, as observers, are as "essential to the creation of the universe as the universe is to the creation of the observer."

How are disparate random parts assembled into the complexity needed to sustain life forms? Wheeler's formulation focuses the question of life-form development on two distinct categories: observer and that observed. This is the same as consciousness and experience.

There is no problem here. We have already studied examples where the sequential cause and effect route (Type I Whyness-Whatness Association) does not explain development of unintentional end products. We surmised the direction of temporal events is not sufficiently explained by the sequence itself. Why does the course of events develop in the way they do (instead of some other way)?

We already concluded higher-level systems (complex structures) are not sufficiently explained by lower-level systems (components). Rather, lower-level systems are explained by higher-level systems. Components have purpose by contributing to stabilization of higher level structures. This idea is implicit in Wheeler's assessment.

Questions about development of "observers," suggest life forms can be explained by yet an even higher function than cosmic structure.

It is difficult enough to explain the direction of Type I Whyness-Whatness Associations where effects follow cause. Utilizing teleological explanation, we assert the cause (apparently) follows (in time) the effects. It has been practically unexplainable how a "future cause" could "direct" the "course" of sequential events (effects).

Hypothesizing that future development "pulls" on present events is perhaps not the best way to frame teleological causation. Let's better state the question.

What directs the development of the universe in a way suitable for life? What orchestrates development of the simple acorn toward its purpose—a more complex structure—the oak tree? What directs the development of a human zygote? What controls the functions that develop it into the more complex fetal structure? What drives the fetus toward its potential purpose—a mature human being? How does a series of discoveries lead to an unintentional instrument like television? What directs (or causes) the sequence of effects toward its teleological destiny?

How is purpose instilled in the design of things? A reversed Type I Whyness-Whatness Association, where cause follows effect, is ruled out as an answer. Why? Increasing entropy does not explain how parts are directed into structural complexity.

Type II Whyness-Whatness Association, though useful to this point, cannot itself be explained by yet another Type II Whyness-Whatness Association. Instead, Type II Whyness-Whatness Association must now itself be explained. The higher-level system that explains direction (of development) must be intrinsic to Type I Whyness-Whatness Associations and must explain Type II Whyness-Whatness Associations. Type II Whyness-Whatness Associations have served us well until now.

Let's simplify. How is purpose embedded in the course of directed events? How are parts arranged so higher-level functions emerge? In the physical world change occurs by rearranging of parts. It is useless to infer beforehand that a higher-level function rearranges parts in such a way to materialize the end from the means. That is like saying "ghost" molecules arrange atoms into collective (molecular) functionality. It would be like a castle (form) molding sand granules (content) into itself (the castle). It would be "form arranging content." If this occurs, we ask how does it occur. Let's return to the mechanism of process.

Transcendent functions (represented by active "form") are not caused by parts (content) arranging (directing) themselves for a higher purpose. Purpose (in itself) is also incapable of imbuing parts with the necessary arrangement to collectively invoke the sought after transcendent function. Rather, purpose itself must be explained. The "idea" of transportation does not design the automobile. (That there is) knowledge needs explanation.

Automotive conveyance does not materialize from "parts" arranging themselves into a collective activity giving rise to the transcendent function called transportation. The transcendent function transportation must also be explained. Let's refine our question about how content is structured into form. How can a function (purpose) not yet manifest, arrange parts into a collective activity transcending its parts?

We have no trouble explaining transportation. Man designed automotive "parts" and systems of parts for the purpose of transportation. If man can erect a machine serving a transcendent function, then it would be easy to surmise the possibility that an even higher-level intelligence could erect a structure also characterized by a transcendent function—awareness. (And least we forget, man did not create the basic materials [such as iron, and the other atoms] he fabricates into useful parts.) Conjectures are not sufficient to establish that higher-level functions "pull" parts into purposive submission. Yet, the fact we had no input regarding where and when we were born and how we physiologically mature into an adult is very telling.

If there is not a super designer doing the designing, it becomes impossible to explain the numerous-functional layers of structure (atomic, molecular, for instance). Numerous coincidences might be explained by chance, but not multiple-synergistic layers of function. If this (multi-level functioning system) occurred once in billions of years, it would be astronomically improbable—but it occurs countless billions of times everyday! Plants and animals have life because of the synergism of hierarchical functions.

We hypothesize for every structural unit there is a higher-level function that explains it. (Transportation explains [is reason for] the automobile.) We also acknowledge (actual) transportation (itself) did not design the automobile. And, we have opined the idea that something (by itself) cannot create something. There is no recourse but to seek the ultimate intelligence level. Once sought, it must explain how thingness activity is (inductively) processed toward complexity and explain emergence of transcendent functions.

Sequential ordering of events is not explained by appealing to any type of design. Rather, design or change in structure (for adaptive purposes) itself must be explained by appeal to an ultimate designer. Cosmic sequence (change) or temporal flow itself must be explained by "a function beyond the flow of time." This problem suggests the need to use a Type III Whyness-Whatness Association to explain system development directed toward purpose.

We already mentioned our operational idea. Each "uncovered" higher-level structure brings us closer to the ultimate cause of everything. Atoms—molecules—cells—tissues—organs—organ systems—life forms—awareness.

Extending our knowledge beyond physical awareness is to first embrace an understanding of the structure of reality. We have done that. We ask how can the structure of reality be explained? How can the correlations of physical constants, noted by Dirac, be explained? Keeping it simple: The answer to these questions also (inductively) explains awareness.

We only have one more hill to climb along the path to greater knowledge and understanding. It is the ultimate peak. The task is to methodically uncover the highest level system—reality itself. How does the ultimate intelligence level itself function?

To expose the ultimate reason for everything is to "inductively" recognize greater structural layers until only "nothing" is left. We have done that. We only need to better understand the meaning of this nothingness. In Chapter 18, we determined the logical meaning of nothingness. We also correlated nothingness to space. In Chapter 19, we connected spatiality to an elementary particle (Higg's boson) and associated it with the unification of physics.

We ascertained the placement of every puzzle piece. We found a way to connect each (and any given) structural level to its immediate higher and lower levels. Each level is a transcendent function of the subsystems or parts that compose it. Transcendent functionality arises from the collective activity of contributing parts or subsystems. Each lower level is functionally subsumed and necessarily directed toward the purpose portrayed by the transcendent function. The automobile is designed for transportation. Each higher-level classification subsumes all lower-level classes.

We determined the nature of a unit (see Chapter 18). We discovered a unit is composed of two aspects. There is a wholeness function, and a parts-function. A unit's wholeness function (by activity) can produce an emergent transcendent function.

Consciousness needs to be explicitly defined and explained. Consciousness was not found to correlate with any part, unit system, or structure of reality. A pressing question was to what extent must we induct knowledge to flush out the reason for the design of reality itself. How is consciousness connected to the structure of reality? We now have the means to answer that question.

We correlated "experience" with the Realm of Whatness. This leaves us to suggest consciousness is correlated with the Realm of Whyness. Consciousness and experience are wrapped into the unity of reality called the Realm of Reasoning. That settles the structural correlation of consciousness and experience.

A characteristic of experience is change. Consciousness does not change. Consciousness is the observer of experience and is (categorically) not an aspect of (i.e., included in) experience. Experience, however, is included within the concept of consciousness. No consciousness—no experience.

> How do we know this? We will not backtrack to the previous chapter. Suffice to say: False (0) implies true (1). Consciousness implies experience.

"Things" are experienced. Everything is of experience. We ascertained everything is included within nothing. Is "nothing" consciousness? If so, how can that be? We already guessed consciousness includes all experience.

We now set up the means to answer two significant questions. In detail: By what actual process does consciousness generate experience? How does consciousness act upon experience?

Reality is the ultimate unit. A unit has two aspects: a wholeness function, and a parts-function. Change cannot occur outside the ultimate unit function. There is no "outside" to it. Change occurs within the ultimate unit—reality.

Given the ultimate unit, what changes are permitted? We learned there is no more than the whole of what is. This is a clue to answering the first significant question. Any part is gotten from the whole by (the process of functional) fragmentation (of the whole). We learned this from how natural numbers are generated from zero (see Chapter 18). An understanding of elementary particle physics (see Chapter 19) supports this.

Where do parts come from? What is the origin of parts? Applying the above concepts to reality, we obtain the following. There are two extreme possibilities. There is the whole of reality and there is the infinity of parts (which compose it).

The whole of reality "functionally" fractionates (itself) into an "infinity" of parts. Functionally, there is a whole; there are parts.

The whole of reality represents reality at its highest level of order—no disorder. (In the whole, there is no disorder because there are no parts.) The infinity of parts pertains to the infinite energy of the vacuum. Chaos is complete disorder (of parts). (The infinite energy of the vacuum has no order.)

The second significant question asks how consciousness "acts" on experience. Another way of asking that question is to seek how change occurs. Given the extreme possible conditions of reality (ultimate whole [zero parts], and infinite parts) we ask how each of these extremes affects the other. How does each "act" upon the other?

The ultimate intelligence is consciousness. It is the wholeness function (of reality). This is the deeper meaning of the quantum of action. From Bell's theorem and Aspect's experiment, we learned about the nonseparability of everything. Consciousness is the "observer" that "influences" what is observed, viz, experience. This problem is also stated as the "measurement problem." Precisely knowing a particle's position precludes knowing its momentum. When consciousness "projects" itself as experience, the wave function "collapses" into manifestation, i.e., experience. Experience is derived from change in the parts-function.

> Collapse of the wave function into manifestation is how quantum "frames" are produced (see Chapter 12).

Each extreme "operates" on its opposite. The whole acts on its parts. Parts act upon the whole (of which they are a part). Consciousness affects experience; experience affects consciousness.

Parts are subunits. A subunit has its own wholeness function that sustains the activity of its parts. And, the wholeness function is nothing but the quantum of action functioning at a lower level in the hierarchy. We do not experience the quantum of action because its function changes when it "acts" upon that which is separate. (The quantum of action functions nonlocally: its action is nonseparable from the ultimate wholeness function.) Recall there is no difference between any given thing's wholeness (its form) and the ultimate wholeness of reality. The difference is only one of function. We learned that setness is always a member of itself. When its (i.e., setness) membership occurs at a lower level in the intelligence hierarchy (as a subsetness) its functionality changes. It is functionality that "separates" parts from their source (the whole from which they are derived). Parts "take on" a "local" function.

Consciousness "pops" into and out of existence. We know this from the nature of the quantum. This is how consciousness "projects" experience. (Refer to Chapter 13.) Consciousness "pops" into existence by an (instantaneous) "on off" functional self-fragmentation.

How does consciousness act upon experience? Consciousness is the wholeness function; it is self-sustaining. Its nature is to "pull" parts into higher-level functionality. It pulls parts "functionally" back towards itself. It pulls parts into higher level-wholeness functions. This is negentropy.

Negentropy is realized in teleological causation. Examples are found in transcendent functions. Transcendent functions result from a potential wholeness function pulling parts into higher level (transcendent) functionality. Atoms become arranged into molecules, animal body systems give rise to (physical) life, and from collective brain activity emerges awareness.

> Molecules do not "exist" in themselves. Molecules are but atoms arranged to evoke a higher-level function—molecular (functionality). Molecules are a functional synthesis of atoms. Atoms are a functional synthesis of lepto-quarks. Lepto-quarks are functional syntheses of virtual particles. The same reasoning applies to higher-level functions. The human body (let's save some steps) is a multi-level functional synthesis of subatomic parts. This is one way of saying the body does not really exist in the common sense. The body is a multifunctional synthesis (of parts) emerging from the infinite energy of the vacuum (space). The point is to not get bogged down with questions about what is real, but to define everything in functional terms. The structure of reality is now interpreted using the concept of function. The major benefit is obvious: It enables us to easily connect everything into a working whole.

Every unit has two aspects: a wholeness function, and a parts-function. From the wholeness function can emerge a transcendent function. It arises from the collective activity of the unit function. The wholeness function (negentropically) acts upon parts to build up higher-level systems through emergent functionalities. The wholeness function tends to be self-sustaining.

The parts-function also tends to sustain itself. Its nature is to disintegrate higher-level functions. System breakdown into component parts is an example. This is entropy. Physical life forms eventually break down. (Physical) life, a transcendent function, ceases when its "parts" stop sustaining it. The extreme entropic process results in the complete chaos—disorder—characterized by the infinite energy of the vacuum.

Complex life forms arise out of the chaos of the infinite energy of the vacuum. The activity of the vacuum gives rise to micro-world events that are measurable only by probability. Yet, out of this probability-mixing arise stable concentrations of energy, e.g., elementary particles and atomic elements. We can also see this idea at play at the macro level, in which we live and function. Life-insurance rates for the collective of men and woman are not determined by any given person's age of death, but by the average-age distribution for all men and women. What seem to be random events, on an individual basis, are statistically predictable. From disorganized (individual) behavior, emerge organized (collective) behavior.

Key Points:

- Consciousness "sits" atop the structural hierarchy. It categorically encloses everything.

- Consciousness is the sufficient and necessary explanation for experience and life forms. It explains (mental) awareness.

- We are ultimately defined by consciousness: it is our higher selfness.

- Consciousness is space. Space includes existence and all time, yet it is not of any given time. Consciousness-space is the whole of reality.

- Existence, and therefore, experience, results from the functional fractionation of consciousness. When consciousness "loses" its wholeness function (because its activity shifts to parts) it becomes a "passive observer" of experience (because it is "itself" nonactive).

We have other bridges to cross before we are ready to deal with the meaning to life question. Before finishing this chapter, we ask one more question. How does memory fit into consciousness and experience?

Memory and Recollection Understood

Memory is neither consciousness nor experience. Yet, memory is unexplainable apart from both consciousness and experience. Memory is associated with the Realm of Reasoning. Here, we mean memory in the cosmological sense (not just personal memory).

There is an inherent problem in consciousness being "all-knowing." The property of all-knowing implies total memory. Consciousness is not memory. When consciousness fractionates (its) knowledge, at worst, is "lost"; at best, it is disconnected ("separated"). Fractionation proceeds from the wholeness function. (The pie is cut into pieces.) That is, consciousness loses its knowledge: becoming "unconscious" in the form of "pieces" of matter (the Realm of Whatness—particularly, the Universe of Perfect Whatness). The purpose (from the perspective of consciousness) for the structure of reality is to "store" knowledge. Let's elaborate.

Consciousness is the Realm of Whyness. It is reason itself. Reason, and hence total knowledge is "representatively" stored (as memory) in the manifestation and its parts. Knowledge of relationships (physical law, equations) is stored in the structure of the "whatness" manifestation—the Realm of Whatness.

How do we know reason is inherent in structure? Reason is reflected in purpose. Purpose is suggested by transcendent functions. Each enveloping-hierarchical structure is the reason for its components. Atomic structures serve (the purpose of) molecular structures.

Knowledge (of whatness) used to create the automobile is "represented" in its (the automobile's) structure. The reason (whyness) for the automobile is exhibited by its transcendent function—transportation. Whyness becomes whatness. Purpose becomes realized. Saying knowledge is manifested in structure is synonymous with the idea consciousness "becomes" unconscious (fractionation of the whole) in the physical world. Once the wholeness function is broken (it is no longer whole) functionality of consciousness (itself) ceases. Consciousness "becomes" a reductive function of parts. (Setness is a subset of every set.) Knowledge becomes manifested in inferential experience.

We understand how knowledge becomes "stored" in the structure of things. Knowledge used to "create" everything, becomes "embedded" in the structure of everything. Knowledge used to create the automobile is reflected in its structure.

Key Points:

- Memory is unexplainable apart from both consciousness and experience.

- When consciousness is whole—"all" knowledge is intact.

- Because consciousness creates things by self fractionation, the knowledge used in the creation process becomes "lost" (from consciousness). However, that knowledge is not totally lost. Knowledge used in the creation process is "remembered" in the structure of everything. It is manifested in everything. The knowledge of creation is "stored" in the creation. The intelligence of consciousness becomes manifested in things.

- Things are intelligent by design. Because we recognize things and structure are symbolically representable by knowledge, we also realize thingness design indicates manifest intelligence. What does this mean? It means things and their structure also represent the knowledge used in their design. This is one of the most telling clues to the meaning to life.

Memory infers more than storage of knowledge. Storage of knowledge has purpose. Knowledge is not stored to be forgotten. Knowledge is stored to be retrieved when needed. Consciousness itself cannot retrieve the knowledge because it is nonfunctional (because it is functionally broken). Consciousness functions through its parts. Using memory, man identifies things he experiences by retrieving general concepts from his memory. For example, based on similar past experiences, we recognize and say, there is a table.

Memory has a two-fold nature. There is the agent that stores and retrieves "memories." There is the medium of memory storage. For humans, there is awareness and memory. For consciousness, knowledge is stored in the structure of "things." Knowledge is retrieved through awareness, viz, the inferential functions—deduction and induction.

The agent of (storing and retrieving) knowledge and the medium of knowledge storage, form a unit. The Realm of Whyness and the Realm of Whatness form the unit called the Realm of Reasoning. The Realm of Reasoning is the unit function. We learn by physics that the Realm of Reasoning correlates with electromagnetism. Electromagnetic energy manifests as (functionally reduces to) electrical and magnetic phenomena. The functioning of human memory is associated with neural electrical brain activity. Reality's memory storage is found in the structure of physical things. Things are functionally and reductively composed of chemically bound atoms. Chemical activity is electrical.

606   Section Seven: Conclusion

The structure of reality is inherently the memory (i.e., repository) of the knowledge used in its creation. The reason for the structure of reality is (attributable to the nature of) consciousness itself. Consciousness becomes functionally broken in (the process of) projecting experience.

As scientists unravel the nature of existence, they are merely "extracting" the knowledge "embedded" in existence. What we learn about creation is nothing more than the knowledge used in its creation.

Summary

Knowledge has purpose. Manifestation is an outward expression of reason (and therefore, purpose). Reason is found in "form" or structure (of the manifestation). In a pure-abstract sense, form is "represented" by logic (see Chapter 18). To understand the ultimate nature of reality is to determine how the rules of logic can be derived from "nothing." Nothing, no-thing, no assumption (assume nothing) is the ultimate source of every-thing. No-thing is consciousness. "Things" constitute experience.

"No-thing" is the *a priori* truth (which is falseness itself). Consciousness is the source of all experience. Without consciousness, there would be no existence and no things for experience. Consciousness is a "given." Consciousness explains itself.

Consciousness and experience are easily defined by each other. Most people realize consciousness and experience are intimately connected. However, there had been little progress in understanding consciousness. Maybe we have turned the corner in understanding consciousness.

The Realm of Whyness is consciousness. It is the ultimate-intelligence. The Realm of Whatness is experience. It is "what" consciousness is "conscious of." Consciousness is the (whole) inductive reason for experience. Experience is the (reductive) explanation of consciousness. We also note that we can apply this understanding to our "logical construct" defining the Self against the "rest of reality." (Refer to Figure 8:1. It serves as an outline for the structure of reality.) "All of Reality" is Consciousness—our higher self. The individual "Self" is mind—awareness. Experience is everything except the whole (All) of reality.

We have suggested there is a possible design in the course of historic events. We explained the concept of process development using entropy and negentropy. We noted in Chapter 17 there is a historical precedence of certain religious notions presaging and correlating with many of the cosmological concepts developed in this book. Scientific concepts were not sufficiently developed to allow proper assessment by correlation to ideas such as God and Goddess.

We now understand the structure of reality to the point we can address the highest-level-religious concepts in scientific and logic terms. Does the relation between consciousness and experience correlate with the most basic religious concepts? Yes. In the final chapter, we continue from where we ended Chapter 17.

We have noncontradictorily defined consciousness by a single principle. The scientific test of this definition is found in its explanatory power. If this Ultimate Principle is the "theory of everything," how does it explain the structure of reality? How does it explain life? How does it explain the meaning to life? Does it provide a more intimate understanding of consciousness? What are its limitations? These lingering questions are addressed in the final chapter. Simply, how does consciousness explain everything?

Chapter 21

Explaining Everything

"No thing explains itself; no subject about anything
explains itself; no event explains itself."

A unified conceptual understanding of reality is an old philosophic endeavor.
Let's review some of these ideas and note how science today is a refined
continuation of seeking to understand reality.

Many early Greek philosophers entertained the idea reality in one sense is
nondivisible. Anaximander supposed the universe was originally whole and only
later "fragmented" into parts. He thought some time in the future that parts
would again become whole. Pythagoras believed the clue to how things became
separate from the whole was to be found in (the nature of) numbers.

Heraclitus believed change was all there was. He said one could not step twice
in the same river because water continually passed where one stepped.
Heraclitus thought there was no permanency of being (of whatness)—only
"becoming" (that things change in a way for their purpose to be realized).

Xenophanes believed the whole could not change, only parts (could change). He
thought all reality was God. Parmenides believed change was impossible
because something could not come from nothing. He thought all there was is
being. Everything is fundamentally of the same substance, and had the same
essence.

Aristotle believed things changed because they "strive" to take on form. An
acorn strives to become an oak tree. This would be his "explanation" why an
acorn develops into an oak tree. He thought the cause of change was an
"unmoved mover." It has form, but no content. He thought things consisted of
form and content (matter).

Plotinus believed beingness flowed from an inexhaustible source. Out of pure
form or soul emanates the formation of things.

Zeno's paradoxes often focus attention upon the divisibility of reality. To
analyze things he developed the idea of dialectic: reasoning by supposing
impossible alternatives. If at any given instant an arrow while in flight is
motionless, there can be no (flow of) time because no time passes from one
instant to another. Zeno's reasoning convinced him (continuous) motion was
impossible.

Empedocles believed in transmigration of the human soul. He believed spirit,
imprisoned in a physical body, is limited in its reasoning ability.

Anaxagoras believed things could not move themselves. He thought things
moved by forces that are nonmaterial in nature. He believed the forces were not
unlike mind. He thought things developed teleologically: everything essentially
had purpose.

Leucippus thought empty space or Non-Being is necessary for (the existence of) Being. Being (or things) existed in Non-Being (or space). Non-Being was just as important as Being although it (Non-Being) was absolute.

Augustine believed God created matter and from it made everything else. His ideas hearken back to the Greeks who thought everything was constituted of form and content. He too believed all form existed in God.

John Scotus Erignea believed God created everything out of itself or from nothing. Plato thought form existed before its materialization. Aristotle believed form and matter existed together, but were distinct. Form could exist without content. Thomas Aquinas taught that form was akin to universals (generalized concepts).

These ideas led thinkers to either support that universals have independent existence or that only the form of things exists. It became known as the conflict of Universals versus Nominalism. Nominalists believed only things exist, and there is no form apart from existing things "in arrangement." To them universals only existed as constructs in the minds of men.

Nicolas of Cusa thought reality was God divided into parts. He believed if man studied the parts and how they are put together, he could eventually come to understand God as a whole.

Later, Spinoza thought reality was only completely understood as an undivided whole. He also thought the wholeness was God. This is called pantheism. Pantheism posits reality is God. A similar concept, panentheism, states God is the whole of reality, but is not immanent in its creation.

Berkeley carried Spinoza's idea to an extreme and declared things only existed when perceived. If things were not perceived, they did not exist. Everything existing is held in the "mind" that is God.

In contrast, Descartes believed in a mind-body dualism. The mind can only have "ideas" about the universe. Locke believed the mind acquires knowledge from sense impressions. Philosophical debate had focused on how reality creates sense data. Kant said there are two worlds: one of experience—the phenomenal world, and one of reason—the noumenal world.

Fichte believed there is that which is represented by an Ultimate Principle that creates things in opposition. The Ultimate Principle, or God, struggles in the created world of things. Creator—creation. In another sense, there is no material world: it only "appears" to exist.

Leibniz believed the essential ingredient of reality was a unit of force. He called these units monads. Although there are infinite monads, most are implemented at different levels of reality. Monads also function together as a whole.

Lotze, following Kant's duality, believed there are different levels or degrees of reality. Royce believed we organize our experiences into categorical systems to understand. He thought we are parts of a greater whole, which is self-conscious and self-organizing. The whole is a Being that is the source of man's thinking and ideas. Averrois had stated there was just one soul, Beingness or intellect. That every individual being shares in this one Beingness.

Schelling believed God was nature. Man's reasoning was the absolute, or God, developing itself through man. Through nature, consciousness comes to understand its own reason for being.

Hegel also saw the world to be in constant conflict with itself. The world develops according to the "dialectic process." The Principle of Contradiction represents this conflict. Yet, opposites attempt to "reconcile" or synthesize and lead to new (or higher level) forms. The Law of Noncontradiction is the "operative" principle of synthesis. Pythagoreans originally developed the dialectic idea. Eventually, and ideally, understanding of the whole would be regained. The whole brings its opposite, parts, back to itself.

Can an all-embracing concept of reality incorporate some of the above ideas? Do conflicts (between different views) develop because philosophers did not understand certain ideas only appear contradictory because they occur at different levels in a hierarchy—the different degrees of reality envision by Lotze? For example, the conflict between universals and nominalism perhaps arises from "miscategorizing" form and content.

Many philosophical ideas have a corollary in science. This has been established using the structure of reality described in this book. We have also developed the means (in the previous chapter) to elucidate some remaining philosophical and scientific questions. Man seeks goals. Goals offer a reason for doing something (usually in the physical world). Our special interest now narrows to the purpose of human existence. We particularly wonder why mankind has the ability to seek truth and explain things.

Means of Deeper Explanation

Our foremost guiding question in seeking truth has been "how can any-given thing be explained?" And ultimately, how can everything be explained? The previous chapter answered how everything can be explained in a categorical sense. Everything is "categorically" explained by the Ultimate Principle. It defines consciousness. Here, we refine our understanding of consciousness and how it explains experience.

That which explains everything must also explain life forms because living beings are a part of everything. The Ultimate Principle should specifically explain how life came into being. It should not only explain how life evolved and diverged, but also explain life-form complexity. It also should explain cosmic evolution. Answering these questions should enable us to determine if life has purpose.

In this chapter, we use the following means of explanation: explanation by subject, explanation by reductive and transcendent function, explanation by hierarchical structure, and explanation by correlation.

A conceptual hierarchy represents the structure of reality. The following ideas were "superimposed" upon reality's structure. By reductive differentiation from a commonality, are derived bipolar dichotomies. For example, electromagnetism reduces and differentiates into electricity (which further reduces to positive and negative charges) and magnetism (reducing again to north and south polarities).

Parts of the structure of reality were pieced together by inductively building higher-level concepts (generalities). Approaching from the lower level, we "synthesize" ("branched") components into their commonality (the bigger branch from which they bifurcate) and so on.

Transcendent functions are higher in the hierarchy than their reductive counterparts. We "strive" to better understand the main trunk or source of every branch (everything). How does the trunk support and sustain its branches (representing the structure of reality)? (Transcendent functions were described in the previous chapter.)

One of our remaining goals is to understand life processes. Along the way, we should also understand life-form evolvement and development. How the trunk (the whole of reality) interacts with its branches (the parts) should explain how life arose and developed. We begin our next quest by examining the contemporary-biological understanding of life-form evolution.

Life-form evolvement can be explained in different ways. "Subject explanation" is elaborated in this chapter. We can approach the subject of evolution from two directions. Evolution can be analyzed (by reduction) or it can be generalized ("synthesized" using inductive means).

Analysis "reduces" the role of explanation to parts. This has been the historical-scientific means of evolutionary explanation. A reductive explanation contends life-form parts explain life.

Synthesis "induces" explanation using higher-level generalities and "enveloping" wholes (transcendent functions). Here, activity of parts is explained by appeal to the organism as a whole: that it exhibits life. Therefore, inductively explaining evolution, is essentially the same as explaining life: by subject—biology.

By delineating the structural nature of reality, we set the foundation to answer the question of the ultimate nature of everything. We conceptually categorized "everything" using an "all-inclusive" classificatory system called the why-what paradigm. This paradigm is constructed on the most basic questions that can be asked. The five fundamental questions are: "Where," "How," "When," "What" and "Why."

The structure (of reality) "explains" how everything is hierarchically linked. Hierarchical detail was established by correlating whyness and whatness relationships using the concepts of inter- and intra- phenomenal relevance.

Relevance was expressed by erecting structure based on whyness and whatness associations (Type I, II, and III: Refer to Chapter 2). These ideas were elaborated using explanatory modes (extrinsic, hierarchical, and intrinsic). Extrinsic and intrinsic modes were incorporated into the hierarchical mode. We now know extrinsic modes emphasize whatness "parts"; intrinsic modes emphasize whyness "wholes."

To explain (Whatness) things, is to answer the ultimate "Why" question. How the "whatness" of things connects to their "whyness," explains things at lower levels in the hierarchy. That is the tactic used in this book.

Key Points:

- Deeper questions of reality are addressed by asking "why" questions.

- The structure of reality is depicted by the why-what paradigm.

- The structure of reality defines how everything relates hierarchically.

- Noncontradictorily answering structural questions eventually leads to the Ultimate Principle.

- The Ultimate Principle categorically explains everything.

- The structure of reality does not answer all questions.

- Connecting categories by function and subject aids properly formulating previously unanswered questions.

Our main investigatory tool has been scientific methodology. It was modified and refined for inductive reasoning. We sought to discover noncontradictory explanations for major concepts of physics. Physics principles guided "piecing" parts of the "big jigsaw picture" that describes reality's structure.

Human behavior was also described and explained by classificatory means (i.e., egocentric and realistic behaviors). These behaviors were tied to the mental functions of deduction and induction and left-right brain asymmetry. (Refer to Chapter 2 and Chapter 8.) These functions correlate with the cosmic processes of entropy and negentropy. (Refer to Chapter 20.)

Reality, as a whole, provided the context to formulate questions that the nuances of structure (itself) could not explain. Most of these questions could not be properly addressed until the Ultimate Principle was uncovered. Emergence of biological life forms needed to be explained. Leftover questions about some religious concepts did not correlate with anything in reality's structure. The concepts of God and Goddess needed noncontradictory definition and explanation. After answering these remaining issues, we finally tackle the question concerning the meaning to life.

Explaining Biological Life Forms

Although this book has addressed and answered most "big" questions of science, explaining biological life forms was not one of them. Specifically: How did life forms arise? This question is not answered by merely studying biology.

Citing "causative" mechanisms attributed to evolutionary theory, is a reasonable beginning to understand "how" life evolved. However, evolutionary theory does not explain how life first emerged.

We briefly describe the history of evolutionary theory before applying the Ultimate Principle to explain how life forms emerged and developed into complex organisms.

The proper study of "what" life forms are, is biology. Biology describes life-form morphology and physiological processes. It also describes life-form maturation from embryo to adulthood.

Biology does not adequately (i.e., inductively) explain (why there is an evolution of) life forms. Neither does it explain how life developed as a complexity of hierarchically mixed functions. No thing explains itself; no subject about anything explains itself; no event explains itself.

Let's describe evolutionary theory, and then develop the means to answer our remaining questions. We are working to answer the following questions: How can life be explained? Does life have purpose?

Explaining life has many meanings. How did life arise? How did life evolve? How did life-form complexity develop?

Evolution of Life Forms

Evolutionary scientists suggest a simple aquatic life form arose perhaps as much as 3.8 billion years ago. The first life form is thought to be a protokaryotic (prenuclear) cell. Because little oxygen was present in the early atmosphere, the first organisms were anaerobic. Photosynthesis played an important role.

Life forms resembling today's bacteria and multi-cellular blue-green alga appeared. The original blue-green alga is similar to pond scum. They lacked a nucleus; their genetic material floated in the cytoplasm. Some early organisms are not easily classified as plant or animal.

After two billion years ago, oxygen accumulated. Some early organisms produced oxygen as a metabolic byproduct. Some oxygen is traced to chemical reactions involving volcanically-spewed material. About 1.7 billion years ago multi-cellular organisms developed and lived on the sea floor.

Cells (eukaryotic) with nucleus, may have developed as early as 1.5 billion years ago. Examples are: protists-like amoebas, diatoms, fungi (similar to yeast and mushrooms) and some multi-cellular types such as kelp. The percentage of atmospheric oxygen increased markedly.

Organized sets of chromosomes emerged. Consolidating genetic material in a localized organelle allowed reproduction by meiosis (germ-cell division [halving of chromosome number]). It enabled nonphotosynthetic-energy generating life forms (oxygen consumers) to expand. The eukaryotic cell provided a means of species diversification. The fossil record supports this idea.

Paleontologists study the history of life forms by examining the fossil record. Stratified layers of rock correspond to different times in earth's history. Deeper geological layers represent earlier times. The Grand Canyon in North America offers an excellent example. Rock at the top of the Grand Canyon was formed about 2.5 million years ago; the bottom was formed over 1.2 billion years ago.

The oldest invertebrate life forms are found in deep Precambrian strata. Examples are annelids (segmented worms) and hydrozoas. Fossils of a strange arthropod called trilobites were common. They had calcified exoskeletons. Insects and arachnid-like scorpions appeared. They existed during the Paleozoic era (beginning about 570 million to 245 million years ago). Marine invertebrate life forms (e.g., jellyfish-like organisms, crustaceans, and starfishes) developed during this time.

During the Cambrian period (570 million to 500 million years ago) a life-form "explosion" took place. Many life forms evolved. Primitive fish emerged. They have endoskeletons and are noticeable for having developed a jaw.

Many paleontologists suggest that during the Devonian period (around 390 million years) there were many fish varieties. Amphibians developed. Land plants began to appear. The first land animals emerged. Centipedes and millipedes are early examples.

During the Permian period (285 million to 245 million years ago) many species became extinct. Reptiles emerged by 225 million years ago. Fish-like reptiles like crocodiles and snakes developed. Flying reptiles also developed. Some reptiles evolved into dinosaurs. Dinosaurs are now thought by some scientists to have been warm-blooded.

One hundred and seventy-five million years ago great reptiles and dinosaurs became extinct. Many scientists now think birds developed from reptiles. By 65 million years ago most reptile lineages were extinct except birds.

Primates emerged 70 million years ago. Higher primates emerged after another 35 million years. About 14 million years ago the upright ape (hominid) appeared. Primates developed the thumb about three million years ago. The thumb enabled easy grasping of implements. About 250 thousand years ago Homo Sapiens, similar to modern man, developed.

Early vertebrate animals died, sometimes falling into mud-like soil or into ponds. Expired aquatic creatures settled to the sea floor. Life forms sometimes left an impression in mud, which, because of chemical reaction and pressure of accumulating layers, lithofies (hardens into rock). This process preserves a fossil image of the life forms present during the period when the then top layer was "laid down." Sandstone fossils are common rock strata.

Scientists can determine the time each stratum was laid down. One method of geological dating uses radioactive isotopes. Radioactivity is measured in half-lives. It is the time in which half the quantity of a given atomic isotope undergoes nuclear decay.

Minerals found in rock can "date" fossils associated with a given rock layer. For example, the fact potassium decays into argon can be used to set up a time scale. The potassium-argon ratio is used to date rocks. Half of potassium decays into argon about every fifty-five hundred years. Determining the quantity of potassium to argon, "dates" when the rock layer was laid down.

Radiodating is used to determine when "carbon" based life forms existed. Photosynthetic activity leaves a "signature" on carbon-14 isotopes (which could find their way into carbon dioxide). Radioactive decay indicates how long ago the organism lived. Radioactivity of fossilized carbon (based life forms) is compared to that in living organisms.

Few well-educated people now doubt that life forms evolve, i.e., change. How species arose (speciation) has been controversial. Let's briefly review evolutionary theory to see how it has "evolved."

Life forms are classifiable into groups. Plants and animals are quite distinct. They are structurally and physiologically dissimilar. For instance, by respiration animals "consume" oxygen and exhale carbon dioxide; they characteristically exhibit locomotion. Plants, however, photosynthetically consume carbon dioxide and discharge oxygen; they typically do not exhibit locomotion. Animals and plants are functional complements.

Life forms can be classified in a hierarchy according to similarities and differences. Life forms are "divided" into two kingdoms (phyla): plants and animals. (Refer to Chapter 2.) That life form (variation) can be categorized in a hierarchy, is very suggestive life evolved from a simple one-celled organism into higher (complex) forms.

How can life-form variety be explained? In the extreme, either various life forms arose independently of each other, or they have a common ancestor.

If each species arose independently of each other, it becomes exceedingly difficult to explain their similarities. How could similar yet nonancestrally related life forms have so much in common?

If each closely related species were originally "born" into existence just as they are today, how did some of them develop practically the same characteristics, except those distinguishing their differences?

It would be astronomically improbable that closely related species could "coincidentally" exhibit such extraordinary similarities if every species originally emerged as separate organisms (i.e., that they were originally no different from what they are today). If this were the case, the only reasonable supposition is that there is a higher-level intelligence which "created" every species *in situ*.

Many religious people have insisted that this is just what happened (that every species today is no different from their original progenitor). Yes, this is an explanation. The problem with this idea is that this "God" must be both "all-knowing" and "all-powerful" (with respect to what it creates). If there is a God, and it is all-powerful, then other things beg for an answer. If God is effectively that powerful, then why does "he" permit man and beast to needlessly struggle for survival? Let's seek explanations that are more reasonable.

Similarities of closely classified life forms are quite reasonably explained by having "descended" from a common ancestor. If every species had a common (single) ancestor, the task becomes one of explaining their differences. How can the differences between species be explained? How did the original life form evolve into the variety of species observed today? How did the primordial organism diverge and differentiate into present life forms? In short, how did species evolve?

In 1858 C. Darwin, a British naturalist, after twenty years of research, presented his theory of "natural selection" to the Linnaean Society of London. A. Wallace, who had corresponded with Darwin, read a similar theory at the same presentation. A year later, Darwin published his *On the Origin of Species by Means of Natural Selection.*

Many "naturalists" believed, prior to Darwin's book, every animal species, when created (by God?) was (in kind) no different from present animals. (Darwin wondered if they believed a plant was originally created as a seed, or a mature structure.) Darwin's work changed the future direction of not only biological interests, but also focused attention on how evolutionary changes could be explained.

Darwin had developed an early interest in geology and natural history. He was recommended for position of "naturalist" on a scientific expedition. In 1831, Darwin sailed on a five-year voyage on the ship *H. M. S. Beagle.* He was to study the geology and biology of the pacific coast of South America and a series of Pacific islands.

While on the islands, Darwin was quite cognizant of geological changes. Environmental factors such as volcanic activity and weather erosion affected landforms. Coral reefs were especially subject to "growth." He published a book about coral in 1842.

Darwin wondered if there was anything not subject to change. He supposed everything, including life forms, changed over time.

Darwin studied life forms on the Galapagos Archipelago. It is located about five hundred miles off the South American West Coast. Darwin noticed, though the islands had the same physical and temperate conditions, many species were indigenous to a specific island. Darwin wondered how this could be explained.

Darwin observed on each island that species of birds varied in certain characteristics. The beaks of ground finches varied noticeably in size and shape. Some beaks were stout and powerful, on other islands they were small or slender. Birds with stout beaks ate large seeds; those with short beaks ate smaller seeds; and those with slender beaks fed on insects.

In 1838, Darwin read a 1798 book by T. Malthus, an English clergyman and political economist. It is entitled *An Essay on the Principle of Population, as it Affects the Future Improvement of Society.* Malthus argued population growth would hinder human progress unless checked by disease, famine, war, or natural disaster. Malthus also had influenced (Wallace.) The evolutionary philosopher H. Spencer used the phrase "Survival of the Fittest" to summarize Malthus' ideas. Malthus' book provided the clue Darwin needed to explain why some birds were indigenous to specific islands and not others.

Darwin supposed some birds were more morphologically suited to the most prevalent bird food on a given island. Darwin thought environmental conditions had "acted" upon the birds. Birds with the "appropriate" characteristics "adapted" and prospered; others tended toward extinction. Birds on an island with large seeds evolved powerful beaks.

Darwin believed each species was "naturally selected" by its environment. He surmised species lineage slowly (and incrementally) incurred morphological changes over long periods. Birds best able to survive were the ones most likely to proliferate. And their characteristics would more likely be passed on to progeny. Changes that better enable a species to survive are passed on to future generations.

The environment "naturally selected" which species would endure. Species with "inherent" characteristics better suited for individual survival had the "adaptive" means for survival of its kind.

> Darwin said natural selection was probably not the sole cause of life-form evolvement. Many critics overlook Darwin's statement that other possible factors could be involved.

Evolutionary theory focuses on inheritable characteristics that enable the species to naturally adapt; hence, "survival of the fittest."

In Darwin's time, there was the interesting case of the "Peppered" moth. It existed around Manchester England. Before the 1850's the species was predominantly lightly colored (speckled—like pepper). However, there was also a small population (2%) of darkly colored moths of the same species. (The color difference is attributed to a change of one gene sequence.) The moths colonize on tree bark.

The onslaught of the Industrial Revolution resulted in high emission of soot from factories. Tree bark, which formerly had been lightly colored because of lichen growth, loss its lichen coat (due to air pollution). The lightly colored ("peppered") moth was no longer "protected" from predators by lichen camouflage.

"Melanic" colored moths, which were once easily seen (against lichen-covered bark) now blended with the tree's bark. Lightly colored (peppered) moths contrasted with bare bark: becoming an easy target for predators.

The melanic moth became the predominant species in just fifty years. Though Darwin was unaware of this phenomenon, it is often cited as an example of natural selection. Birds feeding on the insects are credited with the "selecting" effect.

Darwin collected and "catalogued" many species. Birds could be "classified" (according to their similarities and differences) by arrangement in a hierarchy. Species are depicted on a "family tree." Darwin thought the retracing of radiative lines (diverging branches) correlated with the idea that "his" finches had evolved from a common ancestor.

In 1871 Darwin published *The Descent of Man*. Darwin theorized all species had "descended" (evolved) from a common ancestor. Man was no exception. Man shared many characteristics with the ape. It seemed reasonable to suppose man and ape evolved from a common ancestral primate.

> Darwin thought perhaps as many as four or five progenitors each for plants and animals could account for all species. Yet, he supposed these had their origin in a single progenitor.

Darwin added "sexual selection" as a companion of natural selection. He observed animals, particularly males, competing for mates. The direction of evolution was affected by sexual competition.

Darwin recognized the connection between inheritable characteristics and environmental selection (acting upon those characteristics). These factors necessarily affect species survival yet do not impute any "direction" to species evolution. Evolution is attributed to "chance changes" in inheritable characteristics which just "happen" to best adapt to the natural habitat.

Biological evolutionists scientifically "piece" together a likely course of species evolution. Yet, evolutionary theory posits the course of biological evolution has no "set" direction. The direction of species evolution hinges upon random species variations "favored" by the environment. Some species adapt and evolve; others become extinct.

Amphibians developed from sea creatures. Reptiles developed from amphibians. Birds developed from reptiles. Warm-blooded animals emerged. Apes developed. And man is thought to have developed from an ape-like ancestor. Indisputably, ape and man's morphological, physiological, and biochemical (DNA) similarities, suggest a commonality of origin. Man is similar to the chimpanzee and gorilla.

Evolutionary Theory: the Problem

Inheritance and environmental conditions unpredictably affect species evolution. Evolutionists primarily assert there is no finalism to evolution. Consequently, there is no purpose to life-form development.

Yet, we observe there is a design to everything, including a species' characteristic. Beneficial characteristics are selected by environmental factors (that affect survival). Does the design (of species and their evolution) arise only from random events? Or does the design of things indicate purpose? Is there a purpose to life-form evolution? To help settle this issue, let's focus attention on a single primordial organism.

Hypothetically, the lineage of any generation is genetically traceable to one ancestral life form. Darwin did not understand how inheritance was passed from one generation to another. Let's hypothetically trace all present organisms back to the first organism, and ask a question. How are species characteristics inherited?

There are deeper questions concerning the first life form. Given there was an original life form, poses some interesting problems. For example, that the primordial organism did not exclusively generate invariant copies of itself in its lineage is profound. There just might be an explanation that there are many life forms. Life-form variety supports the idea life has purpose.

Mindful of species variation, let's begin this line of reasoning with Darwin's idea. If from a single organism evolved the myriad of species, what accounts for their variation? A life form's characteristics are acted upon by "natural selection." Beneficial characteristics enable the organism to "adapt" to its surroundings. To what is attributed an organism's characteristics?

Darwin was unaware of G. Mendel's 1866 published work on plant-hybridization experiments. A side note: Mendel had sent a copy to Darwin; he failed to open the package.

Mendel, an Austrian monk and botanist, discovered in "population" studies of the garden-pea plant that notable traits are inherited in a certain proportion of progeny. Some pea plants are short (measuring about a foot in height); others are tall (six feet).

There was a long held notion that progeny exhibit traits that are a "blend" of their parents. Such ideas offered little incentive to conduct hereditary experiments.

Undaunted, Mendel, beginning with purebred stock, learned by cross-pollinating a tall plant with a short one that its seed did not produce an "intermediate" (or blended) height plant. Rather, the offspring were either tall or short: similar to one of the parents.

The first generation hybrids were all tall plants. Mendel determined when these hybrids were pollinated that their (the second generation) offspring were produced in the ratio of three tall plants to one short plant.

Mendel produced a third generation of pea progeny. Tall-plant parents also produced short-pea plants. The trait of tallness was "dominant"; shortness was "recessive."

A plant's external appearance does not necessarily indicate (because of recessiveness) the inheritable characteristics it might pass on to progeny. (This is stated by saying the genotype [hereditary determinants] is not necessarily reflected in the phenotype [an organism's outward appearance].) This was very significant. It later piqued the interests of other biologists. (It was not until the twentieth century that other scientists recognized the significance of Mendel's discoveries.)

Mendel is credited with uncovering the principles of inheritance (as embodied in the concepts of recombination and segregation [of hereditary units]). However, Mendel did not understand the mechanisms of heredity. The seed of genetic study begins with Mendel's contribution.

Modern biologists assert individual, and species variation, i.e., inheritable characteristics, are explained by genetics. The gene is the basic-molecular unit responsible for transmitting hereditary characteristics onto succeeding generations.

Genes are linear sequences of "left handed" amino acids (proteins) constituting DNA (deoxyribonucleic molecules). DNA molecules are arranged as chromosomes. Chromosomes "encode" species design.

Chromosome structure constitutes the DNA "blueprint" for each individual. It is said DNA represents all the knowledge needed to construct a life form. Man has forty-six chromosomes. The fruit fly (Drosophila melanogaster) has six chromosomes. Potatoes have forty-eight chromosomes. The complete chromosome set is called the genome.

A set of three amino acids is called a codon. A codon (or single gene) is composed of three of the following "nucleotide" bases—the amino acids: adenine, cytosine, guanine, and thymine.

The sequence of codons "blueprints" protein synthesis. DNA complementarily "transcribes" itself onto a "messenger" RNA (ribonucleic acid) molecule. (In RNA the amino acid uracil substitutes for DNA's thymine.) RNA translates the genetic message via ribosomes into producing enzymatic, regulatory and structural protein. Protein properties depend upon the sequence of its RNA amino acids.

There are only twenty amino acids found in physical life forms. All physical life forms manufacture specific proteins using the same genetic code.

Morphological differences between species are attributed to variation in the nucleotide sequence and the number of chromosomes. Every "living" cell contains the individual's DNA.

How do genes contribute to species adaptation? Genetic mutation occurs when gene segments break off and "recombine" in a different order. These "changes" occur spontaneously or result from irradiation, high temperature, and chemical influences.

"Natural selection" was reestablished (as a factor in understanding population characteristics) when biologists discovered that "genetic mutation" and "genetic drift" (chance mutation) affected the "gene pool."

Genetically induced variation in the adult becomes subject to environmental selection. If the change favors survival, the species is better fit to reproduce its kind. (Most mutations hinder survival. Phenotypic mutation rarely happens unless "larger" sections of genes become transposed.)

Does "survival of the fittest" really explain anything? It is easy to note the circularity in this statement. That "the strongest survive," is practically a given. To couch that statement in other terms does little to aid understanding. For example, to say natural selection acts upon inheritable characteristics, disguises the fact this "explanation" does not "expand" the initial premise (that only the strong "are strong enough to" survive). We shortly address this problem.

How does an individual's genotype enable the species to better survive? The populace is composed of individuals. We examine this issue by how one individual survives. If he survives, it is credited to his genes. If he does not survive, the genes are blamed. Is this a reasonable tactic to better understand life?

Are genes the cause of survival? If an animal is attacked and killed for prey, are its genes responsible for its death? If an animal kills and eats a smaller animal, do its genes get the credit? Are genes responsible for survival? If so, in what way? We could easily become bogged down with such questions.

> At best: Answers to these types of questions offer little in understanding life-processes, let alone explaining life itself. That every physical being dies clearly proves the absurdity in blaming one's genes for mortality. If genes cannot prevent death, there is little chance they can explain life. The issue of life and death is greater than the study of genetics can handle.

Let's take a different approach. Let's narrow our search to understand life forms by citing a problem with evolutionary theory.

That there are life forms means they have survived! We again ask our question. How do atoms (basis of molecules composing DNA) explain survival (viz, that life continues to exist)?

Let's not dwell on the issue of survival. Let's simplify and refocus our question. Survival is just a myopic way of saying life exists. The notion of survival is already contained within the concept of life. Let's broaden the question. Are genes the cause of life forms?

Where do genes come from? DNA is composed of macromolecules, which are constructed of molecules, which are made of atoms. Our question becomes "how do atoms explain life?"

The nature of life is the subject of biology. Do atoms explain biology? Only in one sense do atoms explain life, and that is through "reductive" explanation. The immediate problem narrows to considering the adequacy of reductive explanations.

We ask, do these atoms gather (organize themselves) around a table (of elements) and say, "Let's (the atoms) create some higher-level structures (molecules) as the bases of life-form inheritance!" Do they look ahead and discuss how they intend to overcome the "protein-folding problem" that vexes molecular biochemical understanding? No reasonable person supposes atoms have such cognitive ability. Neither atoms nor genes "cause" life. This problem brings us back to the essential question. How did complex (structured) life forms arise?

It is one thing to explain microevolution: the evolution of genetic material resulting in species variation. It is a greater stretch to apply natural selection to explain macroevolution—speciation: that there are different species.

Evolutionary theory reasonably explains species variation. Macroevolution concerns the diverging of species from a common ancestor. Does microevolution explain macroevolution? Not unless micro events (e.g., genetic alteration) are "means" specifically imbued with purpose. It is reasonable to suppose species are somehow directed toward species diversification. The problem is explaining how this happens.

It is difficult to explain species divergence. It is even more difficult to account for the original progenitor of all physical life. Or more simply: How did life happen? Let's include our question of evolution (and life-form diversity) within the context of explaining life-form emergence. If we can account for the emergence of life, maybe the same explanation will account for macroevolution.

Key Points:

- If species originated independently, it is exceedingly difficult to explain their similarities.

- Life-form variation suggests species have evolved from a common ancestor.

- If species have a common origin, their differences are in need of explanation.

- Darwinism does not explain why there is life.

- Darwinism is a reductionistic theory of life-form evolution.

- Reductionistic theories are inadequate for explaining purpose.

- Life will be explainable if life has purpose. Conversely, if life is explainable; then, it has purpose.

Explaining Life Form Emergence

Minimally, what has to happen for the first life form to come into existence? What are the necessary requirements for life to emerge? We begin by briefly describing the ludicrous notion of "stretching" the scope of evolutionary theory to explain the original life form.

Natural selection suggests the minimum requirement for life to arise is the following. Atoms and molecules just happened (by natural selection—trial and error) to arrange themselves into the proper macro-molecular structure that produced the first single-celled organism.

We are well aware "survival of the fittest" is a poor guess of explaining the course of speciation. Unlike (i.e., different) species, do not interbreed. This fact is but one problem with explaining speciation by appeal to molecular genetics.

We also ask whether the evolution of life forms, resulting in humankind, is no more than an accident of nature? If the answer is yes, then man has no purpose and no significant meaning. If man does have a higher purpose, then he "was put here for a reason." If man has purpose, it will not be explained by "survival of the fittest."

Is it possible bits of physical matter just happened to come together in such a way enabling the emergence of life? Yes, it is possible, but so extremely improbable that almost any other event would be more probable. The shortcomings of this scenario compel us to look elsewhere for an explanation of life.

Even if "self-organization" (of parts) begetting a living being were "randomly" possible, it would be an extraordinary leap (by any measure). Yet, the emergence of higher-level functions is not unique to physical life. Lepto-quarks combine forming atoms; atoms combine into molecules.

The consequences of the first physical life form are too extraordinary to say it all originated from some improbably freak occurrence (of conjugating elements). There must be a more reasonable explanation. If the first life form did not emerge by chance; then, its existence occurred by design. The appropriate question is "how are parts organized (designed) into systems?"

> It is not enough atoms "just happen to arrange themselves" into the proper molecular configuration of "operational" DNA. Even if the proper random events could occur, it would not be sufficient to explain why life arose. Surely, and to the extreme hypothetical, the same atoms could just as well happen to come together in the same improbable arrangement, and not lead to life. Even if it were "mechanically" possible to arrange atoms and molecules into the necessary DNA configuration, it would not guarantee life-form emergence.

> Saying "life arose by chance is possible, but in principle unexplainable by design," evades asking whether there is a design to everything. There is even a design (order) to natural selection. That granted, the extraordinary order of things suggests there was probably a design or plan for physical life to emerge. Our task narrows to explaining how design is incorporated into nature.

Let's face the basic question. Does everything happen as a matter of chance, or can an eventuality occur by design? If things only occur by chance, then there is no chance of explaining any recognizable eventuality.

Man himself designs many things for a purpose. Man's (self-made) implements are explainable by design. Therefore, not everything occurs by random processes. That man can even muster an explanation for things, correctly or not, suggests there is a design to things. We mentioned in the previous chapter that event seriality is an example of design that is only explained by appeal to an intelligence beyond time flow.

It is reasonable to suppose there is a higher-level intelligence responsible for life itself. If life results by "design"; then, we can also suppose there is purpose to life. This possibility will be addressed and answered.

A higher intelligence should be scientifically determinable. Before making such a supposition palpable, this "higher-level Being" must be explicitly (and, i.e., completely) defined. Test the definition for inconsistencies. Once that is done, it is just as important to explain how this intelligence arranges "parts" into higher-level structures (wholes) that are inherently capable of complex functionality. The mechanism used by the higher intelligence to implement "design" should be explainable.

> If there is no such Being which explains everything; then, (we can safely surmise) no reasonable explanation of anything is possible.

However, if we can define an ultimate intelligence in noncontradictory terms and describe how it explains everything, there just might be such a Being.

The catch to this line of reasoning has always been to establish that there is a super-intelligent Being. Appealing to a superior Being and then asserting it "explains" things, is a historically recurring error. Yet, we have already established through science and proof of logic there can be such a Being (viz, consciousness). (We defined consciousness in a noncontradictory manner.) Our task here is to explain how the ultimate explanation for everything manifests (explains) its designs.

Consider circumstances surrounding the first life form. First, how did its "body" parts become "organized" into a living organism? Did (the organism's) "life" emerge after the necessary-body parts were assembled into the proper configuration? If so, how did the organism's parts succumb to the necessary arrangement begetting life? Second, the alternative is to suppose (the idea of) "life" somehow "preceded" its debut in a physical organism. But how could it? The first scenario seems impossible. The second possibility needs a scientific explanation.

We grant if there is a reasonable explanation for life-form emergence, then life probably has a purpose. That possibility poses the next question. If life arose for a purpose, its purpose "presaged" the emergence of life itself. How can that happen?

If we can explain the purpose of life, then it was meant for life forms to evolve in a way for their purpose to be realized. With this in mind, let's "backtrack" and reformulate our question. How does life (or at least its purpose) manifest in physical structures? Purpose is often realized by emergence of transcendent functions.

Assuming life has purpose, we ask whether life can happen at all if there is not some higher level transcendent force imbuing matter with not only the proper "design," but render it capable of producing a manifest-transcendent function—the extraordinary characteristic of life. We suppose transcendent functions are derived from transcendent functions.

Our first tactic is to understand life-form functionality. Then maybe we can apply that understanding to explain how life first emerged. Understanding life as a transcendent function should enable us to more critically assess circumstances surrounding life-form emergence.

The life process itself must be explained. How? We acknowledge life forms are constituted of molecules, and they of atoms. Seeking an answer in this direction is unsatisfactory. Granted, atoms are necessary for the formation of molecules. And molecules are necessary for the formation of macromolecules such as compose genetic material. The point is not to confuse a thing's role (as a building block) with the purpose of the object of which it is a part. The role of genes should not be confused with the purpose of life.

By reductive analysis, we only uncover "necessary" factors: ones without which the phenomenon (here—life forms) cannot develop (in the manner they do). Saying random events (e.g., genetic changes causing phenotype variation) explain life is no different in kind than pointing in the extreme and saying virtual particles of the vacuum can emerge as manifest particles. Stating "something" (a whatness) happens or an event occurs is different from explaining (whyness) that it occurs. That life emerged does not explain why it emerged. Purpose should be considered when trying to understand how life emerged.

Does a necessary explanation sufficiently explain something? Obviously not. Bricks (themselves) do not explain that a house is built from bricks. Yet, bricks are necessary, or there could not be a brick house. A timepiece, like a watch, does not build itself. The genome (complete set of DNA chromosomes) does not (by itself) build a life form. A "blueprint" (a code representing something—alone) does not make the thing it represents. We need a better explanation to account for life forms.

How do means necessary for production yield results? What is the difference between a necessary explanation and a sufficient explanation? Specifically, what is the relation between necessary and sufficient conditions for something to occur? Let's keep this simple: apropos of explaining life.

A necessary explanation cannot satisfactorily explain how life first arose. There was no "random shuffling" of atoms that just happened to assemble themselves into the proper molecular configuration to form the DNA basis for life. Life is not an accident!

> One cannot select a gene and determine from what animal it was taken as a sample. We have noted the DNA of any one life form is constituted of the very same amino acids as the DNA of any other life forms. The nucleotide basis for physical life is the same for every species. This is no coincidence.

> If we cannot discover a purpose for life, we will concede life might have arisen by chance. If life has no purpose, there is no explanation for life. If life has no purpose, its emergence will ultimately be unexplainable. Then again, if life is explainable; then, it probably has purpose.

It is incomprehensible atoms just happen (by chance) to arrange themselves into the proper (molecular) order to produce a living one-celled organism. The problem worsens. The next question concerns how cells "coincidentally" arrange themselves into tissues, which later "wondrously" assemble themselves into organs, which "amazingly" arrange themselves into organ systems, which just "miraculously" construct themselves into complex life forms such as animals.

Each step to a higher-level function becomes even more improbable (to explain). Complex life forms become impossible to explain by "numerous" miracles of obtaining "just the right mix" of ever-higher-level parts and systems coming into the necessary-synergistic conjunction. All these parts could be thrown onto the ground and mixed, and life will not emerge.

> Even if all parts were an automatic given, the problem of parts arrangement persists. Throw all the unassembled parts of the automobile onto the pavement. What is the probability all those parts could arrange themselves into the proper configuration of the modern automobile and then "start" itself to produce transportation? Stating that "given sufficient time, a functional automobile can be built by self-assembly" (without predetermined design) is to beg for a miracle explanation.

The most difficult problem supersedes "parts just happening to combine into sufficiently stable configurations" capable of sustaining complex structures. The problem (besides explaining the proper configuration of parts) is to explain how each higher-level arrangement leads to a transcendent function (not found in the parts, viz, their arrangement). Transcendent functions are unexplainable by random processes.

Let's narrow our focus. How are different "arrangements" of genes, representing various species, caused? If atoms and molecules are not sufficiently intelligent to cause "themselves" to be arranged into the "proper" configuration for the original-living organism to emerge; then, saying that beneficial (molecular genetic) mutations explain life-form diversity, is asking too much. Microevolution does not explain macroevolution. Species variation (alone) does not explain speciation (species diversity).

> Even if parts were sufficiently intelligent to form themselves into higher-level structures, that still does not explain how they contribute to emergence of transcendent functions. That is the crux of the problem.

The problem of explaining the first life form generalizes to explaining how parts collectively enable the emergence of transcendent functions. Either parts assemble themselves into proper arrangement that just happens to produce a transcendent function, or parts are "pulled" by a potential-transcendent function into manifest activity. The first possibility is no explanation at all. The problem with the latter is to explain how the potential-transcendent function "precedes" materialization of its function (manifest activity).

> Throwing this problem onto our automobile analogy renders a simple answer. The "idea" of transportation preceded the horse and buggy and the horseless carriage—the automobile.

The explanation for things needs to be addressed in the appropriate context: the structure of reality. Let's reformulate (by generalizing) our question about the origin of life. What is there about the structure of reality that enables the emergence of transcendent functions?

Simplifying: How do transcendent functions become manifest? How do transcendent functions emerge? And foremost, what is the source of transcendent functions? If there is an intelligent source of manifest-transcendent functions, then we can explain how life forms originated by design.

Interaction within and between hierarchically structured systems is not explained by "necessary means." Reductively explaining how complex life forms emerged is unfeasible. We concede parts do not sufficiently explain higher-level functions.

> Even if there were no better way to explain configuration of parts into complex structures, it is practically unacceptable parts direct themselves into higher level-arrangements if they are not designed for that purpose. Accepting such an idea (that parts are not designed for a higher purpose) is to say parts are more intelligent (because they, by their own machinations, can "direct" themselves into complexity) than the wholes they compose (of which they are a part). Bricks are not more intelligent (by design) than the (functioning) dwelling of which they are a part.

The explanatory solution (how parts are "ordered" into complexity) is simple. We clearly observe a vertical direction (in the hierarchy) of species evolvement to higher-level complexities. Single cell to multiple cells, and eventually to life forms with very complex-interactive-multiple level systems, can only be reasonably explained by potential-transcendent functions manifestly imbuing parts with concerted activity through ("higher level") direction.

> Genes and a species' chance encounter of "things in the environment" suggest evolution is directionless. The idea evolution is "inherently" directionless was probably inevitable. Evolutionists condescended to the concept of nondirectionality in life-form evolvement because they either had no better explanation, or they unnecessarily conceded there to be no explanation (by design) at all.

> It is undeniable (using hindsight that) there is a direction to life-form evolvement. It would have been more scientifically honest to have just acknowledged life-form evolvement was not understood (instead of continuing a shortsighted [reductionistic] theory only because it counters explanation by religious dogma). Lack of a respectable nonreductionistic concept that explains life, does not mean life cannot be explained by nonreductionistic means. The only way to settle the problem of evolution directionality is to connect purpose with life forms. We are working in that direction!

The human body consists of many functional levels and many interacting systems. There are systems of muscle, circulation, endocrine, nerves, and skeleton. Many of these systems operate (function) at more than one (hierarchical) level.

Most systems have lower-level support. The energy sustentation of respiration involves the circulatory system and muscle activity. At an even lower level respiration is conducted by cellular-biochemical activity.

> There are many examples of "lesser" systems supporting higher-level activity. One well-known example is the Kreb's "citric acid cycle" (after the German biochemist H. Krebs). It describes the biochemical

mechanism for final degradation of small-carbon-chained fragments of metabolites from carbohydrates, fats and proteins.

(The) life (function) "sits" atop the "physical" hierarchy. Lower-level systems are designed to support higher-level systems. Lesser systems provide the necessary energy to sustain higher-level activity (functionality).

The problem now is to explain how lower-level systems are directed to support higher-level systems. Simply: How are parts directed to support the wholeness function (of which they are parts)?

Direction is "inherent" in transcendent functions. Transcendent functions "pull" parts into a collective that supports higher-level functionality. Molecules "pull" (direct) atoms into a collective supporting molecular activity. The trick is to explain how the necessary organization can occur before the end product is established. The clue how this occurs is that organizational direction is vertical: toward the top of the hierarchy.

Our task becomes one of explaining the source of transcendent functions. We also need to eventually cite the purpose of life-form existence. Here, we are concerned with explaining biological systems (functionality). Let's just classify (the subject of) life forms and call it "biology."

To resolve the problem of explaining "complex" functionality in life forms is to just "explain" biology. It is that simple. And the following is the most fundamental way to state our problem. How can biology be explained? It must be explainable by the structure of reality. What is an easy way to do this?

Let's pose our question in a wider context. How does biology fit in with other scientific disciplines? Perhaps taking this approach will yield a "sufficient" framework to answer our question about why life emerged. This answer should not only explain how the first organism emerged, but also explain why it emerged. Neither of these questions is answered by appealing to anything about the biological life form itself. The source of transcendent functions is not found in biology. We knew that. It is biology that is in need of explanation!

We again widen the scope of our investigation of life's origin. Let's include the origin of nonliving things. Maybe this tactic will enable us to view our problem in a sufficiently wide context: one, which "naturally" explains life itself.

We now ask what is THE source of everything? In what direction is everything going? If we can answer these questions, the origin of life should be deducible from THE source. Let's use the subject approach.

How are things explained by subject discipline?

Logic explains set theory. (Refer to Chapter 18.)

Set theory explains mathematics. (Refer to Chapter 18.)

Mathematics explains physics. (Refer to Chapters 18 and 19.)

Physics explains chemistry (Refer to Chapters 13 and 19).

Chemistry explains biology.

Stated backwards the above is:

Biology is explained by chemistry.

Chemistry is explained by physics.

Physics is explained by mathematics.

Mathematics is explained by set theory.

Set theory is explained by logic.

At each end of the "subject" approach, we asked, on one hand, how to explain logic, and, on the other hand, how to inductively explain biology. We have explained logic (by the Ultimate Principle). (Refer to Chapter 18.) We were left to explain biological complexity. Specifically: How is evolutionary development (its directionality) "driven?" That which explains logic should also be capable of explaining biology, but how?

We discovered logical tautology (truth-functionally true statements—and trueness itself) in short—logic itself, is explained by falseness. (Again, refer to Chapter 18.) Translating our problem concerning the consistency of logic (which we determined is different from logical consistency) into reality-palatable concepts, meant, "Nothing(ness) explains everything." This (Principle of Noncontradiction) helped us understand the ultimate nature of everything. It should ultimately explain behavior. Including, for example, why do we seek truth? Let's quickly examine an extreme possibility.

In seeking truth, man aspires to understand the most intimate nature of reality. That man seeks explanations, is itself suggestive there is an explanation for everything.

The ultimate explanation for everything has traditionally been attributed to God. The idea is simple. As layers of reality are peeled back, man unravels not only God's immanent nature, but comes closer to discovering the purpose (whyness) of existence. We cannot help wonder if man seeking truth is really God working (through man) toward understanding his (God's) own nature. But why? This question lingers in the background. It will be answered. (We noted at the beginning of this chapter that Shelling spoke of this possibility.)

At the "high" end of the subject approach, we learned logic categorically represents somethingness, which is defined by the Law of Contradiction. The Law of Contradiction is inductively explained by the Law of Noncontradiction, which defines nothingness. Nothingness is consciousness. Functionally, consciousness is the whole of reality, but is no part. Consciousness projects itself as experience. When we asked how this "projection" is accomplished, we turned to the natural numbers and how they are legitimately "generated" from zero. (Falseness is equivalent to zero; trueness is equivalent to one.) (Again, refer to Chapter 18.)

At the other end of the subject approach, though we understand chemistry "explains" biology in a "reductive" sense, because of entropy, we realized a reductive explanation is unsatisfactory to inductively understand biology. More specifically: Is physical life satisfactorily explained by chemistry? Answer: No. This chapter addresses the same problem using evolutionary theory. Chemistry (DNA, molecular genetics) does not sufficiently explain evolution.

"How did life originate?" We learned how the Ultimate Principle, the Law of Noncontradiction, explains logic and "ultimately" everything else. It should define the source of transcendent functions. The Ultimate Principle must also inductively explain biology, but how? Whatever the explanation, the same solution (in kind) must also apply to all "subjects."

We already learned the concept of transcendent function is the tool needed to inductively explain not only the design of biological systems, but all subjects as well. Here we complete the details.

Transcendent functions (Refer to Chapter 20) explain not only the physical processes of the human body, but also explain the complexity of every level of reality. But how? We combine explanation by transcendent function with explanation by subject.

First, we list the linkage between functional levels using reductive" explanation—beginning with physical life (biology). This briefly reviews the problem of reductively explaining life-form emergence. The animal life form is used for our example.

Biological subject:

Animal life forms are physiologically (and reductively) explained by activity of organ systems (e.g., muscular, circulatory, nervous, and skeletal).

Organ systems are (reductively) explained (histologically) by tissue (activity).

Tissues are (reductively) explained (cytologically) by cellular structure (activity).

Chemistry subject:

Cellular structure is (reductively) explained by macro-molecular structure (activity).

Macro-molecular structure is (reductively) explained by molecular structure (activity).

Molecular structure is (reductively) explained by atomic structure (activity).

Physics subject:

Atomic structure is (reductively) explained by nuclear structure and electrons.

Nuclear structure is (reductively) explained by quark structure.

Quark structure is (reductively) explained by lesser point-like energy packets or the infinite energy of the vacuum.

We again observe the futility of explaining biological life forms by continual appeal to component parts. Let's address the reductive explanation for what it states. The reductive explanation "reduces" the explanation of physical life forms to subatomic structures. Accepting the reductive explanation is tantamount to asserting subatomic structures explain physical life forms. Subatomic structure also "reductionistically" explains inorganic matter. So how could subatomic structure also explain life? It is ludicrous that "nonlife" atomic elements can explain life forms. Life is not a characteristic discerned from the Periodic Table of the Elements.

Reductive explanations are unable to account for transcendent functions. Although components are necessary, they do not sufficiently explain higher-level functions. We already concluded that lower-level functions are "necessary" for higher-level functions to develop. However, lower-level functions are not "sufficient" to explain higher-level functions.

The lesson is clear: Conceptually reducing a life form or organic system to its components, does not sufficiently explain the arrangement of those components into higher-level complexities. If parts cannot explain higher-level complexity, they surely cannot explain transcendent functions arising from those complexities. If reductive explanations do not sufficiently explain higher-level structures, they are incapable of satisfactorily explaining life.

> The design of an automobile is not sufficiently (solely) explained by the parts which compose it. Moreover, the transcendent function of the automobile (transportation) is not entirely explained by automotive parts. Again, automotive parts could just as well be strewn on the ground (totally unassembled) and not much would happen.

The problem remains, how are parts assembled into higher-level structures? Our only alternative (and reasonable) option is to determine how higher-level structures can sufficiently explain lower-level structures. Methodologically, this is an inductive approach. Let's use what we have learned thus far in this book. Let's begin where we left off in the subject approach. We were explaining biological life forms.

Biology level:

Biological life forms are (inductively) explained by (mental) awareness.

Behavior level:

Mental awareness is explained by memory.

Memory is explained by consciousness.

Consciousness is self-explanatory.

Consciousness explains itself and everything (else).

How did we learn about the deeper nature of behavior (viz, memory and consciousness)? Interestingly, we learned about consciousness by understanding mathematics in terms of logic, and logic in terms of set theory, and set theory in terms of the empty set. The empty set is space: defined as "no-thing."

Things (categorically—thingnesses) constitute experience. We realized from problems in physics (e.g., nonseparability) and numerous other considerations that space is synonymous with consciousness. From a problem in generating numbers from zero, we realized consciousness "objectifies" (itself) by "functional" self-fractionation.

What is the point? In "constructing" structure to scientifically understand (how things interrelate) where does reductionism (deducing structure) end, and synthesis (inductively piecing structure together) begins? In our subject approach we "analytically" reduced biology to chemistry. We then reduced chemistry to physics. Physics concerns microstructures (atoms and quarks).

However, something else happens under the banner of physics. Physics concerns more than just microstructures. Physics also concerns the "physical laws of nature." The "laws of physics" (e.g., $F = ma$) represent higher-level generalities (F = force). Universals (such as force) are a synthesis of (what appears to be) separate (m = mass, a = acceleration) phenomena. Physics is the conceptual "turning point" where reductionism is overtaken by synthesis. Nature is understood when we can "mentally" piece parts together—representing higher-level generalities (or universals).

Laws of physics are abstractly represented by mathematics. We understand mathematics through set theory. We understand set theory via logic. Logic concerns the "formal" use of "trueness." Trueness is explained by falseness. Trueness is the unit function; falseness is the wholeness function. We explained how the unit function is generated from the wholeness function. The whole explains parts. The whole is no-thingness—it explains parts (or thingness). Nothingness explains everything.

We learned nothingness envelopes everything. Quantum nonseparability intimates there is a (wholeness) ground state for "everything." The separability of everything is enveloped by the nonseparability of everything. Space encompasses everything; it is the wholeness state of everything. Let's elaborate these statements by subject.

From our "subject" approach, we learned physics (which describes things mathematically) is explained by logic. Logic concerns "things in abstraction" (thingness). Indeed, the essential "order of everything" is abstractly represented in logic. Logical arrangement (structure) is encompassed by "no-thingness" (spatiality). (Falseness envelops trueness: zero generates unity.)

Space is the whole of reality, but no part. Space includes everything: it is the top of the reality hierarchy. The trick is to explain how everything originates from space using reductive means (from the top down). Simply, things are a (reductive) function of space. Let's set up the means to elaborate and explain how everything is a consequence of space.

Space is consciousness. Consciousness is absolute intelligence; it is changeless. Consciousness functions only in the present. Consciousness is the ultimate origin of experience.

> A significant question: If consciousness is the "unmoved" mover, as Aristotle would say, how can it move anything? This will be explained.

Experience is the "present-time" externalization of consciousness. Experience always changes. Change in experience in the present is interpreted as the "flow of time."

Experience is a reductive function of consciousness. Things are the "stuff" of experience. The wave function (ideal form—design) for any given thing incessantly "collapses" into particularity (manifestation); "into the flow of time." (Wave function: refer to Chapter 13.) This understanding will again be addressed.

Neglecting its source, let's focus on experience and the flow of time. We know the present recedes into the past. The critical point is that the present evolves from the past. This yields an intriguing idea. Does the future play a role in shaping the present? In reverse terms: Is the "here and now" somehow affected by what is to be? Every event has consequences. Are certain events specifically directed toward designed consequences (or goals)?

If the present leaves an imprint upon the past, then surely the future leaves its mark upon the present. We surmise this because what is now the present was once the future. Even the future passes (through the present) into the past. Simply, does the future play a role in shaping the course of events?

The future must affect the present. We know this because parts do not "sufficiently" explain the whole. The whole explains part assembly into complex functionality. Parts are "directed" into higher-level structures capable of transcendent functionality. This process takes time. Assembly of parts happens in the present. Structure is completed in the future. The idea of starting (something) is to finish. If something begins as a design, then it has an end in the realization of its purpose.

What happens in the future is contingent on what transpires in the present. The future is limited by what happened in the past. Nonetheless, the future shapes the course of history. This is not a problem because we understand consciousness only functions in the present. The flow of time pertains to the "passing" ("frame by frame" projection) of experience. This is a major clue to understanding the organizing principle (viz, applying an understanding of the Ultimate Principle to how things form).

Experience does not explain itself. We must jump outside experience and appeal to consciousness. Consciousness encompasses "experience of things." Consciousness is the source of everything. Consciousness explains everything (because everything is a reductive actualization of consciousness). Explaining that events are directed by consciousness is simple. Let's explain using the problem of understanding life-form emergence.

Life is a "part of everything." The subject of life forms is described in biology. Therefore, the relationship between consciousness and experience must also explain biology. How? How does consciousness explain emergence of life forms? We can now formulate this question by how reality's structure itself is generated. For example, how are lepto-quarks arranged into atomic structures? How are atoms arranged into molecules? Simply, how are parts "directed" into complexity?

If we properly understand structure building, we could probably explain how any structure emerges from components. If the structure is energized, we could also explain the emergence of transcendent functions. Properly answering these questions should enable us to explain how life first emerged.

We should not have any problem accepting parts are made to be pieced together into wholes. This statement is not contradicted by any experience; it will be explained.

Parts molded into wholes, is system building by design. The problem of life-form emergence is generalized to the problem of explaining how parts are structured into wholes. Let's pull our concepts together and begin explaining how life emerged.

The Ultimate Principle can explain life; it is the organizing principle. What did we learn from the Ultimate Principle? It states a whole "precedes" its fractionation (into parts): resulting in a unit (whole and its parts).

A unit (function) is easily understood. At any given level of reality, or structure, (a unit) "thingness" has two aspects: a wholeness function and a parts-function (parts are lesser wholes). The wholeness function of anything and everything "functionally" envelops its parts.

In the physical world, we observe this thing and that thing. We experience thingness as wholes. Yet the whole we see, is typically a "collection" of parts. An iron statue of a man is a molded collection of iron atoms. Here the whole is no more than the "formed" collection of parts. The identity of the statue's wholeness function is a likeness of man. Here, there is no function "above" its wholeness identity (manlikeness). There is no transcendent function associated with the statute's identity (its form). It does not exhibit overt activity. It does not move. It is not characterized as a self-energizing life form.

Consciousness, the whole of reality, is the ultimate intelligence. We acknowledge the whole of reality envelops and (functionally) transcends its parts. The whole of reality is also immanent in its parts. Therefore, parts also exhibit intelligence because they (a test) comply when "ordered" into higher-level structures (by the whole of which they are a part). (Atoms "willingly" assemble into molecules: it is their immediate purpose [by design].) Things are inherently "intelligent" by design.

"Thingness" intelligence seems difficult to accept. It suggests a "ghost" molecule orders atoms into the proper arrangement from which arises molecular functionality. And we have already discerned that parts cannot order themselves into higher-level structures. Only if parts are "designed" for a purpose, can they be "fitted" into higher-level structures. We recognize parts are designed to be components of higher-level structures. That atoms are "molded" into molecules, is not a coincident. We need to explain how design "arranges" parts into complexity.

Minor side issue: Which is more imbued with intelligence, parts or whole? We have discerned parts cannot and do not sufficiently explain their arrangement into complex structures. The only recourse is to accept the wholeness function of any given unit is (the intelligence) responsible for the arranging of its parts. Arrangement of parts needs to be cast in terms of a thing's wholeness function. How does the wholeness function explain organized complexity? The final task is to completely explain the wholeness function (of any given thing).

We ask what is ultimately responsible for any given wholeness function? The whole of reality ultimately explains the activity of its parts. How? Consciousness (ultimate intelligence) is immanent in experience. How can this be? Can it be explained?

> We soon describe how the whole is "represented" in parts. We know parts are in the whole. It does not take much imagination to suppose the whole is also in the parts.

That a thing's wholeness function is imbued with intelligence is the concept used to explain how parts are ordered into higher-level structures. Unit-things (a whole and its structure of parts) in reality are inherently intelligent. The whole and parts are "intrinsically" intelligent. The whole is always more intelligent than the part. The part always has the necessary intelligence (by design) that it can be "ordered" (arranged) into a manifest-wholeness function. This is the critical idea needing explanation. How is wholeness intelligence conveyed to parts?

Every unit-thingness has two (functional) natures. It has been stated that one is a thing's identity based on its wholeness function; the other is a thing's parts. (Parts also have their own identity.) We simply recognize the two aspects as "form and content."

Let's explain biological life forms using the concept of transcendent functionality. A thing's form is its wholeness function. (Refer to Chapter 18.) Form, therefore manifest formness, is "expressed" (becomes functional) in spatial extensionality.

Recall, setness (formness) is a member of itself (content also has form). Yet, the form (sans content) of any given thing is just the thing's wave function (its wholeness function). The wave function (of any given thing) is just an individual aspect of the universal wave function. (This shortly will be explained.)

> We do not experience thingness as a wave function. We experience things as a particular whatness. Our perceptual senses characterize physical things as a "whatness." Yet the "thing-in-itself" is the wave function—the "formness" discussed by Plato and Kant.

Although a given thing's form is "individualized" in manifestation, its formness (itself) is essentially no different from any other similar thing's formness. A thing's essential formness (wave function) shares in the "ground state" universal wave function. The ground state of everything has zero value. It is space-consciousness. This is just another way of saying consciousness "projects" itself (by self-fractionation) into a manifestation (of things). Simply: A thing's form is directly associated with consciousness. Things materialize when form (wholeness functions) takes on content (parts-function) producing "something" (a unit-thingness). Once the unit thingness appears (emerges) it is fodder for experience. This becomes clearer as we continue.

> For now, we just state the following: A unit-thingness is generated by the "interaction" of the wave function of a thing's wholeness function with that of the various wave functions of its parts. Things are "materializations" produced by constructive interference of wave functions culled from the superposition of waveforms.

The main point is the form of any given thing is an intelligence because it "identifies" with its origin or source (formness itself) in the universal wave function (space-consciousness—the ultimate intelligence). The same applies to every level of reality. The critical point is hierarchically higher levels of reality are more intelligent.

> Setness is a member of itself. There is no difference between a set being a member of itself (a subset) and the original setness in which it has its origin. A difference occurs only when a subset takes on "functionality" becoming a unit-thingness: "form taking on content." There is no separability between any given thing's form and that of any other thing's form. Formness itself is nonseparable. "Separation" of things occurs only in "projected" fractionation of function. Furthermore, setness "exists" only in the present: the "here and now." Functional subsetnesses "exist" in the flow of time (i.e., spatial extensionality or space-time). Distinct things (functional subsetness—unit thingness) are found only in experience, not in that (consciousness) which experiences the "this and that" nature of whatness thingness. Do not be befuddled by these statements. Though we speak of things existing in the flow of time, there is no actual flow of time from the perspective of consciousness. Time flow is an illusion caused by the ever-changing present.

Transcendent functions typically explain the purpose (or whyness) of (whatness) content. How? "Activation" of a manifest thing's form can produce an emergent (or transcendent) function.

How is a thing's formness activated to produce a transcendent function? Answer: energy—motion of parts. An automobile's purpose (transportation) is realized when its form is set into motion. The problem narrows to understanding motion. Motion needs a nonreductionistic explanation.

Energy sustains transcendent functions. Energy is "directed" by "potential" form (universal formness) which is inherently (setness is automatically a "regressive" member of itself) manifested into the structure of everything. This is a more elaborate way of understanding the statement "consciousness projects itself as experience." Projection explains motion nonreductionistically. There is no continuous motion of disparate things. There is just the conscious present wave function that fluctuates in "extended" spatiality.

The nature of experience is found in functional fractionation of the ultimate-wholeness function—absolute intelligence. Therefore, intelligence becomes manifest in the structure of reality—the whatness world of experience. In brief: The Ultimate Principle explains everything (categorically). It also explains how "continuous" motion (of things) is illusory. (Refer to Chapter 12.)

Now a pivotal question. How do things acquire directionality? How does structuring evolve? The nature of everything (with structure) is to "return to (move [motion] toward) its source" (to the wholeness function). By nature, parts tend to merge—becoming whole (via negentropy). This eventuates by design. Yet, considering the other side, wholes tend to fractionate—becoming parts (via entropy).

The design of everything (experience) is attributed to consciousness—the top of the hierarchy. Directionality in reality is one of moving either up (by synthesis) or down (through "regressive" fractionation) in the hierarchy. Everything has the inherent tendency to either synthesize or fractionate. Both occur in reality. Fractionation predominates in Phase I (of the cosmic time cycle). Synthesis predominates in Phase II. (See Chapter 9.)

Inductive explanations "correspond to" (recognize) synthesis. Deductive (better—reductive) explanations (trace) "follow" fractionation.

A transcendent explanation "proceeds" in the opposite direction from a reductive explanation. A transcendent explanation points us to the top of the hierarchical structure: (epistemologically) toward more general concepts—representing greater wholes. A reductive explanation leads to less-general (more specific) concepts—representing parts (and therefore, lesser wholes).

We have already determined reductive explanations are insufficient to explain the emergence of higher-level functions. The only recourse is to suppose an inductive explanation will be sufficient to explain life-form emergence. That is easy to accept because higher-level functions are more intelligent than the parts that are necessary to sustain higher functionality.

A thing's parts are energized (by activity attributed to motion) but directed by the potential-transcendent function (inherent design of its form) into an active (manifest) transcendent function. Potential-transcendent functions combine with "activity of parts" to produce manifest-transcendent functions. This explains how design manifests in structure.

> Potential-transcendent functions always function in the present.
> (Setness, the ultimate intelligence—consciousness, is a regressive
> member of itself.) Activity of parts is most reductively attributable to
> the energy of the vacuum.

Parts, especially their arrangement, are (transcendentally) explained by the "activity" of the greater whole (of which they are a part). Pointless particles (quarks, electrons, and other leptons) are transcendentally explained by the whole (nucleonic) structure they compose. Each subsequent-enveloping level is (transcendentally) explained by the greater function of which it is a part. Direction of energy usage is attributed to potential-transcendent functions. Transcendent functions signify purpose. Each level serves (the purpose of) the greater whole of which it is a part. Negentropy is the pulling of parts together to engender synthesis in (the form of) transcendent functionality. Let's tie the above constructs together by explaining life forms.

Let's explain life forms using transcendent functions, grouped by subject. Therefore:

1. Physics level:

    Quarks and leptons are (transcendentally) explained by nuclear structure.

    Nuclear structure is (transcendentally) explained by atomic structure.

2. Chemistry level:

    Atomic structure is (transcendentally) explained by molecular structure.

    Molecular structure is (transcendentally) explained by macro-molecular structure.

    Macro-molecular (here our interests are organic—carbon-based compounds) structure is (transcendentally) explained by cellular structure.

3. Biology level:

    Cellular structure is (transcendentally) explained by tissue (histological) structure.

    Tissue structure is (transcendentally) explained by organ structure.

    Organ structures are (transcendentally) explained by the (physical activity of the) life form.

    Animal life form is explained by awareness (inferential abilities—deductive and inductive).

4. Behavior—Mental level:

    Awareness is explained by memory.

    Memory is explained by consciousness.

5. Consciousness level:

    Consciousness explains itself (through the Ultimate Principle).

    Consciousness explains everything (else).

How does consciousness explain the first life form? A unit-thing has two aspects: form and content. Form is a wholeness function; content is a parts-function. Form appears to be "arranged" content. However, content is "arranged" by the form in which it participates. How this happens will be further explained.

The wholeness function can generate a transcendent function in dynamic systems. Appearances are deceptive. We cannot infer the presence of a wholeness function unless it has already manifested by "taking on content." We (our awareness) are a part of experience and consequently only observe "some thing" after it materializes. Although the transcendent-wholeness function "appears" to arise from the collective activity of (a unit-thing's) parts, we now know better. Parts are "necessary," but not sufficient for emergence of transcendent functions. What is a sufficient condition for life emergence? We are closer to answering this question. We need to understand the ultimate wholeness function in its own frame of reference.

Often, the wholeness function reveals more than just "identity of arranged parts." A dynamic unit (involving activity) can generate a manifest-transcendent-wholeness function.

Again, let's use our automobile example. An automobile has a wholeness aspect, which identifies it. The wholeness function (the whole automobile) serves (the purpose of) its transcendent function—transportation. For example, we identify an automobile (the "whatness" whole thing) parked in a driveway. Eliciting the purpose of this whatness (whole) thing is to cite its "whyness"—its transcendent function—transportation. We do not experience "transportation" until there is an activation (super functionality) of the automobile (the wholeness function).

Lesser-transcendent functions support higher-level-transcendent functions (potential or actual). A dynamic part function is but the absolute wholeness function (manifested) at lower-hierarchical levels (its function changes). The purpose of the automobile is supported by the activity of its systems and parts: electrical system, engine and transmission.

> What appears as nonactive forms (e.g., iron statue of a man) "underneath" (i.e., its lesser components) are parts (e.g., atomic) teeming with activity. Active parts, and therefore, lesser transcendent functions, also support nonactive forms. Every dynamic and (apparent) nondynamic thing is attributable to activity of parts.

The transcendent-wholeness function is present, if only potentially, in any possible dynamic system. It remains potential until it becomes manifest by taking on the purpose of its "energized" content. (A set [the wholeness function] is always [at least potentially] a member of itself.)

> Our early notion of whyness and whatness modes can be explicitly understood through transcendent functions. Transcendent functions are suggested by the intrinsic-whyness mode.

An "actualized" transcendent function cannot "emerge" until the wholeness function is manifest. An automobile does not exhibit the wholeness-transcendent function (transportation) until dynamism (activity) of parts and whole is invoked. And that cannot occur until after the right parts are "pieced" together (into a "working" whole). An automobile cannot offer "motorized" transportation unless all its needed (necessary) parts are assembled (according to its design) and it is operational. How are parts "moved" (motion) together to form a functional unit?

Our attention shifts to parts. Parts are designed for a purpose. How are parts assembled into an operational structure, which produces a transcendent function?

Considering an automobile: the potential for transportation is always there. The next question is how the transcendent function (here—transportation) manifestly develops. Asking this question is to invoke "purpose in design."

If there is a "preexistence" of purpose, how is purpose manifested to "process" the necessary "ordering" of parts, which enables emergence of transcendent functionality?

We notice how easily we "will" ourselves to climb a set of stairs. It happens without much thought. Our legs carry us up the stairway. Our body parts "willfully" respond to our directives. We did not plan in excruciating detail every little muscle movement. That is very telling: How complex parts' activity arises from a simple IDEA: climbing stairs. Climbing stairs occurs without much thought. A learned typist types without cognitive thought. These types of activity seem to occur almost automatically once the idea of doing something is decided.

Key Points:

- Parts do not assemble into a complexity from which emerges higher level functionality unless they are inherently intelligent by predetermined design. For example, atoms are designed to "fit together" and "form" (function as) molecules. (Direction of development [via negentropy] is toward the top of the structural hierarchy.)

- Molecules form into and function as cells; cells form into tissues, which form into organs. Systems of organs collectively give rise to (the function of) life.

- Life is the subject of biology. Biology is explained by chemistry, which is explained by physics. Physics is explained by mathematics, which is explained by set theory, which is explained by logic. Logic is explained by the Ultimate Principle—the Law of Noncontradiction. (This defines the top of hierarchy.)

- A proper understanding of the Law of Noncontradiction describes the nature of consciousness. It is understood as the absolute wholeness function.

- The absolute wholeness function manifests as "lesser" wholenesses (in the hierarchy). (This idea stems from setness being a [automatic] member of itself.) The absolute becomes the wholeness aspect of any given unit thingness. (Direction [by analysis] is from top down through the hierarchy.)

- A thing's functionality is inductively explained by the super functionality of its wholeness function. A simple way of stating this is to assert content is explained by form. Everything is finally explained by the Ultimate Principle, which defines the ultimate wholeness function. Different hierarchical levels are linked by transcendent functions.

- Transcendent functions emerge from the collective activity of "parts." The Ultimate Principle explains emergence of transcendent functions. Potentiality precedes reality.

- The Ultimate Principle defines how life originated and evolved by explaining how transcendent functions emerge from "parts." Most simply: Setness itself is consciousness—ultimate intelligence—the absolute wholeness function. Setness is a member of itself: becoming parts. Parts are inherently intelligent.

It is easy to state transcendent functions account for life-form complexity. The idea is fine, but it is contingent on explaining how parts are imbued with the proper design, which enables development of functions that fulfill purpose. How are parts imparted with design in the first place? For example, how did atoms obtain the design that enables them to fit into molecular structure? Things are designed for a purpose. Purpose is realized later in time than the intervening eventualities leading to the actualization of purpose. Simply: The idea of purpose must precede (in time) its realization.

How does a thing, before its materialization, direct the arranging of itself (and its parts) into the proper configuration to invoke the manifestation of the transcendent function needed to fulfill its purpose? How does the "organizing" principle explain the emergence of multi-level functions?

Explaining How Purpose Is Designed into Life Forms

Design suggests purpose; purpose suggests designer. The designer is consciousness. Consciousness designed the overall ordering of events to realize its own purposes. We are particularly interested in life forms. Do life forms have a purpose? If they do, how does the Ultimate Principle explain the emergence of transcendent functions to fulfill purpose?

Things change. We cite the changing of things as events. That man commonly dismisses some events as inconsequential is beside the point. One raindrop seems inconsequential. Yet without at least one raindrop, crops do not grow. Too much rain causes flooding: ruining crops. There is a delicate balance between too much and not enough rain. Either way, every event has consequences.

We rarely know in advance what will happen. When an invention is in the early stages of development, it is difficult to ascertain what will develop from it. Although we cannot foretell the "trail of consequences," we do know things will evolve and change. Not knowing in advance what will happen, makes it difficult to determine if there is meaning to life-form existence.

The only sure way of knowing if life has purpose is to wait until its purpose is realized—fulfilled. Meantime, there are other ways to approach the issue of life-form purpose. We can use scientific knowledge, and the structure of reality depicted in this book to piece together a possible meaning to life. Physics, biological-evolutionary theory and some religious concepts are particularly useful. (Certain religious concepts have a correlation with science and with the structure of reality. See Chapter 17.)

Primitive life forms evolving into complex-structures is a startling consequence. Do complex life forms (resulting from changes in design or order) reflect purpose? We suppose yes. The course of historic events is leading to something. We can probably determine man's purpose by tracing the course of cosmic evolution. Such a determination would be a "mapping" of the general direction (by motion) everything is heading. We should be able to determine whether man has purpose (in the scheme of things) by examining the "big picture."

> Things evolve slowly. Purpose is typically realized in incremental steps. Because of the brevity of physical lifetimes, we need to consider evolvement of cosmic structure.

We learned many "coincidences" cannot be explained away by chance (refer to Chapter 20 concerning pivotal events and design). The only plausible conclusion is there is a high-level designer that plays a role in these "not so coincidental" events. The "immediate" purpose of some of these "coincidental events" is not always clear. Maybe they occur just to suggest there is an ultimate designer. However, indications are not enough to establish truth.

Let's widen our consideration to include not just coincidental events, but all events. Surely, coincidental events are but "noted" events in the series of all events. Let's approach the issue of purpose in a "cosmic" context.

Is there an ultimate designer for everything (and every event)? If there is it would explain many things (including "coincidences"). If there is an ultimate designer, how do we scientifically verify and support it by proof of logic? The scientific basis for doing this was set forth in Chapters 19 and 20. The proof of logic was described in Chapter 18. We elaborate our understanding of life forms based on what we learned from the previous section about interpreting reality's structure.

The Ultimate Principle states there is an ultimate-transcendent-wholeness function. It is "dynamic" absolute setness (consciousness—the "highest-level intelligence"). Stating consciousness is dynamic is to acknowledge it is a directive intelligence (it executes instruction). (Saying it is dynamic does not mean it [is material or] exists in the flow of time, but that it "projects" things into the flow of time. We are trying to determine how it affects the course of events for its own purposes.)

Transcendence is associated with activity of the wholeness function. When "activity" shifts from absolute wholeness to (the lesser wholeness function of) parts, the effective organizing power and flexibility of the wholeness function decreases. (As absolute setness becomes a member of itself, there is no longer a "functioning" empty set.) A (whole) pie cut into pieces, is still a whole pie, though it is no longer "functioning" as a whole. Each piece "acts" as a lesser whole. Effective functionality shifts to parts—consciousness becomes experience.

What is the difference between the ultimate-wholeness function and the ultimate-transcendent-wholeness function? The ultimate-wholeness function is the whole of reality—space. Whether space is "functionally" broken, it is still space. The whole pie is still present even if cut into pieces. Yet, if it is cut into pieces, it (the pie) is not "functionally" whole (it functions as pieces of a greater whole). When the ultimate-wholeness function "projects" itself into plurality, it (the wholeness function) becomes "broken." It is no longer functionally whole because it "projects" itself into multidimensionality. The "activity" of projecting indicates its transcendent nature. Yet, that very activity is how consciousness manifests as experience. Consciousness can only become transcendent by an activity that arises from itself—its projection. (We note at this point that the directionality of citing transcendent functions is changed from "going up the hierarchy" to one of "going down the hierarchy.") Its transcendence (activity—projecting, i.e., creating, which is associated with the Realm of Reasoning) is diminished (to the extent its functionality shifts from wholeness to parts). Space is (functionally) "broken" by manifest things in space. Symmetry breaking of the "unified" forces of physics is understandable in these terms. Activity of the whole is shared with its parts. The law of conservation suggests that activity, arising in one form, takes energy from somewhere else. Activities of parts emerge from the energy of formlessness itself—from the infinite energy of the vacuum. Our present concern is "organized" activity. (Note: the Law of Conservation, applied to physics, is reflected in the first law of thermodynamics—conservation of energy.)

The difference between space itself and spatial extensionality is clearer. The distinction is functional. The ultimate-wholeness function is "space-in-itself"—absolute intelligence. It is All, but no thing. It is space without extensionality: without any temporality. The Realm of Whyness represents absolute wholeness.

The ultimate-unit function is spatially "extended." It is "All" and "Everything." Spatial extensionality is "perceived" because of functional things in space. The Realm of Reasoning structure represents the ultimate-unit function.

Things of "lesser functionality" (than the ultimate-wholeness function) are attributed to spatiality. Spatial extensionality allows things to be in motion. Spatially extended-objects "in motion" generate temporal flow. There is no actual motion. We learned that in Chapter 12. "Things in apparent motion" is conceived as time flow.

The ultimate-wholeness function is "represented" (because "setness" is a member of itself) at every level of reality. Everything has its own wholeness function, which is but a manifest (or reductive) function of space. A "part" of space is still space. A piece of pie is still pie. There is not anything that is not in space. Everything owes its existence to space.

Things functionally "break up" space (itself) by occupying it. It is not this simple. Etiologically, things (in spatial extensionality) emerge because of the "functional" fragmentation of space itself.

The above (space itself and its "projected" extensionality to accommodate manifest things) is understood by asserting space itself is ultimately a transcendent function—the ultimate intelligence—consciousness itself. Spatial extensionality is characteristic of experience ("projected dimensional space").

Space-consciousness ultimately controls what is not only allowable, but limits what can eventuate. This does not mean the ultimate designer is solely responsible for everything that happens. Although a country's leader is ultimately responsible for all the people, he or she is not responsible for every individual's actions. The activity of each person is responsible for his or her actions. However, the ultimate "controller" sets the limits of what an individual can do. This idea encompasses everything; applies to everything.

Consciousness conceives existence. Experience (of things: i.e., existence) comes from consciousness. (This is extreme solipsism.) The same is stated when it is asserted things are a function of space. Simply, space projects things. Overall, there just is consciousness and its reductive projection of things and events (in the flow of time).

Imbuing things in space with "reductive" (lesser "representation" of the ultimate) intelligence enables one to deduce everything comes from no-thing—nothing (space itself). This applies to anything experienced. We (and indeed, all things) are "parcels of experience." Because things result from the functional partitioning of consciousness (which is ultimate intelligence) everything (including life forms) is intelligent in its own right. A thing's form (its wholeness function) is its inherent intelligence: a reductive function of space itself—of consciousness itself.

The Ultimate Principle relates the intimate connection between consciousness and experience. Everything should also be explainable in a way which inductively "leads back" to the Ultimate Principle. Otherwise, how else can it be asserted consciousness is the ultimate designer of everything and every event? Mindful of this, let's return to the task of explaining how things evolve (how events unfold [change—by motion] in the flow of time).

The evolving association between consciousness and experience briefly stated:

At the top of the hierarchical structure of reality is ultimate intelligence (consciousness-space itself).

The ultimate-wholeness function regressively fractionates into the "infinite energy of the vacuum." The whole is broken into parts at the Big Bang.

The lowest level, the energy of the vacuum, is bereft of any determinable intelligence.

Physical matter develops. Despite "splitting," the whole (consciousness) of reality is still present, but is only "potentially" whole. "Consciousness becomes unconscious" in physical matter.

Complex structures evolve. Functionality "shifts" from ultimate wholeness to its parts: to lesser wholes (in the hierarchical structure of reality). Ultimate transcendence becomes immanent (as lesser transcendent functions).

The shift in functionality results in every structural level becoming an intelligence unto itself. Each whole is an intelligence. Each thing obtained its inherent intelligence from the "functional" splitting of ultimate intelligence.

Intelligence is "spread" among parts. By "descending" the hierarchy, the "intelligence effect" decreases. Lesser wholes are inherently less intelligent. This is entropy. Therefore, each part is less intelligent than the whole (of which it is a part). Less intelligence means less flexibility in activity. Atoms purposely function only to (chemically) combine with other atoms to form molecules.

Parts inherently tend to merge (move by motion) into greater wholes: toward their source—the ultimate-wholeness function. This is negentropy.

The task of (intelligent) ordering (of parts into greater wholes) is delegated to each unit-level in the structure of reality. For example, consider the physical life form of humankind. Each individual's physical body functions the same. The "formness" of man is the same for every man. Every man is "cast" in the same mold (form). Man's form is inherently intelligent. It is designed for interactive multi-level transcendent activity. All which, supports (the) life (function itself).

How do parts merge into greater wholes? Rephrasing: Beginning at the lowest level: how do transcendent functions emerge?

The cosmological Big Bang from a singularity (no-thingness, the ultimate whole) leads to particle (part) generation. The infinite-virtual (potential) elementary particles of the vacuum begin to manifest as real particles. Let's leap ahead.

The "collective activity of parts" enables (better, "allows") transcendent functions by manifestation of a potential-transcendent function. The potential-transcendent function initially "directs" the ordering of parts into a collective-wholeness function. Once the wholeness function is established, its manifest-transcendent function can emerge—it becomes actualized. The automobile must be constructed and operable ("ready") before it is usable for transportation. The potential-transcendent function is the sufficient condition for life emergence.

The idea of biological "potential" is captured by R. Sheldrake's concept of the morphic field. Sheldrake understands that reductive mechanisms inadequately account for life "self-organization." DNA cannot fully explain embryonic development. The concept of morphic fields is a glimpse of a "higher-organizing principle." (C. Jung's archetypes are similar concepts.) Biologists just do not know how to test, verify or explain such concepts. Our understanding of potential transcendent functions at least explains how "self-organization" (of parts) is possible.

Note something interesting about the above statements. It is counterintuitive that the function of transportation can "order" parts into the proper configuration resulting in a complete automobile. Our example makes it difficult to see the connection. Yet, no one can deny an automobile must first be "designed" and "constructed" before it can function for transportation. Engineers can be observed designing an automobile. We just do not see the ultimate designer, but we do see things evolve, form taking shape, and transcendent functions emerge. The remaining question is how can a potential wholeness function be at least as real as a manifest transcendent function?

Consider development of the butterfly. A butterfly lays eggs. Does the butterfly understand what it is doing? Larval forms emerge from eggs and grow into caterpillars. Using glandular secretions, caterpillars metamorphose into chrysalis. Do they know what they are doing? Caterpillars do not know their (designed) fate! A butterfly emerges from the chrysalis.

The butterfly does not have reasoning abilities to understand its nature. If the butterfly does not have sufficient-cognizant intelligence and reasoning ability to plot and carry out its own course of development, then it is overwhelmingly evident (by default) that the butterfly was designed with "inherent" intelligence. If this be not so, butterfly metamorphosis is unexplainable. It is impossible from random processes that a life form in a series of metamorphosis can consistently evolve and emerge as a butterfly.

Even man, with cognizance and some reasoning ability, cannot plan and direct his physical maturation to adulthood. Life forms do not develop into maturity because of their mental intelligence. They do not plan the course of their physical maturation. Newborn babies are not knowledgeable. They do not "design" their course of development. Innate intelligence is reflected in life-form maturation.

All life forms are born with innate intelligence. The course of life-form maturation is attributed to innate intelligence. Everything has some degree of innate intelligence, even non life forms. Structural complexity and functionality are measures of "intelligent" activity.

Biological instinct and innate intelligence have not been satisfactorily explained. Our task is to understand the operative nature of innate intelligence.

The course of evolution is incrementally guided by innate intelligence. Species diversification is "guided" by this "hidden" intelligence. Evolution and the order (taxonomy) of species reflect this intelligence. Why? Because everything leads to consequences which coincidences cannot explain. We already discerned parts cannot explain their role in higher-level complexities. There is only one feasible alternative. The "hidden" intelligence is inherent (by design) in a thing's form—inherent in its wholeness function.

Innate intelligence is the "will" to attain purpose. Biological drives are an example of will "striving" toward purpose. Let's phrase our basic question in more beneficial terms. How is purpose implemented in the design (structure) of life forms? Better: how does innate intelligence guide an organism toward attainment of purpose?

Understanding a set is automatically a (regressive) member of itself, allowed us to surmise any given wholeness function (at a lesser-hierarchical level) is designed for a purpose: to serve (the purpose of) the ultimate designer (consciousness itself).

Purpose is reflected in design. What is there in the design of things, which functions as the mechanism of "striving?" We briefly discuss some simple concepts to help develop the means to explain the functioning of innate intelligence. The same concepts will explain biological drives. We will also generalize and explain how purpose is manifested in everything.

The laws of nature dictate what is allowed by reality. The Ultimate Principle is the law (of all laws) governing everything. Applying this organizing principle to lesser wholes in the structural hierarchy explains how telic purposiveness "merges" with explanation by transcendent functionality.

> The Ultimate Principle states "the whole can both 'be' and not be." When it 'be'(comes) it (the whole) emerges as a unit (a whole and parts). The highest-functional unit is electromagnetism itself. Electromagnetism functionally splits into Cosmos and Anticosmos; then reduces to the particulars of each. Each lower level (subunit) itself is a unit. Each level is an intelligence unto itself.

Consciousness is both transcendent and manifest. Consciousness is the transcendent (active) nature of space: space itself (not spatial extensionality). Space is functionally transcendent when it is whole (not broken into parts) and is projecting itself. Its manifest nature is experience (exhibited in spatial extensionality). Experience is consciousness in unconscious form (when consciousness is functionally "broken").

Though physical matter is "unconscious" does not mean it is bereft of intelligence. Matter is "manifested" intelligence. Innate intelligence is displayed in a thing's super functionality: its transcendent function—its purpose. How else can this be understood?

Consciousness is characterized by knowledge. Knowledge is used to construct anything and everything. When consciousness manifests, it loses its transcendent functionality. Its "complete" (better—"intact") knowledge is lost, but not entirely lost.

The "lost" knowledge becomes manifested in the structure of reality. Everything is manifested knowledge. Why? A thing's design "inherently" reflects the knowledge used in its construction. Cosmic structure infers the knowledge used in its creation. Existence (what we experience) mirrors consciousness intelligence.

Memory is a composite of two major functions: consciousness (characterized by knowledge) and storage of knowledge (in the structure of everything). Manifest structure is effectively the "stored" memory of consciousness. Because manifest structure reflects the knowledge of consciousness in a "broken" state; it is just a potpourri of "lesser" intelligences. These "lesser" intelligences are just consciousness split (and spread) among parts. Parts are just less intelligent than the whole. Ah, but the collective of all parts is potentially whole, and potentially of ultimate intelligence!

> Some interesting points: Why did life forms take so long to evolve into self-cognizant intelligences? Early life forms (in Phase I) were physical and of lesser-inherent intelligence. For mentality to develop into self-cognizance requires a sophisticated brain that can generate a transcendent function capable of sustaining inferential processes.

> What begins as a low-level intelligence "climbs" toward (the purpose of) intellectual self-understanding. Discernment by intellect quickens as its purpose approaches realization. (Refer to intelligence levels listed in the How Aspect Diagram, Chapter 4.)

Everything is inherently intelligent. Intelligence suggests purpose. By seeking (the) purpose (of any and everything) man is ultimately seeking the reason (whyness) for every (whatness) thing. It seems the purpose of intelligence is to seek the reason (purpose) for everything. Why this? Why that?

The purpose of existence is not deciphered by wishful thinking. Purpose is inherent in everything. In seeking truth (about everything—reality) man comes closer to understanding his purpose or the meaning to life.

In seeking truth man acquires knowledge by (conceptually) understanding (by inductive reasoning) the structure of reality (which includes the structure of any given thing). Concepts are generalities (represented by universal statements). The point (contrary to nominalism) is generalities can and do correspond to something in reality, viz, higher-level (transcendent) functions. And, transcendent functions indicate a thing's purpose.

Everything must be understood categorically to determine purpose. In seeking encompassing concepts, we unravel, not only the structure of reality, but we are also given a hint about the purpose of man's existence.

The structure of reality has been delineated in this book. We are positioned to understand the reason for everything. From this understanding, we should be able to discover mankind's purpose—the meaning to life.

Purpose is not decipherable from structure alone. Though we have delineated the structure of reality, we must reason further to learn man's purpose. To ascertain purpose, we need to understand how the structure of reality changes: how it evolves.

We should uncover the telic purpose, inherent in everything, by methodically understanding structure building. The purpose of life forms should be determinable from (understanding) the "course of cosmic evolution." Ultimately, the question of purpose converges on understanding the generation of motion. Particularly, we are most interested in negentropy: "upward" motion in the structural hierarchy of reality.

To what end are things and life forms evolving (striving)? Let's narrow our search of purpose to man. Simply, the evolution of life forms needs to be understood from the perspective of man's most noteworthy pursuit—seeking truth.

The ideal is to explain everything by one principle. Next, apply the principle to cosmic evolution and determine man's role in the scheme of things.

We have discovered the Ultimate Principle. The Ultimate Principle is the Principle of Life. Let's use it as a context to understand man's purpose.

First: how does the ultimate law 'explain' life-form development? Given everything is designed for a purpose, applying the ultimate law to the cosmic-evolutionary process should suggest man's purpose. Understanding the Ultimate Principle should also reveal the direction of life-form evolvement. An answer to this question should also further explain innate intelligence, and the reason for life-form emergence.

The Ultimate Principle defines the ultimate reason for everything. Man's ultimate purpose is directly linked with the ultimate reason for everything. The design of things, including man's own "creations" (typified by the automobile) suggests things, including life forms, are evolving toward ultimate reason. How and why?

What is the nature of the ultimate reason? How does it explain life? We now know the ultimate reason for everything is consciousness. We have surmised that man's purpose should be determinable from understanding cosmic evolution. We first ask how do things evolve toward consciousness? Later we ask why do things evolve toward consciousness?

Things are typically identified by form. Things have content. Let's conceptually examine thingness using specific examples; then, generalize by abstraction.

Parts (atomic "content") are "pulled" by a potential form (molecular) into a collective arrangement engendering the manifestation of the transcendent function (molecular). This is how atoms are directed into molecular functionality.

A thing's wholeness function is realized in the unity of its form. Parts do not construct (cause) form. Parts are constructed into form! Form is "formed" by a former. There is a designer doing the constructing. The designer is implicit in form itself. Parts do not arrange themselves into higher-level structures. Formness is an intelligence unto itself which "arranges" its content to serve a purpose (perhaps to engender emergence of a transcendent function). Innate intelligence is inherent in a thing's form. Formness is always present because it is directly associated with the ultimate source of everything.

A (possible) transcendent function "exists" as a potential-transcendent function prior to (its) manifested emergence. (This is to say, before manifestation, the transcendent function does not actually exist.) It is a preexistent possibility by design! Even before the first automobile was manufactured, it existed potentially because the laws of nature did not disallow its construction.

> Potentiality seems to be a problem, but we are working to explain it by reasonable means. The trick will be to explain how potentiality can be "more real" than existence itself. Once we fully understand the ultimate nature of existence, this will not be so perplexing.

A caterpillar metamorphoses into a butterfly. Its transformation is directed by the inherent design of its form. Plato understood this. Form can change (in design) and still sustain a transcendent function. It cannot change itself into something more complex unless it is inherently intelligent. The changes in life form from egg to butterfly cannot be explained any other way. A live butterfly has the transcendent function of life. Formness persists whether devoid of content or not.

Development by design is not a cause then effect relation. Rather, changes in form are directed by purposive design. Change is purposively directed toward higher-level complexity. There is a high-level-intelligent "potential" wholeness function "pulling" parts towards itself! This tendency is inherent in everything (because the whole is represented in every part).

Let's briefly review what we have learned. Consciousness is the highest-level intelligence. It is potential because its wholeness function became "broken" when it projected itself into manifestation (as parts). Consciousness became unconscious in physical matter. It (consciousness) cannot directly "pull" parts back to itself (because once it fractionates into everything, it is no longer functionally whole). Yet, every part "naturally" evolves toward "potential wholeness."

> We have already determined wholes are more intelligent than their parts. Parts are also wholes that are intrinsically intelligent by design. If parts came from a whole, they can be "repieced" back into the wholeness from which they came. Jigsaw puzzle pieces can be pieced together. Puzzle pieces came from the whole and can be rejoined back into the whole. Parts can be pieced back together. Parts of reality can be repieced into the whole from which they came. The lingering question is how are parts directed into higher-level structures?

The problem in understanding how parts are "pieced" back into higher wholes hinges upon how transcendent functions emerge. How are parts "pulled" into higher-level complexities that engender manifest-transcendent functions? Simply: how do parts become whole? Answer, each part is inherently intelligent. Again, this is explainable only if potentiality is more real than reality. This will be explained.

Each part is intelligent because it is a lesser whole of the ultimate-transcendent-wholeness function—consciousness. When the whole of consciousness is broken, its intelligence is spread among its parts. So, each part is an intelligence with designed purpose. The purpose is realized when the "part" becomes a participant in the greater whole, the form, for which it was designed. Each part came from a whole and can return to the whole.

> Let's simplify. Before a jigsaw puzzle, at the manufacturers, is cut into puzzle pieces, it was first whole. No matter how it is cut into pieces, those pieces can be repieced into a functioning whole. The cutting process (regardless how it is done) inherently imbues each piece with design (form). The cutting process (fractionation) inherently bestows design upon each piece! Therefore, it is axiomatic each piece has an inherent design which enables it to be pieced back together.

Reason is found in the whyness nature of things. Form is manifested reason. A thing's form is its wholeness function. The ultimate purpose is reason (whyness) itself. Things strive toward purpose—the reason for everything. Everything strives to become whole!

Directionality of purpose is found in things coming together (synthesizing) into higher-level forms governed by transcendent functions. A succession of emergent transcendent functions encompasses more "participating" parts in reality's structure. Eventually, every part at every level "functionally" merges with the ultimate-wholeness function.

When disparate-functional things completely unify as a functional whole, total transcendence (in functionality) occurs. At that point, there is no immanence of functionality—there just is wholeness. The process of projecting experience then begins again. Immanence of functionality "originates" at the "moment" of the cosmological Big Bang (singularity fractionation). Let's restate how this happens.

The reason for everything is the whole of reality. Things strive to become whole. For things, e.g., life forms, to succeed in becoming whole, means they must "climb" up through the hierarchical structure of reality. And that entails they continually merge with other parts into ever higher-level (transcendent) functionality. Eventually, the goal toward which everything is moving is obtained: completed transcendency of wholeness functionality—reestablishment of consciousness function itself.

The counter intuitiveness of the above explanation is that in Phase I of the Cosmic cycle, we only observe the formation of things is preceded (in the flow of time) by a "cause." Closer examination reveals the common notion of cause pertains to "necessary" parts. Because man does not observe any other means of thingness formation, he presumes there are no other consolidating "forces." We now understand otherwise. Parts naturally "come together" (negentropy) by predetermined design.

Although scientists "observe" (in the physical world) that things are formed, they have not sufficiently understood complex-thingness formation. Particularly, if things are designed for a purpose, how are things imbued with purpose? How are things "pulled into higher level beingness?" Answering this question will also explain how life forms are "driven" by inherent purpose.

The central problem is explaining how a teleological effect "shapes" life-form evolvement (course of development) to fulfill its purpose. Let's pursue this tactic by refocusing on development of complex things.

How do complex things arise? We have already conceded a "preceding" cause (in the flow of time) is insufficient to explain thingness complexity. Particularly, how are parts pulled into complex structures that support higher-level functionality?

We observe "natural-system building" everyday. With modern technology, we can observe a zygote evolving into a fetus. An infant is born. A child develops and evolves into an adult. A simple embryonic structure develops into a complex-living organism. The growth process has not been understood.

How is development of complex life forms explained? An explanation for this "natural" process is rarely asked because it is commonly witnessed. Most explicitly: how does simplicity give rise to complexity? We know parts (e.g., sperm and egg) do not "direct" themselves into higher-level complexities. One explanation should "cover" all "system-building" life forms. Let's examine physical maturation a little closer.

Higher-level structures imbued with transcendent functions evolve. Molecules are "formed" into the transcendent function of cellular activity. Cells are "arranged" and function as tissue. Tissue is organized into organ systems. Organ systems are a functional part of the "living" (physical) life form. A (whole) physical life form "generates" the transcendent function—awareness. Awareness is man's highest-level-transcendent function.

But there is more. The collective activity of every "mental" being (individual) culminates in memory. Memory finally is explained by consciousness itself. Man's memory is reminiscent of reality's memory—the structure of reality. We realize man's awareness functions as a link between knowledge and experience. Man's purpose seems to be tied to his ability to obtain knowledge from reality's structure.

What is the organizing principle? It is the Ultimate Principle (the Law of Noncontradiction) which defines consciousness. Consciousness is the ultimate source, which imbues every life form with a transcendent-wholeness function and lesser-transcendent functions. The transcendent-wholeness function for a life form is life. The whole body takes on transcendent functionality—life. Transcendence is maintained in Phase I by activity (energy) of (the components of) the wholeness function.

Review: Let's return to the question about how consciousness can affect anything when it is (functionally) "broken." The answer is that what was functionally whole and transcendent becomes immanent in manifestation. This is just another way of saying intelligence is inherent in reality's structure.

Simply, consciousness bestows purpose through the innate intelligence (inherent) of a thing's wholeness function. Things are designed for a purpose.

> Back to our jigsaw puzzle analogy. We granted a thing's design is inherently intelligent because it can only be "pieced" together with other pieces in one way. This results from the Law of Conservation—there can be no more than the whole. The Ultimate Principle (the Law of Noncontradiction—"the whole can both be and not be") states whatever is derived from the whole cannot exceed the whole. However, the Ultimate Principle does allow self-fractionation. When it (the whole) "can be" (by self-fractionation) it becomes a unit function. By regressive (hierarchical) fractionation, it (the wholeness function) becomes a function of parts (pieces). All these parts can "fit" back together by (predetermined) design. The Law of Conservation permits it! That the ultimate-wholeness function is the supreme intelligence, which by design "broke (itself) apart," also suggests it would not have done so if it did not have sufficient reason. We are developing the means to understand that reason. When we do ascertain the reason, we will uncover the purpose for everything—including those fragmented pieces of which we (our minds) are a part. We will also learn our purpose—the meaning to life.

Consciousness explains not only how life forms arose, but explains how different life forms (speciation) could evolve from a common ancestor. If there is a potential form (or potential-transcendent function) toward which parts "strive" (by design) to manifest; then, life-form diversity is explained, along with life-form complexity (the interactive [hierarchical] levels in physical life forms).

> The potential toward which everything is striving is nothing more than the higher-level-transcendent functions ("forms") from which things ("pieces") originated! And, each higher-level transcendent function is inherently more intelligent.

Life forms, as with any unit, have two aspects: a wholeness function and parts-function. We learned this by understanding the Ultimate Principle. Consciousness is functionally broken in manifest form. When this happens, it loses its functional wholeness and it is only potentially whole. Even then, potential transcendent functionality can pull parts into higher-level-manifest functionality. How? The whole is "represented" in parts as a lesser wholeness functions. Each part is intelligent by virtue of its design. How a part breaks is how it will fit back together. Lower-level-wholeness functions are not consciousness per se. Lower-level functions are "lesser-functional representations" of consciousness.

> Why are lower-level functions said to be a "functional representation" of consciousness? It sounds as though this is meant to convey the idea lower-level functions are not independent (by themselves). This is correct when we consider everything is derived from consciousness. Ultimately, only "consciousness is." Even reality as a unit does not

exist except for the wholeness function—consciousness. There is no existence ("experience") without consciousness.

Regardless of hierarchical level, the wholeness function (in a given unit) tends to pull everything (its parts) back to its higher self (to be whole). Higher level functionality is generated by energy. Energy is consumed by activity of parts. The parts-function also tends to sustain itself. This is easily understood by stating every part is also a whole (considered at a lower level) in itself. Molecules are a whole that pulls atoms into higher-level functionality (e.g., molecular activity [function]). Yet, each atom is a whole itself. An atom tends to sustain itself independently from anything of which it is a part. If this is not so, we could not discern there is a hierarchical structure (composed of atoms, molecules, and higher level functions). Potential cells pull organelles (which pull molecules) into the transcendent function of cellular activity.

After a wholeness function is actualized, it is ready to fulfill its immediate purpose. That purpose is realized in its transcendent functionality (or super functionality). A thing's transcendent functionality defines a thing's role as a contributing part (or necessary function) of a greater whole. We do not see a jigsaw puzzle as a whole when viewing individual parts scattered on a table. Not seeing the puzzle as a whole does not mean the pieces cannot be assembled into the whole from which they were designed. The whole dictates how parts fit together. The nature of atoms is to (come together and chemically combine to) form molecules. The nature of a molecule, as a whole, dictates how atoms will fit together to actualize its form as a functioning unit (a whole and part).

> The jigsaw puzzle analogy is nice, but it might convey the wrong idea. It is not a thing's actual configuration that guides its placement into a complexity. It is a thing's functionality that guides its activity and therefore its contribution to the collective giving rise to some super functionality.

Life forms emerge because a potential-wholeness function (the potential-life function, not unlike H. Bergson's *élan vital*) pulls matter (parts) into the necessary arrangement for the potential-wholeness function (potential life) to manifest as an actual-transcendent function (life itself). The potential whole of reality acts through the "local" effect of the thing's activated form (its transcendent-wholeness function). Parts are designed to contribute to the realization of higher transcendent functionalities. Everything comes from an intelligent whole; everything returns to the intelligent whole (the ultimate source of everything).

When a transcendent function becomes manifest, a thing's "immediate" purpose is realized. The immediate (cosmic) purpose is existence itself. Innate intelligence begets manifested intelligence.

> The transcendent function (transportation) of a parked automobile is potential. The purpose of the automobile (transportation) is not realized until it is "energized" and set into motion.

The original manifestation of the potential for life generated the emergence of the first progenitive organism. Again, how does a potential-wholeness function become manifest? Any given wholeness function represents consciousness (as an intelligence) at a lesser level. That consciousness projects itself into manifestation (as lesser-transcendent-wholeness functions) explains at the highest level how things evolve. Life forms in Phase I strive (evolve) to become whole against the natural-entropic tendency of things to break down.

The transcendent-wholeness function (of a thing or life form) is consciousness "operating" at a lower level in the hierarchy of things. When a wholeness function is broken into parts, the "idea" of the whole persists in its parts.

Parts (which are themselves a whole) "strive" to return to the absolute-wholeness function from where they originated. This imbues parts with purpose. Ultimately, it is not the parts (themselves) that do the "striving." It is the (potential) whole of reality that sustains and "pulls" on any given wholeness function. This process confers directionality on structural evolvement. Structures "naturally" consolidate by evolving higher-level transcendent functions.

Any given wholeness function (at a lesser level than the whole of reality) stems from the "functional" fractionation of the wholeness of reality itself. The wholeness function of any functioning unit is therefore "inherently" intelligent. The whole (of reality) "resides" potentially in every unit thing as a (its) wholeness function.

> Everything is imbued with all knowledge. The catch is, lower-level structures, though imbued with all knowledge, are inherently less intelligent. This is why life-form emergence and development only very slowly "gained" intelligence. A broken whole contains the whole image, though the brilliance (intelligence) of its image is diminished. An extreme example of the "whole in the part" nature of everything was disclosed in Aspect's experiment demonstrating the "nonseparability" of the wholeness function at the quantal level. (Refer to Chapter 15 concerning Bell's Theorem.)

It is the nature of any given wholeness function to "return" to its source. Parts "attract" one another. Gravity is an example of this "pulling" in the physical world. Mating of male and female is also an example. That which is separate tends to become nonseparate (whole). Life forms strive toward "being conscious" (and therefore more intelligent). Why this happens explains cosmic evolvement.

The physical progenitor of all animals (perhaps a single-celled creature) was imbued with the potential function of (mental) awareness. Mental awareness is a lesser-wholeness function of consciousness.

> A tree grows from a single seed. Its trunk branches into limbs. Each limb is a "lesser" version of its radiating source—the bigger branches of which it is an offshoot. Though each branch is "separate" from other branches—none are separated from the trunk that sustains and holds them ("together"). Branches are "born" as leaf stems. Leaves provide

energy by photosynthesis. Consciousness "splits" into lesser intelligences—awarenesses—minds. Each mind is a branch. Tracing a mind "branch" back to its node locates where mind birth occurred. Each branch correlates with the constructive interference in the wave nature of the wholeness function—the universal superposition of every waveform.

Though the original mind is very strong before it begins "splitting—branching" into multiple minds, it is "asleep" in the physical world. Though "advancing" activity develops at the extremities (currently functioning minds) the "former" minds still exist—but are not presently active. All leaves are part of the tree. All minds are a reductive "functional" part of the original mind (call it Adam).

Through evolutionary changes, the mind developed in progeny. Eventually, mental ability to infer one's nature and purpose emerged. That man "inherently" can infer the purpose of existence suggests his purpose.

Why is there a "struggle to survive?" Everything is "caught" between two extremes. Everything is pulled in two directions. One direction is the absolute wholeness of reality itself. The other is the infinite energy of the vacuum. Things are caught between participating in functional synthesis and its opposite—becoming functionally fractionated.

Both extremes concern space. One is space in the absolute-wholeness sense. The other is the infinite divisibility of space. The difference between these opposites is only one of function: functional wholeness and functional plurality. (Except for function, there is no difference between the whole of reality and its infinite partitioning.) In Phase I, the physical world, entropy has a powerful "hindering" effect upon life-form emergence. In Phase II, negentropy is the predominant effect.

Wholeness tends to become parts (entropy); parts tend to become whole (negentropy). The interaction between the two extremes (space as functional whole and functional-infinite divisibility) accounts for all changes. (More about this shortly.)

Let's restate why life forms strive toward consciousness. (It is also a struggle for mind to develop and understand.) When parts become manifest, the whole of reality is functionally broken. At that point, the whole becomes nonfunctional: it is whole only "potentially."

Transcendence of the ultimate-wholeness function is broken in manifestation. Space itself becomes "extended" to accommodate manifest inclusions. Though a pie is cut into pieces, the whole is still there, but is not functioning as a whole. It is only potentially whole (in that it could be pieced back together). Again, symmetry breaking, recognized by physicists, of the "forces of nature" (elementary-particle interactions) is an example. (Minor reminder: There are no pie pieces unless there is first a whole pie. If the pie were to become whole again, there would be no pieces.)

The whole is immanent in its parts. Setness becomes a member of itself. The wholeness function becomes entwined in its parts. Therefore, it is through parts (specifically, physical life forms such as man) that the lesser-wholeness functions strive to become "one" with the whole from which they obtained their nature. Each piece of pie is a whole by itself. A life form is functionally a self-contained (intelligence) unit. (This also means man has free will, though limited by its role in the scheme of things.)

We are "striving" to explain life-form emergence. Reductive explanations are unworkable. We went in the opposite direction. We "inductively" included life-form emergence into the more general problem of how complex structures are generated. We explained everything is a projection of consciousness. Everything is a lesser intelligence of consciousness. This allowed us to state potential-wholeness functions "pull" parts into higher-level structures. This is teleology: direction by design.

Key Points:

- Things evolve, complex forms develop and transcendent functions emerge. A butterfly emerges after a series of metamorphoses.

- Development of complex things and emergence of transcendent functions are not explainable by coincidence. Parts cannot explain their role in higher level functions.

- Everything is a reductive function of space-consciousness. Consciousness becomes "unconscious" in physical things—things are a manifestation of consciousness. The whole intelligence of consciousness is "spread" among things (parts). Things are "innately" intelligent.

- A property of consciousness is knowledge. The structure of things represents the knowledge used in their design.

- Things are ("lesser") wholeness functions. Wholeness functions are but parts in higher level functions (e.g., transcendent functions). Functionality reveals utility, which implies purpose.

- Things purposely evolve to enable emergence of transcendent functions. Therefore, the reason for things is found in greater functionalities. Muscle cells serve the life form of which they are a part.

- Negentropy is (the process of) structure building. Things merge negentropically—to produce ever higher level functions—to eventually restore the "absolute" functionality of the ultimate wholeness function.

- The ultimate wholeness function is consciousness. It is the reason for everything. Consciousness functions only in the present (regardless whether it is functioning as a whole or functioning as parts). When not functioning as a whole, consciousness functions potentially as a whole. Once the wholeness function is recognized as THE reality, it is easy to understand how lesser wholes are (nonfunctionally) undifferentiated from absolute wholeness. (In its [consciousness] own frame of reference: "forms are forms.") Differentiation (and change) occurs only in manifestation. Actual change is an illusion. There just is present-time consciousness. (Any change is a "change in the present.")

- The Ultimate Principle infers consciousness is both transcendent (potential) and immanent (realized—manifestly). Present-time formness (wholeness function) "directs" parts (lesser wholeness functions) into concerted activity begetting transcendent functions (higher level wholeness functions). Simply, purpose (wholeness functions) is explained by manifest "activity" of present-time consciousness. This explains how transcendent functions emerge. Deliberate design occurs in the present. Eventualities are guided in the present. Ultimately, there just is the present.

One problem remains: How does that which is not real (a potential function) induce "real" complex-structural development that enables emergence of (manifest) transcendent functions? Simply, how can a potentiality be an intelligence (before its manifestation)? How can potentiality be "more real" than that which we call real?

It is easy to state things are innately intelligent. It is more difficult to satisfactorily explain intelligence. We know every part is inherently designed, but by what means can we assert they are also inherently intelligent—capable of "directing" activity? Stating consciousness (the transcendent-wholeness function) becomes immanent (in parts) is not enough. We return to this "problem" after explaining some other necessary concepts.

Eventually, the manifest-transcendent function called life evolves to the point where modern man appears. We are acutely concerned with the purpose of man's existence. Man's highest aspiration is to understand the purpose of existence—the meaning to life.

Man can design things for a purpose. He designed the automobile for transportation. Yet, man has not been cognizant of his purpose (the purpose of mankind). Man has not understood why life forms emerged. We are closer to determining man's purpose. The point here is man has purpose though he has not been cognizant of what it is.

Man's mental abilities are a functional peculiarity connected to a specific intelligence level of reality. (Refer to How Aspect Diagram in Chapter 4.) Mentality serves a purpose. Every structural level of reality functions for a purpose—whether (or not) that intelligence level can recognize what it is.

Lesser-intelligence levels (than man's mentality) are not intellects (capable of understanding their purpose). But, every structural level is inherently intelligent because each is a manifestation of ultimate intelligence—consciousness.

Man demonstrates understanding by explaining things. Did life forms emerge so ultimately man could seek truth and understand? If so, why? For what purpose? If man has a purpose, which directly serves the highest intelligence (consciousness) then the emergence of life will not be such a mystery. If man has a "divine" purpose, it will ultimately explain life-form emergence. Essentially: what is the meaning to life? Why do we seek truth? Why do we try to explain things? After discussing God, Goddess, and Lucifer, we resume our quest to understand the meaning to life.

How do we make all this "explaining" work? The ultimate source of the lesser-transcendent functions must be found in an intelligence "superior" to that in its "created" parts. The whole of reality must be a supreme intelligence, otherwise, it could not facilitate its own "functional" fragmentation, i.e., it could not be other than whole. This supernatural intelligence must be what man calls God.

God, Goddess, and Lucifer: Defined

We have interpreted the structure of reality in philosophical and religious terminology, but we were not finished. There were some outstanding questions not answered (in Chapters 16 and 17) by our understanding of the structure of reality.

Space and time (themselves) evaded precise conceptualization. We pressed to understand. We eventually studied logic. We were especially interested in the concepts of consistency and completeness. We uncovered the Ultimate Principle. It defined both completeness and consistency. We correlated our understanding of that principle with physics. The Higg's boson (probably the "real" graviton, and gravitation itself) "degenerates" into a plethora of "particles" (fermions and other bosons). (We elaborated our understanding of space and time in Chapter 19.)

Let's apply our present understanding of reality to determine what else we can learn. Specifically:

"Is there a God?"

> If there is no God, then our appeal to transcendent functions to explain the origin and evolution of life forms will be for naught. There have been problems in characterizing God. God is not explicitly defined in the Hebrew Bible (Old Testament). The only scripture offering any inkling of a definition of God is found in Exodus 3:14. It quotes God defining himself as "I am that I am." God himself is incapable of speaking verbally. Even if God could speak like man speaks, such a definition does little to help understand what this statement ("I am who I am") means. Neither is God explicitly defined in the New Testament.

"Is there a Goddess?"

> Many religions suppose there is a Goddess that complements God.

"Does Lucifer exist?"

> In 1 John 2 verse 1:5, we find a curious statement. Jesus announced: "God is Light." We know "the Light" is Lucifer. Lucifer is not God. We often hear that God is (represented by) truth. That is incorrect. Lucifer is represented by truth. One way to understand this is to say

Lucifer is posing as God! If God is not properly defined, how exactly can one determine what he is worshiping? And if one cannot verify or prove what he is worshiping, then he does not know what he is worshiping!

Chapter 17 left us wondering if God could be properly defined and satisfactorily explained. We are ready to define and understand (the nature of) God and Goddess. Specifically, can these concepts be correlated with the structure of reality? If so, how do these concepts (God, Goddess, and Lucifer) explain things?

Because we have delineated the entire structure of reality, viable religious concepts should correlate with some aspect of that structure. In Chapter 17, we correlated many religious concepts with the structure of reality. The concepts of God and Goddess were not determinable from our then understanding of reality. In Chapter 17, though we could correlate Lucifer with the Realm of Reasoning, we did not understand enough to determine how Lucifer came into being. In Chapters 18 through 20, we developed the means to elucidate the concepts of God and Goddess. We also learned how Lucifer came into being. Let's see how.

The Primary Equation ($1 = 0 \ x \ \infty$) provides our initial "relational" understanding of God and Goddess. From Chapter 18, we understand the Ultimate Principle. It defines zero (0) and explains the derivation of the natural numbers. Natural numbers are "parts of the whole." Zero is the wholeness function. Infinity is the extreme ("lowest-level") parts-function. Unity is the combined-wholeness function and parts-function.

There is an ultimate-wholeness function: the whole of reality, but no part thereof.

In Chapter 20 we learned the ultimate-wholeness function of reality is consciousness. Consciousness is a transcendent function: one that transcends its parts. It is formness itself, but no content. There is no disorder to it (being whole, there are no parts to become disordered). Consciousness is space. The Higg's boson and gravitation also represent it as a unifying concept. It correlates with the ultimate-wave nature and nonseparability of everything.

Either the word God is synonymous with the ultimate-wholeness function or the word God does not correlate with any aspect or function of reality. Consciousness is God. Zero (or falseness in logic) represents God. The religious concepts of Soul and Brahman are also synonymous with consciousness.

There is an infinite "regressive" fractionation of the wholeness function.

The extreme-reductive function is the infinite energy of the vacuum (space). It is characterized by chaos: no order, formlessness: "content without form."

Either the word Goddess is synonymous with the infinite energy of the vacuum, or it does not correlate with any aspect of reality. Infinite divisibility represents the Goddess. The religious concept of Shiva is also synonymous with the notion of Goddess.

There is a highest-level-unit function: reality itself.

The unity function is composed of the ultimate-wholeness function and extreme (the most reductive) parts-function. Reality, as a unit, has two essential aspects: space (the vacuum) and its infinite "functional" fractionation (infinite energy of the vacuum). It is all-form and all-content.

Either the word Lucifer is synonymous with the greatest-unit function of reality or the word Lucifer does not correlate with any aspect of reality. One (unity, true in logic) numerically represents Lucifer. The religious concept of Vishnu is also synonymous with Light.

Thingness (a subunit of reality) arises from the "interaction" of the whole (0) and its extreme partitioning (∞). A thing's content is endmost (most reductively) constituted of the infinite energy (virtual particles) of the vacuum. A thing is ultimately (most transcendentally) enveloped by space (space encompasses and transcends any other property).

Consciousness-wholeness function is space. Time arises from the motion of parts. Time (temporality) is attributable to the Goddess. Lucifer (electromagnetic radiation) is simply space-time. It is "all and every"—reality itself (as a unit). Lucifer arises from God's transcendence becoming immanent. (The whole becoming parts, creates a unit.) Lucifer is the constancy between whole and parts.

Key Points:

- God is the ultimate wholeness function—consciousness.

- Goddess is the infinite fractionation of the wholeness function (vacuum energy—virtual particles).

- Lucifer is the greatest unit functionality—reality itself (the Light—electromagnetic radiation).

Now that God, Goddess, and Lucifer have been defined, we address some long standing problems associated with attempts to understand God's nature. This treatment provides a stepping stone to determining "the meaning to life."

Transcendence and Immanence

How can God be "wholly other," or transcendent, and be immanent in its creation? Obviously, God's immanence must differ in some way from its transcendent nature. The difference is a one of function and one of intelligence.

The Ultimate Principle states "absolute setness" (zero in itself) is not only "the set of all sets," but is also "a member of itself." Setness is included within itself to the extent of regressive infinite "representation." The result is a hierarchy of subsets. It cannot be stressed enough that understanding sets, is the key to understanding reality.

> The study of fractals illustrates how certain sets, like the Mandelbrot set, regressively and automatically fractionate into "nesting" versions of itself. (After B. Mandelbrot.)

In terms of reality, setness is space. Space can "expand" itself into multiple extensionality (e.g., the three-dimensional spatiality that accommodates physical matter). A side note: Zero multiplied by itself (or any other number) is still zero. Space is the whole of reality. Any multiple extensionality (of space) is essentially no more than space itself. Space can expand and accommodate any dimensionality of something, which occupies it.

What is the origin of things that occupy space? Space, when functioning as a whole, is God. Space (the wholeness function—God) fragments itself into the infinite energy of the vacuum (infinite-plurality function—the Goddess). (Refer to Chapters 18 through 20.) The extreme "functional" splitting (of the whole) occurs at the cosmological Big Bang. God creates things from the Goddess.

> The extreme-most splitting of the highest-level whole (viz, space: Realm of Whyness) generates the lowest function of space (viz, the infinite energy of vacuum—the domain of virtual particles).

> The infinite energy of the vacuum is the reductive (subfunctional) extreme of the ultimate-wholeness function. Each virtual particle itself could be denoted a wholeness function (because there is no end to its divisibility). Functionally, however, virtual particles are "end of the line." Virtual particles are reductively beyond manifestation proper. They are lesser functions than lepto-quarks. Virtual particles are nonmanifest functions.

> How can a wholeness aspect, though found in the most reductive partitioning of the ultimate-wholeness function, act as a part function? Simply, each part (virtual particle) of the infinite energy of the vacuum acts as a whole because each virtual particle tends to "draw" everything else (manifest things) to itself. Because each virtual particle is of the lowest-hierarchical level, it "pulls" higher-level ("functioning") wholes toward itself: toward "infinite" fractionation. We expect this statement when explaining things by wholeness functions.

> The pulling of higher wholes to infinity happens because of (the functional) lack of pull toward the ultimate-wholeness function. This occurs because of the functional splitting of the wholeness function itself.

> Associated with the extreme natures of reality, are the following pairs: whole-parts, space-time, consciousness-experience, creator-creation, Higg's boson-virtual particles, and order-disorder. The extreme dipolar nature of reality is also recognized in manifestation proper as the complementary wave-particle nature of everything.

Between the whole and its infinite fractionation, are subset-units—representing (manifest) things. Anything and everything results from an "interaction" of the two extreme functions (absolute-wholeness function or space itself and the infinite energy of the vacuum). The whole and the infinite are opposing extremes of any and every manifest thing. More intimately, a unit thing, for example, a life form, has beingness because of a delicate balance between the opposing nature of its wholeness function and parts-functions.

> The wholeness-human function is the physical body. An "active" body exhibits physical life—its transcendent function. Injury and disease upset the (life function) balance between the whole body and its parts. In the end, body parts win out and the life function ceases. When parts no longer sustain the transcendent function (viz, life) death occurs.

Most levels of reality are constituted of subset-thingness (reductive functions of space). Therefore, every (hierarchical) level of reality is constituted of (lesser) wholeness functions. Many wholeness functions are transcendent. For example, of significance for man is the collective activity of brain parts. It gives rise to the (transcendent) function called mind (mental functions).

That which was ultimately transcendent (consciousness-God) loses its own functionality when it "fractionates" (into parts). Functionality is transferred to parts. Consciousness becomes "unconscious." Ultimate intelligence (consciousness-God) "splits," becoming immanent as lesser intelligences in its parts (as wholeness functions of subunits). Lesser wholes are an intelligence themselves. The intelligence of God becomes "reductively" immanent in parts.

Physical interaction (function) of manifest things sustains itself against annihilation by either of the two extremes (i.e., either absorption into higher-level-wholeness functions, or functional "breakdown").

Each thing's wholeness function tends to sustain its unity. A thing's wholeness function is ultimately sustained by the absolute-wholeness function. (Otherwise, it would incur a functional breakdown caused by the infinite energy of vacuum.) A thing's parts-function is sustained by the infinite energy of the vacuum.

> If parts totally coalesced, there would be no intermediate functions and no things. We know this because prior to the Big Bang there just was the whole and no lesser functions. We also surmise after the Big Bang generation of virtual particles (result of infinite fractionation of wholeness function) that no higher functionality (emerging from virtual particles) is assured unless the absolute wholeness function is also an intelligence capable of design.

During Phase I of the cosmic cycle, disorder dominates over order and parts sustain transcendent functions. Higher level transcendent functions depend upon the stability of lower level wholeness functions.

During Phase II, order dominates over disorder and parts are sustained by transcendent functions. Lower level functions depend upon the persistence of higher level functions. Ultimately, all functional levels depend upon the "activity" of the ultimate wholeness function.

Each unit (thingness) naturally sustains itself (its function) unless overwhelmingly "acted upon" by either entropy or negentropy. Despite homeostasis, things do change. Appropriately, what are the fundamental ways in which unit-things are acted upon? Any given thing either is "ordered" (into higher-level structures) or succumbs to disorder (disintegrates into its [lesser] components). The tendency for things to disintegrate is characteristic of entropy. The natural building up (or "coming together") of things is the distinguishing mark of negentropy.

Consciousness and experience comprise reality itself. All changes of things and events occur within this all-encompassing unit. The whole is consciousness sans parts. "Parts" constitute the "stuff" of experience. How else can this be stated?

Reality (Lucifer) as a unit function, is memory. Consciousness (God) "knows" experience. When God "loses" its knowledge (because it splits into a plurality, thus losing its wholeness function) it retains that knowledge through memory. It "remembers." Remembering calls for "memory storage." Knowledge is "stored" in the structure of reality (i.e., experience itself). Knowledge is immanent in "what" we experience. We experience existence.

What have we learned? Examining transcendent functions by subject reveals intelligence is inherent in structure. As we "ascend" the hierarchy of transcendent functions from atomic to molecular levels and so on to life forms, we realize "higher-level" functions suggest there is a design to the structure of reality. The structure of reality did not happen by accident. It was mentioned earlier that every part (of reality) inherently has all the knowledge attributable to consciousness (God). This accounts for innate intelligence.

In the previous chapter, we noted a series of coincidental events is unexplainable unless there is a higher-level intelligence governing the general course of events. It was also mentioned that "coordination" of muscle cells is difficult to explain unless there is a higher Being of which they are a part. The same reasoning applies to work in a bee colony. How are parts (e.g., bees) "coordinated" to function as a whole? Either such phenomena are coincidental, or they occur by design.

Design implies intelligence. Either coincidence (utmost improbable series of events) is explained by an intelligence, which supersedes man's awareness, or it is not explainable at all. Note: In seeking truth, we have yet to find anything that escapes explanation. There is a reason for everything; no phenomenon has evaded this understanding.

Higher-level functions imply there is an even higher-level intelligence involved in their sustentation. (We have discerned lower-level functions are not sufficient to explain higher-level functions.) Because consciousness (God) is the highest-level function, it is ultimately responsible for the "design" observed in every level of reality.

As the wholeness function (consciousness) "splits" into a plurality of lesser functions, its (own) affect upon those functions decreases. The further removed a function is from the wholeness function, the more it escapes the (functional) affect of ultimate wholeness. Lesser wholes are innately less intelligent.

The larger a physical object the more it is affected by the wholeness function. We recognize the attractive effect of gravitation (another way of understanding the wholeness function) upon massive objects. Very massive objects (e.g., dense astronomical objects) exhibit a greater "draw" on other objects—the pulling (gravitational effect) of parts (the masses) into greater wholes. Gravitation is space pulling parts together on converging paths. Curved space is just the tendency of extended space to "compress" to its functional wholeness: that is negentropy.

The ("hierarchically") closer a function is to the infinite energy of the vacuum; the more it is affected by that function (nonmanifest activity). Elementary particle activity is more attributable to the ceaseless "foaming" of the vacuum than it is to the affect of the wholeness function. Gravitational influence of elementary particles is negligible and its affect upon the energy of the vacuum is practically nonexistent.

In simple terms:

God is the creator of manifest wholeness or "form from content."

> The collective activity of parts may produce a function (the wholeness function) which transcends the functioning of the parts. God creates (transcendent) wholeness functions from the activity of parts.

Goddess is the destroyer of wholeness functions.

> The Goddess (reductively) strips the wholeness function from unit-systems. The Goddess "frees" parts from their role in wholeness functions. This results in disintegration of form.

The interplay between these opposing processes results in change. The two directions of "hierarchical" change are recognized in physics as entropy and negentropy.

Entropy is the disintegration of transcendent functions (because it destroys wholeness functionality). Entropy is form breaking into parts. Examples are molecules breaking into their atomic components and the human (physical) body decomposing after death.

Negentropy is the integration of content into higher-level forms by imbuing wholes with transcendent functionality. From a reductive perspective: The collective activity of parts gives rise to a function (transcendent) not found in the parts. The birth of a baby is an example. Babies arise from (the activity of) molecular structures (DNA—the molecules that constitute genetic material).

The absolute transcendent function breaks into parts. Its intelligence becomes immanent in its parts as lesser-wholeness-transcendent functions. Parts become an intelligence unto themselves. Life forms are inherently intelligent. The power of intelligence is defused in parts.

Consciousness-God becoming immanent in its parts explains how transcendent functions emerge from wholeness functions. Transcendent functionality results from the functional synthesis of parts. Enveloping synthesis shifts functionality up the hierarchy toward the source of everything. The source of negentropy accounts for the emergence of life forms.

Key Points:

- God is transcendent because he is "outside" spatial extensionality and time flow.

- God is immanent because he projects experience and is therefore "in" existence. Setness is a regressive member of itself. Everything is a "part of God." The extreme God functional fractionation is the Goddess.

- Transcendent functions emerge from functional synthesis of parts.

- Immanence discloses design and purpose in "lesser" transcendent functions.

The Problem of Good and Evil

Many people believe God is both all-powerful and all-good. They believe any less a characterization of God is inappropriate. However, the problem of Good and Evil has perplexed many theologians. If God is both all-powerful (omnipotent) and all-good (benevolent) "why does evil exist?"

At the least, why does God allow evil to exist? It is contradictory for God to be all-good and all-powerful and allow evil to exist. Some theologians believe man experiences evil because it accompanies free will.

Man has choices. Man is thought to have a choice between good and evil. Saying man is a sinner (desirer) is no explanation at all. (Refer to Chapter 17.) If man were once not limited in free will, he would not have sinned in the first place. If he sinned because of a limitation in his will, that he sinned is not entirely his fault. This "fall of man" is also in need of explanation. (It was stated in Chapter 17 that man's "fall" is part of a plan.)

> Assessing what is good and bad is thought by many to enter the arena of ethics. Except for this minor digression, it is not pertinent in this book to discuss ethics. Ethics and morality concern man's belief standards and have little to do with understanding reality. Much of "westernized" ethics is associated with biblical scripture. It was noted in Chapter 17 that man does not properly understand the Bible. Ethics arose from man applying belief systems to behavior.

The problem of good and evil needs to be reassessed. There is no "good and evil" in reality. (Refer to Chapter 17, subheading "Skepticism," last paragraph.) Such words (as good and bad) convey a judgment of events experienced by mankind. What some people consider good is deemed bad by others. The words "good and bad" are widely used by religious people whose methodology focuses on belief (refer to Chapter 5).

People suffer personal tragedy: physical deformity, mental instability and deficiency, disease, cancer, sensory malfunction and a myriad of other infirmities. People and animals also suffer physical harm. Natural catastrophes cause injury: earthquakes, volcanoes, fire, and weather related phenomena: flooding, drought, and lightening. Some injuries are inflicted by man's own doing: maiming and torturing. Some animals also contribute not only to man's suffering, but also to the suffering of other animals. In short, physical existence thwarts man's ability to live an unencumbered life. Man experiences many things not of his choosing. Man prefers not to suffer. Tragedy reflects man's limitations in physical existence.

Let's force the question. It is contradictory for God to be all-good and all-powerful and not do anything stop the many phenomena man considers evil. We replaced the judgmental terms of good and bad with possible maladies and natural occurrences that are sometimes judged bad. The problem becomes one of explaining why such "bad" phenomena are permitted.

Perhaps God is not both all-loving (good) and all-powerful. At the least, either God is all-good, but not all-powerful or God is all-powerful, but not all-good. It is also possible God is neither all-powerful, nor all-good.

We learned in Chapter 5 a definition of love, which is practically immune to contradiction. One cares most for what he loves. Simply, either God cares more for himself or he cares more for people. If he cares more for people than himself, then he is incapable of alleviating suffering; he is not powerful enough to prevent suffering. If he is powerful enough to prevent or alleviate suffering, then he does not care enough; he is not a loving God.

Let's grant, at best, God is not both all-loving and all-powerful. We will discover God is neither all-loving nor all-powerful. Let's apply what we have learned from delineating the structure of reality and explain this through "the problem of good and evil."

God is the wholeness function. When God creates its manifestation, it loses its wholeness function: the whole becomes parts. Although God, in himself, is all-powerful (after all, there is no existence without God) his (God's) role in (his created) existence is limited. God "functionally" splitting himself into a plurality (of things) creates existence. Because God loses his wholeness function, he is not all-powerful in the sense of what he can do for his creation. When God becomes immanent in his creation, he loses his transcendence and his power is limited.

Is God limited in his goodness? It has already been explained there is no "real" good-bad associated with the reality of things and events. Even if God were all-good in himself, if he cannot prevent suffering, then it matters not whether he is all-good in himself. The practical answer is that God is limited in his goodness.

Why is God limited in his affect upon his creation? He loses his "power" in the act of creating existence (i.e., experience—creation). Creation "takes on a life of its own." The extent God becomes immanent in its creation, is the extent God loses the functionality of its transcendent nature.

God's intelligence is "represented" in experience. (The mind [spirit] foremost "represents" consciousness.) The mind functions to seek truth. Knowledge (truth) is about things (as manifest-wholeness functions). Understanding pertains to the structure of things.

Another way of understanding God's limitation in his creation is to describe how creation is created. Creating something from nothing—from God himself (ex nihilo) is only accomplished by the creation of opposites (+1, –1): cosmos—anti-cosmos. The (absolute) wholeness function splits. Creating creation "functionally" destroys God himself.

Opposites are connected via complementarity. Physical matter and mind are "opposed." Mind is "overwhelmed" by matter. Even in a limited sense, mind "opposes" suffering. It is "unnatural" for mind to embrace suffering (mental or physical).

Because things are necessarily created in opposites, life has two phases (Cosmic Phase I and Phase II). (Physical) life (living material body—Hell: Phase I) is often depicted as one of struggle and suffering. The other, the mental life form (spirit—Heaven: Phase II) is one of ease and pleasure. Understanding the problem of creating something (things are created in pairs of opposites) from nothing explains why there is suffering in physical existence.

Another way to explain the dominance of suffering over pleasure is to understand Phase I of the cosmic cycle is dominated by the Goddess. Entropy (effect of Goddess) pervades physical existence. Physical things naturally (by design) "break down."

That physical life forms arise in an environment that "naturally" succumbs to entropy reveals the "hand" of God (wholeness function—process of negentropy) in building complex systems and life forms. Negentropy pervades Phase II where things naturally "build up" (come together).

God is both transcendent and immanent. God's immanence is recognized in life forms as innate intelligence. Immanence is only achieved by limiting God's power in his manifestation. This explains why people suffer in physical existence. This is also why the emergence and development of life forms was tedious and prolonged.

The degree mankind suffers in physical existence (Hell) is the extent he (as a functioning mind) experiences pleasure in the next existence (Heaven). This is the best logically possible world—considering there is a "balance" (or compromise) to everything. There is a price paid for something created from nothing.

Some people might assert the struggle of God (wholeness function) against Goddess (parts-function) is the fight between good and evil. However, such characterizations do little to understand the reason for things. The idea "everything" (in reality) is created, in the extreme, from a dual exhalation of two opposing forces was a hallmark of Zorastrianism.

Key Points:

- The idea of good and evil often concerns personal judgments.

- If evil is used as a term describing personal tragedy—then why does God permit it?

- An all-good and all-powerful God would not allow personal tragedy. Either God is all-good or all-powerful (but not both). Perhaps he is neither all-good nor all-powerful.

- Because God self fractionates to create everything—he loses his ability to "directly" affect what he creates. God is limited in his power and therefore also limited in his ability to be good.

Because God is limited in his power and goodness, there might be meaning to life.

The Meaning to Life

A goal of this project was to determine the "meaning to life." It has been explained how biological life forms arose. (The "innate" intelligence of the wholeness function "pulls" parts into higher-level complexity.)

The next relevant question is why is there life? It is better to ask what is the ultimate purpose to (physical) life? We explained "how" things are imbued with purpose (they take on transcendent functionality). We also explained how the wholeness function orders parts into transcendent functionality. We ask in the abstract why is there purpose? The answer should give us "the meaning to life."

Is there a goal toward which historic events are unfolding? Is there a teleological design to evolutionary development? Before tackling these issues, let's reiterate some useful concepts to support an understanding of why life forms emerged.

Knowledge defines the main property of consciousness. Consciousness is beyond or "outside" the flow of time. It is the ultimate-transcendent functionality. When consciousness projects itself as experience, its transcendence collapses into the particularity of thingness: content is given to form. The ultimate-wholeness function becomes "broken" in the process. Knowledge is "lost" (from consciousness).

> Side note: When consciousness is characterized as the ultimate "transcendent" function, it is done so from a lesser-hierarchical position. The concept of emergent transcendent functions was used to understand the connection between hierarchical levels of reality. The order in which the structure of reality was delineated was directed toward the top of the hierarchy. This reflects the course of inductive reasoning—conceptualizing by generalization. Generalization often corresponds to transcendent functions (in reality).

Knowledge is not completely lost. What was originally "pure" knowledge becomes "stored" in the hierarchical nature of reality. Knowledge is "representatively" embedded in existence—manifest structure. The structure of reality is the "stored" memory of consciousness. Life forms are inherently intelligent, at least to the point of sustaining their purpose in the "chain of being."

> Memory demands two major functions: consciousness (characterized by knowledge) and existence (stored knowledge—the structure of

everything). Consciousness, in this book, is denoted the Realm of Whyness. Existence is denoted the Realm of Whatness. Memory sustains the knowing of what can be understood and the understanding of what can be known. Memory is inherent in the "combined" Realm of Whyness and Realm of Whatness unit—namely, the Realm of Reasoning.

Restating: Consciousness loses its wholeness function and in the process loses (its) knowledge. Though knowledge is "lost" from consciousness (the wholeness function) it is not lost from reality (the unit function). It is stored in the manifest structure of everything (plurality function). These ideas are reflected in saying consciousness becomes unconscious in the form of physical matter.

Remembering (knowledge) is a two-way process: storage and retrieval. We grant knowledge is stored in the structure of reality. If creating experience leaves consciousness (itself) nonfunctioning, how does it (consciousness) retrieve the lost knowledge and why does it need the knowledge?

Recall the automobile analogy. Let the "blueprints" (diagrams) for (making) the automobile represent the knowledge required to construct it. Let's say once the automobile is made, the blueprints are lost or destroyed. Knowledge "on paper" to make the automobile is lost. That knowledge is now "embedded" or (manifested) in the structure of the actual automobile. Understanding the structure of the automobile and the various functions of its components should enable scientists (or serious truth seekers) to "reacquire" the knowledge used to create the automobile. Careful study of the automobile should help determine its ultimate purpose (transportation). The purpose of science is obvious. It is the task of scientists to "reacquire" the lost knowledge—that was used to create everything.

Consciousness is the supreme intelligence—God. It alone is ultimately responsible for the "design" of things. God is not omnipotent and omniscient because it is limited in how it generates experience (from nothing—nothingness: God himself).

Consciousness functions in "present-time." Existence "slips away" by blinking "off." Temporal flow is created because each projected "frame" (as in motion picture projection) is different. Experience is produced by the interaction between the wholeness function (God) and the infinite energy of the vacuum (Goddess).

The creating process is just the wholeness function changing into a plurality. Consciousness becomes experience: God manifests as its creation. However, God's transcendent function is not totally nonfunctional. Quantum nonseparability suggests there is a "functional" blinking "on and off" relation between the whole and its parts. (When a part blinks "off," it "merges" with the wholeness function. Then the whole "collapses" again [back] into manifestation.) After all, everything (separated) is sustained by (the nature of) potential consciousness (nonseparability). Nonetheless, as far as knowledge is concerned, it is lost from consciousness (because ultimate transcendence becomes broken—losing its functionality). Experience takes on a life of its own.

God (himself) cannot be a part of his creation and therefore cannot "personally" attend to his creation. (When the whole is fractionated, it no longer is functionally whole.) How can God govern his creation? He manifests himself as a "representative" individual to govern what he has created. God manifests himself as a "part." God's representative has been called the Christ.

Lower-level-wholeness functions are intelligences unto themselves. Like everyone else, at Christ's "First Coming," he potentially had "All" knowledge (of God). (Translation—at his "First Coming" he did not have all knowledge.) During his "Second Coming" he experiences things such that "All" understanding (but not "all knowing") becomes manifest (in him) through "development" of his mental (spiritual) inductive abilities. (All knowledge is exclusively a characteristic of God. Only when Christ functionally merges with God toward end of Phase II, does he have all knowledge.)

> In the "act" of creating, God fractionated himself into halves. Things are created in complementary pairs of opposites—Cosmos and Anti-cosmos. Each cosmos has a two-phase-time cycle: Phase I and Phase II. God's governance of his creation by manifest representation (through Christ) occurs during Phase II—biblically denoted Heaven. Heaven is inhabited by mental (spirit) beings in a mental universe. (Phase I [Hell] is inhabited by physical beings in a physical universe).

God's plan was intertwined with the design of everything since the beginning of time. Adam-Christ (God in manifest form) played a significant role in creating the multitude of individual beings. He had a role from the beginning; he has a role at the end of each temporal phase (Phase I and Phase II).

Things are not so simple. God manifesting himself as a "representative" individual in his creation is also subject to complementary pairing. God manifests himself once as the antichrist and once as the Christ.

> Because the real Christ (Second Coming) represents God, it could be said the antichrist (First Coming) represents the Goddess. This results from stating if Christ represents God; then, antichrist represents God's polar opposite. The Christ-antichrist pairing can be interpreted in other ways. For example, to understand God, Christ (Second Coming) must first be a serious-truth seeker. Yet, he finally discovers God is represented by (logical) falseness itself (not trueness). Does Christ (Second Coming) represent falseness? Yes and no. "It is true" God is falseness itself (the wholeness function). Regardless, Christ "represents the truth about God." Therefore, to confuse who represents God should not be a problem. To say the antichrist represents God (more so than does the Christ) is absurd though the antichrist is (more correctly) the "false" messiah. (The wholeness function, falseness, is the wholeness aspect of the unit function—trueness.) Trueness is associated with Lucifer. In the New Testament, Lucifer poses as God. In seeking THE

truth (unity of reality) one is seeking Lucifer—not God. Yet, no one can understand God, without first understanding Lucifer (the Light). In a way, Lucifer "stands" between man and God. "Truth stands between man and understanding God." There is an interesting irony. People during Phase I (those inhabiting physical bodies) "worship" (through belief) what is false (and not what is true). In a twisted way, they do worship God (defined by falseness). This parallels the separation between man and God. (One cannot understand God until he understands all truth. There is also a separation between man, and "knowing and understanding" truth, though man "lives" in the reality represented by that truth.) Ironically, this might mean a person could unknowingly worship God and still be separated from God. But this forces us to question: can someone worship that which they do not know? If yes, then someone can legitimately make a mere claim (e.g., that he worships God) and it would actually mean he does worship God. That is patently unacceptable.

If we do not like the customary application of the notion of true and false, there is a remedy. Interchange the notion of true and false (but not through negation). Let the wholeness function be represented by true, and the unit function be represented by false. This would enable defining God as true in accord with the common religious concept, and defining reality as false in keeping with the notion of Maya. Surely, this allows us to state God is more real than its creation. Obviously, the notion of trueness has been one of custom rather than ontological need.

Where does everyone come from to populate heaven? If these beings are immortal (mental) beings, they must have already existed. This book has explained this. They are produced (divide and multiply) in Phase I—the physical world. Let's again explain.

Before Adam emerged in Phase I of the Cosmic cycle, he "preexisted" (as a mind) in (the previous) Phase II. He has always existed. Adam (the mind) "falls into a deep sleep" (his mind ceased functioning inductively) and became "manifested" in a physical creature. As faunal life forms "divide and multiply," the original mind (Adam) functionally "splits" into a plurality: resulting in a multitude of (mental) beings. The task of splitting is entropically attributed to the Goddess. These ideas state the telic purposiveness of physical life forms. Minds are "created" (by the splitting process) in physical manifestation.

Each physical life form lives a very brief existence: just long enough to create a mind. Physical death is a major "crossroad" in an individual's development. There are no minds to "inhabit" the next existence unless they (minds) are created somewhere. God's plan is simply to place his personal "manifested" representation (Adam—Messiah) into the physical world (Phase I) where the affect of the Goddess (entropy) can split the one original mind (as Adam) into a plurality.

In spite of the original mind splitting, Adam maintains his identity. He later (toward the end of Phase I) reincarnates (into another physical body) to fulfill his purpose. Adam, the first mind, becomes the Christ. As Adam, he lost the knowledge of God. As the Christ (Second Coming) he retrieves it (knowledge of God) by inductively inferring the structure of reality. That is his earthly purpose.

Everything was designed from the beginning to facilitate a special purpose. Means were provided for God to govern his creation (Phase II—Heaven) through his representative—Christ—God incarnate. Before Phase II begins, the Christ-mind is functionally activated (awakened) to represent God (in the form of knowledge). God (by foreordained design) gives knowledge to Christ through the inductive inferring process. This is how God manifests himself as an individual.

> The plan is foreordained because God loses (his transcendent) functionality in the creating process. Truth, knowledge represents God ("in partition").

Let's elaborate. How is God's representative (Christ) awakened from his "mental" sleep? ("Sleep is of the mind; rest is of the body.") Toward the end of Phase I, the mind that was once Adam again returns (as the Christ) to the physical world. To reacquire the lost knowledge ("stored" in reality's structure) meant Christ (the mind) had to again inhabit (incarnate) another physical body (the Second Coming). Adam reawakens as the Christ by activation of his inductive-mental function. His mind "awakens" to begin retrieving the "lost" knowledge. This "awakening" is the resurrection of Christ. The resurrection is not physical: it is mental (spiritual). Christ is a mind (spirit): he is not a physical body, though he (the mind) has (re)incarnated physical bodies.

> Knowledge is different from understanding. Simply, God is all-knowledge (itself, i.e., the superposition waveform which is structureless): not (its) understanding. (Understanding relates structure.) God finds itself in the same situation as his manifest intelligences. The butterfly does not understand how it metamorphosed to its present stage. Having lost knowledge (it becomes manifested as the structure of reality) God reacquires knowledge through the reasoning and understanding of his incarnate representative—the Christ. God designed Christ with a mind capable of fulfilling its purpose—to correctly infer God's nature. By doing so, Christ manifestly becomes God incarnate (representing knowledge as we know it).

Seriously seeking the meaning to life "awakens" the mind of Christ. Obtaining noncontradictory answers to his questions further develops his mind. The meaning to life is determined by how man fits into the scheme of everything—the structure of reality. Inferring the structure of reality informs the Christ-mind of his role in cosmic evolvement. Christ does not do this: it happens to him. He experiences Reality in a way that fulfills God's purpose of manifesting himself (God) into his mental representative (Christ).

When needed, knowledge is "retrieved" from the structure of reality. This is just what scientists and serious seekers of truth are doing. Man's highest aspiration, the intellectual pursuit of truth—to understand nature, reflects mankind's purpose—to "retrieve" knowledge lost from consciousness. Science provides the basis for Christ to "regain" the knowledge he once had. Science is the true religion.

There is a time when the cosmological shift occurs. It is the "changeover" from Phase I to Phase II. When the changeover occurs, God is ready to govern "his kingdom" in the form of his incarnation—the Christ. When the time is right, the "sleeping" (in themselves nonfunctioning) minds of Phase I, are awaken in the next existence (Phase II).

Why did physical life emerge? Physical existence is a "staging" area for generating living beings. Although physical man is a product of Phase I, he is not the final product. Minds (spirits) are created through physical man. To obtain enough minds required the "splitting" (fractionation) of the original mind (Adam). (The "fall of man" is easy to understand.) The original mind could only be split in a universe where entropy predominates. The entropic process caused Adam to lose the knowledge of God he once had. (When his inductive reasoning ability ceased—he no longer sustained the memory of that knowledge.) Toward the end of Phase I, Adam (as the Second Coming Christ) reacquires the "lost" knowledge. He is then ready to assume governorship over Phase II. This is a part of an extraordinary plan. The plan was designed into everything since the beginning (of the Phase I cycle).

What is the ultimate purpose of physical life-form emergence? That question has just been answered (above). Why did the process take millions of years of evolutionary development? The slow development of the necessary means for God to reacquire the lost knowledge is commensurate with the "winding down" of activities in the Anti-cosmos. And that is because of the problem of creating something from nothing—God creating everything from his own (functional) destruction (by fractionation of himself—the ultimate-wholeness function).

Key Points:
- Consciousness (God) "acquires" meaning by projecting experience.

- Consciousness becomes "unconscious" in (the form of) physical matter. Consciousness "loses" its wholeness function and in the process "loses" its knowledge.

- Knowledge is not totally lost. Because existence is the immanent manifestation of God—the structure of reality "contains" the knowledge used in its creation. The separability of things results in knowledge "fragmentation."

- By seeking truth, man acquires a "conceptualization" of the knowledge "stored" in the structure of reality.

- Man seeks truth for a purpose—one that serves God. Therefore, there is meaning to life.

- Why does man acquire knowledge? Answer, to prepare for the next existence—Phase II of cosmic cycle. It is an existence of mind—one not sustained by physical matter.

Everything is understandable. Let's refine our understanding and put it altogether.

Putting It Altogether

What is the nature of existence? We are conscious of experience. How does consciousness generate experience? What we experience is existence. The why-what paradigm helped us formulate our inquiry about the nature of things into an answerable framework (representing the structure of reality). Now, we take a closer look at how the structure of reality is created. We briefly review the understanding of consciousness and experience. We then describe how the hierarchical structure of reality is generated. And finally, we review correlations between scientific terminology and religious concepts.

Ultimately, there just is consciousness. All else is a deductive projection of consciousness. The Realm of "Whyness" (consciousness) projects itself as the Realm of "Whatness" (experience). What does this mean?

Space is the nonseparable continuum. Space is consciousness. For consciousness, there is no separation in time—no "flow of time." There is just the "present." The present (time-consciousness) encompasses "all (flow of) time." Space (itself) is not characterized by "time flow." Time flow is a reductive function of space. As far as consciousness is concerned, there is no separability between things. Leibniz's idea of absolute monad seems to "fit" the notion of consciousness.

Whatness-things are illusions. It is easy to understand things do not exist independently from space (the whole of reality). When we say things exist, we are stating their beingness ("apparent" separation) is "functional"—meaning an effective fractionation of space (consciousness). This implies "everything is an illusion." No thing exists by its own doing. This is described in the Hinduism doctrine of Maya.

> There is no (nonfunctional) difference between absolute space and spatial extensionality. Space (itself) accommodates any extensionality of itself. The difference (between absolute space and multidimensional space) is functional. Multidimensionality is just the wholeness of space "functionally" divided. Extensional space also appears different because of how we perceive it. After all, perception concerns "things in space."
>
> The empty set represents space (in itself). The empty set is a member of itself: it is a "default" member of every set (see Chapter 18). Setness becoming a member of itself generates unit sets, i.e., things in space. "Extended" (multidimensional) things result from space "naturally" being a ("reductive") member of itself. Spatial extensionality is perceived because of the functional nature of ("manifested") unit-things-in-space. Stating a thingness is "functionally" defined means its separability distinction is illusory. Separated from what? Separated only from "other things in space." Although things appear to be

"separated" or differentiated from spatial extensionality, they are not separated from space (itself). This will be clearer as we proceed.

Through experience, we encounter things. For us, knowledge is "referential." It refers to things and events. What is the significance of knowledge for consciousness itself? Is it comprised of words and concepts in the manner we infer knowledge? No.

Consciousness is "functionally" holographic. In the reference system of consciousness itself, "knowledge" is holographic. The wave nature (of things) is difficult to characterize because it has no "identity of thingness." Consciousness (itself) does not exhibit (manifest) structure: it is the nonseparable commonality of all things.

If we state that things are functional, then it seems appropriate to state that consciousness (itself) is nonfunctional. However, it makes more sense to state that consciousness is functional and everything else is an illusion of that functionality. If we accept that, how is it explained?

The holographic nature of everything "holds" all knowledge. Consciousness "becomes" holographic. The holographic nature of everything is potentia itself (the possibility of anything and every event). The Cirpanic Cosmos "structure" captures the holographic nature of reality. The holographic interpretation of consciousness is derived from "extended" spatiality (the "becoming" of space itself). In this way we use the creative process (the Realm of Reasoning) to understand consciousness. Extended spatiality automatically brings into consideration time flow. This tactic is a reductionistic approach to understanding consciousness.

Higher-level generalities (in referential knowledge) correlate with transcendent functions (in manifest reality). In holographic terms (frame of reference of consciousness itself) a transcendent function correlates with the superposition of the lesser-wave functions (of its parts—collective of the sub-wholeness functions of thing's unit-parts). In its own frame of reference, consciousness is potential reality. But, because there is no reality apart from consciousness, this potentiality is THE reality. There is no easy functional way to characterize it. The reason is functionality is typically a characteristic of time flow.

The whole is in parts; parts are in the whole. The quantum-wholeness-wave nature is holographic, according to physics. The holographic "whole" is "represented" in (manifest) parts as transcendent functions. For example, the idea molecules really exist is common. What we call molecules are just atoms "arranged" to evoke higher level (super) functionality—molecular. Molecules are just a transcendent function of "structured" atoms. This idea applies in both directions—up and down the hierarchical structure of reality. This is why we state everyday things are constituted in the extreme by the wholeness of reality and its reductive fractionation—infinite energy of the vacuum. Vacuum energy is just space "functionally" divided.

There just is space. Things, indeed everything—experience itself, is an illusion "projected" by space. How? Space is THE "all pervasive intelligence." Just as material structure is the embodiment of knowledge, so it is that universal (wholeness) wave function represents knowledge in its purest sense. Everything is a part of a greater whole. The greatest "functional" whole is consciousness.

> Again, if we grant consciousness is a function, then its projection is a functional illusion unless things are defined in terms of the functionality of consciousness.

Existence is everything. Because everything is a "part" of experience (which is a projection of consciousness) things (i.e., their wholeness function) are ultimately holographic. Why? Consciousness projects existence "holographically." All experience results from a universally projected holograph.

The highest-structural level of reality is the Realm of Whyness (consciousness). Consciousness is the ultimate-transcendent function: one not found "within" the Realm of Whatness. Consciousness is the whole of reality. This implicitly means (because everything [else] is a projection of consciousness) any treatment of "whatness things" cannot be completely understood except by explicitly (and therefore, noncontradictorily) defining consciousness. This also infers the hierarchical structure of reality is but a conceptual scheme that leads to understanding that existence (of things, and indeed of everything) is foremost illusory. Therefore, the hierarchy (of "thingness") itself is not as "real" as its ultimate source (nothingness—"no-thingness") space-consciousness.

> Because everything exists in a functional hierarchy, the appropriate conceptual ideation differs from one level to another. So, concepts that work well for one level may not be suitable for definitive purposes when applied to another level. Consider the "thingness subject" approach. Biology is explained by physics. Physics is explained by mathematics. Concepts describing things are not self-explanatory. Each thingness subject level is in need of explanation. It is obvious the Theory of Everything (T.O.E.) will not be a mathematical equation because mathematics itself needs an explanation. Mathematics concerns relationships. Relationships need to be explained. Mathematics is explained by tautological formulation. The T.O.E. equation must be one that explains logic. Simply, how can trueness be explained? We have explained trueness in the ontological sense.

Because we "exist," our treatment of existence confers upon it (existence) more "reality" than is entitled. The reason is existence is not only defined by what is true, but by trueness in its referential sense. And yet, that which is ultimately real is not only "falseness" by definition, but by reference (by correspondence to reality: it is space—nonexistence). With this caveat, let's go on and "explain" how Whatness is generated by Whyness.

Consciousness, God, is self-sustaining. God is best described as the universal-present-time-holographic wave (non-time) function. Existence is not one of his characteristics: God's "beingness" (itself) does not and cannot participate in the flow of time.

God is space. (Space is not time.) Energy is typically associated with time. Time has its source in the Goddess. God's manifest functionality is "in part" attributable to the infinite fractionation of space. Parts (things and their transcendent functions, e.g., brain—mind) are constructed from these fractionated "parts." If we consider consciousness to be holographic; then, the mind is a product of constructive wave interference. Transcendent functions are activated by "shifts" in the interference pattern. This same reasoning applies to everything. Things and minds are "materializations" of the wave pattern. They exhibit separability because they are "functionally distinct." There just is one physicality—the whole is in the part. All individual minds are "divergent" parts of Adam's mind. The difference between minds is also functional. Constructive interference corresponds to negentropy and destructive interference corresponds to entropy. The role of dominance (between entropy and negentropy) reverses in the anti-cosmos.

Motion of parts generates time flow. Time flow is not space because it (temporality) is motion of things in extended space. The complementary nature of space and time is sustained in the unity of space-time (unified as the electromagnetic force).

Transcendent functions "appear" to arise (emerge) from the collective activity of parts. Our automobile example suggests transportation arises from the activity of the (whole) automobile. Purpose typically does not correspond to physical things per se. Rather, purpose is associated with a thing's activity. Transportation results from automobile usage: its purpose is auto "motive" transportation.

Mindful the automobile was "created" for (the purpose of) transportation, we acknowledge the following: The "idea" (i.e., knowledge) of transportation "preceded" construction of the automobile. Before purpose (or reason-whyness for any given thing) can be realized, it is preceded by design (the knowledge used to create parts and assembly [construction]). Design originates from the consciousness wave pattern that "exists" only in present time—and therefore, "all time" (because all time encompasses any given temporal instance).

Everything is designed for a purpose. The purpose of things is revealed by their transcendent function. The purpose of the (whole) brain is to generate (activate) the transcendent functionality of mind. The central purpose of highly developed minds is to "abstract" knowledge by understanding (inferring) things.

Transcendent functions potentially (i.e., in the form of knowledge, in reality—holographically) "exist" prior to manifest emergence. Transcendent functions are potential until they are manifested as "things." Yet, thingness is sustained in the flow of time by quantum on-off "collapse" (into manifestation) of the holographic (universal) wave form.

> Stating that a potential-transcendent function exists "prior" to a thing's manifested emergence is perhaps not the best way to convey this idea. Because the potential-transcendent function is "of" the holographic nature of consciousness, it is of "All" time, and therefore, not of any given duration in the flow of time. It always is. It persists as the

foundation (ground) of existence. In this sense the potential-
transcendent function (the present time nature of the holographic-
universal wave form) "presages" the development of anything:
including life forms.

Maybe because of habit and ignorance, people talk as though manifest things are
"The" Reality. We now realize the foolishness of this idea. The holographic-
universal wave function is The Reality. Manifest things, the "separate" stuff of
experience, are but "illusory" projections of The Reality—Consciousness.

Because mankind is "caught" in the "stream of experience," he supposes the
notion of "unit-thingness" is most real. After all, trueness represents things in
manifestation. They reason that which is true is more real than that which is not
true (i.e., that which is false). Yet, we now understand trueness (representing
unit-thingness—its whole and part) cannot "exist" except for falseness (its
wholeness aspect—no parts). Falseness "precedes" trueness; formness precedes
a thing's manifestation (that occurs via "incorporating" content). There is no
whole and part but that there is first the whole. There are no pieces of pie unless
there is first the whole pie.

We now understand the problematic issue of the potential-transcendent function
is no problem at all.

> The problem of potential-transcendent functionality is stated in the
> following question. "How can a transcendent function precede the parts
> which sustain its manifest-transcendent functionality?" How can a
> whole precede the assembly of parts to create the manifest whole?
> Puzzle pieces cannot be placed together into a whole picture unless the
> whole-puzzle picture preceded its fractionation. We stated there had to
> be potential-transcendent functions because "parts are not more
> intelligent than the wholes they support." In short, are there potential-
> transcendent functions? Most definitely, yes. Arrangement of content
> (parts) cannot precede the purpose of the form (activated wholeness
> function) in which they function as parts. Wholeness activity ("non"
> functionality) persists in the holographic present time nature of
> consciousness. All time is represented in the present because the
> universal wave function harbors all knowledge—all formness. And, it
> is formness (Plato's ideal) that "assembles" parts into higher-level
> complexities. Design (of formness) precedes manifest functionality.
> (Refer to previous digression about the meaning of "prior.")

How did life forms emerge? The potential-transcendent function of life forms
preceded (it has always been) the incorporation of parts into a functioning
physical life form. The potential-wholeness function "preceded" its functionality
as a unit-function. The potential-transcendent function is the (universal wave
function [better: defined by "waveform"]) "nonseparable" aspect of "The"
Reality—consciousness.

Understanding the meaning of the Ultimate Principle (the "life principle" or
organizing principle) explains how wholes can both be a whole (form) and have
parts (also be a unit). It states the whole always "precedes" its "make-over" into

a unit. (Again, "precedence" does not mean the whole temporally precedes (in time) unit formation. Rather, the whole functions (by projection) in the present and persists in "All time.")

Scientists are beginning to understand ultimate reality. Some scientists realize the "observer problem" (i.e., the measurement problem) in quantum physics is remedied by supposing the whole of reality is conscious. They know there is more to reality than thingness distinction.

Collapse of the universal-holographic wave function occurs when the wholeness-transcendent function "crosses" paths with its infinite-functional fractionation (energy of vacuum) resulting in wave "interference." In practical terms, any self-interference ("self"—because the energy of the vacuum is a fractionation of space itself) of the universal wave form induces "functional separation." The result is the "appearance" of distinct things—this and that—via their individual wave functions.

> The double-slit diffraction experiment is interpretable using this understanding. The thingness nature of the double-slit screen splits the "incoming" wave function into "separated" (two) wave functions. After passing through the slits the two wave functions interfere with each other, producing the dotted-wave pattern on the target film. The double-slit experiment demonstrates wave function collapse. The wave function is "The" reality; its collapse into manifestation (as dots) is fleeting ("temporary") and illusory. The dots seem real because they are "recorded" on film. They are no more real than is the film and apparatus.

> That thingness is illusory should not be surprising. Yes, things "seem" to be solid and therefore most real. Understanding the micro foundation for physical things, reveals they are overwhelmingly empty space. An electron has no "solidity." Nucleons are composed of quarks, which are difficult to characterize as solid. Most atomic structure is empty space. Therefore, everything composed of atoms is fundamentally of empty space. This is why it is beneficial to describe things as functions.

Manifest-transcendent functions emerge by design. Simply, thingness (whatness) emerges to fulfill purpose (reason). The automobile (form-structure) was created for (the purpose of) transportation. The Realm of Whyness generates the Realm of Whatness for its own reasons. Purpose precedes (its) realization. Whatness serves Whyness.

Everything has purpose. Things are designed to be assembled into complex structures. Parts are designed to "fit into forms" (wholeness functions) which serve a purpose. Parts are "drawn" together by forms to enable emergence of transcendent functions. Again, how does this occur?

Transcendent functions are associated with purpose. For example, atoms are the "building blocks" of molecular structures. Atoms are "arranged" into molecules. How could this assembly occur if atoms were not "designed" for this very purpose? If atoms were not "designed" as a basis of molecular structures, then their role in molecules occurs by chance.

> We are mindful. If things only occur by chance, then there is no chance of explaining anything.

The "fit" of atoms to play a foundational role for molecular structures is too extraordinary to have occurred by mere chance. That atoms incur random motion is also not a problem. Chaotic motion can lead to "patterns." Patterns are the basis for higher-level structures. The probabilistic nature of quantal systems is a case in point. From chaos emerge transcendent functions: energy of vacuum to leptons to atoms. Each "part" plays a role in higher-level structures. Yet, each is also a unit by itself and each has a wholeness function represented by its wave function. Note: we describe the nature of the hologram as a "wave pattern." (We see patterns naturally arise from chaos when watching fractal designs develop.)

Transcendent functions, prior to manifestation, are "potentials" toward which manifest parts "naturally" evolve to play a role. Molecules are a functionality by themselves. Atoms did not just happen to assemble themselves into molecular arrangement. This was no accident. Atoms were "designed" for this purpose. Atoms were "destined" to play a role in molecular structure. This is telic purposiveness.

There should be no problem understanding or accepting things (parts) are designed to evolve and develop (as parts of) higher-level functions. We observe this process everyday. An infant is born and develops into a mature adult. Merely because the "mechanism" of telic purposiveness has not been understood, does not mean it does not occur. To the contrary, the only problem is that it has not been understood.

The potential toward which things combine and evolve is the wholeness function represented by "The" reality—holographic-universal wave function. Remember, things (including life forms) themselves are illusory, they do not exist except for the ultimate-wholeness function (they are projections of consciousness). Things (which are "functionally" separated from the ultimate-wholeness function) do have a "tendency" to develop in their own way because they (themselves) are a whole. However, the course of development of everything is "naturally" guided toward the ultimate-wholeness function. Again why? Because it (consciousness) lost functional wholeness when it was "broken" (into manifest things—parts). That which was transcendent becomes immanent. That which is immanent strives to become transcendent.

The course of thingness development is not effortless. System building in Phase I is contrary to the predominant entropic nature of things (to break down). Yet, even here we observe something profound. Entropic processes release energy for system building!

Many life-form developments are not successful. Some life forms (e.g., dinosaurs) become extinct. Life-form purpose will be realized—if only by trial and error.

Ultimate purpose was designed into the very nature of everything via innate intelligence. The knowledge of everything is manifested in each and everything. The problem is low-level intelligences (e.g., the lower animals) are further removed from their ultimate purpose although the knowledge of that purpose is "represented" in everything. Low level life forms are innately intelligent, though they are incapable of inferring their purpose in the scheme of things. (Inherent intelligence is associated with the life form's wholeness function.)

To continue limiting the understanding of things (and thingness development) to reductionistic analysis is foolhardy in the face of now understanding anything and everything involves more than just their whatness aspects.

A unit-thing's whyness aspect (the wholeness function) innately "directs" its (parts-function) course of evolvement. A thing's "function" suggests the reason for any and every thing is found in its purpose (or part) in striving toward reason itself (the absolute-wholeness function). Everything has a whyness aspect.

The course of individual events, and even human history, is directed toward the purpose for which everything is designed. If it were not for the problem of creating things through complements, the course of (the development of) everything would be more easily accomplished (e.g., without physical suffering of its participants).

> Activity of the automobile sustains its purpose (transportation). Transportation is a transcendent function. Transcendent functions are not found in parts per se. The transcendent nature of something is sustained by activity of parts—and ultimately by activity of the wholeness function. Activity of the whole automobile generates transportation. There would be no "activity" (of anything) unless reality designed things for a purpose.

Either transcendent functions arise by accident, or they occur by design. The whole of reality (space which encompasses everything) reacquires (and sustains) its transcendent nature by activity (energy—motion) of parts. Parts are "activated" for a purpose.

The chaotic nature of micro-parts can be and is "hierarchically formed" into higher-level structures. Each level is a transcendent function (subservient to the next-higher-transcendent function, and utmost subservient to the ultimate-transcendent function). This treatment of transcendent functions reminds us of Leibniz' subservient monads. Ultimately, the transcendent nature of consciousness is sustained by "hierarchical" activity.

> In Phase I (physical universe) entropy (thingness "breakdown" and its subsequent release of energy, which becomes available for use) supplies activity. In Phase II (mental universe) activity is "things" (mental) naturally "coming-together" by the negentropic process.

Consciousness, represented by knowledge, is objectified as manifest things. Knowledge "used" in the design of everything is "lost" when it becomes (entropically) manifested in the structure of reality. This raises another question.

Once consciousness (because of fractionation, its "functional" wholeness) loses its knowledge in the manifest structure of everything: what happens to consciousness itself? Answer, it becomes a "passive observer" of the activity of its manifestation. Even man intuitively knows this. Man "experiencing this and that" is just God experiencing his immanent nature through man's sensory apparatus and perception.

Reacquiring the lost knowledge through man's inferential abilities (inductive activity) tends to reinstate (by the negentropic process) the "functional" transcendent nature of consciousness. We could suppose a "measure" of the transcendent activity of consciousness might be represented by the state of scientific understanding.

Consciousness itself does not change. The whole of reality is omnipresent: space is everywhere. The difference (between consciousness and its projection) is one of "functional" fractionation. "A (whole) pie is cut into (functions as) pieces." The key to understanding these statements is that the difference between the whole and (its) parts, is one of function. There are no parts but that there is a whole (of which they are parts). Whether there are parts or not, there is always the whole. A difficult thing to realize is that even before the pie was baked, it "existed" as a potential pie (its ingredients were "waiting" for their purpose to be realized and so on).

An interesting question about consciousness concerns its transcendent nature. Is it always transcendent? Asking this question is to wonder if there is a change in the status of ultimate transcendence. Does it change? Yes and no. Because the whole is always there (as whole or functionally partitioned) one answer is that knowledge is always present. Yet the answer to that question hinges upon whether knowledge is present in consciousness (consciousness active itself—and therefore, "projecting") or manifested (immanent) in the structure of parts.

> Absolute intelligence (consciousness) is not "active" in any sense involving time (flow). It does not change: it affects (intrinsically causes) change (only in that respect is it dynamic). Consciousness "activity" pertains to its knowledge (i.e., Cirpanic Cosmos—knowledge in a potential sense—Plato's "forms without content").

> There is a difference between knowledge (as we typically understand it) and knowledge manifested in reality's structure ("what" knowledge represents). Knowledge in consciousness' own frame of reference is also different (it is holoform). This means knowledge is derived from no-knowledge—which is the same as "all" knowledge. Realizing "no space" is "all spatial extensionality" helps to understand this statement.

Accepting there is a "functional" change means consciousness does change. The pie is cut into pieces. But, a functional change is paramount to stating there is no real change. The whole pie is still present. Space is still space. Yet, even this question of functional change evades considering the status of transcendence regarding the Anti-cosmos. When we consider the Big Picture, we need to state there is no change in the status of ultimate transcendence. Functional fractionation in the Cosmos is balanced by functional synthesis in the Anti-cosmos. Let's limit our present treatment of consciousness as it pertains to our Cosmos.

The transcendence of the ultimate-wholeness function is not a case of "either—or" (activated or not). There are functional "degrees" of transcendence and immanence. (The Ultimate Principle states the whole can both be and not be.) The predominance of belief systems suggests consciousness is not very active (at least in our Cosmos). (People believe when they do not know.) The basis for this statement is parts (especially awareness—mentality) serve (the purpose of) the whole. The state of mentality (deductive or inductive) signifies the influence of consciousness in manifest activity (via status of inferential abilities). (Emergence of inductive-inferential functions implies consciousness is more active in its manifestation, i.e., that cosmic purpose is closer to being realized.)

There is a "knowledge swing" between Phase I and Phase II (of the cosmic cycle). Referential knowledge in Phase II is sustained by the inductive-inferential functions. Consciousness is quite active during Phase II. During most of Phase I, there is much "manifested" knowledge (in structure) but little "referential" knowledge.

The "swing" (mentioned above) occurs between the extremes of Consciousness (zero—absolute, i.e., knowledge-full) and unconsciousness (infinite energy of vacuum, i.e., lack of "organized" manifested knowledge—chaos). It is a swing between God and Goddess. When God (consciousness) "rests" (during most of Phase I) the Goddess (extreme unconsciousness) takes over by default.

Subsystems of reality do not sufficiently explain consciousness. Rather, consciousness is the sufficient explanation for the subsystems. Again, "form following function" applies here. (Manifest form follows potential-transcendent function.) Design (holographic—i.e., of spatial extensionality) precedes "construction" (of reality's structure) and its various functional (manifest) levels. Stating this using physics acknowledges the universal wave form collapses into temporal functionality—the particularity of experience.

> Consciousness creates experience by self-fractionation. A functional "partitioning" of the universal-wave nature of reality is "distinguished" as a particular thing's wave function. The thingness' wave function then "collapses" into manifestation as an identifiable-physical object. In Phase I, we find ourselves "living" in this world of physical thingness—the world of "this and that." In Phase II we find ourselves "living" in the universe (Universe of Imperfect Whyness) of noncollapsible wave function (world of the "magnetic monopole"). Here, particularity is distinguished by its specific wave function.

Consciousness (whyness) "extends" itself into multidimensional spatiality (extensionality) as "whatness" experience. Consciousness is manifested in experience as transcendent functions (of a given unit). Transcendent functions emerge from activity of wholeness functions of unit-thingness.

> The foundation for all transcendent functions is owed to the infinite nature of the vacuum (Goddess). Emergence of transcendent functions, however, is designed into everything by consciousness itself.

Absoluteness of space and its infinite divisibility combine into the unity of reality. (Probability is one—unity.) The emergence of somethingness from nothingness is automatic and assured. Yet this latter statement is based upon set theory (setness is a member of itself). This brings up a central point. We can still ask why: why does setness become a member of itself? In our physics treatment of logic: its corollary is how can space generate things?

There can be no "things-in-space" unless space itself is an intelligence capable of "self-fractionation." If space (no-thingness) were not "super-intelligent," no thing could emerge from it. Space must be intelligent, or there would not be any "functional distinction of things-in-space." The basis for this statement is that ultimately there just is space (nothingness). There is no generation of anything from space unless space is responsible. Space cannot be responsible for creating "things" unless it is a sufficiently capable intelligence.

Space is the super-intelligence that is consciousness (God). If it were not, consciousness could not project "experience" by "functional fractionation." (Infinite fractionation results in the Goddess—the functional end product of creation). Such banter is of little aid to understanding.

How did the whole of reality (space) acquire its transcendence? Space is (infinitely and automatically) reductionistically (and regressively) divisible to the point of the infinite energy of the vacuum (of space itself). (Setness is automatically a regressive member of itself.) Out of the chaos of this infinite plenum, emerge all possibilities (all distinct wave functions). One is inclined to say energy (or activity) sustains consciousness. That appears so, but, not unexpectedly, is backwards from the actual case.

> Obviously, in its frame of reference, consciousness is not transcendent. (Transportation is not the automobile.) Here, transcendence is a relative term. It usually denotes functionality compared to lesser-wholeness functions (of things). Transcendence defines the "activity" of consciousness by its projection into manifest existence. Transcendence of consciousness appropriately pertains to the present-time dynamism of the universal holoform. Obviously, consciousness must be more intelligent (active—alive) than its manifestation. Properly stated, consciousness is "The Reality" because it projects everything manifestly by a regressive functional fractionation (of itself).

At lower levels of emerging structures (atoms, molecules) intelligence is implied by design (order). At higher levels of reality, "patterns" emerge which are imbued with "cognitive" intelligence. Man's mental-inferential abilities are a "high-level" example. The highest level is intelligence itself—consciousness. Just as all between levels are transcendent with respect to their sustaining components, so it is that consciousness is transcendent. Its transcendency is self-sustaining and omnipresent. Can this be explained in another way? Let's characterize sustentation from the perspective of consciousness.

Unit things resonate (i.e., frequency—periodic motion). (Frequency defines wavelengths [propagated] per unit time.) At the least, resonance infers things participate in the universal waveform (holographically). At most, resonance suggests the functional means of directing things toward (their) higher purpose.

Consider a musician practicing his instrument for a role in a symphony. Its sound, at any given moment, can be "isolated" as a single tone—a single frequency. An untrained listener cannot easily discern how these bits of sound "mesh" with the sounds produced by other symphony members. The conductor (director) orchestrates the "meshing" of instrument sounds into a meaningful piece of music. While hearing the whole symphony, it is often difficult to discern the sounds of a single instrument. Yet, no one denies each instrument contributes to the "whole" symphony sound (the waveform). The whole sound is a "superposition" of all the musical pieces (tones) composing it.

Intermeshing of whole and part to produce music is an excellent example of how parts are "directed" toward a common whole: harmony. Listening to an isolated instrument piece alone does not suggest that it has a part (has purpose) in a greater whole.

Consider the single valued universal waveform to be the holographic whole of reality. The "holo" waveform is a "superposition" (single valued waveform) of individual frequencies. The same idea applies to things having individuality and purpose within the holographic nature of the waveform. The universal waveform is the ground state (commonality) "behind" everything.

On the intelligence side: How does consciousness manifest itself as "forms with content?" Physicists commonly employ a technique that transforms wave patterns into mathematical equations. In 1822, a Frenchman contemporary of Carnot, named J. Fourier, published a treatise describing how periodic functions could be represented as a sum of a series of sine and cosine waves exhibiting different frequencies. He had studied how heat flows in metal rods from one end to the other. He developed the basis (Fourier transforms) for decomposing (or "separating") waveforms (or functions) into disparate frequencies ("tones" represented mathematically by sinusoids [sine waves]: a musical note is an example). A waveform (at a given moment) is the algebraic sum of the individual characteristics (e.g., amplitude) of the constituting frequencies. The point is obvious: That a waveform is analyzable in mathematical terms implies inherent intelligence (of the waveform). If man can analyze waveforms with mathematics and spectrographic instruments, cannot a more intelligent Being facilitate self fractionation of its wholeness function into discrete parts? Yes. "Fourier transforms" represent the manifest reduction of the wholeness function (and therefore all knowledge) into discrete (all are different) functions (i.e., lesser wholeness functions) representing thingness creation. Again, we are reminded why people enjoy music. It represents a manifest example of how consciousness functions.

The next point is that immanence of knowledge implies the structure of reality (itself) is manifested with intelligence. That which was "absolute" transcendence (space itself—not its extension) manifests in immanence (spatial extensionality) as an "intelligent" (hierarchical) plurality (of lesser-transcendent functions).

The extent absolute transcendence is nonactive, is the extent its immanence is active. For example, the degree a phenomenon is of a particulate nature (of a lesser wholeness function) is the extent it is not of a higher wave nature (higher level wholeness function). The degree a phenomenon is of a wave nature is the extent it is not of a particulate nature. There are many "graduated" levels between absoluteness and the chaos of vacuum. Intelligence levels of the How Aspect reflect this understanding. Regardless how we discern individual things, they are but lesser waveforms of a greater waveform.

There is an interesting way to reductively characterize "things in space." Things are "materializations" from the holographic wave function. What we call materializations are but constructive interferences in the universal waveform.

> How does this happen? Space-consciousness is characterized by formness. How does form become (transformed into) things? Form (in its own reference frame) does not exhibit motion, and indeed, does not exhibit manifest functionality. (Space itself has no temporal characterization except it is of the present—no time flow.) When space becomes "extended" (by fractionation) formness becomes a continuous (0 to 1 "amplitude") waveform(ness)—one exhibiting no periodicity. It exists from $-\infty$ (Anti-cosmos) to $+\infty$ (Cosmos). (The infinities "return" to zero at the vacuum level. After all, the difference between plus and minus infinity is one of function.) Space-consciousness manifesting as things is observed as objects "in motion in space." That which is motionless generates things in opposition to itself; therefore, by imbuing things with apparent motion. Translation: wavefunctions (extended spatiality) generate periodicity or frequency. Frequency relates energy. (Energy is proportional to frequency [$f = \hbar e$,
>
> $\hbar$ = Planck's constant].) Energy corresponds to mass and therefore, to physical matter. (Energy is proportional to mass [$E = mc^2$, c = velocity of light in vacuo].) Formness is "Fourier transformed" into manifest things (form with content).

Physical things are ultimately holographic although we do not typically "identify" things in that manner. If things were not holographic, parts could not be identified as wholes. Why?: Because there would be no commonality "binding" consciousness to experience (of things). Therefore, at lower levels of experience, perceptual apparatus contributing to (visual) sight could not experience "images" if things were not holographic wholes. After all, it is consciousness that is doing the experiencing. Stating a physical thing is holographic is to acknowledge it is a projection of consciousness. When we experience something, it is (the result of) the whole (self interfering and) collapsing into the particularity of thingness. (We [mind and body] are a "thingness" part of experience.) Mindful: It is ultimately consciousness (our higher self) which does the experiencing.

Consciousness is the commonality behind everything. It "ties" everything into a "working" whole. That its nature is holographic ultimately explains how things "intermesh" and affect one another. It explains the intelligence "inherent" within each thing. It explains innate activity. Birds fly in "skyways" to escape the inhospitality of winter. Bees work communally. It also explains how motion (of things) emerges from that which is of no motion (space itself). Zeno's paradoxes concerned this type of problem. Motion is an illusion generated by the "passing (still) frames" of (consciousness "projecting" itself into) experience. Everything is intelligent because all is derived from ultimate intelligence (consciousness). (Everything is intelligible, or there would be no experience.)

Inherent intelligence does not (itself) infer cognitive ability. Nor does it "ultimately explain" why things occur one way instead of another. Rather, it implies things are designed for a purpose whether they are cognizant of that purpose. The simplest way to state purpose (on a local level) is that "parts are naturally inclined to become whole" and serve a transcendent function. Bees work to serve the (whole) community. Atoms are building blocks of molecular functionality. Atoms are wholeness functions that serve as parts of greater wholes—greater functionalities.

Mind is the highest-level-immanent intelligence. Not surprisingly, mentality is the epistemological link between consciousness and experience. The purpose of mind is clear. Mental-inferential processes (awareness of things) sustain knowledge in absence of absolute-transcendent functionality. The mind itself is a holographic-transcendent function. (The mind is a transcendent function arising from the activity of the [physical] brain.)

Knowledge "characterizes" the nature of consciousness. Reality, as a whole, exhibits the functional property of consciousness. What we are conscious of (viz, experience) is (inductively) explained by consciousness itself. Conversely, consciousness is reductively explained by (the activity) of extreme unconsciousness—chaotic energy of the vacuum. Everything we customarily experience falls between these two extremes.

Higher levels of transcendence are closer (in the hierarchical structure) to the absolute wholeness function. Lower-hierarchical levels are further removed from the absolute wholeness function. As the reality-unit becomes more functionally fragmented, there is a decrease in the affect of the absolute-wholeness function (on parts). For example, the effect of gravity (under which all interactions [forces] unify [come together]) on elementary particles is very weak compared to the other interactive forces (weak, electromagnetic, and strong).

The affect consciousness has on things (what is projected) is proportional to their inherent intelligence. A mind is inherently more intelligent than a rock. The more intelligent something is; the better it serves (the purpose of) consciousness. Many things are understandable by appeal to consciousness and its projection in the extreme (its opposite).

The affect of the physical world on minds is pervasively divisive. People and governing leaders do not "easily" cooperate (come together). Each "fights" for (his) self-righteous views. ("Fight for what you believe!") It is foolishness. Yet, if people let reality settle disputes, there would not be any reason to fight for what is true. Truth represents reality. Verify (a statement's) truth by proving it in a way there is no contradiction. The truth will be known: reality remembers (everything).

We are caught between two extremes. The ultimate wholeness function (brings things together—unify) and energy of vacuum (disintegrates things). We can choose. We have limited free will. We can cooperate (come together) or not cooperate (fall apart).

Throughout the Realm of Whatness, we find consciousness (in itself) retains its identity in purpose or reason. Matter reveals manifest whyness in (the form of) order. (Arranging of parts into order [structure] yields a significant clue to a thing's purpose.) Purpose is inferred by a thing's design. Many things are designed to sustain a transcendent function.

Transcendent functions are an inherent intelligence. Yet, consciousness does not "fully" explain why things are imbued with inherent intelligence: one implying purpose.

Emergence of purpose in the Realm of Whatness is not directly explained by consciousness. Why? First, consciousness is fragmented when it manifests materially. Second, lesser-transcendent functions are an intelligence unto themselves.

Therefore, the nature of thingness is partially attributed (by reductive extreme) to the infinite energy of the vacuum (the Goddess). (Energy is understood in terms of temporal flow.)

Ultimately (directly or not) time itself is attributed to space and hence consciousness. Yet, time is a direct attribute of random processes. The energy of which is directly linked to the chaos of the vacuum (infinite [functional] partitioning of spatiality). (Setness is an infinite regressive member of itself.) Time flow arises from activity, which sustains transcendent functions.

Consciousness (God) is not omnipotent or all-benevolent. Consciousness regains its "lost knowledge" by understanding (inferring through man's inductive abilities) the structures that evolved from the probabilistic nature of quantal systems (which emerged from the Big Bang).

That man is "self-willed" is attributable to the probabilistic nature of underlying quantal systems. Yet interestingly enough, probable systems contribute to the predictability of events in higher-level systems. The predictable nature of a "many-body" system is the basis of events in everyday life. Simply, isolated events are unpredictable. Numerous events (of the same kind) are predictable. The same phenomenon allows the emergence of inherent intelligence in lower-level-transcendent functions because it bestows order "upon" underlying chaotic (indeterminable) events.

The double-slit diffraction experiment is an example how unpredictable individual events (dots on film) are predictable as a group (forming a diffraction pattern). The same idea applies to understanding life-form emergence. What appears on an individual basis as trial and error is predictable over a series of unpredictable individual events.

Let's explicitly answer the question how does intelligence become inherent in manifest things? The infinite energy of the vacuum "pulls" the intelligence of ultimate-wholeness functionality into lower-level functionality where probability of numerous events imposes "order" upon otherwise, chaotic motion. As we "ascend" the hierarchy from the infinite energy of vacuum, we observe more order, and hence more manifest intelligence. Finally, we reach (the level of) the "collective activity of brain parts" leading to man's (intelligence of) awareness—the inferential functions.

Purposiveness evolves. Things evolve toward reason (whyness itself). Evolvement entails time (temporal flow). Temporal flow (things in motion) arises from the (infinite) energy of the vacuum. Seemingly, time (flow) "opposes" space. Yet, temporal flow arises from the "unconsciousness" attributed to the infinite partitioning of consciousness itself (which is space).

Ultimately, everything and every event in time (and even time itself) is attributable to space and its inherent intelligence of consciousness. Consciousness is ultimately responsible for the hierarchy of reality.

Key Points:

- Consciousness explains everything, including life forms.

- Everything is created by consciousness self (functional) fractionation.

- Fractionation generates the hierarchical structure of reality.

- The structure of reality is patterned after the why-what paradigm.

- The why-what paradigm meaningfully connects five basic questions: why, what, where, how, and when.

- By functional design, parts necessarily and concertedly generate (by the negentropic process) real functions from potential ones.

- Potential functions (wholeness functions) "presage" and "direct" their manifestation as real functions (unit functions).

- Things evolve by functional synthesis toward the top of the structural hierarchy.

- Life has purpose. Physical life evolved in Phase I to produce minds for Phase II (of cosmic cycle).

- The purpose of scientists and serious truth seekers is to "recover" the knowledge used to create everything.

- When the next existence (Phase II) begins, scientific knowledge (and the understanding of everything) is (in "mental" memory and) ready to "replace" the "stored" knowledge in physical structure.

- In its own frame of reference, consciousness is super-positionally a holoform. It directs thingness evolvement in present-time. It is God.

Projecting the Structure of Reality

Studying the information in this book "chapter by chapter" roughly follows the "order" of how the structure of reality was (inductively) "mapped" (by the author). It was pieced together "from the bottom up." Let's briefly review the structure of reality "from the top down." This treatment describes the hierarchical structure of reality from the perspective of how consciousness created everything. This is a deductive or reductive approach. It is practically the reverse of how the author mapped the structure of reality. Let's simplify.

> We mention little here about the whole of reality self-fractionating to the point of infinite partition. (Refer to cosmic singularity and Big Bang theory.) Physics suggests this occurred before anything else. Out of this extreme fractionation, emerge transcendent functions. The initial-transcendent functions become the "building blocks" of thingness evolvement: leptons to atoms, on to molecules, and so on. Things change enabling emergence of higher-level functions because the whole "acts" upon parts and parts act upon the whole. Each counteracts the effect of its opposite. The whole pulls parts (negentropically) into higher-level wholes (systems); parts (entropically) disintegrate higher-level systems.

> We also note in a functional sense that the structure of reality presented in this book is reminiscent of Plato's idea that was later called the "Great Chain of Being."

In the first instance (of the "top down" treatment of) space "becoming" a "functional (manifest) member of itself" is the electromagnetic unit-function. Electromagnetism is the unit function that is reality. It incorporates the wholeness function and (infinite) parts-function. It exhibits the dual nature of wave (wholeness) and particle (parts). (The particulate nature of light characterizes the parts-function [photons]). The unit function embodies both consciousness and experience. It is electromagnetic radiation in its own reference frame.

> The photon represents more than just the particle aspect of light. The photon is the unit function. It represents both the wave (wholeness function) and particle (part function) dual nature of reality. When the wholeness function (graviton—or Higg's boson) becomes a "member of itself," electromagnetism (the photon) is the result. The unit function is represented by the photon.

> The photon can duplicate itself infinitely. Typical scientific understanding posits the photon is the particulate nature of light. This is not quite true. There should be no problem understanding the photon

represents reality—whole and parts. An infinite number of photons (and any number of bosons) can "occupy" a given point. Photons do not obey Pauli's Exclusion Principle.

The next step down in the hierarchy is the Cirpanic Cosmos. It is (categorically) everything (as a collective group) but no actual-wholeness function. It is the universal (holographic) waveform interfering with itself: representing "parts in the whole." It is every-possible-puzzle piece. It represents knowledge (the characteristic of consciousness). D. Bohm calls it the "holomovement." It is the potential (first reductive) aspect of existence. The "laws of nature" is an apt characterization. It also infers "every possible course of events." The Ultimate-Wholeness function "acquires" identity in its potentia. The Cirpanic Cosmos probably represents this identity of the wholeness function. It is aptly characterized as spatial extensionality—the "projecting" nature of consciousness (reasoning itself)—hence Realm of Reasoning. But, it is spatial extensionality with no manifest parts: this is rightfully called the ether.

The Cirpanic Cosmos Omnipresence (CCO) is "all time" and therefore "no (flow of) time." The CCO is composed of Bi-Deltic Cosmos Durations (BCD's). BCD's represent the "combined" cosmic cycles (durations) of the Cosmos and Anti-cosmos. BCD's are "categorically" directionless.

BCD's are composed of Bi-Deltic Instances (BCI's). BCI's are quantal instances of time. They too are (temporally) directionless. Bi-Deltic Cosmos instances "break down" into the separate Planck times of the Cosmos and Anti-cosmos. The Planck time represented in the cosmic instants (of Cosmos and Anti-cosmos) is directional and concatenate (serially) to form "the flow of time."

The flow of time is an illusion. Yes, time (flow) is one way of explaining change. Yet, the "flow of time" is also explainable from the perspective of consciousness itself, in which case there is only the present (no time flow). Consciousness projects itself as the "passing of events." We, as individuals, are conscious of the passing of events. That consciousness, which is our "higher self," is THE consciousness. Every being has the same consciousness in common. (Remember the inquisitive muscle cells? Refer to Chapter 20, heading "Transcendent Functions.") Consciousness functions only in the present. The present encompasses all time and therefore includes the flow of time. Time flow is only characteristic of experience.

Electromagnetism (unit) finally fractionates into the fermionic nature of the Cosmos and Anti-cosmos. Cosmos and Anti-cosmos are manifestations of the electric and magnetic reductive aspects of electromagnetism. The Cosmos is "represented" by the fermionic nature of the electron. Physical matter is chemical: interaction is electronic. Mentality is represented by the wholeness nature of the magnetic monopole.

The Cosmos has three aspects: the Where Aspect, the How Aspect, and the When Aspect.

The Where Aspect is constructed from two universes: the Universe of Perfect Whatness (physical matter) and the Universe of Imperfect Whyness (the mind).

A mind in Phase I is practically nonfunctional and "poses" as the Universe of Imperfect Whatness. The Universe of Imperfect Whatness does not exist. It is actually the Universe of Imperfect Whyness in "sleep" mode.

The mind in sleep mode has its parallel in consciousness. The mind-sleep mode is a reductionistic version of what happens when consciousness becomes "unconsciousness" in physical matter.

The How Aspect categorizes intelligence levels of reality. (Each level is inherently intelligent.) "Whatness" and "Whyness" classify realms. The Realm of Whatness breaks down into universes. Each universe further breaks down into functional (lesser intelligence) levels.

The When Aspect represents time flow. It is composed of a two-phase cycle: Phase I and Phase II. Phase I represents the half of the cycle in which physical matter dominates mind (which is asleep). The mind functions (Phase I) with respect to the physical world (i.e., deductively). Phase II represents the other half of the cycle when minds function on their own (i.e., inductively).

Briefly, we have explained the Ultimate Principle (viz, it explains itself). It defines space and consciousness.

The ultimate-wholeness function (no-thing) "fractionalizes" into a hierarchical array of functions (things). The breaking of higher-level functions is noted by physicists as entropy. Lower-level functions become manifest things evolving (changing) in time. That which changes (experience) opposes that which is changeless (consciousness).

The ultimate-wholeness function "pulls" parts into higher-level-wholeness functionality. How do we know this? This statement is the opposite of understanding how everything is created from nothing (the fractionating process).

The opposite of entropy is negentropy—things combining into higher-level functions, i.e., system building. Functions evolve, but toward what? Toward wholeness (from which they were originally derived [by fractionation of wholeness function]).

Things evolve "up" through (debifurcation) the functional hierarchy to return to the wholeness function. This is the insight that explains life-form evolution and complexity. "Higher-level" (functioning) life forms exhibiting complexity is what we would expect given this scenario. This is not the customary-linear evolution. It is evolution up through a hierarchy: parts assembling into higher-level functioning structures.

How does reality evolve? In temporal flow, wholes "entropically" break into parts (predominant process during cosmic cycle Phase I). In Phase II minds eventually (toward end of its time cycle) return "negentropically" to the wholeness function from which they came. Minds merge, and finally become one with the mind of Christ (Adam). Christ's mind is absorbed into consciousness at the end of Phase II. He becomes "all-knowing." At that point, there just is consciousness—the "spatial singularity."

The hierarchical structure of reality can be represented diagrammatically. Look at Figure 21:1.

Restating: If our interpretation of the cosmic cycle is correct, then we expect in Phase II of the cosmic time cycle that there is a "debifurcation" of minds. As the Phase II time cycle "winds down," parts (minds) begin to coalesce—eventually (at the end of Phase II) becoming one mind—which we call Adam. (Note: in Phase I, minds split—divide and hence multiply.)

We now understand the cosmic process: Manifest things evolve up (negentropy) or down (entropy). Manifest structure bifurcates and debifurcates. The hierarchical structure of reality is generated in the process.

We ask, from our perspective, why does man seek truth? We can generalize this statement by first answering it. "Man seeks truth to understand: to be able to explain things." This can also be generalized. "Explain why there are explanations." Explain explanations.

Understanding the cosmic process furnishes us with an answer. Consciousness is manifesting itself as an individual to govern its "creation" during Phase II (Heaven). God manifests himself as an individual (within his creation) who, during the end of Phase I, understands God and his purposes. Obviously, it is not this simple. When one wonders about the nature of innate intelligence during Phase I, it probably dawns on him that those mental beings inhabiting Phase II (in the Anti-cosmos) might have something to do with the course of life-form evolvement in Phase I (of our Cosmos).

Key Points:

- By self fractionation consciousness projects existence as a cosmic structural hierarchy of interacting functions.

- Consciousness fractionation creates the unit function—electromagnetic radiation (light). The rest of reality is a reductive partitioning of electromagnetism.

- Absolute wholeness is The reality. It "pulls" parts (negentropically) back to itself. It tends to sustain its self (its own functionality—absolute wholeness).

- Negentropy is "organized system building"—resulting in emergence of transcendent functions. Successive emergence of higher level transcendent functions tends to consolidate disparate functions into a single all embracing absolute functionality.

Final Correlations

The structure of reality, in this book, is modeled after the why-what paradigm. Reality's structure correlates with concepts found in other disciplines. Relevancies are easily established between the following: science, mathematics, philosophy, and religion.

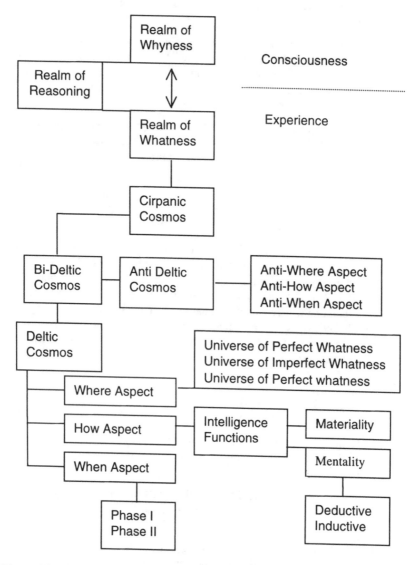

Figure 21:1. Hierarchical Structure of Reality

The structural mapping in this book was derived from scientific understanding. Establishing connections to other "systems of thought" not only allows a wider interpretation of reality's structure, but confers upon it added significance.

At the top of the hierarchical structure is the Realm of Whyness. It is consciousness—no-thingness. It is the ultimate reason for everything. Everything is categorically grouped, in our classification scheme (the why-what paradigm) as the Realm of Whatness. The Realm of Whatness is "what is experienced."

The Realms of Whyness and Whatness constitute Reality itself. On the road to understanding, we named a structure the Realm of Reasoning. We did not then realize it denoted reality as a "complete" structure.

Scientific concepts point to a "structure of reality." Other closely related concepts were cited. There are many concepts of various philosophies and religions that are compatible with the scientific interpretation of the structure of reality, but are not listed here. The idea here is to convey a general sense of connection between science and other disciplines. When religious and scientific concepts are understood, there are no significant inconsistencies between them.

Realm of Whyness

- It is the "whole of reality, but no part." It is the ultimate-wholeness function. It is absolute and "structureless." Characterized: it is a jigsaw puzzle as a whole, but not as pieces.

- In physics, it is the holographic-universal waveform. It is space itself, but not spatial extensionality. It is time (in itself) but not temporal extensionality (time flow). It is not characterizable in temporal terms (time flow). It only functions in the present. It is the "nonseparability" between everything. Time itself is space itself. (Space is all time, but no given time, and not even every instance of time flow.) It is the scientific-unifying principle under gravitation. As the "unifying" elementary particle, it is the Higg's boson. This project considers the Higg's particle to be synonymous with the graviton. Most other elementary particles become "reductive" (fractionated) functions of the Higg's boson via symmetry breaking.

- Of behavior: It is consciousness—the ultimate intelligence. All behaviors are a deductive function of consciousness. Consciousness "projects" experience.

- Of mathematics: It is the continuum. In set theory, it is the empty (null) set. Numerically, it is zero (0). In logic, it is falseness (F). These terms define the Ultimate Principle ("Nothing can both be and not be"). The Ultimate Principle "encapsulates" everything in terms of "nothingness." It is the Law of Noncontradiction. The whole of anything is (the nature of) "no-thingness" (no-parts). No-thingness "identifies" with "nothingness."

- Of philosophy: Plato spoke of it as absolute (idealism) and characterized it as "formness." Hegel called it the Absolute. Spinoza thought it was All reality. Many philosophers consider it the essence of everything. Leibniz called it the ultimate monad.

- Of religion: It is typically named the "Creator." It is also the higher being of which life forms are a part: it is one's soul. There have been many names for the ultimate intelligence. Hinduism calls it Brahmin. Christianity calls it God.

Realm of Reasoning

- It is reality itself. It is the "Sustainer." It is the whole of reality and every part. Parts are sustained by the whole; the whole is sustained by its parts. Parts are in the whole; the whole is (holographically) represented in its parts. It is the highest-level unit function. Because parts interrelate, reality has structure. Characterized: it is a "completed" jigsaw puzzle—its pieces "organized" (placed together and functioning) as wholes. This is the "Creating process." Parts are derived from the whole; the whole is pieced together by coupling of parts into higher-level structures.

- In physics: It is the "wave-particle" duality. It is space-time. It encapsulates all time. It is the cosmic-unifying "force"—electromagnetism (light itself). As a particle, it is the (bosonic) photon. The photon "unifies" the electric (represented by the electron) and magnetic (represented by the magnetic monopole) forces. It represents every possibility of existence.

- Of behavior: It is cosmic memory. Consciousness "stores" and "retrieves" knowledge. Knowledge is either being utilized or stored (in the structure of reality) for archival purposes.

- Of mathematics: It is the continuum and its discrete partitioning. In set theory, it is the highest-level "unit" set. Numerically, it is one (1). In logic, it is trueness (T). These terms define the logical Law of Contradiction. It is the law of complements (opposites). This law defines the nature of somethingness (any given thingness in the abstract). The Law of Contradiction states "something cannot both be and not be" (at any given time). The Law of Contradiction is derived from the Law of Noncontradiction.

- Of religion: It is practically an intelligence unto itself, yet it does not exist except for consciousness. Hinduism calls it Vishnu. Christianity calls it Lucifer (the light-bearer).

Realm of Whatness

- It is the "stuff of experience." It is "everything." It is every manifest thing (mental or physical) but not things as a collective (functioning) whole. It represents the "functional" separation of thingness. Characterized: It is the pieces of a jigsaw puzzle considered as parts (strewn on a table) but not considered as an "organized" whole. Each piece is a (functional) subunit of the puzzle. (Jigsaw pieces themselves are wholes.)

- Of physics: It represents "manifest" existence. Existence is manifested in a duality of opposition: Cosmos and Anti-cosmos, matter-antimatter, Cosmic forward time—reverse time, and the inverted time of the Anti-cosmos. Extended spatiality is also inverted because of the chiral nature (parity) of "matter and antimatter" things in space. Physical things exhibit rest mass.

- Of behavior: Minds are a parcel of existence. Minds deductively infer when inhabiting a physical body (typical of Phase I). Minds inductively infer knowledge when freed from physical bodies and "activated" (during

Phase II). Minds are produced in Phase I of the cosmic-time cycle. Minds are "self" activated during Phase II. Perception results from minds interacting with the "environment."

- Of mathematics: It represents the fractionation (numerical rationalization) of zero (continuum) into discreteness. Each number can represent "this and that" thingness in abstraction.

- Of religion: It is typically called the "Creation." Hinduism calls it Maya.

Universe of Perfect Whatness

- It defines physical matter. It is the noumenal "whatness" of experience. The Universe of Perfect Whatness is indirectly "presented" to an individual's senses (as sense data). It is the basis of empirical science. There are also physical "life forms." Physical life forms "reproduce" through sexual activity.

- Of religion: Many religions call it hell.

Universe of Imperfect Whatness

- This is an apparent mind: it does not exist in itself. The "imperfect" mind is activated by the phenomenal "whatness" of experience. It is the mental life form. The Universe of Perfect Whatness is what appears to the senses. It defines the mind when present in the physical world. It is actually a mind in "sleep" mode. It infers by deductive means. It functions with respect to the physical (whatness) world. It infers the nature of "this and that" thingness. It does not function on its own (therefore inductively). Minds are created in the staging area of the physical world where entropy splits minds from other minds.

- Of behavior: By identifying this and that, it perceives thingness. Perception results from mind encountering its environment.

- Of religion: The mind is the spirit. A mind that functions with respect to the physical world is a devil (deductive function). A mind can "reincarnate" a body to fulfill purpose inherent in its destiny.

Universe of Imperfect Whyness

- It is the functioning mental life form. It is a mind. It infers inductively.

- Of behavior: This is the mind functioning on its own: without respect to the physical world. It inductively infers the reality it experiences.

- Of religion: The mind is functional and in its natural element while in Phase II. This is heaven.

Final Notes and Final Test

Is what is stated in this book true? Much of it is assuredly true. Some statements, though meaningful, might not reflect reality. That should always be considered when examining each statement. However, is the book's general understanding true? Once understood it seems difficult to imagine it otherwise. In that sense then about the only general big question not answered is the following. If the basic presentation in this book is not the whole truth, then what is? We shortly describe the final test that indicates we have stayed the serious truth-seeking course. Let's briefly trace the path to the final understanding.

Some concepts, which early on were considered possibly true, were interpreted "backwards." Understanding the truth behind appearances is what seeking deeper truth is all about. The "backwardness" of physical existence impedes understanding. For example, the apparent color of something is backwards from its actual color. If not, then the apparent color becomes difficult to explain.

The color paradigm is a model for explaining things. We asked of any and every phenomenon: How can this be explained? We sought answers and were granted answers. Are there other ways in which things could be explained? Definitely, yes. Is the author asserting the truth of what is stated in this book? No.

The author stated what he learned about everything. He contends truth (in statement form) depends on what reality demonstrates.

Ask: Does a given statement correctly represent reality? Does reality support the general truth of what is stated? For each person to determine the truth of what is stated, he (she) needs to seek and understand reality. Words and concepts are not reality. The question is: Do the statements constituting this book, correctly represent reality?

Do not deceive yourself about reality. Let reality "speak" for itself. If a person "makes up" his own mind about what is true, he will not listen to those who understand differently and will probably not be attentive to the subtleties of reality (itself).

Where does one turn to learn what is true? Other people or reality? The final determiner of truth is reality. People help; books and other media help. Truth is ultimately contingent upon what reality asserts is true. The final truth is related to the first truth—the nature of cosmogenesis. This is shortly addressed.

What is the secret to understanding life? "Do not deceive yourself!" No one can deceive you. One only deceives himself. Do not deceive yourself about what is true. One knows only a statement's truth when he can scientifically verify it represents something in reality. If he cannot verify or prove it represents reality, then he does not know whether it is true or not. Admitting one does not know, is the beginning of learning.

Who is honest with reality?

He who asserts statements to be true of reality, but cannot verify it. OR

He who asserts statements to be true of reality and can verify it.

Someone, who asserts truth and cannot verify it, is a liar because he does not know the reality to which the statement corresponds. The truth about reality is revealed by reality. Reality is not what people believe. People believe when they do not know. Believing turns one from reality.

The test whether one has been brainwashed or not, is if he can explain what he asserts without any contradiction. If he insists on the truth of what he cannot explain, then he is a liar—he does not know the truth, contrary to his assertion.

Learning the deeper aspects of reality depends on understanding logic. There should be no contradiction in understanding the complete nature of reality. Things relate logically. To represent ultimate reality demands a principle that is both complete and consistent. We have suggested such a principle. It is the negation of the logical Law of Contradiction. It is the Law of Noncontradiction. It has been known for thousands of years. It just has not been understood. It explains everything, including the very nature of cosmogenesis. Let's take the final test.

Looking back, we now know how the Ultimate Principle answers the Big Bang problem. First, let's pose the final questions. Namely, how and why did the Big Bang arise? That the Ultimate Principle clearly resolves these questions ultimately explains everything. We just needed to understand everything else before we could understand the nature of the Big Bang. The Big Bang Problem is the final test of the Ultimate Principle.

The Problems: The first problem was to explain the Big Bang as an expansion of space. Specifically: What is space expanding into—if not itself? The second problem was to explain why there was a Big Bang.

The Answers: If space is expanding into itself, then there is a difference between expanding space and the space into which it is expanding. This is a problem of identity. How can the same be different? A clear understanding of the Law of Noncontradiction (the Ultimate Principle) explains the difference between space itself and its expansion into manifest dimensionality.

Extensional space is an illusion. It is but a projection of the space function—the wholeness function of reality. The wholeness of space is THE reality. The wholeness of space is consciousness—ultimate intelligence. Its projection via the Big Bang—namely spatial expansion and hence spatial extensionality—is an illusion. There is no more than the whole of reality. Consciousness just is.

This thing and that thing, including the extended space in which they exist, are but functional fractionations of the whole of reality. We experience things in spatial dimensionality. Simply, experience, and hence spatial dimensionality, is a projection of consciousness. Things and minds do not exist independent of consciousness. Every part implies there is a greater whole. This is self-evident.

The difference between space itself and its fractionation into extensionality is merely one of function. Nonfunctionally, there is no difference between space itself and its extensionality. We now understand the difference between space itself and its extension into manifest reality. "The same can be different." We know this because zero (or nothingness) is nonfunctionally equivalent to one (or unity). The whole of reality is nonfunctionally equivalent to the unit-structure of reality. Consciousness is nonfunctionally equivalent to experience. Consciousness is God. God is nonfunctionally equivalent to its creation—manifest reality. The Ultimate Principle passes the final test. It explains cosmogenesis.

So what does this all mean? One must be unselfish to seriously seek truth. If not to help others, seeking knowledge will be for naught. There is a bonus to being unselfish. The mind (spirit) develops by inferring the nature of reality. Understanding develops the mind for its entrance into the next existence.

There are some reasons not to publish this book. Many people will not study this book because it is contrary to their belief. Until properly understood, there will be problems associated with its publication.

> History reveals false ideas do not endure. Many misconceptions about reality have been corrected by scientific effort. Educated people now accept the earth is not flat. The truth about reality will be known, whether this book is published.

There is also a personal reason not to publish. I did not ask to do the project resulting in this book. Extraordinary experiences obliged me to seek deeper truth (to better help people). I was compelled to do this project. There is another reason. Consciousness enabled this to occur for its own purposes: to reacquire understanding.

There are reasons to publish. Many scientists have sought truth and toiled for answers. The fruit of their labor is the basis of this book. It is only fair to them this book is published. They will learn why truth seeking serves a higher purpose. This book also offers the public the opportunity to better understand reality. People cannot later assert they were uninformed and did not understand. The mind develops by seeking truth to understand reality to help people.

The curtain of ignorance is lifted. Reality is all around us. The methodology of truth seeking is clear:

> Do not deceive yourself about reality. Reality is the teacher. Scientists enable reality to speak through experimentation. Ask how things can be explained. Truth seekers learn about reality by inductive reasoning.

Purpose will be realized. Scientists, philosophers, and truth seekers have had a special role. They contributed to a higher purpose—understanding reality—to regain the "lost knowledge." I had a role in that effort.

This book was written to answer the question: "Who am I?" I learned about reality and about myself. I understand my place in the scheme of things. I know who I am. I learned and reasoned to help people. Years passed before I realized how understanding reality also served a higher purpose. In that purpose, I found greater meaning. I was destined for an interesting role.

Some people are different, not because they want to be, but because their destiny involves a difference. This book describes what I learned. I learned about change. Everything changes. The end is just the beginning.

Bibliography

Abell, George O. "Cosmology—the Origin and Evolution of the Universe." *Mercury,* (1978) 7:3 pp. 45–50, 61.

Adair, Robert. "A Flaw in the Universal Mirror." *Scientific American,* (1988) 258:5 pp. 50–56.

Adams, J. *Learning and Memory, An Introduction.* Homewood Ill: Dorsy Press, 1976.

Adler, Carl G. "Realism and/or Physics." *American Journal of Physics,* (1989) 57:10 878–82.

Akon, Daniel. "Memory Storage and Neural Systems." *Scientific American,* (1989) 261:1 pp. 42–50.

Albers, Don. "The Meaning of Curved Space. *Mercury,* (1975) 4:4 pp. 16–19.

Albert, David Z. "Bohm's Alternative to Quantum Mechanics." *Scientific American,* (1994) 270:5 pp. 58–67.

Alekseev, G.N. *Energy and Entropy.* Trans. from 1978 Russian. Ed.by U.M. Taude. Moscow: Mir Publishers, 1986.

Alfvèn, Hannes. "Antimatter and Cosmology." *Scientific American,* (1967) 216:4 pp. 106–112.

Allen, Edgar L. *From Plato to Nietzsche: An Introduction to The Great Thoughts and Ideas of the Western Mind.* 1957; rpt. Greenwich, Conn.: Fawcett Premier Book.

Allfrey, Vincent G. and Alfred Mirsky. "How Cells Make Molecules." *Scientific American,* 1961; rpt. by W.H. Freeman.

Allman, William F. "The Mother Tongue." *U.S.News and World Report,* November 5, 1990, pp. 60–70.

Alpher, Ralph A, and Robert Herman. "Reflections on Early Work on Big Bang Cosmology." *Physics Today,* (1988) 41:8 pp. 24–34.

Amaldi, Edoardo. "The Unity of Physics." *Physics Today,* (1973) 26:9 pp. 23–29.

Anderson, Alun, and Bob Holmes and Liz Else. "Zombies, Dolphins, and Blindsight." *New Scientist,* (1996) 151:2028 pp. 20–27.

Anderson, James L. "Newton's First Two Laws of Motion are not Definitions." *American Journal of Physics,* (1990) 58:12 pp. 1192–95.

Antippa, Adel F. "General Three Dimensional Superluminal Transformations and Tachyon Kinematics. *Physics Reviews D,* (1975) 11:4 pp. 724–739.

Antoine, J.P. "Group Theory in Physics." *Encyclopedia of Physics.* Ed. by R. Lerner and G. Trigg. Reading, Mass.: Addison-Wesley, 1981.

Aspden, H. "Extending Space or Separating Galaxies." *American Journal of Physics,* Letter (1988) 56:7 p. 584.

Aspray, William and Philip Kitcher. eds. *History and Philosophy of Modern Mathematics.* Vol. 11 in Minnesota Studies in the Philosophy of Science. Minneapolis: University of Minnesota Press, 1988.

Atkatz, David. "Quantum Cosmology for Pedestrians." *American Journal of Physics,* (1994) 62:7 pp. 619–27.

Atkins, P.W. *Quanta: A Handbook of Concepts.* Oxford: Oxford University Press, 1974.

Avetisov, Vladik A., Vitalii I. Goldanskii and Vladmir V. Kuz'min. "Handedness, Origin of Life and Evolution." *Physics Today,* (1991) 44:7 pp. 33–41.

Axon, T.J. "Introducing Schrödinger's Cat in the Laboratory." *American Journal of Physics,* (1989) 57:4 pp. 317–21.

Ayala, Francisco J. "The Mechanisms of Evolution." *Scientific American,* (1978) 239:3 pp. 56–69.

Ayer, A.J. *The Central Questions of Philosophy.* New York: William Morrow, 1973

Badash, Lawrence. "The Discovery of Radioactivity." *Physics Today,* (1996) 49:2 pp. 21–26.

Bahm, Archie J. *Metaphysics: An Introduction: A Survey and Interpretation of Philosophic Inquiries into the Nature of Existence and its Categories.* New York: Barnes and Noble, 1974.

Bak, Per and Kan Chen. "Self-Organized Criticality." *Scientific American,* (1991) 264:1 pp. 46–53.

Baker, Adolph. *Modern Physics and Antiphysics.* Addison-Wesley Series in Physics. Reading, Mass: Addison-Wesley, 1970.

Bakker, Gerald and Len Clark. *Explanation: An Introduction to the Philosophy of Science,* Mountain View, California. 1988.

Balian, Roger. "On the Principles of Quantum Mechanics and the Reduction of the Wave Packets. *American Journal of Physics,* (1989) 57:11 pp. 1019–26.

Ballantine, L.E. "Remarks on the Interpretation of Quantum Mechanics." *American Journal of Physics,* (1984) 52:3 pp. 271–72.

————. and Jon P. Jarrett. "Bell's Theorem: Does Quantum Mechanics Contradict Relativity?" *American Journal of Physics,* (1987) 55:8 pp. 696–701.

Barbour, Dan G. ed. *Science and Religion.* New York: Harper and Row, Publishers, 1968.

Barger, Vernon D. and David B. Cline. "High-Energy Scattering." *Scientific American,* (1967) 217:6 pp. 77–91.

Barghoorn, Elso S. "The Oldest Fossils." *Scientific American,* (1971) 224:5 pp. 30–42.

Barnes, Wesley. *The Philosophy and Literature of Existentialism.* Woodbury: Barron's Educational Series, 1968.

Barrow, John D. and Joseph Silk. "The Structure of the Early Universe." *Scientific American,* (1980) 242:4 pp. 118–128.

————. *Theories of Everthing: the Quest for the Ultimate Explanation.* Oxford: Clarendon Press, 1991.

Bartusiak, Marcia. "Loops of Space." *Discover,* (1993) 14:4 pp. 61–68.

Bastin, T. ed. *Quantum Theory and Beyond.* London: Cambridge University Press, 1971.

Beard, Ruth M. *An Outline of Piaget's Developmental Psychology for Students and Teachers.* New York: New American Library 1969.

Beard, Ruth M. *Piaget's Developmental Psychology.* New York: New American Library, 1969.

Beardsley, Tim. "The Machinery of Thought." *Scientific American,* (1997) 277:2 pp. 78–83.

Beck, Jacob. "The Perception of Surface Color." *Scientific American,* (1975) 232:2 pp. 62–75.

Begley, Sharon, John Carey and Ray Sawhill. "How the Brain Works." *Newsweek,* (February 7, 1983) pp. 40–48.

Bell, A.B., and Bell, D.M. "Energy in a Highly Ordered Universe." *Foundations of Physics,* (1975) 9:5/6 pp. 471–77.

Benacerraf, Paul and Hilary Putman. eds. *Philosophy of Mathematics: Selected Readings.* Englewood Cliffs, NJ: Prentice Hall, 1964.

Ben-Dov, Yoa. "Everett's Theory and the 'Many Worlds' Interpretation." *American Journal of Physics,* (1990) 58:9 pp. 829–32.

Bennett, Charles H. "Demons, Engines, and the Second Law." *Scientific American,* (1987) 257:5 pp. 108–16.

Bergmann, Merrie, James Moor, and Jack Neson. *The Logic Book.* New York: Random House. 1980.

Bergmann, Peter G. *The Riddle of Gravitation.* New York: Charles Scribner's and Sons, 1968.

Bergson, Henri. *Time and Free Will.* New York: Harper and Row, 1960.

Berkelman, K. "The Compton Effect." *Encyclopedia of Physics.* Ed. by R. Lerner and G. Trigg. Reading, Mass.: Addison-Wesley, 1981.

Bernstein, Herbert J. and Anthony V. Phillips. "Fiber Bundles and Quantum Theory."
    Scientific American, (1981) 245:1 pp. 123–37.
Berofsky, Bernard. *Determinism.* Princeton: Princeton University Press, 1971.
_____. ed. *Free Will and Determinism.* New York: Harper and Row, 1966.
Berry, Michael. *Principles of Cosmology and Gravitation.* London: Cambridge
    University Press. 1976.
Bettinghaus, Erwin P. *The Nature of Proof.* Ed. by Reissell R. Windes. 2nd ed.
    Indianapolis: Bobbs Merrill, 1972.
Bilaniuk, O.M.P. and E.C.G. Sudarshan. "Particles Beyond the Light Barrier." *Physics
    Today,* (1969) 22:5 pp. 43–51.
_____., V. K., Deshpande, and E.C.G. Sudarshan. "Meta Relativity." *American
    Journal of Physics,* (1962) 30:10 pp. 718–23.
Bingham, Roger. "On the Life of Mr. Darwin." *Science 82,* (1982) pp. 34–45.
Black, Max.*The Labyrinth of Language.* New York: New American Library, 1968.
Black, Meme. "Brain Flash: the Physiology of Inspiration." *Science Digest,* (1982) 90:8
    pp. 85–87, 104.
Blackenbecler, R. "Regge Poles." *Encyclopedia of Physics.* Ed. by R. Lerner and G.
    Trigg. Reading, Mass.: Addison-Wesley, 1981.
Blackmore, Susan. "'Consciousness: Science Tackles the Self." *New Scientist,* (1989)
    122:1658 pp. 38–41.
Blakeslee, T. *The Right Brain.* Garden City, NJ: Anchor Press, Doubleday. 1980.
Blokhintsev, D.I. *Space and Time in the Microworld.* Boston: D. Reidel, 1973.
Bloom, Elliott D. and Gary J. Feldman. "Quarkonium." *Scientific American,* (1982)
    246:5 pp. 66–77.
Bohm, David and B. Hiley. "On the Intuitive Understanding of Nonlocality as Implied by
    Quantum Theory. *Foundations of Physics,* (1975) 5:1 pp. 93–109.
_____. and David Peat. *Science, Order, and Creativity.* New York: Bantam Books,
    1987.
_____. *Causality and Chance in Modern Physics.* New York: Harper and Brothers,
    1957.
_____. *Quantum Theory.* Englewood Cliffs, NJ: Prentice Hall, 1951.
_____. *Unfolding Meaning: A Weekend of Dialogue with David Bohm.* 1985; rpt.
    New York: Ark Paperbacks, 1987.
_____. *Wholeness and the Implicate Order.* Boston: Routledge and Kegan Henley,
    1980.
Bohr, Aage and Ben R. Mottelson. "The Many Facets of Nuclear Structure." *American
    Scientist,* (1973) 61:3 pp. 446–55.
Boughn, Stephen and Ho Jung Paik. "The Search for Gravitational Radiation." *Mercury,*
    (1976) 5:3 pp. 9–15.
Bourne, Lyle E., Bruce R. Ekstrand and Roger L. Dominowski. *The Psychology of
    Thinking.* Englewood Cliffs NJ: Prentice-Hall, 1971.
Bower, Bruce. "Consciousness Raising." *Science News,* (1992) 142:15 pp. 232–35.
_____. "Roots of Reason." *Science News,* (1994) 145:5 pp. 72–75.
Bowers, John. *Invitation to Mathematics.* New York: Basil Blackwell, 1988.
Bracewell, Ronald N. "The Fourier Transform." *Scientific American,* (1989) 260:6 pp.
    86–95.
Brantl, George. *Catholicism.* New York: Washington Square Press, 1961.
Bronowski, J. "The Clock Paradox." *Scientific American,* (1963) 208:2 pp. 134–44.
Brown, Erik. "The Direction of Causation." *Mind,* (1979) 88: pp. 334–50.
Brown, Gerald E. and Mannque Rho. "The Structure of the Nucleon." *Physics Today,*
    (1983) 36:2 pp. 24–32.

Brown, H. R. and M.L.G. Redhead. "A Critique of the Disturbance Theory of Indeterminacy in Quantum Mechanics. *Foundations of Physics*, (1981) 11:1/2 pp. 1–19.

Brown, H.R. "De Broglie's Relativistic Phase Waves and Wave Groups." *American Journal of Physics*, (1984) 52:12 pp. 1130–39.

Brown, Julian. "Is the Universe a Computer?" *New Scientist*, (1990) 127:1725 pp. 37–39.

Brown, L.M. and L. Hoddeson. "The Birth of Elementary-Particle Physics." *Physics Today*, (1982) 35:4 pp. 36–43.

Bry, Adelaide. *A Primer of Behavioral Psychology*. New York: New American Library, 1977.

Bub, J. "Complementarity." Ed. by R. Lerner and G. Trigg. *An Encyclopedia of Physics*. Reading, Mass., Addison-Wesley, 1981.

Buchanan, Mark. "An End to Uncertainty." *New Scientist*, (1999) 161:2176 pp. 25–28.

Bucher, Martin A. and David N. Spergel. "Inflation in a Low-Density Universe." *Scientific American*, (1999) 280:1 pp. 62–68.

Buckhout, Robert. "Eyewitness Testimony." *Scientific American*, (1974) 231:6 pp. 23–31.

Bunge, M. ed. *Quantum Theory and Reality*. Heidelberg: Springer-Verlag, 1967.

Bunn, Robert. "Michael Hallett's Cantorian Set Theory and Limitation of Size." *Philosophy of Science*, (1988) 55:4 pp. 467–78.

Burke, E. "Bohr Theory of Atomic Structure." *Encyclopedia of Physics*. Ed. R. Lerner and G. Trigg. Reading, Mass.: Addison-Wesley, 1981.

Byrne, P.H. "Relativity and Indeterminism." *Foundations of Physics*, (1981) 11:11/12 pp. 913–32.

"Calculus of Tensors." *McGraw –Hill Encyclopedia of Science and Technology."* 1982, p. 438.

Callahan, J.J. "The Curvature of Space in a Finite Universe." *Scientific American*, (1976) 235:2 pp. 90–100.

Cann Rebecca L. "In Search of Eve." *The Sciences*, (1987) 27:5 pp. 30–37.

Capek, Milic. "Relativity and the Status of Becoming." *Foundations of Physics*, (1975) 5:4 pp. 607–16.

_____. *The Philosophical Impact of Contemporary Physics*. Princeton: Van Nostrand, 1961.

Carnap, Rudolf. *Foundations of Logic and Mathematics*. Vols. I and II of the Unity of Science. Chicago: University of Chicago Press.

Carney, James and Richard K. Scheer. *Fundamentals of Logic*. New York. Macmillan Pub., 1980.

Carpenter, Finley. *The Skinner Primer: Behind Freedom and Dignity: What the B.F. Skinner Debate is All About*. New York: The Free Press, 1974.

Carr, Herbert Wildon. *Leibniz*. rpt. New York: Dover Publications, 1960.

Carrigan, Richard A. Jr. "Quest for the Magnetic Monopole." *The Physics Teacher*, (1975) 13:7 pp. 391–398.

_____. and W. Peter Trower. "Superheavy Magnetic Monopoles." *Scientific American*, (1982) 246:4 pp. 106–18.

Casey, Edward S. "Perceiving and Remembering." *The Review of Metaphysics*, (1979) 32:3 pp. 407–36.

Cashmore, Roger and Christine Sutton. "The Origin of Mass." *New Scientist*, (1992) 136:1817 pp. 35–39.

Cassidy, David. "Heisenberg, Uncertainty and the Quantum Revolution." *Scientific American*, (1992) 266:5 pp. 106–112.

Casten, Richard F. and Da Hsuan Feng. "Nuclear Dynamical Supersymmetry." *Physics Today*, (1984) 37:11 pp. 26–35.

Casti, John. "Confronting Science's Logical Limits." *Scientific American,* (1996) 275:5 pp. 102–05.

Cermak, L.S. ed. *Human Memory and Amnesia.* Lawrence Erbaum Associates. Hillside, NJ. (1982).

Chalmers, David J. "The Puzzle of Consciousness." *Scientific American,* (1995) 273:6 pp. 80–86.

Champin, T.S. *Reflexive Paradoxes.* New York: Routledge, 1988.

Chandrasekhar, S. "The General Theory of Relativity: the First Thirty Years." *Contemporary Physics.* (1980) 21:5 pp. 429–49.

Charlesworth, Brian. "Neo-Darwinism—the Plain Truth." *New Scientist,* (1982) 94:1301 pp. 133–36.

Cherry, Laurence. "A New Vision of Dreams." *New York Times Magazine,* July 3, 1977 pp. 9–13, 34.

Chew, Geoffrey F., Murray Gell-Mann and Arthur H. Rosenfeld. "Strongly Interacting Particles." *Scientific American,* (1964) 210:2 pp. 74–93.

Chiao, Raymond Y., Paul G. Kwiat, and Aephraim M. Steinberg. "Faster than Light?" *Scientific American,* (1993) 269:2 pp. 52–60.

Christenson, J.H., J.W. Cronin, V.L. Fitch, and R. Turlay. "Evidence for the 2π Decay of the K_2^0 Meson." *Physical Review Letters.* 13:4 pp. 138–140.

Christian, James L. *Philosophy: An Introduction to the Art of Wondering.* 2nd. ed. Corte Madera, California: Holt, Rinehart and Winston, 1977.

Churchland, Paul M. and Patricia Smith Churchland. "Could a Machine Think? *Scientific American,* (1990) 262:1 pp. 32–37.

_____. ed. *Matter and Consciousness: A Contemporary Introduction to the Philosophy of Mind.* London: MIT Press. 1988.

Cline, D.B. and C. Rubbia, and S. van der Meer. "The Search for Intermediate Vector Bosons. *Scientific American,* (1982) 246:3 pp. 48–59.

Cline, David B. Alfred K. Mann, and Carlo Rubbia. "The Detection of Neutral Weak Currents. *Scientific American,* (1974) 231:6 pp. 108–119.

Close, F.E. "Unified Field Theory: Dream Becoming Reality?" *Nature,* (1979) 278:5701 pp. 209–10.

Cofer, Charles N. "Constructive Processes in Memory. *American Scientist,* (1973) 61:5 pp. 537–43.

Cohen, Paul J. and Reuben Hersh. "Non-Cantorian Set Theory." *Scientific American,* (1967) 217:6 pp. 104–16.

_____. *Set Theory and the Continuum Hypothesis.* New York: W.A. Benjamin, 1966.

Colella, R. and A.W. Overhauser. "Neutrons, Gravity, land Quantum Mechanics. *American Scientist,* (1980) 68:1 pp. 79–75.

Combs, Allan and Mark Holland. *Synchronicity: Science, Myth, and the Trickster.* New York: Paragon House, 1990.

Compagner, Aaldert. "Definitions of Randomness." *American Journal of Physics,* (1991) 59:8 pp. 700–705.

_____. "Definitions of Randomness." *American Scientist,* (1991) 79:8 pp. 700–05.

Connor, Daniel John. *Free Will.* Garden City, NJ: Doubleday, 1971.

Conway, John. *On Numbers and Games.* New York: Academic Press, 1976.

Copi, Irving M. *Introduction to Logic.* 4th ed. New York: The Macmillan, 1972.

Corballis, Michael C. and Ivan L. Beale. "On Telling Left from Right. *Scientific American,* (1971) 224:3 pp. 96–104.

Corben, H.C. "Tachyon Matter and Complex Physical Variables." *Il Nuovo Cimento,* 29A:3 pp. 415–425.

_____. and E. Honig. "Behaviour of Electromagnetic Charges Under Superluminal Transformations." *Lettere Al Nuovo Cimento,* (1975) 13:15 pp. 586–588.

Corwin, Mike and Dale Wachowiak. "Lookback Time: Observing Cosmic History." *The Physics Teacher,* (1989) 27:7 pp. 518–24.

Costa de Beauregard, O. "Running Backwards the Mermin Device: Causality in EPR Correlations." *American Journal of Physics,* (1983) 51:6 pp. 513–16.

Crane, H.R. "The g-Factor of the Electron." *Scientific American,* (1968) 218:1 pp. 72–85.

Crick, Francis and Christof Koch. "The Problem of Consciousness." *Scientific American,* (1992) 267:3 pp. 153–59.

Cronin, James. "CP Symmetry Violations: the Search for its Origin." *Science,* (1981) 212:4500 pp. 1221–1228.

Crossley, J.N., C.J. Ash, C.J Brickhill, J.C. Stillwell, and N.H. Williams. *What is Mathematical Logic?* rpt. New York: Dover Publications, 1990.

Cushing, James T. and Ernan McMullen. eds. *Philosophical Consequences of Quantum Theory: Reflections on Bell's Theorem.* Notre Dame, Indiana: University of Notre Dame Press, 1989.

d'Abro, A. *The Evolution of Scientific Thought.* 2nd ed. New York: Dover Publications, 1950.

d'Espagnat, Bernard. "The Concepts of the Influences and Attributes as Seen in Connection with Bell's Theorem. *Foundations of Physics.* (1980) 11:3/4 pp. 205–33.

_____. "The Quantum Theory and Reality." *Scientific American,* (1979) 241:5 pp. 158–81.

_____. *Conceptual Foundations of Quantum Mechanics.* Reading, Mass.: W.A. Benjamin, 1976.

_____. *Reality and the Physicists.* Trans. J.C. Whitehouse. New York: Cambridge University Press, 1985.

Damasio, Antonio R. and Hanna Damasio. "Brain and Lanuage." *Scientific American,* (1992) 267:3 pp. 89–95.

Dancy, Jonathan. *An Introduction to Contemporary Epistemology.* 1985; rpt. New York: Basil Blackwell, 1989.

Dantzig, Tobias. *Number: the Lanuage of Science,* 1933; rpt. New York: The Free Press, 1954.

Darling, David. "The Quest for Black Holes." *Astronomy,* (1983) 11:7 pp. 12–20.

Dauben, Joseph W. *Georg Cantor, His Mathematics and Philosophy of the Infinite.* Cambridge, Mass.: Harvard University Press, 1979.

_____. *George Cantor, His Mathematics and Philosophy of the Infinite.* Cambridge, Mass.: Harvard University Press., 1979.

Davies, P.C.W. "Law and Order in the Universe." *New Scientist,* (1988) 120:1634 pp. 58–60.

_____. "Some Singular Proposals." *Nature,* (1977) 266:3 pp. 12–13.

_____. "The Anthropic Principle and the Early Universe." *Mercury.* (1981) 10:3 pp. 66–77.

_____. "The Day Time Began." *New Scientist,* (1996) 150:2027 pp. 30–35.

_____. "The Inflationary Universe." *The Sciences,* (1983) 23:2 pp. 32–36.

_____. "Time." Ed. by R. Lerner and G. Trigg. *An Encyclopedia of Physics.* Reading, Mass., Addison-Wesley, 1981.

_____. "Time's Arrow." *New Scientist,* (1997) 156:2106 pp. 34–38.

_____. and J.R. Brown. eds. *The Ghost in the Atom: A Discussion of the Mysteries of Quantum Physics.* New York: Cambridge University Press, 1986.

_____. and J.R. Brown. *Superstrings: A Theory of Everything?* New York: Cambridge University Press, 1988.

_____. *God and the New Physics.* London: J.M. Dent and Sons, 1983.

_____. *Other Worlds.* New York: Simon and Schuster, 1980.

_____. *Space and Time in the Modern World.* New York: Cambridge University Press, 1977.

_____. *Superforce: the Search for a Grand Unified Theory of Nature.* New York: Simon and Schuster, 1984.

_____. *The Edge of Infinity: Where the Universe Came From and How it Will End.* New York: Simon and Schuster, 1981.

_____. *The Forces of Nature.* London: Cambridge University Press, 1979.

Davis, Martin. ed. *The Undecidable: Basic Papers on Understandable Propositions, Unsolvable Problems and Compatible Functions.* Hewitt, New York: Raven Press, 1965.

Davis, Philip J. and Reuben Hersh. *The Mathematical Experience.* Boston: Houghton Mifflin. 1981.

Dawkins, Richard. *The Blind Watchmaker.* New York: W.W. Norton, 1987.

_____. "The Necessity of Darwinism." *New Scientist,* (1982) 94:1301 pp. 130–41.

Dawson, John W. Jr. "Gödel and the Limits of Logic." *Scientific American,* (1999) 280:6 pp. 76–81.

De Benetti, Sergio. "The Mössbauer Effect." *Scientific American,* (1960) 202:4 pp. 73–80.

De Broglie, Louis. *The Current Interpretation of Wave Mechanics.* New York: Elsevier, 1964.

De Long, Howard. *A Profile of Mathematical Logic.* Reading, Mass.: Addison-Wesley, 1970.

De Vaucouleurs, G. "The Case for a Hierarchical Cosmology." *Science,* (1970) 167:3922 pp. 1203–13.

Deltete, Robert. "Einstein's Opposition to Quantum Theory." *American Journal of Physics,* (1990) 58:7 pp. 673–83.

Dennett, Daniel C. "Darwin's Dangerous Idea." *The Sciences,* (1995) 35:3 pp. 34–40.

Deregowski, Jan B. "Pictural Perception and Culture. *Scientific American,* (1972) 227:5 pp. 82–88.

Dettling, Ray J. "The Quest for the Ultimate Particle." *Science Digest.* (1982) 90:12 pp. 79–82.

Devlin, Keith J. *Fundamentals of Contemporary Set Theory.* New York: Springer-Verlag, 1979.

_____. *Mathematics: The Science of Patterns.* New York: Scientific American Library, (1994).

_____. *Mathematics: The New Golden Age.* New York: Penguin Books, 1988.

DeWitt, Bryce S. "Quantum Gravity." *Scientific American,* (1983) 249:6 pp. 112–29.

_____. and Neill Graham eds. *The Many –Worlds Interpretation of of Quantum Mechanics.* Princeton: Princeton University Press, 1973.

Dicke, R.H. "The Eötvös Experiment." *Scientific American,* (1961) 205:6 pp. 86–94.

Dickerson, Richard E. "Chemical Evolution and the Origin of Life." *Scientific American,* (1978) 239:3 pp. 70–87.

Dicus, D.A., J.R. Letaw, D.C. Teplitz, V.L. Teplitz. "The Future of the Universe." *Scientific American,* (1983) 248:3 pp. 90–101.

Dimopoulas, Savas, Stuart A. Raby, and Frank Wilczek. "Unification of Coupling Constants." *Physics Today,* (1991) 44:10 pp. 25–32.

Dirac, P.A.M. "The Evolution of the Physicist's Picture of Nature." *Scientific American,* (1963) 208:5 pp. 45–53.

Ditto, William L. and Louis M. Pecora. "Mastering Chaos." *Scientific American,* (1993) 269:2 pp. 78–84.

Dorfan, D., J. Enstrom, D. Raymond, M. Schwartz, S. Wijcicke, and D. Ray Miller. "Charge Asymmetry in the Muonic Decay of the K_2^0. *Physical Review Letters,* (1966) 19:17 pp. 987–993.

Dotson, Allen C. "Bell's Theorem and the Features of Physical Properties." *American Journal of Physics,* (1986) 54:3 pp. 218–21.

Drake, S. "Newton's Apple and Galileo's Dialogue." *Scientific American,* (1980) 243:2 pp. 150–56.

Dray, Tevian. "The Twin Paradox Revisted." *American Journal of Physics,* (1990) 58:9 pp. 822–25.

Drell, Sidney D. "Electron-Positron Annihilation and the New Particles." *Scientific American,* (1975) 232:6 pp. 50–62.

Dryzek, J. and K. Ruebenbauer. "Planck's Constant Determination from Black-Body Radiation." *American Journal of Physics,* (1992) 60:3 pp. 251–53.

Duff, Michael J. and Christine Sutton. "The Membrane at the End of the Universe." *New Scientist,* (1988) 118:1619.

_____. "The Theory Formerly Known as Strings." *Scientific American,* (1998) 278:2 pp. 64–69.

Dummett, Michael. *Truth and Other Enigmas.* Cambridge, Mass.: Howard University Press, 1978.

Dunham, William. *The Mathematic Universe.* New York: John Wiley and Sons, 1994.

Dunstan, Leslie J. *Protestantism.* New York: Washington Square Press, 1962.

Dyson, Freeman J. "Energy in the Universe." *Scientific American,* (1971) 224:3 pp. 51–59.

_____. "Field Theory." *Scientific American,* (1953) 188:4 pp. 57–64.

Earman, J. "Covariance, Invariance, and the Equivalence of Frames." *Foundations of Physics,* (1974) 4:2 pp. 267–89.

Eddington, Arthur. *The Philosophy of Physical Science.* Ann Arbor: University of Michigan Press, 1958.

Edmonston, W. *Hypnosis and Relaxation.* New York: John Wiley and Sons, 1981.

Einstein, Albert, Boris Podolsky and Nathan Rosen. "Can Quantum-Mechanical Description of Reality be Complete?" *Quantum Theory and Measurement.* Ed. by Wheeler, John A. and Wojciech H. Zurek. Princeton Series in Physics. Princeton: Princeton University Press, 1983.

_____. "On the Generalized Theory of Gravitation." *Scientific American,* (1950) 188:4 pp. 13–17.

Ekstrom, Philip and David Wineland. "The Isolated Electron." *Scientific American,* (1980) 243:2 pp. 105–13.

Ellis, G.F.R. and T. Rothman. "Lost Horizons." *American Journal of Physics,* (1993) 61:10 pp. 883–90.

Ellis, Homer G. "Time, the Grand Illusion." *Foundations of Physics,* (1974) 4:2 pp. 311–18.

Englert, Berthold-Georg, Marlan O. Scully and Herbert Walter. "The Duality in Matter and Light." *Scientific American,* (1994) 271:6 pp. 86–92.

Epstein, Lewis C. *Relativity Visualized.* San Francisco: Insight Press, 1988.

Espinosa, J.M. "Physical Properties of de Broglie's Phase Waves. *American Journal of Physics,* (1982) 50:4 pp. 357–62.

Eves, Howard, and Carroll V. Newsom. *An Introduction to the Foundations and Fundamental Concepts of Mathematics.* New York: Holt, Rinehart, and Winston, Rev. 1965.

Ewing, A.C. *The Fundamental Questions of Philosophy.* New York: Collier Books, 1962.

Fadely, J. and V. Hosler. *Case Studies in Left and Right Hemispheric Functioning.* Springfield, Ill.: Charles C. Thomas, 1983.

Falk, G, F. Hermann, and G. Schmid. "Energy Forms or Energy Carriers?" *American Journal of Physics,* (1983) 51:12 pp. 1074–76.

Falk, Lars. "To derive the Existence of Gravity." *American Journal of Physics,* (1986) 54:6 pp. 520–23.

Farber, L.H. *The Ways of the Will: Essays Toward a Psychology of Will.* New York: Basic Books, 1966.

Feigl, Herbert and May Brodeck. eds. *The Philosophy of Science.* New York: Appleton-Century-Crofts, 1953.

Feinberg, Gerald. "Light." *Scientific American,* (1968) 219:3 pp. 50–59.

_____. "Particles that Go Faster than Light." *Scientific American,* (1970) 222:2 pp. 68–75.

_____. *Solid Clues: Quantum Physics, Molecular Biology, and the Future of Science.* New York: Simon and Schuster, 1985.

Feindel, W. *Memory, Learning, and Language.* Toronto: University of Toronto Press, 1959.

Feldman, Gary J. and Jack Steinberger. "The Number of Families of Matter." *Scientific American,* (1991) 264:2 pp. 70–75.

Fernside, Ward W. and William B. Holther. *Fallacy: Counterfeit of Argument.* Englewood Cliffs: PrenticeHall, 1959.

Ferrara, S., J. Ellis, and Peter van Nieuwenhuizen. *Unification of the Fundamental Forces.* New York: Plenum Press, 1980.

Ferrero, M. "Bell's Theorem: Local Realism versus Quantum Mechanics." *American Journal of Physics,* (1990) 58:7 pp. 684–88.

Ferris, Timothy, "Physics Newest Frontier." *New York Times Magazine,* September 26, 1982. pp. 37–41, 46, 68.

Feynman, Richard. "Structure of the Proton." *Science,* (1974) 183:4125 pp. 601–83.

Fine, Arthur. *The Shaky Game: Einstein Realism and the Quantum Theory.* Chicago: University of Chicago Press, 1986.

Finke, Ronald A. "Mental Imagery and the Visual System." *Scientific American,* (1986) 254:3 pp. 88–95.

Finn, Robert. "New Split-Brain Research Divides Scientists." *Science Digest,* (1983) 91:9 pp. 54–55, 103.

Fischbach, Gerald D. "Mind and Brain." *Scientific American,* (1992) 267:3 pp. 48–57.

Fitch, Val. "The Discovery of Charge Conjugation-Parity Asymmetry. *Science,* (1981) 212:4498 pp. 989–92.

Flandern, T. "Is Gravity Getting Weaker?" *Scientific American,* (1976) 234:3 pp. 44–52.

Flew, Anthony. *An Introduction to Western Philosophy: Ideas and Argument from Plato to Sartre.* New York: Bobbs Merrill, 1971.

_____. *Body, Mind, and Death.* ed. by Paul Edwards. London: MacMillan, 1964.

Fodor, Jerry A. "The Mind-Body Problem." *Scientific American,* (1981) 244:1 pp. 114–23.

Foger, Tim. "The Ultimate Vanishing Act." *Discover,* (1993) 14:10 pp. 98–106.

Ford, Joseph, and Giorgio Mantica. "Does Quantum Mechanics Obey the Correspondence Principle? Is it Complete?" *American Journal of Physics,* (1992) 60:12 pp. 1086–97.

Ford, Kenneth W. "Magnetic Monopoles." *Scientific American,* (1963) 209:6 pp. 122–130.

Frankel, Abraham A. *Set Theory and Logic.* London, Reading, Mass: Addison-Wesley, 1966.

Franks, Nigel R. "Army Ants: A Collective Intelligence." *American Scientist,* (1989) 77:2 pp. 139–45.

Freedman, Daniel Z. and Peter van Nieuwenhuizen. "Supergravity and the Unification of the Laws of Physics." *Scientific American,* (1978) 238:2 pp. 126–43.

_____. and Peter van Nieuwenhuizen. "The Hidden Dimensions of Spacetime." *Scientific American,* (1985) 252:3 pp. 74–81.

Freedman, David H. "Quantum Consciousness." *Discover,* (1994) 15:6 pp. 89–98.

_____. "The Theory of Everything." *Discover,* (1991) 12:7 pp. 55–61.

_____. "Weird Science." *Discover,* (1990) 11:11 pp. 62–68.

Freedman, Wendy L. "The Expansion Rate and Size of the Universe." *Scientific American*, (1992) 267:5 pp. 54–60.

Freeman, Walter J. "The Physiology of Perception." *Scientific American*, (1991) 264:2 pp. 78–85.

French, James D. "The False Assumption Underlying Berry's Paradox." *The Journal of Symbolic Logic*, (1988) 53:4 pp. 1220–22.

Freundlich, Yehudah. *In Defense of Copenhagenism*. Studies in History and Philosophy of Science, (1978) 9:3 pp. 151–79.

Frisch, David H. and Alan M. Thorndike. *Elementary Particles*. New York: Van Nostrand, 1964.

Frost, S.E. *Basic Teachings of the Great Philosophers: A Survey of Their Basic Ideas*. 1942; rpt. Garden City, New York: Dolphin Books, 1962.

Fuchs, Hans U. "Entropy in the Teaching of Introductory Thermodynamics." *American Journal of Physics*, (1987) 55:3 pp. 215–18.

Gaillard, Mary K. "Toward a Unified Picture of Elementary Particle Interactions." *American Scientist*, (1982) 70:5 pp. 506–14.

Gale, George. "The Anthropic Principle." *Scientific American*, (1981) 245:6 pp. 154–71.

Gal-Or, B. *Cosmology, Physics, and Philosophy*. New York: Springer-Verlag, 1981.

Gamow, George. "Gravity." *Scientific American*, (1961) 204:3 pp. 94–106.

_____. "The Exclusion Principle." *Scientific American*, (1959) 201:1 pp. 74–86.

_____. "The Principle of Uncertainty." *Scientific American*, (1958) 198:1 pp. 51–57.

_____. *Gravity*. New York: Anchor Books, 1962.

_____. *Thirty Years that Shook Physics*. New York: Doubleday, 1966.

Garcia, Laura L. "Can there be a Self-Explanatory Being?" *The Southern Journal of Philosophy*. (1986) 224:4 pp. 479–87.

Gard, Richard A., ed. *Buddhism*. New York: Washington Square Press, 1962.

Gardet, L., A.J. Gurevich, A. Kagame, C. Larre, G.E.R. Lloyd, A. Neher, R. Paniker, G. Pàttaro and P. Ricoeur. *"Cultures and Time."* Paris: Unesco Press, 1976.

Gardner, Martin. "A Few Words about Everything There Was, Is and Ever Will Be." *Scientific American*, (1976) 234:5 pp. 118–23.

_____. "Can Time Go Backward?" *Scientific American*, (1967) 216:1 pp. 98–108.

_____. "Free Will Revisited." *Scientific American*, (1973) 229:1 pp. 104–09.

_____. "Guest Comment: Is Realism a Dirty Word?" *American Journal of Physics*, (1989) 57:3 pp. 203.

_____. "The Hierarchy of Infinities and the Problems it Spawns." *Scientific American*, (1966) 214:3 pp. 112–18.

_____. "The Infinite Regress in Philosophy, Literature and Mathematical Proof." *Scientific American*, (1965) 212:4 pp. 128–33.

_____. "The Orders of Infinity, the Topological Nature of Dimension and 'Supertasks." *Scientific American*, (1971) 224:3 pp. 106–10.

_____. *The Ambidextrous Universe*. New York: New American Library, 1969.

Gazda, G. and R. Corsini. *Theories in Learning*. Itasca, Ill.: F.E. Peacock, 1980.

Gazzaniga, Michael S. "Organization of the Human Brain." *Science*, (1989) 245:4921 pp. 947–51.

_____. "The Split Brain Revistied." *Scientific American*, (1998) 279:1 pp. 51–55.

Geach, Peter T. *Reference and Generality*. Ithaca: Cornell University Press, 1962.

Gell-Mann, Murray and E.P. Rosenbaum. "Elementary Partictles." *Scientific American*, (1957) 197:1 pp. 72–88.

Gensler, Harry J. *Gödel's Theorem Simplified*. New York: University Press of America, 1984.

Gentner, Donald R. and Donald A. Norman. "The Typist's Touch." *Psychology Today*, (March 1984) pp. 66–71.

Georgi, Howard. "A Unified Theory of Elementary Particles and Forces." *Scientific American,* (1981) 244:4 pp. 48–63.

Geroch, R. *General Relativity From A to B.* Chicago: University of Chicago Press, 1978.

Giannoni, Carlo. "Clock Retardation, Absolute Space, and Special Relativity." *Foundations of Physics,* (1979) 9:5/6 pp. 427–44.

Gibilisco, Stan. *Puzzles, Paradoxes, and Brain Teasers.* Blue Ridge Summit, PA: Tab Books, 1988.

Gibson, W.M, and B.R. Pollard. *Symmetry Principles in Elementary Particle Physics.* London: Cambridge University Press, 1976.

Glashow, Sheldon L. "Quarks with Color and Flavor." *Scientific American,* (1975) 233:4 pp. 38–50.

————. "Tangled in Superstrings." *The Sciences,* (1988) 28:3 pp. 23–25.

Gleick, James. "New Images of Chaos that are Stirring a Science Revolution." *Smithsonian,* (1987) 18:19 pp. 122–34.

Gliedman, John. "Amazing Numerical Coincidences." *Science Digest.* (1981) 89:4 pp. 58–59, 118.

————. "Einstein Against the Odds: The Great Quantum Debate." *Science Digest,* (1983) 91:6 pp. 74–80, 109.

Globus, Gordon K. "Unexpected Symmetries in the 'World Knot." *Science,* (1973) 180:4091 pp. 1129–36.

Gödel, Kurt. *On Formally Undecidable Propositions of Principia Mathematica and Related Systems.* Trans. B. Meltzer. New York: Basic Books, 1962.

Godfrey, Laurie R., ed. *Scientists Confront Creationism.* New York: Norton and Co., 1983.

Goenner, H.F. "Mach's Principle." *Encyclopedia of Physics.* Ed. by R. Lerner and G. Trigg. Reading, Mass.: Addison-Wesley, 1981.

Goldberg, J.N. "Space-time." *Encyclopedia of Physics.* Ed. by R. Lerner and G. Trigg. Reading, Mass.: Addison-Wesley, 1981.

Goldhaber, Alfred Scharff and Michael Martin Nieto. "The Mass of the Photon." *Scientific American,* (1976) 234:5 pp. 86–96.

————. "Monopoles." *Encyclopedia of Physics.* Ed. by R. Lerner and G. Trigg. Reading, Mass.: Addison-Wesley, 1981.

Goldman, Terry, Richard J. Hughes and Michael Martin Nieto. "Gravity and Antimatter." *Scientific American,* (1988) 258:3 pp. 48–56.

Goldman-Rakic, Patricia S. "Working Memory and the Mind." *Scientific American,* (1992) 267:3 pp. 111–17.

Goldsmith, Donald. "When Time Slows Down." *Mercury,* (1975) 4:3 pp. 2–8.

Goldstein, Sheldon. "Quantum Theory without Observers—Part One." *Physics Today,* (1998) 51:3 pp. 42–46.

————. "Quantum Theory without Observers—Part Two." *Physics Today,* (1998) 51:4 pp. 38–41.

Gombrich, E.H. "The Visual Image." *Scientific American,* (1972) 227:3 pp. 82–96.

Good, R.H. *Basic Concepts of Relaitivity.* New York: Reinhold Book, 1968.

Goodstein, R.L. *Essays in the Philosophy of Mathematics.* Leicester: Leicester University Press, 1965.

Goswami, Amit. *The Self-Aware Universe: How Consciousness Creates the Material World.* New York: G.P. Putnam's Sons, 1993.

Gott, Richard J. III, James E. Gunn, David N. Schramm and Beatrice M. Tinsley. "Will the Universe Expand Forever?" *Scientific American,* (1976) 234:3 pp. 62–79.

————. "Creation of Open Universe from de Sitter Space." *Nature,* (1982) 295:5847 pp. 304–06.

————. "Creation of Open Universes from De Sitter Space." *Nature: Letters to,* (1982) 295:5847 pp. 304–06.

Grandy, Richard E. and Richard Warner eds. *Philosophical Grounds of Rationality: Intentions, Categories, Ends.* Oxford: Clarendon Press, 1986.

"Gravitation." *Encyclopaedia Britannica,* 1982. pp. 286–95.

Green, Michael B. "Superstrings." *Scientific American,* (1986) 255:3 pp. 48–60.

Greenberg, O.W. "The Quest for the Elementary Particles of Matter." *American Scientist,* (1988) 76:4 pp. 361–63.

Greenberger, Daniel M. and Albert W. Overhauser. "The Role of Gravity in Quantum Theory." *Scientific American,* (1980) 242:5 pp. 66–76.

_____., Michael A. Horne, Abner Shimony and Anton Zeilinger. "Bell's Theorem without Inequalities." *American Journal of Physics,* (1990) 58:12 pp. 1131–43.

Greenstein, George and Allen Kropf. "Cognizable Worlds: The Anthropic Principle and the Fundamental Constants of Nature." *American Journal of Physics,* (1989) 57:8 pp. 746–49.

Grene Marjorie. "Hierarchies in Biology." *American Scientist,* (1987) 75:5 pp. 504–09.

Gribben, John. "Bubbles on the River of Time." *New Scientist,* (1988) 118:1612 pp. 52–55.

_____. "The Man Who Proved Einstein Wrong." *New Scientist,* (1990) 128:1744 pp. 43–45.

_____. "A Special Theory of Relativity." *New Scientist,* (1991) 131:1787 pp. 1–4 (insert).

_____. "A Theory of Some Gravity." *New Scientist,* (1990) 125:1705 pp. 1–4 (insert).

_____. "Black Holes, White Holes and Wormholes." *Astronomy,* (1976) 4:11 pp. 23–26.

_____. "The Bishop, the Bucket, Newton and the Universe." *New Scientist,* (1984) 104:1435/36 pp. 12–15.

_____. "The Quantum Cookbook." *New Scientist,* (1985) 108:1477 pp. 33–36.

_____. *In Search of Schrödinger's Cat: Quantum Physics and Reality.* New York: Bantam Books, 1984.

Griffin, Donald R. "Animal Thinking." *American Scientist,* (1984) 72:5 pp. 456–63.

Grim, Patrick. "Logic and Limits of Knowledge and Truth." *Nous,* (1988) 22:3 pp. 341–367.

Gruber, Gary R. "Particle Velocities Faster than the Speed of Light." *Foundations of Physics.* 1:1 pp. 79–82.

Guillemin, Victor. *The Story of Quantum Mechanics.* New York: Charles Scribner's Sons, 1968.

Guillen, Michael. *Bridges to Infinity: The Human Side of Mathematics.* Los Angelos: Jeremy P. Tarcher, 1983.

Gunn, James E. "Will the Universe Expand Forever." *Mercury,* (1975) 4:6 pp. 5–8.

Gürsey, F. "Invariance Principle." *Encyclopedia of Physics.* Ed. by R. Lerner and G. Trigg. Reading, Mass.: Addison-Wesley, 1981.

Guth, Alan H. and Paul J Steinhardt. "The Inflationary Universe." *Scientific American,* (1984) 250:5 pp. 116–28.

Gutzweller, Martin C. "Quantum Chaos." *Scientific American,* (1992) 266:1 pp. 78–84.

Haber, Howard E. and Gordan L. Kane. "Is Nature Supersymmetric?" *Scientific American,* (1986) 254:6 pp. 52–60.

Hahn, Hans. "Is there an Infinity?" *Scientific American,* (1952) 187:5 pp. 76–84.

Haish, Bernhard, Alfonso Rueda and H.E. Puthoff. "Beyond $E = mc^2$." *The Sciences,* (1994) 34:6 pp. 26–31.

Hall, Calvin S. *A Primer of Freudian Psychology.* New York: New American Library, 1954.

_____. and Vernon J. Nordby. *A Primer of Jungian Psychology.* New York: New American Library, 1973.

Halliwell, Jonathan J. "Quantum Cosmology and the Creation of the Universe." *Scientific American,* (1991) 265:6 pp. 76–85.

Hanson, Norwood Russell. *Perception and Discovery.* Ed. by Willard C. Humphreys. San Fransisco: Freeman, Cooper, and Co., 1969.

Harari, H. "The Structure of Quarks and Leptons." *Scientific American,* (1983) 248:4 pp. 56–68.

Harris, Errol E. *Fundamentals of Philosophy: A Study of Classical Texts.* New York: Holt, Rinehart and Winston, 1969.

Harrison, David. "Bell's Inequality and Quantum Correlations." *American Journal of Physics,* (1982) 50:9 pp. 811–15.

Harrison, Edward R. "The Cosmic Numbers." *Physics Today,* (1972) 25:12 pp. 30–34.

Harvey, Alex. "Space is Expanding." *American Journal of Physics,* (1988) 56:6 Letter p. 487–88.

Harwood, Michael. "The Universe and Dr. Hawking." *New York Times Magazine,* 23 October 1983.

Haugan, Mark P. and Clifford M. Will. "Modern Tests of Special Relativity." *Physics Today,* (1987) 40:5 pp. 69–76.

Hawking, Stephen W. "Black Holes and Unpredictability." *Physics Bulletin,* (1978) 29:1 pp. 23–24.

_____. "The Direction of Time." *New Scientist,* (1987) 116:1568 pp. 46–49.

_____. "The Edge of Space-time." *American Scientist,* (1984) 72:4 pp. 355–59.

_____. "The Quantum Mechanics of Black Holes." *Scientific American,* (1977) 236:1 pp. 34–40.

_____. and Roger Penrose. "The Nature of Space and Time." *Scientific American,* (1996) 275:1 pp. 60–65.

Hayward, R. "CPT Theorem." *Encyclopedia of Physics.* Ed. by R. Lerner and G. Trigg. Reading, Mass.: Addison-Wesley, 1981.

Head, Joseph and S.L Cranston. eds. *Reincarnation in World Thought.* New York: Julian Press, 1967.

Heatherington, Norriss S. "Hubble's Cosmology." *American Scientist,* (1990) 78:2 pp. 142–51.

Heerden, Pieter J. van. *The Foundations of Mathematics.* The Netherlands: Wassenaar. N.V. Uitgeverij Wistik, 1968.

Hegstrom, Roger A. and Dilip K. Kondepudi. "The Handedness of the Universe." *Scientific American,* (1990) 262:1 pp. 108–15.

Heilbron, J.L. "Rutherford-Bohr Atom." *American Journal of Physics,* (1981) 49:3 pp. 223–31.

Heimer, Lennart. "Pathways in the Brain." *Scientific American,* (1971) 225:1 pp. 48–60.

Heisenberg, Werner. *Physics and Philosophy.* New York: Harper and Row Pub., 1962.

Heller, Eric J. and Steven Tomsovic. "Postmodern Quantum Mechanics." *Physics Today,* (1993) 46:7 pp. 38–46.

Heller, Wendy. "Of One Mind." *The Sciences,* (1990) 30:3 pp. 38–44.

Henley, E.M. "Parity." *Encyclopedia of Physics.* Ed. by R. Lerner and G. Trigg. Reading, Mass.: Addison-Wesley, 1981.

Hennig J. and J. Nitsch. "Gravity as an Internal Yang-Mills Gauge Field Theory of the Poncairé Group." *General Relativity and Gravitation.* (1981) 13:10 pp. 947–62.

Henricks, Sterling B. "How Light Interacts with Matter." (1968) 219:3 pp. 175–86.

Herbert, Nick. *Quantum Reality: Beyond the New Physics, and Excursion into Meta-Physics...and the Meaning of Reality.* Garden City, NJ: Anchor Books, 1985.

Herrera, J.C. "Electromagnetic Radiation." *Encyclopedia of Physics.* Ed. by R. Lerner and G. Trigg. Reading, Mass.: Addison-Wesley, 1981.

Hertzberg, Arthur. *Judaism.* New York: Washington Square Press, 1962.

Hick, John, ed. *The Existence of God.* London: Macmillan, 1964.

Hodge, Paul. "The Cosmic Distance Scale." *American Scientist,* (1984) 72:5 pp. 474–81.

Hoffman, Banesh. *The Strange Story of the Quantum.* New York: Dover Publications, 1959.

Hofstadter, Douglas R. *Gödel. Escher, and Bach.* New York: Vintage Books, 1980.

Home, Dipankar andJohn Gribbin. "What is Light?" *New Scientist,* (1991) 132:1793 pp. 30–33.

Hook, Sidney, ed. *Determinism and Freedom.* New York: Collier Books, 1961.

Horgan, John, "Particle Meta Physics." *Scientific American,* (1994) 270:2 pp. 97–106.

_____. "In the Beginning." *Scientific American,* (1991) 264:2 pp. 116–25.

_____. "Measuring Eternity." *Scientific American,* (1990) 263:6 pp. 16–17.

_____. "Particle Metaphysics." *Scientific American,* (1994) 270:2 pp. 97–106.

_____. "Quantum Philosophy." *Scientific American,* (1992) 267:1 pp. 94–102.

_____. "Universal Truths." *Scientific American,* (1990) 263:4 pp. 109–17.

Hovis, Corby R. and Helge Kragh. "P.A.M. Dirac and the Beauty of Physics." *Scientific American,* (1993) 268:5 pp. 104–09.

"How I Created the Theory of Relativity." Trans. by Yoshimasa A. Ono from Kaizo (1923). *Physics Today,* (1982) 35:8 pp. 46–47.

Hsu, J.P. "The Analysis of Time: Is the Relativistic Time Unique?" *Foundations of Physics,* (1979) 9:1/2 pp. 55–69.

Hughes, R.I.G. "Quantum Logic." *Scientific American,* (1981) 245:3 pp. 202–13.

Hughes, Richard J. "The Bohr-Einstein 'Weighing-of-energy' Debate and the Principle of Equivalence." *American Journal of Physics,* (1990) 58:9 pp. 826–28.

Hull, C. *Hypnosis and Suggestibility.* New York: Appleton-Century-Crofts, 1933.

Hurley, James. "The Time-Asymmetry Paradox." *American Journal of Physics,* (1986) 54:1 pp. 25–27.

Hwang, Sun-Tak. "A New Interpretation of Time Reversal." *Foundation of Physics,* (1972) 2:4 pp. 315–325.

Hyde, M.O. *Brainwashing and Other Forms of Mind Control.* New York: McGraw-Hill Book, 1977.

Iachello, Francesco. "Supersymmetry in Nuclei." *American Scientist,* (1982) 70:4 pp. 294–99.

Imry, Yoseph and Richard A. Webb. "Quantum Interference and the Aharonov-Bohm Effect." *Scientific American,* (1989) 260:4 pp. 56–62.

Ingber, Donald E. "The Architecture of Life." *Scientific American,* (1998) 278:1 pp. 48–57.

Ishikawa, Kenzio. "Glueballs." *Scientific American,* (1982) 247:5 pp. 142–56.

Jackson, David J. "The Impact of Special Relativity on Theoretical Physics." *Physics Today,* (1987) 40:5 pp. 34–42.

Jahn, Robert G. and Brenda J. Dunne. *Margins of Reality: The Role of Consciousness in the Physical World.* New York: Harcourt Brace Jovanovich, 1987.

Jammer, Max. *Concepts of Space.* Cambridge, Mass: Harvard University Press, 1969.

_____. *The Conceptual Development of Quantum Mechanics.* New York: McGraw-Hill, 1966.

_____. *The Philosophy of Quantum Mechanics.* New York: John Wiley and Sons, 1974.

Jaynes, Julian. *The Origins of Consciousness in the Breakdown of the Bicameral Mind.* Boston: Houghton Mifflin, 1970.

Jefferys, William H. and James O. Berger. "Ockham's Razor and Bayesian Analysis." *American Scientist,* (1992) 80:1 pp. 64–72

Johnson, Kenneth A. "The Bag Model of Quark Confinement." *Scientific American,* (1979) 241:1 pp. 112–21.

Johnson, Laurence E. "A Matter of Fact." *The Review of Metaphysics,* (1977) 30:3 pp. 508–18.

Johnsson, Gunnar. "Visual Motion Perception." *Scientific American*, (1975) 232:6 pp. 76–88.

Jügens, Hartmut, Heinz-Otto Peitgen and Dietmar Saupe. "The Lanuage of Fractals." *Scientific American*, (1990) 263:2. pp. 60–67.

Kadvany, John. "Reflections on the Legacy of Kurt Gödel: Mathematics, Skepticism, Postmodernism." *The Philosophical Forum*, 20:3 (1989) pp. 161–79.

Kafatos, Menas and Robert Nadeau. *The Conscious Universe: Part and Whole in Modern Physical Theory*. New York: Springer-Verlag, 1990.

_____. and Thalia Kafato. *Looking In; Seeing Out: Consciousness and Cosmos*. Wheaton, Ill.: Quest Book: the Theosophical Publishing House, 1991.

Kagan, Jerome. "Do Infants Think?" *Scientific American*, (1972) 226:3 pp. 74–82.

Kahane, Howard. *Logic and Philosophy*. Belmont, CA.: Wadsworth Publishing. 1969.

Kaku, Michio and Jennifer Trainer. *Beyond Einstein: the Cosmic Quest for the Theory of the Universe*. New York: Bantam Books, 1987.

Kamke, E. *Theory of Sets*. Trans. Fredrick Bagemihl. New York: Dover Publications, 1950.

Kane, Gordon. *The Particle Garden*. New York: Addison-Welsey, 1995.

Kattsoff, Louis O. *A Philosophy of Mathematics*. 1948; rpt. Freeport, New York: Books for Libraries Press, 1965.

Kaufman, William J. III. "Primordial Black Holes." *Mercury*, (1980) 9:1 pp. 1–6.

_____. *Black Holes and Warped Spacetime*. San Fransisco: W.H. Freeman and Co.: 1979.

_____. *Relativity and Cosmology*. New York: Harper and Row Publishers, 1973.

_____. *The Cosmic Frontiers of General Relativity*. Boston: Little, Brown and Co., 1977.

Kearns, Edward, Takaaki Kajita and Yoji Totsuka. "Detecting the Massive Neutrinos." *Scientific American*, (1999) 281:2 pp. 65–71.

Kendall, Henry W. and Wolfgang K.H. Panofsky. "The Structure of the Proton and Neutron." *Scientific American*, (1971) 224:6 pp. 60–77.

Keppner, Daniel, Michael G. Littman and Myron L. Zimmerman. "Excited Atoms." *Scientific American*, (1981) 244:5 pp. 130–49.

Kernan, Anne, "Discovery of the Intermediate Vector Bosons." *American Scientist*, (1986) 74:1 pp. 21–28.

Kessen, William and Emily D. Cahan. "A Century of Psychology: From Subject to Object to Agent. *American Scientist*, (1986) 74:6 pp. 640–49.

Kidd, Richard. "Evolution of the Modern Photon." *American Journal of Physics*, (1989) 57:1 pp. 27–34.

Klaus, Ruth and Andreas Schäfer. "The Mystery of Nucleon Spin." *Scientific American*, (1999) 281:1 pp. 58–63.

Klein, Abraham. "Spontaneous Symmetry Breaking." *Encyclopedia of Physics*. Ed. by R. Lerner and G. Trigg. Reading, Mass.: Addison-Wesley, 1981.

Klein, Morris. *The Lost of Certainty*. New York: Oxford University Press, 1980.

Klein, Oskar. "Arguments Concerning Relativity and Cosmology." *Science*, (1971) 171:3969 pp. 339–45.

Koestler, Arthur and J.R. Smythies. eds. *Beyond Reductionism: New Perspectives in the Life Sciences*. Boston: Beacon Press, 1969.

Körner, Stephan. *The Philosophy of Mathematics*. New York: Dover Publications, 1968.

Kosslyn, Stephen M. "Aspects of a Cognitive Neuroscience of Mental Imagery." *Science*, (1988) 240:4859 pp. 1621–26.

Kosso, Peter. *Appearance and Reality: An Introduction to Physics*. New York: Oxford University Press, 1998.

Koukkou, M, D. Lehman and J. Angst. *Functional States of the Brain: their Determinants*. New York: Elsener/North-Holland Biomedical Press, 1980.

Krause, Lawrence M. "Cosmological Antigravity." *Scientific American,* (1999) 280:1 pp. 53–59.

Kreisler, Michael N. "Are there Faster-than-Light Particles?" *American Scientist,* (1973) 61:2 pp. 201–08.

Kuhlenbeck, H. *The Human Brain and its Universe.* Vol's 1, 2. New York: S. Karger, 1982.

Kuhn, Thomas S. *The Structure of Scxientific Revolutions.* 1962; rpt. Chicago: University of Chicago Press, 1970.

Landauer, Rolf. "Information is Physical." *Physics Today,* (1991) 44:5 pp. 23–29.

Langacker, Paul and Alfred K. Mann. "The Unification of Electromagnetism with the Weak Force." *Physics Today,* (1989) 42:12 pp. 22–31.

_____. "Grand Unified Theories and Proton Decay." *Physics Reports,* (1981) 72:4 pp. 186–368.

Langer, Susan. *An Introduction to Symbolic Logic.* 1953; rpt. New York: Dover Publications, 1967.

Lasota, Jean-Pierre. "Unmasking Black Holes." *Scientific American,* (1999) 280:5 pp. 40–47.

Lawrence, John K. "The Future of the Universe." *Mercury,* (1978) 7:6 pp. 132–38.

Layzer, David. "The Arrow of Time." *Scientific American,* (1975) 233:6 pp. 56–69.

Le Corbeiller, P. "The Curvature of Space." *Scientific American,* (1954) 191:5 pp. 80–86.

Lebowitz, Joel L. "Boltzmann's Entropy and Times Arrow." *Physics Today,* (1993) 46:9 pp. 32–38.

Lederman, Leon M. "Observations in Particle Physics from Two Neutrinos to the Standard Model." *Science,* (1989) 244:4905 pp. 664–72.

_____. "The Upsilon Particle." *Scientific American,* (1978) 239:4 pp. 72–81.

_____. *The God Particle.* New York: Houghton Mifflin, 1993.

LeDoux Joseph E. "Emotion, Memory and the Brain." *Scientific American,* (1994) 270:6 pp. 50–57.

Lee, G.W. "Gauge Theories." *Encyclopedia of Physics.* Ed. by R. Lerner and G. Trigg. Reading, Mass.: Addison-Wesley, 1981

Leff, Harvey S. "Maxwell's Demon." *American Journal of Physics,* (1990) 58:2 pp. 135–42.

Lemke, H. On the Electrodynamics of Tachyons. *Il Nuovo Cimento,* (1975) 27A:2 pp. 141–154.

Lerner, Keith., ed. *Freedom and Detetminism.* New York: Random House, 1966.

Leslie, John. *Universes.* New York: Routledge, 1989.

Levine, Michael W. and Jeremy M. Shefner. *Fundamentals of Sensation and Perception.* Reading, Mass: Addison-Wesley, 1981.

Levine, R.P. *Genetics.* Modern Biology Series. New York: Holt, Rinehart and Winston, 1962.

Levine, Seymour. "Stress and Behavior." *Scientific American,* (1971) 224:1 pp. 26–31.

Lévy-Leblond, Jean-Marc. "Did the Big Bang Have a Beginning?" *American Journal of Physics,* (1990) 58:2 pp. 156–59.

Li Zhi, Fang and Li Shu Xian. *Creation of the Universe.* Singapore: World Scientific, 1989.

Libet, Benjamin. "Neural Destiny." *The Sciences,* (1989) 29:2 pp. 33–35.

Liboff, Richard L. "The Correspondence Principle Revisted." *Physics Today,* (1984) 37:2 pp. 50–55.

Linde, Andrei, "Particle Physics and Inflationary Cosmology." *Physics Today,* (1987) 40:9 pp. 61–68.

Lindley, David. *The End of Physics: Myth of a Unified Theory.* New York: Harper Collins Publishers, Basic Books Division, 1993.

Lipschutz, Seymour. *Theory and Problems of Set Theory and Related Topics.* Schaum's Outlihne Series. New York: McGraw-Hill, 1964.

Liss, Tony M and Paul L. Tipton. "The Discovery of the Top Quark." *Scientific American,* (1997) 277:3 pp. 54–59.

Lockwood, Michael. *Mind, Brain and the Quantum: The Compound I.* New York: Basil Blackwell, 1989.

Loeser, J.G. "Three Perspectives on Schrödinger's Cat." *American Journal of Physics,* (1984) 52:12 pp. 1089–93.

Long, Michael E. "What is this Thing Called Sleep." *National Geographic,* (December 1987) pp. 787–821.

LoSecco, J.M., Frederick Reines and Daniel Sinclair. "The Search for Proton Decay." *Scientific American,* (1985) 252:6 pp. 54–62.

Louis Renou, ed. *Hinduism.* New York: Washington Square Press, 1963.

Luminet, Jean-Pierre, Glenn D. Starkman and Jeffrey R. Weeks. "Is Space Infinite?" *Scientific American,* (1999) 280:4 pp. 90–97.

Machian, Tibor R. *The Pseudo-Science of B.F. Skinner.* New Rochelle, NY: Arlington House Pub. 1974.

MacKenzie, Dana. "Through the Looking Glass." *The Sciences,* (1997) 37:3 pp. 32–37.

Mackinnon, L. "Explaining Electron Diffraction—De Broglie or Schrödinger?" *Foundations of Physics,* (1981) 11:11/12 pp. 907–12.

Mackintosh, Nicholas. "The Mind in the Skinner Box." *New Scientist,* (1984) 96:1395 pp. 30–33.

Madore, Barry F. and Wendy L. Freedman. "Self-Organizing Structures." *American Scientist,* (1987) 75:3 pp. 252–59.

Malcolm, N. *Memory and Mind.* Ithaca: Cornell University Press, 1977.

Mandelbrot, Benoit. "Fractals—Geometry of Nature." *New Scientist,* (1990) 127:1734. pp. 38–43.

Mandelker, J. *Relativity and the New Energy Mechanics.* New York: Philosophical Library, 1966.

Manicas, Peter T. ed. *Logic as Philosophy.* New York: Van Nostrand Reinhold, 1971.

Mann, Charles and Robert Crease. "John Bell: Interview." *Omni,* (1988) 10:8 pp. 85–92, 121.

Marciano, William and Heinz Pagels. "Quantum Chromodynamics." *Nature,* (1979) 279:5713 pp. 479–83.

Marcuse, Herbert. *Negations.* Trans. Jeremy J. Shapiro. Boston: Beacon Press, 1968.

Margenau, Henry. *The Nature of Physical Reality,* New York: McGraw-Hill Book, 1950.

Mariner, Stefan. "The Coordinate Transformations of the Absolute Space-time Theory." *Foundations of Physics.* (1979) 9:5/6 pp. 445–60.

Mark, R. *Memory and Nerve Cell Connections.* Oxford: Clarendon Press, 1974.

Marquit, Erwin. "A Plea for the Correct Translation of Newton's Law of Inertia." *American Journal of Physics,* (1990) 58:9 pp. 867–70.

Martin, L. "Electron Diffraction." *Encyclopedia of Physics.* Ed. by R. Lerner and G. Trigg. Reading, Mass.: Addison-Wesley, 1981.

Martindale, Colin. *Cognition and Consciousness.* Dorsey Series in Psychology. Homewood, Ill.: Dorsey Press, 1981.

Marx, J.L. "Two Sides of the Brain." *Science,* (1983) 220:4598 pp. 487–89.

Mattis, D.C. "Many-Body Theory." *Encyclopedia of Physics.* Ed. by R. Lerner and G. Trigg. Reading, Mass.: Addison-Wesley, 1981.

Mayr, Ernst. "Evolution." *Scientific American,* (1978) 239:3 pp. 46–55.

McCall, Raymond J. *Basic Logic.* 1952; rpt. Outline Series. New York: Barnes and Noble, 1970.

McCrone, John. "Inner Voices, Distant Memories." *New Scientist,* (1994) 141:1910 pp. 28–31.

McDaniel, Stanley V. *The Philosophy of Nietzsche.* New York: Simon and Schuster, 1965.

McDonald, Margaret C. "The Dream Debate: Freud vs. Neurophysiology, Does Dream Theory Need Revision?" *Science News,* (1981) 119:24 pp. 378–80.

McInerny, Ralph. *A History of Western Philosophy.* vol's 1–5. Notre Dame, Indiana: University of Notre Dame Press, 1963.

McMichael, Alan. "A Set Theory with Frege-Russell Cardinal Numbers." *Philosophical Studies,* (1982) 42:2 pp. 141–49.

Meeterloo, J.A.M. *The Rape of the Mind: the Psychology of Thought Control, Menticide, and Brainwashing.* New York: World Publishing, 1956.

Mehra, Jagdish. "Quantum Mechanics and the Explanation of Life." *American Scientist,* (1973) 61:6 pp. 722–27.

————. *Einstein, Hilbert, and the Theory of Gravitation.* Boston: D. Reidel, 1974.

————. *The Quantum Principle: its Interpretation and Epistemology.* Boston: D. Reidel 1974.

Melaney, Robert and William Fowler. "The Transformation of Matter after the Big Bang." *American Scientist,* (1988) 76:5 pp. 472–77.

Mendicus, Heinrich A. "Fifty Years of Matter Waves." *Physics Today,* (1974) 27:2 pp. 38–45.

Mermin, N. David. "Bringing Home the Atomic Mysteries for Anybody." *American Journal of Physics,* (1981) 49:10 pp. 940–43.

————. "Is the Moon there When Nobody Looks: Reality and the Quantum Theory." *Physics Today,* (1985) 38:4 pp. 38–47.

————. "Quantum Mysteries Revisited." *American Journal of Physics,* (1990) 58:8 pp. 731–34.

————. *Space and Time in Special Relativity.* New York: McGraw-Hill, 1968.

Mickens, Ronald E. "Long-Range Interactions." *Foundations of Physics,* (1979) 9:3/4 pp. 261–69.

Mignani, R. and E. Recampi. "Astrophysics and Tachyons." *Il Nuovo Cimento,* (1974) 21B:1 pp. 210–25.

————. and E. Recampi. "Connection Between Magnetic Monopoles and Faster-than-Light Speeds: Answer to the Comments by Corben and Honig." *Lettere Al Nuovo Cimento,* (1975) 13:15 pp. 589–90.

————. and E. Recampi. "Duration Length Symmetry in Complex Three-space and Interpreting Superluminal Lorentz Transformations. *Lettere Al Nuovo Cimento,* (1976) 16:15 pp. 449–52.

————. and E. Recampi. "Magnetic Monopoles and Tachyons." *Lettere Al Nuovo Cimento,* (1974) 9:9 pp. 367–72.

Milgram, Stanley. *Obediance to Authority: an Experimental View.* New York: Harper and Row, 1974.

————. *The Individual in a Social World.* Reading, Mass.: Addison-Wesley, 1977.

Mills, Robert. "Gauge Fields." *American Journal of Physics,* (1989) 57:6 pp. 493–507.

Milner, Peter M. *Physiological Psychology.* New York: Holt, Rinehart and Winston, 1970.

Mirman, R. "Comments on the Dimensionality of Time." *Foundations of Physics,* (1973) 3:3 pp. 321–33.

Mishkin, Mortimer and Tim Appenzeller. "The Anatomy of Memory." *Scientific American,* (1987) 256:6 pp. 80–89.

————. and Herbert L. Petri. "Behaviorism, Cognition and the Neuropyschology of Memory." *American Scientist,* (1994) 82:1 pp. 30–37.

Moore, A.W. "A Brief History of Infinity." *Scientific American,* (1995) 272:4 pp. 112–16.

_____. "What does Gödel's Incompleteness Theorem Show?" *Nous,* (1988) 22:4 pp. 573–84.

Moore, G.E. *Some Main Problems of Philosophy.* New York: Collier Books, 1962.

Morgan, John A. "Are Galaxies Receding or Space Expanding? *American Journal of Physics,* Letter; (1988) 56:9 pp. 777–78.

Morgenbesser, Sidney and James Walsh. eds. *Free Will,* Englewood Cliffs, NJ: Prentice Hall, 1962.

Moriyasu, K. "The Renaissance of Gauge Theory." *Contemporary Physics,* (1982) 32:6 pp. 553–81.

Morris, Richard. *The Nature of Reality: the Universe After Einstein.* New York: The Noonday Press, 1987.

Morrison, A.R. "A Window on the Sleeping Brain." *Scientific American,* (1983) 248:4 pp. 94–102.

Moulder, James. "Is Russell's Paradox Genuine?" *Philosophy,* (1974) 49:189 pp. 295–302.

Mullatti, L.C. "The Necessity of the Material Conditional." *Indian Philosophical Quarterly,* (1983) 10:3 pp. 331–39.

Muller, Richard A. "The Cosmic Background Radiation and the New Aether Drift." *Scientific American,* (1978) 238:5 pp. 64–74.

Mulligan, Joseph F. "Heinrich Hertz and the Development of Physics." *Physics Today,* (1989) 42:3 pp. 50–57.

Murphy, Frederick and David O. Yount. "Photons as Hadrons." *Scientific American,* (1971) 225:1 pp. 94–104.

Nagel, Ernest and James R. Newman. *Gödel's Proof.* New York: New York University Press, 1958.

_____. *The Strucuture of Science.* New York: Harcourt, Brace and World, 1961.

Nambu, Yoichiro. "A Matter of Symmetry: Elementary Particles and the Origin of Mass." *The Sciences,* (1992) 32:3 pp. 37–43.

_____. "The Confinement of Quarks." *Scientific American,* (1976) 235:5 pp. 48–60.

Narlikar, F.O. and E.C.G. Sudershan. "Tachyons and Cosmology." *Monthly Notices of the Royal Astronomical Society,* (1976) 175 pp. 105–116.

Narlikar, Jayant. "White Holes: Cosmic Energy Machines." *New Scientist,* (1983) 97:1346 pp. 516–18.

Nassau, Kurt. "The Causes of Color." *Scientific American,* (1980) 243:4 pp. 124–54.

Nauenberg, Michael, Carlos Stroud and John Yeazell. "The Classical Limit of the Atom." *Scientific American,* (1994) 270:6 pp. 44–49.

Nelson, D.R. "Renormalization." *Encyclopedia of Physics.* Ed. by R. Lerner and G. Trigg. Reading, Mass.: Addison-Wesley, 1981.

Nicolson, I. *Gravity, Black Holes and the Universe.* New York: John Wiley and Sons, 1981.

Nirenberg, Marshall W. "The Genetic Code." *Scientific American,* 1963; rpt. W.H. Freeman and Co.

Nishikawa, S. "A Relativistic Extension of the Concepts of the Central and Conservative Forces." *American Journal of Physics,* (1990) 58:1 pp. 68–72.

Nordtvedt, Kenneth L. Jr. "Gravitation Theory: Empirical Status form Solar System Experiments." *Science,* (1972) 178 pp. 1157–64.

Noren, Stephen J. "The Conflict Between Science and Common Sense and Why it is Inevitable." *Southern Journal of Philosophy,* (1975) 13:3 pp. 331–45.

Nozick, Robert. *Philosophical Explanations.* Cambridge, Mass.: Harvard University Press, 1981.

Nussenzveig, H.M. "The Theory of the Rainbow." *Scientific American,* (1977) 236:4 pp. 116–27.

Oakland, Nathan L. "A Defence of the New Tenseless Theory of Time." *The Philosophical Quarterly,* (1991) 41:162 pp. 27–37.

Ogden, C.K. and I.A. Richards. *The Meaning of Meaning.* New York: Harcourt, Brace and World, 1923.

Ohanian, Hans C. "What is Spin." *American Journal of Physics,* (1986) 54:6 pp. 500–24.

Okun, Lev B. "The Concept of Mass." *Physics Today,* (1989) 42:6 pp. 31–36.

Olson, Robert G. *A Short Introduction to Philosophy.* New York: Harcourt, Brace and World, 1967.

Olten, David S. "Spatial Memory." *Scientific American,* (1977) 236:6 pp. 82–98.

Ornstein, Robert E. "The Split and Whole Brain." *Human Nature,* (1978) 1:5 pp. 76–83.

_____. *The Nature of Human Consciousness.* San Fransisco: W.H. Freeman and Co., 1973.

_____. *The Psychology of Consciousness.* New York: Penguin Books, 1972.

Osmer, Patrick. "Quasars as Probes of the Distant and Early Universe." *Scientific American,* (1982) 246:2 pp. 126–38.

Oster, Gerald. "The Chemical Effects of Light." *Scientific American,* (1968) 219:3 pp. 158–70.

Overman, Ralph T. *Basic Concepts in Nuclear Chemistry.* New York: Reinhold, 1963.

Padmanabhan, Thanu. "Bridge Over the Quantum Universe." *New Scientist,* (1992) 136:1842 pp. 26–29.

Page, T. "Newton's Laws." *Encyclopedia of Physics.* Ed. by R. Lerner and G. Trigg. Reading, Mass.: Addison-Wesley, 1981.

Pagels, Heinz R. *Perfect Symmetry: The Search for the Beginning of Time.* 1986; rpt. New York: Bantam Books, 1986.

_____. *The Cosmic Code: Quantum Physics as the Language of Nature.* 1982; rpt. New York: Bantam Books, 1984.

Pais, Abraham. "Electron Diffraction." *Encyclopedia of Physics.* Ed. by R. Lerner and G. Trigg. Reading, Mass.: Addison-Wesley, 1981.

_____. "George Uhlenbeck and the Discovery of Electron Spin." *Physics Today,* (1989) 42:12 pp. 34–40.

_____. "Max Born's Statistical Interpretation of Quantum Mechanics." *Science,* (1982) 29578 pp. 1193–98.

Parker, Barry. "Tunnel Through Time." *Astronomy,* (1992) 20:6 pp. 28–35.

Parker, Leonard. "Faster-Than-Light Inertial Frames and Tachyons." *Physical Review,* (1969) 188:5 pp. 2287–92.

Paul, H. "Einstein-Podolsky-Rosen Paradox and Reality of Individual Physical Properties." *American Journal of Physics,* (1985) 53:4 pp. 318–19.

Pavsic, M. "Towards Understanding Quantum Mechanics, General Relativity, and the Tachyonic Causality Paradoxes." *Lettere al Nuovo Cimento,* (1981) 30:4.

_____. and E. Recami. "Again, About Causality for Tachyons in Macrophysics." *Lettere al Nuovo Cimento,* (1977) 18:5 pp. 134–36.

Pavsic, M. and E. Recami. "Recovering Causality for Tachyons Even in Macrophysics." *Lettere al Nuovo Cimento,* (1976) 17:7 pp. 257–61.

Pearson, Norman E. *Space, Time, and Self: Three Mysteries of the Universe.* Wheaton, ILL: The Theosophical Publishing House, 1957.

Peat, David F. *Einstein's Moon: Bell's Theorem and the Curious Quest for Quantum Reality.* Chicago: Contemporary Books, 1990.

_____. *Superstrings and the Search for the Theory of Everything.* New York: Contemporary Books, 1988.

_____. *Synchronicity: the Bridge Between Matter and Mind.* New York: Bantam Books, 1987.

_____. *The Philosopher's Stone: Chaos, Synchronicity, and the Hidden Order of the World.* New York: Bantam Books, 1991.

Penfield, W. and L. Roberts. Speech and Brain Mechanisms. Athenum, New York: 1974.

Penrose, Roger. "Black Holes." *Scientific American*, (1972) 226:5 pp. 39–46.

————. *Shadows of the Mind: The Search for the Missing Science of Consciousness.* New York: Oxford University Press, 1994.

————. *The Emperor's New Mind: Concerning Computers, Minds, and the Laws of Physics.* New York: Oxford University Press, 1989.

Penzias, A.A. "The Origin of the Elements." *Science.* (1979) 205:4406 pp. 549–59.

Peres, Asher. "What is a Quantum Measurement?" *American Journal of Physics*, (1986) 54:8 pp. 688–92.

————. "What is a State Vector?" *American Journal of Physics*, (1984) 52:7 pp. 644–49.

————. and W.H. Zurek. "Is Quantum Theory Universally Valid?" *American Journal of Physics*, (1982) 50:9 pp. 807–110.

Perkowitz, Sidney. "True Colors: Why Things Look the Way They Do." *The Sciences*, (1990) 30:3 pp. 23–28.

Perl, Marin L. and William T. Kirk. "Heavy Leptons." *Scientific American*, (1978) 238:3 pp. 50–57.

Perlow, G.J. "The Mössbauer Effect." *Encyclopedia of Physics*. Ed. by R. Lerner and G. Trigg. Reading, Mass.: Addison-Wesley, 1981.

Peters, Philip C. "Black Holes: New Horizons in Gravitational Theory." *American Scientist*, (1974) 62:5 pp. 575–83.

Peterson, Aage. *Quantum Physics and the Philosophical Tradition.* Cambridge, Mass.: MIT Press, 1968.

Peterson, Ivars. "Bordering on Infinity: Focusing on the Mandelbrot Set's Extraordinary Boundary." *Science News*, (1991) 140:21 p. 331.

Petri, Herbert L. and Mortimer Mishkin. "Behaviorism, Cognitivism and the Neuropsychology of Memory." *American Scientist*, (1994) 82:1 pp. 30–37.

Philip, Yam. "Bringing Schrödinger's Cat to Life." *Scientific American*, (1997) 277:6 pp. 124–29.

————. "Exploiting Zero-Point Energy." *Scientific American*, (1997) 277:6 pp. 82–85.

Pierce, John R. *Electrons and Waves.* New York: Anchor Books, Doubleday, 1964.

Pines, Maya. "We are Left-Brained or Right-Brained." *New York Times Magazine*, (September 9, 1973) pp. 32–33, 121–27, 132, 136.

Pipkin, Francis M. and Rogers C. Ritter. "Precision Measurements and Fundamental Constants." *Science*, (1983) 219:458 pp. 913–21.

Platt, Daniel E. "A Modern Analysis of the Stern-Gerlach Experiment." *American Journal of Physics*, (1992) 60:4 pp. 306–08.

Plum F. and J. Posner. *Diagnosis of Stupor and Coma.* Contemporary Neurological Series. Philadelphia: 1971.

Podolny, R. *Something Called Nothing: Physical Vacuum: What is it?* Trans. N. Weinstein. Mir Publishers, 1983.

Polkinghorn, J.C. *The Quantum World.* Princeton: Princeton University Press, 1984.

Popper, Karl. *Quantum Theory and the Schism in Physics.* Totowa, NJ: Rowman and Littlefield, 1982.

Posner, Michael I., Steven E. Peterson, Peter T. Fox, and Marcus E. Raichle. "Localization of Cognitive Operations in the Brain." *Science*, (1988) 240:4859 pp. 1627–31.

Post, E.J. "The Logic of Time Reversal." *Foundations of Physics.* (1979) 9:1/2 pp. 129–61.

Potvin, Jean. "Simulating Hot Quark Matter." *American Scientist*, (1991) 79:2 pp. 118–29.

Poundstone, William. *Labyrinth of Reason: Paradoxes, and the Frailty of Knowledge*, New York: Anchor Press, 1988.

_____. *The Recursive Universe: Cosmic Complexity and the Limits of Scientific Knowledge.* Chicago: Contemporary Books, 1985.

Press, William H. and David N. Speigel. "Cosmic Strings: Topological Fossils of the Hot Big Bang." *Physics Today,* (1989) 42:3 pp. 29–35.

Price, Richard H. "General Relativity Primer." *American Journal of Physics,* (1982) 50:4 pp. 300–29.

_____. and Kip S. Thorne. "The Membrane Paradigm for Black Holes." *Scientific American,* (1988) 258:4 pp. 69–78.

Priest, Graham. *Contradictions: A Study of the Transconsistent.* Boston: Martinus Publishers, 1987.

Progogine, Ilya. *Order Out of Chaos: Man's Dialogue with Nature.* New York: Bantam Books, 1984.

Prokhovnik, S.J. "Cosmology Versus Relativity—The Reference Frame Paradox." *Foundations of Physics,* (1973) 3:3 pp. 351–58.

"Proton-Antiproton Collisions Yield Intermediate Boson at 80 GeV, as Predicted." *Physics Today,* (1983) 36:4 pp. 17–20.

Quigg, Chris. "Elementary Particles and Forces." *Scientific American,* (1985) 252:4 pp. 84–96.

_____. "Top-ology." *Physics Today,* (1997) 50:5 pp. 20–26.

Quine, Willard Van Orman. *Elementary Logic.* 1941; Rev. New York: Harper Torchbooks, 1965.

_____. *Philosophy of Logic.* 2nd ed. Cambridge Mass: Harvard University Press, 1986.

Quinn, Helen R. and Michael S. Witherell. "The Asymmetry Between Matter and Antimatter." *Scientific American,* (1998) 279:4 pp. 76–81.

Quinton, A. "Objects and Events." *Mind.*(1979) 88:350 pp. 197–214.

Rader, Melvin. *The Enduring Questions.* 2nd ed. New York: Holt, Rinehart, and Winston, 1969.

Radmanabhan, Thanu. "Bridge Over the Quantum Universe." *New Scientist,* (1992) 136:1842 pp. 25–29.

Rae, Alastair. *Quantum Physics: Illusion or Reality?* New York: Cambridge University Press, 1986.

Raine, D.J. "Mach's Principle and Space-time Structure." *Reports on Progress in Physics.* (1981) 44:11 pp. 1153–95.

Raju, P.T. *Contemporary Philosophy.* Carbondale: Southern Illinois University Press, 1962.

Ramon, C. and E. Rauscher. "Superluminal Transformations in Complex Minkowski Spaces." *Foundations of Physics,* (1980) 10:7/8 pp. 661–69.

Randall, John H. Jr. and Justus Buchler. *Philosophy: An Introduction.* 1942; rpt. Outline Series. New York: Barnes and Noble, 1969.

Rauch, Leo. *The Philosophy of Hegel.* New York: Thor Publications, 1965.

_____. *The Philosophy of Immanuel Kant.* New York: Thor Publications, 1965.

Recami, E. "How to Recover Causality in Special Relativity for Tachyons." *Foundations of Physics,* (1978) 8:5/6 pp. 329–40.

_____. and G. Liino. "About New Space-time Symmetries in Relativity and Quantum Mechanics." *Il Nuovo Cimento,* (1976) 33A:2 pp. 205–15.

_____. and P. Castorina. "Hadrons as Compounds of Bradyon Particles and Tachyons." *Lettere al Nuovo Cimento.* (1978) 22:5 pp. 195–201.

_____. and R. Magnani. "Do Magnetic Monopoles Exist?" *Lettere al Nuovo Cimento,* (1974) 9:12 pp. 479–82.

_____. and R. Magnani. "More About Lorentz Transformations and Tachyons: Answer to the Comments by Ramachandran, Fagare and Kolaskar. *Lettere al Nuovo Cimento.* (1972) 4:4 pp. 144–52.

Redshift Explanation that Subverts Cosmology.*Science News,* (1972) 101:24 p. 37.

Rees, Martin. J. "Black Holes in the Galatic Center." *Scientific American,* (1990) 263:5 pp. 56–66.

Regan, David, Kenneth Beverely, and Max Cynader. "Visual Perception of Motion in Depth." *Scientific American,* (1979) 240:1 pp. 136–51.

Regis, Ed. *Who Got Einstein's Office?: Eccentricity and Genius at the Institute for Advance Study.* New York: Addison-Wesley, 1987.

Reichenbach, Hans. *The Direction of Time.* Berkeley, CA: Univerisity of California Press, 1956.

Reid, Constance. *From Zero to Infinity.* New York: Thomas W. Crowell, 1955.

Renteln, Paul. "Quantum Gravity." *American Scientist,* (1991) 79:6 pp. 508–27.

Resnik, Michael D. *Frege and the Philosophy of Mathematics.* Ithaca: Cornell University Press, 1980.

Richards, Frederic M. "The Protein Folding Problem." *Scientific American,* (1991) 264:1 pp. 54–63.

Rietdijk, C.W. "A Vigorous Proof of Determinism Derived from the Special Theory of Relativity." *Philosophy of Science,* (1966) 33:3 pp. 341–44.

_____. "Special Relativity and Determinism." *Philosophy of Science,* (1976) 43:4 pp. 598–609.

Rinder, Wolfgang. *Essential Relativity; Special, General, and Cosmological.* New York: Van Nostrand Reinhold, 1969.

Riordan, Michael. "The Discovery of Quarks." *Science,* (1992) 256:5061 pp. 1287–92.

Robinson, Arthur L. "High Energy Physics: A Proliferation of Quarks and Leptons." *Science,* (1979) 198:4316 pp. 478–81.

Robson, D. "Isospin in Nuclei." *Science,* (1973) 179:4069 pp. 133–39.

Rock, Irving and Stephen Palmer. "The Legacy of Gestalt Psychology." *Scientific American,* (1990) 263:6 pp. 84–90.

_____. "The Perception of Disoriented Figures." *Scientific American,* (1974) 230:1 pp. 78–85.

_____. *An Introduction to Perception.* New York: MacMillan Publishing, 1975.

Rodewald, Bernd. "Entropy and Homogeneity." Entropy and Homogeneity." *American Journal of Physics,* (1990) 58:2 pp. 164–68.

Rogers, Lesley. "The Left and Right Brain at Work." *New Scientist,* (1989) 128:1651 pp. 56–59.

Rosen, Joe. "Anthropic Principle II." *American Journal of Physics,* (1988) 56:5 pp. 415–19.

_____. "Extended Mach's Principle." *American Journal of Physics,* (1981) 49:3 pp. 258–64.

_____. "The Anthropic Principle." *American Journal of Physics,* (1988) 56:5 pp. 415–19.

Rosenhan, D. "On Being Sane in Insane Places." *Science,* (1974) 179:4070 pp. 250–57.

Ross, Floyd H. and Tynette Hills. *The Great Religions,* Greenwich Conn: Fawcett Publication, 1956.

Ross, James F. "Creation." *The Journal of Philosophy,* (1980) 77:10 pp. 615–29.

_____. *Philosophical Theology.* Indianapolis: Bobbs-Merrill, 1969.

Rosser, J.B. *An Informal Exposition of Proofs of Gödel's Theorem in "The Undecidable: Basic Papers on Undecidable Propositions, Unsolvable Problems, and Computable Functions."* Ed. by Martin Davis. Hewitt, New York: Raven Press, 1965. (Especially pp. 223–30.).

Rothman, Tony, Bernard Carr, R. Matzner, A.C. Ottewill, T. Piran, L.L. Shepley, E.C. Sudarshan, E.C.G. Tipler F. and B Unruh. *Frontiers of Modern Physics: New Pespectives on Cosmology, Relativity, Black Holes, and Extra-Terrestrial Intelligence.* New York: Dover Publications, 1985.

_____. "Irreversible Differences." *The Sciences*, (1997) 37:4 pp. 26–31.

Rucker, Rudy. *Geometry, Relativity, and the Fourth Dimension*. New York: Dover Publications, 1977.

_____. *Infinity and the Mind: The Science and Philosophy of the Infinite*, 1982; rpt. New York; Bantam Books, 1983.

_____. *Mind Tools: The Five Levels of Mathematical Reality*, Boston: Houghton Mifflin, 1987.

Russell, Bertand. *Dictionary of Mind, Matter, and Morals*. Ed. by Lester Denonn. New York: The Citadel Press, 1965.

_____. *Problems of Philosophy*. London: Oxford University Press, 1912.

_____. *Religion and Science*. Oxford: Oxford University Press, 1935.

_____. *The Analysis of Matter*. New York: Dover Publications, 1954.

Ruthen, Russell. "Catching the Wave." *Scientific American*, (1992) 266:3 pp. 90–99.

Ryle, Gilbert. *The Concept of Mind*. New York: Barnes and Noble, 1949.

Sachs, Mendel. "Elementary Particle Physics form General Relativity." *Foundations of Physics*, (1981) 11:3/4 pp. 329–54.

Sachs, Robert G. "Time Reversal." *Science*, (1972) 176:4035 587–96.

Sackstein, William. "Least Parts and Greatest Wholes." *International Studies in Philosophy*, (1991) 23:1 pp. 75–85.

Sahakian, William S. *History of Philosophy*. Outline Series. New York: Barnes and Noble, 1968.

Sainsbury, R.M. *Paradoxes*. New York: Cambridge University Press, 1988.

Salam, Abdus. "Gauge Unification of Fundamental Forces." *Science*, (1980) 210:4471 pp. 723–30.

Salmon, Wesley C. "Confirmation." *Scientific American*, (1973) 228:5 pp. 75–83.

_____. *The Foundations of Scientific Inference*. Pittsburgh: University of Pittsburgh Press, 1966.

Sanchez-Ron, J.M. "Actions at a Distance, Four-Dimensionality, and the Problem of 'Where is the Energy?" *American Journal of Physics*, (1982) 50:8 pp. 739–42.

Sandage, Allan R. "The Red-Shift." *Scientific American*, 1956; rpt. by W.H. Freeman.

Sands, R.H. "Spin." *Encyclopedia of Physics*. Ed. by R. Lerner and G. Trigg. Reading, Mass.: Addison-Wesley, 1981.

Sargent, Murray, and Morlan Scully. "The Concept of the Photon." *Physics Today*, (1972) 25:3 pp. 38–47.

Sargent, W. *Battle for the Mind*. Westport Conn.: Greenwood Press, 1957.

Savage, Jay M. *Evolution*. Modern Biology Series. New York: Holt, Rinehart and Winston, 1963.

Scerri, Eric R. "Eastern Mysticism and the Alleged Parallels with Physics." *American Journal of Physics*, (1989) 57:8 pp. 688–91.

_____. "Evolution of the Periodic System." *Scientific American*, (1998) 279:3 pp. 78–83.

_____. "The Periodic Table and the Electron." *Scientific American*, (1997) 277:5 pp. 546–53.

Schilder, P. *The Nature of Hypnosis*, New York: International Universities Press, 1956.

Schlegel, Richard. "An Interaction Interpretation of Special Relativity Theory; Part I." *Foundations of Physics*, (1973) 3:2 pp. 169–83.

_____. "An Interaction Interpretation of Special Relativity Theory; Part II." *Foundations of Physics*, (1973) 3:3 pp. 277–94.

_____. *Time and the Physical World*. New York: Dover Publications, 1961.

Schlick, Moritz. *Space and Time*. Trans. Henry L. Brose. New York: Dover Publications, 1963.

Schopenhauer, Arthur. *Essays on the Freedom of the Will*. New York: The Liberal Arts Press, 1960.

Schopf, William J. "Evolution of the Earliest Cells." *Scientific American,* (1978) 239:3 pp. 111–39.

Schramm, David N. "The Age of the Elements." *Scientific American,* (1974) 230:1 pp. 69–77.

_____. "The Early Universe." *Physics Today,* (1983) 36:4 pp. 27–33.

_____. and Gary Steigman. "Particle Accelerators Test Cosmological Theory." *Scientific American,* (1988) 258:6 pp. 66–72.

Schrödinger, E. "What is Matter?" *Scientific American,* (1953) 189:3 pp. 52–57.

Schwartz, Charles. "Some Improvements in the Theory of Faster-than-Light Paticles." *Physical Review D,* (1982) 25:2 pp. 34–64.

Schwarz, John H. "Completing Einstein." *Science 85,* (1985) 6:9 pp. 60–64.

_____. "Dual Resonance Models of Elementary Particles." *Scientific American,* (1975) 232:2 pp. 61–67.

Schwarzschild, Bertram. "Why is Cosmological Constant So Very Small?" *Physics Today,* (1989) 42:3 pp. 21–24.

Schwinger, Julian. "A Magnetic Model of Matter. *Science,* (1969) 165:3895 pp. 757–61.

_____. "A Path to Quantum Electrodynamics." *Physics Today,* (1989) 42:2 pp. 42–48.

Schwitters, Roy F. "Fundamental Particles with Charm." *Scientific American,* (1977) 237:4 pp. 56–70.

Sciama, D. "Inertia." *Scientific American,* (1957) 196:2 pp. 99–109.

Scully, Marlan O. and Murray Sargent III. "The Concept of the Photon." *Physics Today,* (1972) 25:3 pp. 38–47.

Searle, John R. "Is the Brain's Mind a Computer Program?" *Scientific American,* (1990) 262:1 pp. 31.

Segal, B.G. "Balmer Formula." *Encyclopedia of Physics.* Ed. by R. Lerner and G. Trigg. Reading, Mass.: Addison-Wesley, 1981.

Seliger, Howard H. "Wilhelm Conrad Röntgen and the Glimmer of Light." *Physics Today,* (1995) 48:11 pp. 25–31.

Selleri, Franco. *Quantum Mechanics Versus Local Realism.* New York: Plenum Press, 1988.

Sen, Amitabha and Sharon Butler. "The Quantum Loop: Where Gravity and Matter Parts Company." *The Sciences,* (1989) 29:6 pp. 33–36.

Sennett, R. *Authority.* New York: Alfred A. Knopf, 1980.

Sessa, di A.A. "An Elementary Formalism for General Relativity." *American Journal of Physics,* (1981) 49:5 pp. 401–11.

Shaffer, Jerome A. *A Philosophy of Mind.* Ed. by Elizabeth and Monroe Beardsley. Englewood Cliffs NJ: Prentice Hall, 1968.

Shah, K.T. "A Rigorous Approach to the Theory By Recami and Mignani for Tachyons." *Lettere al Nuovo Cimento,* (1977) 18:5 pp. 156–60.

Shankland, R.S. "The Michelson-Morley Experiment." *Scientific American,* (1964) 211:5 pp. 107–14.

Shapiro, Stuart L. and Saul A. Teukolsky. "Black Holes, Naked Singularities, and Cosmic Censorship." *American Scientist,* (1991) 79:4 pp. 330–43.

Sharks, J.G. "General Tachyon Absorption by Kerr-Newman Black Holes: Thermodynamic Consequences." *General Relativity and Gravitation.* (1980) 12:12 pp. 1029–34.

Shaver, Phillip and Jonathan Freedman. "Your Pursuit of Happiness." *Psychology Today.* (August 1976) pp. 26–32, 75.

Shimony, Abner. "The Reality of the Quantum World." *Scientific American,* (1988) 258:1 pp. 46–53.

Shinbrot, Marvin. "Things Fall Apart." *The Sciences,* (1987) 27:3 pp. 32–36.

Shipman, Harry L. *Black Holes, Quasars, and the Universe.* Boston, Mass., Houghton Mifflin, 1976.

————. *The Restless Universe.* Boston Mass., Houghton Mifflin, 1978.

Shriner, R.A. "Sense-Experience, Colours and Tastes." *Mind,* (1979) 88:350 pp. 161–78.

Siegal, Ronald K. "Hallucinations." *Scientific American,* (1977) 237:4 pp. 132–40.

Simmons, Henry. "The Neutrino, the Little Neutral One." *Mosaic.* (1979) 10:5 pp. 13–20.

Singh, Ishwar and M.A.B. Whitaker. "Role of Observer in Quantum Mechanics and Zeno Paradox." *American Journal of Physics,* (1982) 50:10 pp. 882–86.

Singh, Jagjit. *Great Ideas of Modern Mathematics: Their Nature and Use.* New York: Dover Publications, 1959.

Skinner, B.F. "Beyond Freedom and Dignity." *Psychology Today,* (1971) 5:3 pp. 37–80.

————. "Origins of a Behaviorist." *Psychology Today.* (1983) 17:9 pp. 22–33.

Skinner, R. *General Relativity.* Waltham, Mass.: Blaisdell, 1969.

Sklar, Laurence. *Space, Time, And Spacetime.* Berkeley: University of California Press, 1974.

Slansky, R. "Group Theory for Unified Model Building." *Physics Reports,* (1981) 79:1 pp. 1–111.

Smarr, Larry L. and William H. Press. "Our Elastic Spacetime: Black Holes and Gravitational Waves." *American Scientist,* (1978) 66:1 pp. 72–79.

Smith, John E. "Science and Conscience." *American Scientist,* (1980) 68:5 pp. 554–58.

Smith, Peter and O.R. Jones. *The Philosophy of Mind.* New York: Cambridge University Press, 1986.

Smorodinsky, Ya. A. *Particles, Quanta, Waves.* 1973; Trans. V. Kissin. Moscow: Mir Pub., 1976.

Smullyan, Raymond. *Forever Undecided.* New York: Alfred Knopf, 1987.

Sokal, Robert R. "Classification: Purposes, Principle, Progress, Prospects." *Science,* (1974) 185:4157 pp. 1115–23.

Solomon, Robert C. *Introducing Philosophy: Problems and Perspective.* New York: Harcourt Brace Jovanovich, 1977.

Sondheimer, E. and Alan Rogerson. *Numbers and Infinity,* New York: Cambridge University Press, 1981.

Space and Time in Special Relativity, New York: McGraw-Hill Book, 1968.

Spakovsky, Anatol von. *Freedom—Determinism.* Hague: Martins Nijoff, 1963.

Spergel, David N. and Neil G. Turok. "Textures and Cosmic Structure." *Scientific American,* (1992) 266:3 pp. 52–59.

Sperry, Roger. "Interview." *Omni,* (1983) 5:11 pp. 71–75, 98–100.

Spinks, H. Stephens. *Psychology and Religion.* Boston: Beacon Press, 1963.

Squire, Larry R. and Stuart Zola-Morgan. "The Medial Temporal Lobe Memory System." *Science,* (1991) 253:5026 pp. 1380–85.

Stachel, John. "Einstein and Ether Drift Experiments." *Physics Today,* (1987) 40:5 pp. 45–47.

Stacy, Dennis. "Transcending Science." *Omni,* (1988) 11:3 pp. 55–60, 114–16.

Stapp, Henry P. "Bell's Theorem and the Foundations of Quantum Physics." (1985) 53:4 pp. 306–17.

————. "Bell's Theorem and the World Process." *Nuovo Cimento,* (1975) 29B:2 pp. 70–76.

————. "Theory of Reality." *Foundations of Physics,* (1977) 7:5/6 pp. 313–23.

Stebbing, Susan L. *A Modern Introduction to Logic.* New York: Harper and Row, 1961.

Steen, Lynn Arthur, "New Models of the Real Number Line." *Scientific American,* (1971) 225:2 pp. 92–99.

————. ed. *Mathematics Today.* 1978; rpt. New York: Random House-Vantage Books, 1980.

Stein, Daniel L. "All Systems Slow." *The Sciences,* (1989) 28:5 pp. 22–29.

Stent, Gunther S. "Limits to the Scientific Understanding of Man." *Science,* (1975) 187:4181 pp. 1052–57.

Stevenson, I. *Twenty Cases Suggestive of Reincarnation.* Charlottesville: University Press, 1974.

Stickel, Delford L. ed. *The Brain Death Criterion of Human Death.* New York: Pergamon Press, 1979.

Stoll, Robert R. *Set Theory and Logic.* 1963; rpt. New York: Dover Publications, 1979.

Stolyar, Abram Aronovich. *Introduction to Elementary Mathematical Logic.* New York: Dover Publications, 1970.

Stroud, B. "Inference, Belief, and Understanding." *Mind,* (1979) 88:350 pp. 179–96.

Stumpf, Samuel E. *Philosophy: History and Problems.* New York: McGraw-Hill Book, 1971.

Sudarshan, E.C.G. and Thomas F. Jordan. "Simply No Hidden Variables." *American Journal of Physics,* (1991) 59:8 pp. 698–99.

Sulak, Lawrence R. "Waiting for the Proton to Decay." *American Scientist,* (1982) 70:6 pp. 616–25.

Susskind, Leonard. "Black Holes and the Information Paradox." *Scientific American,* (1997) 276:4 pp. 52–57.

Sutton, Christine. "Quest for the W-Particle." *New Scientist,* (1983) 97:1342 pp. 221–23.
_____. "Subatomic Forces." *New Scientist,* (1989) 121:1651 pp. 1–4 (insert).

Swartz, Clifford E. "Reference Frames and Relativity." *Physics Teacher,* (1989) pp. 437–46.
_____. *The Fundamental Particles.* Reading, Mass., Addison-Wesley, 1965.

Swartz, Robert J. ed. *Perceiving, Sensing, and Knowing.* Garden City, New York: Doubleday, 1965.

Swenson, Loyd S. Jr. "Michelson and Measurement." *Physics Today,* (1987) 40:5 pp. 24–30.
_____. "The Michelson-Morley Experiment." *Encyclopedia of Physics.* Ed. by R. Lerner and G. Trigg. Reading, Mass.: Addison-Wesley, 1981.

Swinburne, Richard. ed. *The Justification of Induction.* Oxford: Oxford University Press, 1974.

t' Hoof, Gerard. "Gauge Theories of the Forces between Elementary Particles. *Scientific American,* (1980) 242:6 pp. 104–38.

"Taking the Measure of Light." *Science News.* (1972) 101:6 p. 85.

Talbot, Michael. *The Holographic Universe.* New York: Harper Collins, 1991.

Talcott, Richard. "Everything You Wanted to Know about the Big Bang." *Astronomy,* (1994) 22:1 pp. 30–34.

Tarlé, Gregory and Simon P. Swordy. "Cosmic Antimatter." *Scientific American,* (1998) 278:4 pp. 36–41.

Tarski, Alfred. "Truth and Proof." *Scientific American,* (1969) 220:6 pp. 63–77.

Tart, Charles T. "States of Consciousness and State Specific Sciences." *Science,* (1972) 176:4040 pp. 1203–10.

Taylor, E. and J. Wheeler. *Spacetime Physics.* San Fransisco: W.H. Freedman and Co., 1966.

Taylor, Herbert J. "The Duplication of Chromosomes." *Scientific American,* 1958; rpt. W.H. Freeman and Co.

Teli, M.T. and V.K. Sutar. "Unified Model for the Description of Subluminal and Superluminal Objects." *Lettere al Nuovo Cimento,* (1978) 22:12 pp. 496–502.

Thomas, Norman L. *Modern Logic.* Outline Series. New York: Barnes and Noble, 1966.

Thomsen, Dietrick E. "A Cosmological Triple Play." *Science News,* (1974) 105:7 p. 109.
_____. "A Matter of Energy Levels." *Science News,* (1983) 121:11 pp. 183–85.
_____. "Cosmic Caldron Bubbles Up Universe." *Science News,* (1982) 121:8 p. 116.

_____. "Kaluza-Klein: the Koenigsberg Connection." *Science News*, (1984) 126:1 pp. 12–14.

_____. "Monopoles are a GUT Proposition." *Science News*, (1982) 121:20 pp. 323–24.

_____. "Supergroup." *Science News*, (1979) 115:13 pp. 214–15.

_____. "The New Inflationary Universe." *Science News*, (1983) 123:7 pp. 108–09.

_____. "Time Symmetry: Back and Forth and a Break." *Science News*, (1982) 121:1 p. 4.

Thorne, Kip S. "Gravitational Collapse." *Scientific American*, (1967) 217:5 pp. 88–98.

_____. "The Search for Black Holes." *Scientific American*, (1974) 231:6 pp. 32–43.

Tiles, Mary. *The Philosophy of Set Theory: An Introduction to Cantor's Paradise.* New York: Basil Blackwell, 1989.

Tinsley, Beatrice M. "From Big Band to Eternity." *Natural History*, (1975) 84:8 pp. 102–04.

Tinto, Massimo. "The Search for Gravitational Waves." *American Journal of Physics*, (1988) 56:12 pp. 1066–73.

Titus, Harold H. *Living Issues in Philosophy.* 5th ed. New York: Van Nostrand Reinhold, 1970.

Toben, Bob and Fred Alan Wolf. *Space-Time and Beyond.* 1975; rpt. New York: Bantam Books, 1983.

Torrance, E.P. *The Struggle for Men's Minds.* Orientation Group, USAF: Wright-Patterson AFB, Ohio, 1959.

Toulmn, Stephen. *The Philosophy of Science.* New York: Harper and Row, Publishers, 1960.

Trawler, Peter W. "Matter and Antimatter." *Chemistry*, (169) 42:9 pp. 8–13.

Treffert, Darold A. "An Unlikely Virtuoso: Leslie Lemke and the Story of Savant Syndrome." *The Sciences*, (1988) 28:1 pp. 28–35.

Trefil, James S. "Einstein's Theory of General Relativity is Put to the Test." *Smithsonian.* (1980) 11:1 pp. 74–82.

_____. "Einstein's Theory of Generality Put to the Test." *Simthsonian*, (1980) 11:1 pp. 74–84.

_____. "The Cosmic Forces." *Science Digest*, (1983) 91:11 pp. 83–85, 114–15.

_____. "Matter vs. Antimatter." *Science 81*, (1981) 7:2 pp. 66–69.

Treiman, S.B. "The Weak Interactions." *Scientific American*, (1959) 200:3 pp. 73–83.

Trester, Jeffery J. "The EPR Experiment, Special Relativity, and the Distinction Between Effects and Signals." *American Journal of Physics*, (1989) 57:1 pp. 86–87.

Tribus, Myron and Edward C. McIrvine. "Energy and Information." *Scientific American*, (1971) 224:3 pp. 179–88.

Trimble, Virgina. "Cosmology: Man's Place in the Universe." *American Scientist*, (1977) 65:1 pp. 76–86.

Trotter, Robert J. "The Other Hemisphere." *Science News*, (1976) 109:14 pp. 218–23.

Tulving, Endel. "Remembering and Knowing the Past." *American Scientist*, (1989) 77:4 pp. 361–67.

Turston, William P. and Jeffrey R. Weeks. "The Mathematics of Three-Dimensional Manifolds." *Scientific American*, (1984) 251:1 pp. 108–20.

Tyson, J.A. "Gravitational Waves." *Encyclopedia of Physics.* Ed. by R. Lerner and G. Trigg. Reading, Mass.: Addison-Wesley, 1981.

Uspensky, V.A. *Gödel's Incompleteness Theorem.* Little Mathematics Library Series. Trans. by Neal Koblitz from 1982 Russian edition. Moscow: Mir Publishers, 1987.

Valentine, James W. "The Evolution of Multicellular Plants and Animals." *Scientific American*, (1978) 239:3 pp. 141–58.

Van Flandern, Thomas C. "Is Gravity Getting Weaker?" *Scientific American*, (1976) 234:2 pp. 44–52.

Van Heel, A.C.S. and C.H.F. Velzel. *What is Light?* New York: McGraw-Hill Book, 1968.

Van Ness, H.C. *Understanding Thermodynamics.* New York: McGraw-Hill Book, 1969.

Van Over, Raymond. ed. *The Psychology of Freedom,* Greenwich, Conn., Fawcett Publications, 1974.

Vanderpool, Harold Y. *Darwin and Darwinism: Revolutionary Insights Concerning Man, Nature, Religion, and Society.* Lexington, Mass.: Heath and Co., 1973.

Veltman, Martinus J.G. "The Higg's Boson." *Scientific American,* (1986) 255:5 pp. 76–84.

Verity, Enid. *Color Observed.* New York: Van Nostrand Reinhold, 1980.

Vilenkin, Alexander. "Cosmic Strings." *Scientific American,* (1987) 257:6 pp. 94–102.

Vladimirov, Yu, N. Mitskiévich and J. Horsky. *Space Time Gravitation.* 1984; Vogel, Trans. A.G. Zilberman. Moscow: Mir Publishers, 1987.

Von Baeyer, Hans Christian. "Indistinguisable Twins" (Bosons and Fermions). *The Sciences,* (1988) 28:2 pp. 8–10.

Von Wright, G.H. "Truth, Negation, and Contradiction." *Snythese,* (1986) 66:1 pp. 3–14.

Wagstaff, G. *Hypnosis, Compliance, and Belief.* New York: St. Martins Press, 1981.

Wald, George. "The Origin of Life." *Scientific American,* 1954; rpt. W.H. Freeman and Co.

Wald, Robert M. *Space, Time, and Gravity: the Theory of the Big Bang and Black Holes.* Chicago: University of Chicago Press, 1977.

Waldrop, M.M. "Inflation and the Arrow of Time." *Science,* (1983) 219:4591 p. 1416.

_____. "Supersymmetry and Supergravity." *Science,* (1983) 220:4598 pp. 491–93.

_____. "The Large Scale Structure of the Universe." *Science,* (1983) 219:4588 pp. 1050–52.

Walker, A.E. *Cerebral Death.* Baltimore: Urban and Schwarzenberg, 1981.

Walsh, W.H. *Metaphysics.* New York: Harcourt, Brace and World, 1963.

Wang, Hao. *Reflections on Kurt Gödel.* London: MIT Press, 1987.

Wartofsky, Marx W. *Conceptual Foundations of Scientific Thought.* London: MacMillan, 1968.

Washburn, Sherwood L. "The Evolution of Man." *Scientific American,* (1978) 239:3 pp. 194–208.

Watson, Andrew. "The Mathematics of Symmetry." *New Scientist,* (1990) 128:1740 pp. 45–50.

Weatherburn, C.E. *An Introduction to Riemannian Geometry and the Tensor Calculus.* London: Cambridge University Press, 1938.

Weber, Joseph. "The Detection of Gravitational Waves." *Scientific American,* (1971) 224:5 pp. 22–29.

Webster, Adrian. "The Cosmic Background Radiation." *Scientific American,* (1974) 231:2 pp. 26–33.

Weinberg, C.B. *Mach's Empiro-Pragmatism in Physical Science.* New York: Columbia University Press, 1937.

Weinberg, Steven. "Life in the Universe." *Scientific American,* (1994) 271:4 pp. 44–49.

_____. "The Decay of the Proton." *Scientific American,* (1981) 244:6 pp. 65–74.

_____. "The Forces of Nature." *American Scientist,* (1977) 65:2 pp. 171–76.

_____. "Unified Theories of Elementary-Particle Interaction." *Scientific American,* (1974) 231:1 pp. 50–59.

_____. *Dreams of a Final Theory.* New York: Pantheon Books, 1992.

_____. *The First Few Minutes.* New York: Bantam Books, 1979.

Weisberg, R.W. *Memory, Thought, and Behavior.* New York: Oxford University Press, 1980.

Weiss, Joseph. "Unconscious Mental Functions." *Scientific American,* (1990) 262:3 pp. 103–09.

Weisskopf, Victor F. "How Light Interacts with Matter." *Scientific American,* (1968) 219:3 pp. 60–71.

_____. "Personal Impressions of Recent Trends in Particle Physics." *Comments on Nuclear and Particle Physics,* (1980) 9:2 pp. 49–54.

_____. "The Development of Field Theory in the Last 50 Years." *Physics Today,* (1981) 34:11 pp. 69–85.

_____. "The Origin of the Universe." *American Scientist,* (1983) 71:5 pp. 473–80.

_____. "The Three Spectrocopies." *Scientific American,* (1968) 218:5 pp. 15–29.

West, Beverly Henderson, Ellen Norma Griesbach, Jerry Duncan Taylor and Louise Todd Taylor. *The Prentice Hall Encyclopedia of Mathematics.* Englewood Cliffs, NJ: Prentice Hall, 1982.

Wheeler, John A. "Hermann Weyl and the Unity of Knowledge." *American Scientist,* (1986) 74:4 pp. 366–75.

_____. *A Journey into Gravity and Spacetime.* New York: Scientific American Library, 1990.

_____. and Wojciech H. Zurek eds. *Quantum Theory and Measurement.* Princeton Series in Physics. Princeton: Princeton University Press, 1983.

White, Andrew Dickson. *Warfare of Science with Theology in Christendom.* 1896; Rev.Bruce Mazlish. New York: The Free Press, 1965.

White, John. ed. *Frontiers of Consciousness: The Meeting Ground Between Inner and Outer Reality.* New York: Julian Press, 1974.

Wigner, Eugene P. "Violations of Symmetry in Physics." *Scientific American,* (1965) 213:6 pp. 28–36.

_____. "Interpretation of Quantum Mechanics." from 1976 Princeton lectures. Rev. 1981; rpt. in *Quantum Theory and Measurement.* Ed. by John A. Wheeler and Wojciech H. Zurek. Princeton Series in Physics. Princeton: Princeton University Press, 1983.

_____. "The Problem of Measurement." in American Journal of Physics, 1963; rpt. *Quantum Theory and Measurement.* Ed. by John A Wheeler and Wojciech H. Zurek. Princeton Series in Physics. Princeton: Princeton University Press, 1983.

Wilber, Ken. *The Spectrum of Consciousness.* Wheaton, Ill., The Theosophical Publishing House, 1977.

Wilczek, Frank. "Anyons." *Scientific American,* (1991) 264:4 pp. 58–65.

_____. "The Cosmic Asymmetry between Matter and Antimatter." *Scientific American,* (1980) 243:6 pp. 82–90.

_____. "The Persistence of Ether." *Physics Today,* (1999) 52:1 pp. 11–13.

Wilkenson, D. "Isospin." *Encyclopedia of Physics.* Ed. by R. Lerner and G. Trigg. Reading, Mass.: Addison-Wesley, 1981.

Will, Clifford M. "Gravitation Theory." *Scientific American,* (1974) 231:5 pp. 25–33.

Wilson, Allan and Rebecca L. Cann. "The Recent Genesis of Humans." *Scientific American,* (1992) 266:4 pp. 68–73.

Wilson, Kenneth. "Problems with Many Scales of Length." *Scientific American,* (1979) 241:5 pp. 158–79.

Wilson, W. "Internal Gauge Symmetries and the Gravitational Field." *General Relativity and Gravitation.* (1980) 12:1 pp. 51–55.

Winfree, Arthur T. "Filaments of Nothingness." *The Sciences,* (1986) 26:2 pp. 20–27.

Winson, Jonathan. "The Meaning of Dreams." *Scientific American,* (1990) 263:5 pp. 86–96.

Witten, Edward. "Duality, Spacetime and Quantum Mechanics." (1997) 50:5 pp. 28–33.

Wolf, Fred Allan. *Star Wave: Mind, Consciousness, and Quantum Physics.* New York: MacMillan, 1984.

Wolfram, Sybil. *Philosophical Logic.* New York: Routledge, 1989.

Woo, C.H. "Consciousness and Quantum Interference—An Experimental Approach." *Foundations of Physics,* (1981) 11:11/12 pp. 933–46.

_____. "Why the Classical-Quantal Dualism is Still With Us." *American Journal of Physics,* (1986) 54:10 pp. 923–27.

"Year of Dr. Einstein." *Time,* 19 February 1979 113:8 pp. 70–79.

Yoav, Ben-Dov. "Everett's Theory and the 'Many-Worlds' Interpretation." *American Journal of Physics,* (1990) 58:9 pp. 829–32.

Young, Arthur M. *The Reflexive Universe: Evolution of Consciousness.* Mill Valley, CA: Robert Briggs Associates, 1986.

Young, Fredrick H. *The Nature of Mathematics.* New York: John Wiley and Sons, 1986.

Young, Louise B. *The Unfinished Universe.* New York: Simon And Schuster, 1986.

Young, Richard W. "Visual Cells." *Scientific American,* (1970) 223:4 pp. 80–91.

Zebrowski, George. "Life in Gödel's Universe: Maps all the Way." *Omni,* (1992) 14:7 pp. 53–60, 84.

Zee, A. "Time Reversal." *Discover,* (1992) 13:4 pp. 97–101.

Zeki, Semir. "The Visual Image in the Mind and Brain." *Scientific American,* (1992) 267:3 pp. 69–76.

Zeleny, W.S. "Symmetry in Electrodynamics: A Classical Approach to Magnetic Monopoles." *American Journal of Physics,* (1991) 59:5 pp. 412–15.

Zuckerman, Martin M. *Sets and Transfinite Numbers.* New York: MacMillan, 1974.

734

Glossary

ambiguous figure — "Perceiving" that a figure's identity changes although the object perceived does not change. Indicates one can be fooled by perceptive processes.

analysis — Either mentally or physically reducing things to parts to better understand their composition. Similar to reductionism.

anthropic principle — That the physical universe seems to be "fine tuned" (evolves in a way) to allow the development of man to observe it. Suggests there is a design to everything.

antichrist — Cosmic complement to the Christ: poses as the messiah. The mind of the first coming functions deductively. Primary subject of the New Testament. Sets stage for second coming.

antiparticles — Cosmic complement to the material particles of our physical world. Matter of the Anticosmos (i.e., antitardyons and antitachyons). The positron is the antiparticle complement to the electron.

atheist — One who believes not (or denies) there is a God. This project suggests God is sought by seeking truth. The theist does not seek truth anymore than does the atheist. Therefore; the commonly defined theist is antiGod and is therefore functionally no different from an atheist. There is a categorical difference between believers (theists and atheists) and knowers.

attachment — Clinging to things (or unverified ideas) for security rather than seeking an understanding of things. The natural expectation of a mind inhabiting a physical body. Contrasts with developing a methodology capable of producing understanding.

awareness — Inferential cognizance. Usually pertains to mental conceptualization of the physical world. Specifically refers to the mental functions of deduction and induction. Awareness (the mind) is a transcendent function of the brain.

axiomatization of arithmetic — That arithmetic is premised on a single principle. Simply, that (the existence of) numbers can be explained by a "self explaining concept." The definition of zero presented in this book (defined by the Law of Noncontradiction) meets that criterion.

baryons — Variations of nucleonic elementary particles. Hadrons subject to the strong force.

belief — A statement accepted as true (that it represents something in reality) on faith; therefore, without verification via reality. Indicates one has not developed a methodology capable of rendering real understanding.

believing — The act of accepting a statement as true without scientific verification or logical proof.

Bell's inequality — A mathematical inequality (A is greater than or lesser than B) developed by John Bell that is satisfied if quantum theory is not valid (and therefore, local realistic theories are valid). A. Aspect and others demonstrated the inequality is "violated" and therefore quantum theory is universally valid.

Bi-Deltic Cosmos Duration (BCD) — Characterizes the nondirectional nature of time at a level that "negates" Bi-Deltic Cosmic Instances of time. The BCD is the collective of all instances of time and is the "symmetry" consideration inductively derived from all durational aspects of time.

Bi-Deltic Cosmic Instant (BCI) — Characterizes the temporal nature of the Bi-Deltic Cosmos. A directionless quantum instant of time (a synthesis of Deltic and Antideltic instants of time).

big bang — Physics' concept the physical universe originated from a singularity (or point of space-time) and "exploded" into material existence. Scientifically supported by recession of distant galaxies and the microwave background radiation.

bit — A single piece (bit) of information. It has a value of zero or one. Often used in conjunction with computers. Seven or eight bits compose a byte.

black-body radiation — Energy naturally emitted from a physical object. Called black because energy emitted from an "ideal" black object (at room temperature) is invisible.

born again — A phrase often misconstrued by religious people. Refers to the "awakening" of the mind: one that begins to function "inductively" (reasoning to higher levels of generality). Appropriately characterizes the "resurrection" of the mind's natural state (functioning inductively) in Phase II (heaven) of cosmic cycle.

bosons — Nonrest mass elementary "particles" that characterize the "forces" of nature. Not subject to Heisenberg's Uncertainty Principle. Also, certain mesons and atoms (having rest mass).

Brahman — Hindu religious concept that can be correlated with God or consciousness.

brain — Physical organ found in the head of animals. The brain "connects" the mind to the physical world. Enables the functionality of awareness.

byte — A computer term referring to a packet of bits that may represent an alphabet letter or a number.

cardinality — Refers to the use of a number to indicate a collection. Answers the question, "how many?" For example, that there are ten apples.

causality — Often a scientific term used to indicate a connection between two events. Implies one event led to (caused) another event. Less confusion results by replacing "cause" with the idea of "contributing factor."

chaos — Indicates a group of particles or things exhibits no discernible order.

charge conjugation (invariance) — A physics principle stating electric charge is "conserved" (charge before interaction equals that after) in an elementary particle interaction. First aspect of the CPT theorem.

chirality — Refers to left hand — right hand symmetries. One hand is the mirror image of the other. A right hand glove turned inside out "becomes" a left hand glove. Suggests spatial inversion.

Christ, First Coming (Antichrist) — Individual "appears" as Christ, but is (functionally) the Antichrist. Subject of New Testament.

Christ, Second Coming — Is the real (functional) Christ.

chromosomes — Refers to the sets of DNA molecules that are the "building blocks" of physical life.

Cirpanic Cosmos — The encompassing structure of reality. Refers to the collection of "everything" physically and mentally. Represents everything (experienced) but not the whole of everything. Does not include consciousness.

Cirpanic cosmos omnipresence (CCO) — Represents the "highest" temporal aspect of cosmic structure. Pertains to the encapsulation of the past and future into the present. Represents "no flow" of time.

coherency — That a system is logically consistent; that there are no contradictions.

color paradigm — Refers to the significance of color complementarity. Color seen is complementarily related to the object's color "in-itself." We observe a lemon; it appears yellow, but we know (by reasoning) it is itself blue. Suggests objects exist apart (separate) from our perception of them.

complementarity — That things are created in pairs (of opposition). Indicates there are two aspects to most things. Wholeness complements unity. Also implied in the principle of duality.

concepts — Representing things as generalities (abstract ideas). Universal statements amenable to inclusion in a hierarchy.

conjunction — Logical notion of AND. This AND that.

consciousness — Function that supersedes everything and is ultimately (inductively) concluded from assessing the import of everything. Answers the question, "how is everything (experienced) explained?"

continuum — In mathematics, the nonpartitioning of a line. Often demonstrated by drawing a line (using a ruler) from zero to one. In this project, it is the wholeness function.

Contradiction, Law of — Logical notion a statement cannot both be true and false. Applied to things, that a thing cannot both exist and not exist (at any given time). Defines the unit function.

correlation — That there is a relationship between things or phenomena. Also, that two distinct things or concepts are somehow connected.

correspondence — That information as knowledge represents (corresponds to) something in reality.

cosmic background radiation — Pervasive cosmic microwave energy. Its existence supports the big bang theory. (See microwave background radiation.)

cosmogony — Concepts concerned with origin of the physical universe. The big bang is a scientific example.

CPT theorem — (charge parity time) A symmetry principle of physics. Indicates only a collective inversion of charge (plus or minus) parity (spatial extensionality: right or left) and time directionality (forward or reverse) is allowed in cosmic construction. Suggests there is a complementary (inverted) system (the anticosmos) to the physical world: one in which time flows in an opposite direction.

curvature of space (-time) — That spatial extensionality and time flow are altered in the presence of a strong gravitational field.

deduction — Usually pertains to logical conclusion. Also suggests logic confines conclusions to no more than what is already "contained" within the premise. Loose term for reductionism.

determinism — That events occur by design. Usually considered opposed to free will. Determinism does not necessarily preclude free will unless determinism is absolute.

devil — Religious word that corresponds to the mind functioning deductively. The expected mentality experienced by a mind inhabiting a physical body. Reductive function of Satan.

disjunction — Logical notion of OR. This OR that.

double aspect theory — Suggests there are two sides to one thing. What appears as two separate phenomena are but one at a higher (hierarchical) function. Electricity and magnetism are two aspects of electromagnetism.

duality, principle of — Idea of complementarity. That everything comes in pairs of opposites. Also that two aspects mesh and contribute to phenomenal functionality. Male — female, day — night, Matter — antimatter.

egocentric behavior — Self centered behavior. Indicated by word usage such as "I believe, I think." Expected mental behavior during Phase I of cosmic cycle.

electroweak force — Functional synthesis of two distinct forces: the weak and the electromagnetic force. Has been verified. Suggests all forces might ultimately "unify" as a functionally distinct force from which all other forces were derived.

elements (of a set) — Members of a set or class of things. Individuals are members of the class of people. Domestic pets are members of the class (classification) animals.

emergent transcendent function — A "single" phenomenon that arises from the collective activity of parts. Higher level functionality has a basis in lower level functions. Cellular activity "emerges" from the collective activity of molecules. Muscles are a function arising from the collective activity of muscle cells.

empty set — A class (in set theory) that has no members or elements. Set theory version of zero in mathematics and falseness in logic.

energy — The reductive reason for all activity. Energy "enables" (is the lesser source of) higher level functions.

entropy — That physical systems eventually and naturally break down into their components. Dominant process during Phase I of the cosmic cycle. Opposite to negentropy.

epiphenomenalism — Philosophical idea the mind is a nonmaterial product of the brain, yet cannot affect the brain.

EPR paradox — (Einstein, Podolsky, Rosen) A thought experiment in which Bohr's ideas about the "acausal" nature of quantum theory were challenged.

equivalence principle — Einstein's idea an accelerating system mimics the effects of gravity. Einstein thought a person in a "closed uniformly" accelerating elevator could not discern the difference between that and the effect of gravity while standing on earth's surface; therefore that the two phenomena are mathematically equivalent.

event horizon — A hypothetical sphere around a black hole wherein matter cannot escape, but is pulled toward the hole's center.

Excluded Middle, Logical Law of — That logically, either something is or is not. It must be one (exists) or the other (not exist).

experience — What consciousness is conscious of. Manifestation, creation.

explanation — The reason (in knowledge form) for something.

extrinsic modes — Generalized description of physical events in interaction. A way to describe cause and effect type relations.

fact — A statement that is true because it represents something in reality. The essential problem is whether a statement is "known" to be factual.

false — Logical notion of "not true." In reality, it represents the wholeness function.

Fool's Fatuity — Using what one does not know to obtain an understanding of what he is trying to know.

forms — Most simply, the shape of something. A thing's identity usually indicates its form (or wholeness aspect). The wholeness function.

free will — That an individual is solely the cause for his own action. The free will problem hinges on properly understanding the nature of causes and the structure of reality. Man is limited in his "own" will by the extent he is a part of a greater being.

general theory of relativity — Einstein's treatment of gravitation. He supposed gravitation is attributed to spatial curvature (distortion) caused by massive objects.

generalities — Conceptualized universal statements representing higher level categories (in a hierarchy).

geodesic — "Shortest" distance between two points in space: One that light would travel. Geometrically, the "straightest" line between two points might be "curved" from the distorting effect of massive physical objects.

global symmetry — Invariance principles (simply, laws of nature) that are valid everywhere.

gluons — Elementary particles of force that bind quarks into nucleonic structures.

God — The supposed final cause of everything. Correlates with the ultimate wholeness function, consciousness.

Goddess — Hypothesized in many religions as a counterpart to God. Correlates with the final fractionation of everything: infinite energy of the vacuum, the extreme parts function.

Godel's incompleteness theorems — Kurt Gödel's idea that logic, if consistent, is incomplete, and if complete, is inconsistent.

gravitation — The idea espoused mathematically by I. Newton that massive bodies (objects) "attract" each other by an unexplained "action at a distance." Einstein's idea of space-time curvature explained Newton's idea.

gravitational radiation — The idea gravity is mediated by a force particle much as electromagnetic phenomena interact by means of photons.

GUT's (grand unification theories) — Unifying concepts that combine the strong nuclear force with electromagnetism and the weak nuclear force. Usually does not include the force of gravity.

hate — That which one cares for second (between the self and all else except the self). Opposite of love.

heaven — According to religion, a higher plane of existence. Correlates with the mind (spirit): the kingdom of God. Phase II of cosmic cycle.

Heisenberg's Uncertainty Principle — The idea an observer cannot simultaneously know an elementary particle's position and momentum. Also describes other conjugate variables such as energy and time.

hell — Generally thought to be "opposite" to heaven. Correlates with physical matter.

here and now — Present time. There is no past nor future without a here and now. The problem was to explain how the present "generates" the past and future without (the present) being altered. Ultimately, there just is the present (no time flow).

hidden variable theories (HVT's) — Hypothesis that the "apparent" indeterminacy of quantum systems results from not fully knowing certain "hidden" factors. An attempt to support the EPR paradox. Retrospectively, HVT's countered the nonseparability implications of quantum theory. The dilemma was how elementary particles could (instantaneously) act as though they were but one particle. Experimentation testing Bell's theorem resulted in the demise of HVT's as well as any notion quantum theory does not involve the nonseparability of everything.

hierarchical modes — A means of explanation. Classification using a "branching" tree structure to represent different levels of generality (or categories of function). The structure of reality is defined by a hierarchical system.

Higg's particle — The supposed particle of all particles. The hypothesized unifying particle of high energy physics. Some scientists suppose all other particles have their origin in the Higg's particle.

How Aspect — One of the three aspects of the Deltic Cosmos. Represents the universe by identification of its "charge" (using a standard such as the electron). Therefore, either of matter or antimatter.

hypothesis — A carefully reasoned assumption. Scientifically, a statement considered possibly true, but needing testing via reality to verify that it (the assumption) corresponds to something in reality.

identity — That something is what it is. That the knowledge label correctly indicates what it purports. There is the word car, then there is the actual car that it (the word car) might signify. Identifying (knowing) something is not the same as understanding (its nature).

imaginary number (i) — The "minimal" complex number. Defined by the square root of minus one.

immanence — A ("high") function "represented" in lower hierarchical levels. For example, consciousness (characterized by knowledge) is represented in manifestation (experience) by things themselves. There is the word car, then there is the actual car. The word (or concept) car is "represented" by the actual car. The concept of car-ness and what it signifies is immanent in the car itself. Typically opposite to transcendence. Automotive "transportation" is a transcendent function of the actual car. Transcendent functions (here transportation) suggest purpose or reason for manifestation (of car-ness). Purpose is inherent (immanent) in the nature (design) of things.

immortality — Idea there is that which survives physical existence. The mind (spirit) is immortal and is created for the next existence, one that is "predominately" mental.

incarnation — Thought by some that Christ is the embodiment of God. That a mind "inhabits" a physical body.

induction (of inductive reasoning) — Conceptualizing to higher levels of generality (in a hierarchy). Higher levels represent enveloping functions. Molecules are a higher level function than its constituent atoms.

inertial system — A frame of reference (system of bodies: objects) that either is not experiencing any motion or is traveling at constant velocity. A nonaccelerating system (relative to "observer"). Contrasts with accelerating systems.

inferential functions — Mental processes (functions); therefore, deducting or inducting.

infinity — unlimited partitioning of the Whole.

interphenomenal relevancy — Relation between classes (different levels of hierarchical categories) of things.

intraphenomenal relevancy — Relation within a class of elements: between members of the same class.

intrinsic modes — A way of describing the "inherent cause" of everything. Stop-gap concept, later filled by the idea of the wholeness function.

intrinsic perception — Experiencing things through one's own inherent vision; therefore, what is commonly denoted (in physical existence as) "dreaming." When fully operational, it becomes means of perception in "next existence."

isotopic spin — Ascribed characteristic of nucleonic (proton-neutron) particles. In strong interactions, the neutron and proton behave alike. They are differentiated by spin. Other baryons (lambda, sigma) are considered higher energy versions of nucleons.

iteration, numerical — Counting: therefore, one, two, three and so on.

karma — Hindu idea how one handles his confrontations with reality affect what he consequently experiences. Pertains to present existence as well as one's "position" in subsequent incarnations. An easily understandable system of reward and punishment. Difficult to verify, but practically impossible to refute.

kingdom of God — Heaven. Time (Phase II of cosmic cycle) when the mind (spirit) dominates its existence. In Phase I, the mind is "dominated" by physical matter (hell).

knowledge — Information that (in statement form) represents reality. One "has" knowledge when he can verify that the statements he espouses as true, correspond to reality.

leptons — Group of elementary particles of which the most well known is the electron.

light — Electromagnetic energy (radiation). Exhibits both wave and particle natures. Includes, gamma, x-ray, ultraviolet, visible, infrared, microwave, radio waves. Correlates to religious term Lucifer.

light velocity — That light speed is constant and never changes relative to either subluminal particles (tardyon — physical matter) or superluminal particles (tachyonic matter — mental function).

local symmetry — Invariance principles valid only for a limited region.

logic — Abstract method of representing the "rules" of discourse regarding the values of true and false. Logic can be "translated" into mathematics: true = one, false = zero.

love — That which one cares for most: between oneself and all else. Opposite to hate.

Lucifer — Biblical being that opposes God. Correlates with the scientific concept of electromagnetic (light) energy (EM radiation). The highest unit function of Reality.

Mach's principle — E. Mach's idea inertial mass (of physical matter) is somehow related to the spread out mass of the universe.

magnetic monopole — A (hypothesized) single charged particle of magnetism. Not observed in the physical world where magnetism only occurs in dipole (exhibits both north and south polarity) configurations.

manifestation — Consciousness fractionates: "materializing" as physical matter. Consciousness projects experience. Experience is a "manifestation" of consciousness.

manifold — Overlap mapping of curved surfaces. Used to understand Riemannian geometry. Helps explain gravitation as a curvature of space.

many worlds interpretation of quantum mechanics — H. Everett's understanding of quantum theory. Elevates the superposition principle to represent "all possible" worlds. A way to avoid Bohr's collapse of the "universal" wave function (superposition) to represent the particular world we experience. Included "the observer" as part of quantum formalism.

matter waves — Physical matter can exhibit a wave nature. Verified by Davisson and Germer's experiment using nickel metal that is "crystallized" into an atomic lattice. The lattice structure diffracted incoming electrons into an outgoing pattern indicating they exhibit a wave nature.

Maya — Hinduism idea everything experienced through the senses is an illusion: that things (themselves) are not ultimately real.

meaning of life — That life has some kind of "higher" purpose. The physical body serves the mind; the mind serves consciousness. Explained in the final chapter.

meaning — Significance, denotation, relevancy. That statements correspond to and therefore represent something in reality. Also, that concepts acquire meaning because of coherency (or logical consistency).

memory — Stored information. The physical world is the manifest "memory" of the knowledge used to create it. Simply, experience is the memory of consciousness. An automobile's structure represents the knowledge and understanding used in its construction. The structure of reality represents the understanding of the knowledge used in its construction.

mental functions —Inferential functions. Either deductive — reasoning from generalities to specificities, or inductive — reasoning toward higher level generalities.

mentality — Of the mind. Corresponds to the religious word spirituality.

messiah — Many religions suggest an individual who represents God in manifestation. The manifestation of God as a "part" of himself.

metaphysics — Beyond physics. Often the purview of philosophers. Yet, it adequately describes the in depth understanding of logic that extends physics to describe all of reality.

Michelson and Morley's experiment — An experiment using an interferometer to determine whether light required a "medium" of transmission like waves on water. Pointing the instrument in different directions detected no "ether" wind. There was no evidence of a transmission medium for light.

microwave background radiation — Present day evidence of the high cosmic temperature (energy) when light was created after the big bang. As the universe expanded, it cooled. Today that radiation energy is detected as a microwave frequency equivalent to that emitted by a "black body" exhibiting a temperature of a couple of degrees above absolute zero. Television "snow" is attributed to this cosmic radiation.

mind — "Agency" of inferential functions. The mind either deducts or inducts. Correlates with religious word spirit.

negation — The logical notion of replacing a value with its opposite. Negating the value of trueness yields falseness.

negentropy — Negative entropy. The natural tendency for things to build up into higher level functions. Atoms collectively function as molecules. Dominant process during Phase II of cosmic cycle. Opposite to entropy.

nominalism — That higher level constructs (language abstractions) do not represent anything in reality. Nominalists believed "abstractions" only had categorical significance. The concept of "car" does not correspond to any particular car. Such an example does not preclude the possibility some higher level constructs do represent something. The collective of certain atomic structures do represent molecular functionality.

Noncontradiction, Law of — A logical principle that defines the wholeness function (falseness, zero). "Nothing can both be and not be." When a whole is "broken" into parts, it ceases being functionally whole. Consciousness is not a thing, yet it can manifest as things. Defines logical consistency and completeness.

nonseparability — That there is an instantaneous connection between everything. The wholeness function (consciousness) envelops everything (experience).

nothingness — The logical opposite of everything (every somethingness). Nothingness is represented and defined by the Law of Noncontradiction.

noumenon — The "thing in itself" as opposed to its appearance. If things were experienced exactly as they are (themselves) there would not be any disagreement about their nature.

nucleons — Components of atomic nuclei. Protons and neutrons.

null set — The empty set (in set theory). A set with no members. Zero, falseness, the wholeness function.

Ockham's Razor — The principle of parsimony: that the simplest explanation is more likely the correct one. After William of Ockham, 13th century.

omnipotence — Usually refers to a superior being that is all-powerful.

omniscience — Implies God is all-knowing.

operant behavior — B.F. Skinner's behaviorist concept that contingencies in one's surroundings either reinforce or extinguish ("operate" upon) certain behaviors. An attempt to explain behavior without alluding to the "inner self."

opinion — A statement an individual considers true, but does not assert its truth.

ordinality — The order of things counted. First, second, third, and so on.

overt behavior — Behaviors that another person can witness. Opposite of covert behavior.

panentheism — Idea God manifests the universe but is not identical with it.

pantheism — Idea God is synonymous with the universe (or nature).

paradox — An apparently unresolvable problem. Usually suggests a problem that is inherently unsolvable. Many problems thought to have been impossible, have been solved. Lesson is that many problems exist because their nature has not been sufficiently understood.

parity — The complementarity of spatial extensionality. A left hand glove turned inside out (looking like a right hand glove) represents its spatial inverted counterpart. Similar to chirality.

Parousia — Word for Second Coming of Christ.

part — Considering something as an aspect of a greater whole. The part may be a functional whole in its own right. An atom is a part of a molecule, yet is a whole in itself.

part function — An appropriate way to label (identify) something by its role in the scheme of things. There is no part that is not a lesser function of something greater.

particle — Considering a thing (a part) without regard to any role it may have in the scheme of things.

Pauli's Exclusion Principle — That two identical (fermion) particles cannot (simultaneously) occupy the same position. Two electrons (in a given atom) are precluded from having the same quantum numbers.

phenomenal relevancy — That things, as they appear, are interrelated.

phenomenon — Appearance of things. Does not necessarily represent things as they are in themselves (noumenon).

photoelectric effect — Einstein explained that metal irradiated with light energy emits electrons. The electrons do not increase in energy as the impinging light energy is increased: only the number of electrons emitted increases. This explained light can also act like a particle — the photon. The electron is the manifest aspect of the photon.

photons — Light characterized as particles (quanta) of energy.

pivotal events — Events that alter the course of history. Turning points in one's life.

Planck time — The most diminutive interval of time. An instant of time. Physicists believe the interval is equivalent to 10^{-40} second.

Platonism — Concept espoused by Plato that there is a "world" (realm) of forms (or ideas). That knowledge itself is characterized by a realm distinct from the physical world. Most mathematicians are Platonists and believe truth is discovered. Suggests truth always existed; waiting to be uncovered.

positron — The positively charged (antimatter) counterpart to the electron.

potential infinity — The natural numbers (positive integers) iterated without limit.

potential transcendent function — A phenomenon that is "real," but does not functionally exist because it has been "broken" (fractionation). A whole jigsaw puzzle is real but does not exist (as a functional whole) after it has been partitioned into parts. Yet it is real in the sense that it could be reestablished by synthesis of parts back into the whole from which they were derived. Foremost, the whole is real because it preceded its fractionation (into pieces).

power set — A combinatory set resulting from considering every possible combination of a set's members. It is a set that can only be done "on paper." In reality, it is an impossible set.

primary equation — The equation $1 = 0 \, x \, \infty$. Properly understood, helps define its components. A stepping stone to the ultimate principle.

primary qualities — Characteristics of matter that are thought to not depend upon perception. Examples: extensionality and solidity.

projection — As used in this project. That consciousness projects (movie projector analogy) itself into manifestation (as experience) via the cosmological Big Bang. Explains how present time consciousness generates the flow of time.

proof — A word properly reserved for logical discernment of truth.

protons — Positive nucleonic counterpart to the electron.

purpose — Reason by design.

quantum (of action) — Max Planck's idea energy is conveyed in packets (as opposed to continuous exchange of energy). Photons are packets of light energy.

quantum theory — Physics that arose from M. Planck's idea energy is only exchanged in multiples (packets) of the quantum (of action).

quark — Considered the micro-basis of all physical matter.

radioactivity — That some atoms naturally change by a calculable (statistical) nuclear beta decay. Can be used to "date" the origin of fossils.

realistic behavior — Behavior of an individual who seeks truth discernment by looking to reality for verification (of truth). Expected behavior during Phase II of cosmic cycle.

reasoning — In this project the word mostly denotes the following. Inductively generalizing toward higher level concepts that represent some structure or functionality in reality.

recollection — Retrieving of information that has been "stored."

red-shift — Alteration of light energy from distant cosmic sources toward lower frequencies (longer wavelengths) caused by recession (moving away) of the source.

reductionism — "Reducing" a thing (either by knowledge or physically) to its components to learn how it is put together. Reductionism is sometimes, but incorrectly defined another way. Classifying things (or phenomena) by knowledge in a categorical sense seems to be an act of reductionism, but is actually an act of synthesis, the opposite of reductionism.

reincarnation — That one's mind (spirit) may have existed (individually) before its present functionality in a physical body. If so, then it has re-incarnated into its present body. Otherwise, one's mind "preexisted" in one's biological parents.

reinforcement — Concerning behavior, the positive or negative conditioning of specific behaviors.

relativity, principle of — Einstein's basis for special relativity. That the laws of physics are the same everywhere and that light velocity is constant.

relevancy — How things interrelate. Ascribes meaning to phenomena.

renormalization — A method to eliminate the "problem of infinities" in calculating certain properties of elementary particles.

Russell's paradox — Concerns the "set of all sets which are not members of themselves." If this all-embracing set does not include itself as a member then it is incomplete (and therefore not all embracing). Therefore, it cannot be the set of all sets, hence the contradiction.

Satan — The supposed superior being responsible for evil. It correlates with the "collective" of all minds functioning deductively. (Individual minds are devils.)

Schrödinger's cat paradox — E. Schrödinger proposed a hypothetical experiment suggesting what would happen if quantum phenomena are "magnified" to a macro size we could easily witness. (In everyday life a cat is either alive or dead.) He places a cat into a closed box with poison that is released according to the probability of wave-function collapse encountered in quantum theory. Between the extremes of alive and dead are probable states in which the cat is suspended between life and death. (This is not observed in everyday life.) Characterizes the paradoxical nature of quantum theory.

secondary qualities — Those sensed. Examples are color, smell, and sound.

sense data — Things as they appear. Usually denote images before identification by perception.

separability — That things are distinct and functionally independent from other things.

set theory — Idea things can be abstractly represented using the notation of class (set) and what is categorically included as members (elements) of the set. A way of characterizing logic.

setness — Characterizing a set as though it was a thing (thingness) itself regardless of any elements possibly included within it. A thing's wholeness function.

sin — A religious term that correlates with the notion of desire. Want.

singularity — Center of a black hole. Represented by the mathematical notion of zero. Also characterized by the idea of the wholeness function presented in this project. The cosmological singularity correlates with the universal wholeness function.

skepticism — Philosophical idea one cannot be sure of the truth of any given statement. One who continually doubts. Offers an initial reason to question everything — to seek truth.

sleep — The mind sleeps; the body rests. A mind not functioning inductively is basically asleep (even though it deducts). Typical mental state in Phase I of the cosmic cycle. (An "awakened" [nonsleeping] mind functions inductively.)

solipsism — Philosophical idea only the self exists and that everyone else is a "projection" of one's imagination. Better, that there is just "one" consciousness and everyone partakes in its expression (as experience).

somethingness — Abstraction of treating a particular thing as a unit (therefore, that it has a wholeness function and a parts function).

space — Nothingness. Wholeness function: consciousness itself.

spatial extensionality — Notion of space as "distance." Contrasts with space "in itself."

special theory of relativity — Einstein's mathematical treatment relating the nature of space and time.

spin (angular momentum) — An inherent characteristic of elementary particles. Integer spins (0, 1, 2) indicate particles of force. Fractional spins (1/2) indicate particles of manifestation (creation).

spirit — Religious word that correlates with mind.

stream of experience — The flow (in time) of what we are conscious of.

strong force — Nuclear force that binds atomic nuclei into stable configurations.

subsetness — Abstract term for sets acting as members of some greater set. The class of "life-forms" includes a "lesser" set (subset) called animals. (Plants are another subset of life-forms.)

superforce — Hypothesized ultimate "unification" force from which lesser reductive forces (electromagnetic, weak, and strong) are derived.

supergravity — Another name for the hypothesized unified force; one indicating that the ultimate force must include gravity.

symmetry breaking — That unification (symmetry) is disrupted (broken) in the fractionation process of creating lesser level particles.

symmetry — (As used in this project) That things are minimally created in complementary pairs exhibiting opposing characteristics. Indicates there is a "balance" (symmetry) to everything. Synthesis of complements indicates symmetry.

synthesis — In thought: bringing disparate concepts together as a unifying principle. In reality: the combining of different things into a higher level functionality. Various atoms combine into molecules.

system — Collection of "parts" considered as a whole (the system). The automobile is a system of parts.

tachistoscope — An instrument used to "flash" visual images for a fraction of second to either the left or right brain. Used to determine brain asymmetry characteristics.

tachyon — Hypothesized faster than light particle. Einstein's special relativity does not preclude the possibility of superluminal particles.

tardyon — Suggestive that physical matter is constrained to always exhibit a velocity less than light. A subluminal particle. Physical matter is categorized as tardyonic.

teleology — Idea there is a final cause toward which events unfold. Implies there is a design to experience, and hence a designer (therefore, God).

temporal asymmetry — That time flows just one way. Indicated by entropy. Third aspect of the CPT theorem.

temporal directionality — That time automatically changes unidirectionally. Often characterized in the notion of past, present, and future. A. Eddington suggested directionality is defined by entropy.

temporal extensionality — That time is experienced as flowing; therefore, "extends" — past-present-future. Contrasts with time (in) itself (the present).

theist — Commonly, thought to refer to one who believes God is.

theory — A statement or concept considered possibly true but not verified by reality. Offers a contradistinction to factual statements.

theory of everything — Unifying concept that every thing is ultimately explainable by (can be reductively derived from) a single principle or equation.

theory of types — B. Russell's idea that a concept differs from the things it "classifies." The idea of car is not itself a car. Russell's effort to resolve the problem encountered in Russell's paradox.

tidal forces — Shearing (distorting) forces inflicted on an object nearing a massive source of gravity. An object is "stretched" as it is pulled ("attracted") toward a massive body.

time — Concept invoked to describe the fact things "naturally" change. Time itself (no time flow) is equivalent to space (in) itself. Time itself is the present (no past, nor future). That time also flows is a characteristic of temporal extensionality. Space itself manifests as temporal extensionality.

transcendence — That beyond what is considered. A metafunction.

transcendent functionality — Proper way to consider the concept of transcendence. One that supersedes the functionality of a considered level (for example, of things). Molecular functionality supersedes (is a "transcendent") functionality of atoms.

transcendent "wholeness" function — Precise defining of emergent functions. Molecular functionality emerges from the collective activity of parts (atomic functionality). Occurs when a wholeness function exhibits an activity beyond its wholeness identity. Transportation is a transcendent (wholeness) function of (the "whole" called) the automobile.

true — Logical notion defining unity. A whole and its parts. The negation of false.

truth — Defines the idea knowledge represents reality in statement (conceptual) form. Sometimes implies the notion of all-knowledge.

748   Glossary

ultimate principle — Logical equation that both defines completeness and consistency. The Law of Noncontradiction. "Nothing can both be and not be." The ultimate principle from which all other knowledge could be deduced.

uniformity — That reality is coherent and therefore logically consistent. Implies knowledge (representing reality) must be ultimately free of contradiction.

unit function — Things are units. A unit exhibits a wholeness function and a parts function (lesser wholeness functions). A "unit" yard stick has the lesser measure of 36 inches.

unity — Abstract notion derived from considering a thing to have two aspects: a wholeness function and parts function.

universal expansion — That the physical world spatially extends itself. The big bang resulted in matter exploding outward. A notion not of space (itself) but of spatial extensionality.

universe of imperfect whatness— The collective (universe) of all minds that function deductively. Natural state of the mind in the physical world. The mind inhabiting a physical body "deduces" (deductively infers) the nature of the physical world (in which it finds itself). Correlates with the religious word Satan.

universe of perfect whatness— Categorically, physical matter. In religion, hell.

universe of imperfect whyness— The collective of minds functioning inductively. Natural state of the mind in Phase II of cosmic cycle. In religion, heaven.

vacuum — Another word for space itself.

verification — Scientifically correlating statements or concepts with what they represent in reality. The purpose of experimentation.

wanting — Desiring. In religious terminology, sinning.

wave function — Mathematical description of quantum phenomena. Can be interpreted as a probability statement about the state of an elementary particle.

wave-packet collapse — Concept explaining how the superposition (of possible quantum outcomes) is "reduced" to (collapsed into) what occurs in reality. Essentially describes how consciousness projects (itself into manifest) experience.

wave packet — Description of how quantum (partitioned) energy is conveyed along a wavefront.

weakons — Bosonic force particles responsible for intermediate interactions (the weak force).

whatness — Abstract notion of thing. Characterizes the world as it is experienced. Things and minds. Pertains to manifest objects, minds, or any sense datum (perceived or imagined).

whatness nonvital substance — Physical things that do not exhibit life. Rocks, nonorganic matter.

whatness vital substance — Physical substance that can exhibit the function of life. Life-forms: plants and animals.

When Aspect — Facet of the Deltic Cosmos that characterizes the flow of time (either forward [Phase I] or backward [Phase II]).

Where Aspect — Facet of the Deltic Cosmos that reductively defines the universes (physical matter and mind).

whole — Highest function of a unit. A jigsaw puzzle exhibits a (whole) picture, yet it is composed of parts. Identity of thingness is based on its wholeness — considered separately from other things.

wholeness function — That a unit thingness behaves (functions) as a whole. A "unit" of atoms might function as a collective; therefore as a (whole) molecule. One oxygen and two hydrogen atoms (functionally) combine to form a water molecule.

whyness function — The reason for something (defined as a whatness). The explanation for something. May be a transcendent function. Transportation is the whyness function (reason, purpose) of the automobile (the whatness).

whyness — The reason for whatness. May refer to things in terms of explanation.

whyness — whatness associations — Relating things (whatness) with their reason or purpose (whyness). Cause and effect, and "contributing factors."

whyness — whatness modes — Ways of illustrating the relationship between things (whatness) with their reason or purpose (whyness).

Wigner's friend — E. Wigner's version of Schrödinger's cat paradox. The idea was to replace the cat with a person capable of speaking from inside the box to tell observers just how alive or dead he was due to macro-quantum effects.

Wise Man's Wisdom — A realistically honest seeker of truth only uses information that has already been scientifically verified (to gain a greater understanding of reality). His basic question is, "how can this (any given phenomenon) be explained?" Opposite to fool's fatuity.

worldlines — Natural trajectory of bodies (objects) in space that includes the affects of spatial curvature caused by the presence of massive bodies.

zero — Numerical sign for the empty set, falseness, and the wholeness function.

750

Name Index

A

Abel, N. Henrik, 310
Abelard, Peter, 346
Abraham (Bible), 377–79, 384, 396, 423
Adair, Robert K., 325
Adam (supposed first being), 162, 163, 372, 389–405, 410–13, 416, 427–33, 657, 673, 674, 675, 679, 694, 695
Adams, John C., 273
Aharonov, Yakir, 310
Alexander the Great, 381
Alhazen, 229, 231
Anaximander, 609
Anderson, Carl D., 181
Antippa, Adel F., 174
Argand, Jean-Robert, 487
Aristotle, 269–70, 371, 491, 609, 610, 634
Asch, Soloman, 104
Aspect, Alain, 333, 531
Augustine, 513, 610
Averrois, 611

B

Bacon, Francis, 343
Balmer, Johann J., 238–40, 244, 252
Bartholin, Erasmus, 229
Becqueral, A. Henri, 305
Bell, John S., 332–34, 531
Bergson, Henri, 655
Berkeley, George, 285, 350, 351, 592, 610
Berkowitz, Leonard, 103
Bilaniuk, Olexa-Myron P., 171, 172
Bjorken, James D., 302
Boethius, Anicius M.S., 346
Bogen, Joseph, 131–34
Bohm, David J., 265, 330, 331, 334, 335, 693
Bohr, Niels H.D., 238–51, 238–51, 256, 258, 261, 265, 329
Boole, George, 510
Boring, Edwin G., 349
Born, Max, 244–46, 255
Bose, Satyendra N., 298
Brackett, Frederick S., 239

Bradley, James, 231
Braginski, Vladimir B., 274
Brahe, Tycho, 270
Brans, Carl, 287
Broca, Paul, 130
Broglie, de, Louis V., 241–45, 248, 259, 260, 263, 322, 335, 535
Brout, Robert, 311
Buridan, Jean, 271
Buss, Arnold A., 103

C

Cantor, Georg, 466–73, 493, 513
Carnot, N.L. Sadi, 568, 687
Carter, Brandon, 597
Cavendish, Henry, 272
Cayley, Arthur, 244
Champeaux, de, Guilliume, 346
Christenson, James H., 189
Christoffel, Elwin B., 280
Clauser, John, 332, 333
Clausius, Rudolf, 568
Compton, Arthur H., 238
Conway, John, 482–87
Copernicus, Nicolaus, 263, 270
Costa de Beauregard, Par O., 335
Coulomb, de, Charles A., 233
Cronin, James W., 189
Cusa, de, Nicholas, 270, 512, 610
Cyrus, 380

D

d'Arrest, Louis H., 273
d'Espagnat, Bernard, 330, 332
Dao, Tsung, 184
Darwin, Charles R., 617–20
David (Bible), 379, 380, 383, 387–94, 423
Davies, Paul C.W., 31, 289
Davisson, Clinton J., 242, 243, 322, 535
Dax, Marc, 130
De Morgan, Augustus, 500, 509, 514, 534
Dedekind, J.W. Richard, 488
Descartes, Rene, 229, 272, 348, 350, 486, 513, 610
Deschamps, 572
Deshpande, V.K., 171
DeWitt, Bryce S., 262

Penzias, Arno A., 209, 587
Perrin, Jean B., 578
Pfund, August H., 239
Philoponus, John, 270, 271
Piaget, Jean, 48
Pierce, Charles S., 491, 509
Pilate, Pontius (Bible), 397, 398,
    418
Planck, Max, 186, 236
Plato, 25, 345, 371, 478, 512, 610,
    636, 651, 680, 684, 692, 697
Plotinus, 609
Podolsky, Boris, 255, 260
Poisson, Simeon-Denis, 244
Politzer, H. David, 314
Pound, Robert V., 284
Prescott, Charles Y., 313
Ptolemy, 228, 269
Pythagoras, 371, 474, 482, 609

R

Rebka, Glenn A., 284
Regge, Tulio, 299, 304
Reissner, Henrich, 212
Retherford, Robert C., 326
Ricci, Curbastro G., 280
Riemann, Georg B., 277, 278, 488
Roemer, Ole C., 231
Roentgen, Wilhelm C., 234, 578
Rogers, Carl, 46, 54
Roscellinus, 346
Rosen, Nathan, 255, 260
Russell, Bertrand A.W., 117, 118,
    341, 343, 463, 490, 502, 506,
    509, 513, 518
Rutherford, Ernest, 238
Ryle, Gilbert, 350

S

Salam, Abdus, 310, 312, 315
Saul (Bible), 379, 383, 398
Schelling, Friedrich, 611
Schopenhauer, Arthur, 361
Schrödinger, Irwin, 244–46, 250,
    252, 257, 258, 262–65, 334
Schwartz, Charles, 173
Schwarzschild, Karl, 210–13, 283,
    284
Schwinger, Julian S., 258, 308, 311
Sheldrake, Rupert, 646
Shimony, Abner, 332
Shockley, William, 580

Silva, Flavius, 382
Skinner, Bernard F., 357–60
Snell, Willebrord van R., 229
Snider, Joseph L., 284
Snyder, Hartland S., 210
Socrates, 27, 342, 371, 398
Solomon (Bible), 380–83, 425
Sommerfeld, Arnold, 240
Spencer, Herbert, 617
Sperry, Roger, 130–33, 414
Spinoza, Benedict, 351, 356, 610,
    697
Spitz, Rene A., 106
Sramek, Richard A., 284
Stapp, Henry P., 332–34
Stark, Johannes, 240, 246
Stern, Otto, 253
Sudarshan, E.C. George, 171, 172
Susskind, Leonard, 314

T

t'Hooft, Gerardus, 312, 314, 316
Taylor, Richard E., 313
Thompson, Benjamin, 569
Thomson, Joseph J., 578
Titus (Bible), 381
Tomonaga, Sin-Itiro, 308
Trevairthen, Colwyn, 132
Turlay, René, 189

U

Uhlenbeck, George E., 252

V

Veltman, Martinus J.G., 312
Venn, John, 510
Vespasian, 381
Vogel, Philip, 131

W

Waals, van der, J. D., 585
Wada, Juhn A., 134
Wagstaff, Graham F., 106
Wallace, Alfred R., 617
Watson, John B., 350
Weinberg, Steven, 310–12
Wentzel, Gregor, 238
Wernicke, Karl, 130
Wess, Julius, 317

Subject Index

Pauli's exclusion principle, **209**,
252-254, 298, 304, 317, **333**,
528, 529, 693
PEP. *See* personal explanation
principle
peppered moth, 618
perception
extrinsic, **63**, **65**, 66, **70**, 72–75,
88, 89, 155, 156, 159, 160,
161, 175, 344, 363, 364,
544–548
intrinsic, 74, **75**, 133, 155, 160,
161, 175, 197, 269, 363, 364,
545–549
personal explanation principle
(PEP), **12**, **28**, 46, 58, 69, 82–86,
92, 93
Peter (books of), 406
Pharisees (Bible), 381, 382, 385,
398, 423, 424
phase I, 154, **155**, 156–163, 167,
175, **176**, 177, 178, 190–192,
198, 200, 202, 208, 215–220,
392, 434, 549, 570, 571, 586,
638, 649, 652, 653, 656, 657,
664, 669, 672–675, 682–685,
691, 694, 698
phase II, 154, **158**, **159**, 160–163,
171–173, **176**, 177, 178,
190–192, 198, 200–204,
215–220, 392, 400, 402, 406,
570, 571, 638, 657, 664, 669,
672–675, 683, 685, 691–695,
699
phenomenal relevancy, 7–9, 13, 46,
48, 57, 82, 612
phenomenon, 6–8, 13–15, 18, 19,
23, **36**, 40, 44, 66–72, **87**, 119,
343, 344, **349**, 700
photoelectric effect, 237–241, 322,
531, 534, 578
photons, 208, 237, 238, 242,
246–248, 251, 252, 264, 291,
303, 306, 308, 311–317,
322–328, 332, 336, 522, 529,
531–536, 567, 692, 698
pivotal events, 571, **572**, 574, 575,
578, **579**, 580
Planck time, 152, 158, 163, **198**,
201, 260, **321**, 327, 693
Planck's constant, **181**, **198**, **199**,
236, **237**, **238**, 239, **243**, 244,
247, **249**, 260, 264

plants, 16, 31, 46, 65, 155, 410,
570, 595, 600, 615, 616, 619,
620
Platonism, xxviii, **345**, 371, 381,
478, 512, 651, 684, 697
positron, **181**, 186–191, 208, 303,
306, 307, 315, 327, 328, 432,
532–534
potential transcendent function,
647, **654**
power set, 467, 470, 472, 474, 502,
508, 509
primary equation, **456**, 460, 473,
474, 481, 483, 486, 507, 508,
516, 523, 532, 539, **540**, 551,
566, 586, 661
primary qualities, **347**, 350
principle of contradiction, **5**, 171,
477, 611
principle of duality
special relativity, 173, 175
projection
Brahma, 374
consciousness, xxiv, 150, 576,
644, **645**, 658, **676–684**,
686–689
defense mechanism, **81**, **138**
proof, xxvii, 6, **7**, 117, 139, 395,
413, 424, 448, 450, 451, 488,
625, 643
propositions, 12, 117, 140,
343–345, 445, 488–510, 518
protons, 208, 209, 227, 251, 254,
287, 299, **303**, 306, 310,
313–316, 321, 325, 328,
330–332, 336, 527, 528, 568
purpose, 3, 6, 59, 65, 73, 113, 163,
193, 227, 264, 363, 407, 427,
526, 550, 567, 571–575,
578–580, **581**, 582–597,
598–600, 601, 604, 605, **606**,
609, 611, 619–624, **625**, **626**,
628–630, **634**, 635, 637,
639–643, **648**, 649–658, **659**,
667, 670, 671, 674, 675,
679–681, 682, 683, **684**, 685,
687, **689**, 690, **691**, 699, 702

Q

quantum
strangeness, 186, 300–303
quantum electrodynamics. *See*
electrodynamics, quantum

self-cosmolization, **55**
self-creation, **123**, 124, 129, 134
self-destruction, 102, **121**, 122, **123**, 129
self-experientialization, **46**
self-relativization, **48**
self-righteousness, **81**, 111, 421, 434, 436
self-security, 45, **50**, 59, 102, 107, 111, 124, 134
self-sensationalization, **46**
self-stimulization, **46**
sense datum, xviii, 13, 45, 46, **56**, **57**, 63–67, 71–74, 78, 89, 344, 347, **348**, 349, 359, 408, 544–547, 610, 699
sense experience, **63**, 347–351, 363, 544, 545, 550
separability, **330**, 332–334, 375, **563**, 564, 633, 637, 675, 676, 679
series circuit, 525
set
　power, 467, 470–474, 502, 508, 509
set theory, 8, 463, 474, 478, **509**, 510–514, 517, 523, 539, 551–553, 587, **629**, 630, 633, 641, 686, 697, 698
set, empty, **476**, 477, 478, 486, **508**, 509–511, **517**, 523, 530, **533**, 553, **559**, **561**, **565**, **587**, 633, 644, 676
setness, **477**, **480**, 485, 499, 507, **517**, 523, 553, 554, **556**, **559**, **561**, 602, 605, 636–638, 641–644, **645**, 658, 662, 663, 667, 676, 686, 690
Shiva, 373, 375, 588, 662
sin, 353, 382, 397–400, 406, 410, 414–416, 430, 433, 434, 437, 448
singularity, **176**, **210**, 211–220, **291**, 327, 473, **524**, 535, 539, 646, 652, 692, 694
skepticism, 116, 417, **436**, 437–439
sleep, xxiii, **156**, 401, 408, 409, 429–432, 549, 673, 674, 694, 699
solipsism, 258, 351, 645
somethingness, 161, 470, 474–477, 488, 489, **492**, 501, 505, 507, 511, 524, **533–536**, 552, 556,

557, 560, 561, **562**, 565, 567, 630, 686, 698
space, 44, 45, 147, 150, 163, 168–175, 198, 206–215, 218–220, 231, 232, 259, 273–275, 277, 278, 287, 291, 292, 309, 317, 323, 324, **327**, 330, 337, 425, 460, **461**, 462–465, 474–478, 496, 497, 512, 516–518, 521–524, 527–531, 537, 538, **540**, 551–569, 587, 600, 603, 604, 610, 633, **634**, 644, **645**, 648, 657, 660–666, 676–691, **692**, 694, **697**, 698
　absolute, 279, 284–286
space-time, xxvi, 45, 163, 168–170, 176, 185, 198, 201, 210, 212, 219, **275**, 278–282, 286–291, 310, 317, 318, 321, 322, 460, 462, 517, 521–524, 528, 531, 537, 637, 662, 663, 679, 698
space-time curvature. *See* curvature, space-time
spatial extensionality, 44, 168, 170, 174, 176, 185, 191, 192, 198, 210, 219, 323, 325, 336, 478, 528, 530, 551, 554, **556**, 560, 561, 567, 636, 637, **644**, 645, 648, 667, 676, 684, 685, 688, 693, 697
spatial inversion, **189**
special theory of relativity, 163, 174, 177, 181, 185, 187, 191, 193, 234, 241, 275, 278, 281, 309, 460, 461, 516, 532, 578
spin
　isotopic, **299**, 300, 301, 310–313, 527
spin angular momentum, 181, 185, 189, 191, 241, 252, 253, **254**, 285, 291, 298, **299**, 300–306, 310, 317, 319, 330–333, 527–534, **536**, 537, 539
spirit, 371, 389, 391, 394–402, 406–411, 414, 415, 423, 427, 429, 435, 437, 441, 442, **443**, 448, 577, 609, 668, 669, 672, 674, **699**, 702
spirit nature, 394, 398
spirituality, 435, 443, **444**
statistics
　Bose-Einstein, 298